Analytical Methods for Coal and Coal Products

Volume II

CONTRIBUTORS

HARVEY I. ABELSON
KEITH D. BARTLE
ROBERT W. FREEDMAN
WALTER W. FOWKES
P. H. GIVEN
JOHN S. GORDON
RYOICHI HAYATSU
R. G. JENKINS
DERRY W. JONES
LAWRENCE F. KING
THOMAS A. LINK
WILLIAM A. LÖWENBACH
O. P. MAHAJAN
HARRY MARSH
DUNCAN G. MURCHISON
C. S. B. NAIR
HOOSHANG PAKDEL
JOHN W. PATRICK
HERBERT L. RETCOFSKY
JANET SMITH
JAMES G. SPEIGHT
MARTIN H. STUDIER
D. N. TODOR
N. I. VOINA
P. L. WALKER, JR.
HERBERT C. WILKINSON
RANDALL E. WINANS
RICHARD F. YARZAB

Analytical Methods for Coal and Coal Products

Edited by CLARENCE KARR, JR.
Department of Energy
Morgantown Energy Technology Center
Morgantown, West Virginia

Volume II

ACADEMIC PRESS New York San Francisco London 1978
A Subsidiary of Harcourt Brace Jovanovich, Publishers

ACADEMIC PRESS, INC.
111 Fifth Avenue, New York, New York 10003

United Kingdom Edition published by
ACADEMIC PRESS, INC. (LONDON) LTD.
24/28 Oval Road, London NW1 7DX

Library of Congress Cataloging in Publication Data

Main entry under title:

Analytical methods for coal and coal products.

Includes bibliographies.
1. Coal--Analysis. I. Karr, Clarence.
TP325.A58 662'.622 78-4928
ISBN 0-12-399902-2 (v. 2)

PRINTED IN THE UNITED STATES OF AMERICA
81 82 9 8 7 6 5 4 3 2

Contents

Part VI MINERALS IN COAL

Part VII COAL CARBONIZATION PRODUCTS: COKE, PITCH

List of Contributors

Numbers in parentheses indicate the pages on which the authors' contributions begin.

HARVEY I. ABELSON (563, 583), The MITRE Corporation, METREK Division, McLean, Virginia 22101

KEITH D. BARTLE (103, 209), School of Chemistry, University of Leeds, Leeds, England

ROBERT W. FREEDMAN* (315), Pittsburgh Mining and Safety Research Center, Bureau of Mines, U.S. Department of the Inteior, Pittsburgh, Pennsylvania 15213

WALTER W. FOWKES† (293), Grand Forks Energy Research Center, U.S. Department of Energy, Grand Forks, North Dakota 58201

P. H. GIVEN (3), Fuel Science Section, The Pennsylvania State University, University Park, Pennsylvania 16802

JOHN S. GORDON (563,583), The MITRE Corporation, METREK Division, McLean, Virginia 22101

RYOICHI HAYATSU (43), Chemistry Division, Argonne National Laboratory, Argonne, Illinois 60439

R. G. JENKINS (265), Department of Material Sciences and Engineering, The Pennsylvania State University, University Park, Pennsylvania 16802

DERRY W. JONES (103, 209), School of Chemistry, University of Bradford, Bradford, England

LAURENCE F. KING‡ (535), Research Department, Imperial Oil Enterprises, Ltd., Sarnia, Ontario, Canada NON 1MO

THOMAS A. LINK (161), U.S. Department of Energy, Pittsburgh Energy Technology Center, Pittsburgh, Pennsylvania 15213

WILLIAM A. LÖWENBACH (563, 583), Löwenbach and Schlesinger, Associates, Incorporated, McLean, Virginia 22101

* Present address: U.S. Bureau of Mines, 4800 Forbes Avenue, Pittsburgh, Pennsylvania 15213.

† Present address: 422 West Farmer Avenue, Independence, Missouri 64050

‡ Present address: St. Clair Parkway, Mooretown, Ontario, Canada NON 1MO

O. P. MAHAJAN (465), Department of Material Sciences and Engineering, The Pennsylvania State University, University Park, Pennsylvania 16802

HARRY MARSH (371), Northern Carbon Research Laboratories, School of Chemistry, The University of Newcastle upon Tyne, Newcastle upon Tyne, England

DUNCAN G. MURCHISON (415), Organic Geochemistry Unit, Department of Geology, University of Newcastle upon Tyne, Newcastle upon Tyne, England

C. S. B. NAIR* (495), Central Fuel Research Institute, Dhanbad, Bihar, India

HOOSHANG PAKDEL (209), School of Chemistry, University of Bradford, Bradford, England

JOHN W. PATRICK (339), British Carbonization Research Association, Chesterfield, Derbyshire, England

HERBERT L. RETCOFSKY (161), U.S. Department of Energy, Pittsburgh Energy Technology Center, Pittsburgh, Pennsylvania

JANET SMITH (371), Northern Carbon Research Laboratories, School of Chemistry, The University of Newcastle upon Tyne, Newcastle upon Tyne, England

JAMES G. SPEIGHT (75), Fuel Sciences Division, Alberta Research Council, Edmonton, Alberta, Canada T6G 2C2

MARTIN H. STUDIER (43), Chemistry Division, Argonne National Laboratory, Argonne, Illinois 60439

D. N. TODOR (619), Building Materials Laboratory, Institutul de Constructii, Bucharest, Romania

N. I. VOINA (619), Building Materials Laboratory, Institutul de Constructii, Bucharest, Romania

P. L. WALKER, JR. (265, 465), Department of Material Sciences and Engineering, The Pennsylvania State University, University Park, Pennsylvania 16802

HERBERT C. WILKINSON (339), British Carbonization Research Association, Chesterfield, Derbyshire, England

RANDALL E. WINANS (43), Chemistry Division, Argonne National Laboratory, Argonne, Illinois 60439

RICHARD F. YARZAB (3), Fuel Science Section, The Pennsylvania State University, University Park, Pennsylvania 16802

* Present address: Research and Development Division, Fact Engineering and Design Organization, The Fertilisers and Chemicals Travancore Ltd., Udyogamandal 683501, Cochin, Kerala State, India.

Preface

Coal is becoming an increasingly important factor in the energy resource projections for the remainder of this century. The major use in the generation of electricity is expected to expand as more power plants are converted from petroleum fuels and natural gas to coal. A smaller volume use, the production of metallurgical coke, is expected to increase in complexity as the traditional high quality coking coals are gradually depleted. Within the next few decades the production of synthetic fuels from coal is expected to start playing a significant role. All of these uses present analytical problems intensified by increasing reliance on new coal resources, increasing requirements for the protection of the environment, and a growing need to find practical ways to utilize waste products.

These volumes have been devoted exclusively to analytical problems, comprising the first major reference work on the methodology for analysis of coal and coal products. Insofar as practical I have arranged the various chapters in these volumes according to either a specific coal process, such as liquefaction, gasification, combustion, or carbonization, the latter two being in Volume II, or a specific coal use problem, such as environmental problems, waste product utilization, the minerals in coal, and the organic structure of coal (the latter two also being in Volume II).

The general philosophy of this work is to strike a balance between sophisticated analyses based on expensive instrumentation such as mass or nuclear magnetic resonance spectrometers, and the more common, less expensive equipment typically employed in the standard methods. Likewise there is an attempt to strike a balance between the expertise available from within the United States and that found in other countries, offering a broader viewpoint. Altogether, a large number of cross references have been entered in these chapters to enable the reader to make maximum use of pertinent information in all of the chapters. The Index has been prepared so as to offer fully detailed entries. A more detailed description of the purpose and philosophy of these volumes is presented in the Preface to Volume I.

I am grateful to all the authors and their organizations for their generous cooperation and support in the preparation of these chapters. Once again I wish to thank Pamala Kisner Stasia for her untiring assistance with the correspondence that was involved in this venture.

Contents of Volume I

Part V

STRUCTURE OF COAL AND COAL PRODUCTS

Chapter 20

Analysis of the Organic Substance of Coals: Problems Posed by the Presence of Mineral Matter

P. H. Given *Richard F. Yarzab*
FUEL SCIENCE SECTION
THE PENNSYLVANIA STATE UNIVERSITY
UNIVERSITY PARK, PENNSYLVANIA

I. INTRODUCTION

Coal may be defined as an organic rock composed of an assembly of macerals, minerals, and inorganic elements held molecularly by the organic matter. Perhaps because of the basic cleavage between organic and inorganic chemistry in much teaching of chemistry, one tends to think that the two kinds of constituents in coals behave quite independently and that if one is interested primarily in the organic matter,

ISBN 0-12-399902-2

mineral matter (equated, falsely, with ash) is merely a diluent, and inert. This position could hardly be more mistaken. In almost every process to which one might subject the organic matter of a coal, whether it be hydrogenation to oil, determination of the carbon content, or oxidation to humic acids, the mineral matter participates to a greater or lesser extent, a fact that is all too often ignored.

The nature of the mineral matter of coal is a subject of interest in its own right, and this is dealt with in Chapters 26, 27, and 28. For the purposes of this chapter, it is assumed that the reader is interested in the composition of the organic matter, and is concerned with the mineral matter only to the extent that its presence may stand in the way of understanding the organic matter; i.e., it is assumed here that the reader regards mineral matter merely as an interference perturbing the data he/she wishes to obtain. The interferences, in this sense, are of two quite different kinds:

(i) The "diluent" effect must be removed: a dry coal contains 73% carbon and 10% mineral matter, hence its organic substance contains $73 \times 100/(100 - 10) = 81.11\%$ carbon. In order to remove this interference, one needs to know the mineral matter content of the sample, which is *not* the same as the yield of ash on combustion at high temperatures.

(ii) The direct participation effect must be removed. In any analytical determination, one tries to select a measurement that relates only to the organic matter or only to the mineral matter, but this is often impossible. A coal contains 10% mineral matter, of which 1% is calcite; in the conventional combustion analysis for organic carbon, the measured yield of CO_2 corresponds to a total carbon content for the dried coal of 73.00%. The calcite decomposes to CaO and CO_2 during the combustion analysis for C, the CO_2 produced by the calcite containing an amount of carbon equivalent to 0.12% of the dry coal. Hence the content of carbon in the organic substance of the coal is

$$(73.00 - 0.12) \times 100/(100 - 10)\% = 80.98\%$$

The two effects just delineated in a simple way are of quite general occurrence in all sorts of practical uses of coal analyses. This fact, and what one can do about it, is the principal theme of this chapter. To be more specific, this chapter is concerned with (a) obtaining ultimate and proximate analyses of the pure organic substance of coals, (b) how ultimate analyses may be used to describe the effect of chemical reactions on the organic substance of coals (including metamorphism or natural increase in rank), and (c) the ways in which the presence of mineral matter may affect the determination of functional groups in coals.

It will be clear that the objectives of this chapter, as just listed, require chiefly discussion of ultimate analyses and their uses. An analysis that can claim to represent the composition of the organic substance of a coal is said to be on the dry mineral-matter-free (dmmf) basis. It is with dmmf analyses that we are mostly concerned here. There is, unfortunately, widespread ignorance of such analyses in this country, probably because the ASTM standards for coal analysis (ASTM, 1976) contain no standards for the determination of mineral matter contents (except the use of the Parr formula in calculating parameters used in classifying coals by rank), nor for obtaining ultimate analyses on the dmmf basis. The corresponding British standard (British Standards Institution, 1971) is not guilty of these omissions.

An excellent review of analytical methods for coal and the problems caused by the presence of mineral matter was published by Shipley (1962) and it is still worthy of study. The review by Rees (1966) has some valuable points. The problems caused by the presence of mineral matter in coals have been discussed in some detail by Given and Yarzab (1975).

II. REACTIVITY OF COAL MINERAL MATTER

Determinations of C, H, N, S (total), O (direct), ash yield, and volatile matter in coals all depend on treating the sample in various ways at high temperatures (750–950°C). Whether or not the conditions are oxidizing has little effect on the changes undergone by the minerals, except that pyrite is burnt to Fe_2O_3 in air or oxygen, but dissociates to FeS in an inert atmosphere. The principal phenomena and their analytical consequences are summarized in Table I. It can be seen that most or all of the minerals undergo changes during the high temperature analyses.

One important consequence is that the high temperature ash yield is almost always less than the mineral matter content; thus a coal containing 2% sulfur and affording 12% ash might have a true mineral matter content of about 14%. Hence if the coal has 73% C on the dry basis, one would calculate 82.95% C on the dry ash-free (daf) basis or 84.88% dmmf; oxygen by difference will also be affected.

Another consequence is that the raw analytical data for C, H, S, and O (if direct) all need corrections in order to obtain the contents of those elements in the organic substance only. The presence of ion-exchangeable cations in low rank coals causes several complications, which are discussed later.

Determination of functional groups in coals is made by means of specific chemical reactions, which are carried out at relatively low temperatures. Reactions may also be carried out at low temperatures for the

TABLE I *The Behavior of Coal Minerals at High Temperatures*

Inorganic species	Behavior on heating	Consequences for analysis
Clays	Lose structural OH groups with rearrangements of structure and release of H_2O	Ash weighs less than MM. Yield of water increases apparent organic hydrogen, oxygen, and VM
Carbonates	Decompose with loss of CO_2, residual oxides fix some organic and pyritic S as sulfate	Ash weighs less than MM, but this effect partly neutralized by fixation of S as sulfate. CO_2 from carbonates increases apparent VM, organic carbon and organic oxygen
Quartz	Possible reaction with iron oxides from pyrite and organically held Ca in lignites. Otherwise no reaction	None, unless reactions indicated take place
Pyrite	In air, burnt to Fe_2O_3 and SO_2. In VM test, decomposes to FeS	Increases heat of combustion. Ash weighs less than MM. S from FeS_2 contributes to VM
Metal oxides	May react with silicates	None (?)
Metal carboxylates (lignites and subbituminous only)	Decompose, carbon in carboxylate may be retained in residue	Uncertainty about significance of ash. Most of organic sulfur in coal fixed as sulfate in ash

purpose, for example, of elucidating the structure of the organic matter. In these cases, the mineral matter may participate in various ways in the reactions. Thus pyrite may be solubilized by oxidizing or reducing agents, and carbonates will dissolve in acid reagents. The nature, extent, and consequences of such participation will obviously depend on the nature of the reagent and the reaction, and no generalizations can be made. Some specific examples of the effect are discussed subsequently.

III. DEFINITIONS OF ULTIMATE ANALYSES

ASTM Standard Method D 3176 (ASTM, 1976) defines an ultimate analysis,† and annotates the definition, as follows:

3.1 *Ultimate analysis.* In the case of coal and coke, the determination of carbon and hydrogen in the material, as found in the gaseous products of its complete combustion, the determination of sulfur, nitro-

† Ultimate analysis is discussed in Volume I, Chapter 6, Section IV.

gen, and ash in the material as a whole, and the calculation of oxygen by difference.

NOTE 1 [Omitted as not relevant here.]

NOTE 2 Moisture is not by definition a part of the ultimate analysis of coal or coke, but must be determined in order that analytical data may be converted to bases other than that of the analysis sample.

NOTE 3 Inasmuch as some coals contain mineral carbonates, and practically all contain clay or shale containing combined water, a part of the carbon, hydrogen, and oxygen found in the products of combustion may arise from these mineral components.†

Section 5.2 of the same Standard and the description of "Scope" in Standard D 3177, Total Sulfur in the Analysis Sample of Coal and Coke, make it clear that the sulfur to be included in the ultimate analysis is the total content.

The scope of Standard D 3178 is described as follows:

1.1 These methods cover the determination of total carbon and hydrogen in samples of coal or coke. Both the carbon and hydrogen are determined in one operation. This method yields the total percentages of carbon and hydrogen in the coal as analyzed and the results include not only the carbon and hydrogen in the organic matter, but also the carbon present in mineral carbonates and the hydrogen present in the free moisture accompanying the sample as well as the hydrogen present as water of hydration of silicates.

NOTE 1 It is recognized that certain technical applications of the data derived from this test procedure may justify additional corrections. These corrections could involve compensation for the carbon present as carbonates, the hydrogen of free moisture accompanying the sample, and the calculated hydrogen present as water of hydration of silicates.

1.2 When data are converted and reported on the "dry" basis, the hydrogen value is corrected for the hydrogen present in the free moisture accompanying the sample.†

Thus it is clear that the ultimate analysis, as defined, does not even purport to represent the composition of the pure organic matter. A number of interferences by mineral matter are correctly pointed out, but no means of eliminating them are provided. It is in the oxygen-by-difference that the mineral interferences accumulate most seriously. On the dry basis, this oxygen, by the ASTM definitions, is given by

$$O = 100 - (\text{Ash} + C_{\text{tot}} + H_{\text{tot}} + N + S_{\text{tot}}) \quad (1)$$

where the subscript tot signifies total; thus the oxygen accumulates not only analytical errors in the direct determinations, but also terms due to interferences by mineral matter. The data can then be placed on the dry, ash-free basis (daf) by multiplying all terms (except the ash) by 100/(100 − Ash).

The daf analysis, as just described, is to be contrasted with the dry, mineral-matter-free analysis (dmmf). British Standard 1016, Part 16 (British Standards Institution, 1971), discusses in detail the derivation of dmmf analyses, but does not give a succinct definition. However, it does state the following:

> 3.51 *General.* The "dry, mineral-matter-free" basis is a hypothetical condition, corresponding to the concept of a pure coal or coke substance.
>
> Since the "dry, ash-free" basis for coal ignores the changes in mineral matter when coal is burnt, the "dry, mineral-matter-free" basis is preferred whenever mineral matter can be calculated (see 3.52), even if some analytical values have to be assumed in the absence of results determined on the sample.†

The Standard makes clear that a dmmf analysis is defined by the performance of the following steps:

(i) Calculate or determine the mineral matter content directly.
(ii) Compute

$$C_o = C_{tot} - \tfrac{12}{44}CO_2 \qquad (2)$$

where C_0 represents organic carbon and CO_2 the amount of gas released by acid from carbonates.

(iii) Compute

$$H_o = H_{tot} - 0.014\text{Ash} + 0.02S_p + 0.02CO_2 \qquad (3)$$

where S_p is the content of pyritic sulfur.

(iv) Compute

$$S_o = S_{tot} - S_p - S_{SO_4} \qquad (4)$$

(v) Evaluate organic oxygen as

$$O_o = 100 - (MM + C_o + H_o + N + S_o + 0.5Cl) \qquad (5)$$

where MM is the mineral matter content.

(vi) Place on the dmmf basis by multiplying all terms (except MM) by 100/(100 − MM).

† Quoted by permission of the British Standards Institution.

TABLE II *Comparison of daf and dmmf Analysis for a High Sulfur Coal*[a]

Constituent (%)	Analysis (%)	
	daf	dmmf
C	78.3	86.4
H	4.28	4.42
N	1.21	1.34
S	9.37 (total)	4.05 (organic)
O (difference)	6.85	3.79

[a] The coal is from the Big Tebo Seam, Tebo Strip Mine, Calhoun, Missouri.

It should be apparent that oxygen-by-difference can be very different on the daf and dmmf bases. The difference tends to increase with increasing rank, and with a low sulfur anthracite the daf value can be double the dmmf. With high volatile A bituminous coals of high pyritic sulfur content, the daf value can also be too high by a factor of 2. The figures given in Table II show how great the difference between daf and dmmf analyses can be for a coal of very high sulfur content.

It must often happen in practice that one is concerned with a process that alters the composition of both the organic and inorganic matter of a coal, and wishes to use ultimate analyses of starting material and product for ascertaining the gross changes brought about. In this case it would be expedient to express the composition of the samples on such a basis that

$$C_o + H_o + N + S_o + O_o + MM = 100\% \tag{6}$$

It might further be expedient to replace MM by (FeS_2 + other MM). Such a basis has not heretofore been defined. It is proposed to refer to this in this chapter as the "dry, mineral-containing basis." In principle, the expression "dry basis" should be sufficient, but this has already been preempted by ASTM to mean something different.

IV. INFORMATION REQUIRED FOR A dmmf ANALYSIS

Unfortunately, quite a large body of data is needed before a true dmmf analysis can be computed. The nature of this information is described in the following subsections, together with notes on methods of obtaining it. In discussing how the data are used, some approximation methods will be noted, for use when the full set of data is not available.

A. Representative Samples

Before proceeding, it should be stressed that no analyses are worthwhile unless representative samples are obtained for the various determinations. Thus the precision of a single determination is dependent on the precision with which the analytical sample can be analyzed *and* on the magnitude of the sampling error. The latter is of course dependent on the size of the sample whose mean composition is wanted. ASTM standards (ASTM, 1976) include procedures for obtaining a representative sample of, for example, the output of a preparation plant, and for obtaining from this a representative subsample for analysis (see Standards D 2234 and D 2013). From material included in the standards it appears that if the mean ash yield of a 1000-ton sample is 12% and the stated procedures are used, the overall precision of a single determination will be ±1.6%.

It might be thought that when mean values for large samples are required, sampling errors are so large that the difference between daf and dmmf analyses becomes insignificant. This is not so. Thus if the mean ash yield for a large sample is 10 ± 1%, the mineral matter content might have been 11.8 ± 1.2%, a significant difference.

Laboratory workers are unlikely to be confronted with the large samples just considered, but it is still important that every time an analytical sample is taken from a bulk supply of 20 g or 1 kg, riffling or cone-and-quartering is used to ensure that a representative sample is used.

B. Mineral Matter Contents

Determination of good values for mineral matter contents is a very important component of coal analysis; alternative methods are now discussed in some detail.

1. *Calculation of Mineral Matter Contents*

The well-known Parr formula (Parr, 1932) permits the calculation of the mineral matter content from the high temperature ash yield† and the total sulfur content:

$$\mathrm{MM} = 1.08\mathrm{Ash} + 0.55\mathrm{S} \tag{7}$$

The first term on the right-hand side of this equation is commented on shortly; the second term purports to allow for the loss in weight when pyrite burns to hematite. The use of this formula is called for in ASTM standards for putting the volatile matter yield and calorific values on the

† For standard methods of ash determination see Volume I, Chapter 6, Section III,C.

dmmf basis in the classification of coals by rank. It will be noted that this formula (a) treats all sulfur as pyritic, (b) makes no allowance for decomposition of carbonates or fixation of sulfate, and (c) implies that no term should be shown for organic sulfur in the ultimate analysis on the Parr dmmf basis.

A more elaborate formula, which allows for a number of effects, was derived by King *et al.* (1936):

$$MM = 1.13\text{Ash} + 0.8CO_2 + 0.5S_p - 2.8(S_{ash} - S_{SO_4}) + 0.5Cl \quad (8)$$

In this formula, CO_2 represents the yield of the gas obtained when the sample is treated with hot hydrochloric acid, S_p represents the pyritic sulfur content, S_{ash} and S_{SO_4} represent sulfur as sulfate in ash and in the coal, respectively. The formula assumes that one-half the chlorine is attached to organic matter and the other half is mineral (see later).

The terms, 1.08Ash and 1.13Ash in Eqs. (7) and (8), respectively, purport to correct for the loss in weight due to elimination of water in the decomposition of clay minerals at high temperatures. Millott (1958) studied the minerals in 80 British coals and concluded that the term in the King–Maries–Crossley (KMC) formula (King *et al.*, 1936) should be revised upwards to 1.13Ash, as shown in Eq. (8), and this revision was accepted (British Standards Institution, 1971) (as originally proposed, the term was 1.09Ash). In fact, it is the uncertain magnitude of this term that constitutes the most serious objection to any mineral matter formula. The different clay minerals lose quite varying amounts of water on decomposition (quartz is presumably included with the aluminosilicates in the formula, but it does not lose water on heating). Thus, to assume, as the formulas do, that a constant mean value for water loss can be assigned to the mixtures of silicates in coals is most unlikely to be valid.

In fact, it has been observed that for coals from three different basins in Australia, the coefficient of the ash term in the KMC formula should be 1.09, 1.13, and variable (1.06–1.20), respectively (Brown *et al.*, 1960). Given and Yarzab (1973) found the water of decomposition of the silicates from about 20 U.S. coals to be highly variable. Both Australia and the United States contain coals of much more diverse origins than does the United Kingdom, where the KMC formula was developed and extensively tested. Hence one might predict that the formula would be less precise outside Europe.

As will be seen presently, direct determination of mineral matter contents can be tedious and slow. Determination of all the parameters needed in the KMC formula is also tedious and slow. Are there short cuts for use when analyses of the highest precision are not needed?

Given *et al.* (1975) have described a simple modification of the Parr formula for use when fuller data are not available:

$$MM = 1.13Ash + 0.47S_p + 0.5Cl \qquad (9)$$

This is open to most of the objections made earlier to the original Parr formula. It assumes the average water of decomposition of clays recommended by Millott. The term Cl was included because chlorine contents are available in the authors' laboratories, but with most U.S. coals little difference will be seen if it is omitted. Apart from requiring few data, the advantages of the modified formula as as follows:

(i) It is more logical in including only pyritic sulfur.
(ii) Since it does include only pyritic sulfur, organic sulfur can be included in the ultimate analysis.
(iii) It readily lends itself to the derivation of other approximate formulas, as will be seen in Section V.

It should be noted that none of the formulas is applicable to the mineral matter of lignites, for reasons explained in Section IV,D.

2. *Direct Determination of Mineral Matter*

It is generally agreed in the literature that the most reliable means of obtaining the mineral matter content of a coal is by the acid demineralization procedure due originally to Radmacher and Mohrhauer (1955) and later modified by Bishop and Ward (1958). The method depends essentially on measurement of the loss of weight of the sample when treated with 40% HF at 50–60°C. The sample is treated with hydrochloric acid before and after the HF treatment to prevent the retention of the highly insoluble calcium fluoride, whose presence causes various complications. Also, pyrite (not dissolved in the treatments), a small amount of residual ash, and the small amount of retained chloride, assumed present as HCl, must be determined separately (the residual ash includes hematite from the pyrite as well as a little undissolved silicates). The mineral matter content is given by

$$MM = \text{wt loss} + HCl + \tfrac{1}{3}(FeS_2) + \text{residual ash} \qquad (10)$$

This method has been used with coals of all ranks and requires no assumptions about the nature of the mineral matter. However, it is slow and tedious (in our laboratories, determinations take an average of about 1.5 man-days). It has been used with coals from many parts of the world and the results compared with those from the KMC formula (Radmacher and Mohrhauer, 1955; Bishop and Ward, 1958; Ward and Millott, 1960; Brown *et al.*, 1959; Savage, 1967–1968; Tarpley and Ode,

1959; Miller and Given, 1972). In all, comparisons were made for 140 coals, and the whole set of data was analyzed statistically by Given and Yarzab (1975), who found that the mean difference, acid demineralization − KMC, was −0.23. This tendency for KMC to overestimate is significant at the 95% confidence level, and became more marked with coals of high mineral matter content. The standard deviation of the differences between data obtained by the two methods is 0.84%. This rather high value is no doubt largely due to the variability of the water of decomposition of clay minerals.

It should be noted that Kinson and Belcher (1975), without explanation, assert that the acid demineralization procedure is not applicable to low rank coals (though it was in fact so used by, for example, Tarpley and Ode, 1959). This assertion presumably refers back to the statement of Frazer and Belcher (1972) that an amount of organic matter in low rank coal equivalent to 1–2% of the mineral matter content is solubilized by the acid treatment.

Tarpley and Ode (1959), from a study of 21 U.S. coals, concluded that the mean standard deviation of differences recorded in duplicate tests was 0.11. We have not been able to achieve this degree of reproducibility (see later).

An alternative means of direct determination of mineral matter in coals is LTA, i.e., ashing at low temperatures in an oxygen plasma asher (Gluskoter, 1965). Frazer and Belcher (1972) have recommended the procedure for routine use, and prefer it to acid demineralization on the grounds of reliability as well as lower man-hours per determination. However, it appears that the (Australian) coals they work with include neither lignites nor coals of high sulfur content.

An extensive study of the application of the procedure to a series of U.S. coals has recently been completed (Miller *et al.*, 1978). Consolidation of the results of this study with those from the work of Frazer and Belcher leads to the conclusion that closely specified conditions of ashing must be adhered to. These include the radiofrequency power level; if this is too high, pyrite is oxidized to hematite, and if it is too low, an excessive amount of organic sulfur is fixed as sulfate. Small amounts of unburnt carbon must be determined as a correction factor after ashing.† With lignites and subbituminous coals, essentially the whole of the organic sulfur is fixed as sulfate, and ashing is slow. This is apparently due to the presence in low rank coals of carboxylate salts of calcium and sodium. Sulfur fixation is small or zero, and the rate of ashing faster, if the coal is pretreated with dilute hydrochloric acid to remove the ex-

† Using the Leco apparatus, this determination is simple and rapid.

changeable cations. Large sulfate fixation, much mitigated by pretreatment with acid, was also observed with HVC coals from the Rocky Mountain geological province, but not with the older coals from the Interior Province (e.g., Illinois). It is inferred that the younger HVC coals contain carboxylate salts.

The LTA procedure is offered by the authors as a valid routine method for U.S. coals provided (a) the specified conditions are observed, and (b) it is only applied to coals of higher rank than those specified earlier (it may be applicable to the lower rank coals if they are first extracted with acid, but this has not yet been fully tested). The precision, calculated from results for 13 coals run in duplicate, was ±0.10%, compared with ±0.33% for duplicate runs of the acid demineralization procedure on the same coals. The results showed that it is impossible in LTA to eliminate entirely the oxidation of pyrite and the fixation of organic sulfur as sulfate; however, the net effect of these side reactions is a bias by which the LTA mineral matter content may be too high by only 0.05–0.15%.

An unresolved complication arises because of reports of nitrogen in low temperature ash. O'Gorman and Walker (1971) and Karr *et al.* (1968) claimed to have found nitrate ion in the LTA from some lignites and subbituminous coals. The conclusion of O'Gorman and Walker was based on the appearance of a very sharp and intense band found at 1380 cm^{-1} in the infrared spectra of the LTA of low rank coals† but not of bituminous coals or anthracites. It must now be questioned whether this band really was due to vibrations of the nitrate ion because (a) Karr and co-workers give 1360 cm^{-1} as the mean frequency of the nitrate ion, with a range of 30 cm^{-1} ; (b) an associated band at 820 cm^{-1} was not reported; (c) the band was much sharper and more symmetrical than that shown by Karr and co-workers for pure nitrate salts; and (d) the band has recently been observed in these laboratories in spectra of blank KBr pellets and of pellets containing various pure minerals. On the other hand, Karr *et al.* (1968) state that they had found nitrate ion in the LTA of lignites but appear not to have published the evidence. Neither set of authors gives quantitative data.

Both sets of authors raise the question of whether the (putative) nitrate is an artifact of the LTA procedure. Accordingly, they sought evidence of the presence of nitrates in hot-water extracts of lignites and a subbituminous coal. Karr's group carefully studied the infrared spectra

† While this chapter was being written, J. Youtcheff and P. Painter kindly determined the precise frequency of this band in the LTA of a lignite, using a Fourier transform infrared spectrometer equipped with an internal means of automatic standardization of frequency. They observed the band at 1384 ± 0.5 cm^{-1} but did not observe a band at 820 cm^{-1} (see later in the text).

and x-ray diffraction of the residues obtained by evaporation of the extracts, and states that the evidence showed the presence of sodium and calcium nitrates. O'Gorman and Walker claim to have found 1.15% nitrate (dry coal basis) from one lignite, by an unspecified method. Both sets of authors admit that they do not know how much additional nitrate is generated in the LTA procedure.

Subsequently, Fowkes (1972) agreed that the formation of nitrate salts during LTA production was very likely, but, on the basis of evidence presented, emphatically denied that fresh North Dakota lignites contained any nitrate.

A survey by Robert N. Miller of the infrared spectra of the LTA from some 400 coals recorded by Gong (1977) showed that an intense sharp band at 1380 cm^{-1} is common for lignites but is not seen at all for coals of bituminous rank. Hamrin *et al.* (1978)† have determined oxygen and nitrogen by fast neutron activation analysis in six coals (including one lignite) and in the LTA from them. These authors report nitrogen contents of the LTA from the bituminous coals that correspond to 0.04–0.20% N expressed on a whole dry coal basis (cf. 0.29% for the lignite). This is the only work on record that we know of that suggests the fixation of nitrate by low temperature ashing of bituminous coals.

This confusing situation has been dealt with in some detail because it bears on the reliability of low temperature ashing as a means of determining the mineral matter content of coals. There is urgent need for a definitive study in which any alleged nitrate is studied quantitatively.

If nitrate is indeed formed during ashing of coals, it is presumably by a route similar to that discussed earlier for fixation of sulfate from organic sulfur; i.e., oxides of nitrogen are formed, and may be trapped in the LTA if effective trapping agents are present, the products of decomposition of carboxylate salts in lignites being the most efficient of such agents. If, as the data of Hamrin and associates suggest, nitrate is fixed in the LTA from bituminous coals, carbonates are presumably the trapping agents and 1 mole of CO_2 is expelled for every two nitrogen atoms added. On this basis, we have calculated from Hamrin's data that the LTA, from this cause alone, would be higher than the true mineral matter content by amounts ranging from 0.09 to 0.45% of dry coal.

C. Other Data Needed

The data considered in this section are concerned with CO_2 from carbonates, forms of sulfur, sulfate in ash, and chlorine. All of these are

† The authors are indebted to Dr. Hamrin for a prepublication viewing of the text he has submitted for publication in *Fuel*.

needed for the KMC formula for mineral matter content. However, even if this formula is not used as such, it provides the basis from which are derived the best available means of correcting total hydrogen for the hydrogen in the water of decomposition of the clays [Eq. (3) in Section III], and volatile matter for a variety of interferences [Eq. (6) in Section IV,D]. Procedures for determining all four parameters are described in various ASTM standards, and are not discussed extensively here, although some comments are offered.

1. CO_2 *from Carbonates*

This figure is needed to derive organic carbon from total carbon, as well as in the KMC formula. The continuous titration method of determining CO_2 from coals, introduced by Pringle (1963), seems not to have been adopted as a national or international standard, but it has sufficient accuracy, and is simple and rapid.

2. *Forms of Sulfur*

It should by now be obvious that quantitative knowledge of the forms of sulfur in coals is very important for a number of reasons.† Sulfur as sulfate is usually very low in coals (0.01–0.05%, dry basis); if it is higher, natural weathering, or oxidation of pyrite during storage of the sample, should be suspected.

No method is yet established for the direct determination of organic sulfur in coals. In the standard procedure, sulfur as sulfate is determined gravimetrically in a hydrochloric acid extract of the coal. Determination of pyritic sulfur calls for extraction of the coal (a) with hydrochloric acid to remove acid-soluble iron, (b) with nitric acid to oxidize pyrite to a soluble form. In principle one could perform the extractions on separate samples of coal, determine the iron in each, and subtract the former from the latter (the "simultaneous" procedure), or alternatively the washed residue from the HCl extraction can be further extracted with nitric acid, in which case analyses are performed on a single sample (the "nonsimultaneous" procedure). Edwards *et al.* (1958) showed clearly that the first alternative can lead to error with some coals (up to 0.4% difference), so that the second is generally preferable; yet the relevant ASTM standard still offers the two approaches as alternatives, and some of the

† Some procedures for determining forms of sulfur in coal are given in Volume I, Chapter 9, Section V.

controversies in the literature on the validity of sulfur analyses must arise because authors persist in using the simultaneous method. In either case, pyritic iron is determined on the nitric acid extract, and pyritic sulfur is calculated from this assuming the stoichiometry FeS_2 (iron is determined in preference to sulfur, because of the possibility that the nitric acid might solubilize some organic sulfur as sulfate).

Edwards *et al.* (1964) were the first to study petrographically the insoluble residues from the nitric acid extraction of coals and to claim that some small pyrite particles, encased in organic matter, had escaped extraction. These authors studied a number of coals of different rank, and stated that the effect is significant, though small (0.1–0.2%), only with coals of relatively high bituminous rank ground to pass a 72-mesh sieve; if −300-mesh coal is used, the effect was insignificant, with coals of any rank. Since the paper of Edwards *et al.* (1964) was published, the question has become a matter of controversy (Brown *et al.*, 1964; Mayland, 1966; Young and Zawadski, 1967; James and Severn, 1967; Neavel, 1966).

The controversy appeared to have been settled, at least for a set of nine Illinois coals, by the work of Kuhn *et al.* (1973) and Shimp *et al.* (1974). They determined the pyrite content of the coals by the ASTM procedure (D 2492) using −60-mesh coal. The authors determined total Fe and S contents in the samples by x-ray fluorescence, and these were essentially equal to the sums of the data obtained in standard methods. They also treated the HCl-extracted coals with lithium aluminum hydride in an organic solvent; this, plus a treatment of the reduced product with dilute HCl, converts pyrite to ferrous chloride and H_2S. Provided the sample was ground to −400 mesh before treatment, the iron solubilized by reduction was essentially equal to that released by oxidation, so that the material balance was again good. The clear implication is that a negligible amount of pyrite remained unmeasured.

Incidentally, the $LiAlH_4$ approach offers the possibility of checking whether the stoichiometry of pyrite in coals is in fact precisely FeS_2, since both Fe^{2+} and H_2S could be measured in the products. Also, the method offers the nearest approach yet available to a direct determination of organic sulfur. The reduction reaction is not at all likely to solubilize organic sulfur. Therefore the total sulfur content of the residue from the $LiAlH_4$ treatment should closely approximate the organic sulfur content. Neither of these interesting approaches appears to have been explored so far.

In the course of their studies aimed at establishing LTA as a valid routine method for the determination of mineral matter contents of coals, Miller *et al.* (1978), using the standard ASTM procedures but with an atomic absorption finish for pyritic iron, determined the pyritic sulfur contents of four coals (HVC, Colorado; HVB, Illinois; HVA, Illinois; and HVA, Kentucky), and of the low temperature ash from them. Iron was determined in HCl extracts as well as in the HNO_3 extracts. The materials balance of iron between extracts of coal and of ash was remarkably close (the pairs of sums were 0.41, 0.43; 2.36, 2.35; 3.90, 3.90; and 0.50, 0.50%, respectively). Hence no significant amount of iron was extracted from the LTA but not the coal (accessibility of pyrite to the reagent should not be a problem with LTA). With 11 other coals of pyritic sulfur content ranging from 0.02 to 3.4%, pyritic sulfur contents in LTA (expressed on the dry coal basis) were less than pyritic sulfur in the coal by amounts ranging from 0 to 0.16%.

Given and Miller (1971) found that the total sulfur contents of some mangrove peat cores from the coastal plains of the Florida Everglades ranged up to 6% (dry basis). At some levels in the cores, pyritic sulfur accounted for a surprisingly small fraction of total sulfur. C. P. Dolsen (1971) made a petrographic examination of the solid residues from the nitric acid extractions of two samples, and reported that the residual pyrite content appeared to be high. Yet HCl extracts of the ashes from the samples showed iron contents barely detectable by atomic absorption spectroscopy. The experience suggests that those who, from petrographic studies, allege that pyrite is incompletely extracted by nitric acid should check their findings by having the residual iron content determined chemically.

Levinson and Jacobs (1977) again raise this point, but also raise a somewhat different one. On the basis of a Mössbauer spectroscopic study of one coal (described as "Pittsburgh seam West Virginia coal"), they conclude that iron is present only as pyrite. Hence they assume, apparently, that the same is true of three other West Virginia coals. With this assumption the Mössbauer data are interpreted to mean that 4–20% of the pyrite (of "extremely fine particle size") is solubilized by the HCl extraction in the standard ASTM analysis, and 6–11% is left unextracted by the nitric acid treatment.

That interpretation of the ^{57}Fe Mössbauer spectra of coals is less simple than is implied by Levinson and Jacobs is suggested by the conclusions of Gary (1977), who found the spectra of a coal difficult to interpret, at least three forms of iron apparently being present.

Any contribution that Mössbauer spectroscopy can make to the understanding of forms of iron (and hence sulfur) in coals will obviously be welcome. But any such contribution must be based on a study of an adequate sample base, and it must be recognized, in this context as in all others, that coal is not a clean system such that findings with pure substances can readily be extrapolated.

Sutherland (1975) has suggested that organic sulfur could be determined in coals by means of the electron microprobe. The approach is interesting, but it cannot be said that a procedure has yet been set up and validated.

3. *Sulfate in Ash*

It appears that in processes ranging from the combustion of coal in power station boilers (Watt, 1969) to the laboratory determination of either high or low temperature ash, some of the sulfur in coals is retained in the ash as sulfate. Thus the sulfate in ash is invariably greater than the sulfate in the original coal, both contents being expressed as fractions by weight of whole coal. This effect is so large with the lower rank lignites that the ash yield may actually be greater than the mineral matter content. However, the recent study of Kiss and King (1977) shows that the fraction of organic sulfur in Australian brown coals that is retained in the ash as sulfate can range from 0 to 99%.

As noted earlier in connection with low temperature ashing of low rank coals, it is the products of thermal decomposition of carboxylate salts that are particularly efficient in trapping organic sulfur as sulfate in ash. With higher rank coals that do not contain carboxyl groups, it is carbonates, or the oxides formed from them by pyrolysis, that tend to fix sulfur as sulfate (Watt, 1969).

From the foregoing, it should be clear that the amount of sulfate in ash depends on both the sulfur content of the coal and the concentration and nature of the materials capable of fixing it during ashing.

The various national and international standards for coal analysis contain procedures for determining sulfate in ash, and no further comment on experimentation is needed here.

4. *Chlorine*

As a consequence of their rather unusual postdepositional history, the coals of the highly productive East Midlands coal field in the United Kingdom may have chlorine contents as high as 1%. Therefore, coal analysis in the United Kingdom has had to concern itself with this element. Experiment showed (Daybell and Pringle, 1958) that by no means could all of the chlorine be accounted for in minerals such as halite (NaCl) or sylvite (KCl), and that on the average, 50% must be assumed to be associated with organic matter, probably as hydrochlorides of pyridine bases.

Gluskoter and Ruch (1971) have studied the chlorine content of 35 coal samples from 27 mines and 5 seams in the Illinois Basin. The values reported range from 0.006 to 0.36%, many lying in the interval 0.05–0.15%. The fraction of the chlorine extractable by water was mostly in the range 30–35%. Extremely small amounts of water-soluble potassium were observed, so that sylvite (KCl) is not a significant mode of occurrence of chlorine. The mole ratio Na/Cl was found to be widely variable, and in roughly half the samples, chlorine was more, or much more, than equivalent to the sodium found. It was concluded that by no means could all of the chlorine be accounted for in minerals, and therefore that a substantial proportion is associated with the organic matter.†

The mean chlorine content of 444 U.S. coals in the Penn State Data Base‡ is shown in Table III to be 0.058%. Since the coals of the Interior Province, which includes the Illinois Basin, accumulated under saline conditions, and in some areas have subsequently been exposed to saline groundwaters, a mean is shown separately for 72 coals from this province, higher than that for the whole set. As shown in Figs. 1 and 2, distributions are highly skewed, most of the coals from any province having chlorine contents less than 0.05%. The principal difference between the population at large and the set restricted to the Interior Province is that if the latter is excluded, no coals have more than 0.25% Cl, whereas 15% of the population from that province have chlorine contents greater than 0.25%.

Thus chlorine in U.S. coals does not present any major problem in the interpretation of coal analyses, although it certainly should not be ignored. Until more data are available, it will be best to assume a 50–50 distribution between organic and inorganic forms, as with British coals.

TABLE III *Chlorine Contents of Coals in Penn State Data Base*[a]

Samples included	Number of samples in set	Mean Cl (%)	Standard deviation	Skewness of distribution
All samples	444	0.058	0.068	2.2
Interior Province only	72	0.103	0.113	1.4
All samples less Interior Province	372	0.050	0.051	1.4

[a] Dry basis.

† Methods of determining chlorine in coal are given in Volume I, Chapter 10.

‡ Assembled under the direction of Dr. William Spackman with support from the U.S. Department of Energy.

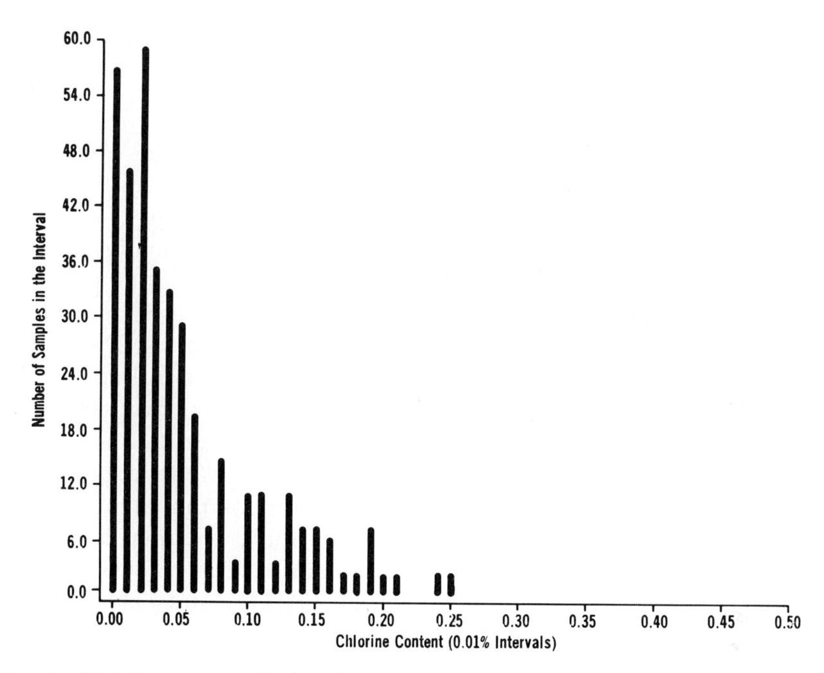

Fig. 1 Distribution of chlorine in 372 U.S. coals excluding those of the Interior Province.

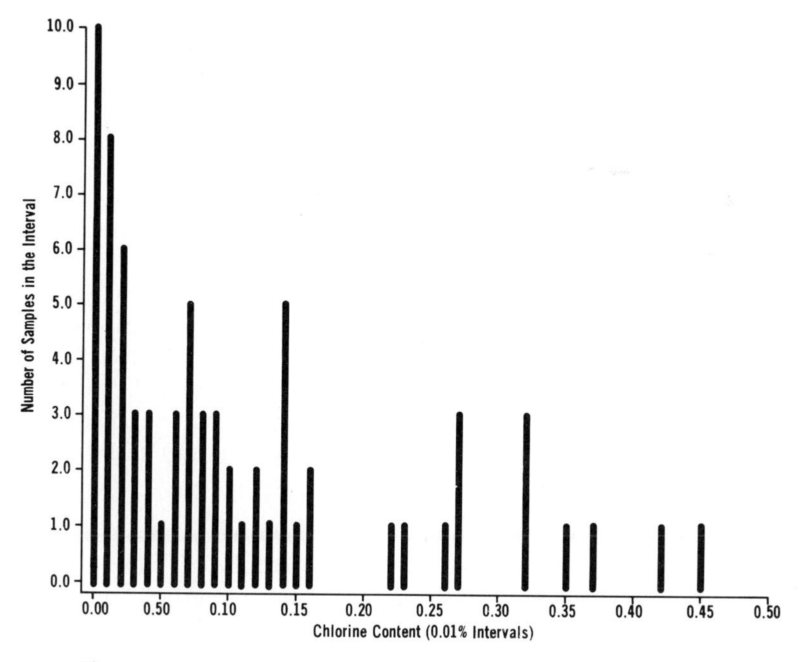

Fig. 2 Distribution of chlorine in coals of the Interior Province.

D. Special Problems Posed by Low Rank Coals

The function of this section is chiefly to gather together points made earlier about lignites and subbituminous coals. These low rank coals contain carboxyl groups, the amounts being in the range 1–4 meq/g. In coals as mined, a variable proportion of the carboxylic acid groups exist as calcium, sodium, or other salts. Hence a substantial proportion (~50%) of the high or low temperature ash from lignites is derived from the cations in these carboxylates rather than from minerals. The pyrolytic decomposition products of these carboxylates are evidently particularly efficient in trapping sulfur as sulfate (and nitrogen as nitrate?).

In these circumstances, the Parr formula, or the modified version of it proposed by the authors, cannot pretend to be able to cope with the situation. The KMC formula for calculating mineral matter content from high temperature ash yield does contain a term designed to allow for fixation of sulfur as sulfate during ashing, and it does assist in obtaining more plausible analyses of low rank coals (Tarpley and Ode, 1959), but the peculiarities of lignites were not considered in its conception for British bituminous coals.

Kinson and Belcher (1975), as noted earlier, assert that the direct determination of mineral matter content by acid demineralization is not applicable to low rank coals. Because of the almost complete fixation of organic sulfur in low rank coals during low temperature ashing, this procedure also is not applicable to the determination of the total inorganic content of low rank coals. The best hope of obtaining valid data for the total inorganic contents of lignites would seem to lie in proceeding as follows:†

(i) Determine the weight loss when the sample is stirred for 1 hr with 1 *N* HCl at room temperature.
(ii) Determine residual chlorine not removed by washing, using the Eschka procedure (whether this step is really necessary will have to be decided by experiment).
(iii) Perform low temperature ashing on the extracted coal using the standard conditions referred to earlier, and note the weight of the LTA.
(iv) Treat the LTA with dilute acid to remove any carbonates, and determine any residual carbon left.

On the evidence now available, this procedure is suggested for use with lignites, subbituminous coals, and coals of high volatile C bituminous rank from the western provinces. Whether fixation of nitrate is a problem must remain an open question for the present.

Kiss and King (1977) have considered how the composition of the pure organic matter in low rank coals can best be expressed. On the

† See also note added in press, p. 41.

grounds that a major part of the ash-forming constituents is bound directly to organic matter rather than in discrete mineral phases, these authors conclude that "the mineral-matter-free basis of expressing the results of ultimate analyses can be just as misleading as expressing results on the dry, ash-free basis." The truth of this obviously depends on the significance attached to the phrase "mineral matter content." Kiss and King evidently assume that it is a quantity derived by calculation from high temperature ash, and, on this assumption, they are correct. In this chapter we tend to treat as synonymous the phrases "total inorganic content" and mineral matter content (i.e., we treat mineral matter as consisting not only of minerals). We finally recommend the procedure just listed, which makes no use of high temperature ash, for determining the best approximation to the total inorganic content, and we still maintain that use of this will give valid dmmf analyses.

Kiss and King proceed to suggest an alternative way of expressing the total inorganic content of low rank coals, which may well be of value for a number of purposes.

V. OPTIMUM AND APPROXIMATE PROCEDURES FOR COMPUTING dmmf ANALYSES

A. Ultimate Analyses

To recapitulate, we have seen that derivation of a dmmf or dry, mineral-containing analysis, all feasible corrections being made, requires that the following data be available: C, H, N, Cl, forms of sulfur, sulfate in ash, mineral matter, and CO_2 from carbonates. The way in which these data are used was set out in Section III as a means of defining a dmmf elemental (or ultimate) analysis, and need not be repeated here. Some of the steps, however, require comment.

Equation (2) is exact as a correction for the contribution of carbonate CO_2 to organic carbon, and the precision is dependent solely on the precision of the experimental measurements. Equation (3), designed to eliminate the effect of the water of decomposition of clays on organic hydrogen, is based on the KMC formula for calculating a mineral matter content from an ash yield, and therefore is not exact and is dependent on the validity of the assumptions on which the formula was based. However, the correction is usually quite small. Equation (3), as quoted earlier from British Standard 1016, Part 16 (British Standards Institution, 1971), stated:

$$H_o = H_{tot} - 0.014\text{Ash} + 0.02S_p + 0.02CO_2 \tag{3}$$

The equation was originally derived by Leighton and Tomlinson (1960) in a somewhat more elaborate form

$$H_o = H_{tot} - 0.014Ash + 0.018S_p + 0.019CO_2 + 0.014SO_3 \quad (3a)$$

The last term represents sulfate in ash, calculated as SO_3 per 100 parts by weight of dry coal. For British coals this rarely exceeds 0.5%, and so the British standard omits the whole term as having a negligible effect on the correction. The ash from some U.S. coals contains sulfate equivalent to considerably more SO_3 than 0.5%, so that the term is not necessarily negligible. Suppose a coal of rather high mineral matter content affords 15% ash, has 2% pyritic sulfur, and yields 4% CO_2, and the ash contains sulfate equivalent to 1.25% SO_3. The algebraic sum of the correction terms in Eq. (3) is −0.09, or −0.08 if one uses Eq. (3a) (−0.098 if the term in SO_3 is omitted).

The hydrogen content of coals is around 5%, so that the corrections just calculated are nearly 2% of the value, and are larger than the precision of the experimental determinations. With many coals, the corrections will be less than those calculated, and the whole matter may appear trivial. Nevertheless, this correction has a considerable influence on the outcome of the Mott–Spooner test (see later).

The validity of the standard methods of determining pyritic sulfur was debated earlier. It should be pointed out that the outcome of this debate has an effect on the validity of organic sulfur determined by difference [Eq. (4)].

Taking account of all these considerations, Kinson and Belcher (1975) have estimated that oxygen-by-difference in a dmmf analysis is accurate to ±0.3% for a coal of mineral matter content of 10–15% (directly determined), whereas with a calculated mineral matter content the value will be accurate to ±0.5%, but with daf analyses there will be a bias of up to +2% (the present authors are convinced that the daf analyses of some coals can show a bias considerable greater than this).

Acquisition of all the data called for in the computation of a dmmf elemental analysis is admittedly a formidable task. The British Standard 1016, Part 16, in the passage already quoted, takes the view that a dmmf analysis involving approximations and assumptions will be a better guide to the composition of the organic matter than a daf analysis, and we are emphatically in agreement with this position. The various approximations for needed data offered in the British standard are well worthy of consideration, but it should be borne in mind that they are based on statistical correlations between the properties of British coals, which are all of Carboniferous age and of bituminous or higher rank. Their value with the greater diversity of U.S. coals needs testing.

The approximate mineral matter content of coals can be calculated by means of the Parr formula, or by use of the modified formula given in Eq. (9). A statistical comparison of the mineral matter contents of 274 coals calculated by means of the formulas showed that the values do not differ significantly, in spite of the more scientific (or less unscientific) basis of the modified formula (Given and Yarzab, 1975). However, the modified formula lent itself readily to some further approximations, which have proved significant.

Suppose that direct determination of the CO_2 yield is not available. Let us assume that carbonate minerals in coals have a mean molecular weight equal to that of calcium carbonate and that they represent 10% of total mineral matter (as indicated by a study of 60 U.S. coals; Miller and Given, 1972). Then a combination of this assumption with Eqs. (2) and (9) affords

$$C_o = C_{tot} - (0.014\text{Ash} + 0.0055S_p) \quad (11)$$

Similarly, Eq. (3) is simplified to

$$H_o = H_{tot} - (0.013\text{Ash} - 0.02S_p) \quad (12)$$

These equations require fewer analytical data than those presented earlier. Are they valid?

The consistency of dmmf elemental analyses of coals, whether obtained with the utmost rigor or with the aid of approximations, can be very usefully tested by means of a semiempirical formula proposed by Mott and Spooner (1938). This formula permits the calculation of the calorific value (CV) of a coal from its elemental analysis, all data being on the dmmf basis. There are two versions of the formula:

For oxygen contents less than 15%:

$$\text{CV} = 144.5\text{C} + 610.2\text{H} + 40.5S_o - 62.46O_{diff} \quad (13a)$$

For oxygen contents greater than 15%:

$$\text{CV} = 144.5\text{C} + 610.2\text{H} + 40.5S_o - (65.88 - 0.31O_{diff})O_{diff} \quad (13b)$$

For a large number of analyses, the mean difference, calculated *minus* the observed CV, is zero (Mott and Spooner, 1938; Neavel, 1974; Given and Yarzab, 1975). It should be noted that a comparison of observed and calculated calorific values involves a major portion of the basic coal analyses: C, H, N, forms of sulfur, moisture, mineral matter content, and calorific value (the value for N affects O by difference, and S_p affects S_o by difference).

Table IV shows that for 265 coals from the Penn State Coal Data Base, the mean difference between observed and Mott–Spooner calculated

TABLE IV *Comparison of Experimental and Mott–Spooner Calorific Values*[a]

Mineral matter basis	Mean Mott–Spooner difference	Standard deviation of differences	95% confidence interval on mean
Parr	−55	206	±25
Modified Parr	+26	182	±27

[a] For 265 coals.

calorific values is significantly less if the modified Parr formula, *plus* Eqs. (11) and (12), is used compared with the situation in which the original Parr formula (alone) is applied. Moreover, in the first case the mean is not significantly different from zero, whereas in the second it is; in addition, the Parr basis shows a significant bias, whereas the modified Parr does not.

That the modified Parr basis gives a better account of the composition of the organic matter is not due to better estimation of the mineral matter content, as we have seen. Nor is it *primarily* due to the inclusion of organic sulfur, with its consequent effect on oxygen-by-difference. Careful examination of the data shows that the better performance of the modified Parr basis is mostly due to its approximate corrections to organic hydrogen and carbon, particularly the former, thus demonstrating the significance of apparently trivial corrections to organic hydrogen.

Therefore, we feel justified in offering Eqs. (9), (11), and (12) as useful and reasonably valid approximations for use when the full set of data for computing dmmf analyses is not available.

B. Volatile Matter

The principal use of the volatile matter yield is in the classification of coals by rank, for which purpose the means of correction for the effects of mineral matter (i.e., the Parr formula) is specified (ASTM, 1976) and has legal force in litigation.

The volatile matter yield from a coal is a function of sample size, particle size, rate of heating, and maximum temperature reached; i.e., it is a parameter to be determined under defined conditions and does not purport to represent composition in any manner recognizable to a chemist. Therefore, meticulous attention to eliminating the effects of mineral matter on the numerical value obtained will usually not be worthwhile.

However, the volatile matter yield is sometimes of value in studies of the systematics of high rank coals, and also in laboratory studies of combustion and gasification mechanisms. For these purposes it may be worthwhile to eliminate as far as possible the effects of the presence of mineral matter on the volatile matter (VM) yield. The best available means of performing this is the equation due to Leighton and Tomlinson (1960):

$$VM_0 = VM_{det} - (0.13Ash + 0.2S_p + 0.7CO_2 + 0.7Cl - 0.20) \quad (14)$$

It should be noted that having applied this correction, the result must be multiplied by 100/(100 − MM) to obtain a VM on the dmmf basis. The correction made by Eq. (14) alone can decrease the apparent VM yield by as much as 4%. Fixed carbon is simply (100 − VM). The ASTM system is unique among national codes for nominally using fixed carbon for classification of coals by rank. It is the volatile matter that contains the decomposition products of the minerals. The use of fixed carbon is, we feel, regrettable because it suggests to the unwary that it is in some sense a chemical form of carbon that exists as such in the coal.

C. Calorific Value

In the ASTM and other systems, the calorific value is used in the classification of coals of rank lower than medium volatile bituminous.† For this purpose the value is put on what is described as the "moist mineral-matter-free basis." The Parr formula is used in the ASTM system to convert ash to mineral matter. "Moist" should mean that the result is calculated on the basis of the coal containing its natural bed moisture or the amount of water determined in the equilibrium moisture test; however, it appears to be determined in most laboratories on the "as received" basis. The rationale for using the calorific value of a moisture-containing coal for classification seems to be that the moisture-holding capacity of a coal is itself a rank parameter, decreasing with increasing rank. Thus calorific values of moist coals depend on two different properties and might therefore be supposed to measure better for this reason the degree of metamorphism. However, for this reasoning to be sound, the sample should be in some standard condition of moistness. At the same time, this may not matter unless a sample is at the lower end of a rank class.

On the other hand, if the calorific value is to be used in research directly as a numerical parameter, for example as a measure of the degree of metamorphism against which another property is to be plot-

† For details on methods for determination of calorific value see Volume I, Chapter 6, Section V.

ted, then all samples should be either dry or in the equilibrium moisture condition, and a better dmmf or mineral-containing basis used.

The data so far discussed are known as "gross calorific values," and refer to the combustion process carried out at constant volume, with products at 25–40°C and water in the liquid state. The "net" calorific value can be calculated as follows:

$$\text{net CV} = \text{gross CV(dmmf)} - \frac{9270 \times H_{dry}}{100 - MM} \tag{15}$$

where H_{dry} is the H content on the dry, mineral-containing basis. The net value refers to the process performed at a constant pressure of 1 atm with products at 25°C and water in the vapor state. Thus the net value is more compatible with thermochemical data as usually recorded.

In calculating the calorific value on a mineral-matter-free basis, the heat of combustion of pyrite must be subtracted. When the Parr formula is used, one subtracts $50S_{total}$ from the determined heat; when another basis is used, in which the presence of organic sulfur is admitted, the correction is $54S_{pyr}$ (in both cases the units are Btu/lb).

VI. THE PROBLEM OF ORGANIC OXYGEN

One of the characteristic differences between coal and petroleum is that most coals contain much more oxygen. Moreover, the oxygen is present in coals in various functional groups which play important roles in the chemistry and geochemistry of the fuel. Thus part of the oxygen is phenolic. As a class phenols are reactive in a number of chemical processes, including oxidation. The oxidation of coals may take place through natural weathering, or because the coals are deliberately treated with an oxidizing agent in structural studies, or as an unwanted side reaction in other reactions. In hydrogenation reactions some of the oxygen will be converted to water or oxides of carbon. The phenolic OH group can form chelate complexes with many metals if another suitable group is adjacent.

Thus for many reasons one often wishes to know the oxygen content of the hypothetical pure organic substance of coals and their reaction products. Yet of all the basic analytical data needed to characterize coals, organic oxygen is by far the most difficult to get satisfactory values for. Procedures for minimizing errors in oxygen-by-difference were discussed earlier. These eliminate various perturbations due to the presence of mineral matter, in a more or less satisfactory way, but the oxygen still accumulates the algebraic sum of the errors in the direct determinations. It is the purpose of this section to discuss methods of determining oxygen directly.

Methods of determining oxygen in organic compounds have been known for many years (e.g., Unterzaücher, 1952). They were applied on a routine basis in the 1950s in the Central Research Laboratory of the Dutch State Mines and in the laboratories of the British Coal Utilisation Research Association. There is a current International Standard (International Organization for Standardization, 1971) for the direct determination of oxygen in coals. However, the Standard specifies that if the ash yield is more than 5%, the coal shall be subjected to acid demineralization before analysis.

The method depends on pyrolysis of the sample to an oxygen-free char in a stream of nitrogen at 900–1100°C. The oxygen is released as CO, which is oxidized to CO_2, and this is determined by titration. Oxygen in CO_2 from carbonates, in the water of decomposition of clay minerals, and in sulfates, is determined as well as organic oxygen, and this causes serious difficulty. Carbonates and sulfate can be determined directly for correcting the determined oxygen. An approximate correction for oxygen in the clay water, similar to that for hydrogen [Eq. (3)], could be worked out, but the magnitude of the correction would be eight times larger, and the necessary approximations in the formula could no longer be tolerated. In principle the clay water could be determined directly for each sample as the water released by pyrolysis of the low temperature ash.

Thus organic oxygen in coals can be determined by the modified Unterzaucher route, but with a very considerable investment of time in each determination, whether one seeks to correct the raw data for interferences or removes the need by prior demineralization of the sample.

Kinson and Belcher (1975) have made an important contribution by considering another approach: the use of radio frequency heating to achieve pyrolysis temperatures of 1950°C. At this temperature, all of the oxygen in the coal, organic and inorganic, is released as CO. Kinson and Belcher then carefully consider how organic oxygen may be derived from total oxygen. One procedure is to perform the determination on demineralized coal. A demineralized sample still contains pyrite and a small amount of residual silicate material, and the authors give an equation by which dmmf organic oxygen can be derived from the oxygen content of demineralized coal.

The authors also consider a number of ways by which dmmf organic oxygen can be derived from total oxygen without prior demineralization. They give preference for reliability to a calculation of inorganic oxygen from "a complete elemental and phase analysis of mineral matter." As second choice they give a formula that requires knowledge of the water of decomposition of the clay minerals. Third, inorganic oxygen may be approximated "for well-known seams of low mineral matter" by assuming it equal to one-half the mineral matter content. Comparative analyses for organic oxygen are shown for six coals. Here again is an approach that can be successful, but at a heavy cost in time and effort.

Two groups of workers have explored the use of fast neutron activation for the determination of oxygen and nitrogen in coals. This method gives the total oxygen content, and again one is confronted with the question, how may organic oxygen be derived from this? Volborth *et al.* (1977a) concede that it would be useful to have the oxygen contents of the low temperature ash of samples studied, but evidently did not have access to the technique. Accordingly, they equated the mean oxygen content of the high temperature ashes of eight coals with the inorganic oxygen contents of each coal. This is totally inadmissible.

Hamrin *et al.* (1975, 1978) and James *et al.* (1976) made creditable efforts to derive organic oxygen from total oxygen as determined by fast neutron activation analysis. They explored two approaches. In the first they determined the oxygen content of the LTA from six coals (using 24 W radio frequency power per channel, at which fixation of organic sulfur as sulfate would have been relatively serious), and equated this to total inorganic oxygen. In the second they determined the oxygen content of the coals demineralized by treatment first with HF and then HCl (without a first extraction with HCl; they are likely to have formed some calcium fluoride not extractable by the final HCl wash, as the authors evidently realized *post facto*). This second approach gave oxygen and nitrogen contents appreciably higher than the first. The authors considered whether this might be due to interference by fluorine (from calcium fluoride) with both determinations, but concluded that it could not.

Among the coals studied by the second approach were five coals provided by the Penn State laboratories. The oxygen contents found by Hamrin and co-workers were higher by 2–5% than those obtained by difference using the procedures discussed in Section V,A. It would follow that the carbon contents in the Penn State analyses are in error by large amounts and that the results of the Mott–Spooner test were totally spurious and fortuitous, both of which propositions appear to us improbable. We are unable to resolve the conflict at this time.

It may well be that fast neutron activation analysis can make important contributions to coal analysis, and in particular to the determination of organic oxygen. In order that it shall do so, more detailed attention to interferences is needed, and also serious and informed consideration, such as that given by Kinson and Belcher (1975) in a parallel situation, must be devoted to the problem of removing the contribution by inorganic oxygen, due attention being given to the differences between coals of low and higher rank.

It would appear that the simplest and most direct way to obtain an inorganic oxygen content of a coal, for use in correcting total oxygen, is to determine the oxygen content of the LTA, and we are puzzled that Kinson and Belcher (1975) do not consider this. It may be that they are concerned with the effect of fixation of sulfate during ashing, of which they are certainly aware. However, we feel that if the optimum conditions for ashing prescribed by Miller *et al.* (1978) are followed, and provided that low rank coals are not in question, this effect will be small.

Any analyst concerned with complex materials is anxious that all determinations should be direct, so that the summation of all terms to 100% can be tested. It cannot be too strongly emphasized that unless or until spectacular advances in such techniques as ESCA (electron spectroscopy for chemical analysis) take place, no such summation is possible with the elemental analyses of coals. References to materials balances in coal analyses and their stoichiometry (Volborth *et al.*, 1977a,b) simply misuse these terms. Organic oxygen contents derived from fast neutron activation are still oxygen-by-difference. Admittedly, the difference involves fewer terms than the procedures discussed in Sections III and IV, and therefore should be more reliable, but this is not the point the authors make.

A materials balance, properly so-called, for a coal analysis would be of the following type:

$$\text{Coal} = C_{tot} + H_{tot} + N + S_{tot} + O_{tot} + Ca + Mg + Fe + Mn + Ti + Si + Al + Na + K + Cl + \cdots \quad (16)$$

Since the *sum* of the minor and trace elements is usually a significant fraction of whole coal, 20–30 elements would have to be included in the summation. The algebraic sum of the errors in each determination would inevitably be such as to vitiate the whole materials balance effort.

In any case the materials balance test that really would be of significance, because it contains the terms the user of analyses needs to know, is

$$C_o + H_o + S_o + N + O_o = 100\% \quad (17)$$

(or the equivalent when mineral matter content is included). But at present there are no means of directly determining C_o, H_o, S_o, or O_o; therefore no valid and useful materials balance can be made for ultimate analyses. There is a way in which a true materials balance for total oxygen can be set up and used in studying liquefaction processes, and this is discussed in Section VII,C.

The principal conclusions of this section must be that (a) there *are* valid methods of determining organic oxygen in coals, (b) these methods involve many determinations of data needed to correct the raw results, and (c) new methods are in sight, but they are not yet established and will not be easy or straightforward. We regard as most significant the finding of Kinson and Belcher (1975) that oxygen-by-difference in a dmmf analysis, obtained by the optimum procedure discussed in Section V,A, with direct determination of mineral matter content, is accurate to $\pm 0.3\%$. It appears to us that this approach is still the least troublesome way to get an acceptable organic oxygen content for coals.

VII. PARTICIPATION BY MINERAL MATTER IN REACTIONS OF THE ORGANIC SUBSTANCE OF COALS

It is alleged that librarians are wont to think, "How efficiently I could run this library if I did not have readers to worry about!" Analytical data on coals are obtained not merely to enable the analyst to exercise his/her professional skill, nor simply to adorn tables, but (at least some of the time) to enable users to solve real problems. In our limited experience, it appears that only too often the research worker sends samples to the analyst, who performs the determinations called for and returns the results without comment, and the user does not take the trouble to find out whether what has been measured is what he really wants to know, or perhaps he is not clear at all as to what he really needs to know. Accordingly, the purpose of this concluding section is to alert users of analyses to hazards that may be anticipated and to indicate how the hazards may be avoided or eliminated.

A. Functional Group Determination

It is not the purpose of this chapter to provide a manual of methods for the determination of functional groups in coals. Methods are briefly described, but exhaustive coverage is not attempted. It should be noted that much of the relevant work was published before 1961. To some extent, comments on interferences by inorganic components are speculative, since very little consideration of the problem can be found in the literature. Methods of determining functional groups in coals were extensively reviewed by van Krevelen (1961), who did not discuss at all the possible role of mineral matter in the selective reactions used for the purpose.

Use of selective chemical reactions for the determination of functional groups in coals is necessarily based on the assumptions that (a) the reaction proceeds *only* by the pathway followed by known organic compounds containing the group in question, and (b) the reaction is essentially complete (in spite of the fact that coals are solids and their pores have very small entrances). Obviously the validity of these extravagant assumptions ought to be tested by study of the products of the "selective" reaction: but this has rarely been done.

Carboxyl contents of low rank coals are determined by barium acetate exchange, whereby all ions associated with the acid groups are replaced by ion exchange. Procedures have been discussed in several publications of which the most recent is due to Shafer (1970; but see also Maher and Shafer, 1976). The coal is demineralized prior to the determination, since clay minerals contain acidic OH groups also capable of ion exchange. Moreover, although Ba^{2+} will replace any group I or group II element that may be present, the exchange with H^+ is more rapid. The ion exchange capacity of clays is of similar magnitude to that of lignites, 1–4 meq/g. Since the content of clay minerals in lignites is usually less than 5%, their effect on ion exchange capacities will be small.

Many different methods of determining OH contents of coals have been tested (van Krevelen, 1961), but the procedure most commonly used has involved acetylation with acetic anhydride in pyridine and determination of the acetyl content of the product (Blom *et al.*, 1957). Blom and co-workers hydrolyzed the acetylated product and titrated the acetic acid released. Hill and Given (1969) replaced this tedious procedure by using ^{14}C-labeled acetic anhydride and determining the acetyl uptake radiochemically. Abdel-Baset *et al.* (1978) have modified this procedure further by combusting the product and absorbing the gas for liquid scintillation counting. Alternatively, the OH groups may be converted to their trimethylsilyl ether derivatives with hexamethyldisilazane in pyridine, and the increase in silicon content determined (Friedman *et al.*, 1961). With both procedures it has been shown that in the infrared spectra of the reaction products, the OH band has been almost completely eliminated and bands to be expected from the groups introduced do indeed appear. There is thus objective evidence that the two assumptions noted previously are valid in this case (Brown and Wyss, 1955; Friedman *et al.*, 1961; Duffy, 1967).

In a recent reexamination of the hydroxyl contents of 35 coals using the acetylation procedure (Abdel-Baset *et al.*, 1978), the potential role of mineral matter in the process was examined. Experiments showed that insignificant amounts of cations were leached from various pure clay minerals by pyridine at 95°C. Both kaolinite and montmorillonite were found to be acetylated to a measurable extent, but, having regard to the range of contents of these minerals in coals, the effect on the determined OH contents of the organic matter was judged to be negligible. On completion of the acetylation reaction, excess acetic anhydride is hydrolyzed; experiment indicated that the dilute acetic acid produced did not solubilize a significant amount of the mineral matter (it would have been a matter of concern if a significant amount had dissolved, since in calculating the OH content of the coal from the acetyl content of the product, one assumes that the mineral matter does not change in weight).

Thus a conclusion of this study was that the mineral matter in the coals has little or no effect on the hydroxyl contents of the organic matter as determined. The importance of the study, however, is not this conclusion, but that for the first time in the history of coal chemistry such questions had even been asked.

The OH groups in silicates can form trimethylsilyl ethers as well as acetates, but this effect also is presumably negligible. Indeed, there is no particular reason to suppose that mineral matter interferes with this determination, other than by forcing a determination by difference.

The content of OH in a coal can be expressed as a fraction of total organic matter or of the total oxygen content (of course, the content of any other oxygen functional group can be expressed in the same two ways, but they have been most frequently used with OH). If expressed on the first of these bases, then comparison of coals or correlations with a rank parameter will contain perturbations due to mineral matter unless expressed on the dmmf basis (g O as OH/100 g dmmf coal). If expressed on the second basis, the whole complex problem of determining the oxygen content of the pure organic substance arises.

In papers by one of the present authors (Given *et al.*, 1960, 1965) oxygen-by-difference was computed by procedures unknown to any national or international standard. Analyses were expressed on the daf basis, but the whole sulfur content was assumed to be organic. In some cases, direct oxygen determined by a modification of the Unterzaucher procedure was used, without correction. Exactly what workers in other European laboratories did cannot now be determined, but they probably did much the same. Workers in the U.S. Bureau of Mines laboratories in Pittsburgh in their classic studies of the organic chemistry of coals used daf analyses, including total sulfur, but they did not use direct oxygen determinations. As excuse for these enormities one can plead that most of the samples studied were handpicked vitrains of high vitrinite content (or concentrates of other macerals). Consequently ash yields were very low (usually 1–2%), and total sulfur was also low (0.6–1%), so that most of it would have been organic. In these circumstances falsification of oxygen-by-difference would have been small, though probably not negligible. Obviously, if one wishes to express the content of a functional group as a fraction of total oxygen, a dmmf value for the latter should be used.

Less work has been done on other oxygen functional groups. Blom *et al.* (1957) determined methoxyl by the Zeisel method (heating with concentrated HI, determination of CH_3I liberated), and carbonyl by reaction with hydroxylamine hydrochloride in pyridine, the result being calculated from nitrogen contents before and after hydrolysis of the oximes. One would guess that interferences from mineral matter were not serious in the Zeisel case (hot concentrated HI would certainly dissolve carbonates and leach elements from clays, and might solubilize pyrite, but these effects should not change the CH_3I yield). One step in the carbonyl determination involves refluxing with mineral acid.

B. Other Reactions

The rank of a coal is no doubt its characteristic of greatest practical importance. Therefore the consequences of the natural processes of metamorphism by which the rank of a coal is increased will always be a topic of interest among research workers. The various major coal basins are of different geological ages, so that different kinds of plant material served as input to coal formation. Moreover, typical temperature–time–pressure histories differ in the different coal basins because of differing rates of various geological processes of subsidence and uplift. A valuable approach to the understanding of the chemical consequences of metamorphism is to study the evolution of coal properties by correlating one property, such as carbon content or reflectance, with another, such as calorific value. Provided enough samples have been studied, one can then determine the probability that coals from different major coal basins constitute distinct populations.

In any such endeavor, it is essential to use dmmf data. If one does not, there will be an additional variance in the data arising from the fact that the effects of the presence of mineral matter on analyses *can vary widely from coal to coal.* The assumption that if data are all on the same (false) basis they will still be comparable is simply wrong, because the degree of falsity is not a constant.

In some chemical processes, carried out by human agency in the laboratory rather than by natural processes in the earth's crust, the situation will be similar: as long as the compositions of reactant and product are on a dmmf basis before comparison, a hazard will have been avoided. A more complicated situation arises when both organic and inorganic matter take part in the reaction. This is most likely to occur when the reaction process is an oxidation or reduction, carried out under acid or alkaline conditions.

Proceeding by induction, let us take a specific example (fictitious but plausible). Suppose a research chemist is studying the oxidation of coals in acidic conditions. The conditions are such that a substantial fraction of the coal is recovered as a solid oxidation product. It is desired to know how much of the organic substance of the coal is recovered in this form, and in particular how much oxygen has been added. (Calculation of the state of confusion of the researcher who tries to use daf analyses is left for any curious reader to make; the authors find the situation too horrific to contemplate.) We shall assume that the researcher is reasonably well informed, and obtains dmmf analyses of reactant and product as follows:

Coal: 82.00% C, 5.50% H, 1.6% N, 1.3% S, 9.6% O
Product: 75.00% C, 5.06% H, 1.61% N, 1.17% S, 16.50% O

These data tell the researcher, correctly, that some C, H, and S have been removed by the oxidation, and much O added. But they give a quite misleading impression of the quantitation, because some of the mineral matter will have been solubilized, and the mass of product will differ from the mass of coal taken for the experiment (a) because some organic matter will have been solubilized, and the mass of product will differ from the mass of coal taken for the experiment, (b) some oxygen has been added, and (c) some mineral matter has been removed. Hence one should compare, not percentage contents, but masses, as shown in Table V, where the figures correspond to the dmmf analyses just displayed and to mineral matter contents of 12.00% in raw coal and 8.26% in product.

The content of mineral matter in the product is decreased primarily because, under the conditions postulated, much of the pyrite and carbonate minerals will be solubilized (for simplicity, it is here assumed that these mineral phases are completely removed from the solid prod-

TABLE V *Comparison of Composition of a Hypothetical Coal and the Product of Its Oxidation in Aqueous Acid*[a]

Mass (g)	Coal	Product	Net loss	Net loss (%)
Total mass	5.000	4.782	0.218	4.4
C	3.608	3.320	0.288	8.0
H	0.242	0.222	0.020	8.3
N	0.071	0.071	0	0
S_{org}	0.057	0.051	0.006	10.5
O_{diff}	0.423	0.724	−0.301	−71.2
FeS_2	0.132	0	0.132	100
$CaCO_3$	0.060	0	0.060	100
Other mineral matter	0.408	0.395	0.013	3.2

[a] For normal dmmf analyses, see text.

uct). One would also expect some ions to be leached from the clay minerals and any hematite removed, hence the loss in weight of "other mineral matter." Thus the simplistic and unspoken assumption that mineral matter is an inert diluent carried through the process unchanged has no foundation in the case postulated. It is submitted that the presentation of data in Table V is the *only* way in which a just representation of the effect of the postulated reaction can be shown.

In practice it would have to be determined experimentally if all the pyrite has indeed been removed. Calculation of the mineral matter content of the coal may be acceptable, but with the reaction product direct determination will almost certainly be needed.

The preceding example was fictitious, but there are real cases in the literature in which the indicated procedure was not followed and the published results must be in error. Thus Brown *et al.* (1958) studied supposedly selective reactions for introducing nitro and sulfonic acid groups into coals. Both the extent of addition of these groups and of unwanted concomitant oxidation were reported, as deduced from a comparison of daf analyses of coal and product, assuming all sulfur to be organic. The fact that carbonates and perhaps pyrite and part of the clays would be solubilized in these processes was ignored, although it is very likely that the relation between mineral matter content and ash yield would thereby differ for coal and product.

The reduction of coals with metallic lithium in ethylene diamine was studied by Reggel *et al.* (1961, 1964). From comparisons of the ultimate analyses (daf, total sulfur included) of original coal and reduction product, it was concluded that some nitrogen had been added (presumably as irreversibly held diamine) and oxygen introduced (oxidation of the product?). Even though the ash yields and sulfur contents of the coals studied were low (1.5–5 and 0.3–1%, respectively), the conclusions can now be seen to be unacceptable, at least in any quantitative sense. Possible solubilization of pyrite by reduction was not considered.

Obviously considerations of the type just discussed will be of very varying significance according to the nature of the reaction process being studied and the objectives of the study. Knowledge of the types of mineral matter in coals and of their chemistry will provide a reliable guide as to whether problems should be anticipated. Trial calculations with estimated numbers will show how seriously the hazard should be taken.

C. Conversion Processes

Analytical problems arising in connection with the gasification and liquefaction of coals are reviewed elsewhere in this work. This section attempts to cover only some specific points related to what has been said in the rest of this chapter.

Many gasification processes produce a char as a by-product. Since this will have already been exposed to high temperatures, its mineral matter will present fewer problems than does that of coals. The discussion of dmmf analyses of cokes in British Standard 1016, Part 16 (British Standards Institution, 1971), will be helpful to those concerned with chars. Kinson and Belcher (1975) discuss direct determination of the oxygen content of cokes and chars.

The products of liquefaction processes present greater problems. Analysis of the liquid products is outside the scope of this chapter, but it is worth remarking that the manifold problems of determining oxygen directly in coals largely disappear in the filtered liquid products, and such determinations should be made. With regard to the solid residue, problems of two different kinds arise.

The matter has not been tested experimentally, but it seems unlikely that any mineral in coals other than pyrite will be changed under the conditions of liquefaction processing. Pyrite is reduced to the nonstoichiometric mineral pyrrhotite, $Fe_{1-x}S$, where $x = 0–0.2$. As a separate crystal phase, this mineral is soluble in acid, yielding H_2S, so that its determination would appear simple. However, a research worker studying kinetics and mechanisms might well have a need to determine pyrrhotite in a solid product formed when the level of conversion has deliberately been held low, so that pyrite may have been incompletely reduced. According to Barnes (1977), pyrite and hexagonal pyrrhotite form a series of solid solutions, some of which are insoluble in acids other than nitric; the composition has to be determined by x-ray diffraction. The tendency of pyrrhotite to be oxidized under the conditions of low temperature ashing is not known, so far as we are aware.

Thus sulfide minerals in solid residues from liquefaction may require special attention, but, apart from this, essentially all that has been said earlier about the effect of mineral matter on the analysis of the organic substance of coals applied also to liquefaction residues.

These remarks are aimed at the chemist seeking a basic understanding of the process. An engineer in process development is confronted with somewhat different problems, in particular with obtaining material balances to ensure that a continuous-flow reactor is properly lined out and that no losses of material are occurring. For these purposes, total contents of C, H, O in coal and solid residue are perfectly acceptable, the O being directly determined, for example, by fast neutron activation analysis. As noted earlier, there are ASTM standards specifying how a representative sample for analysis is to be obtained from the output of a mine or a coal preparation plant, and the procedures should be used for sampling the input to a coal conversion pilot plant. There are as yet no standards for obtaining representative samples of the solid and liquid output, but the importance of obtaining such samples should not be overlooked.

An ash materials balance has been used to check whether a continuous-flow reactor is fully lined out. An assumed high temperature ash materials balance has been used by Neavel (1976) in the calculation of conversion in coal liquefaction experiments; the assumption of no loss of mass is invalid with pyrite, as already seen, though the effect may be small with some coals. However, Shou and Pitts (1977) have objected to Neavel's procedure on the grounds that it must underestimate conversion since ash yield is nearly always less than mineral matter content, so that if the mineral matter content of the residue were 100%, the ash yield would be less, and liquefaction conversion could never achieve the theoretical maximum of 100%. Shou and Pitts recommend the use of LTA instead of high temperature ash.

ACKNOWLEDGMENTS

We are happy to acknowledge financial support for our researches under contracts from the Energy Research and Development Administration (now Department of Energy).

REFERENCES

Abdel-Baset, Z., Given, P. H., and Yarzab, R. F. (1978). *Fuel* **57,** 95.

ASTM (1976) "Annual Book of ASTM Standards," Part 26," "Gaseous Fuels; Coal and Coke; Atmospheric Analysis." Am. Soc. Test. Mater., Philadelphia, Pennsylvania.

Barnes, H. (1977). Personal communication, Pennsylvania State Univ., University Park, Pennsylvania.

Bishop, M., and Ward, D. L. (1958). *Fuel* **37,** 191.

Blom, L., Edelhausen, L., and van Krevelen, D. W. (1957). *Fuel* **36,** 135.

British Standards Institution (1971). "British Standard 1016: Methods for the Analysis and Testing of Coal and Coke," Part 16, "Reporting of Results." Br. Stand. Inst., London.

Brown, H. R., Durie, R. A., and Shafer, H. N. S. (1959). *Fuel* **38,** 295.

Brown, H. R., Durie, R. A., and Shafer, H. N. S. (1960). *Fuel* **39,** 59.

Brown, H. R., Burns, M. S., Durie, R. A., and Swaine, D. J. (1964). *Fuel* **43,** 409.

Brown, J. K., and Wyss, W. F. (1955). *Chem. Ind. (London)* p. 118.

Brown, J. K., Given, P. H., Lupton, V., and Wyss, W. F. (1958). *Proc. Residential Conf. Sci. Use Coal, Sheffield, Eng.* pp. A-43–A-47. Inst. Fuel, London.

Daybell, G. N., and Pringle, W. J. S. (1958). *Fuel* **37,** 283.

Dolsen, C. P. (1971). Unpublished data.

Duffy, L. J. (1967). Ph.D. Thesis, Pennsylvania State Univ., University Park, Pennsylvania.

Edwards, A. H., Daybell, G. N., and Pringle, W. J. S. (1958). *Fuel* **37,** 47.

Edwards, A. H., Jones, J. M., and Newcombe, W. (1964). *Fuel* **43,** 55.
Fowkes, W. W. (1972). *Fuel* **51,** 165.
Frazer, F. W., and Belcher, C. B. (1972). *Fuel* **53,** 41.
Friedman, S., Kaufman, M. L., Steiner, W. A., and Wender, I. (1961). *Fuel* **40,** 33.
Gary, J. H. (1977). Personal communication, Colorado School of Mines, Golden, Colorado.
Given, P. H., and Miller, R. N. (1971). *Annu. Meet. Geol. Soc. Am., Abstr. Programs* **3**(7), 580.
Given, P. H., and Yarzab, R. F. (1975). "Problems and Solutions in the Use of Coal Analyses," Tech. Rep. No. 1. Pennsylvania State Univ., University Park, Pennsylvania. (Report FE-0390-1 submitted to U.S. Energy Research and Development Administration under Contract No. E(49-18)-390.)
Given, P. H., and Yarzab, R. F. (1973). Unpublished data.
Given, P. H., Peover, M. E., and Wyss, W. F. (1960). *Fuel* **39,** 323.
Given, P. H., Peover, M. E., and Wyss, W. F. (1965). *Fuel* **44,** 425.
Given, P. H., Cronauer, D. C., Spackman, W., Lovell, H. L., Davis, A., and Biswas, B. (1975). *Fuel* **54,** 48.
Gluskoter, H. (1965). *Fuel* **44,** 285.
Gluskoter, H., and Ruch, R. R. (1971). *Fuel* **50,** 65.
Gong, H. (1977). Unpublished data.
Hamrin, C. E., Maa, P. S., Chyi, L. L., and Ehman, W. D. (1975). *Fuel* **54,** 70.
Hamrin, C. E., Johannes, A. H., James, W. D., Sun, G. H., and Ehman, W. D. (1978). *Fuel* (in press).
Hill, L. W., and Given, P. H. (1969). *Carbon,* **7,** 649.
International Organization for Standardization (1971). Recommendation R1994.
James, R. G., and Severn, M. I. (1967). *Fuel* **46,** 476.
James, W. D., Ehman, W. D., Hamrin, C. E., and Chyi, L. L. (1976). *J. Radioanal. Chem.* **32,** 195.
Karr, C., Estep, P. A., and Kovach, J. J. (1968). *Am. Chem. Soc., Div. Fuel Chem., Prepr.* **12**(4), 1.
King, J. G., Maries, M. B., and Crossley, H. E. J. (1936). *J. Soc. Chem. Ind., London, Trans. Commun.* **55,** 277.
Kinson, K., and Belcher, C. B. (1975). *Fuel* **54,** 205.
Kiss, M. T., and King, G. N. (1977). *Fuel* **56,** 340.
Kuhn, J. K., Kohlenberger, L. B., and Shimp, N. F. (1973). *Environ. Geol. Notes, Ill. State Geol. Surv.* No. 66.
Leighton, L. H., and Tomlinson, R. C. (1960). *Fuel* **39,** 133.
Levinson, L. M., and Jacobs, I. S. (1977). "Mössbauer Spectroscopic Measurement of Pyrite in Coal," Rep. No. 77CRD121. General Electric Corp., Corporate Res. Dev., Schenectady, New York.
Maher, T. P., and Shafer, H. N. S. (1976). *Fuel* **55,** 138.
Mayland, H. (1966). *Fuel* **45,** 97.
Miller, R. N., and Given, P. H. (1972). Unpublished data.
Miller, R. N., Yarzab, R. F., and Given, P. H. (1978). *Fuel* (in press).
Millott, J. O'N. (1958). *Fuel* **37,** 71.
Mott, R. A., and Spooner, C. E. (1938). *Fuel* **19,** 226, 242.
Neavel, R. C. (1966). Ph.D. Thesis, Pennsylvania State University, University Park, Pennsylvania.
Neavel, R. C. (1974). Personal communication.
Neavel, R. C. (1976). *Fuel* **55,** 237.
O'Gorman, J. V., and Walker, P. L., Jr. (1971). *Fuel* **50,** 135.

Parr, S. W. (1932). "The Analysis of Fuel, Gas, Water, and Lubricants." McGraw-Hill, New York.
Pringle, W. J. S. (1963). *Fuel* **42,** 63.
Radmacher, W., and Mohrhauer, P. (1955). *Brennst.-Chem.* **36,** 236.
Rees, O. W. (1966). *Ill. State Geol. Surv., Rep. Invest.* No. 220.
Reggel, L., Raymond, R., Steiner, W. A., Friedel, R. A., and Wender, I. (1961). *Fuel* **40,** 339.
Reggel, L., Wender, I., and Raymond, R. (1964). *Fuel* **43,** 75.
Savage, W. H. D. (1967–1968). *S. Afr. Chem. Process.* p. 177.
Shafer, H. N. S. (1970). *Fuel* **49,** 197.
Shimp, N. F., Helfinstine, R. J., and Kuhn, J. K. (1974). *Am. Chem. Soc., Div. Fuel Chem., Prepr.* **20**(2), 99.
Shipley, D. E. (1962). *Br. Coal Util. Res. Assoc., Mon. Bull.* **26,** 3.
Shou, J. K., and Pitts, W. S. (1977). *Fuel* **56,** 343.
Sutherland, J. K. (1975). *Fuel* **54,** 132.
Tarpley, E. C., and Ode, W. H. (1959). *U.S. Bur. Mines, Rep. Invest.* No. 5470.
Unterzaücher, J. (1952). *Analyst* **77,** 584.
van Krevelen, D. W. (1961). "Coal," pp. 160–176. Elsevier, Amsterdam.
Volborth, A., Miller, G. E., Garner, C. K., and Jerabek, P. A. (1977a). *Fuel* **56,** 204.
Volborth, A., Miller, G. E., Garner, C. K., and Jerabek, P. A. (1977b). *Fuel* **56,** 208.
Ward, D. L., and Millott, J. O'N. (1960). *Fuel* **39,** 293.
Watt, J. D. (1969). "The Physical and Chemical Behavior of the Mineral Matter in Coal under Conditions Met in Combustion Plant," Part 2. Br. Coal Util. Res. Assoc. Lit. Surv., Leatherhead, Surrey, England.
Young, R. K., and Zawadski, E. A. (1967). *Fuel* **46,** 151.

Note added in press: Recent (unpublished) work by Miller and Given suggests that the procedure suggested on p. 22 for determining the mineral matter content of low rank coals would be considerably improved by removing ion-exchangeable cations with an excess of 1 *N* ammonium acetate (pH 7) rather than with dilute HCl. This would obviate the need for a chlorine determination on the product, and any acetate not removed by washing would be volatilized during low temperature ashing.

Chapter 21

Analysis of Organic Compounds Trapped in Coal, and Coal Oxidation Products

Martin H. Studier *Ryoichi Hayatsu* *Randall E. Winans*
CHEMISTRY DIVISION
ARGONNE NATIONAL LABORATORY
ARGONNE, ILLINOIS

I. INTRODUCTION

Mass spectrometry (MS) is an exceedingly useful analytical tool for the organic chemist. When combined with separation techniques such as solvent extraction, fractional distillation, and various forms of chromatography, the analytical power is greatly increased. With mass spectrometry–gas chromatography combinations (GC/MS) very complex mixtures of organic compounds can be separated, identified, and

ISBN 0-12-399902-2

quantitatively measured. The organic matter in coals is an intricate system of large nonvolatile, insoluble macromolecules with lesser amounts of volatile, solvent-extractable compounds trapped within the macromolecular matrix. We shall describe the isolation of trapped compounds, which bear a relationship to the macromolecular material, by vacuum distillation and solvent extraction, followed by separation by gas chromatography and identification by time-of-flight mass spectrometry (GC/TOFMS). The macromolecules were degraded by pyrolysis and a variety of selective oxidative procedures designed to identify structural units indigenous to coal. The products of the oxidations were chiefly carboxylic acids which were esterified prior to analysis by GC/TOFMS, a TOF variable-temperature solid inlet, and high resolution mass spectrometry (HRMS). In general, the progress of chemical and physical procedures was integrated closely with MS or GC/MS. Unless specified otherwise the work described herein is our own, based on techniques successfully applied for a number of years to a variety of problems and recently used for the study of coals and coal products.

II. ANALYTICAL METHODS

A. Mass Spectrometry

1. *Time-of-Flight Mass Spectrometry*

Our principal tools are Bendix (Model 12) time-of-flight mass spectrometers modified to use a continuous ion source instead of the original pulsed ion source. The modification increased sensitivity by a factor of 300 to a limit of about 10^{-15} Torr with improved resolution (Studier, 1963). A "blanking generator" (Haumann and Studier, 1968) was designed and built to gate the electron multiplier detector of the TOFMS. This device enables one to choose at will the mass region for detection and to blank out superfluous parts of the spectrum. This minimizes saturation and dirtying of the multiplier. The total ion current of the "window" furnished by the blanking generator is used for chromatographic detection with coincident mass spectra obtained from a prior gate.

Another useful adjunct is a variable-temperature solid inlet probe consisting of an electrically heated metal ribbon 1.5 mm wide and 0.05 mm thick placed through a vacuum lock into the source region of the mass spectrometer. The ionizing electron beam grazes across the top of the metal ribbon on which the sample is mounted. With two powerstats in series, and with a 6-V filament transformer, very responsive temperature control results. This system combined with a continuous oscilloscope display of the entire mass spectrum permits prompt observation of a vaporizing species. This often permits detection of readily decomposed compounds not observable in less responsive sources. Since the probe can be heated to the melting point of the metal very nonvolatile species can be studied.

In addition to its extensive use in the study of coals, this system has been used successfully for the analyses of purine and pyrimidine bases (Studier *et al.*, 1968a; Carbon *et al.*, 1968), amino acids (Studier *et al.*, 1970), chlorophylls (Wasielewski *et al.*, 1976), and organic nitrogen compounds and polymeric material from meteorites (Hayatsu *et al.*, 1971, 1972, 1975a, 1977).

An auxiliary metal–glass vacuum line, shown in Fig. 1, connected directly to a TOFMS has proved to be very useful and versatile. With this system small samples can be taken for mass analysis, samples can be manipulated, chemical reactions monitored, or an open tube column loaded for GC analysis.

2. *High Resolution Mass Spectrometer*

Another very useful tool in coal studies has been a high resolution mass spectrometer (AEI MS 902 with PDP-8 computer). With such an

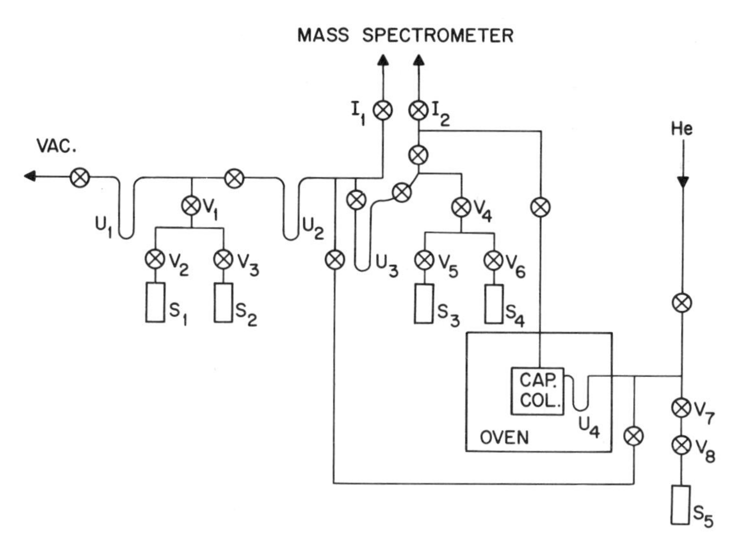

Fig. 1 Diagram of auxiliary vacuum system. See text for identification of the various parts.

instrument the stoichiometric formulas of molecular ions and fragments can be determined. We have found that extensive information can be obtained from analyses of complex mixtures from coals, especially when combined with TOF solid probe analyses and GC/TOFMS data. Details on high resolution mass spectrometry are given in Chapter 16, Section IV,A, and Chapter 17, Section II,B, of Volume I.

3. *General References*

A number of reviews and books on the application of mass spectrometry to organic chemistry and organic geochemistry have been published (Beynon *et al.*, 1968; Biemann, 1972; Budzikiewicz *et al.*, 1967; Burlingame, 1970; Cornu and Massot, 1975; Eglinton and Murphy, 1969; *Mass Spectra*, 1974; Friedel, 1970; Markey *et al.*, 1974; McLafferty, 1963; Porter and Baldas, 1971; Stenhagen *et al.*, 1974; von Kienitz, 1968). It is recommended that these works be consulted for detailed accounts of the technical aspects and applications. In particular, the review by Burlingame *et al.* (1976) and *Mass Spectrometry: A Specialist Periodical Report* (1971, 1973, 1975) provide very extensive bibliographies. Comparisons of mass spectra obtained by electron ionization, chemical ionization, field ionization, and field desorption are made by Fales *et al.* (1975). Field ionization is described in Volume I, Chapter 16.

B. Gas Chromatography—Mass Spectrometry†

1. *Wall-Coated Open Tube–TOFMS*

Wall-coated open tube (WCOT) columns (Ettre, 1965) are particularly useful for analyses of complex mixtures of organic compounds. These "capillary" columns are very efficient separators, improving with decreasing capillary bore. However, sensitive detectors must be used because of the sample size limits imposed by the fine capillaries. The modified TOF with its sensitivity increased by the continuous ion source (see Section II,A,1) has proved to be an excellent detector with one gate for total ion detection and a prior gate for coincident mass spectra. With columns 91.5 m long, an inside diameter of 0.25 mm, and an atmosphere of helium at the inlet, the entire effluent is admitted to the source of the spectrometer without impairment of its performance. Liquid coatings of Apiezon L and OV101 have been found most useful. Columns are loaded using the vacuum line illustrated in Fig. 1. With an isolated mixture at S_5 and the capillary column inlet evacuated, an

† For other details on GC/MS see Volume I, Chapter 17, Section,II,C.

aliquot of the mixture is taken between valves V_7 and V_8 and distilled to U_4, which is part of the column immersed in liquid nitrogen. The inlet is heated to 150°C and He to 1 atm is admitted. The column loading is banded by lowering the liquid nitrogen on U_4 for a few minutes. Then the Dewar is removed and the column is temperature-programmed at 0.5°C/min to give peaks in general about 2 min wide. With this technique of loading without solvent a section of the column can be saturated with sample, and because of banding, good resolution is maintained for sensitive detection of less abundant species. Although an abundant peak may show saturation, less abundant species are symmetrical and the overall dynamic range is increased. Using this technique as many as 200 compounds have been separated and detected in a single gas chromatographic run. Before applying this system to the study of coal products we used the GC/TOFMS technique successfully in numerous analyses of trapped organic compounds in meteorites and Fischer–Tropsch products produced from carbon monoxide and hydrogen (Anders *et al.*, 1973, 1974; Hayatsu *et al.*, 1972; Studier *et al.*, 1968b, 1972). Figure 2 is a relatively simple chromatogram of a synthetic product from reaction of deuterium gas with carbon monoxide in the presence of an iron–nickel catalyst (Studier and Hayatsu, 1968). One hundred perdeutero aliphatic and aromatic hydrocarbons were separated by an Apiezon L wall-coated column and identified by TOFMS. The prominent peaks (numbered 1, 4, 11, 22, 32, 46, 59, 72, 91, and 102) are all normal perdeutero alkanes with 5–14 carbon atoms. Between the prominent peaks are a series of branched alkanes, *n*-alkenes, branched alkenes, and aromatics.

2. *Commercial GC/TOFMS*

The WCOT–TOFMS system described in the preceding section is limited to compounds no less volatile than C_{20} or C_{21} alkanes or trimethylnaphthalene. Times of elution become prohibitively long for compounds much less volatile. A more versatile system with a commercial gas chromatograph (Perkin Elmer 3920B) interfaced with a TOF mass spectrometer is shown in Fig. 3. This system can handle packed as well as open tube columns. The bulk of the carrier gas goes through a flame ionization detector (FID) with a fraction entering the mass spectrometer through a microcapillary valve (No. 1) without a helium separator. The split ratio can be varied continuously, with between 5 and 20% going to the mass spectrometer when using support-coated open tubular (SCOT) columns. The indicated source pressure (Bayard–Alpert gauge) may vary from 3×10^{-6} to 2×10^{-5} without affecting significantly the opera-

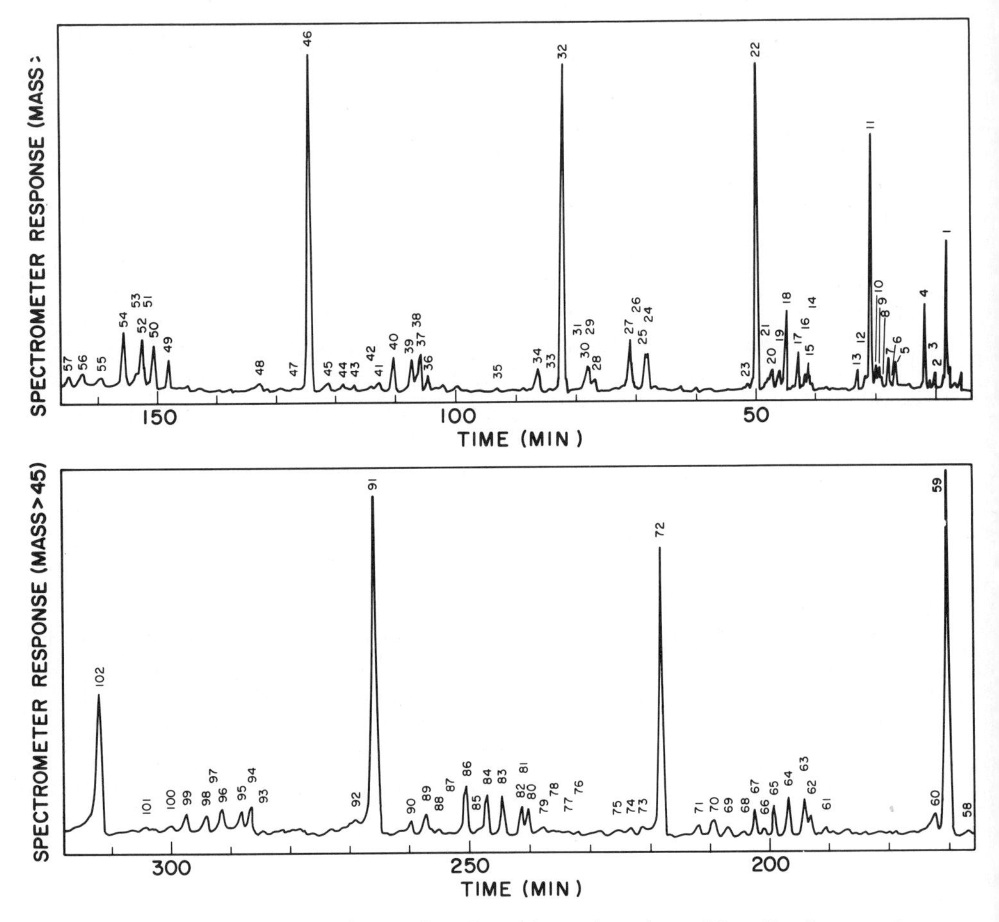

Fig. 2 Chromatogram of a Fischer–Tropsch product from $CO + D_2$. See text for a discussion of the prominent peaks. (For a complete peak identification, see Studier and Hayatsu, 1968.)

tion of the mass spectrometer. If desired, both the FID and the total ion current of this mass spectrometer may be used for GC detection. There is a small time lag between the time a compound is detected by FID and the time it is detected by the mass spectrometer. It is convenient to observe the oscilloscope to determine when a compound is entering the MS and when a mass spectrum should be recorded.

An auxiliary vacuum line for batch sampling is coupled to the interface via valve No. 3. With this arrangement solvent and the more abundant compounds can be bypassed to the cold trap of the vacuum line, thus minimizing memory effects in the mass spectrometer. The sensitiv-

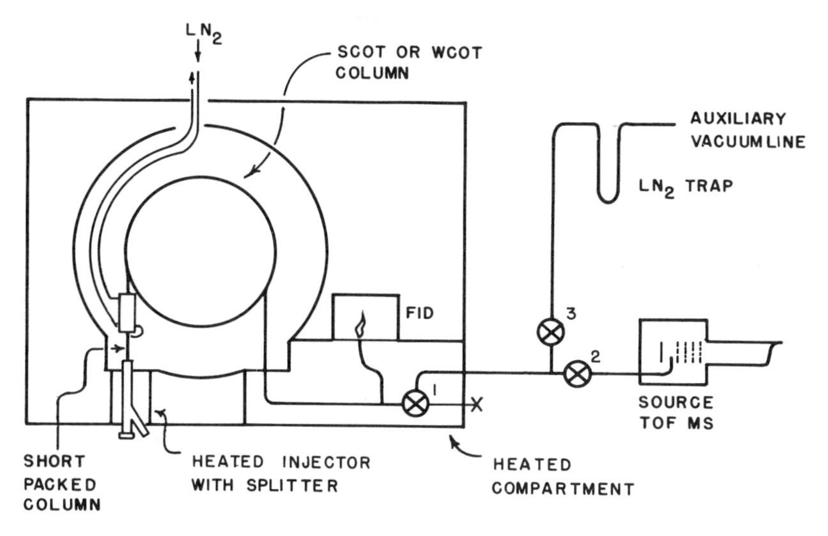

Fig. 3 Diagram of a commercial GC/TOFMS interface. Valve numbers are identified in text.

ity of the system is excellent. A 100-pg sample of methyl stearate injected onto a 15-m OV-101 SCOT column gave a molecular ion peak at $m/e = 298$ with an estimated limit of detection of 10^{-11} g (Winans *et al.*, 1978).

For most mixtures encountered in the study of coals and coal products the SCOT columns are a good compromise between packed columns and WCOT columns. They can be operated at higher temperatures and flow rates than WCOT columns and give better resolution than packed columns.

The readers interested in other approaches may wish to consult several review papers and books which furnish a broad introduction into GC, GC/MS, and GC/MS/computer techniques and applications (*Mass Spectrometry: A Special Periodical Report*, 1975; Biemann, 1972; Cram and Juvet, 1976; Burlingame, 1970; Burlingame *et al.*, 1976; McFadden, 1973).

C. Other Techniques

A high performance liquid chromatograph (HPLC) (Done *et al.*, 1974; Stock and Rice, 1974) promises to be a very useful tool for studying coal and its derivatives (Coleman *et al.*, 1977). It is particularly effective when combined with GC, MS, HRMS, and GC/MS. Fundamental

HPLC/MS or HPLC/MS/computer studies by others (Lovins *et al.*, 1973; Scott *et al.*, 1974; McLafferty *et al.*, 1975) may be of interest to the reader.

III. ANALYSIS OF TRAPPED ORGANIC COMPOUNDS IN COALS AND PYROLYSIS PRODUCTS

A. Vacuum Distillation

1. *Solid Probe Differential Analysis*

Solid samples of raw coal are placed on the metal ribbon (see Section II,A,1) which is then placed into the source of the mass spectrometer. As the sample is heated a variety of organic compounds are volatilized and detected.

The coals used in this study were a Wyoming lignite (Decker Coal Company, Sheridan, Wyoming, maf analysis: 67.3% C, 4.8% H, 1.3% N, 1.2% S, 25.4% O by difference); a high volatile Illinois bituminous (Peabody Coal Company, Seam #2, maf analysis: 77.8% C, 5.4% H, 1.4% N, 2.1% S, 13.3% O by difference); and a Pennsylvania anthracite (Penn State PSOC #85, maf analysis: 91.3% C, 3.9% H, 0.6% N, 1.1% S, 3.1% O by difference). In Fig. 4 are shown mass spectra at 350°C for these three coals. Phenol ($m/e = 94$), methyl phenol (107, 108), and dimethyl phenol (121, 122) are evident in both the lignite (a) and the bituminous coal (b), but are not detected in anthracite (c). Molecular ions or prominent fragments of alkylbenzenes (91, 105) and alkylnaphthalenes (128, 141), biphenylene (152), bifluorene (165), phenanthrene (178), and pyrene (202) are seen in all three. Note that the spectra become simpler with increasing rank from lignite, to bituminous, to anthracite coal.

The variable-temperature solid probe has been used for analysis of a great variety of relatively nonvolatile mixtures. Frequently by carefully programming the temperature, mixtures can be markedly fractionated. In Fig. 5 are shown two mass spectra of a mixture of benzenecarboxylic acid methyl esters obtained from nitric acid oxidation of a char produced by pyrolysis of bituminous coal at 800°C. Mass spectrum was observed (a) at about 50°C and (b) at about 100°C. Base peaks $(M-OCH_3)^+$ and molecular ions of the tri-, tetra-, penta-, and hexacarboxylic acid methyl esters are observed. At the lower temperature the base peak of benzenetetracarboxylate at $m/e = 279$ is most abundant. At the higher temperature the less volatile benzenehexacarboxylate ($m/e = 395$) is most abundant. By recording spectra over a broad temperature range a

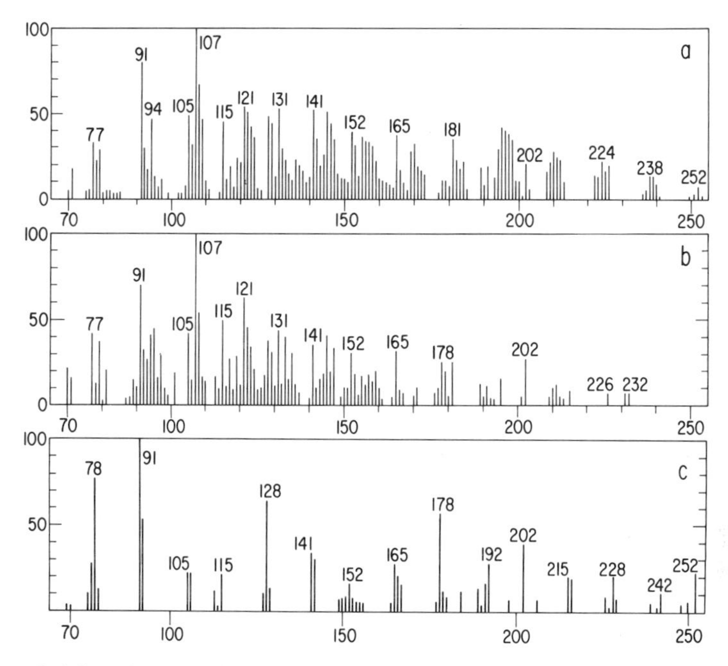

Fig. 4 Solid probe MS of three coals at 360°C: (a) lignite, (b) bituminous, and (c) anthracite. Major peaks are identified in text.

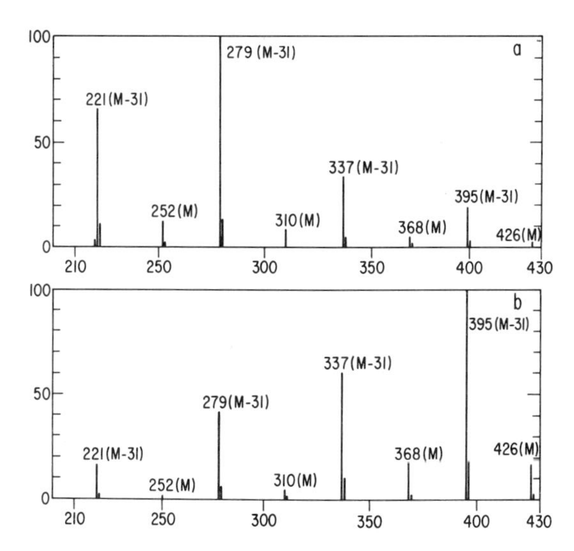

Fig. 5 Fractionation of benzenecarboxylic acid methyl esters from solid probe: (a) 50°C, (b) 100°C. Numbered peaks are identified in text.

rough separation according to vapor pressures can be made so that components in complex mixtures can be identified. By judicious control of probe temperature and an extra bakeout filament, small amounts of nonvolatile compounds can be identified in complex mixtures.

2. *Collection and Fractional Distillation*

When freshly powdered bituminous coal is placed into a vacuum and gradually heated, large numbers of organic compounds are slowly liberated. Figure 1 is a diagram of a vacuum system which has proved to be very versatile and useful. For example, freshly ground bituminous coal is placed in tube S_3, and S_3, U_3, and S_4 are cooled with liquid nitrogen. Air is pumped away through U_3, then V_4 is closed, S_3 is permitted to warm with condensables distilling into S_4, and noncondensables are periodically pumped away through U_3. Samples are admitted to the spectrometer to monitor the evolving gases. After evolution at a given temperature subsides, the entire collection is condensed to the bottom of S_4 and then the entire tube is cooled with liquid nitrogen. With liquid nitrogen on U_3, the cold bath is removed from S_4. As S_4 warms, the sample is fractionally distilled into U_3 while periodically small samples are introduced into the mass spectrometer through I_2. Usually a great deal of semiquantitative data can be obtained by this simple procedure and surprising separations can be made. The system behaves somewhat like a short, temperature-programmed GC column. For example, in Fig. 6 are shown two mass spectra obtained from the product of two hydrocracking experiments with an Illinois coal. Both samples were heated to 400°C with 4 atm of hydrogen. The sample which produced the spectrum shown in Fig. 6b also contained a cracking catalyst (activated montmorillonite clay). The mass spectra shown in Fig. 6 were obtained after the major products, volatile saturated hydrocarbons, had been fractionated away. Figure 6a shows peaks due to typical pyrolysis products when coal is heated above 250°C. The major peaks have been identified as phenol (m/e = 94), methyl phenol (107, 108), and dimethyl phenol (121, 122). In the presence of clay these phenols were converted to hydrocarbons, benzene (78), toluene (91, 92), and xylenes (105, 106). Mass 128 is due to naphthalene.

3. *WCOT–TOFMS*

If a sample is so complex that the simple distillation–mass analysis procedure does not give sufficient information, the sample S_4 is moved to position S_5 for loading onto the capillary column as described in

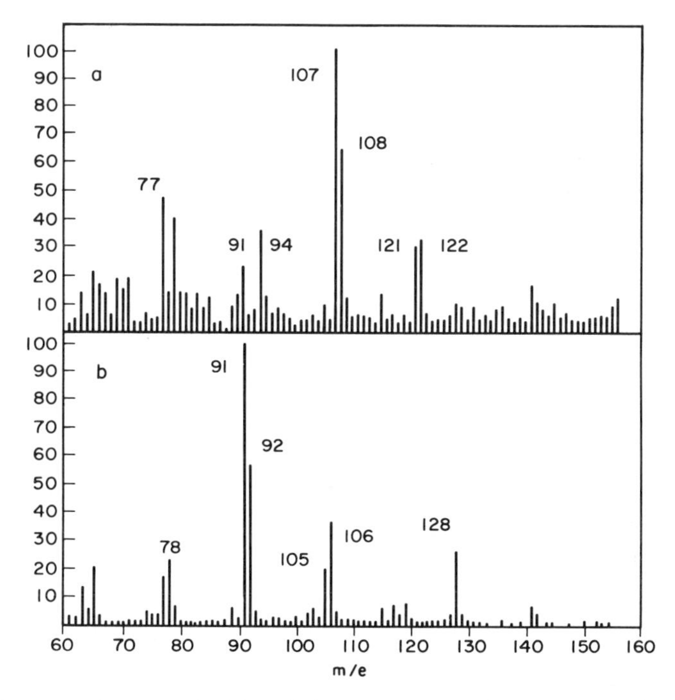

Fig. 6 Mass spectra of product from bituminous coal hydrocracked at 400°C: (a) 4 atm hydrogen only; (b) hydrogen (4 atm) + montmorillonite. Pertinent peaks (see text) are labeled with their mass-to-charge ratios.

Section II,B,1. In Fig. 7a is shown a gas chromatogram of the volatiles isolated by vacuum distillation from a freshly powdered Illinois bituminous coal (Peabody #2) at 150°C. The more volatile components with less than six carbon atoms and comprising about 90% of the volatiles were pumped away from a −78°C bath prior to this run. An aliquot of the remainder was taken at 0°C and loaded on the column as described earlier. Total ion current above $m/e = 45$ was used for detection with coincident mass spectra providing identification. About 200 compounds were detected, most of which are shown in Fig. 7a and identified in Table I. From Fig. 7a it was estimated that 37% were aliphatics (alkanes, alkenes, alkynes, and alkadienes), 29% were alicyclic (cyclohexane, cyclohexene, and numerous methyl derivatives), 29% were hydroaromatic and aromatic (tetralin, indan, benzene, naphthalene, and numerous derivatives), and 5% were thiophene and its derivatives. No nitrogen- or oxygen-containing compounds were detected (Hayatsu *et al.*, 1978).

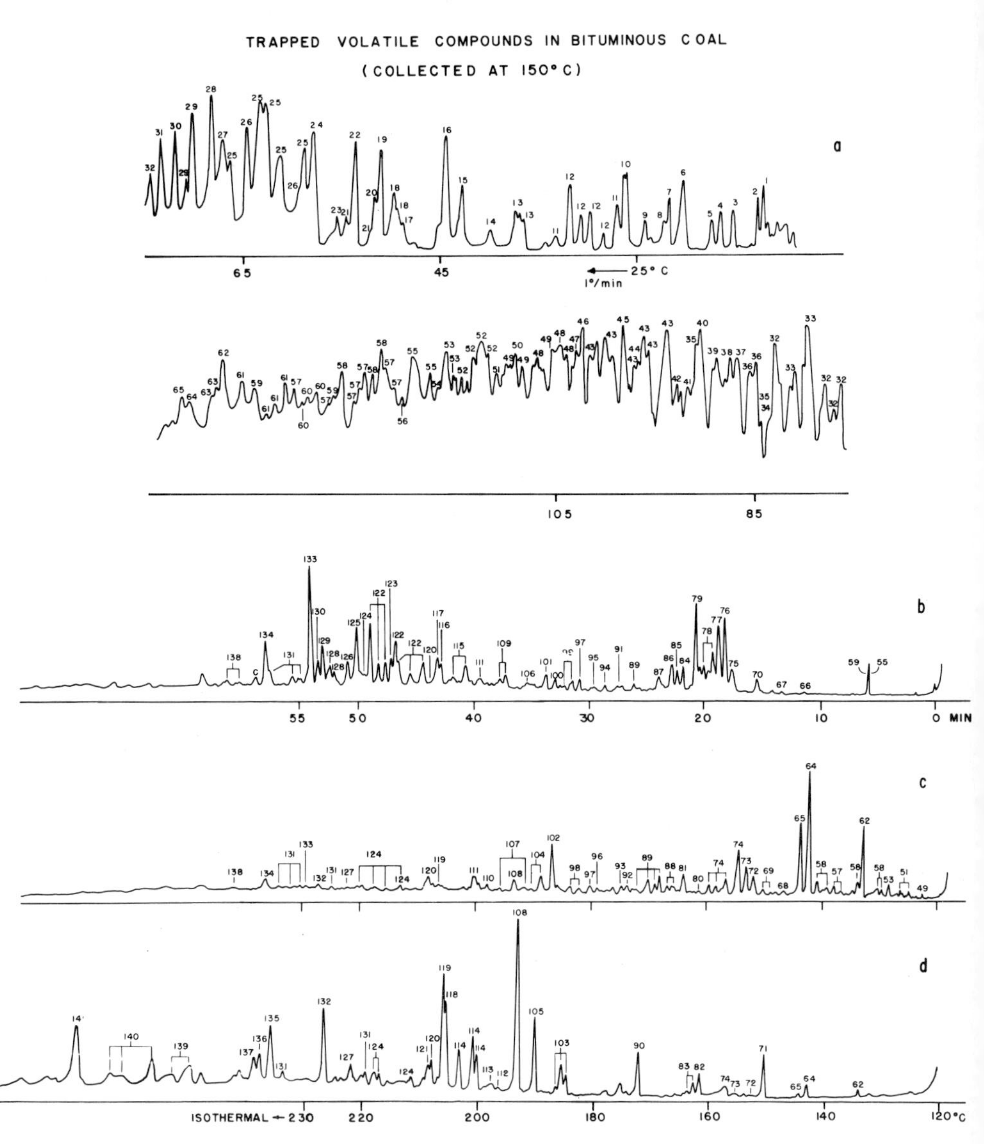

Fig. 7 Gas chromatograms of trapped compounds from coals. Numbered peaks are identified in Table I. (a) Trapped volatiles vacuum-distilled from bituminous coal at 150°C loaded on an open tube column (91.5 m long and 0.25 mm i.d. coated with OV-101) and temperature-programmed from 25°C at 1°C/min. (b) Hydrocarbon-rich fraction (16.3%) of the solvent extract of lignite separated on a 15.2-m SCOT column 0.51 mm i.d. coated with OV-101 and temperature-programmed from 120 to 230°C at 2°C/min. (c) Hydrocarbon-rich fraction (22%) of the solvent extract of bituminous coal run under the same conditions as (b). (d) Total solvent extract of anthracite run under the same conditions as (b).

4. *Discussion of Results*

These trapped compounds are believed to be products and residuals of the original coalification process and are clues to the structure of the macromolecular material which comprises the bulk of coals. This distribution of the trapped volatiles from bituminous coal with numerous isomers and much branching and much alkylation of aromatic rings suggests that during coalification much cracking and re-forming took place. The volatiles (10 carbon atoms or less) from a lignite and an anthracite were much fewer in number. Except for methane, only traces of other light molecules were observed. Presumably these have not yet been produced at the lignite stage of coalification and have been lost or condensed at the anthracite stage of the process.

B. Extraction and Separation

1. *Solvent Extraction*

The organic molecules not readily removed by vacuum distillation were extracted by refluxing with benzene–methanol. Typically 25–30 g of freshly ground fine powdered coal was extracted by refluxing with 150–200 ml of a mixture of benzene–methanol (3 : 1) for 24–48 hr with stirring. Yields as percent of the coal weights are 4.3 for lignite, 6.3 for bituminous, and 0.04 for anthracite coal.

2. *Separation by Liquid Phase Chromatography*

Chromatographic techniques, separation methods based on adsorption, and partition are extensively employed to fractionate complex geological sample extracts. For example, coal extracts, coal products, and oil shale extracts have been separated on alumina (Anders *et al.*, 1975; Imuta and Ouchi, 1973), on silica gel (Spence and Vahrman, 1970; Schweighardt *et al.*, 1976; Farcasiu, 1977), and on Florisil columns (Brooks and Smith, 1969). Paper and thin-layer chromatography (Schweighardt *et al.*, 1976) may be carried out on an analytical or preparative scale.

In our experiment, 1.4 g of the lignite coal extract was chromatographed on an alumina column (2.3 × 16 cm). The column was eluted successively with hexane–benzene mixtures, benzene, and benzene–methanol mixtures. The first fraction (230 mg) consisted primarily of hydrocarbons. A similar separation was made for the bituminous extract. Compare with the procedure described in Volume I, Chapter 16, Section II,C.

TABLE I *Compounds Identified by GC/MS and High Resolution MS*[a]

No.	Compound	No.	Compound
1.	*n*-Hexane	72.	2-Ethylnaphthalene
2.	2-Hexene	73.	1-Ethylnaphthalene
3.	Dimethylbutane, methylcyclopentane	74.	Dimethylnaphthalene
4.	Cyclohexane	75.	Cadinane
5.	C_7-Alkene (B)	76.	Dihydrocadinene (T), C_9-alkylcyclohexane
6.	Benzene	77.	Selinane and eremophilane (?)
7.	Thiophene	78.	Dihydroselinene (T) and/or dihydroeremophilene (T)
8.	C_7-Alkane (B), C_7-alkene (B)	79.	Dihydrocadinene (T)
9.	C_7-Alkadiene (B) or C_7-alkyne (B)	80.	C_5-Alkylindan
10.	Cyclohexene	81.	Methylacenaphthene, 2(?)-isopropylnaphthalene
11.	C_7-Alkane (B)	82.	Diphenylmethane
12.	Dimethylcyclopentane	83.	3- or 4-Methylbiphenyl
13.	2- and 3-Methylhexanes	84.	C_{10}-Alkylbenzene
14.	Heptene	85.	Tetramethylindan
15.	2,3-Dimethyl-2-pentene	86.	$C_{15}H_{28}$-Sesquiterpenoid hydrocarbon (?)
16.	Methylcyclohexane	87.	C_5-Alkyltetralin
17.	Dimethylhexane	88.	Methylethylnaphthalene and/or trimethylnaphthalene
18.	Heptyne	89.	Trimethylnaphthalene
19.	Trimethylpentane	90.	Fluorene
20.	Methylheptane	91.	1,6-Dimethyl-4-isopropyl-1,2-dihydronaphthalene (T)
21.	Methylheptene	92.	Isobutylnaphthalene, trimethylnaphthalene
22.	Trimethylcyclopentane	93.	1-Methyl-4-isopropyl-naphthalene
23.	1-Methylcyclohexene	94.	Eudalene (1-methyl-7-isopropylnaphthalene)
24.	Toluene	95.	C_5-Alkyltetralin (?)
25.	Dimethylcyclohexane	96.	1-Methyl-2-propylnaphthalene (T)
26.	Methylthiophene	97.	Cadalene (1,6-dimethyl-4-isopropyl-naphthalene)
27.	C_9-Alkene (B)	98.	Tetramethylnaphthalene
28.	Ethylcyclohexane	99.	1,4-Dimethyl-6(?)-isopropylnaphthalene, C_6-Alkyltetralin (T)
29.	Trimethylcyclohexane	100.	C_6-Alkyltetralin (T)
30.	*n*-Propyl and/or isopropylcyclohexane	101.	1,2,5,7-Tetramethylnaphthalene
31.	C_4-Alkylcyclopentane (?)	102.	Pristane
32.	C_9-Alkane (B), C_9-alkene	103.	Methylfluorene
33.	C_9-Alkyne (?) and/or C_9-alkadiene (?)	104.	Pentamethylnaphthalene
34.	Ethylbenzene	105.	Dibenzothiophene
35.	Dimethylthiophene	106.	Trimethyloctahydrophenanthrene

36. *m*- and *p*-Xylene
37. *o*-Xylene
38. C_9-Alkene (B)
39. Tetramethylcyclohexane
40. C_{10}-Alkene (B)
41. C_4-Alkylcyclohexane
42. Diethylcyclohexane (?)
43. C_{10}-Alkane (B), C_{10}-alkene (B)
44. C_{10}-Alkene
45. Ethyloctane (?)
46. Trimethylthiophene
47. Propylbenzene
48. Methyl ethylbenzene
49. Trimethylbenzene
50. C_{11}-Alkene (B)
51. C_4-Alkylbenzene
52. C_{11}-Alkene (B) and C_4-alkylbenzene
53. Tetramethylbenzene
54. 1-Methyl-4-isopropyl-3-cyclohexene (?)
55. Methylindan
56. C_{12}-Alkene (b)
57. Dimethylindan
58. C_5-Alkylbenzene
59. Tetralin
60. C_6-Alkylbenzene
61. C_{13}-Alkene, C_6-alkylbenzene
62. Naphthalene
63. C_{13}-Alkane (B), C_{14}-alkene (B)
64. 2-Methylnaphthalene
65. 1-Methylnaphthalene
66. C_3-Alkyldecalin
67. C_4-Alkyldecalin
68. Trimethylindan
69. Tetramethylindan and/or trimethyltetralin
70. C_9-Alkylcyclohexane
71. Biphenylene
107. Methyltetrahydrophenanthrene (T)
108. Phenanthrene
109. $C_{14}H_{23}$(*m/e* 191,base peak), $C_{20}H_{36}(M^+)$ tricyclic terpenoid(?)
110. Dimethyltetrahydrophenanthrene (T)
111. Ethyltetrahydrophenanthrene (T)
112. Anthracene
113. Naphthofuran
114. Methyldibenzothiophene
115. C_9-Alkyltetralin (?)
116. $C_{20}H_{32}$(abietadiene ?)
117. $C_{19}H_{30}$(tricyclic diterpenoid ?)
118. 2- and/or 3-Methylphenanthrene
119. 1- and/or 9-Methylphenanthrene
120. 1,7-Dimethylphenanthrene
121. Dimethyldibenzothiophene
122. C_{10}-Alkyltetralin (or C_{11}-alkylindan) (?)
123. Dehydroabietene
124. Dimethylphenanthrene and/or dimethylanthracene
125. Dehydroabietane
126. $C_{20}H_{32}$(tricyclic diterpenoid) (?)
127. Fluoranthene
128. C_{11}-Alkyltetralin (?)
129. $\Delta^{6,8,11,13}$-Abietatetraene (T)
130. 1,2,3,4-Tetrahydroretene (T)
131. Methylethylphenanthrene and/or trimethylphenanthrene
132. Pyrene
133. Simonellite
134. Retene (1-methyl-7-isopropylphenanthrene)
135. 1,2-Benzofluorene
136. 2,3-Benzofluorene
137. 3,4-Benzofluorene
138. Methylpropylphenanthrene and/or dimethylethylphenanthrene
139. Methylbenzofluorene
140. Tetramethylphenanthrene and/or tetramethylanthracene
141. Chrysene and/or triphenylene

[a] B = branched; T = identification tentative; ? = identification uncertain.

3. *GC/MS, HRMS, Solid Probe*

Figures 7b, 7c, and 7d are gas chromatograms of the hydrocarbon fractions solvent-extracted from the three coals. The data were obtained with a 15-m OV-101 SCOT column using the system described in Section II,B,2. Numbered peaks are identified in Table I. The lignite chromatogram reveals the presence of a large number of sesqui and diterpenoids and related compounds. Not detected by GC/MS but shown to be present by HRMS were steranes (characteristic fragments at *m/e* = 218, 217, 149) and triterpenoids (typical fragment at *m/e* = 191). In a more polar fraction were identified γ-pyrones ($C_6H_6O_2$, $C_7H_8O_2$, $C_6H_6O_3$, and $C_7H_8O_3$), terpenoid-type compounds ($C_{15}H_{24}O$, $C_{19}H_{30}O$, and $C_{20}H_{30}O$), and oxygen-containing pentacyclic triterpenoids (fragments at *m/e* = 207, 189). Terpenoids, steroids, and other biologically interesting compounds found in geological sources are extensively reviewed by Maxwell *et al.* (1971) and Albrecht and Ourisson (1971). The lignite solvent extract also contained an abundant "wax" fraction (67% of the extract) which contained compounds with empirical formulas $C_nH_{2n}O_2$ for n = 24–32 as determined by HRMS. The bituminous chromatogram is dominated by naphthalene derivatives and pristane. The peaks in the anthracite chromatogram are entirely aromatic, showing a high degree of condensation. The solid inlet probe and HRMS revealed the presence of polycyclic aromatic compounds in the anthracite extract too nonvolatile to be separated by GC. These had molecular formulas $C_{21}H_{14}$, $C_{22}H_{14}$, $C_{22}H_{16}$, $C_{24}H_{14}$, $C_{25}H_{18}$, and $C_{26}H_{14}$.

IV. ANALYSIS OF OXIDATION PRODUCTS

A. Background

The bulk of the organic matter in a coal consists of a macromolecular material of complex and variable composition. To study this chemically it is necessary to break it into smaller pieces which can then be identified. In so doing the question always arises as to what changes the chemical procedures have produced. Pyrolysis of coal, for example, yields a large number of identifiable molecules but it is not clear what relationship these have to the original coal since pyrolysis usually yields at least 50% char. A number of oxidizing agents have been used to degrade the macromolecules. In the past, commonly used oxidants have been HNO_3, HNO_3–$K_2Cr_2O_7$, $KMnO_4$, O_2, H_2O_2–O_3, and NaOCl (Bearse *et al.*, 1975; Bitz and Nagy, 1967; Chakrabartty and Kretschmer, 1972,

1974; Lowry, 1963; Tingey and Morrey, 1973; van Krevelen, 1961). In general these are drastic oxidants degrading coal to benzenecarboxylic acids and much of the work depended on classical chemical techniques before modern GC/MS instrumentation was available. We have chosen to use a series of oxidants, one drastic, HNO_3, and several others more selective, $Na_2Cr_2O_7$, air–ultraviolet, and AcOH–H_2O_2 (Hayatsu *et al.*, 1975b, 1978; Winans *et al.*, 1978).

B. Nitric Acid Oxidation

Nitric acid is a drastic oxidizer which will degrade a macromolecular material leaving a residue of benzenecarboxylic acids. The following reactions of organic compounds illustrate the approach:

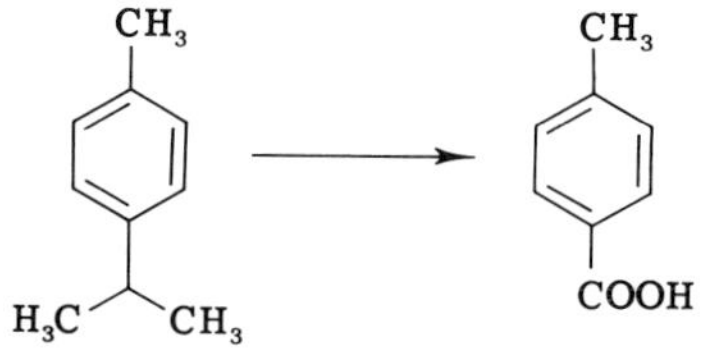

(1)

(Tuley and Marvel, 1955) wherein aliphatic groups larger than methyl are readily oxidized to carboxylic groups, but the methyl group is somewhat more resistant.

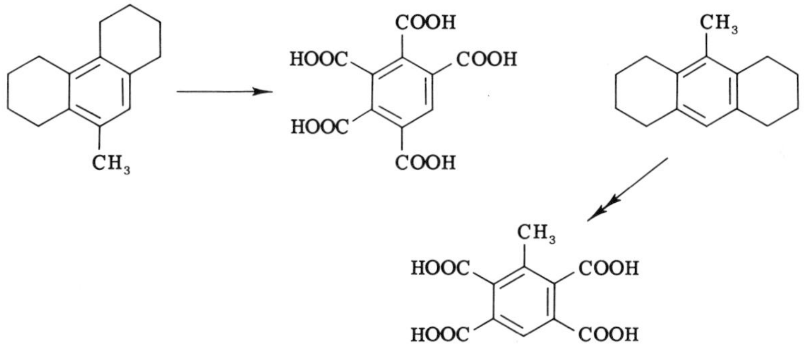

(2)

(Nes and Mosettig, 1954) where the alicyclic groups of hydroaromatic compounds are readily oxidized.

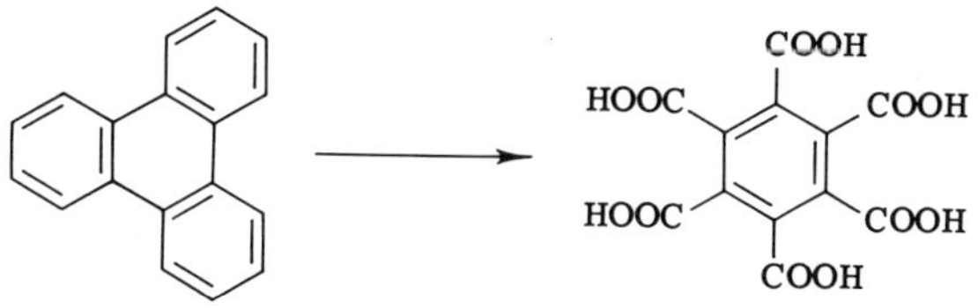

(3)

(van Krevelen, 1961) where complex aromatic groups are oxidized to benzenecarboxylic acids, with the more complex structures yielding more carboxylic acid groups. A series of samples were refluxed with 70% nitric acid for 20 hr yielding a clear orange solution. The acid solutions were evaporated to dryness under reduced pressure and the residue weighed. To increase volatility for GC and MS analysis the acids were methylated with diazomethane. The methylated acids were analyzed by GC/MS, HRMS, and the solid probe. Fragmentation patterns and precise mass determination of both molecular ions and fragments, in particular $(M-OCH_3)^+$, were used for identification. In Fig. 8 are summarized the data for benzenecarboxylic acids (as their methyl ester) from seven samples, including one completely synthetic Fischer–Tropsch product. The synthetic sample had been prepared by heating CO, H_2, and NH_3 with an Fe–Ni catalyst at 200°C for 6 months (Anders *et al.*, 1973; Hayatsu *et al.*, 1977). It was a macromolecular material insoluble in organic solvents, HCl, HF, and KOH. Despite the drastic

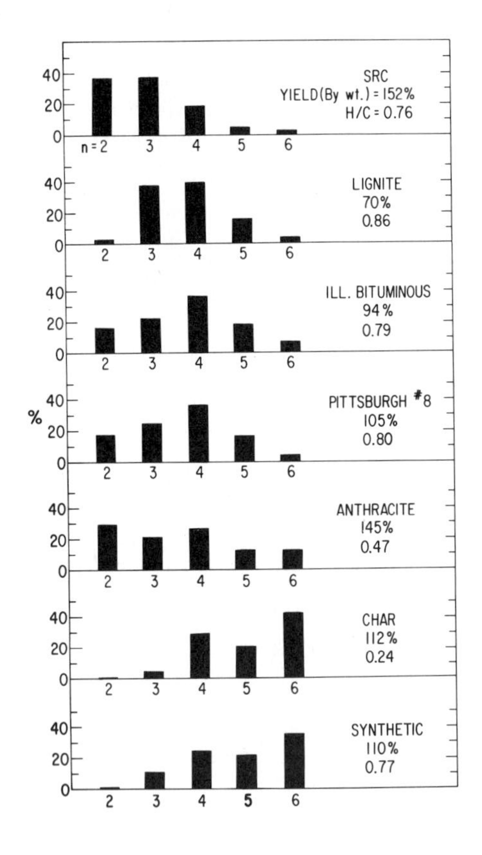

Fig. 8 Abundances (%) of benzenecarboxylic acids determined as their methyl esters, produced by nitric acid oxidation. *n* is the number of $COOCH_3$ groups per benzene ring.

nature of the nitric acid oxidation it appears that useful information can be obtained by the procedure. For example, the yield of total acids and the number of carboxylic acid groups per benzene ring seem correlated with the degree of condensation of the original material. The spectra from the synthetic sample from the Fischer–Tropsch reaction and from char are relatively simple, consisting primarily of the benzenecarboxylic acid esters with from two to six ester groups. The coal samples and coal-derived products (except for char) are more complex and contained nitro-substituted esters and pyridine derivatives (Hayatsu *et al.*, 1977). Note the similarity between the synthetic sample and the char prepared from Illinois bituminous coal by heating under vacuum to 800°C. The abundance of the hexacarboxylic acid for these two suggests a high degree of condensation in the original samples. The similarity in distribution of the oxidation products from the two bituminous coals (the Illinois and Pittsburgh) is striking. The solvent-refined coal (SRC) derived from the Pittsburgh #8 coal is shown in Section IV,C to have a higher degree of aromatic ring condensation than its feed coal. Thus it is surprising to see the shift to fewer acid groups for SRC. This may mean that many aliphatic cross-links were destroyed in the SRC process and evolved as light hydrocarbons.

C. Sodium Dichromate Oxidation

Friedman *et al.* (1965) and Wiberg (1965) have shown that when heated to 250°C with excess sodium dichromate the alkyl or alicyclic substituents of aromatic compounds are oxidized to carboxylic acid or keto groups with a minimum of degradation of the aromatic rings. Methoxy derivatives of alkyl benzenes are also oxidized to corresponding methoxybenzenecarboxylic acid in fair yield in the presence of sodium dihydrogen phosphate. The stoichiometry of oxidation is shown as

$$2Ar(CH_3) + 3Na_2Cr_2O_7 \rightarrow 2ArCO_2Na + 2Na_2CrO_4 + 2Cr_2O_3 + 3H_2O$$

Samples (3–5 g) were shaken in an autoclave for 40 hr with a fivefold by weight excess of $Na_2Cr_2O_7$ in H_2O. The reactor was cooled to 60–80°C and hydrated chromic oxide filtered off.

The clear, alkaline filtrate was extracted with benzene–ether (1 : 1) to remove nonacidic compounds, and then acidified with cold concentrated HCl in an ice bath. The solvent was distilled off under reduced pressure at 40°C. The resulting mushy residue was extracted three times with 30 ml methanol–ether (3 : 1) and twice with 20 ml chloroform. The combined extracts were dried over Na_2SO_4, and then evaporated at room temperature under a nitrogen stream. The resulting acids were

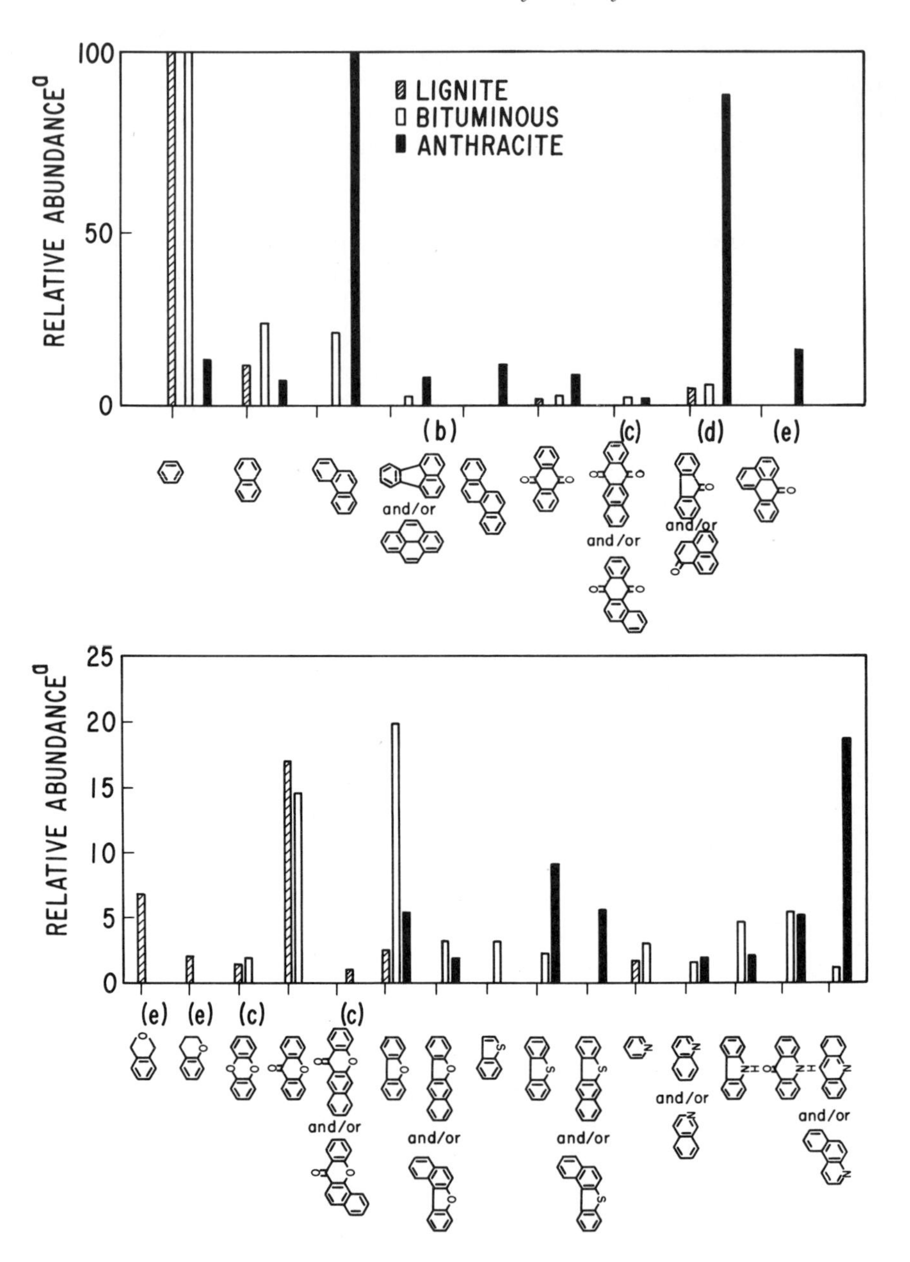

Fig. 9 Relative abundances of aromatic and heteroaromatic units produced by sodium dichromate oxidation of lignite, bituminous, and anthracite coals. Identification based on TOF variable-temperature solid inlet, GC/TOFMS, and high resolution MS. Aromatic units in the upper graph (left to right): benzene, naphthalene, phenanthrene, fluoranthene/

dissolved in methanol–ether (3 : 1), and esterified with diazomethane (Fieser and Fieser, 1968, 1969) for mass spectrometric analysis. Three coal samples, a Wyoming lignite, an Illinois bituminous, and a Pennsylvania anthracite, were oxidized with aqueous sodium dichromate. Essentially all of the lignite and bituminous coals were converted to volatile or soluble compounds but only two-thirds of the anthracite was converted. The weights of water and organic solvent-soluble compounds produced by the oxidations were 51, 70, and 55% of the original coal weights for lignite, bituminous, and anthracite coals, respectively. The anthracite also yielded 17% of a humic acid type of material of high molecular weight soluble only in alkaline aqueous solution. The more water-soluble acids were esterified by the method of Eisenbraun *et al.* (1970). In Fig. 9 are shown the aromatic units identified and their relative abundance in the three coals. As expected the degree of condensation increased in the order lignite, bituminous, and anthracite.

Using NaOCl as an oxidant, which they believed to be selective, Chakrabartty and Kretschmer (1972, 1974) and Chakrabartty and Berkowitz (1974) suggested that coal is largely nonaromatic. They found no evidence for aromatic compounds other than benzene derivatives. They suggested the possibility that major structural rearrangements with pyrolytic formation of polynuclear aromatic compounds occur during $Na_2Cr_2O_7$ oxidations at 250°C (Chakrabartty and Berkowitz, 1976a). The fact that no polynuclear aromatic compounds with more than two fused rings were detected in the oxidation products of lignite and that the degree of condensation increases with increasing rank of coal is internally consistent and suggests that condensation during oxidation with $Na_2Cr_2O_7$ is minimal (Hayatsu *et al.*, 1976, 1978). In a blank control experiment the Illinois bituminous coal (5.3 g) was heated with 60 ml of H_2O in an autoclave at 250°C for 40 hr. An insoluble residue (96.2%) and inorganic salts (2.9%) were obtained. Analysis of the residue by a variable-temperature solid inlet probe (see Sections II,A,1 and III,A,1) suggested that no oxidation, pyrolytic cleavage, or structural rearrangement occurred. The H/C ratio had not been changed. Oxidation of model compounds, tetralin and 9,10-dihydrophenanthrene, with

---

pyrene, chrysene, anthraquinone, naphthacenequinone/benzanthraquinone, fluorenone/phenalenone, benzanthrone. Heteroaromatic units in the lower graph (left to right): phthalan, chroman, dibenzo-*p*-dioxin, xanthone, benzoxanthones, dibenzofuran, benzonaphthofurans, benzothiophene, dibenzothiophene, benzonaphthothiophenes, pyridine, quinoline/isoquinoline, carbazole, acridone, acridine/benzoquinoline. (a) Determined as methyl esters of carboxylic acids and as nonacidic compounds. The benzene unit is normalized to 100 for the lignite and bituminous coals. Phenanthrene is normalized to 100 for anthracite.

sodium dichromate gave phthalic acid and diphenoic acid, respectively, in 85–95% yield as expected. Furthermore, the procedure has been applied successfully to structural determination in meteorites (Hayatsu *et al.*, 1977). The specificity of NaOCl as an oxidant has been questioned (Mayo, 1975; Ghosh *et al.*, 1975; Landolt, 1975) and is still in dispute (Chakrabartty and Berkowitz, 1976b).

The usefulness of the sodium dichromate oxidation procedure for analysis of coal products is illustrated by the comparison of a solvent-refined coal (SRC) with the feed coal from which it was derived. The samples were obtained from the Wilsonville, Alabama pilot plant. The SRC had been produced by heating a bituminous coal (Pittsburgh #8) with a donor solvent at 460°C and 1700 psi of hydrogen. It was separated into three fractions on the basis of solubility: hexane (4.5%); hexane insoluble, benzene–methanol (3 : 1) soluble 82.3%; benzene–methanol insoluble, pyridine soluble 11.8%; and 1.4% residue. The hexane extract consisted primarily of neutral polynuclear aromatics and heteroaromatics such as naphthalenes, phenanthrenes, biphenyl, acenaphthene, fluorenes, pyrenes, fluoranthenes, dibenzofurans, and dibenzothiophenes. The benzene–methanol extract was shown to be primarily a macromolecular material by the variable-temperature solid inlet. Both the benzene–methanol-extracted material and the feed coal itself were oxidized by the $Na_2Cr_2O_7$ procedure. The two gas chromatograms are shown in Fig. 10. The numbered peaks are identified in Table II with calculated relative abundances. Two mass spectra shown in Fig. 11 illustrate that reasonable identifications can be made although authentic mass spectra are not available. The mass spectrum shown in Fig. 11a, and believed to be that of dibenzofuran monocarboxylic acid methyl ester, was obtained from the compound of peak 7, Fig. 10. The peak at $m/e = 226$ is at the molecular weight of methyl dibenzofurancarboxylate. The peak at 195 (M-31) from the loss of —OCH_3 is typical of all aromatic carboxylic acid methyl esters. The peak at 167 represents the loss of a CO group and the peak at 139 represents the loss of a second CO group. The stoichiometric formula was verified by HRMS. Figure 11b represents the mass spectrum of peak 8 in Fig. 10 and is believed to be that of methyl naphthalenedicarboxylate. Again the principal peak is typically parent minus 31 (OCH_3). There is a prominent parent peak with the stoichiometric formula again verified by HRMS. The other peaks can be accounted for by loss of CO to give $m/e = 185$, loss of another OCH_3 to give $m/e = 154$, and then another CO to give $m/e = 126$.

Compounds too nonvolatile or polar to be observed by GC/MS were

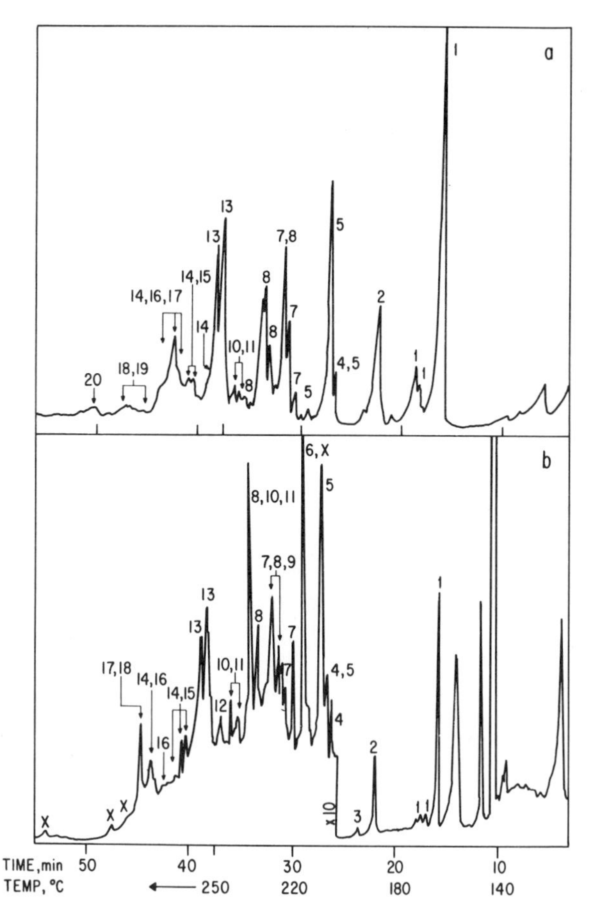

Fig. 10 Gas chromatograms of methylated $Na_2Cr_2O_7$ oxidation products of (a) a solvent-refined coal (SRC) and (b) the Pittsburgh #8 feed coal. The numbered peaks are identified in Table II.

identified by solid probe TOFMS and by HRMS. The greater degree of aromatic condensation of the SRC over that of its feed coal is evident. For example, naphthalene, phenanthrene, pyrene, and anthraquinone units are much more abundant in the SRC product than in the original coal. The predominant heterocyclic units in the coal product, benzo- and dibenzothiophenes and xanthones, are much less abundant in the SRC product and were apparently destroyed by the SRC process. However, *N*-heteroaromatics and dibenzofuran appear to have survived the SRC procedure.

TABLE II *Aromatic Acids Found in the Oxidation of SRC and Its Feed Coals[a] Identified as Methyl Esters*

GC peak number	Nucleus	Number of —$COOCH_3$	Relative abundance (±15%)[b] SRC extract[c]	Coal[c]
1	Benzene	2	100	100
5		3	58	29
10		4	9	10
4	Biphenyl	1	3	9
2	Naphthalene	1	47	30
8		2	82	29
16		3	24	8
		4	7	3
13	Phenanthrene	1	82	28
20		2	8,15	8
		3	5	7
		4	1 (T)	—
19	Pyrene/fluoranthene	1	7	3
		2	4	2
11	Fluorenone	1	4	8
		2	3	2
		3	3	—
14	Anthraquinone	1	26	4
		2	3	—
7	Dibenzofuran	1	32	21
17		2	9	4
		3	10	4
15	Xanthone	1	4	9
		2	6	7
		3	2 (T)	3
3	Benzothiophene	1	—	4
12	Dibenzothiophene	1	2	8
		2	—	5
	Pyridine	3	—	5 (T)
	Carbazole	1	4	3
		2	3	4
18	Benzoquinoline/acridine	1	3	7
		2	5	4
		3	2	2
6[d]			—	11
9[e]			—	22

[a] Ninety-two percent of the extract was oxidized and the yield of total acidic and non-acidic, less volatile compounds, was ~59%. For the feed coal, 84% of the coal was oxidized; yield of totals was ~58%. T means identification tentative; —, not detected; x, peak consists of more than one component which is difficult to identify by GC/MS.

[b] Benzenedicarboxylic acid methyl ester is normalized to 100.

[c] Identification and estimation of relative abundances were made by TOF variable-temperature solid inlet and HRMS, because of difficulty of identification by GC/MS. Yields in italics.

[d] Peak number 6 is tentatively identified as trimethoxyxanthone.

[e] Peak number 9 shows prominent mass ions at 216 and 215. HRMS shows their elemental composition corresponding to $C_{12}H_{10}O_3N$, $C_{12}H_9O_3N$ or $C_9H_{12}O_6$, $C_9H_{11}O_6$.

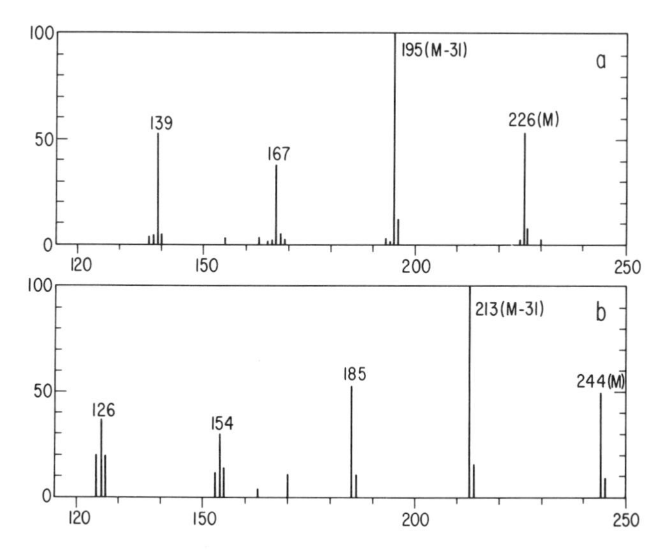

Fig. 11 Mass spectrum of (a) dibenzofuranmonocarboxylic acid methyl ester and (b) naphthalenedicarboxylic acid methyl ester. Numbered peaks are identified in text.

D. Photochemical Oxidation

A number of photochemical cleavage reactions are known (Ranby and Rabek, 1975) but do not appear to have been applied to degradation of complex carbonaceous materials such as coals. We have studied photo-oxidative degradation of lignite, bituminous, and anthracite coals (Hayatsu *et al.*, 1975b, 1978). As a general procedure 0.5–1.0 g of fine powdered coal, extracted with concentrated HCl or with an organic solvent, suspended in 50 ml of 10% HCl aqueous solution, was placed in a quartz flask. The flask was irradiated with ultraviolet light from a high pressure mercury lamp for 8 days while air was bubbled through the HCl solution. The temperature rose to 45–50°C during the reaction. After filtration the nonreacted residue was washed successively with hot water, methanol, and chloroform. The filtrate and washings were evaporated under reduced pressure below 40°C. The resulting carboxylic acids were esterified with diazomethane. Almost no photooxidation was observed for anthracite, whereas lignite and bituminous were oxidized in yields of 25 and 30%, respectively. If the unreacted residue was further oxidized by the same procedure, 22–24% of the residue was oxidized again with the same distribution of products. This procedure is particularly attractive because no nonvolatile reagents are added to complicate the isolation and identification of products.

In Table III are listed the aromatic carboxylic acids isolated from a bituminous coal by photochemical oxidation. The products from lignite were primarily benzenecarboxylic acids with only traces of pyridinetricarboxylic acids and xanthonedi- and tricarboxylic acids. Several aliphatic carboxylic acids (as methyl esters) were positively detected in the photooxidation product of bituminous by TOF variable-temperature solid inlet and high resolution MS. They are methyl esters of malonic acid, succinic acid, glutamic acid, and saturated monocarboxylic acids [CH_3–$(CH_2)_n$–$COOCH_3$, n = 1–7]. The fragments for 3-methyl and 3,3-dimethyl aliphatic carboxylic acid methyl esters were also seen in the mass spectra. A similar observation was made for the oxidation product of lignite. From the photooxidation of model substances (polynuclear aromatic hydrocarbons, heterocyclics, and their alkyl derivatives and polymers), it was found that benzene, pyridine, and condensed *N*-heteroaromatic rings are quite stable (Hayatsu *et al.*, 1977). Five-membered heterocyclic derivatives (Augustine and Trecker, 1971; Buchandt, 1976) and phenolic compounds were found to be easily oxidized. Photooxidation of a poly(2-vinylpyridine)polystyrene copolymer yielded benzoic acid, pyridine-2-carboxylic acid, pyridine-2-aldehyde, malonic acid, and succinic acid as major products.

E. Hydrogen Peroxide–Acetic Acid Oxidation

It is probable that in the oxidation experiments just described aromatic units with phenolic groups would have been destroyed. Since the presence of phenolic groups in coals has been established by Vaughan and Swithenbank (1970), we have been searching for oxidizing agents that do not destroy phenols. Schnitzer and Skinner (1974) have shown that an acetic acid–hydrogen peroxide mixture under mild conditions oxidizes humic acids while preserving phenolic groups. Using this procedure we have oxidized our lignite and bituminous coals with over 80% conversion to methanol-soluble acids and have methylated the acids produced.

The gas chromatogram of the aromatic portion of the methylated product from lignite obtained with an OV-101 SCOT column is shown in Fig. 12. The identification of individual compounds was made by coincident MS and by high resolution MS of the mixture. The methyl esters identified gave the following approximate distribution: 36.1% benzene, 7.6% methylbenzene, 22.1% methoxybenzene, 15.9% furan, and 18.2% dibasic aliphatic. For the bituminous coal the methoxy derivatives were half as abundant. These results suggest that the lignite

TABLE III *Aromatic Acids Found in the Photochemical Oxidation of Bituminous Coal: Identified as Methyl Esters*[a]

Nucleus	Number of —$COOCH_3$	Precise mass $(M\text{–}OCH_3)^+$ Elemental composition	Observed	Dev. × 10^3	Relative abundance[b] ±15–20%
Benzene	2	$C_9H_7O_3$	163.0388	−0.7	42
	3	$C_{11}H_9O_5$	221.0457	0.8	100
	4	$C_{13}H_{11}O_7$	279.0510	0.6	97
	5	$C_{15}H_{13}O_9$	337.0555	−0.3	29
	6	$C_{17}H_{15}O_{11}$	395.0593	−2.0	4
Cl-benzene[c]	2	$C_9H_6O_3Cl$	196.9990	−1.5	24
	3	$C_{11}H_8O_5Cl$	255.0066	0.6	31
	4	$C_{13}H_{10}O_7Cl$	313.0125	1.1	26
	5	$C_{15}H_{12}O_9Cl$	371.0124	−4.4	2
Fluorenone	2	$C_{16}H_9O_4$	265.0494	−0.6	9
Anthraquinone	2	$C_{17}H_9O_5$	293.0451	0.2	3
	3	$C_{19}H_{11}O_7$	351.0520	1.6	1
Phthalan	2(T)	$C_{11}H_9O_4$	205.0478	−2.2	2
	3(T)	$C_{13}H_{11}O_6$	263.0572	1.8	1
Xanthone	2	$C_{16}H_9O_5$	281.0461	1.2	9
	3	$C_{18}H_{11}O_7$	339.0510	0.6	7
Dibenzofuran[d] and/or naphthofuran	2	$C_{15}H_9O_4$	253.0477	−2.3	3
	3	$C_{17}H_{11}O_6$	311.0560	0.6	2
Pyridine	2	$C_8H_5O_3NCl$[e]	197.9974	1.6	3
	3	$C_{10}H_8O_5N$	222.0408	0.5	3
	4	$C_{12}H_{10}O_7N$	280.0488	3.2	3.5
Quinoline and/or isoquinoline	1	$C_{10}H_6ON$	156.0412	−3.7	0.5
	2	$C_{12}H_8O_3N$	214.0468	−3.5	2
Carbazole	1	$C_{13}H_8ON$	194.0598	−0.7	4
	2	$C_{15}H_{10}O_3N$	252.0660	−0.0	2
	3	$C_{17}H_{12}O_5N$	310.0708	−0.6	2
Acridone	1	$C_{14}H_8O_2N$	222.0525	−2.9	5
	2	$C_{16}H_{10}O_4N$	280.0592	−1.8	2
	3	$C_{18}H_{12}O_6N$	338.0635	−2.8	1

[a] Identification based on TOF variable-temperature solid inlet, GC/TOFMS, and high resolution MS. T means identification tentative.

[b] Benzenetricarboxylic acid methyl ester is normalized to 100. Relative abundances were estimated from the GC and an integration of the base peak of each compound during the time that the sample was completely volatilized in the MS.

[c] Monochlorobenzenecarboxylic acids were always obtained when coal or model compounds were oxidized in 10% HCl aqueous solution. Chlorocarboxylic acids of other aromatics were also observed in very low yield.

[d] Relatively large amounts of dibenzofurans were isolated from $Na_2Cr_2O_7$ oxidation, while photooxidation yielded dibenzofurancarboxylic acids in low yield.

[e] For pyridinedicarboxylic acid, only chloro derivatives were found.

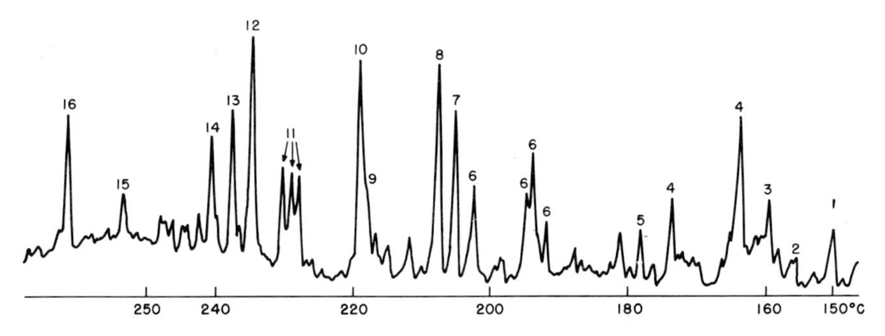

Fig. 12 Methyl esters from hydrogen peroxide–acetic acid oxidation of lignite separated on an OV-101 SCOT column temperature-programmed from 90°C at 4°C/min. Numbered compounds: 1, methyl *o*-methoxybenzoate; 2, methyl *m*- or *p*-methoxybenzoate; 3, methyl methylfurandicarboxylate; 4, methyl benzenedicarboxylate; 5, methyl carbomethoxyphenylacetate; 6, methyl methoxybenzenedicarboxylate; 7, methyl 1,2,3-benzenetricarboxylate; 8, methyl 1,2,4-benzenetricarboxylate; 9, methyl methoxybenzenetricarboxylate; 10, methyl furantetracarboxylate; 11, methyl methoxybenzenetricarboxylate; 12, methyl 1,2,3,4-benzenetetracarboxylate; 13, methyl 1,2,4,5-benzenetetracarboxylate; 14, methyl 1,3,4,5-benzenetetracarboxylate; 15, methyl methoxybenzenetetracarboxylate; and 16, methyl benzenepentacarboxylate.

has twice the phenolic content of the bituminous coal. It is interesting to note that methyl furantetracarboxylate has been identified and other furan derivatives have been tentatively identified.

It appears that the coals are acting as catalysts for this oxidation. If 2,6-dimethylnaphthalene is reacted with H_2O_2–AcOH under the same conditions as used for the coals, only 2,6-dimethyl-1,4-naphthoquinone is isolated in essentially quantitative yield. With the addition of a small amount of lignite to the reaction, 5-methylphthalic acid is obtained as the major product. Also no phenolic carboxylic acids were isolated, which indicates that this procedure does not hydroxylate aromatic rings. Transition metals are known to catalyze reactions of hydrogen peroxide (Augustine and Trecker, 1971).

No polyhydroxyl benzenecarboxylic acids have been observed in the products from either coal. It is expected that these species would undergo ring oxidation and subsequent degradation. However, the concentration in coal of these species is probably small, since these compounds are expected to be very reactive and probably would not survive the coalification process. From the aqueous NaOH extract of the Wyoming lignite, numerous hydroxylated aromatic hydrocarbons and aromatic acids were identified, but no polyhydroxylated species were detected.

F. Conclusions

The identification of the aromatic units indigenous to the coal macromolecules obtained by these selective oxidative degradations is an important step in elucidation of the structures of coals. This information should be very useful for understanding the chemistry in coal conversion processes and in the search for new methods for converting coal. The internal consistency of these results in relation to the rank of the coals supports the claim that these methods are selective and that the aromatic units are not being produced during the coal degradation. This type of argument is more persuasive than any studies on model compounds since the reactivity of coal is usually different and many times unpredictable.

V. DETERMINATION OF AROMATICITY OF COAL BY FLUORINATION

The aromaticity of coal can be determined from the stoichiometry of the reaction of coal with fluorine gas (Huston *et al.*, 1976a,b). By careful addition of fluorine at room temperature a fluorinated coal can be prepared with replacement of most of the hydrogen in coal with fluorine. In addition, fluorine adds to aromatic rings. Since HF is formed when hydrogen is replaced the weight increase due to replacement can be distinguished from that due to the addition by measuring the quantity of HF formed. When both aromatic and nonaromatic carbon are present the two can be distinguished gravimetrically by subtracting from the total fluorination weight increase of the substrate that part of the increase which is produced by hydrogen replacement. Thus, using the equation

$$\underset{13}{\mathrm{CH}} \xrightarrow{F_2} \underset{31}{\mathrm{CF}} + \underset{20}{\mathrm{HF}} \tag{4}$$

the weight increase of the substrate due to hydrogen replacement equals the weight of HF produced times (31 − 13)/20 equals the weight of HF × 0.9. Then the weight of aromatic carbon in the sample equals

$$\tfrac{12}{19}\,[\text{weight increase} - 0.9\,(\text{weight HF})] \tag{5}$$

Dividing the weight of aromatic carbon by the total carbon in the sample gives the aromaticity. The procedure can be illustrated by fluorination on a mole basis of benzene and cyclohexane:

$$\underset{78}{C_6H_6} \xrightarrow{F_2} \underset{300}{C_6F_{12}} + \underset{6\times 20}{6HF}$$

$$\text{weight increase} = 300 - 78 = 222$$

$$\text{weight of HF} = 6 \times 20 = 120$$

$$\tfrac{12}{19}\,(222 - 0.9 \times 120) = \tfrac{12}{19} \times 114 = 72 \text{ g aromatic carbon}$$

Thus benzene is calculated to be 100% aromatic.

$$\underset{84}{C_6H_{12}} \xrightarrow{F_2} \underset{300}{C_6F_{12}} + \underset{12\times 20}{12HF}$$

$$\text{weight increase} = 300 - 84 = 216$$

$$\text{weight of HF} = 12 \times 20 = 240$$

$$\tfrac{12}{19}\,(216 - 0.9 \times 240) = \tfrac{12}{19} \times 0 = 0 = \text{no aromatic carbon in cyclohexane}$$

It should be pointed out that olefinic carbon would contribute to apparent aromaticity while replaceable hydrogen in functional groups would decrease the apparent aromaticity. A high volatile bituminous Illinois coal (Peabody #2) demineralized and dried *in vacuo* at 200°C was used for the fluorination experiments. In four experiments the aromaticity was calculated to be 68.6, 69.8, 68.8, 67.0% with an average of 68.6%.

ACKNOWLEDGMENTS

We are grateful to Leon P. Moore and Robert G. Scott for technical assistance and help in preparing this manuscript. Work was performed under the auspices of the Divisions of Fossil Energy and Physical Research of the U.S. Energy Research and Development Administration.

REFERENCES

Albrecht, P., and Ourisson, G. (1971). *Angew. Chem.* **10,** 209.

Anders, D. E., Doolittle, F. G., and Robinson, W. E. (1975). *Geochim. Cosmochim. Acta* **39,** 1423.

Anders, E., Hayatsu, R., and Studier, M. H. (1973). *Science* **182,** 781.

Anders, E., Hayatsu, R., and Studier, M. H. (1974). *Astrophys. J.* **192,** L101.

Augustine, R. L., and Trecker, D. J., Eds. (1971). "Oxidation," Vol. 2. Dekker, New York.

Bearse, A. E., Cos, J. L., and Hillman, M. (1975). "Production of Chemicals by Oxidation of Coal," Battelle Energy Program Rep. Battelle Memorial Inst., Columbus, Ohio.

Beynon, J. H., Saunders, R. A., and Williams, A. E. (1968). "The Mass Spectra of Organic Molecules." Elsevier, Amsterdam.

Biemann, K. (1972). *In* "Elucidation of Organic Structure by Physical and Chemical Method" (K. W. Bently and G. W. Kirby, eds.), 2nd Ed., Vol. 4, Part 1. Wiley (Interscience), New York.

Bitz, M. C., and Nagy, B. (1967). *Anal. Chem.* **39,** 1310.

Brooks, J. D., and Smith, J. W. (1969). *Geochim. Cosmochim. Acta* **33,** 1183.
Buchandt, O., Ed. (1976). "Photochemistry of Heterocyclic Compounds." Wiley, New York.
Budzikiewicz, H., Djerassi, C., and Williams, D. H. (1967). "Mass Spectrometry of Organic Compounds." Holden-Day, San Francisco, California.
Burlingame, A. L., Ed. (1970). "Topics in Organic Mass Spectrometry." Wiley (Interscience), New York.
Burlingame, A. L., Kimble, B. J., and Derrick, P. J. (1976). *Anal. Chem.* **48,** 368R.
Carbon, J., David, H., and Studier, M. H. (1968). *Science* **161,** 1146.
Chakrabartty, S. K., and Berkowitz, N. (1974). *Fuel* **53,** 240.
Chakrabartty, S. K., and Berkowitz, N. (1976a). *Nature* (*London*) **261,** 76.
Chakrabartty, S. K., and Berkowitz, N. (1976b). *Fuel* **55,** 362.
Chakrabartty, S. K., and Kretschmer, H. O. (1972). *Fuel* **51,** 160.
Chakrabartty, S. K., and Kretschmer, H. O. (1974). *Fuel* **53,** 132.
Coleman, W. M., Woaton, D. L., Dorn, H. C., and Taylor, L. T. (1977). *Anal. Chem.* **49,** 533.
Cornu, A., and Massot, R. (1975). "Compilation of Mass Spectral Data," 2nd Ed., Vols. 1 and 2. Heyden, London.
Cram, S. P., and Juvet, R. S., Jr. (1976). *Anal. Chem.* **48,** 411R.
Done, J. N., Knox, J. H., and Loheac, J. (1974). "Application of High-Speed Liquid Chromatography." Wiley, New York.
Eglinton, G., and Murphy, M. T. J., Eds. (1969). "Organic Geochemistry: Methods and Results." Springer-Verlag, Berlin and New York.
Eisenbraun, E. J., Morris, R. N., and Adolphen, G. (1970). *J. Chem. Educ.* **47,** 710.
Ettre, L. S. (1965). "Open Tubular Columns in Gas Chromatography." Plenum, New York.
Fales, H. M., Milne, W. A., Winkler, H. V., Beckey, H. D., Damico, J. N., and Barron, R. (1975). *Anal. Chem.* **47,** 207.
Farcasiu, M. (1977). *Fuel* **56,** 9.
Fieser, L., and Fieser, M. (1968). "Reagents for Organic Synthesis," Vol. 1, pp. 191, 293, 295. Wiley, New York.
Fieser, L., and Fieser, M. (1969). "Reagents for Organic Synthesis," Vol. 2, p. 102. Wiley, New York.
Friedel, R. A., Ed. (1970). "Spectrometry of Fuels." Plenum, New York.
Friedman, L., Fishel, D. L., and Shechter, H. (1965). *J. Org. Chem.* **30,** 1453.
Ghosh, G., Banerjee, A., and Mazumder, B. K. (1975). *Fuel* **54,** 294.
Haumann, J. R., and Studier, M. H. (1968). *Rev. Sci. Instrum.* **39,** 169.
Hayatsu, R., Studier, M. H., and Anders, E. (1971). *Geochim. Cosmochim. Acta* **35,** 939.
Hayatsu, R., Studier, M. H., Matsuoka, S., and Anders, E. (1972). *Geochim. Cosmochim. Acta* **36,** 555.
Hayatsu, R., Studier, M. H., Moore, L. P., and Anders, E. (1975a). *Geochim. Cosmochim. Acta* **39,** 471.
Hayatsu, R., Scott, R. G., Moore, L. P., and Studier, M. H. (1975b). *Nature* (*London*) **257,** 378.
Hayatsu, R., Scott, R. G., Moore, L. P., and Studier, M. H. (1976). *Nature* (*London*) **261,** 77.
Hayatsu, R., Matsuoka, S., Scott, R. G., Studier, M. H., and Anders, E. (1977). *Geochim. Cosmochim. Acta* **41,** 1325.
Hayatsu, R., Winans, R. E., Scott, R. G., Moore, L. P., and Studier, M. H. (1978). *Fuel* (in press).
Huston, J. L., Scott, R. G., and Studier, M. H. (1976a). *Fuel* **55,** 281.
Huston, J. L., Scott, R. G., and Studier, M. H. (1976b). *Org. Mass Spectrom.* **11,** 383.
Imuta, K., and Ouchi, K. (1973). *Fuel* **52,** 174.
Landolt, R. G. (1975). *Fuel* **54,** 299.

Lovins, R. E., Ellis, S. R., Tolbert, G. D., and McKinney, C. R. (1973). *Anal. Chem.* **45,** 1553.
Lowry, H. H., Ed. (1963). "Chemistry of Coal Utilization," Suppl. Vol. Wiley, New York.
McFadden, W. H. (1973). "Techniques of Combined Gas Chromatography-Mass Spectrometry: Application in Organic Analysis." Wiley (Interscience), New York.
McLafferty, F. W., Ed. (1963). "Mass Spectrometry of Organic Ions." Academic Press, New York.
McLafferty, F. W., Knutti, R., Venkataraghavan, R., Arpino, P. J., and Dawkins, B. G. (1975). *Anal. Chem.* **47,** 1503.
Markey, S. P., Urban, W. G., and Levine, S. P., Eds. (1974). "Mass Spectra of Compounds of Biological Interest," Vols. 1 and 2 (TID-26553-P1,2,3). U.S. ERDA Tech. Inf. Cent., Oak Ridge, Tennessee.
"Mass Spectra: Eight Peak Index" (1974). 2nd Ed., Vol. 1-3. Compiled by Mass Spectrometry Data Center, in collaboration with ICI Ltd. Pendragon House, Palo Alto California.
"Mass Spectrometry: A Specialist Periodical Report" (1971). Chem. Soc., London.
"Mass Spectrometry: A Specialist Periodical Report" (1973). Chem. Soc., London.
"Mass Spectrometry: A Specialist Periodical Report" (1975). Chem. Soc., London.
Maxwell, J. R., Pillinger, C. T., and Eglinton, G. (1971). *Q. Rev., Chem. Soc.* **25,** 571.
Mayo, F. R. (1975). *Fuel* **54,** 274.
Nes, W. R., and Mosettig, E. (1954). *J. Am. Chem. Soc.* **76,** 3186.
Porter, Q. N., and Baldas, J. (1971). "Mass Spectrometry of Heterocyclic Compounds." Wiley (Interscience), New York.
Ranby, B., and Rabek, J. F. (1975). "Photodegradation, Photooxidation and Photostabilization of Polymer." Wiley, New York.
Schnitzer, M., and Skinner, S. I. M. (1974). *Can. J. Chem.* **52,** 1072.
Schweighardt, F. K., Retcofsky, H. L., and Friedel, R. A. (1976). *Fuel* **55,** 313.
Scott, R. P. W., Scott, C. G., Munroe, M., and Hess, J., Jr. (1974). *J. Chromatogr.* **99,** 395.
Spence, J. A., and Vahrman, M. (1970). *Fuel* **49,** 395.
Stenhagen, E., Abrahamsson, S., and McLafferty, F. W. (1974). "Registry of Mass Spectral Data," 4 vols. Wiley, New York.
Stock, R., and Rice, C. B. F. (1974). "Chromatographic Method." Wiley, New York.
Studier, M. H. (1963). *Rev. Sci. Instrum.* **34,** 1367.
Studier, M. H., and Hayatsu, R. (1968). *Anal. Chem.* **40,** 1011.
Studier, M. H., Hayatsu, R., and Fuse, K. (1968a). *Anal. Biochem.* **26,** 320.
Studier, M. H., Hayatsu, R., and Anders, E. (1968b). *Geochim. Cosmochim. Acta* **32,** 151.
Studier, M. H., Moore, L. P., Hayatsu, R., and Matsuoka, S. (1970). *Biochem. Biophys. Res. Commun.* **40,** 894.
Studier, M. H., Hayatsu, R., and Anders, E. (1972). *Geochim. Cosmochim. Acta* **36,** 189.
Tingey, G. L., and Morrey, J. R. (1973). "Coal Structure and Reactivity," Battelle Energy Program Rep. Battelle Pacific Northwest Lab., Richland, Washington.
Tuley, W. F., and Marvel, C. S. (1955). "Organic Synthesis," Collect. Vol. 3, p. 822. Wiley, New York.
van Krevelen, D. W. (1961). "Coal." Elsevier, Amsterdam.
Vaughan, G. A., and Swithenbank, J. J. (1970). *Analyst* **95,** 890.
von Kienitz, H. (1968). "Massenspektrometrie." Verlag Chemie, Weinheim.
Wasielewski, M. R., Studier, M. H., and Katz, J. J. (1976). *Proc. Natl. Acad. Sci. U.S.A.* **73,** 4282.
Wiberg, K. B. (1965). *In* "Oxidation in Organic Chemistry" (K. B. Wiberg, ed.), Part A, p. 69 Academic Press, New York.
Winans, R. E., Hayatsu, R., and Studier, M. H. (1978). In preparation.

Chapter 22

Assessment of Structures in Coal by Spectroscopic Techniques

James G. Speight
FUEL SCIENCES DIVISION
ALBERTA RESEARCH COUNCIL
EDMONTON, ALBERTA, CANADA

I. INTRODUCTION

There is little doubt that coal has been known for many hundreds of years (Galloway, 1969). The advent of the industrial revolution of the nineteenth century caused man to turn from wood to coal as a major source of fuel. With the onset of the twentieth century and the development of a technologically oriented society, further uses were found not only for coal but also for derived products. With the pronounced advances in physicochemical techniques over the last three decades, chemists and technologists have been able to gain a better understanding of the chemical changes occurring during the processing of a fossil fuel. Nevertheless, as composition studies advance to the investigation of the higher molecular weight materials, it becomes impractical to consider the determination of the individual compounds

ISBN 0-12-399902-2

and investigators must be content with information on molecular types of hydrocarbons and on structural groups; it is perhaps in this field that spectroscopy has its most valuable application. Although several reviews are available on the chemistry and physical properties of coal (van Krevelen and Schuyer, 1957; Tschamler and de Ruiter, 1963; Dryden, 1963, 1964), it is the purpose of this chapter to assemble and correlate the spectroscopic techniques which are relevant to the determination of the structural entities present in coal.

II. SPECTROSCOPIC TECHNIQUES

A. Infrared Absorption Spectroscopy

Of all the physical techniques, infrared spectroscopy gives perhaps the most valuable information about the constitution of organic materials. Indeed, qualitative information about specific structural elements can often be surmised even though the spectra are too complex for individual compound analysis.

1. *Carbon–Hydrogen Absorptions*

Infrared spectroscopic studies of coal have been the subject of many publications and attempts have been made to diagnose the functional groups and carbon skeleton of the coals. From the results of these investigations, it was generally established (Table I) that coal contained various aliphatic and aromatic carbon–carbon and carbon–hydrogen functions but few, if any, isolated C=C and C≡C bonds (Roy, 1965). Brown (1955a) reported on the infrared spectra from 650 to 4000 cm^{-1} for a variety of coal fractions (in particular, the optical densities of the two peaks at 3030 and 2920 cm^{-1}) and noted that the ratio of aromatic hydrogen to total hydrogen increased with rank; this phenomenon has also been observed in a series of Japanese coals (Kojima *et al.*, 1956; Osawa *et al.*, 1969). It has also been reported by Murchison (1966) that fractions of bituminous coals display strong absorption in the infrared which arises from nonaromatic carbon–hydrogen functions, intense absorption from aliphatic or alicyclic groups, and that the intensity of the 1600-cm^{-1} absorption typifies lower rank fractions.

The three absorption bands at 760, 814, and 870 cm^{-1} have been assigned (Gourlay, 1950; McMurry and Thornton, 1952; Hadzi, 1954; Brown, 1955a) to out-of-plane vibrations of one isolated, two adjacent, and four adjacent aromatic CH groups, respectively, while the relative intensity of these bands was suggested to give an indication of the degree of condensation of the aromatic clusters.

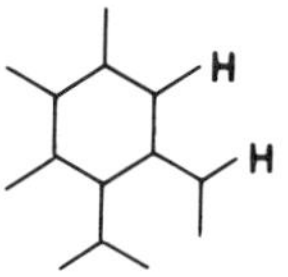

Isolated CH groups

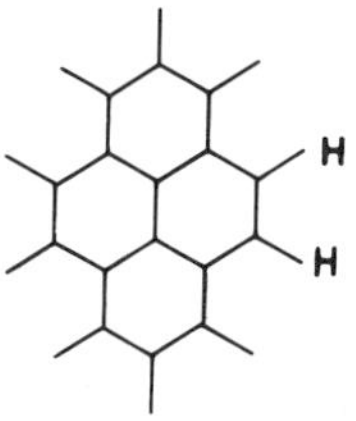

Two adjacent CH groups

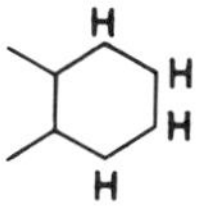

Four adjacent CH groups

More recently, it has been noted (Fujii *et al.*, 1970) that the absorption band at 3030 cm^{-1} (aromatic CH), which appears in a coal with 81% carbon, becomes more pronounced with increasing coal rank, while the band at 2920 cm^{-1} generally increases with rank to 86% carbon but thereafter decreases sharply.

2. *Functional Groups*

The assignment of absorptions in the infrared spectra to various oxygen functions has also received some attention (Table II). In addition, den Hertog and Berkowitz (1958) reported that pyridine extraction of

TABLE I *Infrared Analysis of Coal: The Carbon Skeleton*

Band (cm^{-1})	Assignment	Source
3030	Aromatic CH	Brown (1955b); Roy (1957a)
2978	CH_3	Kojima *et al.* (1956)
2940	Aliphatic CH	Cannon and Sutherland (1945); Orchin *et al.* (1951); Cannon (1953); van Vucht *et al.* (1955); Brown (1955b)
2925 2860	CH_3, CH_2, CH	Cannon (1953); Kojima *et al.* (1956); Roy (1957a)
1600	Aromatic ring C–C	Cannon and Sutherland (1945); Orchin *et al.* (1951); van Vucht *et al.* (1955); Brown (1955b)
1575	Condensed aromatic ring C–C	Cannon and Sutherland (1945)
1460	Aliphatic CH_2 and CH_3 groups	Cannon and Sutherland (1945); Orchin *et al.* (1951); van Vucht *et al.* (1955); Osawa *et al.* (1971)
1370	CH_3 group, cyclic CH_2 group	Cannon and Sutherland (1945); Orchin *et al.* (1951); van Vucht *et al.* (1955)
870 814 760	Hydrogen atoms on substituted benzene rings	Orchin *et al.* (1951); Gordon *et al.* (1952); Cannon (1953); van Vucht *et al.* (1955); Brown (1955b)

TABLE II *Infrared Analysis of Coal: Functional Groups*

Band (cm^{-1})	Assignment	Source
3300	Associated OH and NH	Cannon and Sutherland (1945); Orchin *et al.* (1951); Cannon (1953); van Vucht *et al.* (1955); Brown (1955b); Osawa and Shih (1971); Osawa *et al.* (1971)
1700	C–O	Cannon and Sutherland (1945); Murchison (1966)
1600	Hydrogen-bonded C–O	Roy (1957a); Depp and Neuworth (1958)
1300–1000	C–O (phenols) C aromatic–O–C aromatic C–O (alcohols) C aromatic–O–C aliphatic C aliphatic–O–C aliphatic	Cannon and Sutherland (1945); Cannon (1953); Brown (1955b); Kojima *et al.* (1956); Roy (1957a); Kogan *et al.* (1969); Alaev and Manaenkova (1970); Osawa and Shih (1971)

coal affords a two-component system, one rich in carbonyl and hydroxyl functions and the other in nonaromatic carbon–hydrogen bonds. Another report on the infrared spectral analysis of coal (Roy, 1965) indicates that the spectra are generally similar in higher rank coals but different in the lower rank coals. Hydrogen-bonded hydroxyl and carbonyl, or polycyclic quinones, were prominent in some fractions while others contained higher proportions of aliphatic and alicyclic ethers, epoxides, sulfoxides, and sulfones.

The infrared spectra of weathered coals (humic acids) are characterized by bands at 1600 and 1700 cm^{-1}, but no absorption band in the region for quinones. Ceh and Hadzi (1956) observed differences in the relative intensities of these two absorptions after methylation which they attributed to the presence of hydrogen-bonded hydroxyquinones. The presence of quinones in weathered coals and the origin of the 1600- and 1700-cm^{-1} bands have been discussed by Brooks *et al.* (1960) who observed that changes in the position and intensities of these bands after methylation and demethylation could be explained without any reference to quinone structures. It appears that failure to observe the characteristic quinone absorption was due to hydrogen bonding of the quinone carbonyl with unreacted adjacent, and/or peri, hydroxyl groups in the methylated materials which retained the carbonyl frequency within the 1600-cm^{-1} region (Ceh and Hadzi, 1956; Brooks *et al.*, 1960). Hydroxyquinones having hydroxyl groups in adjacent and peri positions to the carbonyl groups exhibit carbonyl maxima in the 1600-cm^{-1} region (Flett, 1948) which remain unchanged by treatment with diazomethane (Schonberg and Mustafa, 1946; Ceh and Hadzi, 1956; Bom *et al.*, 1957); it seems likely that their contribution to this band would be unaffected by methylation. However, hydroxyl groups of these types can be acetylated (Johnson *et al.*, 1951; Blom *et al.*, 1957), destroying the hydrogen bonding. The position of the carbonyl absorption in the acetoxyquinone is comparable to that in the parent quinone, but an additional band near 1770 cm^{-1}, due to the acetoxy group, appears (Johnson *et al.*, 1951). Subsequent examination of the infrared spectra of the methylated and acetylated humic acids showed two new absorption bands at 1780 and 1660 cm^{-1} which were assigned to acetate esters and quinones, respectively.

Czuchajowski (1961) also discussed the presence of quinones in coal and oxidized coals and the products of the reaction of the latter with bisdiazonium compounds. An absorption band at 1640 cm^{-1}, observed in the spectra of model compounds obtained from the polycondensation of phenanthrene, pyrene, and chrysene with formaldehyde and subsequent pyrolysis and nitric acid oxidation, was also assigned to nonchelated quinoid carbonyl groups. This band was absent from the spectra of coal, oxidized coals, and their products following reaction with bisdiazonium compounds after pyrolysis and oxidation with nitric acid. It is evident that the quinone structures present in weathered coals are relatively simple since x-ray diffraction data for low and medium rank coals exclude polycondensed systems containing more than, say, four to five rings (Francis, 1961, p. 701) and only a few polynuclear quinones absorb at or near 1660 cm^{-1}. On the other hand, polarographic evidence produced by Given and Peover (1960) supports the idea that hydroxyl-substituted naphthaquinones as well as anthraquinones and more complicated systems involving heterocyclic oxygen and polynuclear hydrocarbons are possible.

Durie and Sternhell (1959) made a quantitative study of the changes occurring in the infrared spectra of coals during acetylation and noted that part of the oxygen is in the form of hydroxyl groups which are sterically hindered from acetylation and are thermally stable up to 450°C. In other infrared spectroscopic studies which include the carbonization of coal, Brown (1955a) noted that the product spectra indicate that the most marked structural changes are loss of thermally labile aliphatic CH and phenolic OH. Oelert (1968) reported a study of the chemical characteristics of the thermal decomposition of bituminous coals at temperatures up to 500°C by examining the residues from thermogravimetric experiments by quantitative infrared spectroscopy. The carbon content in methyl groups was calculated from the 1380-cm^{-1} band and carbon in aromatic CH groups from bands in the region

650–910 cm^{-1}. The proportion of carbon-bearing phenolic hydroxyl was deduced from the band at 1750 cm^{-1} in the spectra of acetylated samples while carbon in quinoid locations was estimated from bands in the range 1540–1640 cm^{-1}. In agreement with this study, an infrared investigation of the thermal treatment of a Korean coal by Li and Boo (1964) showed that the ratio of aliphatic to aromatic moieties decreased with increasing temperature while absorbance fluctuated with temperature.

3. *The 1600-cm^{-1} Band*

Many research groups have focused their attention on an assignment of the 1600-cm^{-1} band in the infrared spectra of coals. This band also occurs in the asphaltene fraction of petroleums, even in those with less than 1% oxygen (Speight, 1972), and has been reported to be unaffected either by acetylation or by oxidation of fractions (Bergmann *et al.*, 1958). From the $LiAlH_4$ reduction of coals, Fujii (1963a,b) concluded that this absorption band is due predominantly to the presence of hydrogen-bonded carbonyl groups in the coals. Rao *et al.* (1962) reported that the band at 1600 cm^{-1} in the infrared spectra of coals could not be assigned principally to carbonyl absorption. An interpretation of the infrared spectrum of an Indian coal was made on the basis of a polynuclear condensed aromatic nucleus, with the aliphatic moiety made up principally of methylene-containing rings. However, Elofson and Schulz (1967) contend that the 1600-cm^{-1} band arises through donor–acceptor phenomena between the aromatic sheets in the coal molecules. More recently, Friedel *et al.* (1971) have concluded that noncrystalline graphite-type structures may be responsible for the 1600-cm^{-1} band.

4. *Quantitative Studies*

Quantitative studies arising from the infrared spectra of coals are infrequent due, in no small part, to the inaccuracies which are inherent in the method. Nevertheless, it has been reported that the aromatic hydrogen–total hydrogen ratio increases with increasing rank, and the data indicate that at a carbon content of 94%, coal is completely aromatic (Brown, 1955a). Durie and Sternhell (1959) studied the changes occurring in the infrared spectra of coals arising from acetylation and concluded that the oxygen (5–7% of the total oxygen) which is not detected in the functional group analysis was predominantly in the form of hydroxyl groups resistant to acetylation.

Bent *et al.* (1964) attempted to estimate the methyl group content of coal and coal derivatives by infrared spectroscopy but an absolute measure of the methyl hydrogens was not possible. It appeared that the percentage of the total hydrogen contained in methyl groups probably lay in the range 15–25% and the methyl content decreased with increasing rank of the coal. Oelert (1965) also attempted to estimate the methyl and methylene groups in a variety of coals using the maximum extinction of a band at about 1380 cm^{-1} and the integral extinction of bands between 650 and 910 cm^{-1} which, with elemental analysis, allowed calculation of the carbon distribution in methyl, methylene, and in aromatic locations, as well as aromaticities of coals. However, Friedel (1959a) has claimed that the conventional carbon–hydrogen bending frequencies for methyl and methylene groups, located in the range 1300–1500 cm^{-1}, cannot be used for accurate estimations of methyl and methylene carbon since the infrared spectra of the chars from fully deuterated materials exhibit absorption in this region, and hence these bands are not entirely due to the aforementioned frequencies.

B. Nuclear Magnetic Resonance Spectroscopy

Nuclear magnetic resonance spectroscopy has proved to be of great value in coal research because it allows rapid and nondestructive determination of the total hydrogen content and distribution of hydrogen among the chemical functional groups present in the sample. Hydrogen atoms are liberally scattered throughout most of the molecular entities encountered in coal and, since the peak areas in the nuclear magnetic resonance spectra are directly proportional to the number of contributing hydrogen atoms, a series of structural parameters can be derived. However, since the major part of coal is notoriously insoluble in a variety of solvents, it has been the custom of investigators to examine thermally treated materials and products in which it was hoped that the basic skeletal structure of the coal remained substantially intact.

1. *Aliphatic Structures*

In an early publication on the subject, Bell *et al.* (1958) reported that in a high rank coal 33% of the hydrogen atoms occurred in methylene groups of bridge and alicyclic structures, whereas in a low rank coal 67% of the hydrogens occurred in this form and/or in long chains. In a further communication, Richards and Yorke (1960) concluded that the proportion of perimethyl groups in coal is very small. Brown *et al.* (1960) studied the nuclear magnetic resonance spectra of the products obtained from the vacuum carbonization of coal and concluded that the nonaromatic hydrogen occurred almost exclusively on saturated carbon atoms as aliphatic, alicyclic, or hydroaromatic structures. There was no evidence that methylene bridges existed to any appreciable extent and the proportions of aromatic hydrogen, hydrogen on carbon atoms alpha to aromatic rings, and hydrogen on other saturated carbon atoms were estimated quantitatively. Solvent extracts of Japanese coals have recently been shown to contain naphthene-type compounds having four or five rings (Yokoyama *et al.*, 1970).

Oth and Tschamler (1963) employed magnetic resonance spectroscopy to derive structural parameters for the distribution of aliphatic hydrogen in a Belgian coal and its extraction products. They noted a ratio of four methylene groups per methyl group and that the methyl content was higher in the soluble material and varied with the solvent power of the solvent. It was possible to estimate the methine content of the samples, however, only by employing derived data which were not in contradiction with other structural parameters. Ladner and Stacey (1961) determined the ratio of aromatic hydrogen to aliphatic hydrogen for a series of low temperature vacuum-carbonized coals and found that the results compared favorably with those derived by the infrared method. More recently, Ladner and Stacey (1965) investigated the carbonization of coals at temperatures up to 900°C and found a significant loss of methylene hydrogen. This finding is consistent with the indication from infrared studies that it is essentially the aliphatic structures in coals that undergo thermal decomposition in that particular temperature region. Bartle and Smith (1965, 1967) employed proton magnetic resonance spectroscopy to calculate the relative incidence of the various saturated structures present in refined coal tar. These authors concluded that more than 80% of the tar could be accounted for in terms of alkyl-substituted open-chain polynuclear aromatic molecules with a structure that was a continuous function of molecular weight. The principal alkyl substituents were determined to be methyl with lesser amounts of ethyl, propyl, and butyl groups.

2. *Aromatic Structures*

Friedel (1959a,b) examined the nuclear magnetic resonance spectrum of a fraction obtained by the hydrogenation of coal at 400°C. Friedel concluded that the large molecules of the product contained a copious distribution of tetralin-type structures but that any aromatic protons were benzenoid rather than those of large condensed aromatic structures. Brown and Ladner (1960) compared hydrogen distribution data from the high resolution nuclear magnetic resonance spectra of low temperature vacuum carbonization products and used the data to calculate aromaticity, degree of aromatic substitution, and ring size for different assumed values of the average composition of nonaromatic structures. The results conformed to the independent results of the infrared and x-ray methods as applied to the parent coals indicating the presence of condensed aromatic structures.

Studies of coals of different rank (Heredy *et al.*, 1966) showed that the proportion of aromatic hydrogen varied from 20 to 32% of the total hydrogen and that no simple relationship existed with rank. β-Paraffinic naphthenic hydrogens comprised 38–66% of nonaromatic hydrogens while the most striking variation in the structures of the coals was in the number of methylene bridges. A general coal structure in accord with these results was determined to be large units (molecular weight >4000) and smaller units (molecular weight 300–600) linked by methylene bridges. Given (1960, 1961) estimated the ratio of aromatic carbon to aliphatic carbon for a variety of coals and reached the conclusion that coal is a peri-condensed polymeric structure, which is in contrast to Yen and Erdman's (1962) postulate of coal being predominantly kata-condensed. Japanese workers (Takeya *et al.*, 1963, 1964) studied the constitutions of the pyridine extracts of several Japanese coals and concluded that the average coal consisted of molecules having four to six aromatic rings carrying aliphatic side chains of four or five carbon atoms; the average bituminous coal had molecules with four or five aromatic rings bearing side chains of three or four carbon atoms.

Brooks and Stevens (1964) examined tars produced by mild pyrolysis of a bituminous and a perhydrous coal and concluded that a typical aromatic unit in the tars contained two to four rings with about 50% of the peripheral carbon atoms carrying short alkyl substituents or hydroaromatic structures. There was no indication of methylene bridges or polyalicyclic compounds in the tars. Brooks and Smith (1964) also showed that tars obtained from the fluidized-bed carbonization of a bituminous coal contained an aromatic fraction composed of alkyl- and hydroaromatic hydrocarbons of the naphthalene, anthracene, and phenanthrene series. High resolution proton magnetic resonance spectra have been recorded at 200 MHz for fractions of coal tar (Bartle and Jones, 1969) and signals were well enough resolved to enable single compounds (naphthalene, anthracene, phenanthrene, acenaphthene) to be identified in several of the lower molecular weight samples; olefinic hydrogens were also identified. Other investigations (Brown and Waters, 1966) have shown that the coking of coals produces soluble extracts which contain about 36 carbon atoms per molecule and have a basic structure of up to seven fused rings of which the majority are aromatic and bear short alkyl (up to butyl) substituents.

It has also been reported (Rao *et al.*, 1960) that coal tar pitch is not completely aromatic but that methyl hydrogens comprise up to 10% of total hydrogen, and the ratio of aromatic hydrogen to aromatic carbon implies a high degree of condensation. Distribution of the major part of the aromatic hydrogen in structures containing four neighboring aromatic CH bonds is attributed to ortho-disubstituted terminal rings of the condensed system. About 72% of the nonaromatic hydrogen was situated on carbon alpha to the ring system and it appeared that aliphatic carbon was largely in methyl groups directly attached to the ring as well as in methylene bridges and small alkyl groups. Durie *et al.* (1966) studied the hydrogen distribution in the pyridine and chloroform extracts from seven Australian coals and showed that in the soluble fractions about 50% of the aliphatic hydrogen was present on carbon atoms directly attached to aromatic rings. Indeed, the aromatic nature of coal has, in part, been reaffirmed by Japanese workers who used proton magnetic resonance to examine hydrogen distribution in coal (Yokoyama *et al.*, 1970, 1975; Yoshida *et al.*, 1975). Solvent extracts of Hokkaido coals were shown to contain alkylbenzenes, alkylnaphthalenes, alkylanthracenes, and alkylphenanthrenes (Yokoyama *et al.*, 1970).

3. *Carbon-13 Magnetic Resonance*†

Friedel and Retcofsky (1963, 1966a,b, 1968) have recently applied carbon-13 nuclear magnetic resonance to the elucidation of structures of soluble coal fractions. In particular, they examined a neutral oil obtained by cracking (at 700°C) a carbonization (450°C) product of coal. The data obtained, used in conjunction with those obtained from proton magnetic resonance (Friedel and Retcofsky, 1961) and elemental analysis, indicated an average molecule of the neutral oil to be 70% aromatic and to consist of a naphthalene ring system bearing two or three saturated side chains, each having less than three carbon atoms. Lately, Retcofsky and Friedel (1968) applied carbon-13 magnetic resonance to the carbon disulfide extract of a Pittsburgh coal. These authors determined that the mean structural unit consisted of two to three polynuclear condensed aromatic rings with 40% of the available aromatic carbons bearing alkyl, phenolic, and/or naphthenic groups. Indeed, the mass spectrum of the extract indicated the presence of alkyl aromatic compounds having from 1 to 10 or more alkyl carbons per molecule. The results obtained using this technique support the classical views that coals are highly aromatic materials and that the aro-

† See Chapter 24 for a full discussion of this technique.

maticity of coal increases with increasing rank (Retcofsky and Friedel, 1973; van der Hart and Retcofsky, 1976).

C. X-Ray Diffraction

Formidable though the combination of the various spectroscopic techniques is in structural analysis, there are times when this combination cannot provide a solution and x-ray diffraction studies have been initiated in attempts to resolve these difficulties. However, this method requires that within a given fraction, the larger portions of the molecules contain within and among themselves certain repeated structural features as, for example, sheets of condensed aromatic rings.

1. *Crystallite Structure*

The x-ray scattering from coals was the subject of several early studies (Mahadevan, 1929, 1930; Brusset *et al.*, 1943; Brusset, 1947; Riley, 1944; Nelson, 1952; Mitra, 1953a), notably that of Blayden *et al.* (1944) who postulated that coal contains aromatic layers about 20–30 Å in diameter, aligned parallel to near neighbors at distances of about 3.5 Å. Later development of the method by Hirsch (1954) and Brown and Hirsch (1955) led to the conclusion that the aromatic carbon occurred in layers composed of four to five condensed rings for coal up to 87% carbon, to about 30 condensed rings in coal having 94% carbon. Nelson (1954) proposed that if coal (up to about 90% wt/wt carbon) contained condensed aromatic ring systems, their average number could not exceed about four rings. Other work by Hirsch (1958) indicated that coal (up to about 90% carbon) contained an appreciable proportion of small layers consisting of one to three rings. In addition, Cartz and Hirsch (1960) stated that these small condensed aromatic regions form part of larger units which may themselves be linked to other such units by aliphatic or alicyclic material or by five-membered rings to form large buckled sheets. Above 90% wt/wt carbon, the condensed aromatic layers increase rapidly in size with increasing rank. Ergun and Tiensuu (1959a) concluded that alicyclic structures were present to a considerable degree in coals of medium rank. As a result of further studies (Ergun and Tiensuu, 1959b; Ergun *et al.*, 1959, 1960), considerable differences were noted in the x-ray scattering characteristics of different coal components, and also that certain high rank coals gave rise to the three-dimensional crystalline reflections of graphite. After a further systematic study of several high rank coals (Menster *et al.*, 1961, 1962), McCartney and Ergun (1965) were able to find single graphitic crystals (up to 3 μm in size) in one of the coals and concluded that coal ultimately becomes graphite. Other workers (Dhar and Niyogi, 1942; Mitra, 1953b; Hirsch, 1954; Inouye *et al.*, 1953) noted that the aromatic interlayer distance of coal decreases (3.9 → 3.5 Å) with increasing rank and approaches the minimum theoretical separation (3.44 Å) between aromatic lamellae (Blayden *et al.*, 1944; Franklin, 1951).

2. *Layer Distribution*

In order to gain additional information about coal structure, Diamond (1957, 1958, 1959) developed a method to derive the layer distribution in coal in terms of molecular size histograms. It was difficult to make an accurate estimate of the number of condensed aromatic rings per layer since edge groups may contribute to the diffraction of the layer. The results, however, indicate that all coals with up to 90% carbon may contain only very few condensed rings (Hirsch, 1958). Application of the method to aromatic polymers gave aromatic size distributions, average layer sizes, and mean band lengths which were in good agreement with the theoretical values (Ruland, 1959). Nevertheless, the method has received some criticism (Brooks and Stephens, 1965) and is used only infrequently.

D. Electron Spin Resonance

With the advent of electron spin resonance, also called electron paramagnetic resonance, data of varied types with both direct and indirect bearing on the structure of coal and the location of the heteroatoms, have been obtained.

Early work by Ingram *et al.* (1954) on a series of carbonized coals gave 3×10^{19} free radicals per gram (1 free radical per 1600 carbon atoms). At the same time, Uebersfeld (1954) independently observed a similar phenomenon, and further work by Ingram (1955, 1957), Austen *et al.* (1958), Uebersfeld *et al.* (1954), and others (Smidt, 1958; Honda *et al.*, 1958; Smidt and van Krevelen, 1959; Losev and Bylyna, 1959) established that the free radical content of coal at first increaes slowly (in the range 70–90% carbon), rises markedly (90–94% wt/wt C), and then decreases to limits below those of detectability. Thus, in a coal having 70% carbon there is 1 radical per 50,000 carbon atoms, but this is increased to 1 radical per 1000 carbon atoms in coal with 94% wt/wt carbon. Berkowitz *et al.* (1961) examined the free radical concentration of a series of coal chars and equated this to carbonization temperature. As a result of a study of the free radical concentrations of coal chars,

Brown and Berkowitz (1966) postulated that the free radical sites were formed via charge transfer processes rather than by rupture of a σ bond. Retcofsky *et al.* (1967) examined coals of varying rank and found values typical of those anticipated for neutral radicals of carbon and nitrogen as well as hydrocarbon ions. It was suggested that free electrons in these coals may be associated with a delocalized π system of electrons stabilized by resonance. Toyoda and Honda (1966) confirmed the early work which showed that the radical concentration of coal increases with rank, and Yokokawa (1968a) concluded that the spin centers of coal were aliphatic or alicyclic up to 500°C but became aromatic above 500°C. In further communications, Yokokawa (1968b, 1969) reported variations in the electron spin resonance spectra of coals of different rank at temperatures between 120 and 260°C in the presence of solvents such as ethylenediamine, pyridine, cyclohexanone, and quinoline. In general, a decrease in free radical concentration occurred and was ascribed to the occurrence of spin–spin coupling after strong swelling of the coals. Other workers (Ohuchi *et al.*, 1969) examined radical concentrations of coal before and after exposure to air and noted an irreversible increase in free radical concentration in the air-exposed samples; it was suggested that air oxidation of virgin coal produced free radical sites in either the aliphatic or alicyclic moieties.

Chauvin *et al.* (1969) concluded that the tar produced during coal pyrolysis had a larger proportion of aliphatic moieties than the original coal and progressive reheating of the tars caused a decrease in the radical concentration. de Ruiter (1962, 1965a) examined the intensities of the electron spin resonance signals in the spectra of coal and coke and noted a reduction in intensity in the presence of oxygen. In a further report, de Ruiter (1965b) showed that the electron spin resonance signals of a series of coke samples underwent reversible changes when the samples were brought into contact with oxygen or concentrated sulfuric acid; the changes were related to penetration by oxygen and sulfuric acid into the pore systems of the cokes. Polish workers (Kuczynski *et al.*, 1965) reported that acid treatment of brown coal caused a general increase in free radical concentration, but an increase in temperature during the treatment caused a decrease in the free radical concentration. These authors suggested that higher temperatures accelerated the recombination of the free radicals. Austen *et al.* (1959) followed the progress of the electrolytic reduction of coal by the electron spin resonance method and reported that radicals resulting from the reduction of coal were stable at room temperature. In a similar manner, Reggel *et al.* (1961) noted that treatment of coal fractions with lithium–ethylenediamine resulted in a substantial decrease in free radical concentration. Brooks and Silberman (1962) prepared a series of cokes at temperatures up to 900°C from a high rank coking coal and observed that, whereas chemical reduction diminished the number of free radicals in the low temperature (<650°C) cokes, the radical concentration in the 700–900°C cokes was increased. They concluded that carbon atoms carrying lone pairs of electrons were present in the high temperature cokes. More recently, Russian workers (Chernyshev and Opritov, 1970; Vasil'eva *et al.*, 1972; Tyutyunnikov *et al.*, 1973) have noted that the free radical concentration increased with increasing coal rank and have suggested that the signals are caused by paramagnetic condensed aromatic rings as well as free radicals.

E. Mass Spectroscopy

Mass spectroscopy adds further knowledge to the structural analysis of coal by allowing calculation of ring distribution as well as identification of individual molecular ions. Thus, application of this technique to the identification of methyl esters of the organic acids obtained by the controlled oxidation of bituminous coal (Holly *et al.*, 1956; McLafferty and Gohlke, 1959) allowed the more volatile benzenecarboxylic acid esters to be identified. A detailed example of this application is given in Chapter 21, Section IV. These were esters of benzenetetracarboxylic acid, trimellitic acid, methylphthalic acid, *o*-phthalic acid, isophthalic acid, terephthalic acid, toluic acid, and benzoic acid. Decarboxylation of the total acid mixture was shown to afford benzene, toluene, C_2-benzenes (i.e., ethylbenzene or xylenes), C_3-benzenes, butylbenzenes, C_5-benzenes, C_7-benzenes, naphthalene, methylnaphthalene, C_2-naphthalene, biphenyl, methylbiphenyl, C_3-biphenyl, indane, methylindane, C_2-indane, phenanthrene, and fluorene (Montgomery *et al.*, 1956; Montgomery and Holly, 1957). It was concluded that these nuclei are prodiminant in bituminous coal but are linked by more readily oxidizable structures. Reed (1960) examined solvent extracts of a British coal and found evidence for the presence of short methylene chains which were either part of an aliphatic chain or an alicyclic ring. Holden and Robb (1960) noted that a series of tetralins (or indanes) and higher kata-condensed aromatics were prevalent in the extracts of this same coal.

High resolution mass spectroscopic analyses (Kessler *et al.*, 1969) of pyridine extracts from reduced and untreated American coals support earlier claims (Reggel *et al.*, 1961, 1964) that ether linkages exist in the coal and are split during hydrogenation, and that hydroaromatic compounds can be formed by addition of hydrogen to the aromatic structures. Notable fragmentation ions in the mass spectra of the volatile extracts were phenols, dibenzofurans, benzonaphthothiophenes, anthracenes (and/or phenanthrenes), and hydronaphthalenes. The pyridine extracts of a Pittsburgh seam coal (representing about 20% of the original coal) were shown to contain peri-condensed aromatics, e.g., pyrene and its alkyl derivatives, as well as kata-condensed aromatics, e.g., dibenzochrysene and its alkyl derivatives (Sharkey *et al.*, 1966). Progressive pyridine extraction of the Pittsburgh coal (Friedel *et al.*, 1968) and examination of each fraction identified aromatics from alkylbenzenes and phenols to six-ring peri-condensed aromatics as constituents of the coal.

Schultz *et al.* (1965) analyzed three fractions of coal tar pitch and concluded that the mean structural unit consisted of four to five aromatic rings although individual ring systems, including alkyl derivatives, varied from naphthalenes to dibenzoperylenes. Furthermore, these authors (Sharkey *et al.*, 1961) had previously reported that a vacuum pyrolysis condensate and methanol–benzene extract of coal contained similar major constituents, namely, alkyl-substituted benzenes, tetralins, condensed aromatics up to chrysenes, and hydroxyl-substituted derivatives of the lower molecular weight compounds.

F. Ultraviolet Spectroscopy

The ultraviolet spectra of coals, examined as suspensions in potassium bromide, show an absorption at 2650 Å which becomes more pronounced with increasing rank of the coal (Fujii, 1959). This band has been assigned to aromatic nuclei (Hunphreys-Owen and Gilbert, 1958) although it has been noted that β-diketones have their strongest band in this region, requiring only small amounts to produce a moderately intense band (Friedel and Queiser, 1959). On the basis of data obtained from a comparison of the specific extinction coefficients of coal with those of standard condensed aromatic compounds, Friedel (1957) concluded that the concentration of aromatic systems in coal is lower than had previously been believed.

de Ruiter and Tschamler (1958) examined the ultraviolet spectra of the pyridine extracts of a Pittsburgh coal and were able to reconstruct the spectrum using a mixture of aromatic hydrocarbons varying from naphthalene to 7,8-benzterrylene. Other workers examined the absorption spectra of coal extracts and identified naphthalenes (Barclay and Layton, 1956; Roy, 1957b), xylenols (Roy, 1957b), and a variety of heterocycles (Roy, 1957b). It concluded that coal contains benzene and naphthalene rings, cyclic ethers, heterocyclic nitrogen, hydroxyl, and methylene groups (Roy, 1957b), and is fairly uniform in structure (Barclay and Layton, 1956). de Ruiter and Tschamler (1962) concluded that the position of the maxima in the ultraviolet spectra of coal fractions indicated a mean cluster size comparable with that obtained by other methods. Ergun *et al.* (1961) studied the ultraviolet absorption spectra of coals of varying rank and concluded that low rank coals contained small aromatic nuclei of various kinds with a gradual coalescence of the units to form much larger layers of aromatic nuclei occurring as rank increases.

G. Electron Microscopy

Waisman *et al.* (1954) employed electron microscopy as a means of elucidating coal structure and observed granules as small as 200 Å in diameter, whereas Alpern and Pregermain (1956) reported that the particles in coal fractions varied from 200–300 Å in diameter to about 1000 Å in dimaeter. In a later communication, Pregermain and Guillemot (1960) reported that particles ranged from 250 Å in a low rank coal to 100 Å in a high rank coal, while McCartney *et al.* (1966) observed two general ranges for thc ultrafine structures, one >100 Å and the other <100 Å, with some of the particles in the form of polygonal platelets.

III. ASSESSMENT OF COAL STRUCTURE

There are many postulates of the structural entities present in coal which involve estimations of the hydrocarbon skeleton, but although spectroscopic techniques may appear quite formidable as an aid to structural analysis of coal, the validity of any conclusions may always be suspect. As an illustration, Ruof *et al.* (1956) showed that the infrared and x-ray diffraction spectra of the aromatic acids and their methyl esters formed during oxidation of carbon black were almost identical with those of mellitic acid and its methyl ester, but that the chemical and physical properties were markedly different and the ultraviolet spectra bore very little resemblance to those of the pure materials. Nevertheless, the major drawback to the investigation of coal structure has been the incomplete solubility of the material, which has dictated that, in many cases, structural determinations be carried out on extracted material. Thus, a series of experiments by Rybicka (1959), designed to examine the solubility of an English coal in a variety of solvents, and examination of the infrared spectra of the extracts by Brown (1959) indicated that coal consisted of structures of basically similar chemical type; these studies suggested that coals closely related in rank may be homogeneous in chemical structure. Indeed, there are numerous examples cited in the literature which support this view. It is not surprising that material extracted from coal has been employed as being representative of that particular rank for structural determinations. Differences exist predominantly in the molecular weights (i.e., degree of polymerization) of the structural entities (Oelert and Hemmer, 1970a,b). However, Friedel *et al.* (1968) showed that the pyridine extracts of a Pittsburgh coal differed substantially with increasing extraction time, and different constituents predominated at different stages of the extraction. In spite of this latter observation, the structural properties of coal fractions have been investigated for many years. Early workers believed that coal was composed predominantly of a vast aromatic network that was only slightly removed from carbon and graphite. As work progressed, it became clear that this was not true and that coal contained distinct organic molecular entities.

An early model proposed for coal structure was based on ovalene:

with the nonaromatic part represented as alkyl substituents (Huck and Karweil, 1953a,b). Another early attempt to illustrate coal structure was made by Dryden (1955) who suggested ether linkages as a means of combining alkyl-substituted pyrene and/or coronene nuclei:

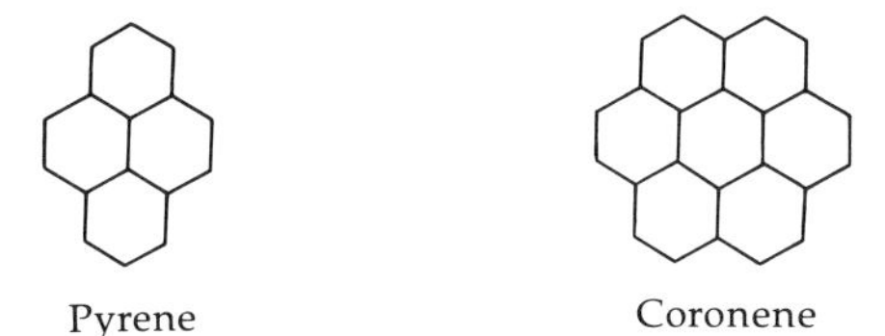

Pyrene Coronene

In a later publication, Dryden (1957) concluded that coal is composed basically of an aromatic cluster unit bearing alkyl, as well as oxygen, substituents, with the low rank coal unit being composed of approximately three fused aromatic rings and high rank coal composed of 10–20 fused rings. Given (1960) also reviewed the evidence available from the infrared and x-ray methods and concluded that coal exists as polymeric structures in which the monomer is based on 9,10-dihydroanthracene:

but the absence of bridging methylene carbons, as determined from high resolution proton magnetic resonance spectroscopy and chemical methods, caused Given (1961) to modify his basic aromatic unit of coal from a dihydroanthracene to a dihydrophenanthrene:

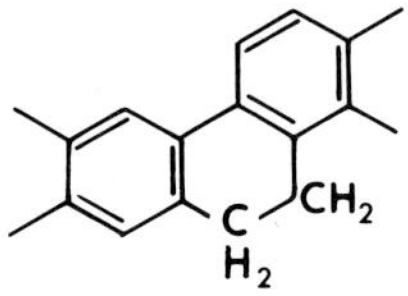

without seriously affecting the overall model for coal structure. Given's postulates of coal structure also contained a novel element insofar as they invoke the concept of an extra dimension. This was achieved by employing a tryptycene moiety in the coal molecule which necessitated that the structure not only be buckled but also that part of the coal molecule project in an additional dimension to the rest (Fig. 1).

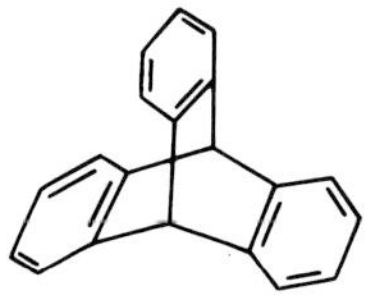

Tryptycene

Cartz and Hirsch (1960) also proposed a structure, consistent with their x-ray data, for coal (84.5% C) wherein the average molecule was represented as a buckled sheet consisting of condensed aromatic and hydroaromatic rings bearing only short alkyl substituents (methyl and

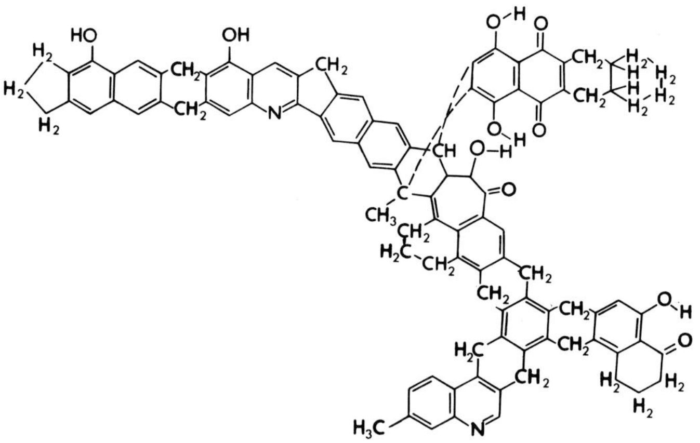

Fig. 1

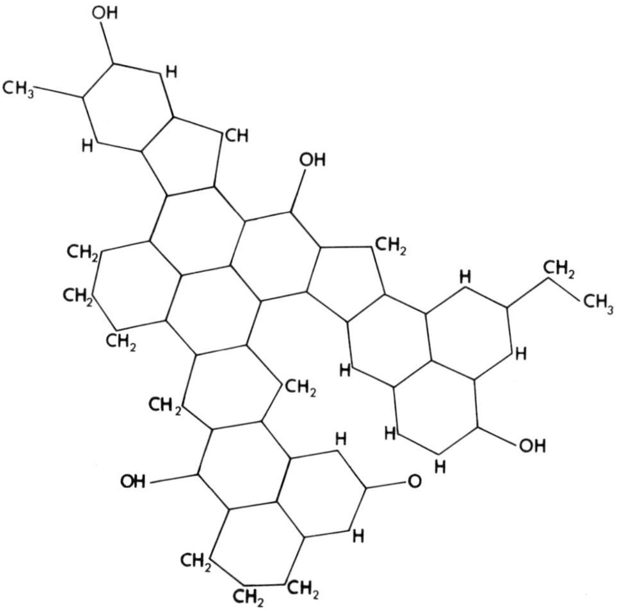

Fig. 2

ethyl) with bridging and nonbridging methylenes adjoining the aromatic rings (Fig. 2). On the other hand, Hill and Lyon (1962) reviewed the chemical and spectroscopic evidence and proposed a vast complex network of condensed aromatic and heteroaromatic nuclei bearing alkyl, as well as heteroatom, substituents. The aromatic centers varied from two to nine rings with the heteroatoms occurring as heterocyclic ring systems or as substituents on the aromatic and naphthenic nuclei (Fig. 3).

Fig. 3

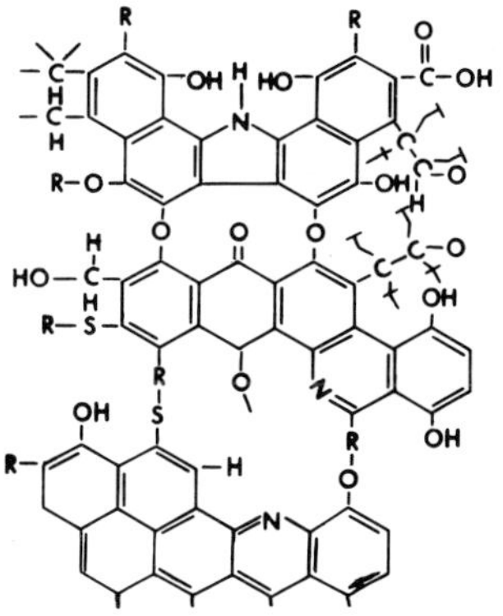

This postulate also invoked the concept that tetrahedral three-dimensional carbon–carbon bonds were present in the coal molecule, thus indicating an even higher degree of complexity than can be illustrated. Mazumdar *et al.* (1962) and Mazumdar and Lahiri (1962) also considered coal to be composed of essentially small, but heterogeneous, condensed aromatic ring systems. They also noted that the sum of the aromaticity and alicyclicity from lignite to high rank bituminous coal was substantially constant and proposed structures to illustrate the transition from lignite to the higher rank coals (Fig. 4).

It is apparent that the theory of coal structure has evolved from those theories which invoke the concepts of highly condensed aromatic clusters. In fact, through the agency of later studies, involving an x-ray diffraction technique, the majority of the investigators deduced that low to medium rank coal consisted of clusters of condensed (three- to six-ring) aromatic nuclei (Hirsch, 1954; Nelson, 1954; Brown and Hirsch, 1955; Hirsch, 1958; Ergun and Tiensuu, 1959b; Cartz and Hirsch, 1960). It was not fully understood what proportion of aliphatic structures occurred in coal and, in the majority of postulates, the aliphatic portion of the molecule was largely ignored. Although all the postulated coal struc-

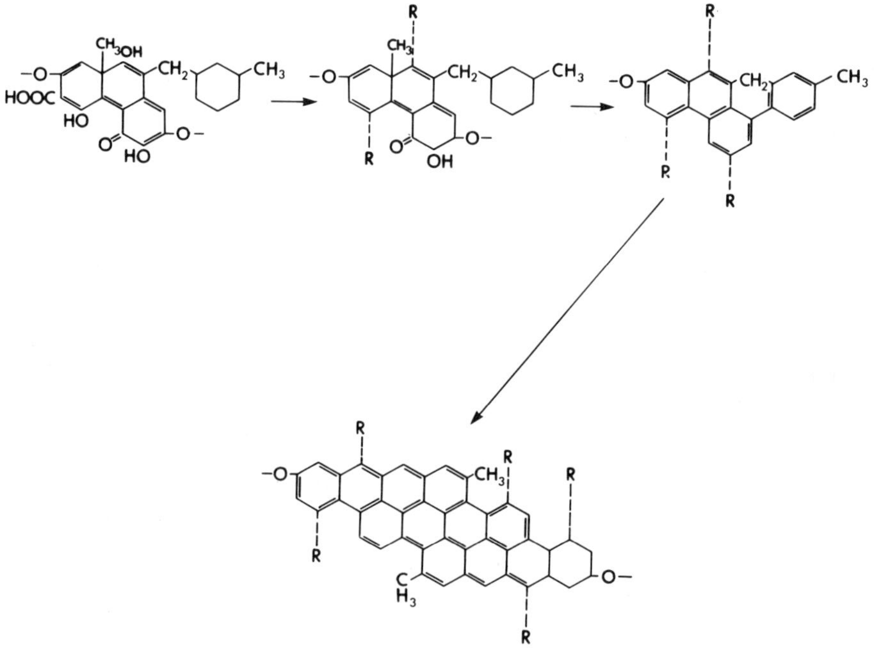

Fig. 4

tures have been found satisfactory to explain the constitution and chemical behavior of coal to some extent (but not entirely), there still remained certain inexplicable facets of the chemical and physical behavior of coal.

In an early attempt to elucidate the role of aliphatic structures in coal (Friedel, 1957, 1959a; Friedel and Queiser, 1959), it was intimated that the principal entities in coal may not be predominantly polynuclear aromatics but could well be composed of significant proportions of tetrahedral carbon structures, including quaternary carbons. In a later report, Friedel *et al.* (1968) showed that progressive extraction of coal with pyridine afforded significant quantities of benzenes, phenols, dihydric and/or alkoxyphenols, and naphthalenes. Kata- and peri-condensed aromatics were obtained in the later stages of the extraction. Similarly, Vahrman (1970) demonstrated that *n*-alkanes, branched alkanes, cycloalkanes, cycloalkenes, alkylated benzenes, naphthalenes, and other condensed aromatics up to picenes, as well as benzfluoranthrenes, can be isolated in substantial quantities by exhaustive extraction. These authors suggested that the saturated moieties of coal may predominate.

Ergun (1958) and Ergun and Tiensuu (1959c) have also questioned the role of the aromatic centers in coal structure and have even shown that clusters of tetrahedrally bonded carbon atoms give rise to x-ray diffraction bands in approximately the same angular region in which the two-dimensional reflections of graphite-like layers occur. In fact, Ergun concluded from the diffuse diffraction peaks that many amorphous carbonaceous materials produced diffraction patterns in these particular regions and that it is very difficult to ascertain if graphite-like or diamond-like structures, or both, are present. Thus, the work of Friedel and Ergun led to the suggestion (Francis, 1961, p. 749) that coal contained significant proportions of adamantane-type structures:

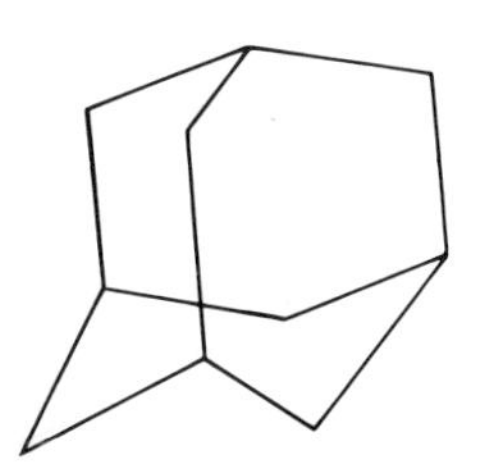

but the theory received little support and was largely ignored. However, the postulate has recently been revived and expanded on the basis of the oxidation of model compounds as well as various coals by sodium hypochlorite (Chakrabartty and Kretschmer, 1972, 1974a,b) which led

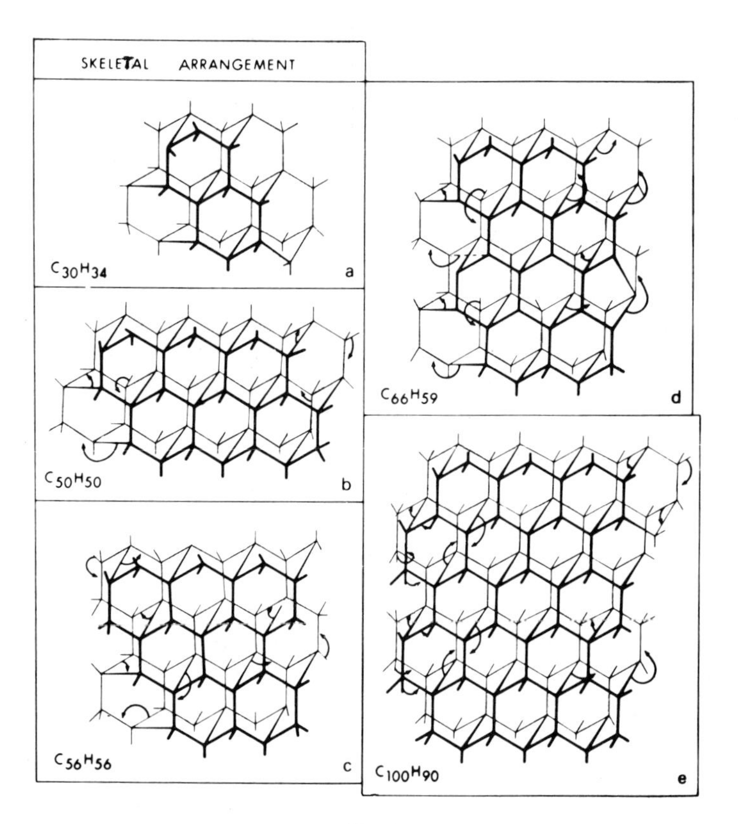

Fig. 5

Chakrabartty and Berkowitz (1974, 1976a,b) to expound that coal is basically a modified bridged tricycloalkane, or polyamantane, system (Fig. 5). Indeed, the isolation of adamantane itself from a Japanese coal (332) can lend credence to this theory. Nevertheless, severe criticism, based on chemical and spectroscopic investigations, has been leveled against the polyamantane system (Mayo, 1975; Ghosh *et al.*, 1975; Landolt, 1975; Hayatsu *et al.*, 1975, 1976; Huston *et al.*, 1976; Retcofsky and Friedel, 1976), and it is still the general consensus of opinion that coal consists of small condensed aromatic systems. Whether these systems are linked by alkyl chains and/or heteroatoms or through the agency of an admantane-type system remains to be proven.

ACKNOWLEDGMENT

The author is indebted to Mr. J. F. Fryer for his comments on the manuscript.

REFERENCES

Alaev, G. P., and Manaenkova, L. N. (1970). *Izv. Nauchno-Issled. Inst. Nefte- Uglekhim. Sin. Irkutsk. Univ.* **12,** 83.

Alpern, B., and Pregermain, S. (1956). *Bull. Microsc. Appl.* **6,** 16.

Austen, D. E. G., Ingram, D. J. E., and Tapley, J. G. (1958). *Trans. Faraday Soc.* **54,** 400.

Austen, D. E. G., Given, P. H., Ingram, D. J. E., and Peover, M. E. (1959). *Fuel* **38,** 309.

Barclay, L. R. C., and Layton, T. M. (1956). *Fuel* **35,** 31.

Bartle, K. D., and Jones, D. W. (1969). *Fuel* **48,** 21.

Bartle, K. D., and Smith, J. A. S. (1965). *Fuel* **44,** 109.

Bartle, K. D., and Smith, J. A. S. (1967). *Fuel* **46,** 29.

Bell, C. L. M., Richards, R. E., and Yorke, R. W. (1958). *Brennst.-Chem.* **39,** 530.

Bent, R., Joy, W. K., and Ladner, W. R. (1964). *Fuel* **43,** 5.

Bergmann, G., Huck, G., Karweil, J., and Luther, H. (1958). *Brennst.-Chem.* **39,** 520.

Berkowitz, N., Cavell, P. A., and Elofson, R. M. (1961). *Fuel* **40,** 279.

Blayden, H. E., Gibson, J., and Riley, H. L. (1944). *Proc. Conf. Ultrafine Struct. Coal Coke* p. 176.

Blom, L., Edelhausen, L., and van Krevelen, D. W. (1957). *Fuel* **36,** 135.

Brooks, J. D., and Silberman, H. (1962). *Fuel* **41,** 67.

Brooks, J. D., and Smith, J. W. (1964). *Fuel* **43,** 125.

Brooks, J. D., and Stephens, J. F. (1965). *Carbon* **2,** 379.

Brooks, J. D., and Stevens, J. R. (1964). *Fuel* **43,** 87.

Brooks, J. D., Durie, R. A., Lynch, B. M., and Sternhell, S. (1960). *Aust. J. Chem.* **13,** 179.

Brown, H. M., and Berkowitz, N. (1966). *Fuel* **45,** 207.

Brown, H. R., and Waters, P. L. (1966). *Fuel* **45,** 17.

Brown, J. K. (1955a). *J. Chem. Soc.* p. 752.

Brown, J. K. (1955b). *J. Chem. Soc.* p. 744.

Brown, J. K. (1959). *Fuel* **38,** 55.

Brown, J. K., and Hirsch, P. B. (1955). *Nature (London)* **175,** 229.

Brown, J. K., and Ladner, W. R. (1960). *Fuel* **39,** 87.

Brown, J. K., Ladner, W. R., and Sheppard, N. (1960). *Fuel* **39,** 79.

Brusset, H. (1947). *C. R. Acad. Sci.* **224,** 1426.

Brusset, H., Devaux, J., and Guinier, A. (1943). *C. R. Acad. Sci.* **216,** 152.

Cannon, C. G. (1953). *Nature (London)* **171,** 308.

Cannon, C. G., and Sutherland, G. B. B. M. (1945). *Trans. Faraday Soc.* **41,** 279.

Cartz, L., and Hirsch, P. B. (1960). *Phil. Trans. R. Soc. London, Ser. A* **252,** 557.

Ceh, M., and Hadzi, D. (1956). *Fuel* **35,** 77.

Chakrabartty, S. K., and Berkowitz, N. (1974). *Fuel* **53,** 240.

Chakrabartty, S. K., and Berkowitz, N. (1976a). *Nature (London)* **261,** 76.

Chakrabartty, S. K., and Berkowitz, N. (1976b). *Fuel* **55,** 362.

Chakrabartty, S. K., and Kretschmer, H. O. (1972). *Fuel* **51,** 160.

Chakrabartty, S. K., and Kretschmer, H. O. (1974a). *Fuel* **53,** 132.

Chakrabartty, S. K., and Kretschmer, H. O. (1974b). *J. Chem. Soc., Perkin Trans. 1* p. 222.

Chauvin, R., Chiche, P., Quinton, M. F., and Uebersfeld, J. (1969). *Carbon* **7,** 307.

Chernyshev, B. N., and Opritov, V. V. (1970). *Khim. Tverd. Topl.* No. 6, p. 147.

Czuchajowski, L. (1961). *Fuel* **40,** 361.

den Hertog, W., and Berkowitz, N. (1958). *Fuel* **37,** 253.

Depp, E. A., and Neuworth, M. B. (1958). *Brennst.-Chem.* **39,** S90.

de Ruiter, E. (1962). *Fuel* **41,** 63.

de Ruiter, E. (1965a). *Fuel* **44,** 49.

de Ruiter, E. (1965b). *Fuel* **44,** 65.

de Ruiter, E., and Tschamler, H. (1958). *Brennst.-Chem.* **39,** 362.
de Ruiter, E., and Tschamler, H. (1962). *Fuel* **41,** 491.
Dhar, J., and Niyogi, B. B. (1942). *Proc. Natl. Inst. Sci. India, Part A* **8,** 127.
Diamond, R. (1957). *Acta Crystallogr.* **10,** 359.
Diamond, R. (1958). *Acta Crystallogr.* **11,** 129.
Diamond, R. (1959). *Phil. Trans. R. Soc. London, Ser. A* **252,** 193.
Dryden, I. G. C. (1955). *Fuel* **34,** 529.
Dryden, I. G. C. (1957). *J. Inst. Fuel* **30,** 193.
Dryden, I. G. C. (1963). *In* "The Chemistry of Coal Utilization" (H. H. Lowry, ed.), Suppl. Vol., p. 232. Wiley, New York.
Dryden, I. G. C. (1964). *In* "Encyclopaedia of Chemical Technology," Vol. 5, p. 606. Wiley (Interscience), New York.
Durie, R. A., and Sternhell, S. (1959). *Aust. J. Chem.* **12,** 205.
Durie, R. A., Shewchyk, Y., and Sternhell, S. (1966). *Fuel* **45,** 99.
Elofson, R. M., and Schulz, K. F. (1967). *Am. Chem. Soc., Div. Fuel Chem., Prepr.* **11,** 513.
Ergun, S. (1958). *Fuel* **37,** 365.
Ergun, S., and Tiensuu, V. H. (1959a). *Nature (London)* **183,** 1669.
Ergun, S., and Tiensuu, V. H. (1959b). *Fuel* **38,** 64.
Ergun, S., and Tiensuu, V. H. (1959c). *Acta Crystallogr.* **12,** 1050.
Ergun, S., McCartney, J. T., and Menster, M. (1959). *Econ. Geol.* **54,** 1068.
Ergun, S., Menster, M., and O'Donnell, H. J. (1960). *Science* **132,** 1314.
Ergun, S., McCartney, J. T., and Walline, R. E. (1961). *Fuel* **40,** 109.
Flett, M. St. C. (1948). *J. Chem. Soc.* p. 1441.
Francis, W. (1961). "Coal: Its Formation and Composition." Arnold, London.
Franklin, R. E. (1951). *Acta Crystallogr.* **4,** 253.
Friedel, R. A. (1957). *Nature (London)* **179,** 1237.
Friedel, R. A. (1959a). *Proc. Conf. Carbon, 4th* p. 321.
Friedel, R. A. (1959b). *J. Chem. Phys.* **31,** 280.
Friedel, R. A., and Queiser, J. A. (1959). *Fuel* **38,** 369.
Friedel, R. A., and Retcofsky, H. L. (1961). *Proc. Conf. Carbon, 5th* **2,** 149.
Friedel, R. A., and Retcofsky, H. L. (1963). *J. Am. Chem. Soc.* **85,** 1300.
Friedel, R. A., and Retcofsky, H. L. (1966a). *In* "Coal Science" (P. H. Given, ed.), p. 503. Am. Chem. Soc., Washington, D.C.
Friedel, R. A., and Retcofsky, H. L. (1966b). *Chem. Ind. (London)* p. 455.
Friedel, R. A., and Retcofsky, H. L. (1968). *Proc. Int. Conf. Coal Sci., 7th, Prague* p. 2.
Friedel, R. A., Schultz, J. L., and Sharkey, A. G. (1968). *Fuel* **47,** 403.
Friedel, R. A., Queiser, J. A., and Carlson, G. L. (1971). *Am. Chem. Soc., Div. Fuel Chem., Prepr.* **15**(1), 123.
Fujii, S. (1959). *Nenryo Kyokai-Shi* **38,** 284.
Fujii, S. (1963a). *Fuel* **42,** 17.
Fujii, S. (1963b). *Fuel* **42,** 341.
Fujii, S., Osawa, Y., and Sugimara, H. (1970). *Fuel* **49,** 68.
Galloway, R. L. (1969). "A History of Coal Mining in Great Britain." David & Charles, Newton Abbott, Devon, England.
Ghosh, G., Banerjee, A., and Mazumdar, B. K. (1975). *Fuel* **54,** 294.
Given, P. H. (1960). *Fuel* **39,** 147.
Given, P. H. (1961). *Fuel* **40,** 427.
Given, P. H., and Peover, M. E. (1960). *J. Chem. Soc.* p. 394.
Gordon, R. R., Adams, W. N., and Jenkins, G. I. (1952). *Nature (London)* **170,** 317.
Gourlay, J. S. (1950). *Research (London)* **3,** 242.

Hadzi, D. (1954). *J. Phys. Radium* **15,** 194.
Hayatsu, R., Scott, R. G., Moore, L. P., and Studier, M. H. (1975). *Nature (London)* **257,** 378.
Hayatsu, R., Scott, R. G., Moore, L. P., and Studier, M. H. (1976). *Nature (London)* **261,** 77.
Heredy, L. A., Kostyo, A. E., and Neuworth, M. B. (1966). *In* "Coal Science" (P. H. Given, ed.), p. 493. Am. Chem. Soc., Washington, D.C.
Hill, C. R., and Lyon, L. B. (1962). *Ind. Eng. Chem.* **54**(6), 36.
Hirsch, P. B. (1954). *Proc. R. Soc., Ser. A* **226,** 143.
Hirsch, P. B. (1958). *Proc. Conf. Sci. Use Coal* A-29.
Holden, H. W., and Robb, J. C. (1960). *Fuel* **39,** 485.
Holly, E. D., Montgomery, R. S., and Gohlke, R. S. (1956). *Fuel* **35,** 56.
Honda, H., Chitoku, K., Yokozawa, Y., and Higasi, K. (1958). *Bull. Chem. Soc. Jpn.* **31,** 890.
Huck, G., and Karweil, J. (1953a). *Brennst.-Chem.* **34,** 97.
Huck, G., and Karweil, J. (1953b). *Brennst.-Chem.* **34,** 129.
Humphreys-Owen, S. P. F., and Gilbert, L. A. (1958). *Proc. Conf. Ind. Carbon Graphite* p. 37.
Huston, J. L., Scott, R. G., and Studier, M. H. (1976). *Fuel* **55,** 281.
Ingram, D. J. E. (1955). *Discuss. Faraday Soc.* p. 179.
Ingram, D. J. E. (1957). *Proc. Conf. Carbon, 3rd* p. 93.
Ingram, D. J. E., Tapley, J. G., Jackson, R., Bond, R. L., and Murnaghan, A. R. (1954). *Nature (London)* **174,** 797.
Inouye, K., Tani, H., and Abiko, M. (1953). *Nenryo Kyokai-Shi* **32,** 292.
Johnson, A. W., Quayle, J. R., Robinson, T. S., Sheppard, N., and Todd, A. R. (1951). *J. Chem. Soc.* p. 2633.
Kessler, T., Raymond, R., and Sharkey, A. G. (1969). *Fuel* **48,** 179.
Kogan, L. A., Popov, V. K., Vetrova, A. K., and Khudyshkina, L. D. (1969). *Podgot. Koksovanie Uglei* No. 8, 3.
Kojima, K., Sakashita, K., and Yoshino, T. (1956). *Nippon Kagaku Zasshi* **77,** 1432.
Kuczynski, W., Stankowski, J., Janiak, J., Dezor, A., and Wieckowski, A. (1965). *Przem. Chem.* **44,** 243.
Ladner, W. R., and Stacey, A. E. (1961). *Fuel* **40,** 295.
Ladner, W. R., and Stacey, A. E. (1965). *Fuel* **44,** 71.
Landolt, R. G. (1975). *Fuel* **54,** 299.
Li, B. S., and Boo, C. K. (1964). *Chosun Kwahakwon Tongbo* p. 34.
Losev, B. I., and Bylyna, E. (1959). *Dokl. Akad. Nauk SSSR* **125,** 184.
McCartney, J. T., and Ergun, S. (1965). *Nature (London)* **205,** 962.
McCartney, J. T., O'Donnell, H. J., and Ergun, S. (1966). *In* "Coal Science" (P. H. Given, ed.), p. 261. Am. Chem. Soc., Washington, D.C.
McLafferty, F. W., and Gohlke, R. S. (1959). *Anal. Chem.* **31,** 2076.
McMurry, H. L., and Thornton, V. (1952). *Anal. Chem.* **24,** 318.
Mahadevan, C. (1929). *Fuel* **8,** 462.
Mahadevan, C. (1930). *Fuel* **9,** 574.
Mayo, F. R. (1975). *Fuel* **54,** 273.
Mazumdar, B. K., and Lahiri, A. (1962). *J. Sci. Ind. Res., Sect. B* **21,** 277.
Mazumdar, B. K., Chakrabartty, S. K., and Lahiri, A. (1962). *Fuel* **41,** 129.
Menster, M., O'Donnell, H. J., and Ergun, S. (1961). *Proc. Conf. Carbon, 5th* **2,** 493.
Menster, M., O'Donnell, H. J., and Ergun, S. (1962). *Fuel* **41,** 153.
Mitra, G. B. (1953a). *Acta Crystallogr.* **6,** 101.
Mitra, G. B. (1953b). *J. Sci. Ind. Res., Sect. B* **12,** 88.

Montgomery, R. S., and Holly, E. D. (1957). *Fuel* **36,** 63.
Montgomery, R. S., Holley, E. D., and Gohlke, R. S. (1956). *Fuel* **35,** 60.
Murchison, D. G. (1966). *In* "Coal Science" (P. H. Given, ed.), p. 307. Am. Chem. Soc., Washington, D.C.
Nelson, J. B. (1952). *Research (London)* **5,** 489.
Nelson, J. B. (1954). *Fuel* **33,** 381.
Oelert, H. H. (1965). *Erdoel Kohle* **18,** 876.
Oelert, H. H. (1968). *Fuel* **47,** 433.
Oelert, H. H., and Hemmer, E. A. (1970a). *Erdoel Kohle* **23,** 87.
Oelert, H. H., and Hemmer, E. A. (1970b). *Erdoel Kohle* **23,** 163.
Ohuchi, H., Shiotani, M., and Sohma, J. (1969). *Fuel* **48,** 187.
Orchin, M., Golumbic, C., Anderson, J. E., and Storch, H. H. (1951). *U.S. Bur. Mines, Bull.* No. 505.
Osawa, Y., and Shih, J. W. (1971). *Fuel* **50,** 53.
Osawa, Y., Sugimara, H., and Fujii, S. (1969). *Nenryo Kyokai-Shi* **48,** 694.
Osawa, Y., Shih, J. W., and Tsunada, T. (1971). *Nenryo Kyokai-Shi* **50,** 31.
Oth, J. F. M., and Tschamler, H. (1963). *Fuel* **42,** 467.
Pregermain, S., and Guillemot, C. (1960). *Int. Conf. Electron Microsc., Proc., 4th, 1958* **1,** 384.
Rao, H. S., Murty, G. S., and Lahiri, A. (1960). *Fuel* **39,** 263.
Rao, H. S., Gupta, P. L., Kaiser, F., and Lahiri, A. (1962). *Fuel* **41,** 417.
Reed, R. I. (1960). *Fuel* **39,** 341.
Reggel, L., Raymond, R., Steiner, W. A., and Friedel, R. A. (1961). *Fuel* **40,** 339.
Reggel, L., Wender, I., and Raymond, R. (1964). *Fuel* **43,** 75.
Retcofsky, H. L., and Friedel, R. A. (1968). *Fuel* **47,** 487.
Retcofsky, H. L., and Friedel, R. A. (1973). *J. Phys. Chem.* **77,** 68.
Retcofsky, H. L., and Friedel, R. A. (1976). *Fuel* **55,** 363.
Retcofsky, H. L., Stark, J. M., and Friedel, R. A. (1967). *Chem. Ind. (London)* p. 1327.
Richards, R. E., and Yorke, R. W. (1960). *J. Chem. Soc.* p. 2489.
Riley, D. P. (1944). *Proc. Conf. Ultra-fine Struct. Coal Coke* p. 232.
Roy, M. M. (1957a). *Fuel* **36,** 249.
Roy, M. M. (1957b). *Fuel* **36,** 344.
Roy, M. M. (1965). *Econ. Geol.* **60,** 972.
Ruland, W. (1959). *Acta Crystallogr.* **12,** 679.
Ruof, C. H., Entel, J., and Howard, H. C. (1956). *Fuel* **35,** 409.
Rybicka, S. M. (1959). *Fuel* **38,** 45.
Schonberg, A., and Mustafa, A. (1946). *J. Chem. Soc.* p. 746.
Schultz, J. L., Friedel, R. A., and Sharkey, A. G. (1965). *Fuel* **44,** 55.
Sharkey, A. G., Schultz, J. L., and Friedel, R. A. (1961). *Fuel* **40,** 423.
Sharkey, A. G., Schultz, J. L., and Friedel, R. A. (1966). *Carbon* **4,** 365.
Smidt, D., and van Krevelen, D. W. (1959). *Fuel* **38,** 355.
Smidt, J. (1958). *Arch. Sci.* **11,** 180.
Speight, J. G. (1972). *Appl. Spectrosc. Rev.* **5,** 211.
Takeya, G., Itoh, M., Suzuki, A., and Yokoyama, S. (1963). *Bull. Chem. Soc. Jpn.* **36,** 1222.
Takeya, G., Itoh, M., Suzuki, A., and Yokoyama, S. (1964). *Nenryo Kyokai-Shi* **43,** 837.
Toyoda, S., and Honda, H. (1966). *Carbon* **3,** 527.
Tschamler, H., and de Ruiter, E. (1963). *In* "The Chemistry of Coal Utilization" (H. H. Lowry, ed.), Suppl. Vol., p. 35. Wiley, New York.
Tyutyunnikov, Y. B., Romadanov, I. S., and Sintserova, L. G. (1973). *Khim. Tverd. Topl.* No. 1, 140.

Uebersfeld, J. (1954). *J. Phys. Radium* **15,** 126.
Uebersfeld, J., Etienne, A., and Combrisson, J. (1954). *Nature (London)* **174,** 614.
Vahrman, M. (1970). *Fuel* **49,** 5.
van der Hart, D. L., and Retcofsky, H. L. (1976). *Fuel* **55,** 202.
van Krevelen, D. W., and Schuyer, J. (1957). "Coal Science." Elsevier, Amsterdam.
van Vucht, H. A., Rietveld, B. J., and van Krevelen, D. W. (1955). *Fuel* **34,** 50.
Vasil'eva, L. M., Anufrienko, V. F., Bochkareva, K. I., Shklyaev, A. A., and Lozbin, V. I. (1972). *Khim. Tverd. Topl.* No. 1, 26.
Waisman, B. A., Krivitzkii, M. D., and Krigman, F. E. (1954). *Dokl. Akad. Nauk SSSR* **97,** 1031.
Yen, T. F., and Erdman, J. G. (1962). *Am. Chem. Soc., Div. Pet. Chem., Prepr.* **7,** 5.
Yokokawa, C. (1968a). *Nenryo Kyokai-Shi* **47,** 872.
Yokokawa, C. (1968b). *Fuel* **47,** 273.
Yokokawa, C. (1969). *Fuel* **48,** 29.
Yokoyama, S., Yamamoto, Y., and Takeya, G. (1970). *Nenryo Kyokai-Shi* **49,** 932.
Yokoyama, S., Itoh, M., and Takeya, G. (1975). *Nenryo Kyokai-Shi* **54,** 340.
Yoshida, R., Maekawa, Y., Yokoyama, S., and Takeya, G. (1975). *Nenryo Kyokai-Shi* **54,** 332.

Chapter 23

Nuclear Magnetic Resonance Spectroscopy

Keith D. Bartle
SCHOOL OF CHEMISTRY
UNIVERSITY OF LEEDS
LEEDS, ENGLAND

Derry W. Jones
SCHOOL OF CHEMISTRY
UNIVERSITY OF BRADFORD
BRADFORD, ENGLAND

ISBN 0-12-399902-2

I. INTRODUCTION

A. Scope

This chapter provides a brief survey of the principles and techniques of nuclear magnetic resonance (NMR) spectroscopy with special reference to the structural applications of NMR to coal and coal products. It thus amplifies and extends Section II,B of Chapter 22 (covering spectroscopic techniques in general) and may be regarded as an introduction to a major part of Chapter 25 in which is described the increasingly significant role NMR plays in the study of paraffinic hydrocarbons extracted from coal, to Chapter 24 (which develops some of the NMR techniques in more detail with regard to their application to Synthoil), and indeed to Section V of Chapter 16 and Section II,F of Chapter 17 in Volume I, which deal with characterization of Syncrudes and related products.

In this chapter, a short summary of some of the relevant theory in Section II and of the range of experimental methods in Section III is followed by a brief reference to spectral interpretation (Section IV) and a series of accounts (Section V) of representative kinds of applications of NMR spectroscopy in coal science and technology.

B. Nuclear Magnetic Resonance Spectroscopy

Nuclear magnetic resonance spectroscopy arises from interaction of the magnetic component of electromagnetic radiation with the very small magnetic moments possessed by atomic nuclei of isotopes with a nonzero spin quantum number I. Discrete nuclear magnetic moment orientation levels have an energy separation, and hence a resonance frequency, proportional to the magnetic field B_0 applied (Section II,A). Assemblies of such nuclear moments can give rise to measurable macroscopic changes. According to the size of B_0 (commonly within 0.5–7 T; note 1 T = 10^4 G) and the particular nuclear species (of which the most favorable is ^{1}H, with a large moment, high natural abundance, and spin $I = \frac{1}{2}$), resonance frequencies are usually in the radio-frequency range 10–300 MHz or so. Thus the quanta are small, so that little disturbance is caused to the system.

On the other hand, the smallness of the separation between ground and excited states, together with the long lifetimes of the excited states (which limit the power that can be applied to induce a transition), gives NMR spectroscopy an inherently low sensitivity. Alleviation of the limitation of this sensitivity may be achieved, as outlined in Section

III, by employment of high magnetic fields, time-averaging computers, and Fourier transform spectrometers.

C. Magnetic Moments of Atomic Nuclei

For certain species of atomic nuclei, including ^{12}C, ^{16}O, and ^{32}S (the predominant isotopes of these elements), for which both mass number (atomic weight) and charge number (atomic number) are even, the spin quantum number $I = 0$. Such nuclei possess zero nuclear angular momentum and zero associated magnetic dipole moment and so cannot undergo NMR. Many other atomic nuclei have nonzero I, either half-integral (odd mass number) or integral (atomic number odd, mass number even). Such $I > 0$ nuclei, with both angular momenta and magnetic moments, will interact with an applied magnetic field, so as to yield quantized energy levels separated by an amount ΔE proportional to both field B_0 and magnetic moment, component μ_z (in many textbooks, B_0 is denoted by H_0); hence, for a transition:

$$\Delta E = |\hbar \gamma B_0| \tag{1}$$

Here γ is the magnetogyric ratio, a constant for any given nuclear species.

Nuclides with $I > \frac{1}{2}$, e.g., ^2H with $I = 1$ or ^{17}O with $I = \frac{5}{2}$, have a nonspherical nuclear charge distribution giving rise to an electric quadrupole moment Q, in addition to the nuclear magnetic moment. The nuclei most used for the NMR of coals and related compounds have been isotopes with $I = \frac{1}{2}$, i.e., ^1H and ^{13}C, which forms only 1.1% of natural carbon. As will be seen later, the low natural abundance of ^{13}C nuclei is responsible for simplifications of high resolution spectra from both ^1H and ^{13}C.

Development of ^{13}C-NMR occurred much later than that of ^1H-NMR because, in a given magnetic field, ^{13}C-NMR has a sensitivity about 6000 times less than that of ^1H for unenriched samples. This is partly because of the low natural abundance of ^{13}C (a factor of nearly 100), and partly because $\gamma(^1H)/\gamma(^{13}C) \approx 4$ and sensitivity is proportional to γ^3 (thus an additional factor of about 60). Although very few measurements have been reported on coal products by NMR with nuclei other than ^1H and ^{13}C, it may be noted that several other kinds of nuclei present in coals possess magnetic moments, and occur with the natural abundances indicated: ^2H, $I = 1$, 0.015%; ^{14}N, $I = 1$, 99.6%; ^{15}N, $I = \frac{1}{2}$, 0.4%; ^{17}O, $I = \frac{5}{2}$, 0.04%; ^{33}S, $I = \frac{3}{2}$, 0.8%. Recently, Schweighardt *et al.* (1976a) reported the application of ^2H-NMR to the study of ^2H-labeled sites in organic structures in coal liquids (see Chapter 24, Section VI,D), and Schweighardt *et al.* (1976b) used a combination of ^1H, ^{13}C, and ^{14}N

resonances (all at a high B_0 field of 5.8 T) to study coal-derived asphaltenes (see Section V,E).

II. THEORY OF NUCLEAR MAGNETIC RESONANCE

The theory of NMR can be treated in two complementary ways, classical and quantum mechanical, of which the rotating-frame classical approach is particularly valuable for pulse experiments.

A. The Nuclear Magnetic Resonance Phenomenon

1. *Quantum Approach to NMR*

For a nucleus with spin quantum number I, the total spin angular momentum vector $\hbar I$ of magnitude $\hbar[I(I+1)]^{1/2}$ (where $\hbar$ is the reduced Planck's constant $h/2\pi$), is related to the magnetic moment

$$\boldsymbol{\mu} = \gamma\hbar\mathbf{I} \tag{2}$$

through the magnetogyric ratio γ. The interaction energy between $\boldsymbol{\mu}$ and the applied magnetic field $\mathbf{B}_0$ is given by the Hamiltonian

$$\mathcal{H} = -\boldsymbol{\mu}\cdot\mathbf{B}_0 = -\gamma\hbar B_0 I_z \tag{3}$$

where I_z is the component of $\mathbf{I}$ resolved along $\mathbf{B}_0$.

Since I_z can assume any of $2I+1$ values, specified by magnetic quantum number $m = +I, (I-1), \ldots, -(I-1), -I$, the possible Zeeman energy levels of the nuclear system are

$$E_m = -\gamma\hbar B_0 m. \tag{4}$$

Thus for $I = \frac{1}{2}$ (and we shall be concerned mainly with such nuclei) there are two levels, for $I = 1$ there are three levels (Fig. 1), and so on, with spacing $\gamma\hbar B_0$ between adjacent levels. Since the selection rule is $\Delta m = \pm 1$,

$$h\nu = \Delta E = -\gamma\hbar B_0 \tag{5}$$

i.e., in terms of angular frequency, $\omega = 2\pi\nu$,

$$\omega = -\gamma B_0 \tag{6}$$

The frequency thus depends on the species of nucleus, which determines γ, and on the magnetic field B_0 of the experiment.

2. *Classical Description of NMR for a Single Nucleus*

An isolated spinning magnetic moment $\boldsymbol{\mu}$, inclined at an angle θ to a static magnetic field $\mathbf{B}_0$, will be subject to a torque $\boldsymbol{\mu}\wedge\mathbf{B}_0$ directed perpendicular to the plane containing $\boldsymbol{\mu}$ and $\mathbf{B}_0$ (Fig. 2). Since this torque is equal to the rate of change of angular momentum, the equation of motion is

$$d(\mathbf{I}\hbar)/dt = \boldsymbol{\mu}\wedge\mathbf{B}_0 \tag{7}$$

Substituting for $\boldsymbol{\mu}$ from Eq. (1), we have

$$d\boldsymbol{\mu}/dt = \gamma\boldsymbol{\mu}\wedge\mathbf{B}_0 = \boldsymbol{\mu}\wedge\gamma\mathbf{B}_0 \tag{8}$$

This corresponds to the precession of a single nuclear magnet of moment $\boldsymbol{\mu}$ about $\mathbf{B}_0$ with an angular velocity $\boldsymbol{\omega}_0$ rad/sec (or $\boldsymbol{\nu}_0 = 2\boldsymbol{\pi}\boldsymbol{\omega}_1$ Hz) so that

$$d\boldsymbol{\mu}/dt = \boldsymbol{\omega}_0\wedge\boldsymbol{\mu} \tag{9}$$

From Eqs. (8) and (9), the so-called Larmor frequency is

$$\boldsymbol{\omega}_0 = -\gamma\mathbf{B}_0 \tag{10}$$

If, from a transmitter, a small additional field $\mathbf{B}_1$ is applied, rotating at a variable angular frequency $\boldsymbol{\omega}$ in the xy plane perpendicular to the direction of the main large field $\mathbf{B}_0$ (which is in the z direction), then resonance will occur when $\boldsymbol{\omega} = \boldsymbol{\omega}_0$ and is in the same sense as the precessing magnetic moment, $\boldsymbol{\mu}$. From the viewpoint of $\boldsymbol{\mu}$, $\mathbf{B}_1$ appears to be constant with time and the torque $\boldsymbol{\mu}\wedge\mathbf{B}_1$ will tend to tip $\boldsymbol{\mu}$ away from $\mathbf{B}_0$, i.e., to increase θ, with consequent absorption of energy from $\mathbf{B}_1$. This is a nuclear magnetic *resonance* since there is no net effect unless $\boldsymbol{\omega} = \boldsymbol{\omega}_0$ and unless $\boldsymbol{\omega}$ is in the correct sense. $\mathbf{B}_1$ is usually derived from a small *oscillating* magnetic field, equivalent to a pair of contrarotating fields, one of which has a negligible effect.

B. Macroscopic Approach: Relaxation Times

In a real NMR experiment, we do not have an isolated nucleus, but rather an assembly of identical nuclei. The Bloch phenomenological formulation follows the net macroscopic magnetization (vector sum per unit volume of the nuclear moments) $\mathbf{M}$ of the assembly of identical nuclei in the sample. In an applied field B_0, parallel to the z direction, a set of $I = \frac{1}{2}$ nuclei precessing about z will give, at thermal equilibrium between the spin system and the surroundings, a small net magnetization in the z direction (more nuclei precess about z than about $-z$), but zero net magnetization in the x and y directions. Application of a $\mathbf{B}_1$ field tips $\mathbf{M}$ away from B_0, so that transverse components M_x and M_y are generated. These decay to zero with a characteristic time T_2, the transverse or *spin–spin relaxation time,* while the logitudinal component M_z is restored to its equilibrium value with a time constant T_1, the longitudinal or *spin–lattice relaxation time.*

Although both parameters can be derived from steady-state experiments, pulse techniques often provide a more satisfactory procedure, especially for measurement of T_1.

1. *Rotating Coordinate System*

Pulse NMR measurements may conveniently be treated in terms of a coordinate frame $x'y'z'$, which rotates at an angular frequency $\boldsymbol{\omega}$ about

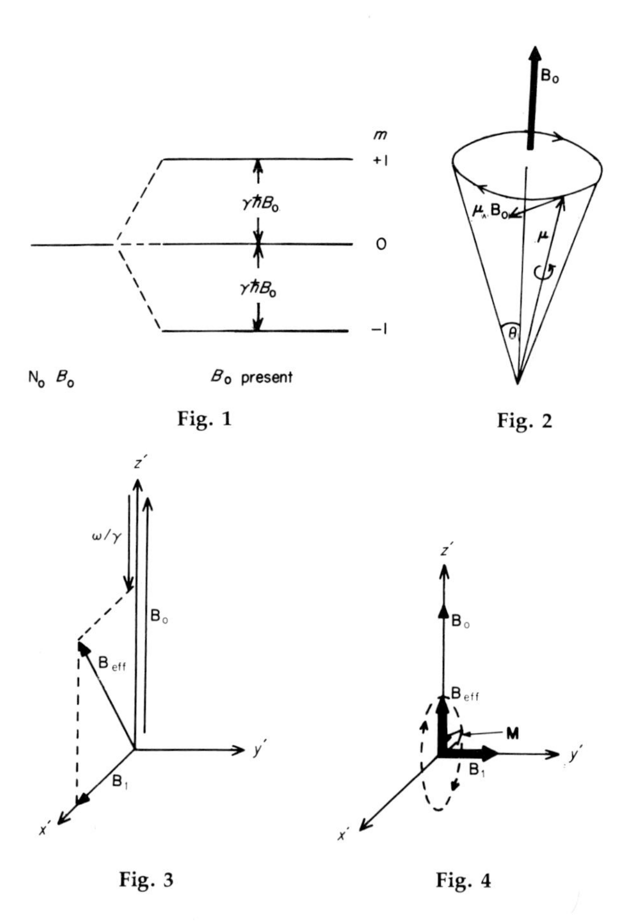

Fig. 1

Fig. 2

Fig. 3

Fig. 4

Fig. 1 Nuclear magnetic energy levels for atomic nucleus with spin $I = 1$.

Fig. 2 Precession of nuclear magnetic moment $\boldsymbol{\mu}$ about fixed magnetic field $\mathbf{B}_0$

Fig. 3 Effective magnetic field $\mathbf{B}_{\text{eff}}$ in rotating reference plane x', y', z', derived from $\mathbf{B}_0$, $\mathbf{B}_1$, and $\boldsymbol{\omega}/\gamma$; the magnetic moment $\boldsymbol{\mu}$ precesses around $\mathbf{B}_{\text{eff}}$.

Fig. 4 Rotation of magnetization **M** through 90° into direction of white arrow by $\mathbf{B}_1$ pulse.

the z direction of the field $\mathbf{B}_0$ instead of in the fixed laboratory frame, xyz; z' and z are parallel. Primes denote quantities as seen by an observer in the rotating frame when these quantities differ from those in the laboratory frame (the rotating field $\mathbf{B}_1$, introduced later, is directed along x'). Reference to this rotating frame can eliminate the time dependence from some of the quantitites.

Transformation of the equation of motion [Eq. (8)] for the nuclear moment $\boldsymbol{\mu}$ precessing in the magnetic field $\mathbf{B}_0$ may be represented (e.g., Pake, 1956) by

$$d\boldsymbol{\mu}/dt = d\boldsymbol{\mu}'/dt + \boldsymbol{\omega}\wedge\boldsymbol{\mu} = \gamma\boldsymbol{\mu}\wedge\mathbf{B}_0 \tag{11}$$

i.e.,

$$d\boldsymbol{\mu}'/dt = \gamma\boldsymbol{\mu}\wedge\mathbf{B}_0 - \boldsymbol{\omega}\wedge\boldsymbol{\mu} = \gamma\boldsymbol{\mu}\wedge(\mathbf{B}_0 + \boldsymbol{\omega}/\gamma) \tag{12}$$

Thus the time rate of change of $\boldsymbol{\mu}$ in the rotating system may be treated as though it were in a fixed system (i.e., the laboratory) in which the effective magnetic filed was

$$\mathbf{B}_{\text{eff}} = \mathbf{B}_0 + \boldsymbol{\omega}/\gamma \tag{13}$$

For the particular case of constant $\mathbf{B}_0$ and $\boldsymbol{\omega} = -\gamma\mathbf{B}_0$, Eqs. (12) and (13) imply that the rotating frame accompanies the Larmor precession; thus in the rotating frame $\boldsymbol{\mu}'$ appears stationary and the nuclei appear to be in zero field.

If the applied magnetic field (Fig. 3) now has a nonzero component $\mathbf{B}_1$ along x', rotating about z' at angular frequency $\boldsymbol{\omega}$ in the region of $-\gamma\mathbf{B}_0$, then $\boldsymbol{\mu}'$ precesses about $\mathbf{B}_{\text{eff}}$, which now includes $\mathbf{B}_1$. With $\boldsymbol{\omega}$ actually at the resonant frequency, the terms on the right-hand side of Eq. (13) are canceled out so that the net $\mathbf{B}_{\text{eff}} = \mathbf{B}_1$. Thus precession (in the rotating frame) is about $\mathbf{B}_1$, the radiofrequency field; i.e., the magnetic moment can be tipped from being parallel to antiparallel to $\mathbf{B}_0$.

Returning to the ensemble of nuclei, the total magnetization $\mathbf{M}$ in the rotating frame has the equation

$$d\mathbf{M}/dt = \gamma\mathbf{M}\wedge\mathbf{B}_{\text{eff}} \tag{14}$$

where the effective field

$$\mathbf{B}_{\text{eff}} = \mathbf{B}_0 + \mathbf{B}_1 + \boldsymbol{\omega}/\gamma \tag{15}$$

Here, $\boldsymbol{\omega}/\gamma$ is the fictitious field arising from the new coordinate system. For $\boldsymbol{\omega} = \boldsymbol{\omega}_0$, the last term in Eq. (15) balances $\mathbf{B}_0$ $(= -\boldsymbol{\omega}_0/\gamma)$, so that $\mathbf{M}$ precesses round B_1 (say, in direction x). It should be noted that the angle α (flip angle in a pulse experiment) through which $\mathbf{M}$ is tipped governs the size of M_y and hence that of the signal observed in the receiver coil:

$$\alpha = \gamma B_1 t_p \tag{16}$$

where t_p is the length of time for which B_1 is applied. The time $\pi/2\gamma B_1$ sec for a 90° rotation of **M** is an important parameter in pulse NMR.

2. *Relaxation Times and Pulse Measurements*

Application of the small radio-frequency field B_1 (rotating about the z axis at the Larmor frequency so that, in the rotating frame $x'y'z'$, it appears fixed in the x' direction) causes **M** to precess in the $x'y'$ plane. **M** tips away from the z' axis, precessing about B_1 at frequency $\omega_1 = \gamma B_1$, so that $M_{z'}$ decreases. In pulse NMR experiments, the rf field B_1 is applied only for times short compared with T_1 and T_2, rather than as a continuous-wave (cw) radio frequency, disturbing the system from equilibrium; at other times the free motion may be described by the equations proposed by Bloch. The pulse duration controls the angle by which **M** is tipped from the z axis.

3. *90° and 180° Pulses*

If B_1 is applied for a time $t_p = \pi/2\gamma B_1$, then **M** will turn by an angle $\alpha = \gamma B t_p = \pi/2$ from the z' direction (Fig. 4). This provides a means of assessing the size of **M** at any instant. After such a *90° pulse,* **M** is turned from z' into the $x'y'$ plane of the rotating frame; in the laboratory frame, its rotation at the resonance frequency in the xy plane changes the magnetic flux linked with the NMR coil so that a maximum signal voltage is induced in the spectrometer receiver. Evidently, the nuclei are no longer precessing about B_0 with random phases but are clustered together. Such a free precession NMR signal (FPS) or *free induction decay* (FID) is the Fourier transform of a normal absorption line; local magnetic fields (much smaller than B_0) and spin exchange cause it to decay with time constant T_2^*, which incorporates contributions from the magnetic field inhomogeneity over the sample volume as well as from the Bloch transverse relaxation time, T_2. Thus, precession of nuclei at a small range of frequencies (because of slightly different local fields) causes them gradually to lose their coherence of phase in the $x'y'$ plane as they exchange energy (spin–spin relaxation) among themselves. The FID will be short in a solid with small T_2, but longer in a solution.

A *180° pulse,* i.e., one of length $t_p = \pi/\gamma B_1$, causes **M** to reverse its direction. Suitable combinations of 90° and 180° pulses enable measurements of T_1 and T_2 to be made (see, e.g., Rushworth and Tunstall,

1973). Thus, T_1 may be measured by 180°–τ–90° (as used by Gerstein *et al.* (1977) for protons in coals, for example) or 90°–τ–90° pulse sequences, where τ is the waiting period between the two short radio-frequency pulses, B_1.

In the 90°–τ–180° *spin echo* pulse sequence for measurement of T_2, the initial 90° pulse turns the magnetization into the $x'y'$ plane (as in Fig. 4), after which there is an FID, with the nuclear magnets gradually fanning out at a rate inversely proportional to T_2. The effect of applying a 180° pulse after an interval τ is to reverse this process, so that after a further τ they will coalesce to yield a large magnetization in the $x'y'$ plane and hence a signal, the so-called spin echo.

To illustrate echo formation, Hahn (1953) drew a parallel between precessing nuclear moments and runners, with different but constant speeds, gradually spacing out along a circular running track after starting together. If at time τ they all reverse direction, then they should converge at time 2τ. Abragam (1961) presents perhaps a better analogy, that of ants crawling at different speeds around the edge of a pancake which is then tossed or "flipped" by the 180° pulse. Provided each ant continues to crawl in the same direction in space, then all the ants will gradually bunch together; subsequently, they will begin to spread out again, just as the nuclear magnets dephase and so the NMR signal decays.

Other sequences used for measuring T_2 include the longer Carr and Purcell (1954) train of pulses 90°–τ–180°–2τ–180°–2τ–180° · · · , and the Meiboom and Gill (1958) spin echo pulse sequence; this differs from the Carr–Purcell method in that the 90° pulse is phase-shifted by 90° relative to the succeeding 180° pulses.

Whereas the T_2 relaxation time can sense moderately low frequencies, e.g., 10–10^2 kHz in a solid, T_1 relaxation is sensitive to large laboratory fields, i.e., to fields fluctuating at frequencies of the order of 10^7–10^8 Hz. Investigation of relatively slow motions in solids requires study of T_1 under a small B_0 field (say, 1–10 G, or 10^{-3}–10^{-4} T), i.e., under conditions that are adverse for sensitive measurements. A third kind of relaxation, characterized by a time $T_{1\rho}$ (where the ρ refers to rotating frame) related to T_1 (measured in the laboratory frame), occurs effectively under the influence of small local magnetic fields and may be applicable to solid coals.

For measurement of $T_{1\rho}$ (Hartmann and Hahn, 1962), the magnetization (initially along B_0) is first aligned along B_1 in the transverse plane $x'y'$ by a strong 90° pulse which is followed immediately by a longer pulse (with the same B_1), of variable duration τ, phase-shifted by 90° from the first pulse. The spins are now "spin-locked" to B_1, and the magnetization M_x decays in this field B_1 with a time constant $T_{1\rho}$. To find the extent of magnetization decay

$$M(\tau) = M_x \exp\{-\tau/T_{1\rho}\} \quad (17)$$

the height of the FID with B_1 switched off is measured. Experiments for various times τ then enable $T_{1\rho}$ to be determined.

C. Broad-Line NMR in Solids: Dipolar Broadening

In a solid sample, the breadth and shape of an NMR line are usually dominated by the magnetic interaction between nuclear spins. The line width is inversely related to the spin–spin relaxation time T_2, which may be as small as 10^{-4}–10^{-6} sec, while T_1 may be as long as several minutes. The extent of direct dipolar interaction (across space) between nearby spins depends on the magnitude and orientation of the spins and on the variation with time of the internuclear vector. Dipolar broadening in solids is much greater than in a liquid; only when it is averaged, as in a solvent extract of a coal or a solution of a single compound, or in certain special experiments with solids (Sections III,E and V,D), can one observe high resolution effects (Section II,D). In a fictitious solid consisting magnetically of a rigid lattice of isolated protons (each so remote from its neighbor as not to sense the magnetic field of the other), nuclei would sense only the applied external field B_0. In an actual solid, the total magnetic field B experienced by a nucleus will also include the (much smaller) local field B_{loc} from neighboring nuclear magnets, so that the resonance will be broadened. If the rigid solid consists of magnetically well-isolated *pairs* of nuclei, the total *static* field B experienced by one nucleus in such a pair, moment μ, would be

$$B = B_0 \pm \tfrac{3}{2}\mu\, r^{-3}(3\cos^2\theta - 1) \quad (18)$$

where r is the internuclear separation and θ is the angle between r and B_0 (Pake, 1956) (the $\pm$ sign allows for the two almost equally probable orientations of a neighboring magnetic dipole with respect to B_0). Such a sample would be expected to yield a spectrum of two sharp lines of equal intensity, flanking the position corresponding to the applied field B_0. Their separation would depend on the crystal orientation in the B_0 field; note that at the "magic angle," $\theta = 54.7°$, $3\cos^2\theta = 1$, and the pairs of lines should merge into one (see Section III,E,2).

Spectra of some crystalline hydrates show such doublets (local fields within a proton pair are of the order of gauss or millitesla). However, the two lines are actually much broader owing to overlap of many weak pairs of lines arising from small dipolar interactions with other pairs of nuclei, and to mutual spin-flipping of nuclei in the rotating component of the magnetic field due to precession of another identical nucleus. It is possible to calculate the shapes of NMR lines for crystals and powders containing simple, closely interacting groups of three identical spins ("three-spin systems," as in CH_3 or H_3O^+ groups) or even for "four-spin systems" as in NH_4^+ groups, and so to calculate the length of r, the interproton vector. Few solids contain sufficiently compact clusters of magnetic nuclei to yield characteristic two- or three-spin NMR spectra; four-spin groups and more complicated cases give experimental curves with undistinguished flat-topped broad single peaks. However, a line of such a shape may be described by means of its *moments*. In terms of a normalized function $g(B)$, representing the line shape, the nth moment (or mean nth-power width) about a point B_0 may be defined by

$$M_n = \langle \Delta B^n \rangle_{\text{av}} = \int_{-\infty}^{\infty} (B - B_0)^n g(B)\, dB \Big/ \int_{-\infty}^{\infty} g(B)\, dB \qquad (19)$$

Thus the second moment, often used for expressing the broadening of a line, has $n = 2$ in Eq. (19), or may be expressed as an analogous mean-square frequency deviation:

$$M_2 = \langle \Delta \nu^2 \rangle_{\text{av}} = \int_{-\infty}^{\infty} (\nu - \nu_0)^2 g(\nu)\, d\nu \qquad (20)$$

Here the line shape function $g(\nu)$ is taken to be normalized. Although knowledge of all moments of a curve is equivalent to knowledge of the line shape function itself, the second and fourth moments are those most used. Van Vleck (1948) derived theoretical expressions for these in terms of the sum over the sample of the contributions to the broadening from all pairs of magnetic nuclei (a summation somewhat analogous to evaluation of a lattice energy).

Motion of molecules, or of parts of molecules, can reduce the time-averaged magnetic interaction between nuclear dipoles, since the local field is influenced by the separation r and orientation θ of neighboring nuclei, both of which can vary with time. Sufficiently fast motion effectively erases the direct dipolar interaction and leads to drastic or so-called extreme line narrowing. Strictly, as Van Vleck (1948) showed, the second moment M_2 of the NMR line is unaffected by motion of the constituent nuclei, although the fourth and higher moments increase (Abragam, 1961). However, the *observable* part of the second moment, i.e., the central part close to the Larmor frequency, is reduced by molecular motion. [Gerstein *et al.* (1977) associate increased M_2 *of FIDs* (not cw absorption signal) in vitrains as proton content decreases with aging, i.e., protons move to more rigid aromatic structures.] It may also be shown (see, e.g., Rushworth and Tunstall, 1973) that

$$T_2 = 1/\Delta\omega = (2\pi\ \Delta\nu)^{-1} = (\gamma\ \Delta B)^{-1} \tag{21}$$

for a Lorentzian line shape; ΔB is the half-width at half-height.

D. High Resolution Effects

In liquids and solutions, with the correlation time (the time for significant molecular movement) τ_0 of the order of 10^{-12} sec, direct magnetic dipolar interactions are averaged out, and relaxation times T_1 and T_2 are much longer (typically of the order of seconds) and more nearly equal than in solids. With line widths of the order of hertz, appropriate apparatus, especially a sufficiently homogeneous magnetic field B_0, then enables much smaller, so-called high resolution effects to be observed. The fine structure of high resolution NMR spectra derives predominantly from two effects: the chemical shift and indirect spin–spin coupling.

1. *The Chemical Shift*

In the presence of the steady laboratory magnetic field B_0, small currents induced in the electron clouds around the nucleus under observation give rise to small local magnetic fields, of magnitude proportional to B_0, and in a sense such as to oppose B_0. In effect, a magnetic nucleus in a molecule is partially shielded from the full applied field:

$$2\pi\nu_0 = \gamma B_0(1 - \sigma) \tag{22}$$

where σ is the screening constant or shielding parameter. Thus the actual frequency ν_0 depends both on the amount of screening (i.e., the chemical surroundings of the nucleus) and on the magnetic field B_0 at which the measurement is made. The (dimensionless) constant σ is independent of B_0 and is very small, of the order of parts per million, for protons but bigger for other magnetic nuclei, e.g., ^{13}C. Since screening causes the magnetic energy levels to approach one another more closely, an increase in magnetic shielding σ will require the B_0 field to be increased slightly, for a fixed resonance frequency ν_0, for the resonance condition to be satisfied. Conventionally, high resolution spectra are represented by the most highly screened nuclei on the right-hand side, and magnetic field strength is regarded as increasing from left to right (i.e., "deshielded" nuclei are on the left-hand side); not all early spectra are in this convention.

Substitution in Eq. (22) for two nuclei with different chemical environments (e.g., methyl ^1H and methylene ^1H) with screening constants σ_1 and σ_2, gives the relative frequency, $(\nu_2 - \nu_1)$:

$$2\pi(\nu_1 - \nu_2) = \gamma B_0(\sigma_2 - \sigma_1) \tag{23}$$

It is usual to take the fractional change $(\nu_1 - \nu_2)/\nu_1$, i.e., to divide the frequency difference by the resonance frequency (or to take the corresponding field expression) and to describe this ratio as the *chemical shift*, δ. Division of the frequency difference in hertz by the NMR frequency in megahertz yields a chemical shift in parts per million, e.g., 1 Hz at 100 MHz is 0.01 ppm.

Instead of referring chemical shifts to a bare or isolated nucleus (i.e., a proton for ^1H resonance or a carbon nucleus for ^{13}C resonance) they are referred to an arbitrary standard; most commonly this is tetramethylsilane (TMS), Me_4Si, for ^1H- and ^{13}C-NMR. TMS is miscible with many organic solvents; it is conveniently volatile, gives a single, narrow ^1H or ^{13}C line, and is, for most purposes, inert. Chemical shifts may thus be expressed as, e.g., 50 Hz downfield from TMS at 100 MHz, or as $(50/100) \times 10^{-6} = 0.5$ ppm downfield from TMS on the δ scale (i.e., with $\delta_{TMS} = 0$); the spectrometer frequency (here 100 MHz) must be specified if the shift is given in hertz but not if it is given in parts per million (δ scale). On the ^1H chemical shift tabulations with respect to the old τ scale, TMS is assigned the arbitrary value 10 (instead of 0) and most coal extracts give ^1H shifts in the range 0–10 τ. The two scales are linked by the relation

$$\tau = 10 - \delta \tag{24}$$

so that, in this example, 0.5 ppm corresponds to 9.5 τ. Variations in nuclear shielding which lead to the differing shifts exhibited by chemical groups arise from many sources, associated both with the molecule or molecules under study, and with the solvent system. However, even if used empirically, chemical shifts provide valuable characterizations of different chemical environments. Tabulations of ^1H shifts may be found in the book by Jackman and Sternhell (1969) and of ^{13}C shifts in the books by Levy and Nelson (1972) and Wehrli and Wirthlin (1976).

Although only a single chemical shift is generally measured in a liquid or solution sample, this is because of averaging by molecular tumbling. The chemical shift screening tensor $\boldsymbol{\sigma}$ is usually anisotropic; if the three components σ_x, σ_y, and σ_z are different, molecular motion will produce a fluctuating magnetic field at the nucleus and will thereby provide a relaxation mechanism. Nuclei in a solid can have several different relaxation mechanisms, each characterized by a relaxation time. Chemical shift anisotropy can sometimes be detected in randomly oriented solids (Sections III,E and V,D).

2. *Nuclear Spin–Spin Coupling*

Although molecular tumbling in liquids and solutions averages out the *direct* (i.e., across space) magnetic dipolar interactions between nearby nuclear spins, magnetic nuclei within the same molecular can still interact *indirectly*, i.e., they can sense each other's magnetic fields *via* the spins of the bonding electrons. Such mutual scalar (non-angular-dependent) nuclear spin–spin coupling is of a magnitude independent of B_0 but depends on the number and kind of magnetic nuclei and on the number and nature of connecting bonds and their stereochemistry. Spin coupling between a given pair of (nonequivalent) magnetic nuclei is specified as the coupling constant J, which may be positive or negative. Couplings between two ^{1}H nuclei are typically of the order of a few hertz but may be much bigger when ^{13}C nuclei are involved.

Spin–spin coupling can give structural information both from the magnitude of J (which is markedly affected by conformational details) and from the multiplicity of spectral lines generated by spin–spin interaction. Chemical shifts are diagnostic for the presence of certain chemically different groups in a compound, while the fine structure (in simple cases, multiplets) consequent on internuclear spin–spin coupling can often reveal which groups are linked together; this represents a specific advantage of high resolution NMR over, for example, infrared spectroscopy in structure elucidation.

III. EXPERIMENTAL METHODS

A. Components of a Continuous-Wave NMR Spectrometer

The minimum requirements for an NMR experiment are (i) a steady magnetic field $\mathbf{B}_0$ as high as possible, to separate the nuclear energy levels; (ii) a radio-frequency field $\mathbf{B}_1$, rotating or oscillating to give a field perpendicular to $\mathbf{B}_0$ and with frequency satisfying the resonant condition of Eq. (6); and (iii) a means of detecting the NMR effect. A continuous-wave (cw) spectrometer must have a facility for slowly varying $\mathbf{B}_0$ (usually by means of auxiliary coils) and provision for recording the output as an NMR spectrum. Detection may be with the same coil as applies the $\mathbf{B}_1$ field to the sample (as is usual for pulsed spectrometers) or by a separate receiver coil, "crossed" or orthogonal to the transmitter coil. For high resolution studies in particular, there are stringent requirements of homogeneity of magnetic field in time and over the region of the sample. Permanent magnets (especially with a "barrel" yoke) with negligible power and water consumption, seem to be most economic for the simplest instruments, at relatively moderate fields, while electromagnets (which require water-cooling) predominate among the medium-field 2.3-T spectrometers. For high resolution spectra of coal products, with many close and overlapping chemical shifts, the availability of very high fields (i.e., above 2.3 T) of high homogeneity not only greatly aids resolution and thus interpretation but also markedly improves sensitivity. In commercial spectrometers, superconducting solenoid magnets, operating at liquid helium temperature, already provide fields appropriate to ^{1}H-NMR at 220 or 300 MHz with resolution comparable (in parts per 10^9 terms) with conventional magnets.

Many spectrometers incorporate a small dedicated computer or can be linked on-line to a computer; this can be utilized for signal averaging, data processing, spectra simulation, and, for pulse spectrometers, fast Fourier transformation (to give the appearance of a conventional spectrum). Facility for variation of the temperature of the sample is commonly available; this can also prove useful in relaxation time measurements on fossil fuels in pulse spectrometers (Miknis and Netzel, 1976). High resolution NMR spectrometers incorporate an electronic integrator for areas of peaks or groups of peaks. The proportionality of peak areas (*not* heights) to numbers of resonating nuclei, characteristic of ^{1}H-NMR spectroscopy, is a particular advantage for quantitative measurements, but there are complications with ^{13}C Fourier transform measurements (Section III,F).

B. Detectability of a Spectrum: Signal-to-Noise Enhancement

The ability of an NMR spectrometer to detect a signal, an essential prerequisite to assignment and interpretation, depends partly on the breadth and structure of the spectral line and partly on the instrumental parameters which define the resolution and sensitivity. Resolution is usually restricted by the variation in magnetic field over the sample. A 100-MHz NMR spectrometer might have a resolution of 0.3 Hz and a resolving power $\nu/\Delta\nu = 10^8/0.3$, so that the magnetic field must stay constant to about 3 parts in 10^9 over the sample during a run, a stringent requirement.

NMR signals are observed against a background of unwanted random fluctuations or "noise." When genuine signals are weak, and further increase in sample size is not feasible, increasing the power of the source may be advantageous. However, excessive power levels can "saturate" the system, or tend to equalize the populations of energy levels, so that the signal strength diminishes, even to zero. The effect of noise at the detection stage may be minimized by lock-in amplification or phase-sensitive detection. The signal is converted into an alternating voltage at a frequency well away from the dominant frequency components (near zero) constituting the noise, and the following amplifier is tuned to the new signal frequency. Modulation of the transmitter frequency (or, by means of subsidiary magnet coils, of the magnetic field) is effected at a few hundred hertz. With a modulation amplitude smaller than the line width, the carrier is modulated in proportion to the difference in absorption over the modulation range, i.e., to the rate of change of absorption with frequency (or field); random signals will tend to be averaged out. Such differential curves appear in the presentation of some broad-line NMR spectra (e.g., see Memory and Parker, 1974).

Lock-in detection and frequency-sensitive amplification imply relatively slow scanning across the spectrum; usually a choice of time constants is available, and the spectrometer must have sufficient long-term stability. Increase of the scanning time permits longer time constants to be used and enables bandwidth, and thus the range of noise frequencies transmitted, to be reduced; signal-to-noise is enhanced in proportion to the square root of the scan time (the bandwidth cannot be reduced with pulsed NMR). When very slow sweep might induce saturation (or when constancy of spectrometer conditions over many hours cannot be guaranteed), a large number n of passes may be made through the spectrum, and the results summed. Whereas the total signal has an amplitude n times that of an individual run, the noise occurs in different parts of the spectrum in successive runs and its total increases only in proportion to $n^{1/2}$. In a computer of average transients (CAT), a digital storage device, the spectrum (total signal and total noise) may be monitored at, say, 10^3 discrete points of frequency (or channels) and the improvement in signal-to-noise is proportional to $n^{1/2}$. With multiscan averaging, it is usual for many fairly fast sweeps to be made, with the B_1 field set to just insufficient to "saturate" the line. The CAT (or multichannel analyzer) can be used both with cw spectrometers and for the FIDs of FT spectrometers.

C. Multiple Resonance Techniques

In a double resonance experiment, a second exciting radio-frequency field B_2 is applied simultaneously to the sample by means of an additional irradiating transmitter (a triple resonance experiment would have a further rf field, but usually "multiple" means "double" here). It may thus be possible to observe the magnetic effect on one group of nuclei in the sample (by means of B_1 at frequency ω_1) while another group of nuclei is perturbed by an rf field B_2 at frequency ω_2. The two sets of nuclei may either be of the same species (homonuclear double resonance, e.g., $^1H-\{^1H\}$) or of different species (heteronuclear double resonance, e.g., $^{13}C-\{^1H\}$, which is the symbol for observations of ^{13}C resonance while ^1H is irradiated). Most double resonance experiments can be performed with the observing nucleus detected either in the cw mode (sweeping field or frequency) or in the FT mode (i.e., with pulsed radiation in the observe channel); in some recent techniques, the decoupler channel is also pulsed. In general, double resonance experiments are concerned with changes in coupling between nuclear spins, whether the coupling is through electrons in connecting chemical bonds, by chemical exchange (as in saturation spin population transfer experiments), or across space via the relaxation field. Multiple resonance experiments (Johannesen and Coyle, 1972) can be classified according to the energy of the perturbing field, $\gamma_{N'}B_2$, applied to nucleus N′. (Normal power levels are applied to the monitoring or measuring field B_1.) If $\gamma B_2 \approx (T_1T_2)_{N'}^{-1/2}$, very low power selective irradiation can be made of the spin systems in what is termed the generalized Overhauser effect. Spin population transfer experiments (SPT) are possible with FT and internuclear double resonance (INDOR) in cw, whereby relative intensities are changed and the presence of resonances under other peaks may be revealed. Thus for INDOR, the observing frequency ν_1 is fixed on some (visible) line; sweeping ν_2 through the *region* in which the coupled nucleus is expected will cause the ν_1 signal to change markedly when ν_2 hits the exact frequency of the second nucleus. In spin tickling experiments, still with low perturbing field $\gamma B_2 \sim (\pi T_2)^{-1} \sim \Delta\nu$, i.e., comparable with the line width, *additional* splitting of lines when frequency ν_1 is swept can help to determine energy level diagrams and signs of coupling constants J.

With $\gamma B_2 \sim J_{N,N'}$, we have selective spin decoupling and, at slightly higher power levels, with $\gamma B_2 > nJ_{N,N'}$, where n depends on the multiplicity of the line, spin decoupling is widely used to simplify spectra. In effect, by inducing rapid transitions between spin states of, say, nucleus X, decoupling can cause the multiplet from nucleus A to collapse and thus provide unequivocal proof that nuclei A and X are electron-coupled together. Such collapse of multiplets will also effectively enhance the signal-to-noise ratio of the collapsed line.

With yet bigger amplitudes, $\gamma B_2 = 10^2$ or 10^3 Hz or so, we have either single-frequency (coherent) or, very commonly with $^{13}C-\{^1H\}$, broad-band noise-modulated irradiation, whereby all coupling to one nuclear species is suppressed. It is also possible to achieve an intensity enhancement due to the nuclear Overhauser effect (NOE), $\eta \leqslant 0.5\ \gamma_{N'}/\gamma_N$, use of which is mentioned in Section III,E. The Overhauser effect is a change in population of spin states and hence in observed intensity arising from relaxation mechanisms, including dipolar coupling which can act across space (rather than along bonds), between or within molecules.

Some heteronuclear multiple resonance pulse sequences can eliminate dipolar broadening to enable high resolution effects to be observed in solids (see Sections III,E and V,D). Fuller accounts of double resonance methods, especially in Fourier transform spectroscopy (see Section III,D), may be found in the book by Shaw (1976). Wehrli and Wirthlin (1976) give a valuable summary of the utility of selective decoupling and off-resonance decoupling (as well as other techniques) in the assignment of ^{13}C resonances. For discussion of *gated decoupling,* useful for quantitative measurements in ^{13}C FT spectroscopy (Section III,F), reference should be made to the works by Müllen and Pregosin (1976) or Shaw (1976).

D. Pulse and Fourier Transform Methods

Spectrometers using short rf pulses for *relaxation time measurements* in solids have rather less rigorous magnetic field homogeneity requirements (about 1 in 10^5) than high resolution cw spectrometers, and somewhat larger samples are permissible. Pulse spectrometers require short (of the order of 1 μsec for solid samples), closely rectangular pulses of high power. The receiver must be able to detect the small FID immediately after the large pulse has been applied, the recovery time of the detection–amplifier system must be short (a few microseconds), and the bandwidth must be wide (in the region of 1 MHz) to minimize distortion of the short rf decay signals; sensitivity enhancement, necessary because of the additional noise thus admitted, is often accomplished with a gated amplifier or "boxcar integrator" (Rushworth and Tunstall, 1973). Pulse equipments also incorporate a programmer to initiate and gate pulse sequences and phases such as those described in Section II,B,2.

In solid coals, with sufficient free radicals to reduce T_1 to the region of 10^{-1} sec, free induction decays (FID) may be accumulated rapidly. Pembleton *et al.* (1977) describe a dc amplifier system with rapid recovery time (so that decays as short as 10^{-5} sec or so can be measured at 60 MHz) in a pulsed spectrometer (Vaughan *et al.*, 1972), with which hydrogen contents in seven coals were determined from the initial value of the FID. Related T_1 measurements on solid coals are reported by Gerstein *et al.* (1977) and Gerstein and Pembleton (1977) who found the effective spin–spin relaxation time (or damping constant) T_2^* too short for useful application of the Carr–Purcell–Meiboom–Gill (Section II,B,2) pulse sequence with the transmitter power available. Pulsed measurements could be made on tar sands and coals similar to those described by Miknis *et al.* (1974) on standard ASTM oil shales; hydrogen contents determined from FID amplitudes sampled 20 μsec after a 90° pulse agreed well with, and were measured much more rapidly than, Fischer assays.

In *Fourier transform (FT) spectroscopy* a wide range of energies (and frequencies) and thus of nuclei is sampled simultaneously. Slow consecutive sweeping through frequencies with time in continuous wave NMR is replaced by short (typically 10^{-5} sec) bursts or pulses of high radiofrequency power. The FT technique is thus a means whereby the advantages to signal-to-noise of monitoring many frequencies at the same time in a multichannel spectrometer may be achieved indirectly (by so-called multiplexing) much more practicably and cheaply than by having 10^4 or 10^5 parallel transmitters and detection systems. The realization of the advantages of pulse excitation, and the availability of stable spectrometers at high magnetic fields, of fast Fourier transformation algorithms, and of (comparatively) cheap small computers have combined to make natural abundance ^{13}C-NMR feasible for solutions of organic compounds such as coal extracts.

By comparison with the spectrum from a conventional single-channel cw instrument in which each frequency is sampled at successive instants, the transient FID observed in FT spectroscopy contains simultaneously excited frequencies corresponding to all frequency separations, acquired moreover in seconds rather than hundreds of seconds. All the information required is present in the FID (since the precession frequencies of the nuclei still reflect their magnetic surroundings and couplings). However, it may be interpreted more readily in terms of chemical shifts and coupling constants if it is decoded or Fourier-transformed (in a computer) from the time domain to the more familiar frequency domain. The usual absorption mode line shape function $g(\nu)_{\mathrm{abs}}$ in the frequency domain is related to the time domain function $g(t)$ by

$$g(\nu)_{\mathrm{abs}} = k \int_0^\infty g(t) \cos(2\pi\nu t)\, dt \tag{25}$$

Usually a series of similar intense pulses is applied, the transient responses of the system are added in a time-averaging computer, and the stored total envelope is then fast-Fourier-transformed by dedicated computer to yield the conventional absorption spectrum with a high signal-to-noise ratio. For a range Δ (in hertz) of chemical shifts, the pulse amplitude must be such that

$$\gamma B_1 \gg 2\pi\Delta \tag{26}$$

In these terms, the time $\pi/2\gamma B_1$ for a 90° rotation of M must be less than $(4\Delta)^{-1}$ where typically the spectral width might be 5000 Hz for ^{13}C or 1000–2000 Hz for ^1H. The time during which the FID is sampled is the acquisition time T_a, which depends on T_1 and may be of the order of 1 sec; for maximum resolution, T_a will be the time T_r between successive pulses. If the 90° pulses are repeated too quickly compared with T_1, the signal intensity is reduced (because the full M_z has not been regained) and so it is usual to have a delay between the end of data acquisition from one pulse and the beginning of the next pulse; in any case, T_r should exceed $5T_1$ (see, for example, Shaw (1976), p. 112).

In order to achieve maximum signal-to-noise with a repetitive sequence of pulses, a compromise is necessary in choice of pulse flip angle, α (often to as low as 40°). This is because $\alpha = 90°$ gives maximum signal (but needs a long delay between pulses) whereas for $\alpha < 90°$ recovery to equilibrium is quicker, so that more pulses can be used in a given time (but for each pulse M_z and M_y are somewhat reduced).

Pulse or Fourier transform spectrometers for high resolution NMR of solutions require a magnet with homogeneity and stability of cw high resolution; electronics (transmitter, probe, receiver) with the stringent rf characteristics of pulsed NMR, including powers of kilowatts rather than watts; and a computer for data accumulation and reduction and for operating the pulse programmer. However, the cost is justified primarily (although there are other advantages) by the immense reduction in time necessary for recording a spectrum of a given signal-to-noise ratio; a factor of 10^2 would be typical for ^1H-NMR and 10^3 for ^{13}C-NMR. The effective improvement in sensitivity (in proportion to the square root of the ratio spectrum-width/line-width) is such that NMR measurements can readily be made on coal extracts with natural abundance ^{13}C, a nucleus for which FT operation is especially advantageous. Also, relaxation times can be measured for individual lines in a spectrum; for example, quaternary carbons, with larger relaxation than other carbons, give much weaker intensities. Among the components of an FT spectrometer, the lock signal, usually effected by means of a nuclear species other than the one under examination, keeps the resonant condition [Eq. (6)] by locking field to frequency. With ^{13}C FT NMR, ^1H–^{13}C couplings are eliminated by noise-decoupling of ^1H over a wide frequency range; this also causes NOE enhancement (see Section III,F).

E. Special Techniques

1. *Solvent Resonances*

Coal extracts often involve rather low concentrations of sample so that, even if a deuterated solvent is used in ^{1}H-NMR, the largest peak may arise from residual ^{1}H in the solvent. Observation of a weak signal in the presence of a strong one raises *dynamic range* problems in FT spectroscopy since information is stored in the computer in words of finite length which could overflow, and since the analog-to-digital converter has limited word length. One way of overcoming overflow in the computer memory is "block averaging," whereby one batch of FIDs is acquired, Fourier-transformed, and stored as a frequency spectrum, then a second batch is acquired to give a second frequency spectrum, and so on, with subsequent addition of these data blocks carried out in the frequency domain.

Although unwanted solvent peaks in ^{1}H resonance when D_2O (containing HOD) is the solvent have received most attention, they can occur with ^{13}C resonance of coal extracts in organic solvents. Several techniques available for their elimination are based on differences in relaxation times T_1 between sample and solvent.

2. *High Resolution Techniques for Solids*

In the solid state, when the static nuclear spin dipole–dipole interactions (Section II,C) are not averaged, chemical shift information is usually obscured in the broad NMR lines. One approach to overcoming this is to spin the solid sample mechanically extremely rapidly (at about 10 kHz) about an axis inclined at $\theta = 54.7°$ [the "magic angle," since it gives $\cos^2\theta = \frac{1}{3}$, so that the angular factor $3\cos^2\theta - 1 = 0$ in Eq. (18)] to B_0. If this can be achieved (Andrew, 1971), it also averages chemical shift anisotropy effects; however, it is difficult experimentally.

Other means of obtaining relatively narrow lines from solids involve double resonance techniques (Pines *et al.*, 1973) and multiple pulse sequences (Vaughan *et al.*, 1976) with large (50 G) rf fields but low (100 W) power which can be designed to remove specific contributions. The multiple pulse series $90°_{x'}$, 2τ, $90°_{-x'}$, τ, $90°_{y'}$, 2τ, $90°_{-y'}$, τ, proposed by Haeberlen and Waugh (1960), causes the nuclear spins to precess about the magic angle and so removes the dipolar broadening. The chemical shift anisotropy is reduced (by $1/\sqrt{3}$) but not to zero unless the multiple pulse technique is combined with macroscopic magic angle spinning, as Pembleton *et al.* (1977) have done, utilizing the REV-8 (Rhim *et al.*, 1973, eight-pulse cycle with overall time of about 10^{-4} sec) sequence.

The double resonance or cross-polarization approach to high resolution

NMR in solids is particularly valuable if the predominant dipolar coupling is heteronuclear, as with ^{13}C-NMR in a sample containing ^{1}H nuclei, so that it may be removed by very high power decoupling ($\gamma B_2 \sim 20$ kHz) at the ^{1}H frequency, since ^{13}C nuclei are rarely close enough to each other to cause more than a few hertz dipolar broadening. Moreover, the sensitivity of these dilute ^{13}C spins can be improved by a modification, proton-enhanced nuclear induction (due to Pines *et al.*, 1972), of the Hartmann and Hahn (1962) solid state spin transfer experiment. The abundant ^{1}H spins are polarized at a low spin temperature with respect to the rotating frame, and the "hot" dilute ^{13}C spins are cooled by contact with the ^{1}H spins in a double resonance experiment; they develop a large transverse magnetization which is observed while they are decoupled from the protons. As usual, successive FIDs of the ^{13}C magnetization are digitized, accumulated, and finally Fourier-transformed. The effective gain in ^{13}C sensitivity by ^{1}H enhancement depends on the ratios of abundances and γ values of the two nuclear species (and increased repetition rate because of the shorter ^{1}H relaxation times) and may be 10^2 compared with conventional ^{13}C FT spectroscopy. Applications of the cross-polarization technique, described in Chapter 24, Section II,C, to whole coals are mentioned in Section V,D of this chapter; VanderHart and Retcofsky (1976) discuss experimental and interpretative problems in the application of the technique to samples of this kind. Shielding anisotropy contributions to the line shape are still present unless the specimen is also macroscopically rotated rapidly at the magic angle, as achieved by Miknis and Netzel (1976) for shales and by Bartuska *et al.* (1977) for coals. A combined double resonance multiple pulse ^{13}C–$\{^{1}H\}$ scheme (Stoll *et al.*, 1976), not yet applied to coals, may be able to yield chemical and geometrical information contained in spin–spin interactions.

F. Quantitative Measurements with ^{13}C-NMR

One of the contributory reasons for the rapid utilization of cw ^{1}H-NMR spectra in structural organic chemistry is the simple relation between peak areas and numbers of nuclei contributing to them, so that quantitative measurements are very straightforward. With Fourier transform operation, however, and especially for ^{13}C-NMR, intensity relations are less simple. A useful summary of the problems of quantitative ^{1}H and ^{13}C FT NMR studies on related hydrocarbon mixtures (crude oils) is given by Vercier *et al.* (1977), and Section V,C of this chapter includes quantitative ^{13}C measurements on coal extracts. In this section, consideration of some of the problems with ^{13}C-NMR (which, in practice, is nearly always by FT) is preceded by a summary of the advantages of ^{13}C-NMR. Sensitivity limitations delayed the application of natural abundance ^{13}C-NMR until recent years. Now, however, with the availability of FT instrumentation, the merits of ^{13}C are being exploited. Not only do ^{13}C chemical shifts cover a much wider range than ^{1}H shifts (300 ppm compared with 12 ppm for most organic compounds relevant to coal) but also dipolar broadening is lower. Absence of spin–spin couplings leads to an enormous simplification of spectra; ^{13}C–^{13}C coupling can be disregarded because proximity of 1% abundant nuclei will be so rare, while ^{13}C–^{1}H coupling can be eliminated (as mentioned in Section III,C) by proton (i.e., heteronuclear) decoupling experiments, ^{13}C–$\{^{1}H\}$. A further characteristic of ^{13}C-NMR is the extent to which spin–lattice relaxation times T_1, generally dominated by ^{13}C–^{1}H dipolar interactions, can be measured for individual nuclei (rather than as averaged values); thus a methylene carbon might have a T_1 of about 40 msec, while the value for a methine carbon might be twice this.

Correlation between the number of carbon atoms contributing to a peak and the peak area is lost in proton-decoupled ^{13}C spectra because of a number of factors (Freeman and Hill, 1971; Stothers, 1974; Gray, 1975). First, irradiation of the proton resonance leads to intensity changes through the nuclear Overhauser effect (Section III,C). Relaxation of carbon nuclei between spin states is often caused largely by dipolar interaction with neighboring protons; saturation of the proton resonance then increases the population difference between the upper and lower energy states of the carbon nuclei. Consequently, the intensity of the carbon signal may be enhanced by as much as a factor of 3 if dipolar relaxation predominates; but, since the strength of the dipolar interaction depends on internuclear separations r, as r^{-6}, selective intensity enhancements are observed, although enhancements close to the theoretical maximum are often observed for C–H.

A second factor which leads to loss of quantitative information is the use of Fourier transform detection. As indicated earlier (Section III,D), this involves the excitation of all the different signals in the spectrum by an intense pulse of rf energy, collection of frequency and amplitude data during the free induction decay (FID) (10^{-1}–10 sec), and, finally, computer calculations of the Fourier transform of the FID so as to recreate the slow passage spectrum.

Owing to the weakness of the ^{13}C signals, it may be necessary to accumulate 10^2–10^4 pulses in order to achieve sufficient improvement (in proportion to the square root of the number of pulses) in the signal-to-noise ratio of the final spectrum. Now, the rate at which the populations of the nuclear spin states are restored to the equilibrium values (characterized by the relaxation time T_1) is determined by the interaction of the ^{13}C magnetic moment with the fluctuating magnetic fields of directly bonded hydrogen nuclei and, to a lesser extent, by other nearby protons. T_1 thus differs greatly for different ^{13}C nuclei; for nonprotonated carbons, it may be long compared with the interval between pulses. Other (less important) factors influencing quantitative ^{13}C measurements in the Fourier transform mode are the effects of limited pulse strength and of adequate digitization of the frequency domain spectrum.

In principle, both the variable nuclear Overhauser effect (NOE) and the effect of long spin–lattice relaxation times may be overcome if an alternative relaxation mechanism for all ^{13}C nuclei is provided in the sample by adding a paramagnetic species. This can quench the NOE and make all T_1 values short compared with the pulse repetition rate (La Mar, 1971; Natusch, 1971; Gansow *et al.*, 1972; Levy and Cargioli, 1973).

A second method of avoiding the adverse effects of NOE and long T_1 makes use of a gated decoupling sequence in which the ^1H irradiation is only switched on during the pulse and acquisition of the interferogram (Freeman *et al.*, 1972). VanderHart and Retcofsky (1976) discuss additional problems in making quantitative ^{13}C-NMR measurements of aromaticity by cross-polarization with protons. By combining ^1H and ^{13}C FT NMR, Dorn and Wooton (1976) used phenanthrene as a model for quantitative H and C analysis of hydrocarbons. Their recommendations and the application of the method to solvent-refined coal are mentioned in Section V,C.

IV. INTERPRETATION OF SPECTRA

Coal and coal products are notoriously complicated materials; indeed, their very complexity underlines both their technological importance and their intrinsic scientific interest. There are two complementary approaches to the study of such materials by NMR spectroscopy. On the one hand, detailed high resolution spectroscopic analyses can be made on solutions of single components believed to be present in coal extracts. On the other hand, more direct (if less comprehensive) compositional and structural indications can be obtained by examination of solutions of complex mixtures extracted from coal and coal products, or even (with appropriate instrumentation) on whole coals. In general, measurements of the first kind add to the fund of relevant knowledge whereby success in the second approach can advance beyond the empirical or fingerprint stage. (Of course, the actual agglomeration of individual molecules into mixtures may perturb their resonances somewhat.)

In this brief section, a summary (Section IV,A) of the interpretation of the shapes of broad NMR lines from coal and coal products precedes an indication (Section IV,B) of the reference sources to interpretation of high resolution studies of solutions of relevant single compounds. Detailed treatment of high resolution applications to coals, extracts, and mixtures follows in Section V.

A. Broad-Line NMR Spectra of Coals

The majority of early NMR studies on fossil fuels involved broad-line ^1H measurements on solids made by cw spectrometers with the output in the form of a differential absorption spectrum. In some solids, as the temperature increases the line widths tend to diminish from the value appropriate to a rigid lattice, and nuclear resonance transitions with fairly sharp decreases in width, corresponding to increased molecular motions, occur. Following the investigations by Bell *et al.* (1958), Richards and Yorke (1960), and others, of the variation in ^1H-NMR line shape with carbon content and maceral type, Tschamler and de Ruiter (1962, 1963) reviewed earlier work and attempted to derive structural units for coals from second moment measurements. Broad-line studies continued (Cunningham *et al.*, 1966) even after the emergence of high resolution NMR. In order to estimate the concentrations of proton types in coal tar pitches, Pearson (1973) has combined high resolution measurements on chloroform extracts with low temperature broad-line second moment measurements on the residues (i.e., after chloroform is removed).

Moist solid samples exhibit a broad-line ^1H-NMR signal with a narrow component, corresponding to protons in rapid motion, superimposed on a broader absorption, associated with protons of shorter T_2. Chapter 7 of Volume I compares many methods of measuring moisture content in coals; Ladner and Stacey (1962) successfully compared the areas of the narrow and broad components of the composite NMR line to measure the moisture content (to $\pm1\%$ moisture for 0–14% moisture content) in a moving stream of powdered solid coal or coke flowing at 100 g/hr in a cw spectrometer. At water contents above 15%, caking prevented free flow of sample under gravity. In a small-scale on-line development (Ladner, 1975), coal was conveyed from the outlet of a supply hopper by a flexible belt which passed through a Perspex tube mounted through the center of the NMR magnet system. Since the belt was about twice as wide as the diameter of the Perspex tube, it was guided from the flat position to almost total enclosure of the conveyed coal during passage through the magnet system, so that the tube remained clean; the belt then returned to the flat conveying condition to discharge the coal into a pivoted weigh hopper. In this semicontinuous sampling system, in which a preset amount of coal was weighed at each cycle, the time taken for this weight to load the weigh hopper was used to gate the integration circuit of the NMR measuring unit. For further discussion of the applications of NMR in monitoring flowing systems, see Jones and Child (1976).

Retcofsky and Friedel (1970b) reported the first derivative-detected broad-line ^{13}C-NMR measurements on whole bituminous coals and anthracites and investigated signal averaging of the first derivative of the dispersion mode signal as a means of overcoming signal-to-noise problems with natural abundance ^{13}C-NMR. Later, Retcofsky and Friedel (1971) used the dispersion mode and about 700 scans on a cw spectrometer at 15 MHz to confirm by ^{13}C-NMR the high aromaticity f_a (see Section V,B) of Dorrance anthracite. With the same technique, Retcofsky and Friedel (1973) found that, for four vitrains, carbon aromaticity increases with coal rank.

B. High Resolution NMR of Relevant Single Compounds

If differences in chemical shift (Section II,D) between nuclei in a molecule are much larger than the corresponding coupling constants, so-called *first-order NMR spectra* of simple multiplets occur. This will be the case for coupled nuclei of different species, e.g., ^1H and ^{13}C in the same molecule (although, as has been mentioned, couplings to natural abundance ^{13}C can generally be neglected in ^1H spectra), and sometimes also for nuclei of the same species, e.g., methyl ^1H and methylene ^1H in the ethyl group of a substituted polynuclear hydrocarbon, provided the B_0 field is high enough. Very often, however, for hydrogen nuclei on one ring of such a compound, for example, the couplings are comparable in magnitude with the chemical shift separations. In such strongly coupled systems, the spectra (while occasionally appearing deceptively simple) become complex and extraction of the exact values of the δ and J terms from such *second-order spectra* can be very difficult.

If two nuclei in a compound happen to possess the same chemical shift, they are sometimes said to be *isochronous*. Symmetry-equivalent (or, more often, just *equivalent*) nuclei, which are more common, give rise to extra lines from spin coupling only when they interact with some other (*nonequivalent*) nucleus or group of nuclei. For a group of nuclei to be *magnetically equivalent*, all must have identical chemical shifts, and must also couple equally to each member of every group of *magnetically equivalent* nuclei in the molecule.

If a group of three equivalent spin-$\frac{1}{2}$ nuclei is loosely coupled to another group of two spin-$\frac{1}{2}$ nuclei, as in the ^{1}H resonance of an ethyl group recorded at sufficiently high field, then first-order multiplets will result. Coupling of three methyl protons causes the methylene protons to sense four values of the total spin: $\Sigma m = +\frac{1}{2} + \frac{1}{2} + \frac{1}{2} = +\frac{3}{2}$, $+\frac{1}{2} + \frac{1}{2} - \frac{1}{2} = +\frac{1}{2}$, $+\frac{1}{2} - \frac{1}{2} - \frac{1}{2} = -\frac{1}{2}$, and $-\frac{1}{2} - \frac{1}{2} - \frac{1}{2} = -\frac{3}{2}$; the probabilities of the corresponding transitions lead to intensity ratios 1 : 3 : 3 : 1 for the lines of the methylene quartet, separated by J. Conversely, the higher field methyl resonance is split into a 1 : 2 : 1 triplet because the methyl protons sense three fields from the methylene protons, with $\Sigma m = +1, 0$, and -1, and twice as many molecules will have $\Sigma m = 0$ as $\Sigma m = \pm 1$; the equally spaced lines are separated by the same coupling constant, J. In general, a group of n spin-$\frac{1}{2}$ nuclei will cause splitting of a nonequivalent nucleus or group of nuclei into $n + 1$ peaks with intensity ratios 1 : 1; 1 : 2 : 1; 1 : 3 : 3 : 1; 1 : 4 : 6 : 4 : 1; 1 : 5 : 10 : 10 : 5 : 1; 1 : 6 : 15 : 20 : 15 : 6 : 1; . . . Coupling of n equivalent spin-I nuclei causes splitting into $2nI + 1$ lines. First-order spectra can be "solved," i.e., chemical shifts and coupling constants extracted, by direct inspection.

In the conventional spin system designations, closely coupled nonequivalent nuclei, with chemical shifts comparable with their coupling constants, are labeled by adjacent letters of the alphabet: A, B, C, . . . ; nuclei with Δ/J very large are given remote letters in the alphabet: . . . , X, Y, Z. A three-spin ABX system would thus be a group of three coupled spins of which the X nucleus is either a different species from the other two or at least has a chemical shift well separated from those of the other two. Nuclei with intermediate shifts are given letters K, L. M. . . . . Subscripts denote the number of equivalent nuclei so that, for example, a closely coupled ethyl group would be an A_3B_2 five-spin system. Primes indicate that chemically equivalent nuclei are not magnetically equivalent.

Since the condition for first-order spectra involves the ratio $\nu_0\delta/J$, in which the numerator depends on B_0 but the denominator does not, increase of the laboratory magnetic field can make a spectrum more nearly first order, or at least will markedly influence its appearance. This is true even for the simplest case of a molecule containing two magnetic nuclei coupled together, AB, if of the same species, or AX, if of different species. The spectral analysis of this and more complicated spin systems, in order to extract chemical shifts and coupling constants, is discussed by Abraham (1971). Typically an iterative computer analysis is involved and the experimental spectrum is compared with one calculated from the parameters deduced.

Elucidation of the ^1H spectra of polycyclic condensed benzenoid hydrocarbons is often difficult because of spectral complexity and overlapping spin systems, so that high field 220-MHz spectra are valuable. Bartle and Jones (1972) have reviewed at length the application of high resolution ^1H-NMR to polycyclic aromatic molecules, many of them relevant to extracts of coal and coal products. More recent examples of detailed ^1H studies include benzofluorenes (Jones *et al.*, 1972), fluoranthene (Bartle *et al.*, 1974), and benzo[*b*]fluoranthene (Jones *et al.*, 1974). Friedel and Retcofsky (1970) measured ^{13}C shifts in pure compounds thought to be present in coal, while Ozubko *et al.* (1974) and Buchanan and Ozubko (1975) have interpreted ^{13}C spectra in other polynuclear hydrocarbons. High resolution 100-MHz ^1H-NMR spectroscopy helped Ouchi and Imuta (1973) to identify an aromatic $C_{30}H_{20}O$ compound containing a diphenylene oxide structure from Yubari coal and aided inference of structures of compounds separated from depolymerized brown coal by Imuta and Ouchi (1973). Ortho-coupling constants in polycyclic hydrocarbons have been reviewed by Bartle *et al.* (1969a) and long-range couplings by Bartle *et al.* (1969b).

V. SELECTED APPLICATIONS OF NMR TO COAL AND COAL PRODUCTS

A. High Resolution ^1H-NMR Examination of Mixtures

In early investigations of coal extracts and other coallike materials by high resolution ^1H-NMR (Friedel, 1959; Brown *et al.*, 1960; Brown and Ladner, 1960; Rao *et al.*, 1960; Ladner and Stacey, 1961; Oth and Tschamler, 1961, 1962; Oth *et al.*, 1961; Takeya *et al.*, 1963; Winniford and Bersohn, 1962), three kinds of hydrogen atoms were distinguished. These were aromatic and phenolic hydrogen (designated H_A), hydrogen on saturated carbon atoms α to an aromatic ring (H_α), and hydrogen on other saturated structures (H_β).

Hydrogens of these types were recognized in carbon disulfide extracts of coke oven (CO) and low temperature (LT) coal tar pitches, as well as the hydrogen present in methylene bridges between aromatic rings ($H_{\alpha,2}$) (De Walt and Morgan, 1962); by addition of pure compounds, it was possible to assign peaks due to CH_2 of acenaphthene and fluorene. Hydrogen types H_A, H_α, $H_{\alpha,2}$, and H_β were also recognized in coal extracts and shock-heated products (Brown and Waters, 1966; Durie *et al.*, 1966; Yokayama *et al.*, 1970; Retcofsky and Friedel, 1970a) and in fractions from low temperature tars (Brooks and Stevens, 1964; Brooks and Smith, 1964) and "depolymerized" coal (Heredy *et al.*, 1966).

These investigations were confined mainly to relatively low-molecular-weight materials (<400). In a study of the pyrolysis of resins from humous coals by ^{1}H-NMR in combination with infrared and ultraviolet spectroscopy, Turenko *et al.* (1972) found that 40% of the higher molecular weight fraction is aromatic, substituted with aliphatic side chains. Bartle and Smith (1965, 1967) extended ^{1}H-NMR studies to fractions of some coal tars with molecular weights up to 1200. More detailed assignments of proton types (Table I) to individual spectral bands were based both on comprehensive compilations of literature chemical shifts and (as indicated in Section IV) on parallel studies on pure compounds (Bartle and Jones, 1972). Proton types in the spectra of bitumen fractions were similarly assigned by Speight (1970).

Although coal-derived materials do not in general give sharply defined resonances, fine structure (Maekawa *et al.*, 1975) in spectra for the

TABLE I *Hydrogen Types Distinguishable by ^{1}H-NMR in Coal Extracts and Derivatives*

δ range[a] (ppm)	Assignment in 60- and 100-MHz spectra	Symbol	Subdivision possible at 220 MHz
5.5–9.0	Aromatic and phenolic	H_A	Sterically hindered aromatic hydrogen $\delta > 8.1$
4.7–5.5	Olefinic		
3.3–4.5	Methylene groups α to two rings (e.g., of fluorene)	$H_{\alpha,2}$ or H_F	
2.0–3.3	Hydrogen on carbon atoms α to ring	H_α	Methylenes of acenaphthenes and indenes, δ 3.0–3.3
1.6–2.0	Naphthenic methylene and methine (other than α to ring)	H_N	
1.0–1.6	Methylene β or more remote from ring; methyl groups β to ring	H_β	
0.5–1.0	Methyl γ or further from ring	H_γ	

[a] Twenty percent solution in $CDCl_3$ or CS_2. Shifts downfield from internal tetramethylsilane.

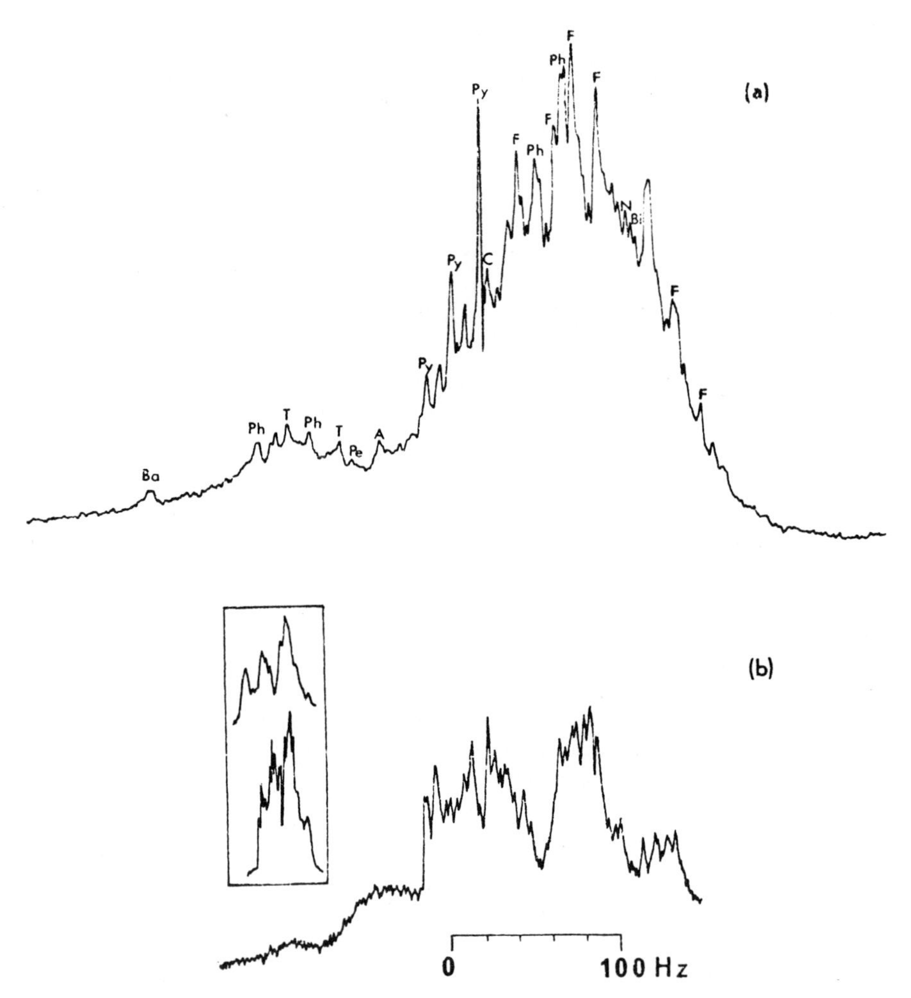

Fig. 5 Fine structure in ^{1}H-NMR spectra of tar fractions (after Bartle, 1972).
(a) Aromatic region of a CO tar fraction (Bartle and Smith, 1965) at 60 MHz. Key: N, naphthalene; Bi, biphenyl; A, anthracene; Ph, phenanthrene; F, fluoranthene; Py, pyrene; Ba, 1,2-benzanthracene; C, chrysene; T, triphenylene; Pe, perylene.
(b) H_α band in 220-MHz spectrum of neutral oil from LT coal tar (Bartle and Jones, 1969). Inset, the same band is compared on a reduced scale at 60 MHz (above) with (below) the much better resolved version at 220 MHz. Reproduced by permission of the publishers, I.P.C. Business Press, Ltd. ©.

lowest molecular weight (~200) fraction may be attributed (Bartle and Smith, 1965; De Walt and Morgan, 1962) to individual compounds (Fig. 5). Limitations of resolution at 60 MHz led Bartle and Jones (1969) to record ^1H spectra of coal-derived materials at 220 MHz. The high resolution and sensitivity available at this frequency are manifest in two ways. First, bands which are the result of the overlap of separate signals with slightly different chemical shifts from protons in many different environments reveal new fine structure. This occurs in the aromatic region [e.g., H-12 of benz[*a*]anthracene in a low-molecular-weight fraction of a blended continous vertical retort (CVR) tar], in the $H_{\alpha,2}$ region (e.g., CH_2 of the <0.5% fluorene in a fraction of a CVR tar), and particularly in the H_α region (e.g., Fig. 5). High field ^1H-NMR affords an alternative to chromatographic and other combinations of separation and spectroscopic methods of identifying the individual products, especially for compounds either present in small amounts or of relatively high molecular weight. [In some cases, the separation procedure may be complementary to high field ^1H-, and ^{13}C-, NMR in the identification of individual compounds present in low concentration; see, e.g., Chapter 25, Section III, and Schweighardt *et al.* (1971).]

A number of bands attributable to additional hydrogen types are also detectable in the 220-MHz ^1H spectra of coal tar fractions (Bartle and Jones, 1969) (see Table I) and the overlap of certain bands may be less marked (Oelert and Köser, 1977). Signals from olefinic protons in a 220-MHz spectrum of a CVR fraction of tar were recognized by Bartle and Jones (1969).

As noted in Section III, Fourier transform ^1H-NMR allows the spectra of much smaller samples to be recorded than with continuous-wave NMR. Submilligram fractions from polynuclear aromatic hydrocarbons of several environmental origins have been analyzed by this method (Bartle *et al.*, 1976b, 1977c) and a number of compounds, especially methyl derivatives, have been identified. Quantitative aspects of the technique have been explained (Thiault and Mersseman, 1975) and application to coal derivatives could utilize the inherent sensitivity and rapidity of the method.

Direct information may be obtained from the distributions of hydrogen between the chemical types in Table I. For example, the aromaticity of coal tar fractions (measured by percent H_A) and the content of alkyl groups (measured by percent H_α, H_β, and H_γ) vary markedly with the carbonization conditions under which the original tars were produced. All fractions show increasing aromaticity and decreasing content of alkyl groups in the sequence LT, CVR, blended (mainly CVR), blended (mainly CO), and CO; CVR and CO tar fractions differ less in percent H_A as molecular weight increases (Bartle, 1972).

B. Average Structural Analysis of Coal Samples from ^{1}H-NMR

Once hydrogen distributions have been determined by ^{1}H-NMR, they may be combined with number-average molecular weights and with elemental, functional group, and other analyses in structural analysis schemes so as to provide a picture of the properties of the average molecule. A method devised by Brown and Ladner (1960) for the analysis of coal carbonization products has been widely applied (e.g., Takeya *et al.*, 1963; Brooks and Stevens, 1964; Brooks and Smith, 1964; De Walt and Morgan, 1962; Oelert, 1967a,b; Retcofsky and Friedel, 1970a; Maekawa *et al.*, 1975), both as originally stated and in modified form; this is despite some criticism (McNeil, 1966) of the arbitrary assumptions of the procedure concerning heterocyclic structures and the nature of alkyl side chains. From ^{1}H-NMR and elemental analyses, Brown and Ladner (1960) calculated the following average structural parameters: f_a, the ratio of aromatic carbon (C_{ar}) to total carbon (C); σ, the fraction of the total available outer edge positions occupied by constituents; and H_{aru}/C_{ar}, the atomic H/C ratio that the average aromatic skeleton would have if each substituent were replaced by a hydrogen atom. These are derived from the following equations, where C/H and O/H are atomic ratios:

$$f_a = \left(\frac{C}{H} - \frac{H_\alpha^*}{x} - \frac{H_\beta^*}{y}\right)\left(\frac{C}{H}\right)^{-1} \qquad (27)$$

$$\sigma = \left(\frac{H_\alpha^*}{x} + \frac{O}{H}\right)\left(\frac{H_\alpha^*}{x} + \frac{O}{H} + H_{ar}^*\right)^{-1} \qquad (28)$$

$$\frac{H_{aru}}{C_{ar}} = \left(\frac{H_\alpha^*}{x} + \frac{O}{H} + H_{ar}^*\right)\left(\frac{C}{H} - \frac{H_\alpha^*}{x} - \frac{H_\beta^*}{y}\right)^{-1} \qquad (29)$$

Here, $H_\alpha^* = H_\alpha/H$, the ratio of "alpha type" hydrogen to total hydrogen, determined by ^{1}H-NMR; $H_\beta^* = H_\beta/H$, the ratio of "beta type" hydrogen to total hydrogen, again from ^{1}H-NMR; x and y are the atomic ratios of hydrogen to carbon in, respectively, alpha and beta structures, both of which are assumed; and $H_{ar}^* = H_{ar}/H$ is the ratio of aromatic to total hydrogen determined by ^{1}H-NMR, with the assumption that 60% of the total oxygen is phenolic.

Brooks and Stevens (1964) successfully applied Eqs. (27)–(29) to fractions of low temperature tar and found that ~50% of each fraction was predominantly aromatic but with some aliphatic and alicyclic substituents. The remainder was divided approximately equally between paraffins and a more highly aromatic analog of the major portion. De Walt and Morgan (1962) applied the Brown and Ladner equations to coke oven pitches (f_a = 0.94–0.99; σ = 0.06–0.12) and also modified them to take into account heterocyclic oxygen and CH_2 bridge hydrogen. The presence of sulfur, nitrogen, and saturated hydrocarbons was ignored. Oelert (1967a,b) combined the Brown and Ladner method with parameters measured by ultraviolet (Oelert, 1967c) and infrared spectroscopy (Oelert, 1971) and applied these to shock-heating products of coal and tar fractions (Oelert, 1967c). De Ruiter and Tschamler (1964) have also applied a combined structural analysis scheme to coal extracts.

Friedel and Retcofsky (1963) applied the Brown and Ladner equation to pitch fractions and later to pyridine and CS_2 extracts of coals (Retcofsky and Friedel, 1970a). More recently, this group has conducted a thorough investigation of the utility of the Brown and Ladner equation in coal research (Retcofsky *et al.*, 1977) (see also Chapter 24). The unambiguous f_a values determined from proton-coupled ^{13}C-NMR spectra of materials derived from coal were compared with values of the same parameter from Eq. (27). The largest difference between f_a determined by the two methods was found to be 0.07, suggesting that published values based on the older ^1H-NMR method are reasonably reliable. In these studies, the arbitrary parameters x and y in Eqs. (27)–(29) have generally been taken as 2.0, although values between 1.6–1.8 (Dryden, 1962) and 2.2 (De Walt and Morgan, 1962) have also been used. Retcofsky *et al.* (1977) discussed the consequences of the various assumed and indirectly determined values of x and y.

During studies of the composition of refined tars, and the chemical changes occurring during their use as road binders (Bartle *et al.*, 1969c), Bartle and Smith (1965, 1967) developed an independent structural scheme. In this, elemental analyses and ^1H-NMR-derived hydrogen distributions in solvent fractions of the tars are modified in order to exclude residual solvent and paraffins, determined by chromatography on silica gel; the authors then refer only to those structures containing an aromatic nucleus. The molecular formula and the numbers of hydrogen atoms per average molecule in each of the categories distinguished by ^1H-NMR are corrected for the presence of atoms other than carbon and hydrogen by a "lift out and replace" method similar to that used by van Krevelen (1961) in studies on coal and applied by Binns (1962) to pitch fractions. These corrections change the numbers measured to the corresponding quantities for the "average equivalent hydrocarbon" (AEH) of the fraction.

The scheme then allows calculation of the numbers and nature of the alkyl substituents. Thus $\frac{1}{3}[H_\gamma]_a$ is the number per AEH molecule of terminal methyl groups of chains three or more carbon atoms long; of course, very many combinations of the length and degree of branching of these chains, and of the number of methyl (m), ethyl (e), and isopropyl (i) groups, would give the same values for $[H_\alpha]_a$ and $[H_\beta]_a$. If it is assumed that the only alkyl groups making significant contributions to the ^{1}H-NMR spectra are methyl, ethyl, propyl, and butyl, then the following equations can be written

$$3(n + b + 2s + c) = [H_\gamma]_a \quad (30)$$

$$3e + 2n + 6i + 4b + 5c = [H_\beta]_a \quad (31)$$

$$3m + 2e + 2n + i + 2b + 3s + c = [H_\alpha]_a \quad (32)$$

Here the numbers of n-propyl, n-butyl, isobutyl, and *sec*-butyl groups per molecule of AEH are n, b, s, and c, respectively. It is then possible to solve these equations to give ranges of possible values for the numbers of different alkyl groups per molecule, provided qualified assumptions are made about the relative distributions of the groups based on analyses of extracts (Table II).

On this basis, Bartle *et al.* (1975b) have given explicit formulas for the numbers of each group type in an application of the Bartle and Smith scheme to a supercritical gas extract of coal (Section V,G). An alkyl-substitution parameter P, analogous to the σ of Brown and Ladner (1960), may also be calculated (Bartle and Smith, 1965, 1967):

$$P = \text{number of alkyl groups}/([H_A]_a + \text{number of alkyl groups}) \quad (33)$$

together with the number of ring carbon atoms C_R, equal to the total number of carbon atoms minus carbon atoms in alkyl groups, and C_J, the number of ring-joining carbon atoms:

$$C_J = C_R - [H_A]_a - \text{number of alkyl groups} - \text{number of fluorene-type } CH_2 \text{ groups} \quad (34)$$

TABLE II *Distribution[a] of Alkyl Groups in Low-Molecular-Weight (~200) Fractions of Refined Tars[b]*

Alkyl	CVR	Blended CVR	Blended CO	CO
Methyl	1.01–1.43	0.89–1.05	0.70–0.79	0.56
Ethyl	0–0.85	0–0.33	0–0.17	0.01
Isopropyl	0–0.42	0–0.16	0–0.08	—
n-Propyl	0.32	0.17	0.04	—
n-Butyl, isobutyl, *sec*-butyl	0.06	0.03	0.01	—

[a] Per molecule of average equivalent hydrocarbon.

[b] After Bantle and Smith (1965) Reproduced by permission of the publishers, I.P.C. Business Press, Ltd. ©.

A further modification (Bartle *et al.*, 1975b) was the incorporation of the extra available parameter H_R, the number of ring hydrogen atoms in the unsubstituted aromatic nucleus. Three-dimensional graphs of C_J, C_R, and H_R were constructed to allow delineation of the structure of the average aromatic nucleus of each fraction for comparison with similar plots for model hydrocarbon types. In two dimensions, the value of C_J at $C_R = 16$ is graphed versus H_R at $C_R = 16$.

The results obtained from the Bartle and Smith scheme applied to a neutral oil from a low temperature tar have been shown (Bartle *et al.*, 1970) to be consistent with values for the comparable parameters of the Brown and Ladner approach.

A third structural analysis scheme, due to Speight (1970, 1971), has found particular application in the analysis of oil and bitumen fractions (Ali, 1971). This approach uses hydrogen distributions determined from ^{1}H-NMR by equations which are modifications of those proposed by Yen and Erdman (1962); in turn, these equations are based on the method of Brown and Ladner (1960). The Yen–Erdman method has been applied to oil asphaltenes (Ferris, 1967) and the original Brown–Ladner equations to other asphaltic materials (Davis and Peterson, 1966). Peterson and co-workers (Ramsay *et al.*, 1967; Helm and Peterson, 1968) combined the Brown and Ladner method with certain empirical equations proposed by Williams (1958) to relate peak heights in the ^{1}H-NMR spectra of petroleum fractions to structural parameters. Further application of this combined method to asphaltic samples has been reported (Dickson *et al.*, 1969). The earlier Williams scheme, which has been extended by Chamberlain (1963) and by Williams and Chamberlain (1963), has also been applied by Gardner *et al.* (1959) to petroleum and asphalt.

For application to petroleum heavy ends, Hirsch and Altgelt (1970) developed a structural analysis scheme in which an iterative computer program combines data from NMR, density, elemental analysis, and molecular weight determinations. This method was modified by Katayama *et al.* (1975) and applied to coal tar pitches. Yamada *et al.* (1976) proposed further modifications, especially in the computational methods employed; they also combined the Brown and Ladner (1960) method with an independent scheme, due to Diamond (1958), which uses x-ray diffraction to yield information about the number of structural units or aromatic rings per molecule. The results for pitch fractions of various origins agreed well with the computer method and with the combined ^{1}H-NMR/x-ray method.

Computer-assisted molecular structure construction (CAMSC) has been developed recently by Oka *et al.* (1977) for the structural elucidation of coal-derived fractions. CAMSC uses elemental analysis, ^{1}H-NMR spectra, and molecular weight data to determine the allowed combinations for functional groups constituting the structure. Results obtained by the method for four sets of experimental data from the literature compared well with those obtained by other methods [e.g., the Brown and Ladner (1960) method and the Bartle and Smith (1965, 1967) scheme as modified by Bartle *et al.* (1975b)].

C. ^{13}C-NMR Spectroscopy of Coal Extracts

Friedel and Retcofsky (1963, 1966) first reported the ^{13}C-NMR spectra of coal derivatives: the neutral oil from the high temperature cracking of a coal carbonization product, three coal tar fractions (pitch, anthracene oil, and heavy creosote), and a CS_2 extract of a bituminous coal (Retcofsky and Friedel, 1968). Use of the dispersion mode with rapid scan gave spectra consisting of broad asymmetric bands, so that approximate integrated intensities could be obtained by planimetry. Two groups of resonances were observed for the neutral oil at about 60 and 170 ppm from CS_2, assigned to aromatic and saturated C, respectively.

A time-averaging computer enabled Hammel and Smith (1967) to accumulate 920 scans of the absorption mode spectrum of a solvent fraction of a CO tar, so that much more precise integration was possible (see Bartle, 1972, Fig. 9). As in the spectra by Friedel and Retcofsky (1963, 1966, and Friedel 1970), the aromatic band is a doublet, and the large peak to low field is thought to contain the low field components of the aromatic CH doublets superimposed on the singlets due to the substituted and ring-joining C atoms. The smaller peak to high field probably contains the high field components of the CH doublets only—an assignment supported by the doublet splitting of 150 Hz [cf. 142–162 Hz in many pure hydrocarbons (Stothers, 1965)] and its chemical shift (62 ppm upfield from $^{13}CS_2$).

Similar spectra were obtained by Knight (1967) and by Clutter *et al.* (1972) from aromatic petroleum fractions. Integrated peak areas were combined with hydrogen distributions from ^{1}H-NMR in structural analysis schemes, both of which incorporated, in part, the scheme of Williams (1958).

Proton-coupled ^{13}C-NMR offers a direct measure of the carbon aromaticity of soluble coal fractions, provided that effects arising from long relaxation times are taken into account (see Section III,F). For the semiquantitative experiments of Friedel and Retcofsky (1963, 1966) and Friedel (1970) mentioned earlier, f_a values were in fair agreement with the estimate obtained indirectly from the ^{1}H-NMR spectra. More recently, however, the same group has carried out a careful comparison of f_a values determined unambiguously from correlation mode (Dadok and Sprecher, 1974) ^{13}C-NMR spectra of a variety of coal extracts and coal-derived materials (see also Chapter 24). The effects of rf power levels, sweep times, and delays between successive scans were optimized to yield reliable intensity data; long time delays between scans were not found to be necessary, presumably because free radicals shortened relaxation times (Retcofsky, 1977; Retcofsky and Friedel, 1976; Retcofsky *et al.*, 1977). (See also Chapter 24, Section III,A.)

Erasure of ^{13}C–^{1}H splittings in proton-decoupled ^{13}C-NMR spectra of complex mixtures such as petroleum fractions (Clutter *et al.*, 1972; Shoolery and Budde, 1976) allows the resolution of signals which overlap in coupled spectra and results in considerable simplification. Preliminary experiments were reported for coal tar pitch and neutral oil (Retcofsky and Friedel, 1966), in which low field signals attributable to bridgehead aromatic carbons and carbons bearing alkyl or naphthenic substituents were discerned. More recently, Fourier transform spectroscopy has allowed many more carbon environments to be recognized by this method. Thus peaks forming part of the aromatic C–H band of the spectrum of a CS_2 extract of the vitrain of an American coal were tentatively assigned to carbons similar to C-4 of phenanthrene, C-7 of fluoranthene, and C-1 of fluorene (Retcofsky, 1977). In the spectra of fractions of a supercritical gas extract of coal (Bartle *et al.*, 1975a) and of a coal hydrogenation product (Maekawa *et al.*, 1975), the aliphatic region could be divided (Table III) into signals from carbons β or further from an aromatic ring, α-C-H and α-CH_3, and chain-ending methyl carbon. A diffuse but unambiguous band, between 37 and 53 ppm downfield from TMS, includes (Bartle *et al.*, 1975a,b) methylene bridge carbons but is much more intense than might be expected from the methylene bridge proton signals in the ^{1}H-NMR spectra; it seems likely that the resonances of certain carbons of naphthenes and of complex alkyl substituents with branched side chains are included. A number of bridgehead carbons of perhydroaromatic hydrocarbons have been shown to resonate in this region (Dalling and Grant, 1974; Balogh *et al.*, 1972). The ^{13}C-NMR spectrum (Clutter *et al.*, 1972) of a mixture of branched chain and cycloparaffins from a petroleum fraction also shows signals between 37 and 53 ppm.

A comparison of the ^{13}C-NMR spectra of coal-derived fractions with those of mixtures of *n*-paraffins from petroleum (Clutter *et al.*, 1972) and from coals and tars (Bartle *et al.*, 1977a) also provides some evidence for the presence of chains longer than butyl, commonly accepted as one of the larger alkyl substituents in coal-derived molecules (Chakrabartty

TABLE III *Carbon Types in Spectra of Coal-Derived Fractions*

	Supercritical gas extract of coal[a]			
	Methanol eluate	Benzene eluate	Coal hydrogenation product[b]	^{13}C shift range[c] (ppm from TMS)
Aromatic ether C–O	2			158–165
Phenol C–O	3			148–158
Aromatic C–C	4	4	Ca_1, Ca_2	129–148
Aromatic C–H[d]	5	5	Ca_3	118–129
Aromatic C–H ortho to ether-O, and OH	6		Ca_4	108–118
Methylene bridge C, bridgehead C of naphthenes and branched alkyls	7	7		37–53
CH α to aromatic ring other than CH_3[e]	8	8	Cp_1, Cp_2, Cp_3	30–37
Certain C β to aromatic ring	9	9	Cp_4, Cp_5, Cp_6	23–30
CH_3 α to aromatic ring	10	10	Cp_7	19–23
CH_3 of ethyl	11	11	Cp_8	17–19
CH_3 γ further from aromatic ring	12	12	Cp_9	13–17

[a] Numbers correspond to those of Fig. 6.
[b] Symbols correspond to those of Fig. 7.
[c] Bartle *et al.* (1975a). Some overlap of shift ranges likely. Only principal carbon types in a given shift range are listed. Shifts downfield from TMS (converted from CS_2 reference).
[d] Fine structure attributed to carbons of following types: C_4 of phenanthene (124 ppm); C_7 of fluoranthene (121 ppm); C_1 of fluorene (120 ppm) (Retcofsky, 1977).
[e] Superposed sharp peak (Cp_3) at 29.7 ppm also assigned to CH_2 of long chains (Bartle *et al.*, 1976a).

and Kretschmer, 1972). Thus there are resemblances in the sharp singlet at 29.7 ppm superimposed on the band assigned to α-C(CH_2 of long chains) and in the signals near 22.8 and 14.0 ppm (CH_2CH_3 and chain-ending CH_3, respectively). These bands are particularly clear in spectra of fractions from a supercritical gas extract of a Turkish lignite (Bartle *et al.*, 1976a). Evidence concerning oxygen functionality was obtained (Bartle *et al.*, 1975a,b) from a comparison of fine structure in the aromatic regions of the ^{13}C-NMR spectra of the silica gel chromatography fractions from the portion soluble in petroleum ether extracted from coal by supercritical gas (Table III). Neither benzene nor methanol eluates displayed detectable signals in the quinone C=O region (178–193

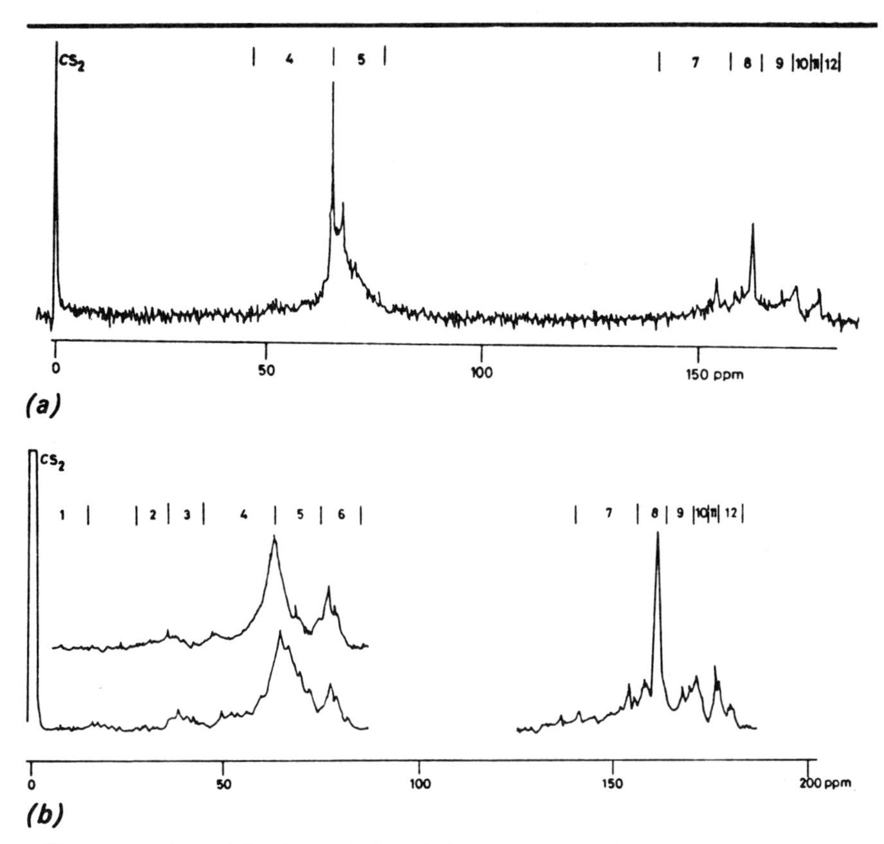

Fig. 6 Broad-band ^{1}H-decoupled FT ^{13}C-NMR spectra of fractions of a supercritical fluid extract of coal: (a) benzene eluate (22.63 MHz; 27,100 scans); (b) methanol eluate [25.14 MHz; 28,000 (upper) and 12,100 (lower) scans]. Numbers on peaks refer to carbon types tabulated in Table III. (After Bartle *et al.*, 1975a.) Shifts upfield from CS_2.

ppm). However, spectra of the methanol eluate contained a band, absent in the oxygen-lean benzene eluate, in the region 158–165 ppm and typical of *C*–OH (148–158 ppm) and *C*–O of heterocyclic and aromatic ether type (158–165 ppm) (Isbrandt *et al.*, 1973). This fraction also showed part of the aromatic C–H band shifted to high field, a feature absent in the benzene eluate spectrum, and attributed in part to aromatic C–H *ortho* to ether C–O (Isbrandt *et al.*, 1973). Aromatic ether oxygen was thought to be a major structural type in this coal extract.

In an attempt to draw some inferences about coal structures by mild dissolution of coal, Franz *et al.* (1977) applied proton-decoupled ^{13}C-NMR (and infrared) spectroscopy to successive extracts of Knife River

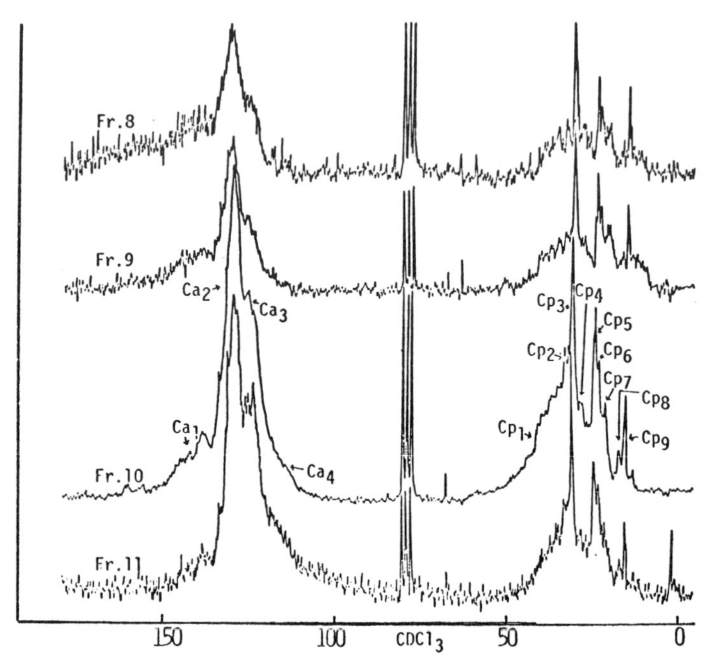

Fig. 7 Broad-band ^{1}H-decoupled FT ^{13}C-NMR spectra (25.14 MHz; 10,000–30,000 scans) from fractions 8–11 extracted from hydrogenated Yubari coal and separated by GPC. The symbols (Ca denotes aromatic carbon; Cp denotes aliphatic carbon) refer to carbon types tabulated in Table III. (After Maekawa *et al.*, 1975.)

coal (a lignite assigned to the ASTM rank of high volatile bituminous C) depolymerized by acid-catalyzed phenol. It appears that up to 74% of the coal structure is contained in methanol extracts. ^{13}C spectra indicate a predominantly aromatic system, with increasing aliphatic CH content and with aromatic spectra becoming simpler as depolymerization proceeds.

Although the use of Fourier transform and random noise decoupling allows dramatic improvement in the signal-to-noise ratios of ^{13}C-NMR spectra, quantitative information is difficult to obtain by these methods. As discussed in Section III,F, the major factors influencing quantitative ^{13}C-NMR measurements are nuclear Overhauser effects and long spin–lattice relaxation times. Promising experiments aimed at overcoming both of these limitations have been reported for hydrocarbon mixtures relevant to coal and petroleum (Thiault and Mersseman, 1976), for phenanthrene (Dorn and Wooton, 1976), and for crude oil (Shoolery and Budde, 1976). In all three studies, the effects of the addition (suggested for coal fractions as early as 1972 by Bartle) of a relaxation agent [*tris*(acetylacetonato)chromium(III), Cr(*acac*)$_3$] was compared with the use of gating of the proton spin decoupler during the delay between pulses. Although Thiault and Mersseman (1976) obtained better precision with gated decoupling (mentioned at the end of Section III,C) than with Cr(*acac*)$_3$, they nonetheless recommended the use of the latter because they regarded the time delays as prohibitive in the electronic method. On the other hand, Shoolery and Budde (1976) showed how both gated decoupling and additions of Cr(*acac*)$_3$ were necessary to obtain reproducible measurements of percentages of aromatic carbon in crude oils. Dorn and Wooton (1976) pointed out that alternative relaxation reagents (Levy *et al.*, 1975) might minimize interaction between the metal chelate and substrate molecules, but also mentioned that such additions may be unnecessary in coal- and petroleum-derived samples if sufficient paramagnetic materials (i.e., free radicals) are already present. Wooton *et al.* (1976) also made quantitative ^{13}C-NMR measurements on solvent-refined coal products by gated decoupling alone. A similar approach was employed by Maekawa *et al.* (1975) to coal hydrogenation products; the presence of free radicals was assumed and ratios of aromatic to aliphatic carbon remained constant at pulse repetition times as short as 0.8 sec, without gated decoupling. For agreement with such ratios determined by ^{1}H-NMR methods (Brown and Ladner, 1960), however, adjustment of x and y values (see Section V,B) was necessary.

D. High Resolution ^{13}C-NMR Spectroscopy of Whole Coals

The applications of high resolution ^{13}C-NMR in coal research outlined in Section V,C were restricted to liquids and soluble materials derived from coals. This is because the strong dipolar interactions,chemical shift anisotropies, and long spin– lattice relaxation times so broaden the lines as to mask chemical shift information.

Furthermore, any inferences about the structure of coal from solvent extract studies by NMR are restricted by the limited solubility of coal in NMR solvents and by the extent to which memory of the coal structure is retained in solution. None of the ^{13}C high resolution NMR methods so far reported allowed the recently expressed view (Chakrabartty and Kretschmer, 1974; Chakrabartty and Berkowitz, 1974) that coal is not a predominantly aromatic material but has a "polyamantane" structure (see Chapter 22, Fig. 5) to be incontrovertibly refuted. Values for f_a estimated on soluble coal derivatives may not represent the aromaticity of the original whole coal, while broad-line ^{13}C-NMR

(Retcofsky and Friedel, 1971, 1973) yielded quantitative information only (Bartuska *et al.*, 1977).

Very recently, a number of solid fossil fuel materials have been subjected to new techniques (Section III,D) which narrow the lines of ^{13}C-NMR signals in solids: cross-polarization or proton-enhanced nuclear induction spectroscopy (Pines *et al.*, 1972, 1973) and magic angle spinning (Lowe, 1959; Kessemeier and Norberg, 1967; Andrew, 1971). Thus, VanderHart and Retcofsky (1976) and Retcofsky (1977) have reported the cross-polarization spectra of solid coals in which dipolar line broadening was eliminated (see also Chapter 24, Sections II,C and V,B). Each spectrum consisted of two overlapping bands because of both the chemical shift anisotropies (~200 ppm for aromatic carbons, 30–60 ppm for nonaromatics) and the chemical shift dispersions (30–60 ppm). Line shape analysis then allowed separation of the aromatic and aliphatic carbon bands and direct measurement of f_a (Pines and Wemmer, 1977). Results from the two high quality vitrains studied support the classical view that coals are highly aromatic materials and that the aromaticity of coals increases with coal rank.

Schaefer and Stejskal (1975a) applied the combination of cross-polarization with magic angle spinning (so that the effects of chemical shift anisotropy but *not* of chemical shift dispersion are eliminated) to synthetic polymers and wood samples (Schaefer and Stejskal, 1975b, 1976). The same group also applied the method to anthracite, lignite, algal coal, and kerogen samples (Bartuska *et al.*, 1977); the coal samples were analyzed as powders in a solid sulfur matrix. The resulting spectra (Fig. 8) without spinning were analogous to those of VanderHart and Retcofsky (1976) but magic angle spinning narrowed the bands sufficiently for complete separation of the aromatic and aliphatic carbon resonances; f_a values of 0.99 (minimum) and 0.72 were estimated for anthracite and lignite, respectively. Spinning is especially important in analytical experiments on samples rich in aromatic carbon: A minor aliphatic carbon signal is obscured by the intense asymmetric aromatic carbon signal.

Miknis and Netzel (1976) have successfully combined magic angle sample spinning with the double resonance (Pines *et al.*, 1973) ^{1}H enhancement technique to obtain ^{13}C spectra of oil shale kerogens (see also Chapter 25). The large aliphatic band is resolved from the weak broad aromatic band, while different shales have mean paraffinic ^{13}C bands differing by several parts per million. Although this section is concerned with ^{13}C, it may be mentioned that Gerstein *et al.* (1977) used multiple pulse ^{1}H-NMR to study ^{1}H chemical environments in Virginia and Iowa vitrains. About 400 Hz of the residual ^{1}H line widths of 700 Hz (reduced

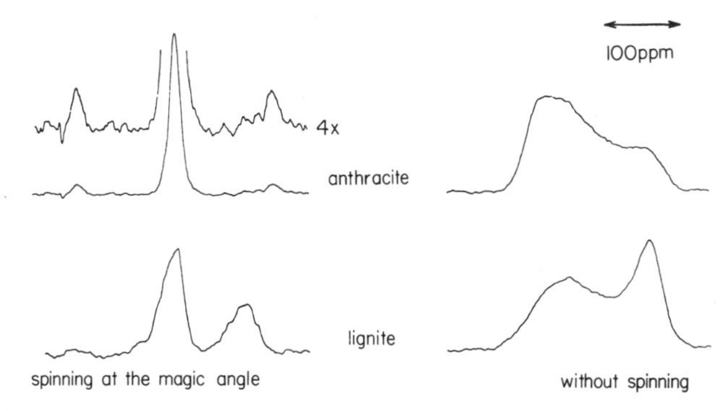

Fig. 8 ^{13}C spectra of anthracite and lignite obtained by the cross-polarization technique, with and without magic angle spinning. (After Bartuska *et al.*, 1977. Reproduced by permission of the publishers, I.P.C. Business Press, Ltd. ©.)

from 30 kHz) are attributed to dipolar coupling between electron spins (from free radicals) and nuclear spins. With interproton dipolar interactions, chemical shift anisotropies (250 Hz or 8 ppm), and electron–proton interactions still present, the REV-8 (Rhim *et al.*, 1973) pulse sequence did not resolve aromatic from aliphatic protons; combination with magic angle spinning would evidently be needed.

E. NMR and an Acid–Base Structure for Coal Asphaltenes

Schweighardt and co-workers have shown how asphaltenes from coal liquefaction products (Synthoil) can be separated into acidic and basic components by dissolving the asphaltenes in toluene and passing dry HCl gas through the solution (see also Chapter 24, Section VI,C,2 for further discussion of the application of NMR to Synthoil products). The basic component precipitates as an insoluble HCl adduct, while the acidic component remains in solution and can be recovered by evaporation of the solvent (Sternberg *et al.*, 1975; Schweighardt *et al.*, 1976b). Similar methods can be used to separate asphaltenes from supercritical gas extracts of coal (Martin and Bartle, 1977) and crude oil (Sternberg *et al.*, 1975). Thin-layer chromatography and electrophoresis of the Synthoil asphaltene acid fractions suggested that these contained polynuclear aromatic phenols and some indole-type nitrogen, whereas for the base fractions the results indicated polynuclear aromatic nitrogen bases with phenolic groups absent (Sternberg *et al.*, 1975; Schweighardt *et al.*, 1976b).

Confirmation came from a series of elegant NMR experiments (Sternberg *et al.*, 1975; Schweighardt *et al.*, 1976b). The 250-MHz

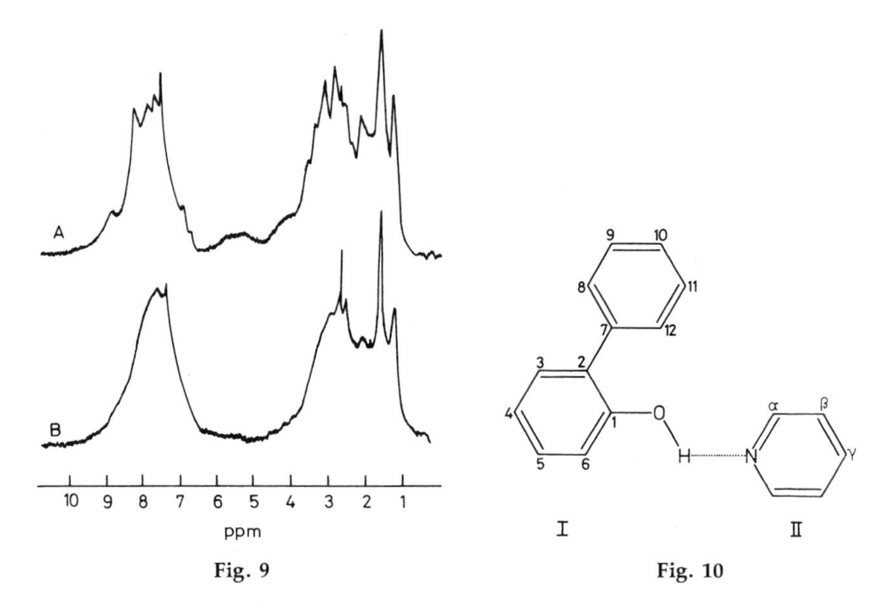

Fig. 9 ¹H-NMR spectra at 250 MHz of acidic (A) and basic (B) component solutions in $CDCl_3$ of asphaltenes derived from Kentucky high volatile bituminous coal. (After Sternberg *et al.*, 1975.)

Fig. 10 Hydrogen-bonded 1:1 complex between pyridine and *ortho*phenylphenol. (After Schweighardt *et al.*, 1976b.)

^{1}H-NMR spectra of the acidic and basic components (Fig. 9) differed particularly in that a broad O*H* resonance at δ 5.35 ppm (which was replaced by a sharp *H*OD signal when D_2O was added) was observed only in the acid fraction; no deuterium-exchangeable protons were detected in spectra of the basic fractions. Moreover, mixing the acidic and basic components yielded a material with an ^{1}H-NMR spectrum rather like that of the original asphaltene (Fig. 10), while incremental addition of the basic component progressively broadened the O*H* resonance and shifted it to low field. The conclusion that a hydrogen-bonded complex is formed between the acidic and basic components was strongly supported by a series of experiments (Schweighardt *et al.*, 1976b) with the model compounds pyridine and *o*-phenylphenol (Fig. 10). When the separate acidic and basic components of the coal-derived asphaltene were added to electron donor and acceptor systems, changes in ^{1}H, ^{13}C, and ^{14}N chemical shift and line width occurred similar to those observed with pure compounds alone.

For asphaltenes from coal supercritical gas extracts (Martin and Bartle, 1977), 100-MHz ^{1}H spectra of the acid and base components were similar to those just reported; the original material, however, gave a spectrum in which an O*H* signal is also detected, so that the presence of uncomplexed phenolic groups may be inferred. Taylor and Li (1978) have applied high field ^{1}H-NMR to acidic and basic components of asphaltenes from bituminous coals. Bartle and Smith (1965) also accounted for the shift of O*H* in the ^{1}H-NMR spectra of coal tar fractions; they further pointed out that a variety of groups other than nitrogen base (e.g., quinone, COOH) could bring about the low field shift. Hydrogen bonding between asphaltene and resin moieties of Athabasca bitumen was also ascribed to interactions between OH and oxygen functions (Moschopedis and Speight, 1976), in part by means of ^{1}H-NMR measurements. Apparent increases in the molecular weight of coal products in $CHCl_3$ have been accounted for by acid/base complexing (Coleman *et al.*, 1977) in much the same way as in a study on asphaltenes a difference was observed in the ability of solvents to break up intermolecular complexes (Haley, 1971). In their solvation studies, with anthracene oil and hydrophenanthrene solvents, of Wyoming subbituminous coals, Ruberto *et al.* (1977) made considerable use of ^{1}H spectra of asphaltenes. Their results favored a coal liquefaction route solid $\rightarrow$ asphaltene $\rightarrow$ resin $\rightarrow$ oil and suggested predominance of groups of about three condensed rings in the coal structure.

F. Electrode Pitches and Tars: Correlation of ^{1}H-NMR Measurements with Other Properties

Analysis of coal tar pitches in general is discussed in Chapter 33 and of coal tar electrode binders in particular in Chapter 34. In this section, possible correlations are considered between ^{1}H-NMR measurements and technologically important macroscopic properties of some electrode pitches and tars.

Following the demonstration by Bartle and Smith (1967) of a linear relation between the densities of coal tar fractions and the percentage of aromatic hydrogen ($\%H_A$) (see Section V,A), a number of pitch properties have been correlated with ^{1}H-NMR measurements. More recently, measurements by Yamada *et al.* (1976) have been shown (Pakdel *et al.*, 1977) to extend the density/$\%H_A$ relation to other carbonaceous materials, including pitches from petroleum and ethylene as well as coal tar pitches. Kini and Murthy (1974) claimed to have correlated the quinoline-insoluble content [related to the compressive strength of electrodes made from the pitch (Montgomery and Godspeed, 1956)] of coal tar pitch with the percentage of aromatic hydrogen determined for the material soluble in CS_2, but scaled up so as to correspond to the whole pitch. Aromaticity in coal tar electrode binders is dealt with in the context of macroscopic measurements in Section III,G of Chapter 34, and NMR determination of aromaticity in coal fractions is discussed in detail in Section III of Chapter 24. The quinoline-insoluble fraction of pitch is the subject of commercial specification (Mason, 1963, 1970; Weiler, 1963; Branscomb, 1966) for electrode binders, because these macromolecular constituents contribute to the binding and agglomeration properties (Greenhow and Sugowdz, 1961; Jurkiewicz and Tengler, 1966).

In further experiments (Pakdel *et al.*, 1977) with electrode pitches and tars, the $\%H_A$ measured from 100-MHz ^1H-NMR spectra for the portion of the pitch or tar soluble in CS_2 was compared with the toluene-insoluble (TI) content of each pitch or tar determined by the STPTC (Standardization of Tar Products Test Committee) method. This diverse range of products [three coal tar pitches, two petroleum pitches, one coal tar (Rexco) obtained by low temperature carbonization, and the two petroleum tars] exhibits, as might be expected, more of a trend than a correlation (Fig. 11) between $\%H_A$ in CS_2 and %TI; at least, low $\%H_A$ seems associated with lower %TI. However, if data for six road tars ($\%H_A$ in the heptane-soluble portion of a dioxan extract, expected to correspond to CS_2-soluble fraction) obtained by Bartle *et al.* (1969c) and

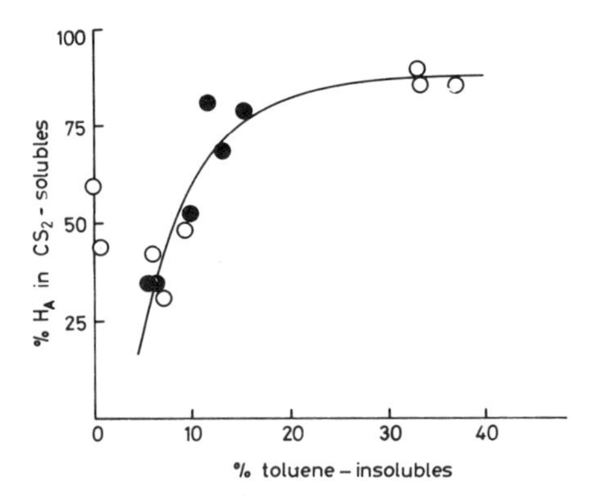

Fig. 11 Graph of percent aromatic hydrogen H_A in soluble fraction versus percent toluene insoluble in whole tar or pitch: ○ indicates data of Pakdel *et al.* (1977) (tars and pitches from coal and petroleum with $\%H_A$ from carbon disulfide solubles); ● indicates data of Bartle *et al.* (1969c) and Wilman (1966) (road tars with $\%H_A$ from heptane-soluble part of dioxane extract).

by Wilman (1966) (%TI) are included, a clearer relation emerges. With the petroleum tars excluded, $\%H_A$ rises steeply with %TI, and then levels off; thus, for $\%H_A > 80$, the $\%H_A$ stays fairly constant for coal tar pitches with high (>30) %TI.

The indication from the results of Pakdel *et al.*, that %TIs increase as H_A rises to near 80% is presumably because the higher molecular weight fractions become less soluble as the aromaticity of the constituent molecules increases. This interpretation lends support to the view that coal (Bartle *et al.*, 1975b) and coal tar (Bartle and Smith, 1967; Bartle, 1972) fractions have structural types which, for a given source, are independent of molecular weight, a conclusion which may extend to some petroleum products. For coal-derived materials, the initial portion of Fig. 11 is, of course, a consequence of the $\%H_A$ versus density relation (Bartle and Smith, 1967; Yamada *et al.*, 1976) already referred to, since McNeil (1961) has shown that the specific gravity of coal tars correlates with the percent insoluble in benzene. However, above a certain degree of aromaticity, the fraction of insoluble high-molecular-weight material seems independent of $\%H_A$, possibly because factors other than aromaticity, and which lead to TIs (thermolysis on distillation, etc.), become important.

The combination of broad-line and high resolution NMR in pitch studies was mentioned in Section IV,A (Pearson, 1973). In a combined mass spectrometric and high resolution NMR study of graphitizable carbons from Gilsonite pitch, Marsh *et al.* (1971) determined the percentage of aromatic protons from 90-MHz ^1H spectra of deuterated pyridine and carbon tetrachloride extracts. The original pitch (supposed by some authors to be mainly aromatic) was only about 5% aromatic, but the percentage of aromatic insoluble material rose appreciably after carbonization.

G. NMR Spectroscopy of Supercritical Gas Extracts of Coals and Lignites

With the decline of carbonization as a method for obtaining chemical feedstocks from coal, newer procedures for deriving these important raw materials are under active investigation. One promising method involves the extraction of coal with supercritical gases (SCE) (Wise, 1970; Whitehead and Williams, 1975; Harrison, 1976). This SCE technique, which uses compressed gases to volatilize high boiling materials and to effect their separation from a nonvolatile residue (Paul and Wise, 1971), is applied to coal undergoing mild pyrolysis in order to extract the liquids which are formed. The liquids are readily recovered by condensation from the gas phase.

Since the extraction takes place at temperatures (300–400°C) below those at which volatile matter is evolved by destructive distillation, extensive thermal degradation of the coal is avoided. The extract may therefore be representative of the lower molecular weight fractions of the coal, which Vahrman (1970, 1972) has suggested may to some extent be retained in the pores of the coal "molecular sieve." With toluene at 400°C and about 10 MPa, as much as one-third of the coal feed can be extracted. The residue from the SCE process is a noncoking porous solid which may find uses in combustion or gasification (Harrison, 1976).

Supercritical gas extracts obtained under different conditions from a British coal (Bartle *et al.*, 1975a,b, 1977d) and from Turkish lignites (Bartle *et al.*, 1976a) have been studied by a range of chromatographic and spectroscopic methods; ^{1}H- and ^{13}C-NMR provided particularly vital information. Thus, a coal extract obtained by semicontinuous extraction of Markham Main coal (NCB Coal Rank Code 802) by means of supercritical toluene at 350°C and 10 MPa amounted to 17% of the daf coal; little gas and water were evolved (Bartle *et al.*, 1975b). The gas extract was subjected to a fractionation scheme based on solubility in petroleum ether and benzene, followed by silica gel adsorption and gel permeation chromatography.

^{1}H-NMR spectra at 100 and 220 MHz were recorded for solutions of the extract in trifluoracetic acid, and 100-MHz spectra were recorded for solutions in benzene-d_6 for some of its fractions. Broad-band proton-decoupled pulse FT ^{13}C-NMR spectra were recorded for solutions in CS_2; the numbers of scans varied between 8000 and 28,000 for pulse intervals between 0.5 and 2 sec (Bartle *et al.*, 1975a).

The ^{1}H-NMR spectrum of the extract showed little extra fine structure at 220 MHz compared with the 100-MHz spectrum and no individual compounds were present in amounts large enough to be discerned in the spectrum. In the spectrum of the nonpolar aromatic fraction, peaks due to bibenzyl (Bovey, 1967) were identified; this compound is thought to arise by thermolysis of a small part of the large excess of toluene used in the extraction process.

Hydrogen distributions determined by ^{1}H-NMR were combined with ultimate and functional group analyses and molecular weight data to calculate the average structural parameters of the extract and its aromatic fractions using a version of the Bartle and Smith (1965) scheme as discussed in Section V,B. The properties of the average molecule (AEH) of each fraction were derived (Table IV). Semiquantitative results from the ^{13}C spectra (Section V,C) provided general support for the assump-

TABLE IV *Numbers of Atoms or Groups in the Average Molecule of Each Fraction of a Supercritical Gas Extract of Coal*[a]

			Petroleum-ether solubles					Petroleum-ether insols/benzene sols	
			Silica gel fractions			Sephadex fractions			
	Property	Whole[b] extract	Petroleum-ether eluate	Benzene eluate	Methanol eluate	Band A[b]	Band B	Band A1	Band B1
Number of atoms	C	28·4	26·0	22·8	22·5	21·5	17·6	44·0	40·5
	H	28·8	51·8	26·0	28·0	20·6	21·1	46·7	40·0
	Phenolic O	1·2	0	0·2	1·1	0·2	1·0	0·9	2·0
	Nonphenolic O	0·8	0	0·3	1·2	0·4	1·2	1·1	2·7
	Basic N	0·05	0	0·0	0·0	0·0	0·0	0·05	0·05
	Nonbasic N	0·35	0	0·0	0·1	0·1	0·1	0·55	0·45
	S	0·1	0	0·1	0·2	0·1	0·1	0·2	0·1
	H_A	7·4	0	7·5	6·5	7·9	4·7	10·4	8·1
	H_{OL}	0·7	0	0·0	0·0	0·3	1·3	0·9	1·1
	H_F	2·1	0	0·8	0·7	0·5	1·7	1·9	3·7
	H_α	9·4	2·6[a]	8·4	8·8	7·9	6·1	9·9	11·6
	H_β	6·3	36·3	6·6	8·8	1·8	5·2	16·2	11·4
	H_γ	2·5	12·9	2·9	3·4	2·3	2·0	7·5	4·0
Average equivalent hydrocarbon	No. of methyls	2·0		1·6	1·4	2·1	1·1	0·3	1·9
	ethyls	0·8		0·8	1·1	0·0	0·6	1·8	1·4
	(normal) propyls	0·4		0·5	0·6	0·4	0·4	1·4	0·8
	isopropyls	0·5		0·4	0·5	0·0	0·3	0·9	0·7
	butyls	0·3		0·3	0·4	0·3	0·2	0·8	0·5
	Total alkyls	3·9		3·6	4·0	2·8	2·7	5·3	5·2
	Olefinic carbon	0·7		0·0	0·0	0·3	1·3	0·9	1·1
	Total C alkyl	7·8		7·1	8·5	4·8	6·7	15·0	12·1
	Ring-joining CH_2 (RJM)	2·3		0·8	1·8	0·9	2·2	2·7	5·1
	Ring carbon atoms	21·9		16·1	15·5	17·4	12·3	30·8	31·8
	Ring-joining carbon atoms	8·3		4·3	3·3	5·8	2·7	12·4	13·3
	Ring-attached hydrogen atoms[c]	16·0		12·6	13·9	12·5	11·9	21·1	23·6
	Degree of alkyl substitution P[d]	0·34 (0·45)		0·32 (0·34)	0·38 (0·49)	0·27 (0·28)	0·36 (0·49)	0·33 (0·39)	0·39 (0·54)

[a] From Bartle *et al.* (1975b) by permission of the publishers, I.P.C. Business Press, Ltd. ©.

[b] Data refer to aromatic components only, a correction having been made for the paraffinic content.

[c] Cycloparaffinic hydrogen.

[d] For unsubstituted nucleus.

[e] Values in parentheses include OH.

tions made in the ^{1}H-NMR-based analyses except in the important respect that the ^{13}C spectra indicated the presence of naphthene and complex branched paraffinic side chains not included in the ^{1}H-based scheme. Model average structures for each fraction were now compared (Table V) with the properties of the average molecule. These represent an extremely wide range of chemical types which contribute to the average. For the higher molecular weight fractions, the average molecules can accommodate most of the structural features required by the chemical types present; for the low-molecular-weight fractions,

TABLE V *Comparison of Properties of the Average Molecule (from NMR-based structural analysis) with a Model Average Structure for Each Fraction of a Supercritical Gas Extract of Coal*[a]

Fraction and model structure	Number of atoms per average molecule								
							Average equivalent hydrocarbon		
	Total alkyl	Alkyl C	Ring-joining CH_2	Non-phenolic O	Phenolic OH	NH	C_R	C_J	H_R
Benzene eluate	3.6	7.1	0.4	0.3	0.2	0.0	16.1	4.3	12.6
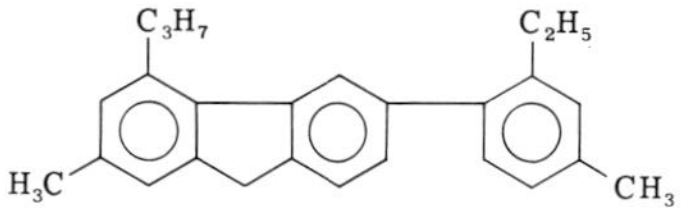	4	7	1[b]	0	0	0	19	6	14

			0.8				16.0	4.4	12.4
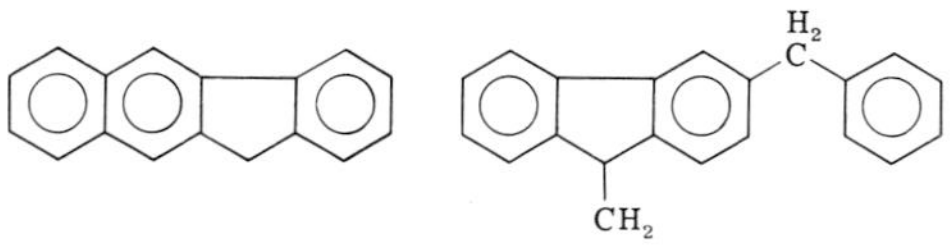									
Methanol eluate	4.0	8.5	0.4	1.2	1.1	0.1	15.5	3.3	13.9
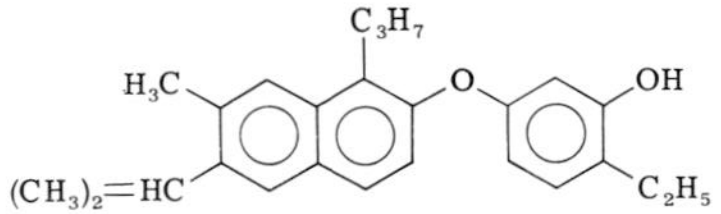	4	9	0	1[b]	1	0	17	4	14

Band A	2.8	4.8	0.3		0.2	0.1	17.4	5.8	12.5
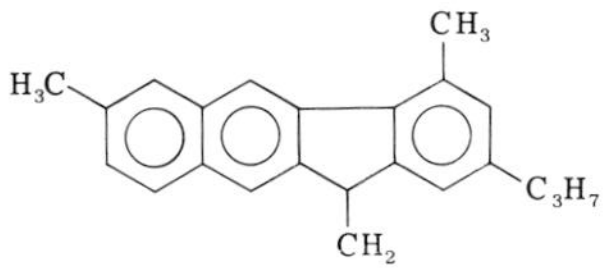	3	5	1[b]	0	0	0	17	6	12
Band B	2.7	6.7	0.9	1.2	1.0	0.1	12.3	2.7	11.9
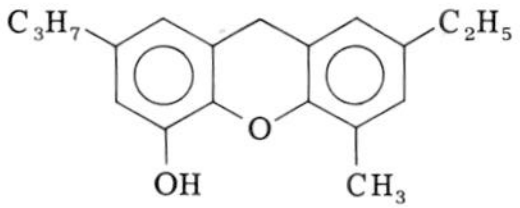	3	6	1	1	1	0	14	4	12
Band A1	5.3	15.0	1.0	1.1	0.9	0.55	30.8	12.4	21.1
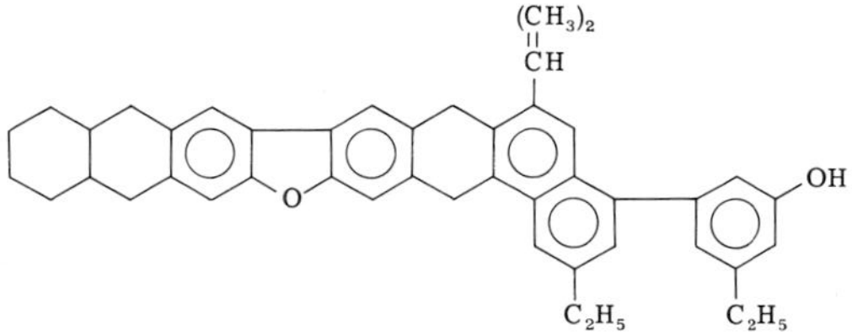	5	15	2[b]	1[b]	1	0	31	12	22
Band B1	5.2	12.1	1.9	2.7	2.0	0.45	31.8	13.3	23.6
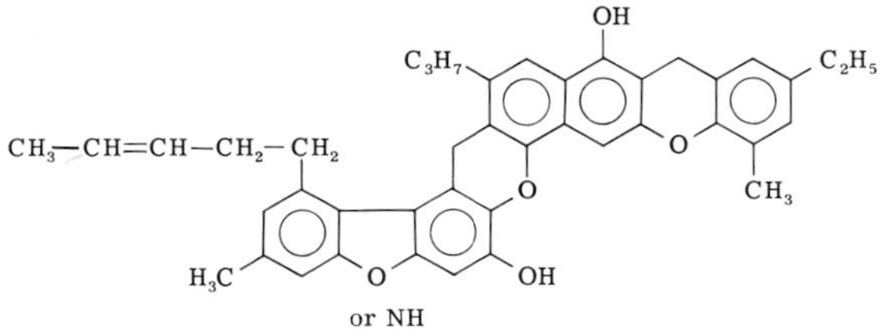	5	12	2	2.5	2	0.5	33	14	24

[a] After Bartle *et al.* (1975b) by permission of the publishers, I.P.C. Business Press, Ltd. ©.

[b] In model structures certain of these CH_2 groups would be replaced by O, S, etc., or vice versa.

however, an *average* of a number of model structures is necessary to account for the parameters. This approach might be applied with profit in more elaborate computer-aided average structure determinations (e.g., Oka *et al.*, 1977).

The low oxygen fractions evidently have a generally open chain polynuclear aromatic average structure with about 33% of the available sites carrying substituents. The most common of these is methyl but there are also substantial numbers of longer chain alkyls (some branched) and naphthenic groups. The fractions rich in oxygen have a similar basic structure and substitution pattern, except for the presence of phenolic hydroxyl groups and ether oxygen links. These conclusions are consistent with the view that the extract has experienced mild conditions. The presence of biphenyl linkages, methylene bridges, and dihydroanthracene-type structures (the latter particularly sensitive to dehydrogenation) points to minimum degradation of the aromatic part.

The application of similar methods to a supercritical toluene extract of the same coal at 400°C in the presence of hydrogen and zinc chloride catalyst showed (Bartle *et al.*, 1977d) that the material consisted of similar molecules incorporating more naphthenic substituents on smaller aromatic clusters with fewer alkyl groups and methylene and biphenyl-type linkages.

Supercritical toluene extracts of lignites (Bartle *et al.*, 1976a) yield fractions for which the ^{1}H-NMR and ^{13}C-NMR spectra are broadly similar to those from corresponding fractions from similar extracts of bituminous coal except for lower aromaticities. The aliphatic carbon region of the ^{13}C spectra of these lignite-derived materials shows fewer features, however, except for the strong CH_2 signal from long chains; more complex aliphatic groups may be present.

Supercritical gas treatment of coals is thought to yield large quantities of material in a state unaltered from that in the coal matrix—the product of coalification of plant material—i.e., without extensive degradation during extraction.

VI. CONCLUSION

The value of high resolution ^{1}H-NMR in providing chemically useful information about individual hydrocarbons, especially cyclic compounds, has been established for many years. Full analyses are often difficult but spectra can generally be solved with the help of high magnetic fields, double resonance experiments, and so on. Remarkably interesting results have been achieved from such superficially unpromising samples as complex mixtures of hydrocarbons extracted from coal and coal products. Over the last decade, ^{13}C has established itself as a complement to ^{1}H-NMR and is becoming almost as widely used for organic compounds. It is of especial value in yielding simpler spectra when, as with solutions extracted from coal-derived materials, the sample contains very many closely similar chemical environments (and thus chemical shifts). Conversely, with such pulsed ^{13}C-NMR experiments, quantitative measurements are less straightforward and involve consideration of more experimental parameters, including relaxation times.

Current instrumentation makes the combination of ^{1}H- and ^{13}C-NMR spectroscopy one of the most valuable nondestructive techniques for studying the composition and structure of solutions of hydrocarbons extracted from coal. In the solid state, early experiments showed that broad-line ^{1}H-NMR could be applied to whole coals to measure the moisture content. Very recently, preliminary high resolution ^{13}C-NMR experiments on solid coal samples with new pulse and magic angle techniques have already given very valuable direct indications of aromatic content in coals.

The power of NMR in organic structural analysis derives from an ability to sense several kinds of parameters. Thus chemical shifts depend on electronic and molecular environments; couplings (when measured) can be indicative of interactions with neighbors; and relaxation times are responsive to motions and other influences on magnetic dipolar interaction. Not all these aspects have been fully exploited yet in the field of coal and coal products. NMR focuses on specific nuclear species, e.g., ^{1}H or ^{13}C, sensing all the nuclei of the species under examination, with consequent complexity of the spectrum. Development of high frequency spectrometers has appreciably advanced the ^{1}H-NMR study of extracts of coal products; superconducting magnets of reasonable homogeneity with even higher fields, and consequently higher NMR operating frequencies, are quite feasible. In organic chemistry, the major instrumental development of the Fourier transform spectrometer has led to an enormous expansion in the exploitation of the potential of ^{13}C-NMR; fossil fuels are now sharing in this.

Finally, it should be recalled that any physical method concentrates attention on certain aspects of structure and composition; thus ^{13}C-NMR probes the regions near carbon nuclei only. In this chapter, we have considered NMR spectroscopy to some extent in isolation. There are many more cases in which NMR has been combined with infrared or other spectroscopic techniques in the study of coal extracts (e.g., Oelert and Hemmer, 1970; Bernhard and Berthold, 1975). Indeed, for a full picture in systems as complicated as coal and coal products, a combined approach is necessary. Chapter 22 puts NMR in perspective as just one of a wide range of spectroscopic techniques for studying coal structures in general, while Chapter 25 illustrates a situation in which NMR is a full partner in a small group of physical methods studying one chemical part of the coal structure, the paraffin hydrocarbons.

REFERENCES

Abragam, A. (1961). "The Principles of Nuclear Magnetism." Oxford Univ. Press, London and New York.

Abraham, R. J. (1971). "Analysis of High Resolution NMR Spectra." Am. Elsevier, New York.

Ali, L. H. (1971). *Fuel* **50,** 298.

Andrew, E. R. (1971). *Prog. Nucl. Magn. Reson. Spectrosc.* **8,** 1.

Balogh, B., Wilson, D. M., and Burlingame, A. L. (1972). *In* "Advances in Organic Geochemistry 1971" (H. R. Von Gaertner and H. Wehner, eds.), p. 163. Pergamon, Oxford.

Bartle, K. D. (1972). *Rev. Pure Appl. Chem.* **22,** 79.

Bartle, K. D., and Jones, D. W. (1969). *Fuel* **48,** 21.

Bartle, K. D., and Jones, D. W. (1972). *Adv. Org. Chem.* **8,** 317.

Bartle, K. D., and Smith, J. A. S. (1965). *Fuel* **44,** 109.

Bartle, K. D., and Smith, J. A. S. (1967). *Fuel* **46,** 29.

Bartle, K. D., Jones, D. W., and Matthews, R. S. (1969a). *J. Mol. Struct.* **4,** 445.

Bartle, K. D., Jones, D. W., and Matthews, R. S. (1969b). *Rev. Pure Appl. Chem.* **19,** 191.

Bartle, K. D., Smith, J. A. S., and Wilman, W. G. (1969c). *J. Appl. Chem.* **19,** 283.

Bartle, K. D., Jones, D. W., Martin, T. G., and Wise, W. S. (1970). *J. Appl. Chem.* **20,** 197.

Bartle, K. D., Mallion, R. B., Jones, D. W., and Pickles, C. K. (1974). *J. Phys. Chem.* **78,** 1330.

Bartle, K. D., Martin, T. G., and Williams, D. F. (1975a). *Chem. Ind. (London)* **54,** 226.

Bartle, K. D., Martin, T. G., and Williams, D. F. (1975b). *Fuel* **54,** 226.

Bartle, K. D., Calimli, A., Jones, D. W., and Pakdel, H. (1976a). Unpublished measurements.

Bartle, K. D., Lee, M. L., and Novotny, M. (1976b). *Proc. Anal. Div., Chem. Soc.* p. 304.

Bartle, K. D., Jones, D. W., and Pakdel, K. (1977a). *In* "Molecular Spectroscopy" (A. R. West, ed.), p. 127. Heyden, London.

Bartle, K. D., Lee, M. L., and Novotny, M. (1977b). *Analyst* **102,** 731.

Bartle, K. D., Ladner, W. R., Martin, T. G., Snape, C., and Williams, D. F., (1977d). *Fuel,* in the press.

Bartuska, V. J., Maciel, G. E., Schaefer, J., and Stejskal, E. O. (1977). *Fuel* **56,** 354.

Bell, C. L. M., Richards, R. E., and Yorke, R. W. (1958). *Brennst.-Chem.* **39,** 530.

Bernhard, V., and Berthold, P. H. (1975). *J. Prakt. Chem.* **317,** 1.

Binns, E. H. (1962). Rep. 0236. Coal Tar Res. Assoc., (now part of BCRA, Chesterfield, England).

Bovey, F. A. (1967). "NMR Data Tables for Organic Chemists," Vol. 1. Wiley (Interscience), New York.

Branscomb, J. A. (1966). *In* "Bituminous Materials" (A. J. Habergel, ed.), Vol. 3, Ch. 12. Wiley (Interscience), New York.

Brooks, J. D., and Smith, J. W. (1964). *Fuel* **43,** 125.

Brooks, J. D., and Stevens, J. R. (1964). *Fuel* **43,** 87, 125.

Brown, H. R., and Waters, P. L. (1966). *Fuel* **45,** 17.

Brown, J. K., and Ladner, W. R. (1960). *Fuel* **39,** 87.
Brown, J. K., Ladner, W. R., and Sheppard, N. (1960). *Fuel* **39,** 79.
Buchanan, G. W., and Ozubko, R. S. (1975). *Can. J. Chem.* **53,** 1829.
Carr, H. Y., and Purcell, E. H. (1954). *Phys. Rev.* **94,** 630.
Chakrabartty, S. K., and Berkowitz, N. (1974). *Fuel* **53,** 240.
Chakrabartty, S. K., and Kretschmer, H. O. (1972). *Fuel* **51,** 160.
Chakrabartty, S. K., and Kretschmer, O. (1974). *Fuel* **53,** 132.
Chamberlain, N. F. (1963). *In* "Treatise on Analytical Chemistry" (I. M. Kolthoff and P. J. Elving, eds.), Part 1, Vol. 4, p. 1885. Wiley, New York.
Clutter, D. R., Petrakis, L., Stenger, R. L., and Jensen, R. K. (1972). *Anal. Chem.* **44,** 1395.
Coleman, W. M., Wooton, D. L., Dorn, M. C., and Taylor, L. T. (1977). *Anal. Chem.* **49,** 533.
Cunningham, A. C., Ladner, W. R., Wheatley, R., and Wyss, W. F. (1966). *Fuel* **45,** 61.
Dadok, J., and Sprecher, R. F. (1974). *J. Magn. Reson.* **13,** 243.
Dalling, D. K., and Grant, D. M. (1974). *J. Am. Chem. Soc.* **96,** 1827.
Davis, T. C., and Peterson, J. C. (1966). *Anal. Chem.* **38,** 240.
De Ruiter, E., and Tschamler, H. (1964). *Brennst.-Chem.* **45,** 15.
De Walt, C. W., Jr., and Morgan, M. S. (1962). *Am. Chem. Soc. Symp. Tars, Pitches, Asphalts, Atlantic City, N.J.* p. 33.
Diamond, R. (1958). *Acta Crystallogr.* **11,** 129.
Dickson, F. E., Davis, B. E., and Wirkkala, R. A. (1969). *Anal. Chem.* **41,** 1335.
Dorn, H. C., and Wooton, D. L. (1976). *Anal. Chem.* **48,** 2146.
Dryden, I. G. C. (1962). *Fuel* **41,** 301.
Durie, R. A., Schewchyk, Y., and Sternhell, S. (1966). *Fuel* **45,** 99.
Ferris, S. W. (1967). *Ind. Eng. Chem., Prod. Res. Dev.* **6,** 127.
Franz, J. A., Morrey, J. R., Campbell, J. A., Tingey, G. L., Pugmire, R. J., and Grant, D. M. (1977). *Fuel* **56,** 367.
Freeman, R., and Hill, H. D. W. (1971). *J. Magn. Reson.* **4,** 366.
Freeman, R., Hill, H. D. W., and Kaptein, R. (1972). *J. Magn. Reson.* **7,** 327.
Friedel, R. A. (1959). *J. Chem. Phys.* **31,** 280.
Friedel, R. A., and Retcofsky, H. L. (1963). *Proc. Carbon Conf., 5th* **2,** 149.
Friedel, R. A., and Retcofsky, H. L. (1966). *Chem. Ind. (London)* p. 455.
Friedel, R. A., and Retcofsky, H. L. (1970). *In* "Spectrometry of Fuels" (R. A. Friedel, ed.), p. 90 Plenum, New York.
Gansow, O. A., Burke, A. R., and La Mar, G. N. (1972). *Chem. Commun.* p. 456.
Gardner, R. A., Hardman, H. F., Jones, A. L., and Williams, R. B. (1959). *J. Chem. Eng. Data* **4,** 155.
Gerstein, B. C., and Pembleton, R. G. (1977). *Anal. Chem.* **49,** 75.
Gerstein, B. C., Chow, C., Pembleton, R. G., and Wilson, R. C. (1977). *J. Phys. Chem.* **81,** 565.
Gray, G. A. (1975). *Anal. Chem.* **47,** 546A.
Greenhow, E. J., and Sugowdz, G. (1961). *Coal Res. CSIRO* **15,** 10.
Haeberlen, U., and Waugh, J. S. (1960). *Phys. Rev.* **175,** 453.
Hahn, E. L. (1953). *Phys. Today* **6,** 4.
Haley, G. A. (1971). *Anal. Chem.* **43,** 371.
Hammel, J., and Smith, J. A. S. (1967). Personal communication.
Harrison, J. S. (1976). *Am. Chem. Soc., Div. Pet. Chem. Prepr.* **21,** 92.
Hartmann, S. R., and Hahn, E. L. (1962). *Phys. Rev.* **128,** 2042.
Helm, R. V., and Peterson, J. C. (1968). *Anal. Chem.* **40,** 110.
Heredy, L. A., Kostyo, A. E., and Neuworth, M. B. (1966). *Adv. Chem. Ser.* **55,** 493.
Hirsch, E., and Altgelt, K. H. (1970). *Anal. Chem.* **42,** 1330.

Imuta, K., and Ouchi, K. (1973). *Fuel* **52,** 174.
Isbrandt, L. R., Jensen, R. K., and Petrakis, L. (1973). *J. Magn. Reson.* **12,** 143.
Jackman, L. M., and Sternhell, S. (1969). "Applications of Nuclear Magnetic Resonance Spectroscopy in Organic Chemistry," 2nd Ed. Pergamon, Oxford.
Johannesen, R. B., and Coyle, T. D. (1972). *Endeavor* **31,** 10.
Jones, D. W., and Child, T. F. (1976). *Adv. Mag. Reson.* **8,** 123.
Jones, D. W., Matthews, R. S., and Bartle, K. D. (1972). *Spectrochim. Acta, Part A* **28,** 2053.
Jones, D. W., Matthews, R. S., and Bartle, K. D. (1974). *Spectrochim. Acta, Part A* **30,** 489.
Jurkiewicz, J., and Tengler, S. (1966). *Koks, Smola, Gaz* **11,** 191.
Katayama, U., Hosoi, T., and Takeya, G. (1975). *Kogyo Kagaku Zasshi* **24,** 127.
Kessemeier, H., and Norberg, R. E. (1967). *Phys. Rev.* **155,** 321.
Kini, K. A., and Murthy, G. S. (1974). *Fuel* **53,** 204.
Knight, S. A. (1967). *Chem. Ind. (London)* p. 1923.
Ladner, W. R. (1975). Personal communication.
Ladner, W. R., and Stacey, A. E. (1961). *Fuel* **40,** 295.
Ladner, W. R., and Stacey, A. E. (1962). *Br. J. Appl. Phys.* **13,** 136.
La Mar, G. N. (1971). *J. Am. Chem. Soc.* **93,** 1040.
Levy, G. C., and Cargioli, J. D. (1973). *J. Magn. Reson.* **10,** 231.
Levy, G. C., and Nelson, G. L. (1972). "Carbon-13 Nuclear Magnetic Resonance for Organic Chemists." Wiley, New York.
Levy, G. C., Edlund, U., and Hexem, J. G. (1975). *J. Magn. Reson.* **19,** 259.
Lowe, I. J. (1959). *Phys. Rev. Lett.* **2,** 285.
McNeil, D. (1961). *J. Appl. Chem.* **11,** 90.
McNeil, D. (1966). *In* "Bituminous Materials" (A. J. Hoiberg, ed.), p. 139. Wiley (Interscience), New York.
Maekawa, Y., Veda, S., Hasegawn, Y., Nakata, Y., Yokoyama, S., and Yoshida, Y. (1975). *Am. Chem. Soc., Div. Fuel Chem., Prepr.* **20,** 1.
Marsh, H., Akitt, J. W., Hurley, J. M., Melvin, J., and Warburton, A. P. (1971). *J. Appl. Chem.* **21,** 251.
Martin, T. G., and Bartle, K. D. (1977). Unpublished measurements.
Mason, C. R. (1963). *Coal Tar Sci.* **16,** 3.
Mason, C. R. (1970). *Fuel* **49,** 165.
Meiboom, S., and Gill, D. (1958). *Rev. Sci. Instrum.* **29,** 688.
Memory, J. D., and Parker, G. W. (1974). *In* "Molecular Physics" (Dudley Williams, ed.), 2nd Ed., Part B, p. 465. Academic, New York.
Miknis, F. P., and Netzel, D. A. (1976). *In* "Magnetic Resonance in Colloid and Interface Science" (H. A. Resing and C. G. Wade, eds.), Symposium Series No. 34, p. 182. Am. Chem. Soc., Washington, D.C.
Miknis, F. P., Decora, A. W., and Cork, G. L. (1974). "Pulsed N.M.R. Studies of Oil Shales: Estimation of Potential Oil Yields." *U.S. Bur. Mines, Rep. Invest.* RI-7984.
Montgomery, D. S., and Godspeed, F. E. (1956). Rep. FRL 213. Fuels Div., Dep. Mines Tech. Surv., Ottawa.
Moschopedis, S. E., and Speight, J. G. (1976). *Fuel* **55,** 187.
Müllen, K., and Pregosin, P. S. (1976). "Fourier Transform NMR Techniques: A Practical Approach." Academic, New York.
Natusch, D. F. S. (1971). *J. Am. Chem. Soc.* **93,** 2566.
Oelert, H. H. (1967a). *Brennst.-Chem.* **48,** 362.
Oelert, H. H. (1967b). *Z. Anal. Chem.* **231,** 105.
Oelert, H. H. (1967c). *Z. Anal. Chem.* **231,** 81.
Oelert, H. H. (1971). *Z. Anal. Chem.* **255,** 177.
Oelert, H. H., and Hemmer, E. A. (1970). *Erdoel Kohle* **23,** 87.

Oelert, H. H., and Köser, H. J. (1977). *In* "Molecular Spectroscopy" (A. R. West, ed.), Ch. 8. Heyden, London.
Oka, M., Chang, H.-C., and Gavalas, G. R. (1977). *Fuel* **56,** 3.
Oth, J. F. M., and Tschamler, H. (1961). *Fuel* **40,** 119, 719.
Oth, J. F. M., and Tschamler, H. (1962). *Brennst.-Chem.* **43,** 177.
Oth, J. F. M., Tschamler, H., and De Ruiter, E. (1961). *Brennst.-Chem.* **42,** 378.
Ouchi, K., and Imuta, K. (1973). *Fuel* **52,** 171.
Ozubko, R. S., Buchanan, G. W., and Smith, I. C. P. (1974). *Can. J. Chem.* **52,** 2493.
Pakdel, H., Jones, D. W., and Bartle, K. D. (1977). Unpublished measurements.
Pake, G. E. (1956). *Solid State Phys.* **2,** 1.
Paul, P. F. M., and Wise, W. S. (1971). "The Principles of Gas Extraction." Mills & Boon, London.
Pearson, R. M. (1973). *Fuel* **52,** 80.
Pembleton, R. G., Ryan, L. M., and Gerstein, B. C. (1977). *Rev. Sci. Instrum.* **48,** 1286.
Pines, A., and Wemmer, D. (1977). Unpublished measurements.
Pines, A., Gibby, M. G., and Waugh, J. S. (1972). *J. Chem. Phys.* **56,** 1776.
Pines, A., Gibby, M. G., and Waugh, J. S. (1973). *J. Chem. Phys.* **59,** 569.
Ramsay, J. W., McDonald, F. R., and Peterson, J. C. (1967). *Ind. Eng. Chem., Prod. Res. Dev.* **6,** 231.
Rao, H. S., Murti, G. S., and Lahiri, A. (1960). *Fuel* **39,** 263.
Retcofsky, H. L. (1977). *Appl. Spectrosc.* **31,** 116.
Retcofsky, H. L., and Friedel, R. A. (1966). *Adv. Chem. Ser.* **55,** 503.
Retcofsky, H. L., and Friedel, R. A. (1968). *Fuel* **47,** 487.
Retcofsky, H. L., and Friedel, R. A. (1970a). *In* "Spectrometry of Fuels" (R. A. Friedel, ed.), p. 70. Plenum, New York.
Retcofsky, H. L., and Friedel, R. A. (1970b). *In* "Spectrometry of Fuels" (R. A. Friedel, ed.), p. 99. Plenum, New York.
Retcofsky, H. L., and Friedel, R. A. (1971). *Anal. Chem.* **43,** 485.
Retcofsky, H. L., and Friedel, R. A. (1973). *J. Phys. Chem.* **77,** 68.
Retcofsky, H. L., and Friedel, R. A. (1976). *Fuel* **55,** 363.
Retcofsky, H. L., Schweighardt, F. K., and Hough, M. (1977). *Anal. Chem.* **49,** 585.
Rhim, W.-K., Elleman, D. D., and Vaughan, R. W. (1973). *J. Chem. Phys.* **58,** 1772.
Richards, R. E., and Yorke, R. W. (1960). *J. Chem. Soc.* p. 2489.
Ruberto, R. G., Cronauer, D. C., Jewell, D. M., and Seshadri, K. S. (1977). *Fuel* **56,** 17, 25.
Rushworth, F. A., and Tunstall, D. P. (1973). "Nuclear Magnetic Resonance." Gordon & Breach, London.
Schaefer, J., and Stejskal, E. O. (1975a). *J. Magn. Reson.* **18,** 560.
Schaefer, J., and Stejskal, E. O. (1975b). *J. Am. Oil Chem. Assoc.* **52,** 366.
Schaefer, J., and Stejskal, E. O. (1976). *J. Am. Chem. Soc.* **98,** 1031.
Schweighardt, F. K., Friedel, R. A., and Retcofsky, H. L. (1971). *Fuel* **55,** 313.
Schweighardt, F. K., Bockrath, B. C., Friedel, R. A., and Retcofsky, H. L. (1976a). *Anal. Chem.* **48,** 1254.
Schweighardt, F. K., Friedel, R. A., and Retcofsky, H. L. (1976b). *Appl. Spectrosc.* **30,** 291.
Shaw, D. (1976). "Fourier Transform NMR Spectroscopy." Elsevier, Amsterdam.
Shoolery, J. N., and Budde, W. L. (1976). *Anal. Chem.* **48,** 1458.
Speight, J. G. (1970). *Fuel* **49,** 76.
Speight, J. G. (1971). *Fuel* **50,** 102.
Sternberg, H. W., Raymond, R., and Schweighardt, F. K. (1975). *Science* **188,** 49.
Stoll, M. E., Vega, A. J., and Vaughan, R. W. (1976). *J. Chem. Phys.* **65,** 4093.
Stothers, J. B. (1965). *Q. Rev., Chem. Soc.* **19,** 144.

Stothers, J. B. (1974). *In* "Topics in Carbon-13 NMR Spectroscopy" (G. C. Levy, ed.), p. 232. Wiley (Interscience), New York.
Takeya, G., Itoh, M., Suzuki, A., and Yokayama, S. (1963). *Bull. Chem. Soc. Jpn.* **36,** 1222.
Taylor, S. R., and Li, N. C. (1978). *Fuel* **57,** 117.
Thiault, B., and Mersseman, M. (1975). *Org. Magn. Reson.* **7,** 575.
Thiault, B., and Mersseman, M. (1976). *Org. Magn. Reson.* **8,** 28.
Tschamler, H., and De Ruiter, E. (1962). *Brennst.-Chem.* **43,** 212.
Tschamler, H., and De Ruiter, E. (1963). *In* "Chemistry of Coal Utilization" (H. H. Lowry, ed.), Suppl. Vol., p. 35. Wiley, New York.
Turenko, F. P., Sakharovskii, V. G., Turenko, L., and Baranski, A. D. (1972). *Fiz.-Khim. Issled. Vzaimodeistviha Solei Shchelochnykh Metzl. Raplavakh Prod. Destruktsii Sapropelitov* p. 106.
Vahrman, M. (1970). *Fuel* **49,** 5.
Vahrman, M. (1972). *Chem. Br.* **8,** 16.
VanderHart, D. L., and Retcofsky, H. L. (1976). *Fuel* **55,** 202.
Van Krevelen, B. W. (1961). "Coal," p. 321. Elsevier, Amsterdam.
Van Vleck, J. H. (1948). *Phys. Rev.* **74,** 1168.
Vaughan, R. W., Ellman, D. D., Stacy, L. M., Rhim, W.-K., and Lee, J. W. (1972). *Rev. Sci. Instrum.* **43,** 1356.
Vaughan, R. W., Schreiber, L. B., and Schwartz, J. A. (1976). *In* "Magnetic Resonance in Colloid and Interface Science" (H. A. Resing and C. G. Wade, eds.), Symposium Series No. 34, p. 275. Am. Chem. Soc., Washington, D.C.
Vercier, P., Thiault, B., and Mersseman, M. (1977). *In* "Molecular Spectroscopy" (A. R. West, ed.), Ch. 10. Heyden, London.
Wehrli, F. W., and Wirthlin, T. (1976). "Interpretation of Carbon-13 Spectra." Heyden, London.
Weiler, J. F. (1963). *In* "Chemistry of Coal Utilization" (H. H. Lowry, ed.), Suppl. Vol., Ch. 14. Wiley, New York.
Whitehead, J. C., and Williams, D. F. (1975). *J. Inst. Fuel* **48,** 182.
Williams, R. B. (1958). "Symposium on the Composition of Petroleum Oils." *Am. Soc. Test. Mater., Spec. Tech. Publ.* No. 224, p. 168.
Williams, R. B., and Chamberlain, N. F. (1963). *World Pet. Congr., Proc., 6th, Frankfurt, 1962* **5,** 217.
Wilman, W. G. (1966). "Chemical Examination of Road Tars from the Hercies Road Experiment," Rep. 0359. Coal Tar Res. Assoc. (now part of B.C.R.A., Chesterfield, England).
Winniford, R. S., and Bersohn, M. (1962). *Am. Chem. Soc. Symp. Tars, Pitches, Asphalts, Atlantic City, N.J.* p. 21.
Wise, W. S. (1970). *Chem. Ind. (London)* p. 950.
Wooton, D. L., Dorn, H. C., Taylor, L. T., and Coleman, W. M. (1976). *Fuel* **55,** 225.
Yamada, Y., Furuta, T., and Sanada, Y. (1976). *Anal. Chem.* **48,** 1637.
Yen, T. F., and Erdman, J. G. (1962). *Am. Chem. Soc., Div. Pet. Chem., Prepr.* **7**(3), 99.
Yokayama, S., Yamamoto, Y., and Takeya, G. (1970). *Nenryo Kyokai-Shi* **49,** 932.

Chapter 24

High Resolution ^{1}H-, ^{2}H-, and ^{13}C-NMR in Coal Research

Applications to Whole Coals, Soluble Fractions, and Liquefaction Products

Herbert L. Retcofsky *Thomas A. Link*
U.S. DEPARTMENT OF ENERGY
PITTSBURGH ENERGY TECHNOLOGY CENTER
PITTSBURGH, PENNSYLVANIA

ISBN 0-12-399902-2

I. INTRODUCTION

As pointed out in Chapters 23 and 25, the uses of high resolution nuclear magnetic resonance (NMR) spectrometry in coal research have expanded considerably during the past few years. This rapid growth was undoubtedly the result of a combination of factors. Two predominant influences were certainly (1) the need for information on the composition of coal-conversion products and on the chemical structure of coal itself, and (2) refinements in instrumentation and the introduction of new techniques which demonstrated the general utility as well as the potential of NMR spectrometry in satisfying a significant portion of that need.

The need for compositional data on coal and environmentally acceptable fuels from coal was dictated primarily by the recent energy shortages experienced by many nations. Compositional data are important not only in evaluating fuels for their intended end use, but also as an aid in the solution of problems associated with the development of economical and efficient coal-conversion processes. Product analysis is also useful in determining proper conditions for the storage of liquid fuels for future use and for the upgrading of fuels to more desirable products. It is likely that this need will continue as new coal-conversion processes are conceived and as tested ones evolve from the bench-scale and process development unit (PDU) stages to the pilot plant and commercial plant stages. Knowledge of the chemical structure of coal and of products and intermediate products from coal conversion, particularly coal liquefaction, is also necessary to elucidate the basic chemical mechanisms involved.

High resolution NMR spectrometry is a proven technique for the determination of the distribution of certain nuclei, particularly protons (^{1}H), among various organic structures in liquids and soluble materials. It was first applied to soluble materials from coal in the late 1950s (Friedel, 1959). A review of this early work is given in Chapter 22, Section II,B. The introduction of new methods for observing NMR resonances, e.g., pulse Fourier transform (FT) techniques and correlation NMR spectrometry, which to a large extent have replaced the more conventional continuous wave (cw) techniques, has extended NMR spectrometry to solutions of very low concentration and to nuclei of low natural abundance such as carbon-13 (^{13}C). Recent developments in NMR spectrometry have provided means by which considerable chemical structure information can be obtained on polycrystalline and amorphous solids. These latter methods, which include cross-polarization and multiple pulse NMR, remove dipole–dipole interactions and thus substantially enhance spectral resolution. The use of "magic angle" spinning adds a new dimension to the quality of NMR spectra of solids by reducing the anisotropic chemical shift contribution to the line width to zero. These techniques, when considered collectively, present the exciting possibility that resolution in the NMR spectra of solids may approach that observed in spectra of solutions.

The scope of the present chapter is somewhat limited in that discussion is confined primarily to NMR studies of whole coals, solvent extracts of coals, and products from a specific coal liquefaction PDU. Also included is an assessment of the potential of NMR as a means for determining the distribution of stable isotopes incorporated into products during coal liquefaction.

II. EXPERIMENTAL METHODS

A. Continuous Wave and Pulse Fourier Transform NMR

The reader is referred to Chapter 23, Section III, for a discussion of the experimental aspects of conventional continuous wave and pulse Fourier transform NMR techniques. Several of the problems associated with the quantitative reliability of ^{13}C FT NMR data for materials derived from coal are discussed in Section III,A.

B. Correlation NMR

Dadok and Sprecher (1972, 1974) showed that the equivalent of a slow passage cw NMR spectrum can be obtained from fast passage spectra by cross-correlation of the spectrum with a calculated or experimental reference line. Typically, fast passage spectra obtained by conventional techniques are quite distorted because of line broadening, shifting of line positions, and the presence of relaxation "wiggles" in the tail of each spectral line. Correlation NMR eliminates these effects mathematically and thus allows the acquisition of cw spectra at a rate comparable to that of pulse FT NMR and, of course, much faster than that of slow passage experiments. Thus, improved speed and sensitivity are two major advantages of correlation mode NMR spectrometry.

The following sequence of operations was proposed by Dadok and Sprecher (1972) as the most practical way to obtain correlation NMR spectra:

(i) Linearly fast-swept NMR spectra (including relaxation "wiggles") are coherently accumulated in the memory of a computer using conventional time-averaging techniques.

(ii) The accumulated spectrum is Fourier-transformed using the fast FT technique.

(iii) The result obtained in (ii) is multiplied point by point by a calculated reference filtering function of the form

$$U_R^*(n) = \left(\frac{ST}{SW}\right)^{1/2} \exp(-j\varphi_0) \exp\left(-\frac{\pi LW}{SW} n\right) \exp\left(-j \frac{\pi}{STSW} n^2\right) \quad (1)$$

(iv) Inverse Fourier transformation of (iii) gives the cw slow passage spectrum.

(v) The resulting spectrum is inspected for proper phasing, filtering, and resolution and, if necessary, steps (iii) and (iv) are repeated with other values of φ_0 (phase) and LW (line width).

Included among the advantages of correlation spectrometry are (1) a weak line may be detected in the presence of a nearby line of much higher intensity, (2) the peak radiofrequency (rf) power needed for excitation is usually smaller than that required for pulse FT NMR by three to five orders of magnitude, and (3) nearly any good high resolution NMR spectrometer capable of fast linear frequency or field sweep can be operated in the correlation mode without major modification.

C. Cross-Polarization ^{13}C-NMR

Cross-polarization (CP) NMR, developed by Pines *et al.* (1972a), is a means for obtaining high resolution NMR spectra of isotopically rare spins such as ^{13}C in solids. In the ^{13}C experiment, abundant spins (usually protons) present in the sample serve as a reservoir of polarization for the ^{13}C nuclei. The method permits accumulation of ^{13}C spectra after each polarization of the ^{1}H spins to equilibrium, thus allowing use of the typically shorter *proton* relaxation time in place of the ^{13}C relaxation time as the waiting time between experiments. High power proton decoupling during the experiment removes the effects of proton dipolar line broadening. The resulting spectrum is termed high resolution in the sense that chemically shifted ^{13}C resonances can ordinarily be resolved or partially resolved. It should be noted that the resulting spectrum is a powder pattern, i.e., anisotropic chemical shifts will *not* be averaged to their isotropic values.

The spin-locking version of the CP method (Pines *et al.*, 1973, and references cited therein) has been used by several investigators (VanderHart and Retcofsky, 1976a,b; Schaefer and Stejskal, 1976; Pines and

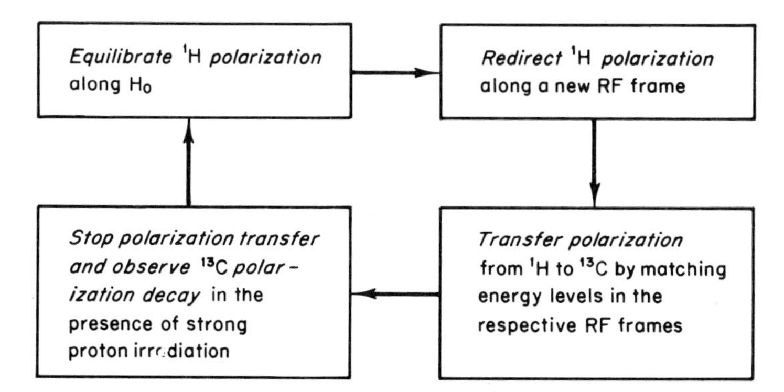

Fig. 1 Block diagram of the spin-locking version of the ^{13}C–^{1}H cross-polarization experiment. (From VanderHart and Retcofsky, 1976b.)

Wemmer, 1976; Bartuska *et al.*, 1976, 1977; Retcofsky and VanderHart, 1978) to obtain ^{13}C-NMR spectra of solid coals. A block diagram describing the method is shown in Fig. 1.

In stage 1, the protons are allowed to build up their polarization over a time on the order of T_1^{H} until equilibrium for that temperature is reached. In the second period, the application of a strong rf pulse shifts this polarization from its original position along the static field H_0 and redirects it along a new axis defined in the rotating frame. Stage 2 is very short, usually lasting only 3–6 μsec.

Polarization transfer from protons to ^{13}C takes place during stage 3. The protons are quantized along the applied rf field H_1^{H} and precess around that axis in direct analogy to the precession of spins around H_0. Meanwhile, in the carbon system, the spins are precessing around a second applied rf field, H_1^{C}. When experimental conditions are such that both precessional frequencies are the same, the Hartmann–Hahn condition (Hartmann and Hahn, 1962) is met and a system of degenerate levels results. The ^{13}C–^{1}H dipolar interaction lifts this degeneracy and new states characterized by strongly coupled protons and carbons are formed. In these newly formed states, energy normally associated with the respective rotating frames is exchanged repeatedly between the two systems. The macroscopic evidence for energy exchange in the form of oscillations in the carbon magnetization damps out rather quickly due to the fact that the protons which interact strongly with a particular carbon also are coupled to many other protons. The many-body character of the proton–proton interaction is in turn responsible for a steady growth of ^{13}C magnetization asymptotically approaching a spin temperature equilibrium between the carbon and the protons.

The last period of the process begins by terminating the CP process by removing the ^{13}C rf field. Simultaneously, the ^{13}C free induction decay is observed in the presence of high power proton irradiation. The resulting Fourier-transformed spectrum is free of line broadening due to proton dipolar interactions.

D. CP NMR with Magic Angle Spinning

The CP NMR experiment as just described removes the adverse effects of dipolar line broadening by protons. The effects of any anisotropy in the ^{13}C chemical shifts, however, remain in the spectra. Although the chemical shift anisotropy may give important information, i.e., chemical shift tensors for simple molecules in the solid state, its presence in CP NMR spectra of coals limits the resolution between the aromatic and nonaromatic carbons.

Andrew (1971) has shown that chemical shift anisotropies in solids are averaged to their isotropic values by spinning the sample at the so-called magic angle of 54.7° (see Chapter 23, Section III,E,2). The spinning must be done at frequencies somewhat greater than the chemical shift dispersion; a spinning rate of 2–3 kHz is generally used in ^{13}C experiments at Larmor frequencies <25 MHz. It should be noted that higher spinning rates (tens of kilohertz) can also be used to remove dipolar interactions; such high speed spinning is difficult but not impossible to achieve. CP ^{13}C-NMR studies of coal samples spinning at the magic angle have been reported by Schaefer and Stejskal (1976) and by Bartuska *et al.* (1976, 1977).

E. Multiple Pulse ^{1}H-NMR

Multiple pulse ^{1}H-NMR techniques can be used to remove homonuclear proton–proton interactions. The REV-8 pulse cycle of Rhim *et al.* (1973) and the phase-altered REV-8 sequence (Dybowski and Vaughan, 1975) have been applied to coals (Gerstein *et al.*, 1977). Although considerable narrowing of resonances was achieved, additional improvements are needed to resolve the resonances of the aromatic and nonaromatic protons in the samples. These techniques will be treated in more detail in Volume III.

III. THE DETERMINATION OF CARBON AROMATICITY

The carbon aromaticity, defined as $C_{aromatic}/C$ and generally designated f_a, has in the past been one of the most elusive chemical structure parameters in coal research. Recent developments in NMR instrumentation and techniques have greatly improved the situation with regard to coals and coal-derived materials. Nevertheless, the aromaticity problem in coal research has not been completely solved and many possible pitfalls await the unwary experimenter.

To discuss the various NMR approaches to determine or to estimate carbon aromaticities, it is convenient to classify coals and coal-derived materials according to their solubilities: (1) those soluble in solvents suitable for direct studies by high resolution ^{13}C-NMR, (2) those soluble in conventional ^{1}H-NMR solvents (but not in ^{13}C-NMR solvents), and (3) those materials which are essentially insoluble.

A. Liquid Products or Materials Soluble in ^{13}C-NMR Solvents

Liquid products such as carbonization oils and materials soluble in ^{13}C-NMR solvents, e.g., carbon disulfide, can be examined directly by ^{13}C-NMR to provide essentially unambiguous f_a values *provided the experimenter takes the necessary precautionary measures to ensure quantitative reliability of his data.* Problems which must be considered include variable nuclear Overhauser enhancements (NOE) which accompany proton decoupling and the rather long spin–lattice relaxation times normally associated with ^{13}C nuclei.

Decoupling of protons in a ^{13}C-NMR experiment to collapse ^{13}C–H spin–spin multiplets sometimes results in an enhanced population in the lower ^{13}C energy level relative to that of the normal Boltzmann distribution. This effect, termed the nuclear Overhauser effect, is frequently different for different carbon atoms in the same sample. Unless due care is taken, these NOE effects frequently lead to NMR peak intensities which are *not* directly proportional to the number of carbons responsible for the resonances.

In early proton-decoupled pulse ^{13}C FT NMR studies of coal-derived liquids, NOE values were estimated and used to correct the observed ^{13}C intensities. Based on data for substituted aromatic compounds (Levy *et al.*, 1973; Johnson and Jankowski, 1972), Bartle *et al.* (1975) assumed that the enhancement of carbons in C–H groupings is three times that of carbons in C–C groupings. For fractions of a supercritical extract of coal, they found good agreement between the ^{13}C f_a values and f_a values estimated from ^{1}H-NMR studies. Maekawa *et al.* (1975), on the other hand, reasoned that NOE effects in coal-derived materials would be eliminated since free radicals are known to be present in such materials. La Mar (1971) had shown earlier that the addition of paramagnetic

TABLE I ^{13}C-NMR *Nuclear Overhauser Enhancements for Selected Fractions from a Coal Liquefaction Product*

	NOE data[a]		
	Aromatics	Hexane solubles	Phenolics
Total aromatic	1.6	1.1	1.2
Total aliphatic	1.7	1.4	1.2
Phenolic C–OH	1.1	1.0	.5
Aromatic C–alkyl C at saturated ring junctions	1.7	1.0	.7
Aromatic bridgehead C	1.3	1.3	1.0
Aromatic C–H	1.7	1.1	1.6
Methylene bridge C α to aromatic rings	1.7	1.3	1.0
C β to ring	1.7	1.6	1.3
CH_3 γ or further from ring	1.7	1.2	1.2

[a] NOE data expressed as $(I_D/I_0) - 1$, where I_D and I_0 refer to the integrated intensities of the decoupled and nondecoupled resonances, respectively.

species in appropriate amounts suppresses nuclear Overhauser polarizations. The present authors have measured NOE values for various types of carbons in soluble fractions from the hydrodesulfurization of coal. The results (Table I) indicate that ^{13}C-NMR intensity data based on estimated NOE values must be viewed with some reservation.

Actually, most modern pulse FT NMR spectrometers have "built-in" capability to suppress Overhauser enhancements. The method (Freeman *et al.*, 1972) involves "gating" the proton-decoupling field on during the data acquisition time, and off for a relative long period of time before the sequence is repeated. The excess ^{13}C spin population will decay during this latter period. Wooten *et al.* (1976) found that time periods of 0.33 and 2.92 sec, respectively, were reasonable ones to use for samples of Solvent Refined Coal (SRC).

The problem of long nuclear relaxation times places severe limitations on the pulse repetition rates that can be used to accumulate ^{13}C-NMR data. One way to ensure quantitatively reliable data is to introduce a delay of several T_1's between pulses. For many materials, this can result in very long accumulation times. The use of pulse delays ranging from 0 sec (Pugmire *et al.*, 1977) to 25 sec (Callen *et al.*, 1976) has been reported for coal-derived materials. The time delay between pulses can be shortened if smaller flip angles are used by the experimenter. It is difficult to use the scientific literature on ^{13}C-NMR studies of coal-derived

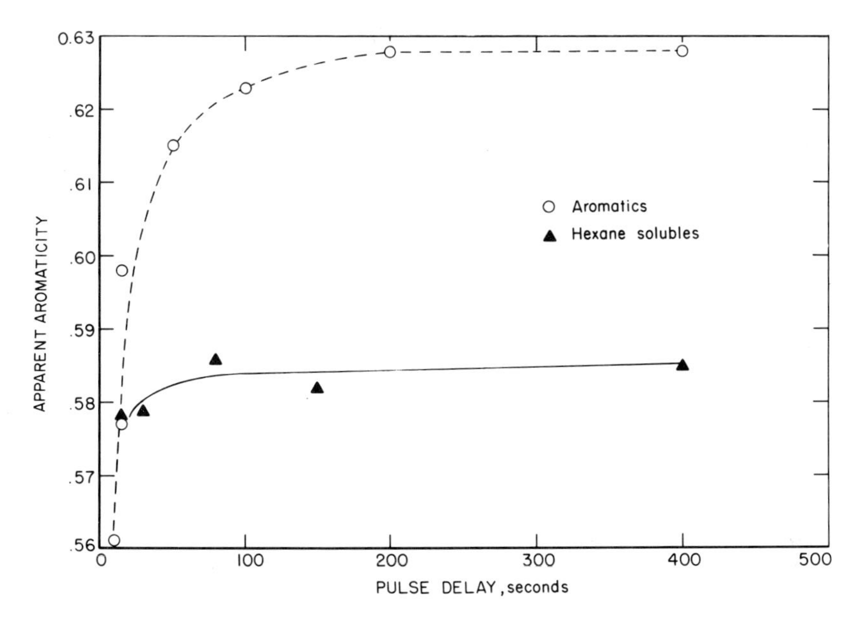

Fig. 2 Dependence of the apparent aromaticity values for materials derived from coal on the time delay between pulses in an FT NMR experiment.

liquids to provide guidance for the proper choice of time delays since most authors do not specify the flip angle used in their work.

Recent work in our laboratory on fractions from a coal-derived liquid show that the choice of time delays between pulses must be given serious consideration. The data depicted graphically in Fig. 2 for the hexane-soluble portion and its aromatic subfraction indicate that pulse delays in excess of 100 sec are necessary to obtain reliable f_a values for *some* coal-derived materials. For materials such as coal extracts and coal-derived asphaltenes, which have reasonably high molecular weights and high concentrations of free radicals, time delays of 25 sec or less can be used with confidence. It should be noted that the data of Fig. 2 were obtained using a 90° flip angle.

B. Materials Soluble in ^{1}H-NMR Solvents (but Not in ^{13}C-NMR Solvents)

Many materials derived from coal are not soluble in solvents suitable for ^{13}C-NMR. For example, pyridine, which is one of the best extraction solvents for coal, cannot be used for ^{13}C-NMR studies. (Perhaps this problem can be overcome by future availability of ^{12}C-enriched, i.e.,

^{13}C-depleted, pyridine.) Many coal-derived materials not soluble in ^{13}C-NMR solvents are soluble in ^{1}H-NMR solvents such as perdeuteropyridine.

Brown and Ladner (1960) in their classic paper showed that f_a values can be estimated from ^{1}H-NMR data provided suitable values for the H/C ratios of the aliphatic groupings can be deduced. The Brown–Ladner method makes use of the quantitative determination of the fractions of the total hydrogen present (1) on aromatic rings and as phenolic OH

$$H^{*}_{ar+\phi} = H_{ar+\phi}/H \tag{2}$$

(2) on benzylic carbons

$$H_{\alpha}^{*} = H_{\alpha}/H \tag{3}$$

and (3) on other nonaromatic carbons

$$H_{0}^{*} = H_{0}/H \tag{4}$$

The aromaticity equation is given as

$$f_a = [C/H - (H_{\alpha}^{*}/X) - (H_{0}^{*}/Y)]/C/H \tag{5}$$

The parameters X and Y refer to the H/C ratios for the benzylic carbon structures and for the other nonaromatic carbon structures, respectively. For materials derived from coal, X and Y are generally taken to be 2 on the assumption that the aliphatic structures are predominantly methylene groups.

Retcofsky *et al.* (1977) reported a comparison of f_a values determined unambiguously from proton-coupled ^{13}C-NMR spectra with those estimated by Brown–Ladner-type treatment of the corresponding ^{1}H-NMR data. Included in the study were the carbon disulfide-soluble materials from five coals of different rank, coal tar pitch, coal carbonization oils, coal-derived asphaltenes, and products or subfractions from five coal liquefaction processes (H-Coal, Solvent Refined Coal, Solvent Refined Lignite, COED, and SYNTHOIL). The results are summarized in Fig. 3. In the figure, f_a refers to the ^{13}C-NMR value whereas f_a' refers to the values estimated from ^{1}H-NMR. For 21 of the 29 samples examined, the differences between the f_a and f_a' values were $\geq |0.03|$. The largest difference found was 0.07. Statistical analysis of the data using the Student t-test indicates that at the 95% confidence limit there is no real difference between the f_a and f_a' values for each of the samples.

The authors conclude that for materials similar to those included in the study, reasonably good estimates of carbon aromaticities can be derived from ^{1}H-NMR data. Furthermore, the results suggest that pre-

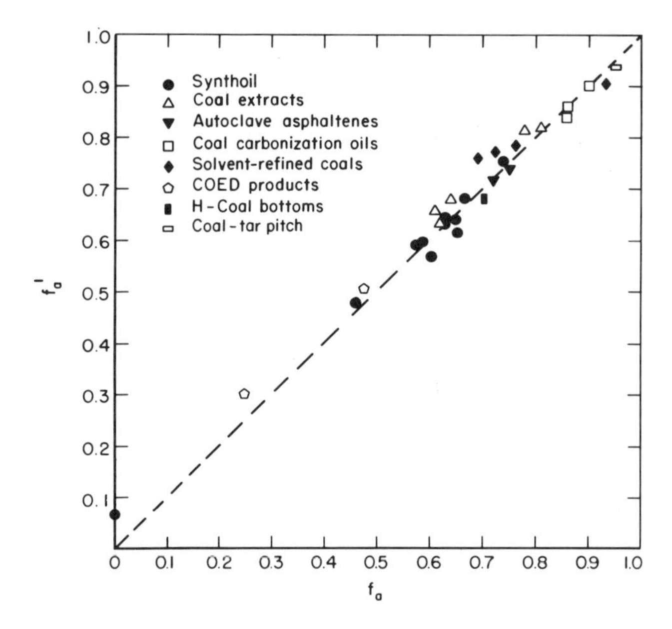

Fig. 3 Comparison of proton (f'_a) and [13]C (f_a) aromaticities for materials derived from coal. (From Retcofsky *et al.*, 1977.)

viously published ^{1}H-NMR f_a values for materials derived from coal can be viewed with a considerable degree of confidence. The authors caution, however, that all f_a values deduced from Brown–Ladner-type calculations must be carefully scrutinized in terms of other information regarding the chemical structure of the material under study. Naturally, there is no substitute for an absolutely unambiguous value requiring no *a priori* assumptions about chemical structure.

C. Insoluble Materials

The high resolution NMR techniques discussed in Sections III,A and III,B are not applicable to solid samples. Such materials can, of course, be examined by the cross-polarization technique described in Section II,C. The question that remains to be answered, however, is whether the results are quantitative. We hasten to emphasize that as of this writing there is no universally accepted answer to the question. The following arguments are based on two recently published papers by VanderHart and Retcofsky (1976a,b).

These authors point out that quantitative determinations by CP NMR depend critically on the distribution of protons throughout the sample.

That is, if a carbon is further than about 0.35 nm from a proton, its contribution to the CP NMR signal will be attenuated and possibly lost. A second requirement is that the protons themselves should not be isolated from each other by more than 0.5 nm. Thus, clusters of carbons in graphite-like or diamond-like structures would not be seen. Another problem which would complicate quantitative determinations is any nonuniformities in cross-polarization times of the carbons in the sample. It is a characteristic of CP NMR that the cross-polarization time depends on the sixth power of the distance between the interacting carbon and proton.

For coals, it was found that the CP intensity was usually less than that expected on the basis of the CP intensity generated by an n-alkane standard. Although a satisfactory explanation for this behavior has not been found, two types of experiments were carried out to investigate the quantitative reliability of the spectra.

One approach to the problem involved comparison of ^{13}C spectra of coals obtained by the CP NMR technique and those obtained by a tedious $(90° - T)_x$ pulse sequence on the carbons (T = 400 sec). The $(90° - T)_x$ experiment should give a truly quantitative line shape provided there are no carbon T_1's in excess of 100 sec. The fact that both experiments gave essentially identical spectra and that full spectral intensity was observed for the $(90° - T)_x$ experiment suggests that the CP NMR spectra have low distortion and most probably include all the carbons in the samples.

The second approach to the quantitative reliability problem was to compare f_a values obtained from CP NMR spectra of solid or frozen coal derivatives with essentially unambiguous values obtained by high resolution ^{13}C solution spectra. The results (Table II) indicated differences in

TABLE II *Comparison of ^{13}C-NMR Aromaticity Values*

	f_a	
Material[a]	Correlation spectroscopy	Cross-polarization NMR
Coal extracts:		
Pittsburgh hvAb	0.62	0.67
Pocahontas No. 4 lvb	0.78	0.80
Coal-conversion products:		
Solvent Refined Coal	0.72	0.79
SYNTHOIL CLP	0.46	0.50[b]

[a] CS_2-soluble portion.
[b] Measurement made on frozen sample.

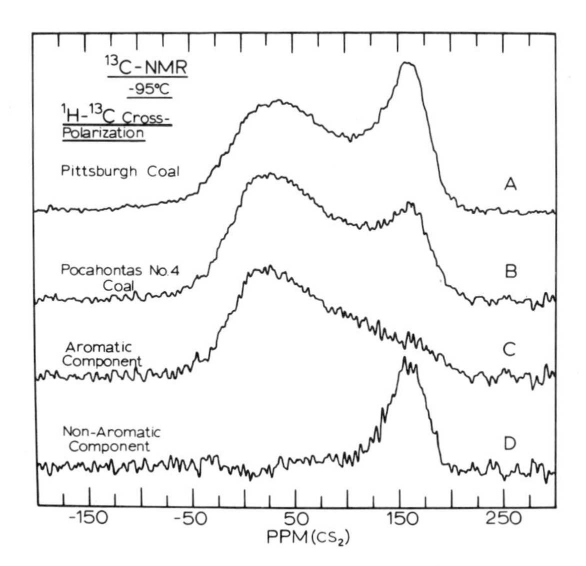

Fig. 4 Cross-polarization ¹³C-NMR spectra of vitrains from selected coals. (From VanderHart and Retcofsky, 1976a.)

aromaticity values of ≤0.1, thus indicating that reasonably reliable f_a values may be obtained by the cross-polarization technique.

The CP NMR spectra discussed in this section were all obtained without magic angle spinning; thus deconvolution of the spectra was necessary to determine f_a values. Representative spectra of two coals are reproduced in Figs. 4A and 4B. By taking appropriate linear combinations, the aromatic and nonaromatic line shapes (Figs. 4C and 4D) are obtained. The criterion used in this deconvolution method is that negative intensity is not allowed. It should be noted that the two component line shapes are consistent with what is known about aromatic and nonaromatic chemical shift tensors (Pines *et al.*, 1972b). The aromatic component is very nearly a linearly decreasing function through the region of overlap between the two components; this linear assumption was used in subsequent analyses of spectra. In more recent work in which the CP NMR experiment was combined with magic angle spinning, Bartuska *et al.* (1977) suggested that the linear extrapolation method used without sample spinning may lead to f_a values somewhat lower than the actual values.

IV. ESTIMATION OF MEAN STRUCTURAL UNITS

The concept of the *mean structural unit,* which van Krevelen (1961) defines as "the assembly of atoms averaged over the number of aromatic clusters," is a useful one in coal research. Mean structural units are usually described by structural parameters, e.g., carbon aromaticity, size of aromatic cluster, ring condensation index. Frequently, coal researchers will propose an average structural formula which incorporates all measurable structural parameters as well as what is known about the chemical reactivity of the material in question. Notable examples include the coal model of Given (1960) and the "asphaltol" model proposed by Farcasiu *et al.* (1976b) which are reproduced in Figs. 5 and 6, respectively. Other coal models are shown in Figs. 2–4 in Chapter 22.

High resolution NMR spectra of coal-derived materials in solution provide an excellent means for determining or estimating three very useful structural parameters. These are the carbon aromaticity f_a, the degree of aromatic ring substitution σ, and the H/C ratio for the hypothetical unsubstituted aromatic nuclei H_{aru}/C_{ar}. The latter parameter is related to the number of aromatic rings that make up the polynuclear condensed aromatic ring systems. For benzenoid structures $H_{aru}/C_{ar} = 1$; naphthalene systems have values of 0.8; three-ring systems, 0.72; and so on.

The means for determining f_a values for coal-derived materials have already been discussed. The direct determination of f_a by high resolution ^{13}C-NMR and the Brown–Ladner method for estimating f_a using ^{1}H-NMR data were described in Sections III,A and III,B, respectively.

To estimate the degree of aromatic ring substitution and the H/C ratio for the hypothetical unsubstituted aromatic nuclei, it is necessary to determine H^*_{ar}. Since the ^{1}H-NMR resonances for aromatic and phenolic

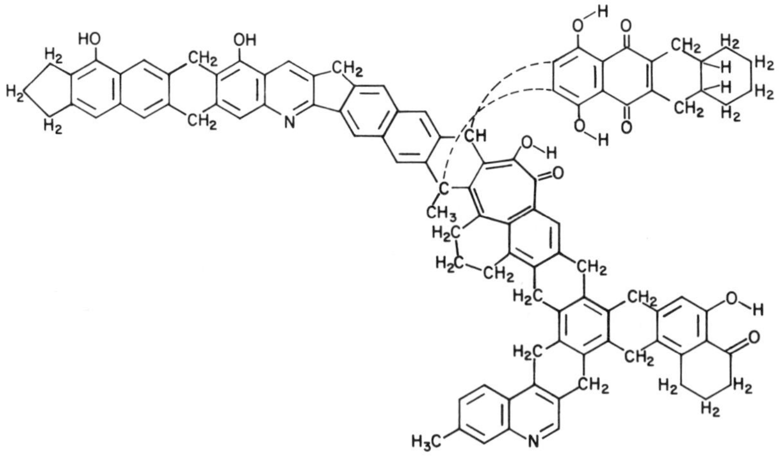

Fig. 5 Proposed molecular model for an 82% carbon coal. (From Given, 1960.)

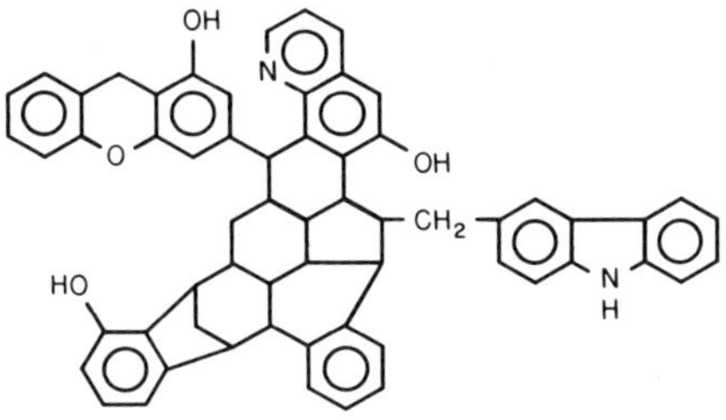

$C_{61}H_{46}N_2O_4$ 73% Aromatic C

900 mol. wt. 60% Aromatic H

Fig. 6 Possible average structure for an "asphaltol" fraction from Solvent Refined Coal. (From Farcasiu *et al.*, 1976b.)

OH protons generally appear in very nearly the same spectral region (for materials derived from coal), the OH content of the sample is frequently determined by an independent method. The experimentally determined phenolic OH content of the sample can then be used to evaluate a constant K and H^*_{ar} can be calculated:

$$H^*_{ar} = H_{ar}/H = (H_{ar+\phi}/H) - K(O/H) \quad (6)$$

Once H^*_{ar} is known, σ and H_{aru}/C_{ar} can be estimated. If only ¹H-NMR data are available, σ can be determined using the second equation of Brown and Ladner (1960):

$$\sigma = [H_\alpha{}^*/X) + (O/H)]/[H_\alpha{}^*/X) + (O/H) + H^*_{ar}] \quad (7)$$

and H_{aru}/C_{ar} may be estimated using their third equation;

$$H_{aru}/C_{ar} = [H_\alpha{}^*/X) + H^*_{ar} + (O/H)]/[C/H) - (H_\alpha{}^*/X) - (H_o{}^*/Y)] \quad (8)$$

In the case of materials for which both ¹H and ¹³C data are available, the following equation can be used to determine H_{aru}/C_{ar}:

$$H_{aru}/C_{ar} = [H_\alpha/X) + H_{ar} + O]/C_{ar} \quad (9)$$

The advantage of Eq. (9) over Eq. (8) is that the denominator is uniquely determined by a direct and unambiguous method.

It should be noted that Eq. (7), (8), and (9) are not without limitations. Their derivations are based on several assumptions, the principal ones being (1) all oxygen atoms are bonded directly to aromatic rings and not shared between them, (2) the aromatic rings are not linked by C–C bonds as in polyaryl systems, and (3) values for X and Y are known or can be estimated. Versions of Eqs. (7) and (8) slightly modified to include methylene bridge protons have been published by DeWalt and Morgan (1962).

V. ^{13}C-NMR—IMPROVEMENTS IN RESOLUTION, 1966–Present

The purpose of this brief section is to illustrate the tremendous advances made in the applications of ^{13}C-NMR in coal research.

A. Solution Spectra

^{13}C-NMR spectra of coal-derived materials were first published by Friedel and Retcofsky (1966). The materials examined were the neutral oil from 700°C cracking of a 450°C coal carbonization product, an anthracene oil, a heavy creosote from coal, and the carbon disulfide-soluble fraction of a coal tar pitch having a softening point of 80–85°C. The aromatic carbon resonance in the proton-coupled ^{13}C-NMR spectrum of the latter (Retcofsky and Friedel, 1966) is shown in Fig. 7A. The intense resonance at low field is due to the solvent. The spectrum is poorly resolved, quite distorted, and exhibits a very poor signal-to-noise ratio since it was obtained in a single scan by observing the dispersion mode under rapid passage conditions (see Lauterbur, 1962, for a discussion of this mode of spectrometer operation). Some improvement in sensitivity and in line width was realized

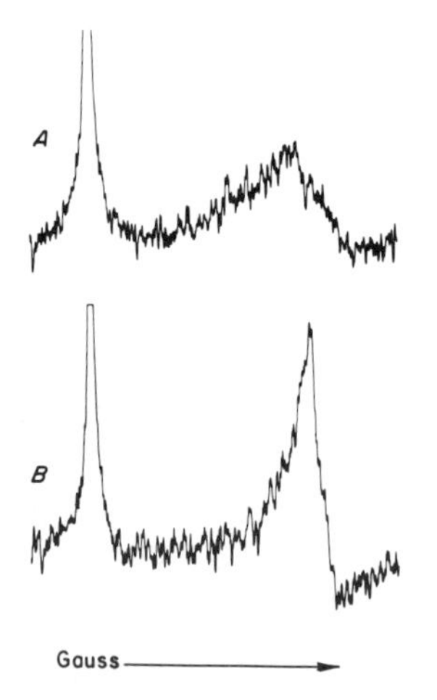

Fig. 7 15.1-MHz rapid passage dispersion mode spectra of the carbon disulfide-soluble material from coal tar pitch: (A) coupled spectrum, (B) single-resonance proton-decoupled spectrum. (From Retcofsky and Friedel, 1966.)

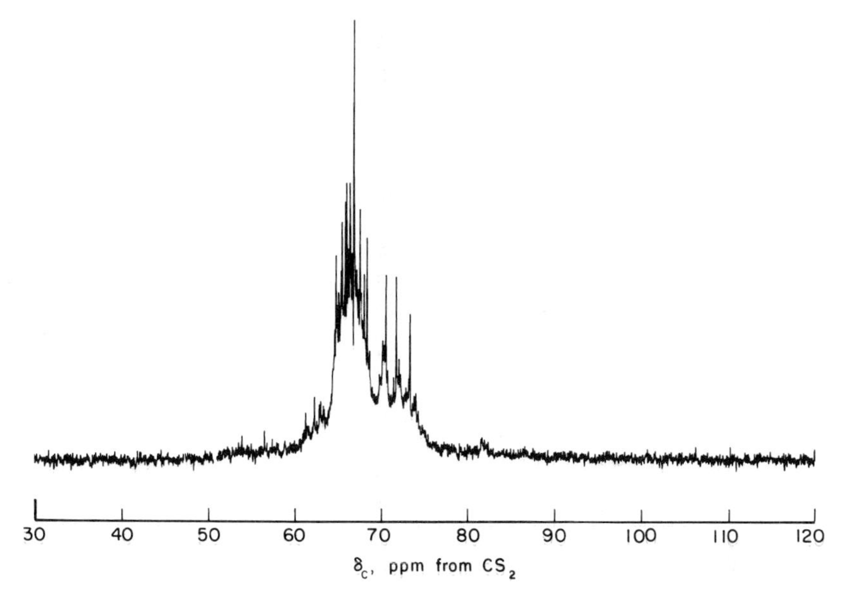

Fig. 8 Proton-decoupled (correlation mode) ^{13}C-NMR spectrum of the carbon disulfide-soluble fraction from coal tar pitch. (From Retcofsky and Schweighardt, 1977.)

upon single-frequency proton decoupling of the aromatic resonances (Fig. 7B). The vast improvement in spectrum resolution using present-day techniques (Sections II,A and II,B) is evident in Fig. 8. Needless to say, the quality of ^{13}C-NMR solution spectra has increased dramatically over the past several years.

B. Spectra of Solids

Quite analogous improvements in the ^{13}C-NMR spectra of solid coals have also taken place. Spectral resolution in CP ^{13}C-NMR spectra was discussed in Section II,C and is illustrated in Figs. 4A and 4B. In contrast to this, the broad-line ^{13}C-NMR spectra of coals (Fig. 9) published just $3\frac{1}{2}$ years earlier (Retcofsky and Friedel, 1973) show little if any resolution between aromatic and nonaromatic carbon resonances. The most highly resolved spectrum of a solid coal, published only recently (Bartuska *et al.*, 1976, 1977; Schaefer and Stejskal, 1976), is reproduced in Fig. 10. The spectrum, that of a lignite, was obtained using the CP NMR technique combined with magic angle spinning (Section II,D). For comparison, the spectrum of the nonspinning sample is also shown.

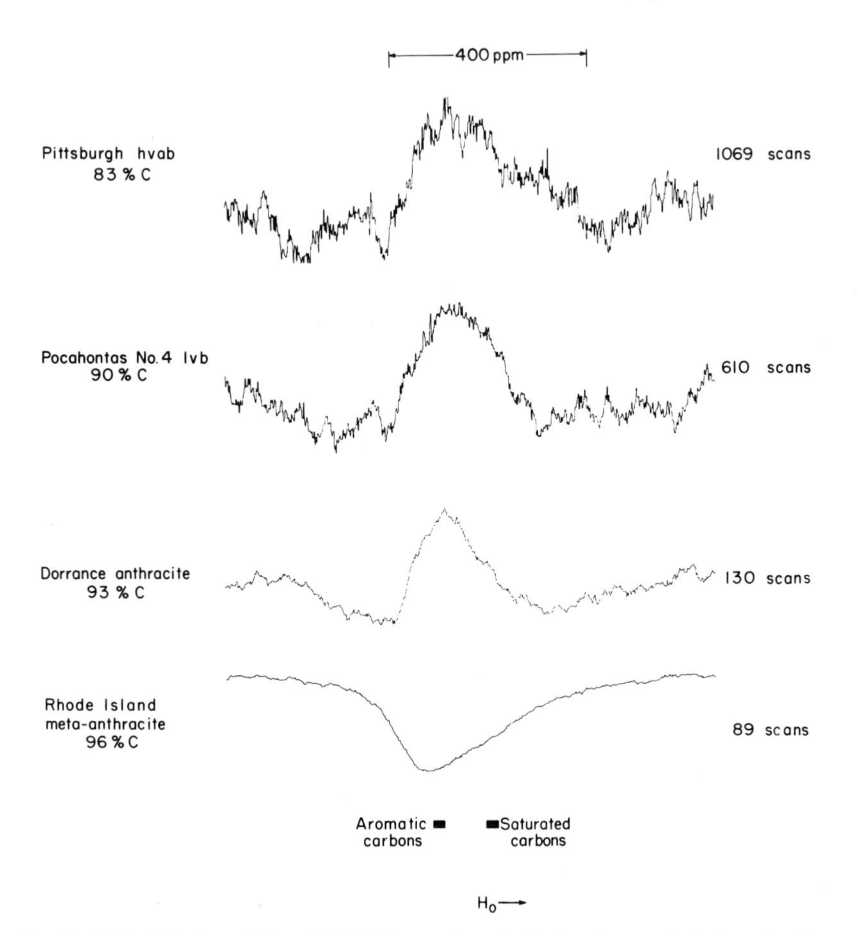

Fig. 9 15.1-MHz broad-line ^{13}C-NMR spectra of vitrains from selected coals. (From Retcofsky and Friedel, 1973.)

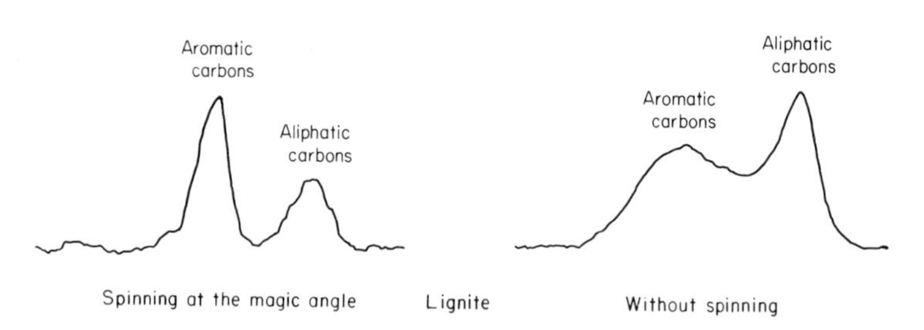

Fig. 10 Cross-polarization ^{13}C-NMR spectra of lignite with and without magic angle spinning. (From Schaefer and Stejskal, 1976 and Bartuska *et al.*, 1977.)

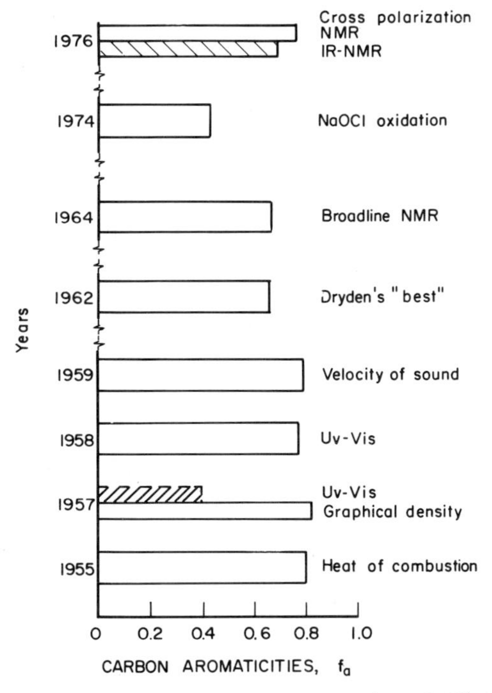

Fig. 11 Estimated carbon aromaticities of an 82.6% C coal.

VI. APPLICATIONS

A. Coals and Their Petrographic Components

As indicated in Section III, the carbon aromaticity has been one of the most elusive structural parameters in coal research. The present state of confusion is illustrated in Fig. 11 which shows the values for the carbon aromaticity of hvAb Pittsburgh coal (82.6% C) which have been published over the past 22 years.† Note that in 1957 and more recently in 1974 the classical view that bituminous coals are highly aromatic substances was challenged. This implies a *qualitative* as opposed to a *quantitative* disagreement over a very fundamental aspect of coal structure. For a more detailed discussion of the work at these earlier dates, see Chapter 22 (Section III and Fig. 5).

The material discussed in Section III,C and the data presented in Table II lead us to believe that reliable carbon aromaticities of bituminous and

† In some cases, f_a values for Pittsburgh coal were interpolated using data for other coals.

TABLE III *Carbon Aromaticities for Vitrains from Selected Coals as Determined by Cross-Polarization* ^{13}*C-NMR*

Source of vitrain	Carbon content (maf %)	f_a
Dorrance anthracite	92.8	>0.96
Pocahontas No. 4 lvb coal	90.4	0.89
Lower Banner mvb coal	86.1	0.81
Powellton hvAb coal	85.1	0.80
Pittsburgh hvAb coal	82.6	0.76
Adaville subbituminous coal	76.3	0.76
Beulah lignite	72.6	0.79(?)
Wisconsin peat	55.0	<0.50

anthracite coals can be determined by CP NMR. VanderHart and Retcofsky (1976a,b) reported CP NMR results for vitrains from nine coals ranging in rank from peat to anthracite. The resulting f_a values (Table III) support the classical views that most coals are highly aromatic and that the aromaticities of coal increase with increasing coal rank. Bartuska *et al.* (1977), using the CP NMR technique combined with magic angle spinning, reported f_a values of 0.99 and 0.72 for an anthracitic and lignitic coal, respectively. These values also support the classical theory of coalification.

CP NMR spectra of macerals from an hvAb coal have also been published (Retcofsky and VanderHart, 1978). The spectra (Fig. 12) indicated that f_a values decreased in the order fusinite (f_a = 0.92–0.96) > micrinite (f_a = 0.85) ≈ vitrinite (f_a = 0.85) > exinite (f_a = 0.66). By assuming that the nonaromatic carbons are predominately methylene and that the oxygen and half the nonaromatic carbons are directly bonded to aromatic rings, these authors estimated that the fusinite contained the largest polynuclear condensed aromatic ring system (> five rings) whereas the mean structural unit of the vitrinite contains three to four condensed rings per aromatic unit.

As indicated in Section II,E, Gerstein *et al.* (1977) applied multiple pulse ^{1}H-NMR techniques to selected coals. These investigators were successful in narrowing the proton line width from ~30 kHz to ~700 Hz. Comparison of the results on coals with those for model compounds indicated that ~400 Hz of the residual proton line width in the coals is due to free radical electron spin–nuclear spin dipolar coupling. They concluded that resolution of the proton resonances of coals into aromatic and aliphatic components will require removal of proton chemical shift anisotropies and electron–proton interactions as well as interproton di-

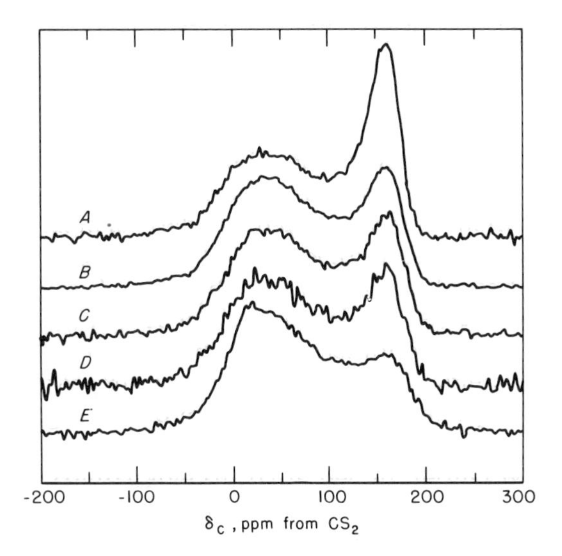

Fig. 12 Cross-polarization ^{13}C-NMR spectra of macerals from Hernshaw coal. A, Exinite; B, vitrinite #2; C, vitrinite #1; D, micrinite; E, fusinite. (From Retcofsky and VanderHart, 1978.)

polar interactions. It was suggested that a combination of multiple pulse line-narrowing experiments and magic angle spinning would produce the desired result.

B. Solvent Extracts of Coal

Solvent extraction is a classical method for studying coal constitution. It is extremely useful in NMR investigations of coal structure since the extracts are amenable to study by conventional high resolution techniques.

Included among the large number of papers reporting high resolution NMR studies of coal extracts are the works of Friedel and Retcofsky (1961, 1963), Oth and Tschamler (1962), Durie *et al.* (1966), Kanai (1966), Oelert (1967), Retcofsky and Friedel (1970), and Retcofsky (1977). The following discussion is based principally on the results reported by Takeya *et al.* (1964), Durie *et al.* (1966), and Retcofsky (1977) since each of these authors investigated extracts from a series of coals of different ranks. Although a number of different extracting solvents have been used in NMR studies, the discussion is limited to pyridine and carbon disulfide extracts. Pyridine was chosen because it is one of the best extracting agents for coal and the resulting extract can be dried and redissolved in completely deuterated pyridine for ^{1}H-NMR measurements. Carbon disulfide, although a much poorer coal solvent, is an excellent solvent for ^{13}C-NMR studies.

Representative ^{1}H-NMR spectra of coal extracts are reproduced in Figs. 13 and 14. Figure 13 shows the spectrum of the pyridine extract of vitrain from Pocahontas No. 3 lvb coal, whereas the spectrum of the carbon disulfide extract of vitrain from Pittsburgh hvAb coal is shown in Fig. 14. The three rather narrow resonances at low field in the spectrum of the pyridine extract are due to residual protons in the deuterated solvent. The small water peak in the spectrum is identified on the figure. The broad envelope of resonances centered near 7.5 ppm is assignable to aromatic and phenolic protons in the extract. The resonances extending from ~0 to 4.2 ppm are assigned to nonaromatic protons; these are further subdivided into benzylic protons (~1.8–4.2 ppm) and protons on nonaromatic carbons beta or further removed from aromatic rings.

The ^{13}C-NMR spectra of the carbon disulfide extracts of Pittsburgh and Pocahontas No. 3 coals are reproduced in Fig. 15. The broad resonances between 50 and 100 ppm (with respect to the carbon disulfide solvent resonance) are assigned to aromatic carbons; those found between 150 and 200 ppm are assigned to aliphatic carbons. It should be

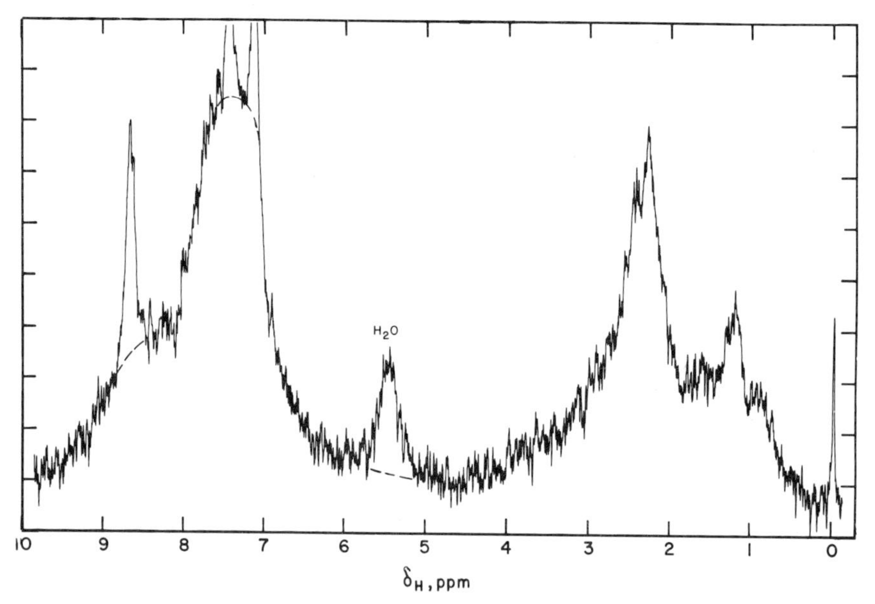

Fig. 13 60-MHz ^{1}H-NMR spectrum (cw) of the pyridine extract of vitrain from Pocahontas No. 3 coal. (From Retcofsky and Friedel, 1970.)

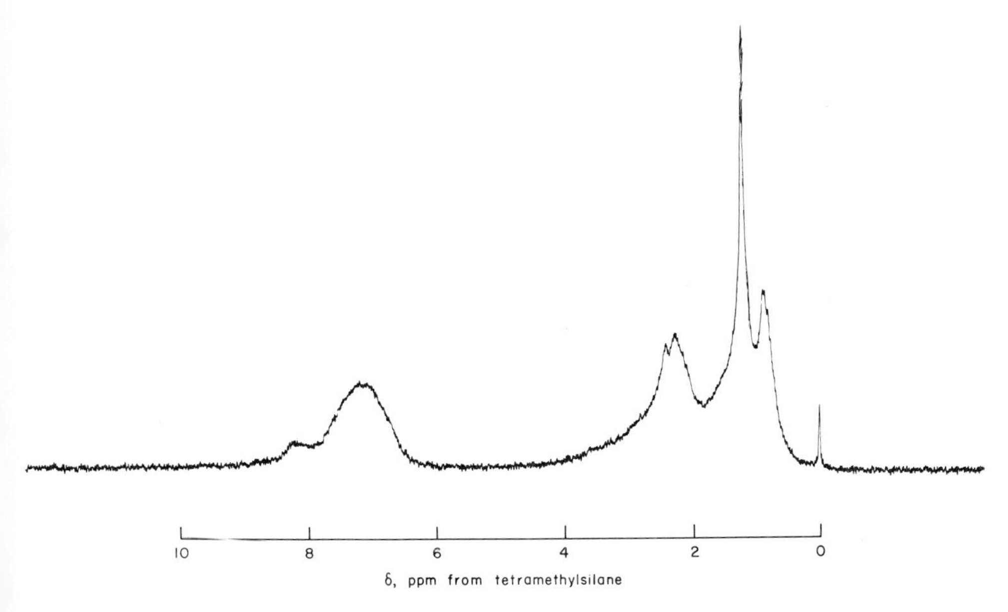

Fig. 14 60-MHz ¹H-NMR spectrum of the carbon disulfide extract of vitrain from Pittsburgh coal. (From Retcofsky and Friedel, 1968.)

noted that these spectra were obtained using correlation mode NMR. Proton decoupling was *not* employed during the quantitative measures to avoid any errors due to variations in the nuclear Overhauser enhancements for the different kinds of carbons in the samples (Section III,A).

The changes in H_{ar}/H, f_a, and H_{aru}/C_{ar} of the pyridine extracts with increasing carbon content of the starting coals are shown graphically in Fig. 16. H_{ar}/H and f_a both increase with an increase in carbon content of the starting coal. The variation in f_a is of particular significance because it supports the classical view that aromatization plays a major role in the coalification process. In contrast to the change in f_a, H_{aru}/C_{ar} decreases with increasing coal rank, indicating an increasing development of polynuclear condensed aromatic ring structures as coalification proceeds, again supporting the classical view of coalification. Although not shown in the figure, the degree of aromatic ring substitution of the pyridine extracts also decreases with increasing rank of the starting coal.

The structural parameters for the pyridine extracts were estimates based on treatment of the ¹H-NMR data in the Brown–Ladner manner (Sections III,B and IV). As indicated in Sections III,A and IV, essentially unambiguous f_a values and improved values of H_{aru}/C_{ar} can be obtained

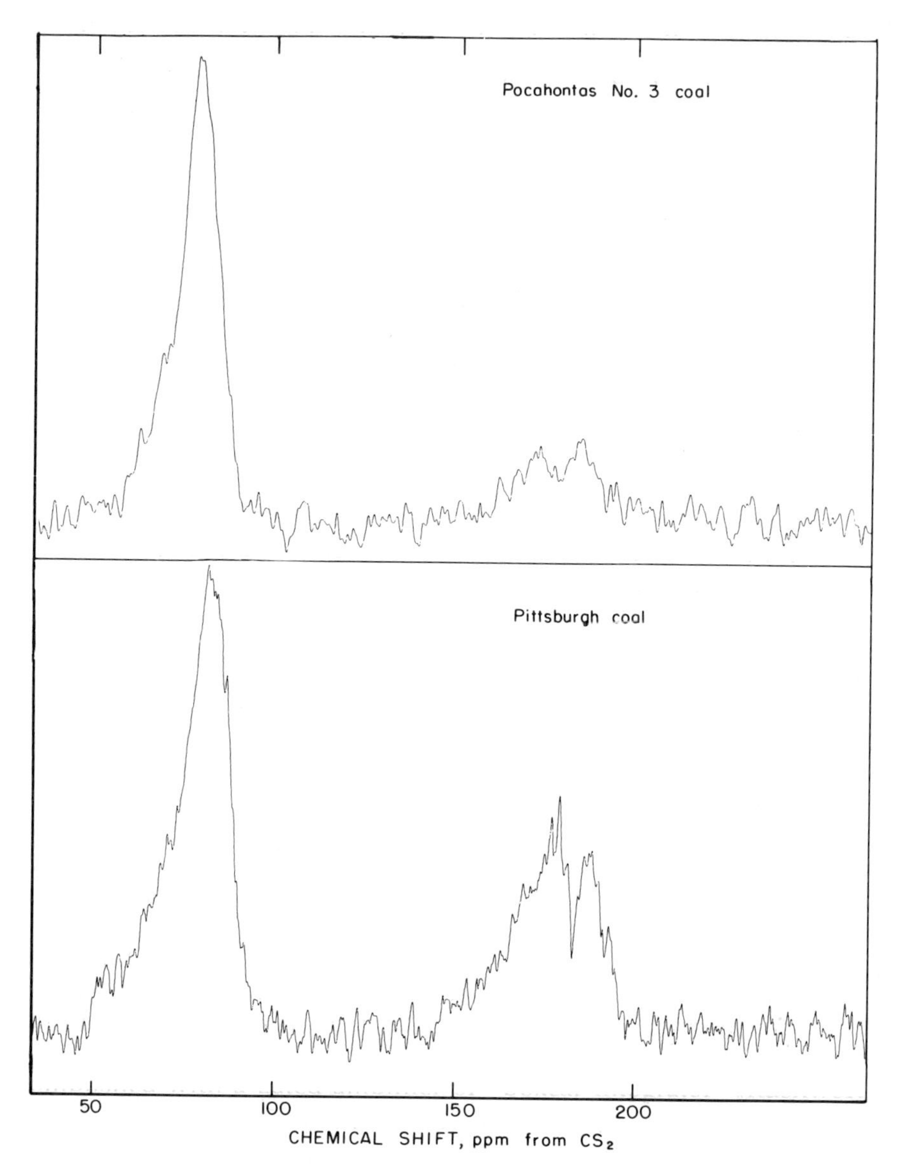

Fig. 15 62.8-MHz (correlation mode) ^{13}C-NMR spectra of the carbon disulfide extracts of vitrains from selected coals. Spectra obtained *without* proton decoupling. (From Retcofsky, 1977.)

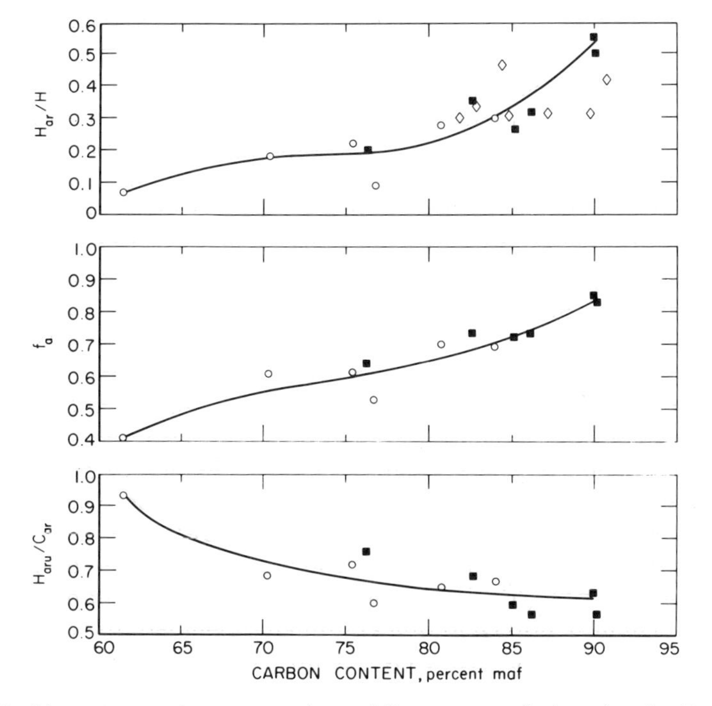

Fig. 16 Mean structural parameters for pyridine extracts of selected coals. (Data taken from Takeya *et al.*, 1964; Durie *et al.*, 1966; Retcofsky, 1977.)

for carbon disulfide extracts of coals by the use of complementary ¹H- and ¹³C-NMR data. Retcofsky (1977) and Retcofsky and Friedel (1970) have reported such data for five coals ranging in rank from subbituminous A to low volatile bituminous. The results, again plotted as a function of carbon content of the starting coals, are shown in Figs. 17 and 18.

The carbon aromaticities of carbon disulfide extracts, pyridine extracts, and whole coals as determined by ¹³C-NMR, ¹H-NMR, and CP ¹³C-NMR, respectively, are plotted in Fig. 17. Inspection of Fig. 17 in light of the discussions of Section III on the determination of carbon aromaticity permits the following conclusions to be drawn: (1) The ¹³C-NMR measurements show unambiguously that carbon disulfide extracts of the coals investigated are highly aromatic substances and, within experimental uncertainty, that their carbon aromaticities increase with rank of the parent coals. (2) The results for the pyridine extracts, which represent higher portions of the whole coals, indicate quite strongly that these extracts are more highly aromatic than the corresponding carbon disulfide extracts. The aromaticities of the

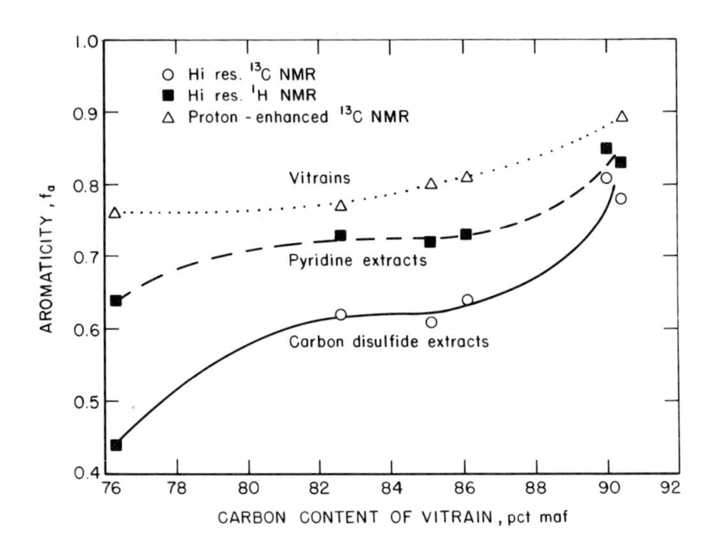

Fig. 17 Aromaticities of coals and coal extracts. (Data from Retcofsky, 1977; VanderHart and Retcofsky, 1976a,b; and unpublished observations.)

pyridine extracts also increase with increasing rank of the starting coals in a fashion analogous to the carbon disulfide extracts. (3) The cross-polarization results for the whole coals also indicate an increase in aromaticity with increasing coal rank.

H_{aru}/C_{ar} values for the coal extracts are shown in Fig. 18. Encircled numbers in the figure refer to pure aromatic hydrocarbons, i.e., ben-

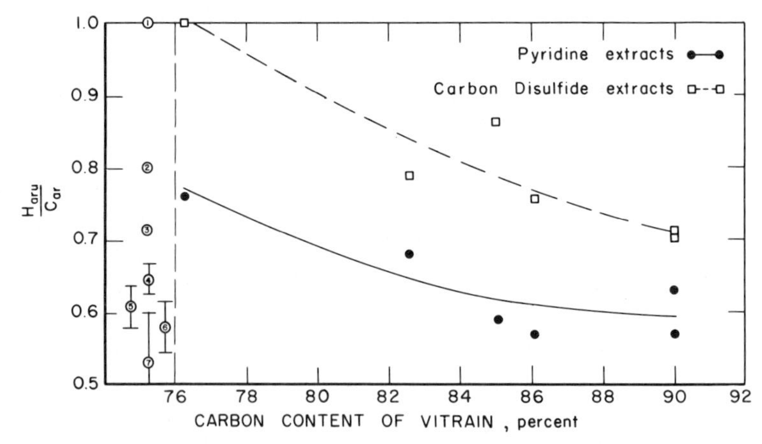

Fig. 18 Atomic H/C ratios for the hypothetical unsubstituted aromatic nuclei for extracts of vitrains from selected coals. Encircled numbers refer to pure aromatic hydrocarbons. (From Retcofsky, 1977.)

zenoid structures are composed of a single ring; naphthalene, two rings; anthracene or phenanthrene, three rings; and so on. The NMR data indicate that the number of aromatic rings per condensed cluster increases (Fig. 18) and the degree of substitution of the rings decreases (not shown) for each set of extracts as coalification proceeds. For the carbon disulfide extracts, the number of rings per condensed aromatic structure increases from one for the case of the subbituminous coal to two to three for the lvb coal. For the pyridine extracts, the ring systems are much larger, ranging from two to three rings for the lowest rank material to five or more for the extracts from the lvb coals. The results for the extracts and the aromaticity plot for the whole coals (Fig. 17) when considered in light of the decreasing hydrogen contents with increasing coal rank suggest that polynuclear condensed aromatic rings also play an important part in the chemical structure of whole coals.

Retcofsky (1977) also reported the proton-decoupled ^{13}C-NMR spectrum of the carbon disulfide extract of an lvb coal. Partially resolved peaks in the aromatic spectral region (not observed in the proton-coupled spectrum) indicated the presence of carbon atoms in chemical environments similar to that of the C-4 carbon in phenanthrene and of carbons in environments similar to those of the C-7 carbons in fluoranthrene and the C-1 carbons in fluorene. It is conceivable that future studies using proton decoupling may yield even more information on coal structure than is presented here.

C. Coal Liquefaction Products

1. *Overview*

The use of NMR for characterizing coal-derived liquids has received considerable attention since 1975. ^{1}H-NMR and, in some cases, ^{13}C-NMR techniques have been applied to Solvent Refined Coal (SRC) by Anderson (1975), Callen *et al.* (1976), Farcasiu (1977), Farcasiu *et al.* (1976a,b), Retcofsky *et al.* (1977), Schwager and Yen (1976), Woolsey *et al.* (1976), and Wooton *et al.* (1976). Studies of Solvent Refined Lignite (SRL) have been reported by Woolsey *et al.* (1976) and Retcofsky *et al.* (1977). Callen *et al.* (1976) and Retcofsky *et al.* (1977) applied both ^{1}H- and ^{13}C-NMR to aid in the characterization of H-Coal. NMR results for the COED product were reported by Schwager and Yen (1976) and Retcofsky *et al.* (1977). A number of investigators, including Sternberg *et al.* (1975), Callen *et al.* (1976), Schwager and Yen (1976), Schweighardt *et al.* (1976a,b, 1978), Retcofsky *et al.* (1977), and Retcofsky and Link (1977), examined materials from the $\frac{1}{2}$-ton/day (TPD) SYNTHOIL PDU by

NMR. NMR studies of materials from other coal hydrogenation or coal hydrogenolysis processes have been reported by Maekawa *et al.* (1975), Yokoyama *et al.* (1976a,b), Yoshida *et al.* (1976), and Pugmire *et al.* (1977).

2. *Coal-Derived Liquids from the ½-TPD SYNTHOIL PDU*

a. Description of PDU. The Department of Energy's ½-TPD SYNTHOIL PDU located at the Pittsburgh Energy Technology Center was designed to convert a high sulfur, high ash bituminous coal into a low sulfur, low ash boiler feed. The process uses a unique reactor system to accomplish a mild hydrogenation of the feed coal. Basically, a feed paste consisting of 35% coal and 65% vehicle oil (recycled product oil) is passed in a rapid, turbulent flow of hydrogen through a preheater and then through a fixed-bed catalytic reactor. The catalyst used is a commercial $CoO–MoO_3–SiO_2–Al_2O_3$ catalyst in the form of ⅛ in. × ⅛ in. cylindrical pellets. The product formed is initially separated from gaseous and low boiling liquid components in liquid receivers. The primary liquid products are separated from solid materials (primarily mineral matter) by centrifugation, yielding the centrifuged liquid product (CLP). More detailed descriptions of the process are given elsewhere (Akhtar *et al.* 1975).

b. Changes in Aromaticity during Coal Liquefaction. In order to elucidate the mechanism of liquefaction and the chemical reactions occurring in different sections of a coal liquefaction PDU, it is important to characterize not only the liquid product but also the process coal and any isolatable intermediate products. NMR spectrometry provides a means by which the changes in at least one important structural parameter, the carbon aromaticity f_a, can be determined. The CP ^{13}C-NMR spectra reproduced in Fig. 19 serve as an example. The spectra are those of the process coal, a sample of the coal–recycle oil mixture removed from the preheater section of the PDU, and the centrifuged liquid product. The decrease in f_a in going from the process coal to the material from the preheater to the CLP is evident from the reduction in the relative intensity of the low field aromatic carbon resonances in the spectra. The quantitative estimates of f_a for each of the samples are given in Table IV. Also included in the table are f_a values for soluble fractions from the process coal and preheater sample. In both cases, the soluble fractions exhibit lower f_a values than the total sample.

Asphaltenes, which are considered by many to be key intermediates in the conversion of coal to oil (Weller *et al.*, 1951a,b), can be isolated from CLPs (Fig. 20) and examined by NMR. The changes in aromaticity

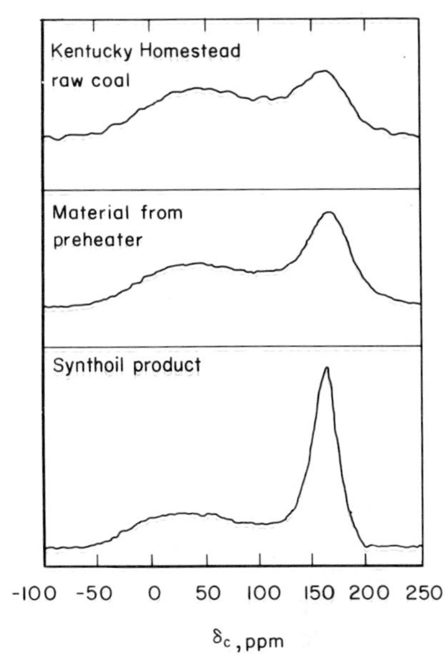

Fig. 19 Cross-polarization ^{13}C-NMR spectra of materials from the Pittsburgh Energy Technology Center's ½-TPD SYNTHOIL process development unit. (From VanderHart and Retcofsky, 1976b.)

during the liquefaction of Homestead, Kentucky coal and the liquefaction of SRC in the ½-TPD SYNTHOIL PDU are shown in Fig. 21. For each run, the f_a values decrease in the order feed material > asphaltene > oil. Thus, decreases in carbon aromaticity accompany the conversion of coal or SRC to oil in the SYNTHOIL process.

TABLE IV *Carbon Aromaticities for Materials from the Pittsburgh Energy Technology Center's ½-TPD SYNTHOIL Process Development Unit Run FB-42*[a]

Material	f_a	NMR method[b]	Yield (wt %)
Process coal	0.80	CP ^{13}C	100
Pyridine extract	0.67	HR ^{1}H	19
Carbon disulfide extract	0.59	HR ^{13}C	1
Preheater sample	0.68	CP ^{13}C	100
Carbon disulfide extract	0.59	HR ^{13}C	74
	0.60	HR ^{1}H	74
Centrifuged liquid product	0.48	HR ^{1}H	100
	0.46	HR ^{13}C	100
	0.50	CP ^{13}C	100

[a] Run conditions: Homestead, Kentucky coal, 450°C, 4000 psig H_2, Co/Mo catalyst.

[b] HR = high-resolution; CP = cross-polarization.

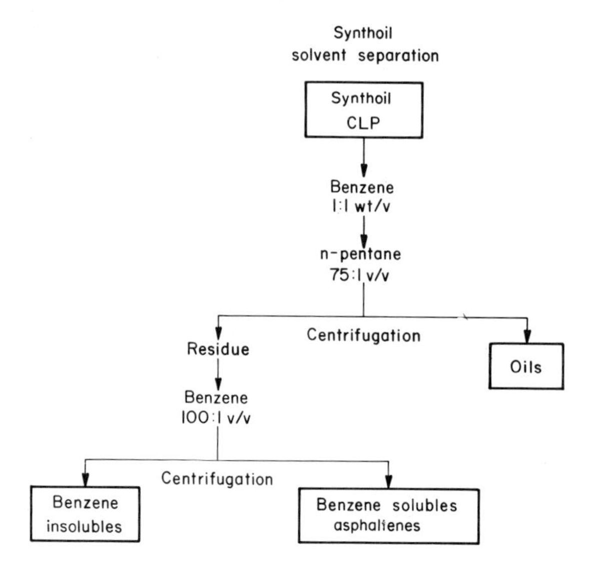

Fig. 20 Scheme for the solvent separation of SYNTHOIL centrifuged liquid products into asphaltenes and oils. (Courtesy of F. K. Schweighardt, Pittsburgh Energy Technology Center.)

c. *Pulse FT NMR Operating Parameters for Coal–Liquid Subfractions.* The remainder of this section deals with the characterization of asphaltenes and various subfractions of hexane-soluble oils by pulse Fourier transform NMR. It is emphasized that the spectrometer operating parameters were determined on the basis of quantitative reliability of the NMR data obtained for materials from the liquefaction of coal in the SYNTHOIL unit just described.

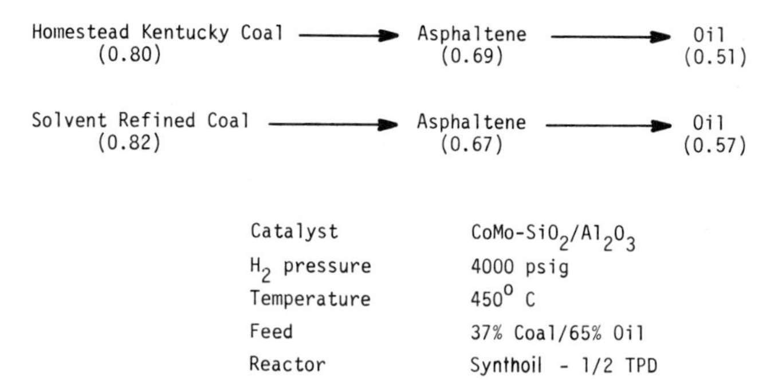

Fig. 21 Changes in carbon aromaticity during the liquefaction of Homestead, Kentucky coal and of Solvent Refined Coal in the Pittsburgh Energy Technology Center's $\frac{1}{2}$-TPD SYNTHOIL process development unit. (From Retcofsky and Link, present work.)

All NMR spectra obtained for this particular investigation were acquired on a Varian Associates† XL-100 NMR spectrometer equipped with a 15-in. magnet, 16K computer, and interactive disk system. Since NMR is usually the first method of analysis applied to these samples and the amount of most samples is limited, careful attention is paid to preparation and handling procedures. No relaxation agents of any kind are added because their presence would prevent analysis by other spectrometric techniques. Only deuterated solvents that could easily be removed from the sample were used.

The ^{13}C spectra were obtained using a flip angle of 90° and pulse delays of 100 sec or more to allow for long relaxation times. The 12-mm sample was irradiated with a 12-μsec pulse, and a 5-kHz spectral region was observed. Proton broad-band noise decoupling was utilized and data accumulated in the suppressed Overhauser mode of operation. An 8K data length allowed for adequate resolution. Internal pulse deuterium lock was employed, allowing very stable operation for long accumulation times.

Proton data were acquired using a flip angle less than 90° and a pulse delay of 7 sec. A 1.5-kHz spectral region and 12K data length allowed for resolution better than 0.2 Hz. Pulse deuterium lock was again employed for stable operation of the spectrometer.

d. Separation Procedures. The CLP was separated into asphaltenes and oil using a procedure similar to that shown in Fig. 20. In some cases, hexane was used in place of pentane in the separation. The oil fraction was further subjected to a functional group separation which is part of the SARA (Saturates-Aromatics-Resins-Asphaltenes) separation (Jewell *et al.*, 1972). The steps in the separation are outlined in Fig. 22. Five subfractions, i.e., nitrogen bases, acids–phenolics, neutral nitrogens, saturates, and aromatics, are collected for further characterization. The results for a SYNTHOIL run with no added CoMo catalyst are summarized in Table V.

e. ^1H- and ^{13}C-NMR Results. Solutions of the asphaltenes and each of the oil subfractions were prepared and examined by high resolution ^1H- and ^{13}C-NMR using the spectrometer operating conditions described earlier. NMR spectra are reproduced in Figs. 23–28. Elemental compositions, average molecular weights, NMR data, and mean structural parameters are summarized in Table VI.

NMR spectra of the asphaltenes are reproduced in Fig. 23. In these and all subsequent ^1H and ^{13}C spectra in this chapter, chemical shifts are

† Reference to brand names is to facilitate understanding and does not imply endorsement by the U.S. Department of Energy.

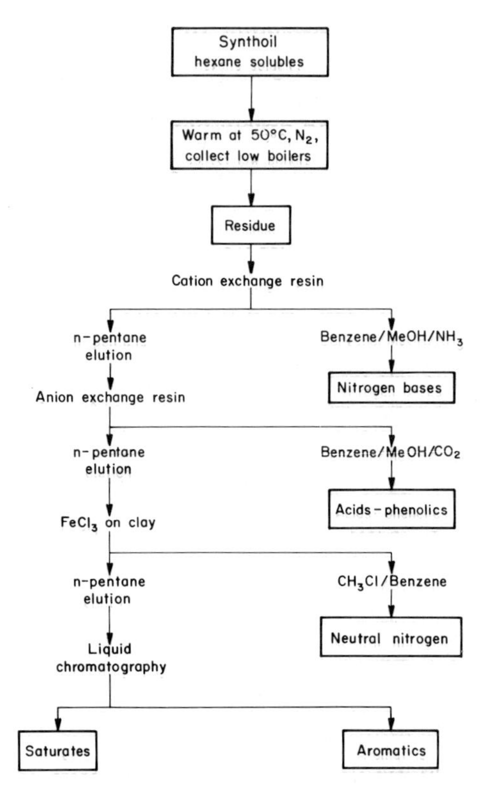

Fig. 22 The functional group separation scheme used for coal-derived oils. (Drawing courtesy of F. K. Schweighardt, Pittsburgh Energy Technology Center, based on method according to Jewell *et al.*, 1972.)

referenced to internal tetramethylsilane. The sharp peak near 54 ppm in the ^{13}C spectrum and that near 5.2 ppm in the ^{1}H spectrum are due to the solvent methylene chloride. The less intense narrow signals symmetrically located about the solvent resonances in the ^{1}H spectrum are assigned to spinning sidebands and ^{13}C satellites. A small amount of residual benzene from the solvent separation is evident from the sharp resonance near 7.5 ppm in the ^{1}H spectrum.

Neither the ^{13}C nor the ^{1}H spectra of the asphaltenes are well-resolved; nevertheless, values for f_a, H_{ar}^*, H_α^*, and H_0^* can be obtained in accordance with the procedures outlined in Section VI,B. The weak ^{13}C resonances in the 155–160 ppm region are part of the aromatic resonances but can be more specifically assigned to phenolic and/or heteroaromatic carbons. ^{1}H NMR studies of the trimethylsilyl derivatives of similar asphaltenes (Schweighardt *et al.*, 1978) confirm the

TABLE V *Solvent Separation Data for a Centrifuged Liquid Product from the Pittsburgh Energy Technology Center's ½-TPD SYNTHOIL Process Development Unit*[a]

Fraction recovered from centrifuged liquid product (wt %)	Solvent-separated fraction[b]	SARA fraction[c]	Fraction recovered from hexane-soluble Oil (wt %)
7.5	Benzene insolubles		
19.8	Asphaltenes		
72.7	Hexane-soluble oil	Nitrogen bases	25.0
		Acids–phenolics	8.3
		Neutral nitrogens	14.0
		Saturates	7.9
		Aromatics	43.4
		Pentane insolubles	0.1
		Losses	1.4

[a] Run conditions: Homestead, Kentucky coal, 450°C, 4000 psig H_2, no added catalyst. Sample taken from batch 20.
[b] Data courtesy of F. K. Schweighardt, Pittsburgh Energy Technology Center. Separation scheme is outlined in Fig. 20.
[c] Separation scheme is outlined in Fig. 22.

presence of phenolic OH groups. The high carbon aromaticity of the asphaltenes (f_a = .74) is evident from visual inspection of the ^{13}C spectrum.

Complementary use of the quantitative ^{1}H and ^{13}C NMR data in conjunction with the elemental analysis (Table VI) gave a value of 2.2 for the H/C ratio for the nonaromatic organic structures. This finding suggests that methylene groups are the predominant aliphatic structures. Assuming that 60% of the oxygen in the asphaltene fraction is phenolic and that the H/C ratio for the *alpha* aliphatic carbons is also 2.2, use of Eqs. (6) and (9) yields an estimated H_{aru}/C_{ar} of 0.65. This is consistent with a mean structural unit consisting of a basic polynuclear aromatic system of 3–4 condensed rings. The high molecular weight (511) and the estimated degree of aromatic substitution (0.40 based on the use of Eq. (7)) requires that the mean structural unit contain two such systems probably in a hydroaromatic network. For additional information on the structure of asphaltenes see Volume I, Chapter 17, Sections II,G and III,F, Fig. 4, and Tables VIII and IX.

Spectra of the nitrogen base fraction (Fig. 24) are only slightly better resolved than those of the asphaltenes. A small amount of chloroform in the sample is responsible for the ^{13}C resonance at 79 ppm and the ^{1}H

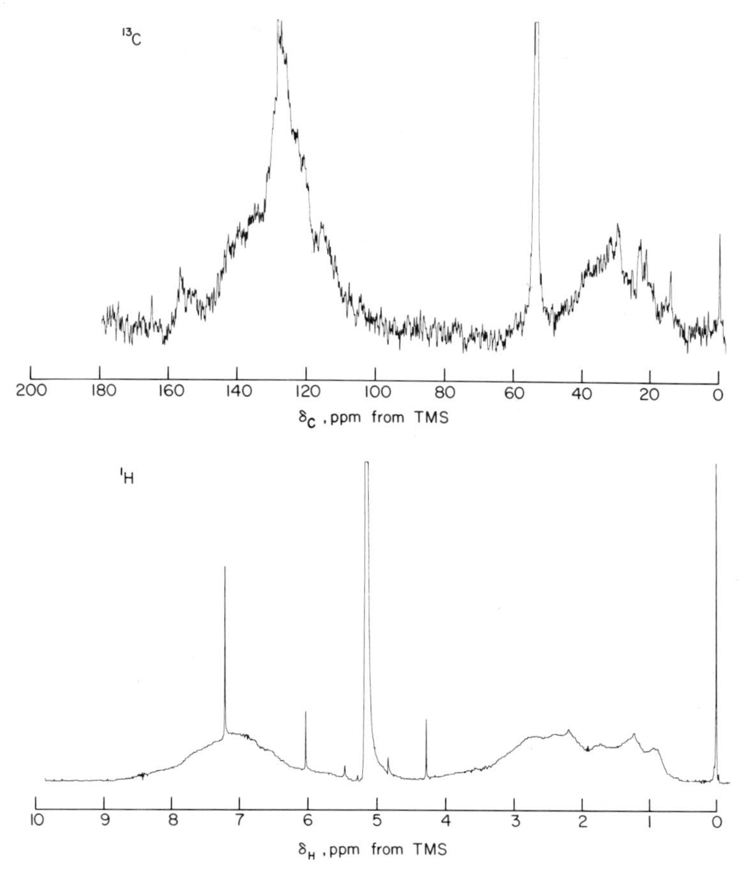

Fig. 23 100-MHz proton and 25.2-MHz ^{13}C pulse Fourier transform magnetic resonance spectra of the asphaltene fraction from SYNTHOIL run FB-57.

resonance at 7.2 ppm. The relatively narrow ^{1}H resonance near 1.2 ppm suggests that the nitrogen bases may contain more long aliphatic chains than the asphaltenes. Treating data in a manner analogous to that of the asphaltenes and assuming that the nitrogen is present only in pyridinelike structures suggest the condensed aromatic units to be larger than those of the asphaltenes. The molecular weight of 372 is less than that of the asphaltenes indicating a smaller mean structural unit.

In contrast to the spectra of the asphaltenes (Fig. 23) and of the nitrogen bases (Fig. 24), both the ^{13}C and ^{1}H spectra of the phenolic fraction (Fig. 25) exhibit a large number of well-resolved peaks. Of particular significance in the ^{1}H spectrum is the OH resonance at 4.9 ppm, which accounts for 4% of the protons and thus 53% of the oxygen in the sample. To obtain additional information about the mean structural unit for this fraction, the aromatic region of the ^{13}C spectrum was treated according to the method proposed by Bartle *et al.* (1975).

Bartle *et al.* (1975) proposed a number of ^{13}C chemical shift–structure correlations which they found useful in studying the composition of a novel, supercritical fluid extract of coal. Of particular interest in the present study of the phenolic fraction separated from a SYNTHOIL coal liquid were their proposed assignments for the aromatic carbon resonances. Briefly, the carbon resonances in the 148–158 ppm region are assigned to phenolic C–OH and to various heteroaromatic carbons, those in the 129–148 ppm region to aromatic C–C resonances, those in the

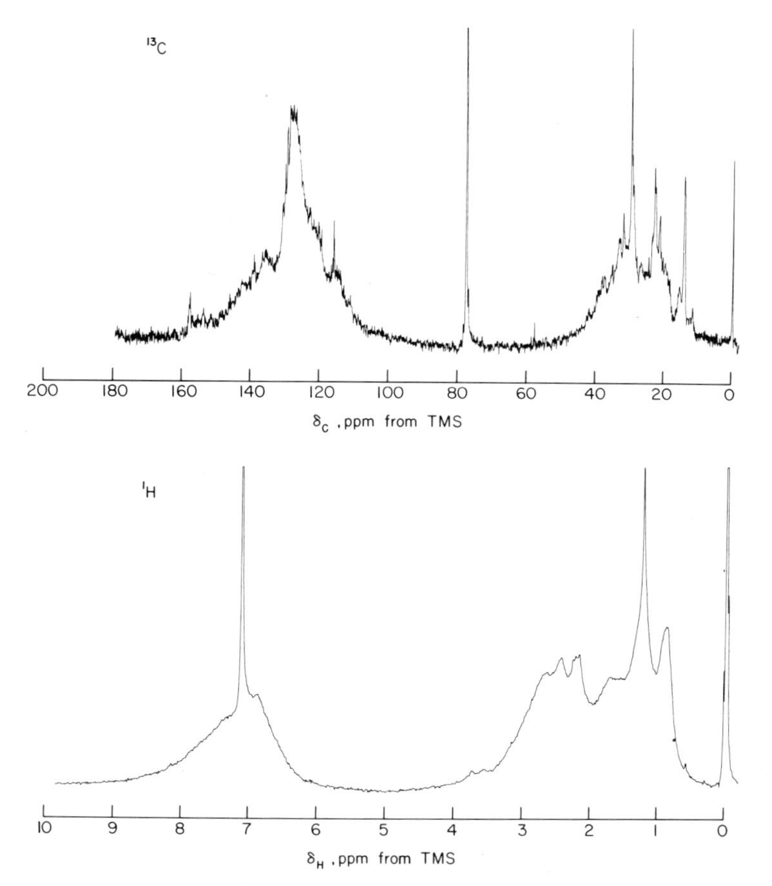

Fig. 24 100-MHz proton and 25.2-MHz ^{13}C pulse Fourier transform magnetic resonance spectra of the nitrogen base fraction from SYNTHOIL run FB-57.

118–129 ppm region to aromatic C–H resonances, and those in the 108–118 ppm region to aromatic C–H ortho to aromatic C–O groups (see Chapter 23, Table III).

Integration of the ^{13}C spectrum of the phenolic fraction (Fig. 25) shows that of the 12.9 carbon atoms, 7.9 are aromatic. Of these 7.9 aromatic carbons, the Bartle treatment of the data suggests that 1.1 are of the phenolic C–OH type, 3 are aromatic C–C carbons, 2 are in aromatic C–H groupings, and 1.8 are in aromatic C–H groupings ortho to aromatic C–O groups. The ^{1}H data indicate that 0.67 protons (and thus 0.67 carbon atoms) are phenolic leaving approximately 0.4 carbons for assignment to heteroatom or possibly aryl ether-type structures. The 1.8 aromatic carbons ortho to phenolic OH suggests that most of the aromatic C–O groupings are isolated, i.e., carbons ortho to these sites do not bear

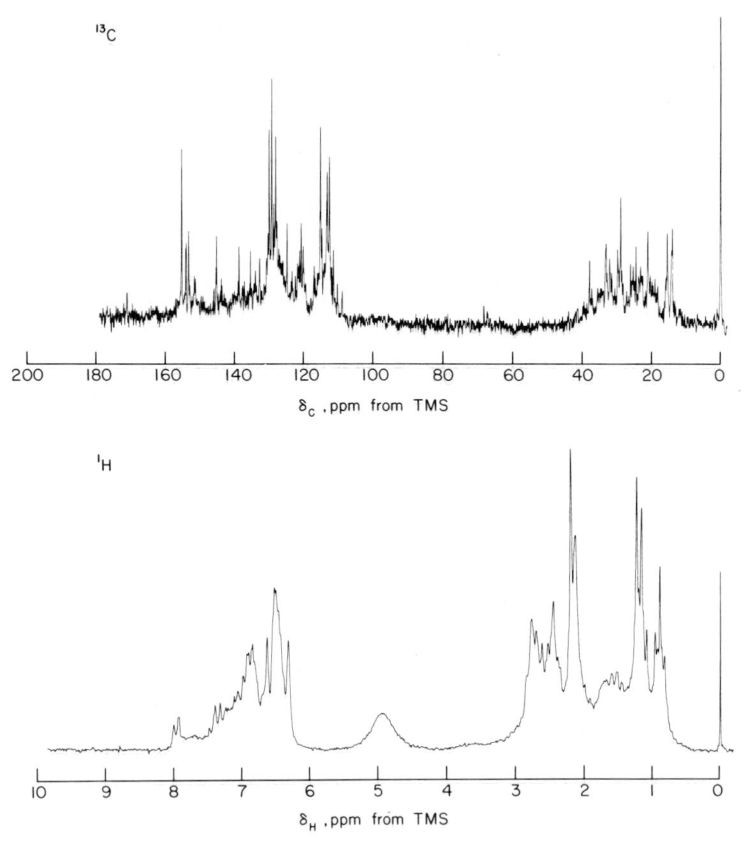

Fig. 25 100-MHz proton and 25.2-MHz ^{13}C pulse Fourier transform magnetic resonance spectra of the acid–phenolic fraction from SYNTHOIL run FB-57.

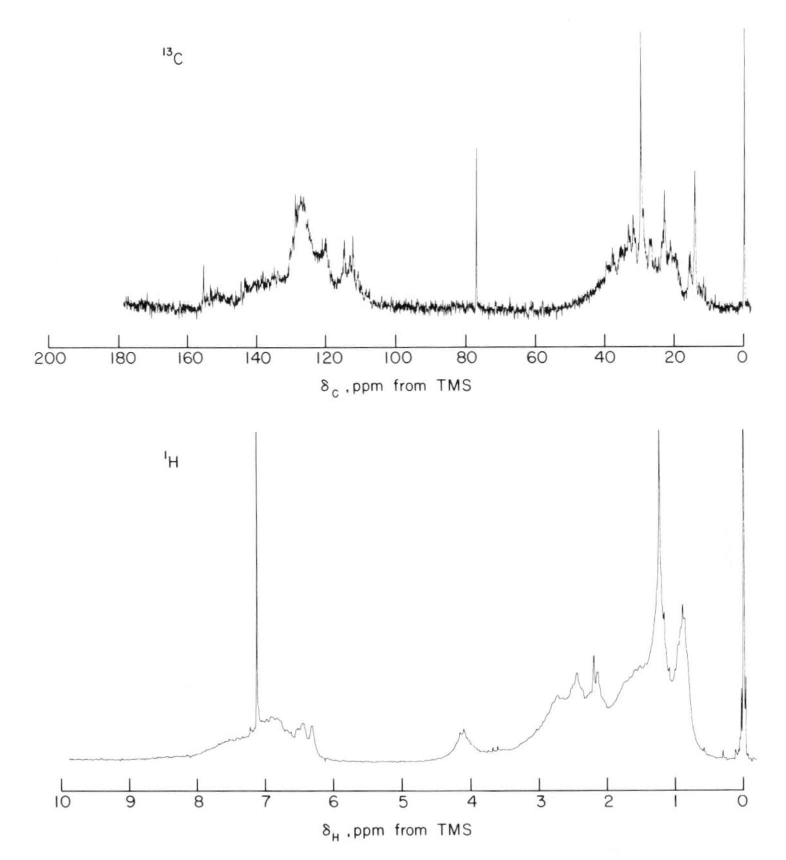

Fig. 26 100-MHz proton and 25.2-MHz ^{13}C pulse Fourier transform magnetic resonance spectra of the neutral nitrogen fraction from SYNTHOIL run FB-57.

substituents other than hydrogen. Treatment of the ^1H data in the manner described in Section IV gives an H_{aru}/C_{ar} value consistent with a benzenoid mean structural unit and a σ value of 0.5. This is in reasonable agreement with the ^{13}C data which indicates that the number of aromatic carbons in the mean structural unit is only slightly more than 6 (7.9) and that 4 of these are in either aromatic C–C or phenolic C–O groups yielding an approximate σ value of 4/7.9 or 0.5. Because of the low molecular weight of the phenolic fraction, it is doubtful that a single mean structural unit of the type shown in Fig. 6 could be drawn that would incorporate all the data obtained from the NMR spectra.

The spectra of the neutral nitrogen fraction (Fig. 26) and the aromatic fraction (Fig. 28) can be treated in a similar manner. Results for these fractions are summarized in Table VI.

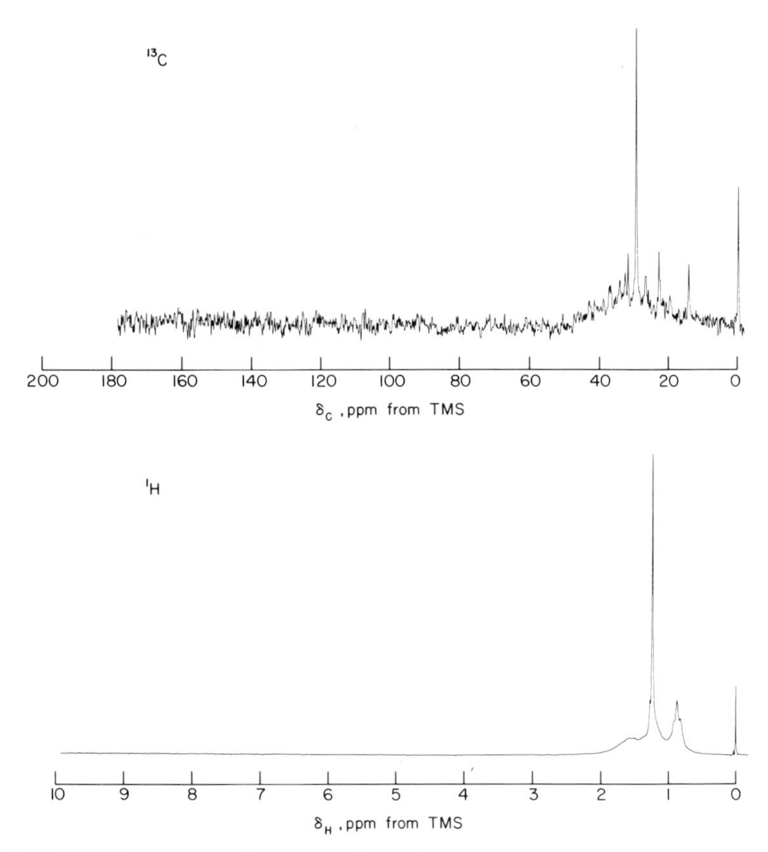

Fig. 27 100-MHz proton and 25.2-MHz ^{13}C pulse Fourier transform magnetic resonance spectra of the saturate fraction from SYNTHOIL run FB-57.

Spectra of the saturate fraction are shown in Fig. 27. The reliability of the separation method is indicated by the absence of resonances in the aromatic regions of both the ^{1}H and ^{13}C spectra. The strongest resonance in the ^{13}C spectrum (~29 ppm) is assigned to the δ^{+} carbons of aliphatic hydrocarbons and strongly suggests the presence of considerable amounts of long chain alkanes. The α, β, and γ carbons of paraffins are responsible for the resonances at 14, 23, and 32 ppm, respectively. Pugmire *et al.* (1977) have used the relative intensities of these bands to estimate the amount of unbranched alkanes and the average chain length in the saturate fraction obtained from products resulting from the hydrogenation of a Utah coal. Unlike the spectrum published by Pugmire *et al.* (1977), the presence of considerable absorption in the 22–27 and 32–40 ppm regions in the spectrum in Fig. 27 suggests the presence

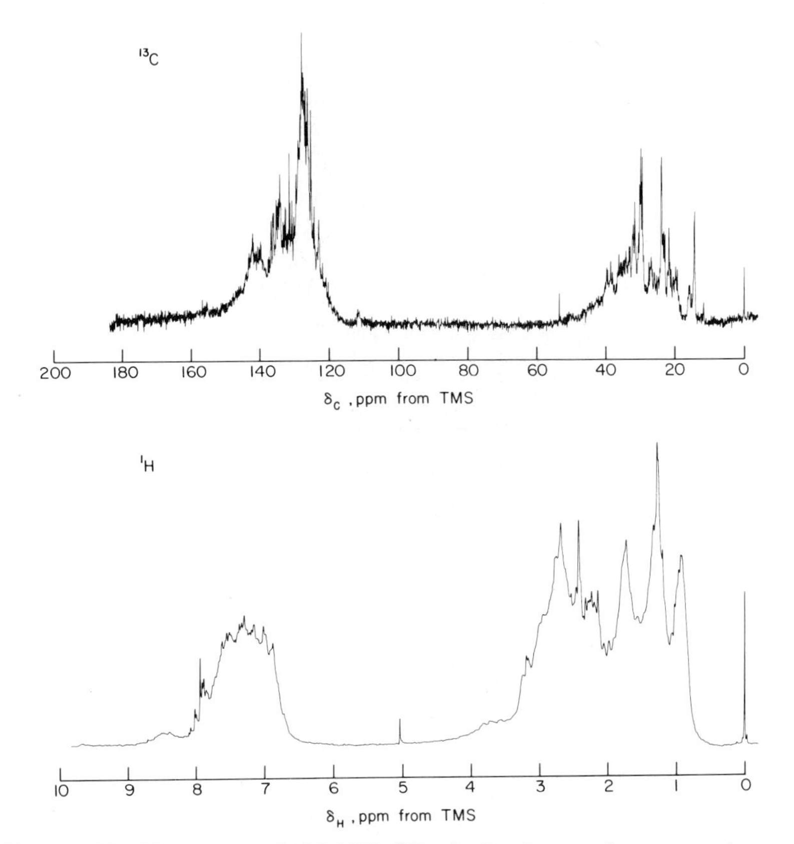

Fig. 28 100-MHz proton and 25.2-MHz ^{13}C pulse Fourier transform magnetic resonance spectra of the aromatic fraction from SYNTHOIL run FB-57.

of considerable amounts of branched or cyclic alkanes. For a detailed discussion of ^{13}C NMR of paraffinic hydrocarbons see Chapter 25, Section V,C and Table II.

D. Deuterium Tracer Studies

Another nucleus which has received some attention in coal research and will undoubtedly receive more in the future is deuterium (^{2}H). This isotope of hydrogen by virtue of its nonzero magnetic moment is amenable to NMR studies. A representative ^{2}H NMR spectrum is shown in Fig. 29. The spectrum is that of a partially deuterated sample of tetralin prepared by Y. C. Fu of the Pittsburgh Energy Technology Center. The strongest peak in the spectrum shows unequivocally that the alpha

TABLE VI *NMR and Other Data for the Solvent-Separated Fractions from a Centrifuged Liquid Product*[a]

	Aromatics	Saturates	Acids–phenolics	Nitrogen bases	Neutral nitrogens	Asphaltenes
Elemental composition (wt %)						
Carbon	90.0	86.1	80.6	83.0	85.3	86.4
Hydrogen	8.5	13.4	8.3	7.6	8.4	6.3
Oxygen	0.8	0.3	10.3	4.9	5.3	4.5
Nitrogen	0.0	0.0	0.4	3.3	0.4	2.1
Sulfur	0.7	0.2	0.5	1.2	0.6	0.6
Molecular weight[b]	260	268	193	372	348	511
Molecular formula	$C_{19.5}H_{21.8}O_{0.1}N_{<0.1}S_{<0.1}$	$C_{19.2}H_{35.5}O_{<0.1}N_{<0.1}S_{<0.1}$	$C_{12.9}H_{15.9}O_{1.2}N_{<0.1}S_{<0.1}$	$C_{25.7}H_{28.0}O_{1.1}N_{0.9}S_{0.1}$	$C_{24.7}H_{28.9}O_{1.2}N_{0.1}S_{<0.1}$	$C_{36.8}H_{32.1}O_{1.4}N_{0.8}S_{0.1}$
Hydrogen distribution						
$H^*_{ar+\phi}$	0.26	0.00	0.34	0.27	0.19	0.36
H_α^*	0.36	0.00	0.36	0.34	0.31	0.38
H_o^*	0.38	1.00	0.31	0.39	0.50	0.27
Aromaticity, f_a	0.65	0.00	0.61	0.73	0.61	0.74
H_{al}/C_{al}[c]	2.4	1.8	2.1	2.9	2.4	2.2
Degree of aromatic substitution (est.)	0.4	—	0.5	N.D.	N.D.	0.4
Number of aromatic rings per PNA unit (est.)	3	—	1	N.D.	N.D.	3–4

[a] Same sample as Table V.
[b] Determined by vapor pressure osmometry/CH_2CL_2 (27°C)
[c] al = aliphatic.

aliphatic position contains the largest amount of incorporated deuterium with a smaller amount in aromatic sites (low field resonance) and essentially negligible amounts in beta aliphatic fractions (highest field peak).

Schweighardt *et al.* (1976c) were the first to use deuterium magnetic resonance (^{2}H NMR) as a means for determining the distribution of incorporated hydrogen among various organic structures in coal liquids. A centrifuged liquid product (CLP) from the U.S. Department of Energy's $\frac{1}{2}$-ton-per-day coal liquefaction pilot plant (SYNTHOIL) and deuterium gas were heated to a maximum of 450°C at 1600 psig D_2 in a stirred autoclave. No catalyst was added and the temperature was maintained above 400°C for 15 min. Mass spectrometric analysis of the product gases showed the presence of H_2, HD, and D_2 and indicated that approximately 4.5% of the hydrogen originally present in the CLP had been replaced by deuterium.

NMR spectra were obtained at nominal operating frequencies of 38.6 MHz and 250 MHz for deuterons and protons, respectively. Acetone-d_6 was used as the internal chemical shift standard for the deuteron measurements and chemical shifts were then converted to ppm from TMS using the published chemical shift value of 2.17 ppm for acetone. Assignments of resonances to specific types of deuterons or protons were made on the basis of previous proton NMR studies of materials derived

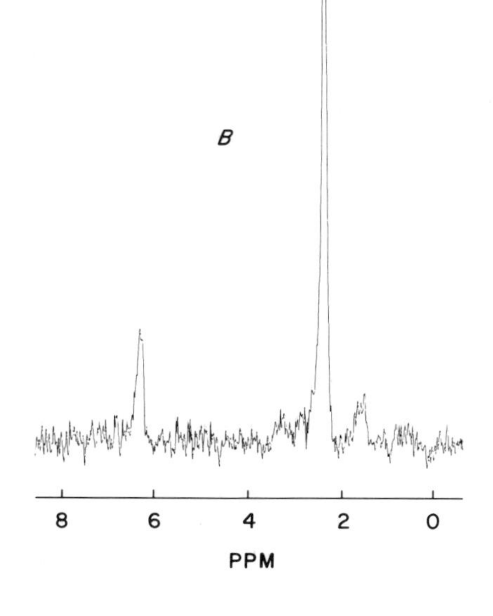

Fig. 29 38.6-MHz deuterium magnetic resonance spectrum (correlation mode) of a partially deuterated tetralin. (Sample courtesy of Y. C. Fu, Pittsburgh Energy Technology Center.)

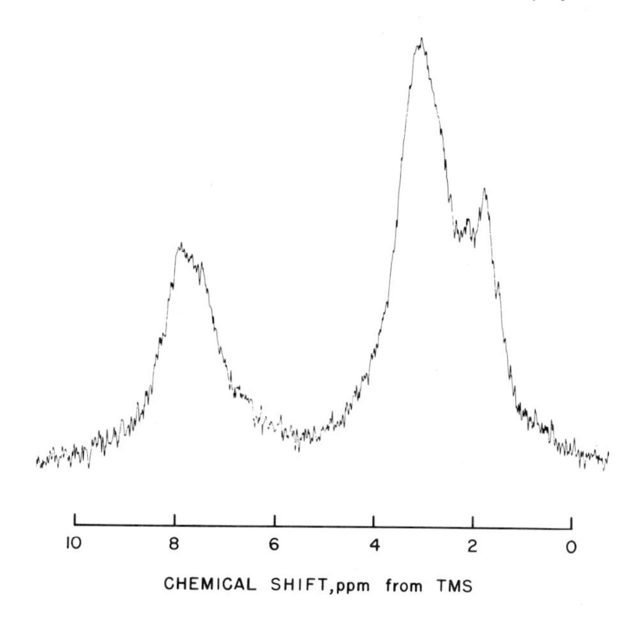

Fig. 30 38.6-MHz deuterium magnetic resonance spectrum (correlation mode) of 140–190°C boiling material (0.1 torr) from a partially deuterated SYNTHOIL centrifuged liquid product. (From Schweighardt *et al.*, 1976c.)

from coal since proton and deuteron chemical shifts measured in the same sample (natural abundance) should be identical. The ^2H NMR spectrum of a distillation fraction from the deuterated CLP is reproduced in Fig. 30. A sample of unlabeled CLP examined under identical instrumental conditions revealed no detectable deuteron NMR signals, showing that only resonances of deuterium present in excess of natural abundance were observed.

The hydrogen and deuterium distributions as determined by integration of the NMR spectra are summarized in Table VII for the deuterated CLP and several fractions thereof. These included (1) distillation cuts covering a wide range of boiling points, (2) the asphaltene components, and (3) hydrocarbon- and phenol-rich fractions of the heavy oil. Labeled material was found in all fractions examined and, for each fraction, deuterium was found at both aromatic and nonaromatic sites. This broad distribution of the label can be best rationalized on the basis of a combination of rapid incorporation and scrambling reactions.

Closer scrutiny of the data in Table VII revealed that a degree of specificity for incorporation at benzylic and aromatic positions prevails. Hydrogens bonded to carbons β or further removed from aromatic rings appear to be the least susceptible to deuterium labeling under the experimental conditions employed. This specificity becomes even more apparent when the deuteron and proton data for the various materials are compared using $(D_x/D)/(H_x/H)$ ratios (x = aromatic, benzylic, or other nonaromatic carbons). In all cases, this ratio lies within the approximate range 1–2 for the aromatic and benzylic sites, whereas the ratio is much smaller, 0.2–0.6, for hydrogens bonded to carbons β or further removed from aromatic rings. The fact that the distribution of deuterium among these three sites is not the same as the corresponding hydrogen distribution for any individual fraction or for the total CLP indicates that thermal isotopic equilibrium has not been achieved.

VII. SUMMARY AND PROGNOSIS

The material presented in this chapter was carefully chosen to provide the reader with an overview of the applications of nuclear magnetic resonance spectrometry in coal research. Both conventional high-resolution NMR techniques and NMR techniques still in their infancy, including those that extend the realm of NMR to the solid state, were discussed. The applications of NMR in coal research cited ranged from basic studies of coal structure to compositional analyses of coal-conversion products.

The horizons for NMR in coal research appear to be exceptionally bright. Although the more or less routine use of conventional NMR in compositional studies of liquid and soluble materials from coal will continue, future emphasis will undoubtedly be on applications of the newer techniques to probe the finer details of coal structure. Certainly, the required instrumentation needed to perform the cross-polarization experiments and special probes to spin solid samples rapidly at the so-called magic angle are becoming commercially available.

A number of new areas of NMR research, however, still require very fundamental exploratory studies to evaluate their potential in coal research. Probing of the nature of heteroatom species in coal and coal-conversion products using NMR of less receptive nuclei such as ^{17}O and ^{15}N has been relatively unexplored and deserves further attention. It is conceivable that ^{17}O and ^{15}N NMR studies might also give insight into intermolecular forces which are thought to play an important role in coal structure. Another area worthy of additional investigation is the use of specialized multiple pulse techniques, which, in theory should permit researchers to obtain proton spectra of solids comparable in quality to the ^{13}C spectra reproduced in this chapter. These techniques, after additional refinement, should aid in the elucidation of the hydrogen distribution in coals and thus further unravel their chemical structure.

TABLE VII *Deuterium and Hydrogen Distribution in Isotopically Labeled SYNTHOIL[a] Product*

Material	Nucleus		Aromatic[b] total	Benzylic total	Other nonaromatic[c] total
Deuterated centrifuged liquid	D		0.35	0.49	0.16
product (CLP)	H		0.24	0.30	0.46
		$(D_x/D)/(H_x/H)$	1.5	1.6	0.35
CLP asphaltenes[d]	D		0.48	0.40	0.12
	H		0.45	0.33	0.22
		$(D_x/D)/(H_x/H)$	1.1	1.2	0.55
CLP heavy oil[d]					
Hydrocarbon fraction[e]	D		0.20	0.66	0.14
	H		0.13	0.35	0.52
		$(D_x/D)/(H_x/H)$	1.5	1.9	0.27
Phenol-rich fraction[e,f]	D		0.41	0.46	0.13
	H		0.30	0.35	0.35
		$(D_x/D)/(H_x/H)$	1.4	1.3	0.37

CLP distillation fractions,[g] boiling range, °C					
49–100	D		0.48	0.42	0.10
	H		0.22	0.29	0.49
		$(D_x/D)/(H_x/H)$	2.2	1.5	0.20
100–138	D		0.32	0.55	0.13
	H		0.17	0.37	0.46
		$(D_x/D)/(H_x/H)$	1.9	1.5	0.28
140–190	D		0.26	0.58	0.16
	H		0.21	0.31	0.48
		$(D_x/D)/(H_x/H)$	1.2	1.9	0.33
190–220	D		0.37	0.46	0.17
	H		0.31	0.29	0.40
		$(D_x/D)/(H_x/H)$	1.2	1.6	0.43
Residue (bp > 258)	D		0.47	0.44	0.09
	H		0.37	0.35	0.28
		$(D_x/D)/(H_x/H)$	1.3	1.3	0.32

[a] See Akhtar *et al.* (1975).
[b] Also includes phenolic protons or deuterons.
[c] Refers to hydrogen or deuterium in aliphatic structures or those bonded to carbon atoms β or further removed from aromatic rings.
[d] Asphaltenes are operationally defined as benzene-soluble, pentane-insoluble material; the deasphaltenated liquids are called heavy oils.
[e] Fractions separated chromatographically; Schweighardt *et al.* (1976b).
[f] Phenols detected by thin-layer chromatography; fraction also contains other polar compounds.
[g] Distillation carried out at 0.1 torr. Distillate accounted for 59.7% of the total CLP.

REFERENCES

Akhtar, S., Mazzocco, N. J., Weintraub, M., and Yavorsky, P. M. (1975). *Energy Commun.* **1,** 21–36.

Anderson, R. P. (1975). *In* "Coal Processing Technology," Vol. 2, pp. 130–132. Am. Inst. Chem. Eng., New York.

Andrew, E. R. (1971). *Prog. Nucl. Magn. Reson. Spectrosc.* **8,** 1–39.

Bartle, K. D., Martin, T. G., and Williams, D. F. (1975). *Chem. Ind. (London)* pp. 313–314.

Bartuska, V. J., Maciel, G. E., Schaefer, J., and Stejskal, E. O. (1976). *Prepr. 1976 Coal Chem. Workshop, Stanford Res. Inst.* pp. 220–228.

Bartuska, V. J., Maciel, G. E., Schaefer, J., and Stejskal, E. O. (1977). *Fuel* **56,** 354–358.

Brown, J. K., and Ladner, W. R. (1960). *Fuel* **39,** 87–96.

Callen, R. B., Bendoraitis, J. G., Simpson, C. A., and Voltz, S. E. (1976). *Ind. Eng. Chem., Prod. Res. Dev.* **15,** 222–233.

Dadok, J., and Sprecher, R. F. (1972). *Exp. NMR Conf., 13th, Asilomar, Calif.* (unpublished).

Dadok, J., and Sprecher, R. F. (1974). *J. Magn. Reson.* **13,** 243–248.

DeWalt, C. W., Jr., and Morgan, M. S. (1962). *Am. Chem. Soc., Div. Fuel Chem. Symp. on Tars, Pitches, and Asphalts Prepr.* pp. 33–45.

Durie, R. A., Shewchyk, Y., and Sternhell, S. (1966). *Fuel* **45,** 99–113.

Dybowski, C. R., and Vaughan, R. W. (1975). *Macromolecules* **8,** 50–54.

Farcasiu, M. (1977). *Fuel* **56,** 9–14.

Farcasiu, M., Mitchell, T. O., and Whitehurst, D. D. (1976a). *Prepr. 1976 Coal Chem. Workshop, Stanford Res. Inst.* pp. 101–124.

Farcasiu, M., Mitchell, T. O., and Whitehurst, D. D. (1976b). *Am. Chem. Soc., Div. Fuel Chem. Prepr.* **21**(7), 11–26.

Freeman, R., Hill, H. D., and Kapten, R. J. (1972). *J. Magn. Reson.* **7,** 327–329.

Friedel, R. A. (1959). *J. Chem. Phys.* **31,** 280–281.

Friedel, R. A., and Retcofsky, H. L. (1961). *Carbon Conf., 5th, Pennsylvania State Univ.*

Friedel, R. A., and Retcofsky, H. L. (1963). *Proc. Carbon Conf., 5th* **2,** 149–165.

Friedel, R. A., and Retcofsky, H. L. (1966). *Chem. Ind. (London)* pp. 455–456.

Gerstein, B. C., Chow, C., Pembleton, R. G., and Wilson, R. C. (1977). *J. Phys. Chem.* **81,** 565–570.

Given, P. H. (1960). *Fuel* **39,** 147–153.

Hartmann, S. R., and Hahn, E. L. (1962). *Phys. Rev.* **128,** 2042–2053.

Jewell, D. M., Weber, J. H., Bunger, J. W., Plancher, H., and Latham, D. R. (1972). *Anal. Chem.* **44,** 1391–1395.

Johnson, L. F., and Jankowski, W. C. (1972). "Carbon-13 NMR Spectra." Wiley (Interscience), New York.

Kanai, H., Suzuki, A., Ito, M., and Takeya, G. (1966). *Nenryo Kyokai-Shi* **45,** 859–866.

La Mar, G. N. (1971). *J. Am. Chem. Soc.* **93,** 1040–1041.

Lauterbur, P. C. (1962). *In* "Determination of Organic Structures by Physical Methods" (F. C. Nachod and W. D. Phillips, eds.), Vol. 2, pp. 472–475. Academic Press, New York.

Levy, G. C., Cargioli, J. D., and Anet, F. A. (1973). *J. Am. Chem. Soc.* **95,** 1527–1535.

Maekawa, Y., Ueda, S., Hasegawa, Y., Nakata, Y., Yokoyama, S., and Yoshida, Y. (1975). *Am. Chem. Soc., Div. Fuel Chem., Prepr.* **20**(3) 1–6.

Oelert, H. H. (1967). *Z. Anal. Chem.* **231,** 105–121.

Oth, J. F. M., and Tschamler, H. (1962). *Brennst.-Chem.* **43,** 35–37.

Pines, A., and Wemmer, D. E. (1976). *Prepr. 1976 Coal Chem. Workshop, Stanford Res. Inst.* pp. 229–232.

Pines, A., Gibby, M. G., and Waugh, J. S. (1972a). *J. Chem. Phys.* **56,** 1776–1777.

Pines, A., Gibby, M. G., and Waugh, J. S. (1972b). *Chem. Phys. Lett.* **15,** 373–376.
Pines, A., Gibby, M. G., and Waugh, J. S. (1973). *J. Chem. Phys.* **59,** 569–590.
Pugmire, R. J., Grant, D. M., Zilm, K. W., Anderson, L. L., Oblad, A. G., and Wood, R. E. (1977). *Fuel* **56,** 295–301.
Retcofsky, H. L. (1977). *Appl. Spectrosc.* **31,** 116–121.
Retcofsky, H. L., and Friedel, R. A. (1966). *In* "Coal Science" (R. F. Gould, ed.), pp. 503–515. Am. Chem. Soc., Washington, D.C.
Retcofsky, H. L., and Friedel, R. A. (1968). *Fuel* **47,** 487–498.
Retcofsky, H. L., and Friedel, R. A. (1970). *In* "Spectrometry of Fuels" (R. A. Friedel, ed.), pp. 70–89. Plenum, New York.
Retcofsky, H. L., and Friedel, R. A. (1973). *J. Phys. Chem.* **77,** 68–71.
Retcofsky, H. L., and Link, T. A. (1977). *Pittsburgh Conf. Anal. Chem. Appl. Spectrosc., Cleveland, Ohio.*
Retcofsky, H. L., and Schweighardt, F. K. (1977). Unpublished results.
Retcofsky, H. L., and VanderHart, D. L. (1978). *Fuel* **57,** 421–423.
Retcofsky, H. L., Schweighardt, F. K., and Hough, M. (1977). *Anal. Chem.* **49,** 585–588.
Rhim, W. K., Elleman, D. P., and Vaughan, R. W. (1973). *J. Chem. Phys.* **58,** 1772–1773.
Schaefer, J., and Stejskal, E. O. (1976). *Poster Pap., Exp. NMR Conf., 17th, Pittsburgh, Pa.*
Schwager, I., and Yen, T. F. (1976). *Am. Chem. Soc., Div. Fuel Chem., Prepr.* **21**(5), 199–206.
Schweighardt, F. K., Friedel, R. A., and Retcofsky, H. L. (1976a). *Appl. Spectrosc.* **30,** 291–295.
Schweighardt, F. K., Retcofsky, H. L., and Friedel, R. A. (1976b). *Fuel* **55,** 313–317.
Schweighardt, F. K., Bockrath, B. C., Friedel, R. A., and Retcofsky, H. L. (1976c). *Anal. Chem.* **48,** 1254–1255.
Schweighardt, F. K., Retcofsky, H. L., Friedman, S., and Hough, M. (1978). *Anal. Chem.* **50,** 368–371.
Sternberg, H. W., Raymond, R., and Schweighardt, F. K. (1975). *Science* **188,** 49–51.
Takeya, G., Itoh, M., Suzuki, A., and Yokoyama, S. (1964). *Nenryo Kyokai-Shi* J. Fuel Soc. (Japan) **43,** 837–848.
VanderHart, D. L., and Retcofsky, H. L. (1976a). *Fuel* **55,** 202–204.
VanderHart, D. L., and Retcofsky, H. L. (1976b). *Prepr. 1976 Coal Chem. Workshop, Stanford Res. Inst.* pp. 202–218.
Van Krevelen, D. W. (1961). "COAL: Typology-Chemistry-Physics-Constitution." Elsevier, Amsterdam.
Weller, S., Pelipetz, M. G., and Friedman, S. (1951a). *Ind. Eng. Chem.* **43,** 1572–1575.
Weller, S., Pelipetz, M. G., and Friedman, S. (1951b). *Ind. Eng. Chem.* **43,** 1575–1579.
Woolsey, N., Baltisberger, R., Klabunde, K., Stenberg, V., and Kaba, R. (1976). *Am. Chem. Soc., Div. Fuel Chem., Prepr.* **21**(7), 33–51.
Wooton, D. L., Dorn, H. C., Taylor, L. T., and Coleman, W. M. (1976). *Fuel* **55,** 224–226.
Yokoyama, S., Bodily, D. M., and Wiser, W. H. (1976a). *Am. Chem. Soc., Div. Fuel Chem., Prepr.* **21**(7), 77–83.
Yokoyama, S., Bodily, D. M., and Wiser, W. H. (1976b). *Am. Chem. Soc., Div. Fuel Chem., Prepr.* **21**(7), 84–89.
Yoshida, R., Maekawa, Y., Ishii, T., and Takeya, G. (1976). *Fuel* **55,** 341–345.

Chapter 25

Separation and Spectroscopy of Paraffinic Hydrocarbons from Coal

Keith D. Bartle
SCHOOL OF CHEMISTRY,
UNIVERSITY OF LEEDS,
LEEDS, ENGLAND

Derry W. Jones *Hooshang Pakdel*
SCHOOL OF CHEMISTRY,
UNIVERSITY OF BRADFORD,
BRADFORD, ENGLAND

ISBN 0-12-399902-2

I. INTRODUCTION

A. Spectroscopy of Coal Hydrocarbons

Investigation of the composition and structure of fuels has long provided one of the major fields of industrial application of molecular spectroscopy (Friedel, 1970). For materials as complex as coals, in particular, preliminary subdivision of extracts is an essential prerequisite for successful exploitation of spectroscopic techniques. Over the past two decades, developments in gas chromatography, mass spectrometry, nuclear magnetic resonance spectroscopy, and other physical techniques have enormously enhanced the ability of chemists to analyze alkane fractions rapidly and quantitatively.

Accordingly, before the discussion of spectroscopic techniques in Section V and the results from them in Section VI, we consider in Section II the extraction of paraffinics from coals, and in Sections III and IV describe fractionation and chromatographic separation into homologous series. Although the applications illustrated are predominantly to materials derived from coals, some reference is inevitably made to the parallels with and differences from paraffinic hydrocarbons derived from petroleum and other fossil fuels. Indeed, in a number of cases, major advances in the techniques utilized have first been made through the stimulus of the analysis requirements for paraffinic hydrocarbons in petroleum.

TABLE I *Alkane Content of Hydrocarbon Minerals and Other Sediments*

Source	Organic matter (%)	Atomic H/C ratio of organic matter	Alkanes in organic matter (%)	Reference
Recent sediments	0.02–1.7	1.7–1.9	0.8–36.9	Smith (1954)
Crude oils	~100	1.5–2.0	>32[a]	Mair (1964)
Oil shales	1–2[b]	1.3–1.6	~6	Cummins and Robinson (1964)
Bituminous coals	~100[c]	0.5–0.7	0.4–0.7	Bartle *et al.* (1975); Vahrman (1970)
Lignites	~100	0.7–1.0	0.01[d]	Brooks and Smith (1967)

[a] Identified alkanes in API Research Project 44 Reference Petroleum.
[b] Extractable by benzene; excludes kerogen.
[c] At least 17% extractable by supercritical gases, 5% by exhaustive solvent extraction.
[d] Nonexhaustive solvent extraction.

B. Proportions of Paraffins in Coals

Most sediments of the earth's crust contain paraffinic hydrocarbons, in the main derived from plant and animal sources. For crude oils, oil shales, and coals, the alkanes are generally present in complex mixtures of paraffinics and naphthenic and aromatic compounds, which show wide variations in both the relative and absolute concentrations of the different classes of organic materials (Table I).

Thus, whereas at least 32% of a reference petroleum has been accounted for in terms of *identified* alkanes (28% as C_1–C_{33} *n*-alkanes and C_4–C_{10} singly branched alkanes) (Mair, 1964), the alkanes of coal are much scarcer—a difference only in part due to the difficulty in extraction and reflected in the atomic H/C ratios of various hydrocarbon minerals (Table I).

C. Classes of Paraffinic Hydrocarbons

The paraffinic hydrocarbons of coal and coal derivatives are of four types: (i) normal alkanes (or paraffins); (ii) singly branched alkanes; (iii) acyclic isoprenoids (with several branches); and (iv) cyclic diterpanes, triterpanes, and steranes (see Table II). Whereas the *n*-alkanes and isoprenoid compounds are prominent constituents, the singly branched and cyclic compounds are much less abundant.

Although the high proportion of paraffins in petroleums is of direct economic importance, the much smaller proportions in coals represent very little direct commercial value. Knowledge of the paraffins of coal tar is of interest in determining the behavior of fuels, coatings, and other derived products. Moreover, the paraffins extracted from coal itself are vitally important in discussions of the nature and origin of coal, indeed quite out of proportion to their small percentage of the total content.

The understanding of the patterns of abundance of paraffins in coals in terms of the composition of the original biological material and of the processes which have led to them represents a continuing goal of organic geochemistry. Thus, the paraffins of coal may be regarded as biological markers, compounds the structure of which can be interpreted in terms of a previous biological origin. Sometimes the compound is originally present in the organism, but more usually a related structure has been derived from the original biolipid by the operation of diagenetic processes in the forming of the coal and in later maturation processes (see Section VI). The carbon skeleton and stereochemistry may thus be preserved and the C–C bonds present in the geolipids are the same ones formed by the living organism.

TABLE II *Paraffinic Hydrocarbons*

	Typical carbon skeleton	Name
(i)		Normal
(ii)		Iso
		Ante-iso
(iii)		Isoprenoid
		Diterpane
(iv)	R	Sterane (C_{27}–C_{29})
		Triterpane (C_{27}–C_{30}) (pentacyclic)

II. EXTRACTION AND SEPARATION OF PARAFFINIC HYDROCARBONS FROM COAL AND COAL PRODUCTS

A. Extraction of Paraffins from Coal

The primary extraction of alkanes from coal has usually been achieved by Soxhlet extraction. While the yields are generally low for this method (Table I), greater quantities may be extracted if the process is continued over a long period. Thus Spence and Vahrman (1970) found that aliphatic compounds were still extracted during the final period of 250 hr of Soxhlet extraction (Fig. 1), up to a total of 0.29% of an orthohydrous coal. Up to 0.70% aliphatic compounds were obtained by autoclave extraction with first benzene and then tetralin at temperatures up to 290°C (Rahman and Vahrman, 1971).

In the Soxhlet extraction procedure, hydrocarbons were found to

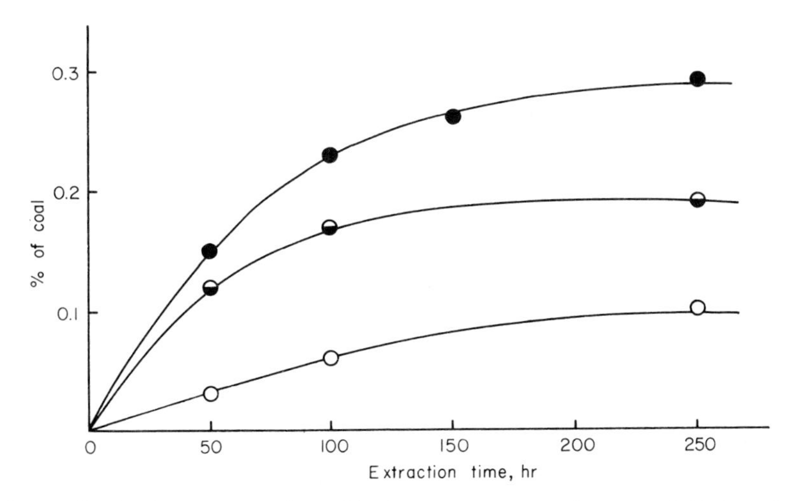

Fig. 1 Graph showing variation with time of extraction of classes of aliphatic hydrocarbons for Soxhlet extraction of coal with 7:3 (vol/vol) benzene–ethanol. Top curve, total aliphatic hydrocarbons; middle curve, branched and cyclics; bottom curve, straight chains. (Data from Spence and Vahrman, 1970.)

emerge in a definite sequence. At first, the branched and cyclic alkanes together predominated over the straight chain (after 50 hr the ratio was 3.3 : 1), but this position was gradually reversed (after 250 hr the ratio was 0.4 : 1) (Spence and Vahrman, 1970). [The alkanes present in low temperature tars were similar in amount to those in extracts (Rahman and Vahrman, 1971)]. Vahrman (1970) has used these data to infer that most of the hydrocarbon molecules are probably inclusions in the micropores which make up a molecular sieve structure for coal.

Solvent extraction of coal with supercritical gases is a new technique for obtaining chemicals from coal (Wise, 1970; Whitehead and Williams, 1975). When coal is heated to about 400°C, the liquids formed are normally too involatile to distill at this temperature, but the presence of a gas near to its critical temperature causes the liquids to volatilize. At least 17% of the coal can be extracted in this way (Bartle *et al.*, 1975); by transferring the gas phase to another vessel at a lower pressure, the extract is precipitated, and thereby separated from the gas solvent. The extract obtained in this way from a British bituminous coal (Bartle *et al.*, 1975) contained alkanes corresponding to 0.4% of the original coal.

B. Separation of Paraffins from Coal-Derived Materials

1. *Introduction*

Coal tar is a multicomponent, highly aromatic system that consists of neutral, basic, and acidic organic molecules. Separation and study of the chemical composition of different tars and pitches from any particular component as groups of compounds are best effected after preliminary separation into narrower fractions.

Isolation of paraffinic-type hydrocarbons from coal products may be achieved by adsorption chromatography or sulfonation of saturated and unsaturated mixtures, from which unattached hydrocarbon types can be separated. To adopt this method to larger scale operation, it would be necessary to apply different preliminary separation methods such as distillation, or solvent extraction followed by distillation (Bartle *et al.*, 1977b). The complex mixtures of hydrocarbons in the coal product must be subdivided into aliphatic, aromatic, and other types of compounds. Methods for this include:

(i) silica gel, or impregnated silica gel column chromatography (Section II,B,3);
(ii) iodine monochloride treatment, followed by purification on a silica gel column (Section II,B,5);
(iii) sulfuric acid treatment, followed by purification on a silica gel column (Section II,B,4);
(iv) gel permeation chromatography (GPC) (Section II,B,6); and
(v) concentrated sulfuric acid treatment (Section II,B,7).

2. *Preliminary Separation*

Among many schemes of solvent analysis or solvent fractionation for coal products, an early method employed by the Coal Tar Research Association (Binns, 1962) consisted of successive extraction with petroleum ether, benzene, and pyridine. This has been extended (Binns, 1962; Bartle and Smith, 1965, 1967; Wilman, 1966; Bartle *et al.*, 1969) to dispersion of pitch in dioxane, followed by precipitation with *n*-heptane to separate the solubles (containing aliphatics) from the insoluble part. As an alternative approach to solvent fractionation schemes based on molecular size and complexity, the Australian group at CSIRO (Smith, 1966; CSIRO, 1966) proposed solvent extraction followed by partition of the soluble fractions between polar and nonpolar solvents. Bartle *et al.* (1970) described a separation scheme, also based on chemical types, which employs γ-butyrolactone as a solvent to extract aromatic species. Possibilities of fractionation of coal tar with novel solvents (e.g., propane) have been explored (Grudzien, 1967); dimethylformamide/heptane has been reported (Quader and Vaidegaswaran, 1966) to be a good solvent for the extraction of aromatics from aromatic–saturated mixtures. Blunt and Vahrman (1960) first separated tars into broad fractions on the basis of their solubility in benzene and light petroleum and then separated phenols, bases, and neutrals within each of these. The light-petroleum-soluble neutrals are then separated by adsorption chromatography into fractions containing predominating hydrocarbon types. To separate the hydrocarbon types in neutral tar, Ahmed and Vahrman (1966) used 70–90°C boiling range petroleum spirit in two procedures: (i) direct extraction with a selective solvent and (ii) partitioning between a solvent pair.

For highly viscous samples, preliminary solution of as much as possible in a favorable solvent (such as chloroform) (Bartle *et al.*, 1977b) can yield a liquid sample with much lower viscosity; this can then be extracted with the original solvent as in (i). Criteria for effective selective solvents include sufficient difference in density and boiling range between solvent and solute; sufficiently low viscosity of solvent (by comparison with the viscous tar oil used); lack of reaction between solvent and tar components; availability of solvents; and, particularly for large-scale operation, the cost.

By elution of the petroleum-soluble fraction on silica gel, four fractions were collected, the first of which consisted of the total aliphatic hydrocarbons. Removal of phenols and bases, found to be a necessary preliminary to the subdivision of hydrocarbons by solvent extraction (Ahmed and Vahrman, 1966), was achieved by 10% sulfuric acid and caustic soda or methanol. Vahrman, and collaborators (Ahmed and Vahrman, 1966; Spence and Vahrman, 1967) also varied temperatures and ratio of furfural (solvent) to oil for separation of aromatic hydrocarbons from aliphatics. They dissolved neutral oil in two parts by volume of 70–90°C light petroleum, and extracted the solution in five stages with one part by volume of furfural.

The residual material in light petroleum was chromatographed on silica gel to obtain the total aliphatic hydrocarbons. Montan wax (Vcelák, 1959) is an extract obtained with organic solvents from some kinds of sapropelic coals; its chemical components were separated into three main fractions on the basis of their different solubilities in isopropanol (Eglinton and Murphy, 1969); asphalts (fraction insoluble in the boiling solvent); waxes (precipitated from solution on cooling); and resins (soluble in cold isopropanol).

3. *Silica-Gel Column Chromatography*

Column adsorption or elution chromatography separates substances according to their relative polarities. Adsorption of molecules on the surface of the adsorbent is related to interaction of adsorbent and adsorbed molecules. Silica gel is the most frequently used adsorbent for separation of unsaturated and saturated hydrocarbons (e.g., Wollrab *et al.*, 1963; Bartle *et al.*, 1977b). When silica gel is heated up to about 150°C, formation of additional free hydroxyl adsorption sites increases the adsorption energy (Snyder, 1968); maximum surface activity appears to be achieved after about 4 hr at 150°C (Watkins, 1967). Paraffinic and naphthenic hydrocarbons are more weakly adsorbed on silica gel than are the unsaturated components. Adsorbents selective for olefinic double bonds can be prepared by impregnating silica or other adsorbent (e.g., Fluorisil) with silver nitrate, which increases the basicity of an adsorbent molecule and generally increases its adsorption energy. The application of these adsorbents has been reviewed (Jurriens, 1965).

Birkofer and Pauly (1969) separated paraffinic hydrocarbons by eluting an alumina column with cyclohexane, although this generally requires several runs for the separation of aliphatics (Bartle *et al.*, 1977b); alumina has a tendency to hydrolyze waxes (esters) and may also be effective for the chromatography of aromatic hydrocarbons (Drake and Jones, 1977). After a preliminary separation, creosote (Watkins, 1967) and coal tars (Bartle *et al.*, 1977b) can be separated into paraffins and other constituents by dissolving them in petroleum spirit (40–60°C) or *n*-hexane, and then allowing them to percolate through a column of silica gel; the paraffins and naphthenes are preferentially eluted and may be estimated gravimetrically.

In the separation of aromatic, olefinic, and paraffinic fractions from a neutral tar oil on silica gel (100–200 mesh), Bartle *et al.* (1970) eluted with *n*-amyl alcohol. For separation of saturated from aromatic hydrocarbons and other polar material by liquid-solid chromatography, Hirsch *et al.* (1972) and Sawatzky *et al.* (1975) used dual silica–alumina columns.

As mentioned in Section VI,A,3, Allan and Douglas (1974) investigated a series of vitrinites and sporinites isolated and concentrated from a range of bituminous coals by solvent extraction and saponification, followed by oxidation and pyrolysis of the extracted residues. Each maceral was pyrolyzed (in an autoclave under a nitrogen atmosphere at 275 and 375°C for 24 hr) and then dichloromethane extracts of the pyrolyzed residues were chromatographed on activated silica gel. Aliphatic hydrocarbons were eluted by petroleum ether. Silica-gel chromatography applied to a wax fraction from Montan wax (Vcelák, 1959) separates the paraffin fraction from esters, alcohols, and acids. Wollrab *et al.* (1965) used silica gel impregnated with 20% silver nitrate for the separation of saturated from unsaturated hydrocarbons.

4. *Sulfuric Acid and Silica-Gel Treatment*

Concentrated H_2SO_4 was used (Ahmed and Vahrman, 1966) to free fractions in light petroleum from olefins and/or aromatics; the fractions were then purified on silica gel columns by removal of the trace of sulfonated materials from the unreacted parts. Olefinic hydrocarbons were extracted with 98% sulfuric acid from a 20% solution of extracted aliphatic hydrocarbons (Spence and Vahrman, 1966) in cyclohexane. For purification of separated paraffins, a 10% solution in cyclohexane was eluted through silica gel.

5. *Iodine Monochloride Addition Compounds*

Olefins may be separated from paraffins by chromatography of iodine monochloride addition compounds. Spence and Vahrman (1967) separated the total aliphatic hydrocarbons into olefins and paraffins by addition of iodine monochloride (Jurriens, 1965) followed by silica-gel chromatography.

6. *Gel-Permeation Chromatography*

Hendrickson and Moore (1966) first described the elution behavior of aliphatic and aromatic hydrocarbons. Edstrom and Petro (1968) concluded that separation occurs, not as a function of a single parameter, but as a complex function of molecular size, shape, and polarity. These authors first suggested characterization of pitches by GPC. Hsieh *et al.* (1969) applied GPC to the fractionation of some coal carbonization and related liquids, separating the mixtures into aromatic, aliphatic, and tar–acid components; within the aliphatic fractions, paraffinic hydrocarbons were then identified by gas chromatography (GC). Following GPC separation of paraffins from mixtures of oil hydrocarbons (Marin-Mudrovčić *et al.*, 1972), preparative-sized fractions (Baldwin *et al.* 1975) containing some paraffins have been derived from solvent-refined coals, prior to NMR characterization (Coleman *et al.*, 1976; Wooton *et al.*, 1976, 1978).

7. *Separation by Concentrated Sulfuric Acid*

Since they are not attacked by sulfuric acid, solid paraffins (Watkins, 1967) may be separated from the coal-tar product crude anthracene by heating the crude with a large excess of pure sulfuric acid. Maher (1966) also applied this method of extraction for neutral oil in two stages; 80% sulfuric acid enabled the olefins to be extracted, and then 98% sulfuric acid was used to separate the aromatic fraction (see also Section IV,B).

III. SUBFRACTIONATION OF COAL PARAFFINS

Despite some earlier reservations about concomitant fractionation (Eglinton and Hamilton, 1963), considerable success in further separation is being achieved by formation of inclusion compounds with urea and thiourea, and by use of molecular sieves to separate according to molecular shape and size. The procedures for these, which are important preliminaries to fuller characterization and isolation of individual alkanes, are discussed in more detail in the following subsections.

A. Methods of Separation by Urea and Thiourea Adduction

Urea adduction as a separating procedure was developed by Benger and Schlenk (1949) and Schlenk (1951) working on milk; subsequently it has been adapted to many fields for recovery and characterization (Swern, 1955). When urea is crystallized in the presence of straight chain hydrocarbon molecules, a channel adduct is formed in which the urea molecules are loosely wrapped in a helix around the hydrocarbon backbone to form an inclusion complex (Smith, 1950). Since the urea helices of the host lattice can accommodate molecules of up to about 5.3 Å in diameter, but exclude branched chains and cyclics with larger diameters, *n*-alkanes and *n*-alkenes (provided they are more than a minimum chain length) can readily be adducted and thus separated. In quantitative studies of the method, Zimmerschied *et al.* (1950) and Johnson and Jones (1968) found that 92.5 ± 0.5% of the *n*-alkanes can be recovered by urea adduction.

Following preliminary separations, Brooks and Stevens (1964) extracted four low temperature tars; higher molecular weight fractions were then separated by solvent extraction and clathration with urea. Although clathration failed to give a clear-cut separation into types, it had the advantage of reproducibility and thus provided a firm basis for comparison; it also simplified the problem of further characterization. Brooks and Stevens mixed the *n*-hexane-soluble part of the tar with urea activated with 10% water and with dichloromethane in the proportion 1:2:2. After being shaken for 12 hr at 25°C, the adduct was separated by vacuum filtration and washed four times with cold dichloromethane. In the further subdivision by type of the main (nonadductable) subfractions, preliminary trials with thiourea failed to yield a significant amount of clathrate, but elution chromatography was more successful.

As one example among many in the petroleum field, Adler (1974) used urea adduction to separate the paraffinics from light gas oil distilled from a Yugoslavian mixture, as a preliminary to infrared and nuclear magnetic resonance (NMR) spectroscopic analysis.

After separating the paraffinic hydrocarbons derived from coal tar, and making a solution in a 1:4 mixture of benzene and acetone, Spence and Vahrman (1967) added freshly ground urea. *n*-Paraffins recovered by the urea adduction process were filtered from the residue (urea dissolved in hot water). Birkofer and Pauly (1969) first applied the urea adduction technique to coals; cyclohexane extracts of Ste. Fontaine vitrinite and gas-flammekohle (gas coal) were further subdivided as a preliminary to a gas chromatography–mass spectrometry GC/MS study which led to isolation of *n*- and branched-chain (isoprenoid) hydrocarbons. Following preliminary extraction of the coal-derived materials, Bartle *et al.* (1970) also applied urea adduction to the extracted fraction and then chromatographed on a silica-gel column with the addition of a trace of a fluorescent indicator dye. The column was eluted with *n*-amyl alcohol. From their fluorescent effect, the bands were traced down the column, from which the paraffinic part was collected first.

The thiourea complexes, which were developed by Angla (1947, 1949), Fetterly (1950, 1951, 1952, 1957), and Swern (1955), form another important class of inclusion complexes. Although urea and thiourea act in closely analogous ways, there is an important distinction in that thiourea can form large channels of diameter 6.1 Å, which can accommodate many branched-chain and cycloaliphatic compounds (while urea complexes with normal hydrocarbons). Methods of recovery of urea and thiourea from complexes, and the conditions for complex formation, were discussed by Fetterley and Angla. Several methods (Mold *et al.*, 1966; Streibl *et al.*, 1964) involving clathration with urea or thiourea are used with long chain paraffins to determine the concentration of the branched and cyclic paraffins in a mixture containing *n*-paraffins.

In order to achieve a perceptible concentration of branched-chain hydrocarbons, which are frequently present in the original mixture in very small concentration, the adduction process is repeated several times. There are few reports available of the use of thiourea with coal products, but recently Sodhi and Vesely (1977) used it with hydrocarbons from petroleum. Thiourea is also commonly applied (Murphy *et al.*, 1967; Wszolek *et al.*, 1972) to branched–cyclic alkane fractions in geochemical studies.

B. Molecular Sieve Separation Methods

A zeolite, or molecular sieve, is a hydrated crystalline aluminosilicate, with a tetrahedrally coordinated framework, that can effect reversible hydration–dehydration, ion exchange, and the sorption of molecules with a smaller critical diameter than the uniform crystalline pores of the structure. Such a definition excludes other crystalline aluminosilicates such as clays, feldspars, and mica. While zeolites appropriate to branched and cyclic hydrocarbons have been developed, molecular sieves are most widely used for separating low-molecular-weight normal paraffinic hydrocarbons; more polar molecules are absorbed more strongly.

Molecular sieves exert strong physical forces on molecules in their vicinity. Molecules with larger diameters than the molecular sieve pore size will be excluded from the active sites. Molecular sieve type 5A ($\frac{1}{16}$-in. pellets, Linde Air Products Company, New York) can accommodate normal aliphatic hydrocarbons in the channels of these macromolecular inclusion compounds (Fetterly, 1964) (with a permanent crystal structure, in contrast to the variable structures in the organic urea and thiourea adducts), while branched-chain and cyclic hydrocarbons are excluded. The rate of adsorption of a solute from solution depends on the temperature, quantity of other adsorbent, and the nature and concentration of the solute. Since normal paraffins are the only solute molecules adsorbed, the rate should decrease with increasing molecular weight. O'Connor *et al.* (1962) demonstrated this by dissolving 2 g each of a series of paraffins in 100 ml of isooctane refluxing with 40 g of type 5A molecular sieve (previously dehydrated for 6 hr at 250°C at 1- to 5-torr pressure in a vacuum oven). Recovery of adsorbed *n*-paraffins was achieved (O'Connor *et al.*, 1962; Chen and Lucki, 1970) by refluxing them in *n*-pentane, followed by extraction with isooctane, benzene, and acetone. Brunnock (1966) suggested destruction of molecular sieves in hydrofluoric acid for recovery of adsorbed *n*-paraffins. Maher (1966) passed 100 ml of the paraffinic fraction of tar through a column of molecular sieve 5A at 236°C to remove *n*-paraffins; this was followed by GLC for quantitative analysis. Jarolimek *et al.* (1965), Eglinton *et al.* (1964), and Bartle *et al.* (1977b) separated *n*-paraffins from branched and cyclic paraffins by a sieving procedure involving reflux of a solution mixture in isooctane with type 5A molecular sieve. Immediately afterwards, in order to remove any residual branched-chain and cyclic molecules adhering to the surface of the sieve, it is advisable either to reflux the molecular sieve in pure benzene or ether or, preferably, to carry out a Soxhlet extraction of the molecular sieve with isooctane. Eglinton *et al.* (1964) isolated homologous series of normal C_{12}–C_{37} and branched-chain paraffins (C_{16}–C_{35}) by the type 5A molecular sieve extraction procedure.

In view of the close similarity between the procedures for separation and identification of the paraffinic content in coal tar and in petroleum, some consideration is now given to the application of molecular sieve extraction methods to paraffinic-type hydrocarbons in petroleum. A precise knowledge of the hydrocarbon types in paraffin waxes is of interest to the petroleum industry in that it enables physical characteristics of petroleum products to be evaluated. O'Connor *et al.* (1962) designed an apparatus for batchwise adsorption of normal paraffins in petroleum waxes by previously activated type 5A molecular sieve. The sample, dissolved in isooctane, was left in contact with the molecular sieve for about 4 hr. The solvent was evaporated from the residue separated from the molecular sieve in order to recover the branched-chain and cyclic paraffinic mixture. Finally, the molecular sieve was transferred to a glass-stoppered flask, filled with *n*-pentane, and allowed to stand at room temperature for at least 15 days for recovery of the adsorbed *n*-paraffins. Applying 5A molecular sieve to the separation of the normal paraffins in gas oil (endpoint 900°F), Chen and Lucki (1970) found that the method of O'Connor and co-workers led to less satisfactory separation in the presence of mono- and polycyclic aromatics or when the size of the *n*-paraffinic molecule exceeded C_{36}. Using 5A molecular sieve activated by dehydrating at 500°C for 6 hr, these authors either refluxed the oil in isooctane or carried out a high temperature molecular sieve adsorption without solvent. Throughout the adsorption process a nitrogen purge of 10 cm^3/min provided a nonoxidative atmosphere for the gas oil.

For adsorption of the *n*-paraffins in the heavy distillate of a petroleum crude, Brunnock (1966) dewaxed the distillate and then selectively adsorbed the normal paraffins present in the wax by refluxing with benzene. More recently, Sista and Srivastava (1976) developed a method, based on the principle of preferential adsorption of *n*-paraffinic hydrocarbons on a molecular sieve, for the direct estimation to ±1 wt % of *n*-paraffins in C_{12}–C_{32} petroleum distillates. A study of the effect of the temperature on the desorption of the *n*-paraffins indicated a minimum furnace temperature of 260°C (at 0.1 torr) for desorption to occur and lower desorption/temperature gradients for higher molecular weight paraffins.

For separation of normals from isoparaffins, Bieser (1977) describes a four-zone system containing 24 beds of adsorbent (e.g., Linde 5A molecular sieve) heated under pressure; the normals are extracted (together with some aromatics in the feed) by use of two desorbants: C_8 aromatics and *n*-pentane.

IV. CHARACTERIZATION OF ALKANES BY GAS CHROMATOGRAPHY

With its ability to operate from very small quantities, gas-liquid chromatography (GLC) (Douglas, 1969) is the technique responsible for most of the identifications of individual alkanes. The principle is the separation of components of a mixture by partitioning them between a liquid phase and a gas phase; the mixture is injected into a stream of carrier gas flowing through a column containing powder coated with the liquid phase. Temperature-programmed operation is usual for the analysis of the wide range of carbon numbers commonly encountered.

GLC can operate in a purely analytical role for the separation of homologs or in a preparative role (partly as a replacement of distillation) and, as is discussed in Section V,A, forms a particularly effective combination with mass spectrometry. Alkane analysis of petrochemicals has provided a stimulus for development of capillary column technology. Some discussion of columns and supports precedes illustrations of the application of GLC to alkane fractions from coals and analogous materials.

A. Gas Chromatographic Columns and Support Materials

In gas chromatography (GC), the column is the crucial component on which the ultimate performance and separation power of the system depends. Within the packed column, a large, inert surface carrying the stationary phase as a thin, uniform film is provided by the solid support, typically some kind of diatomaceous earth, e.g., Chromosorb, Celite, or C-22 firebrick. In the preparation and coating of the support, particular care must be taken in the choice of the appropriate mesh range of support material and in ensuring that the coating is homogeneous. Production of a homogeneous surface is favored by washing the support material with acid and alkali or by modification by means of adsorbed uniform monolayers (Vidal-Madjar and Guiochon, 1974). When, as sometimes for petroleum hydrocarbons (Sawatzky *et al.*, 1976), an inorganic salt such as lithium chloride is used as a support, its low surface area can be enhanced by coating onto diatomaceous silica or porous silica beads. For preparative GLC, support material of 30–60 mesh is commonly used, with 60–120 mesh for high resolution packed columns.

Factors to be taken into account in the selection of the most appropriate liquid phase for a separation include thermal stability, low vapor pressure, chemical inertness, and solubilizing power.

As an example of the influence of the choice of stationary phase, a 5 ft × 0.125 in. column failed to separate a mixture of *n*-heptadecane and 2,6,10,14-tetramethylpentadecane (pristane), with either silicone rubber (SE-30) or Apiezon L liquid phase, but they were separated on a column containing tetracyanoethylated pentaerythritol (TCEPE) (Eglinton *et al.*, 1966). From experience gained in identification of some petroleum products, Cole (1968) suggested that the average packed column can provide sufficient compositional detail of material only up to carbon number 13. For fractions in the higher boiling range, the additional resolution can be achieved with a capillary column (Fig. 2).

Resolving power in a column is influenced by choice of stationary phase, particle size of support, solubility of the solute, length and diameter of column, flow rate, and temperature.

Pyrolysis gas chromatography is a valuable precolumn technique in which the organic materials are pyrolyzed under carefully controlled conditions; the volatile products are swept onto the top of the column and subsequently separated. Pyrolysis of gas chromatographic eluates, followed by further gas chromatography of the pyrolysates to give unequivocal fingerprints, has been demonstrated by Levy and Paul (1967) for fatty acids and for saturated and unsaturated hydrocarbons.

In practice, gas chromatographic separation of a complex mixture, such as a coal extract, commonly gives rise to overlapping peaks. Tentative evidence for the presence of a single component, provided by observation of a single peak on a chromatogram of moderate resolution, would be corroborated by occurrence of a single peak on a high resolution column (packed or preferably capillary); the evidence would be progressively more conclusive if the single peak persisted on several different stationary phases. Injection of an authenticated sample can be a valuable aid to identification of an individual compound.

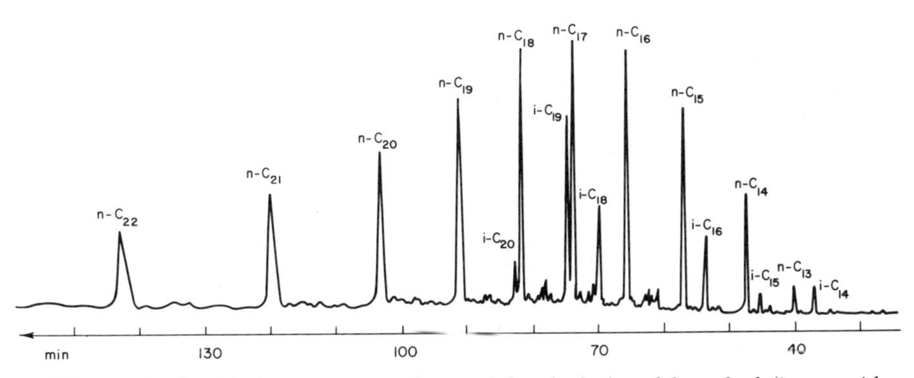

Fig. 2 Gas-liquid chromatogram of normal (marked n) and branched (isoprenoids, marked i) paraffinic hydrocarbons from Rexco coal tar. Recorded with 50 m × 0.3 mm glass capillary column coated with OV 101; nitrogen flow rate 2 ml/min; temperature-programmed at 1.5 K/min from 373 to 473 K; injector temperature 523 K.

B. Gas–Liquid Chromatography of Alkanes

Following qualitative GLC analysis of narrow distillate cuts by Karr and Chang (1959, 1961, 1962), Maher (1966) used Apiezon L as stationary phase to analyze hydrocarbons from a neutral oil produced by low temperature carbonization of bituminous coal (bp 200–335°C) from the Liddel Seam, Foybrook Open Cut, New South Wales. From the relative peak areas, a semiquantitative analysis in terms of normal paraffins, isoparaffins, and naphthenes was achieved for 71 constituents. Previously, GLC analyses had been carried out on a synthetic mixture of 13 hydrocarbons (aromatic, normal, iso-, and cycloparaffins) before and after treatment with sulfuric acid (to extract unsaturateds and, subsequently, aromatics) and 5A molecular sieve (to extract normal paraffins). To obviate structural breakdown of saturated molecules by concentrated sulfuric acid treatment (Section II,B,7), Maher (1966) recommended either very careful extractions (presumably several treatments with small amounts of acid at a low temperature) or the use of selective GC methods for extraction of unsaturated compounds. He also suggested that improved separation of isoprenoids from naphthenes could be effected by dehydrogenation of naphthenic hydrocarbons, e.g., by Rowan's (1961) method. Blytas and Peterson (1967) determined the kerosene range *n*-paraffins by a combination of 5A molecular sieve (see Section IV,D on GSC) and GLC.

Drake and Jones (1977) examined extracts of macerals selected from coals from several English collieries on a 2 m × 6 mm (4% OV 101 on Celite AW, 80–100 mesh) column, N_2 carrier gas 40 ml/min, programmed from 150 to 320°C. Whereas all samples extracted with CS_2 showed appreciable *n*-alkane content (C_{34} down to C_{21} or lower in some cases), extractions with pyridine yielded far fewer *n*-alkanes in most samples. Spence and Vahrman (1967) separated paraffinic hydrocarbons from a low temperature tar (Rexco) and studied their composition by GLC, temperature-programmed between 100 and 300°C, with 15% methyl-type silicone elastomer (SE-30) on Chromosorb-W (60–80 mesh) packed in a 6 ft × 0.25 in. stainless steel column. The peak areas of the *n*-alkane chromatogram indicated a nearly Gaussian distribution of the homologous series from C_{10} up to C_{37}, with a maximum around C_{17}–C_{21}. In addition, the branched-chain and cyclic fraction showed a less well-resolved chromatographic pattern (with boiling range similar to that of the straight chains), from which only pristane and phytane were positively identified.

n-Alkanes and branched–cyclic fractions, previously adduct-separated (Section III,A) from a Saar gas flame coal (80.63% C, 36.6%

volatile matter) and from vitrinite (83% C, 33.1% volatile matter), were analyzed by Birkofer and Pauly (1969) on a 50-ft GC column coated with HI-PAK silicone gum (SE-30) (35 ml He/min carrier gas flow rate), temperature-programmed over 100–300°C at 2°C/min. A Gaussian distribution of *n*-hydrocarbons in the range C_{13}–C_{33} was found for the gas flame coal, with a small proportion of monomethyl branched-chain paraffins from 2-methyltetradecane up to 2-methyltriacontane. While vitrinite showed much the same composition, there was some difference in the abundance of *n*-paraffins in the range C_{11}–C_{17} as well as of isoparaffins in the range C_{22}–C_{30}. For the branched-chain fraction from vitrinite, the mixture was so complex that the column was unable to separate individual compounds. In more detail, four isoprenoids—2,6,10-trimethyltridecane, 2,6,10-trimethylpentadecane, 2,6,10,14-tetramethylpentadecane, and 2,6,10,14-tetramethylhexadecane—were detected in the gas flame coal fraction, with pristane predominant; in addition, four much weaker peaks with higher retention times were attributed to perhydrophenanthrene derivatives (see Section VI,C). Birkofer and Pauly also observed four weak perhydrophenanthrene derivative peaks in the vitrinite branched–cyclic paraffinic fractions; these authors surmised that the one well-resolved intense (pristane) peak was from an impurity maceral rather than from vitrinite.

Pichler *et al.* (1970) analyzed the aliphatic hydrocarbons separated from a low temperature tar on a 100 m × 0.25 mm capillary column at 220°C under isothermal conditions, with polyphenylether as coating material. A Gaussian distribution of *n*-paraffins in the range C_{10}–C_{21} was detected, together with methylparaffins; the following isoprenoids were identified from mass spectra: 2,6,10-trimethylundecane, 2,6,10-trimethyldodecane (farnesane), 2,6,10-trimethyltridecane, 2,6,10-trimethylpentadecane (norpristane), 2,6,10,14-tetramethylpentadecane (pristane), 2,6,10,14-tetramethylpentadecene, and 2,6,10,14-tetramethylhexadecane (phytane). Monomethylparaffins were eluted before the corresponding *n*-paraffins in the sequence 7-, 6-, 5-, 4-, 2-, 3-. The column proved to be effective in separating the strongly branched chains without prior separation of *n*-paraffins.

Comparison of the detailed patterns of capillary gas liquid chromatographic peaks lying between the *n*-paraffin peaks in fractions from different crude oil sources provided Cole (1971) with useful data in the region between C_{13} and C_{14}. The patterns can act as fingerprints for comparison of crude oils, and can aid correlation studies of the origin of crude oil in sedimentary basins (Welte, 1970). Connan (1972) also used this technique for comparing hydrocarbon distributions in studies of natural and artificial diagenesis in sediments, while Grantham (1973) has applied the comparison technique for investigation of diagenetic history (see also Section VI). High resolution gas-liquid chromatographic studies by Allan and Douglas (1974) indicate the probable presence of at least three homologous series in the alkane fraction of sporinite. The vitrinite fraction showed a trend toward a lower average molecular weight with increasing rank.

Peterson and Rodgers (1972) were able to separate (in terms of carbon number) cyclic, iso, and normals in the range C_5–C_{10} from aromatic and other unsaturated compounds by a combination of a 5A, 13X molecular sieve column, and precolumn hydrogenation of unsaturated compounds. Separation was achieved in two stages. In the first stage, the mixture was passed through the molecular sieve columns which retained the paraffins. In the second stage, the 5A molecular sieve column was warmed and the 13X column was cooled; the fraction desorbed in the first column was injected into the 13X molecular sieve column, which was then analyzed by a second temperature-programming cycle. Following pyrolysis of alkanes extracted from Nottinghamshire vitrinite, Allan *et al.* (1975) found that gas chromatographic analysis revealed the effectiveness of pyrolysis. For hydrocarbons in the higher boiling range (300–400°C) from immature source material, pyrolysis can be informative about the composition of the mixture.

For quantitative determination of *n*-paraffins and isoparaffins in mixtures, Denisenko *et al.* (1976) found that use of activated (NaOH-treated) 13X molecular sieve (0.3–0.5 mm grain size) and GC on a 100 × 0.3 cm stainless steel column (temperature increasing at 0.2°/min from 300 to 400°C, He as carrier gas) was simpler and more effective than GLC with a 100-m column and octadecene as fixed phase.

Novotny *et al.* (1972) were able to analyze petroleum waxes for *n*-alkanes up to C_{60} using glass capillary columns coated with highly thermostable poly-*m*-carborane siloxane polymers.

C. The Relation between Retention Characteristics and Structural Properties

1. *Retention Volume and Retention Time*

Two kinds of empirical linear relations have proved useful in the identification of *n*-alkanes in unknown mixtures of paraffins: those between the logarithm of the adjusted retention time R (Groenendijk and Van Kemenade, 1968) or of specific retention volume V_g (Grant, 1971) and the carbon number n_c, and between the retention time and the boiling point (Green *et al.*, 1964; Gouw *et al.*, 1970; Gouw, 1973). Merritt

(1969) found a linear dependence on carbon number for *adjusted* retention volume V' (for temperature-programmed GC conditions) or its logarithm (for isothermal GC conditions) in several homologous series. However, because of pressure drop along the column and the dependence of retention volume on the amount of stationary phase, it would be preferable to use specific retention volume (which is independent of the quantity of stationary phase).

The specific retention volume may be calculated from the equation

$$V_g = \frac{273K}{\sigma_L T_c} \tag{1}$$

where K is the partition coefficient, σ_L the density of the liquid phase, and T_c the column Kelvin temperature.

2. *The Kováts Retention Index in Gas Chromatography*

Since most published GC retention data are relative, difficulties may arise in relating these to one another owing to different choices of standards. Such difficulties may be obviated by means of the retention index scheme proposed by Kováts (1958). In this, the retention of a substance is expressed either as an absolute value or by comparison with a single standard which expresses the retention behavior of the substance in question on a uniform scale determined by a series of closely related standard compounds.

In the original definition with the normal paraffins as fixed points, the retention index I of an unknown substance x in the mixture, under isothermal GC conditions, was calculated from the equation

$$I = 100N + 100n \left(\frac{\log R_x - \log R_N}{\log R_{N+n} - \log R_N} \right) \tag{2}$$

Here R_x is the adjusted retention time of the unknown compound, and R_N and R_{N+n} are the retention times of the two n-alkanes of carbon number N and $N + n$ with retentions immediately below and above that of x (retention times R may, of course, be replaced throughout by retention volumes). Earlier Habgood and Harris (1960) had pointed out that, with linear temperature-programming, the elution temperature of the normal paraffins is approximately (but not exactly) proportional to chain length. Van den Dool and Kratz (1963) showed that in temperature-programmed GC the logarithms of the net retention volumes may be replaced by the retention temperatures θ in calculations of retention indices:

$$I \approx 100N + 100n \left(\frac{\theta_n - \theta_N}{\theta_{N+n} - \theta_N} \right) \tag{3}$$

Ettre (1964) discussed the dependence of the retention indices on the temperature and on the chemical nature of the stationary phase. Reviewing the accuracy of retention indices, which provide a valuable means of relating retention data to the physical and structural properties of the compounds separated, Guardino *et al.* (1976) recommended calculating I from the equation

$$\log R'_x = aI + b \tag{4}$$

where R'_x is the corrected retention time, and the parameters a and b have been determined by least-squares adjustment. Following hydrogenation of sesquiterpenes, Anderson and Falcone (1969) calculated Kováts indices for several sesquiterpanes.

Shlyakhov *et al.* (1975) compared the Kováts retention indices of isoprenoids and monomethylalkanes determined on three stationary phases—polar diethylene-glycol-adipate ester cross-linked with pentaerythritol (DEGA-PE), and nonpolar Apiezon L and methylsilicone rubber SE-30—in copper and stainless steel capillary columns typically 70–90 m long. For both classes of hydrocarbon, I increases slowly with temperature for the nonpolar stationary phases; for the polar DEGA-PE, I decreased over the temperature range 120–190°C, although it was hinted that this might be nonlinear over a wider range. At lower temperature, retention indices derived from the three phases become less dissimilar, i.e., the phases become less selective. It was suggested that the differing temperature dependences of I for different stationary phases might be of some help in the identification of individual compounds. Retention indices for successive members of the three homologous isoprenoid series (2,6-dimethyl-, 2,6,10-trimethyl-, and 2,6,10,14-tetramethylalkanes) do not necessarily differ by 100 units. For $C_{14/15}$, the retention index difference is greater and, as found in other hydrocarbon classes (Sojak *et al.*, 1973), it is subsequently smaller and asymptotically increases toward 100; thus extrapolated values can be obtained for the indices.

For the prediction from known retention indices of the indices for structurally related polymethylalkanes relevant to organic geochemistry, Shlyakhov *et al.* (1975) used two techniques. In the first, I is synthesized from additive components corresponding to the kinds of carbon atom (primary, secondary, and tertiary, but further characterized by the carbons bound to the neighbors), with due allowance made for the dependence of the isoprenoid link increment on molecular size. Thus, equations relating two pairs of branched hydrocarbons differing by one isoprenoid unit may, for example, enable I for one hydrocarbon to be deduced from the known values of I for the other three:

$$
\begin{aligned}
IC_n(2,6,10) - IC_{n-5}(2,6) &= IC_{n+5}(2,6,10,14) - IC_n(2,6,10) \\
&= IC_{n+10}(2,6,10,14,18) - IC_{n+5}(2,6,10,14) \\
&= \cdots \\
&= \text{increment for isoprenoid unit } \text{—C—C(C)—C—C}
\end{aligned}
$$

The second method followed the empirical approach used by Wiener (1947) to evaluate boiling points of paraffins on the basis of the number, kind, and arrangement of the carbon atoms in the molecule; for structural variables, Wiener used path number W (allowing for compactness of the molecule) and a polarity number (pairs of carbon atoms that are separated by three C–C bonds). Here, the retention index difference between an n-alkane and an isoprenoid of the same carbon number is given (for Apiezon L at 100°C, with no polarity number) by

$$\delta I^{n}_{\text{iso}} = 4.638(100\,\Delta W_{\text{i}}/n^2 + 4.57k - 0.3l) - 39.4 \tag{5}$$

where k, l are the numbers of methyl groups at positions 2,3; n is the number of carbon atoms in the molecule; and $\Delta W_{\text{i}} = W_n - W_{\text{iso}}$, where W is the sum of products of the number of C atoms on one side of each C–C bond multiplied by those on the other side. For either method, of course, separate calculations are required for each combination of stationary phase and separation temperature. Agreement is good for both methods, but, as might be expected, divergence between experimental and calculated values increases somewhat at higher molecular weights.

Shlyakhov *et al.* (1975) used retention indices calculated thus to show the presence of 3,7,11-trimethylalkanes and higher (C_{21}–C_{25}) isoprenoids in alkane–naphthenic fractions of Romashkino petroleum. Although this work demonstrates that the presence of such compounds in a complex multicomponent mixture such as a coal or petroleum extract can be established purely from isothermal GC data and calculated logarithmic retention indices, presumably a more realistic application would be in conjunction with mass spectrometry.

Bellas (1975) has described a FORTRAN computer program for calculating both the Kováts logarithmic index I and the linear retention index l (Vigdergauz and Martynov, 1971) when some paraffin reference compounds are missing. For an unknown hydrocarbon with retention time R between two standard ones (subscripts 1 and 2), Bellas deduced the relation $I - l = d - (\exp(d/B) - 1)/(\exp(l/B) - 1)$ where $d = I - I_1$, and B is the coefficient in the rearranged Kováts index equation $I = A + B \ln R$.

D. Gas–Solid Chromatography

In the analysis of higher molecular weight components, gas–liquid chromatography at higher temperatures suffers from baseline drift. This and other problems in the analysis and identification of geometrical isomers, can often be overcome by gas–solid chromatography (GSC) with its greater specificity (Vidal-Madjar and Guiochon, 1974). In GSC, solutes are separated by differential adsorption (rather than by partition) and the stationary phase is an active solid (instead of a liquid). Here, however, tailing of peaks can be troublesome, especially for higher molecular weight materials. Tailing presumably arises from nonlinearity in the adsorption isotherm because of the long average desorption time of the adsorbate from the active site (Giddings, 1963). The problem is being overcome as more homogeneous and specific molecular adsorbents of high chemical purity are synthesized.

Scott and Rowell (1960) used alumina deactivated with 40% sodium hydroxide for separation of high-molecular-weight hydrocarbons from C_{15} to C_{36}. Symmetrically eluted hydrocarbon peaks were obtained by Scott and Phillips (1965) on alumina modified with a variety of inorganic salts. Barrall and Baumann (1964) used a 10-ft-long gas-liquid chromatographic column in series with a 15-ft molecular-sieve 5A gas–solid chromatographic column to separate normal from branched-chain hydrocarbons in the range C_7–C_{20}. Alkanes and cycloalkanes of similar carbon number can be distinguished by GSC from the difference in their retention volumes (Snyder and Fett, 1965; Scott, 1966). Blytas and Peterson (1967) have extended to the kerosine range a subtractive scheme of columns whereby molecular sieve 5A was employed to extract *n*-paraffins from gasoline. McTaggart *et al.* (1971) discussed GSC methods based on the use of 13X, 10X, and 5A molecular sieves for rapid and accurate determination of *n*-paraffins within a narrow boiling range. More recently, Sawatzky *et al.* (1976) analyzed petroleum compounds and distillates on 50% lithium chloride-coated acid-washed diatomaceous silica (60–80 mesh) or porous silica (Porasil 100–80 mesh); a mixture of standard *n*-alkanes (C_{12}–C_{36}) was injected. Tailing was almost eliminated by the improved quality of packing resulting from heating at 700°C; column efficiency for the separation of *n*-alkanes was about 10^3 theoretical plates.

V. SPECTROSCOPIC EXAMINATION OF COAL PARAFFINS

A. Mass Spectroscopy

Mass spectrometry, for many years widely used for examination of petroleum fractions, has for the past decade been one of the major analytical techniques in geochemical research (Burlingame and Schnoes, 1969). Applications to aromatic rings in coal products are discussed in Chapter 22, Section II,E. Mass spectrometry has a major advantage over the techniques discussed here in its ability to enable the integer molecular weight of the compound to be determined with a low resolution instrument, and to establish the molecular formula with a high resolution instrument.

1. *Fragmentation Patterns*

Molecular ion abundance can be a most useful criterion in assessing the type of compound under investigation in that the fraction of ions so collected will be roughly proportional to the activation energy for the most facile fragmentation. Some compounds may not give molecular ions because either (a) the activation energy for decomposition is very low or zero (so that no M^+ ions survive to reach the collector) or (b) the sample decomposes thermally prior to ionization. As an aid to recognition, the loss of 14 mass units is a strong indication for the presence of a homolog differing in formula by a CH_2 unit. Because of the large volumes of data involved, structure elucidation of an unknown compound is often now accomplished with the aid of a computer on-line to a high resolution mass spectrometer.

In branched-chain hydrocarbons, formation of tertiary carbonium ions is favored in primary fragmentation, so that often no molecular ion (M^+) peak occurs. Increasing number of double bonds favors the formation of more intense molecular ion peaks. Thus, in normal paraffinic hydrocarbons, the major ions of interest occur at the parent (molecular weight) mass and at lower mass/charge (m/e) values resulting from cleavage of successive C–C bonds. In methyl-substituted paraffins, the spectra vary widely according to the position of the methyl group: In some instances, a small parent peak is observed, while in others no molecular ion peak is formed at all.

Cycloalkanes give spectra very similar to those of linear alkanes except that the molecular ions are more intense. The spectra for cycloalkanes show an appreciable parent peak but dissociation of the five- and six-membered rings follows different paths. The 1-cyclopentyl types give a strong peak at M-28, and a smaller peak at M-70 corresponding to

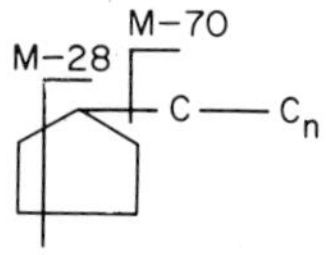

the loss of the ring. Large peaks at M-84 and at *m/e* 83 as a result of cleavage α to the ring are indicative of 1-cyclohexylalkane and methyl-substituted cyclohexylalkanes, respectively.

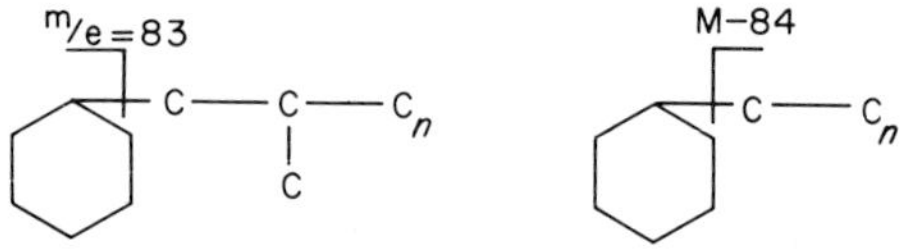

Fragmentation of isoalkanes proceeds with some loss of methyl radical (yielding an ion with mass number M-15) and considerable loss of isopropyl (ion at *m/e* 43). The ante-isoalkanes can be identified mass spectrometrically by a very pronounced loss of an ethyl radical, so that a high intensity fragment at M-29 is observed.

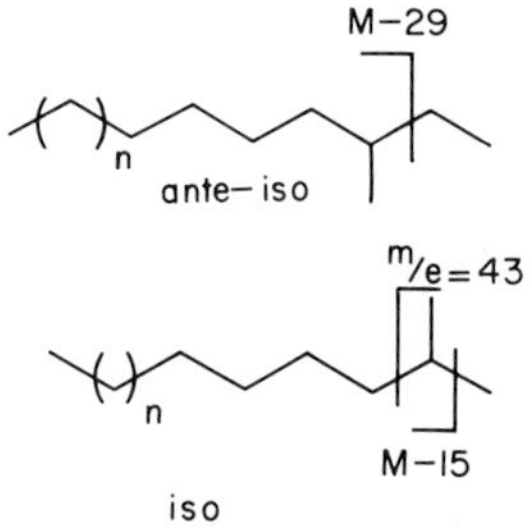

In normal paraffins, olefins, cyclics, and aromatic hydrocarbons, the parent ion intensity is higher than in long and highly branched aliphatic side chains. Thus, the following hold true:

(i) The relative height of the parent peak is greater for straight chain compounds and decreases as the degree of branching increases.
(ii) Loss of a fragment containing a single carbon atom is unlikely unless the compound contains a methyl side chain.
(iii) Fragmentation is most likely at highly branched-chain atoms.
(iv) The energies of bonds and the relative stabilities of positive ions and neutral fragments result in the following sequence in the fragmentation patterns in alkanes and cycloalkanes:

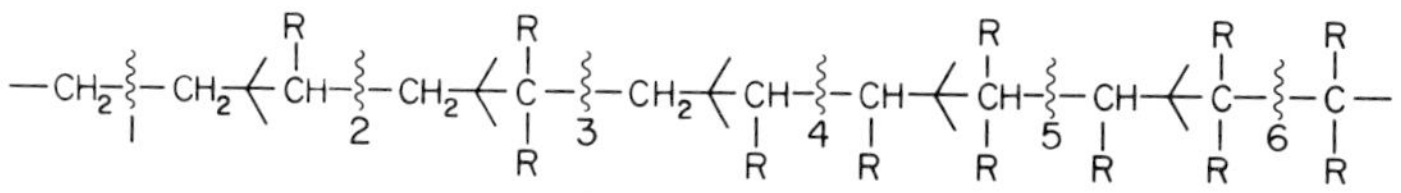

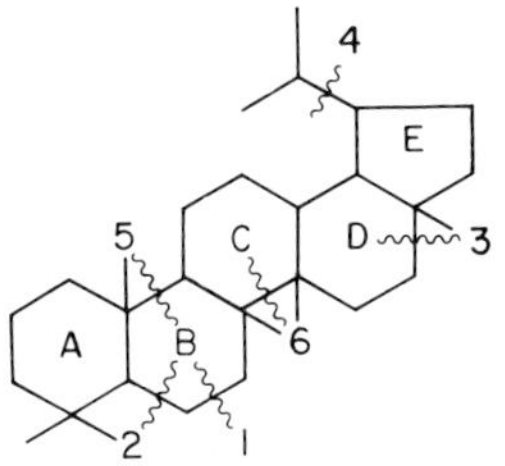

Fragmentation patterns in cyclic structures usually occur at more than one C–C bond. The very intense peak characteristic of steranes at *m/e* 217 allows ready differentiation between steranes and triterpanes; peaks at *m/e* 149 or 151 are also typical. Terpane-type cyclic hydrocarbons generally give rise to a peak at *m/e* 191; however, exceptions are two tetracyclic terpane isomers, $C_{23}H_{40}$, which give peaks at *m/e* 177 and 203 (Gallegos, 1970), and two pentacyclic terpanes, $C_{30}H_{52}$ and $C_{31}H_{54}$, with *m/e* 205 fragments (Whitehead, 1974).

2. *Combination of Mass Spectrometry with GLC*

Independently, gas chromatography and mass spectrometric instrumentation have reached a high degree of sophistication; their combination in a single instrument is now common. This versatile arrangement has considerable speed and sensitivity (Hedfjall and Ryhage, 1975), which are especially required for the low concentration constituents of paraffin fractions.

A complete mass spectrometric analysis can be obtained in essentially the time required for an ordinary GLC run; with computer data processing, each component can be analyzed and the homogeneity of a single GLC peak can be checked by obtaining successive mass spectral scans. Use of a capillary column can overcome difficulties due to a contribution from other hydrocarbon material, e.g., contamination of an isoprenoid alkane with its isoalkane isomer (Bartle *et al.*, 1977b).

Mass spectra (70 eV ionizing voltage) were recorded by Bartle *et al.* (1977a) on an AEI MS902 spectrometer for each compound that emerged from the GLC column (Pye, with a 1.5-m glass column packed with 20% Apiezon L on Celite, temperature-programmed between 110 and 250°C at 6°C/min, N_2 as carrier gas at 40 ml/min flow rate) for the urea-adductable and -nonadductable fractions of a low temperature carbonization coal tar (see Section III for separation and Section V,C for NMR spectroscopy). Combined GLC and mass spectrometry of the urea-adductable fraction showed *n*-alkanes in the range of *n*-C_{11} to *n*-C_{22}, while the nonadductable fraction showed the presence of six acyclic isoprenoid hydrocarbons—$C_{14}H_{30}$, $C_{15}H_{32}$, $C_{16}H_{34}$, $C_{18}H_{38}$, $C_{19}H_{40}$, $C_{20}H_{42}$—as well as normal alkanes from C_{11} to C_{17}. Molecular ion peaks were detected, characteristic of each *n*-alkane.

With normal paraffins, the first fragmentation (CH_3 with *m/e* 15) has only a limited tendency to occur, but the tendency for fragmentation at successive C–C and C–H bonds increases with increasing number of CH_2 groups. The Gaussian distribution of *n*-alkanes in the mixture was also evident from the molecular ion peaks of the mass spectra (Bartle *et al.*, 1977a).

Since acyclic isoprenoids are characterized by the orderly repetition of isoprene units, their mass spectra would be expected to exhibit orderly modes of fragmentation. This was observed in a series of characteristic C_nH_{2n+1} fragment ions which recur at five-carbon increments (Bendoraitis *et al.*, 1963). The regularly condensed isoprenoids usually exhibit prominent $C_8H_{17}^+$ (*m/e* 113) and $C_{13}H_{27}^+$ (*m/e* 183) ions in their mass spectra. Invariably, the mass spectra of fractions which have shown either or both of these as prominent ions have been found to contain acyclic isoprenoid hydrocarbons.

Field desorption mass spectra of some CS_2 extracts of British coals (Drake *et al.*, 1977) provide evidence of *n*-alkanes of surprisingly high molecular weight.

B. ^1H Nuclear Magnetic Resonance Spectroscopy

Detailed high resolution ^1H nuclear magnetic resonance (NMR) spectroscopic analyses have been carried out on solutions of many single compounds from among the many aromatic compounds associated with coal and related materials; see, for example, Bartle and Jones (1972) and Chapter 23, Section IV,B. For these analyses, detailed consideration is required for the relation between chemical shifts and spin–spin coupling constants. For the paraffin hydrocarbons, however, and particularly for mixtures of them, with many closely similar chemical shifts, the emphasis is much more on peak *areas, patterns,* and *profiles* than on shift–coupling relations; indeed, very little use has been made of spin–spin coupling constants in this area.

Two instrumental developments in the past decade have markedly favored the application of NMR to mixtures of closely similar compounds. One is the wider availability of high field superconducting magnets. While the gains for mixtures of paraffins have not yet proved as great as in the aromatic region for mixtures of coal products (see Chapter 23, Section V,A), at 220 MHz ^1H methyl resonances are clear of most methylene and methine protons. [See Fig. 4 in Oelert and Köser (1977), which gives a useful review of recent European ^1H- and ^{13}C-NMR spectroscopy of petroleum and coal.] The second development is the application of Fourier transform techniques (see Chapter 23, Section III,D); the major influence of this has been to ^{13}C-NMR (Section V,C).

Before these advances, Bartz and Chamberlain (1964) had carried out an extensive survey of the reliability of 60-MHz ^1H-NMR for determining the structure of paraffinic hydrocarbons. In normal paraffins, the $—CH_2$ groups appear with a sharp signal at 1.24 ppm and $—CH_3$ groups at 0.84 ppm; the multiplicity of the $—CH_2$ signal decreases with increasing number n of CH_2 groups. Bartz and Chamberlain found that the multiplicity reduces to unity at $n > 5$; the CH_3 pattern also changed with n. Approximate ranges of ^1H chemical shifts in normal and branched-chain paraffins are CH_3, 0.70–1.12 ppm; CH_2, 1.18–1.28 ppm; and $—CH$, 1.50–1.80 ppm. At 220 MHz the enhanced splitting of the $—CH_3$ band can be of diagnostic value (Bartle and Jones, 1969). Bartle *et al.* (1970) combined 60-MHz high resolution ^1H-NMR with other spectroscopic and chromatographic data to derive distributions of hydrogen types and average structural parameters. Thus, for the urea-adductable fraction of a fluidized-bed tar (see Section III,A,1) thought to be branched and/or cyclic paraffins, these authors deduced a high methyl content, in addition to methylene groups. Differences were evident between the methyl group resonances in the straight chain and those in the branched–cyclic fraction in this tar fraction, and Bartle *et al.* (1970) concluded that it was composed of branched-chain rather than cyclic derivatives. From the similarity in area ratios of NMR peaks $CH_3 : CH_2 + CH$, for high- and low-molecular-weight continuous vertical retort and coke oven coal tars, the nature of the paraffins of low temperature tars appears to be fairly independent of carbonization temperature (Bartle, 1972).

Combination of ^1H-NMR study of paraffinic fractions from Australian low temperature tar (Brooks and Stevens, 1964; CSIRO, 1964) with ultraviolet spectroscopy and urea adduction showed them to be equal mixtures of both straight chain aliphatic compounds of structure

$$—C—(\overset{|}{\underset{|}{C}})_n—C— \qquad (n = 15\text{–}20)$$

and analogous compounds with an average of four branches per molecule; Maher (1963) and Waddington (1961), on the other hand, had reported only a small degree of branching.

Recently, Myers *et al.* (1975) developed a 100-MHz ^{1}H-NMR method for quantitative determination of hydrocarbon types and of H/C ratios in gasolines which represents a considerable improvement in speed and precision over the fluorescent indicator analysis (FIA) and combustion methods, respectively. It was suggested that an automated device would be feasible, based on a 100-MHz continuous-wave spectrometer and without sample spinning, in which a gasoline could be analyzed in 2 min.

C. ^{13}C Nuclear Magnetic Resonance Spectroscopy

As a result of instrumental advances in the past few years, particularly in pulse techniques leading to Fourier transform spectrometers, signal-to-noise problems with natural abundance ^{13}C-NMR may be said to have been overcome so that the technique is now widely used for the study of paraffinic hydrocarbons in petroleums and is being increasingly applied to sediments and coals. While quantitative measurements are considerably more difficult than with ^{1}H-NMR, ^{13}C-NMR (Wehrli and Wirthlin, 1976) has several intrinsic advantages over ^{1}H-NMR for qualitative structural analysis, and the two nuclear probes form a powerful combination.

When ^{13}C-NMR spectra are recorded by broad-band noise-modulated proton decoupling, all heteronuclear ^{13}C–^{1}H couplings are removed so that each chemically nonequivalent carbon in the molecule gives rise to a separate absorption line; for natural abundance samples ^{13}C–^{13}C spin–spin couplings are usually absent. Not only are the ^{13}C spectra thus often simpler and easier to analyze than corresponding ^{1}H spectra, but they are spread over a much wider range of chemical shifts (~200 ppm).

Lindeman and Adams (1971) recorded the ^{13}C spectrum of the isoprenoid pristane, $C_{19}H_{40}$, and improved the Grant and Paul (1964) shift prediction parameters so as to embrace isomers of the C_5–C_9 paraffins. The acyclic paraffins' ^{13}C chemical shift range is shown in Table III (Lindeman and Adams, 1971). For linear and branched open-chain alkanes the chemical shift $\delta_c(k)$ of the kth carbon may be calculated from the equation

$$\delta_c(k) = A_n + \sum_{m=0}^{2} N_m{}^{\alpha}\alpha_{nm} + N^{\gamma}\gamma_n + N^{\delta}\delta_n \tag{6}$$

where n is the number of hydrogens at carbon k,

$$\overset{k}{\downarrow}\;\; \overset{\alpha}{} \qquad \beta \quad \gamma \quad \delta$$
$$>\!\!-\text{CH}_n\text{—}\text{CH}_m\text{—C—C—C—}$$

TABLE III *Ranges of* ^{13}*C-NMR Chemical Shifts, Classified on the Basis of* α-, β-, *and* γ-*Carbon Atoms*[a]

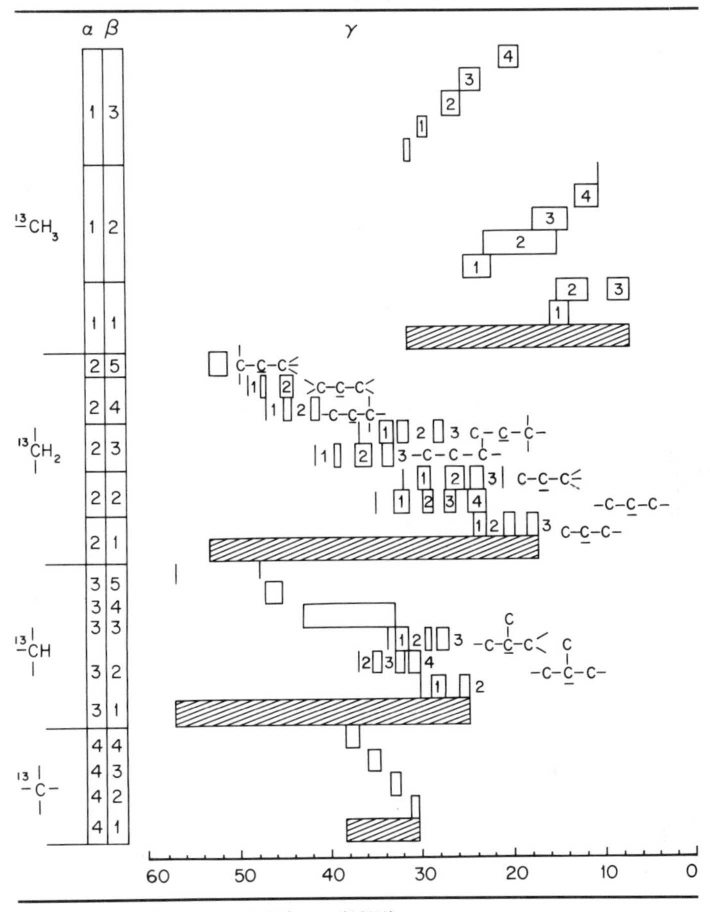

[a] After Lindeman and Adams (1971).

m the number of hydrogens at the α-carbon; $N_m{}^{\alpha}$ the number of CH_m groups at the α position ($m = 0, 1, 2$; α-CH_3 groups are ignored); N^{γ} the number of γ-carbons; N^{δ} the number of δ-carbons; and A_n, α_{nm}, γ_n, and δ_n are tabulated parameters.

Subsequently Carman *et al.* (1973) measured the ^{13}C resonances in a further series of highly branched alkanes up to C_{25}, containing regularly repeating methylene sequences. Magnetic nonequivalence was observed between isopropyl methyl carbons even when the molecule did not contain an asymmetric carbon atom.

The availability of pulsed Fourier transform spectrometers enabled natural abundance proton-decoupled ^{13}C spectroscopy to be applied to mixtures of aliphatic hydrocarbons. In the ^{13}C-NMR spectrum of a C_{15}–C_{20} *n*-paraffin mixture isolated from petroleum, Clutter *et al.* (1972) detected four distinct methylene carbon types. Not only is the greater magnitude of the ^{13}C chemical shifts advantageous in structural determinations, but, from other experiments, the ^{13}C–^{1}H coupling constants can be used to help make unequivocal assignments of particular carbon types.

The average carbon numbers can be determined from the integrated ^{13}C-NMR spectrum of a mixture of normal paraffins if one uses the chemical shifts given by Grant and Paul (1964). Clutter *et al.* (1972) emphasized that the ^{13}C spectrum can be much more informative than the ^{1}H spectrum about the composition of a mixture of isoparaffins and cycloparaffins from crude oil.

In an investigation of terpanes and steranes in shales and sediments, Balogh *et al.* (1973) measured ^{13}C spectra from 18.4 mg of a preparative GLC thiourea-nonadducted fraction extracted from an oil shale bitumen (Green River formation). They were able to establish (Balogh *et al.*, 1972, 1973) unequivocally that the pentacyclic triterpane was (as suggested from GLC and MS) 17-α(H)-hopane rather than any other of 25 possible stereoisomers. By comparison with spectra of pristane and of a pristane–phytol mixture, spin–lattice relaxation time measurements, and use of the single-frequency off-resonance proton-decoupling technique, Goodman *et al.* (1973) have assigned almost all the ^{13}C resonances in phytol (see Section VI,B).

Not only does ^{13}C-NMR enable isoprenoids to be differentiated from *n*-alkanes and from singly branched compounds, but it has also enabled Bartle *et al.* (1977a) to identify [by comparing observed chemical shifts with those calculated from the Lindeman and Adams (1971) additivity scheme] six individual isoprenoids in mixtures separated from low temperature carbonization coal tar oil and petroleum crude extracts.

D. Infrared Spectroscopy

In the study of coals, most of the infrared (ir) spectroscopic measurements (see Chapter 22, Section II,A) employed, for example, by Oelert and collaborators (e.g., Oelert and Hemmer, 1970), have been more concerned with coal constituents other than the paraffins, which form the subject of this section.

For the paraffins extracted from Rexco tar (see Section III), Spence and Vahrman (1967) recorded the ir spectrum of each fraction over the range 3500–700 cm^{-1}. Moderately intense bands at 770 cm^{-1} were shown by

Cummins and Robinson (1964) and by Bendoraitis *et al.* (1963) to be characteristic of terminal ethyl groups. Cummins and Robinson also commented on the appearance of two bands at 1380 and 1370 cm^{-1} as significant evidence for isoprenoid hydrocarbons. After Bellamy (1968) had demonstrated the estimation of chain length and of degree of branching from intensity measurements of the 720- to 740-cm^{-1} bands, the procedure was applied to petroleum fractions (Hughes and Martin, 1957) and to coal-derived materials (Andrzejak and Ignasiak, 1967, 1968; Wachowska *et al.*, 1972; Ignasiak *et al.*, 1975). Bartle *et al.* (1977b) recorded infrared spectra on KBr disks, at frequencies in the range 300–700 cm^{-1}, for coal tar extracts (Roomheat urea-nonadductable and -adductable, and branched–cyclic Rexco paraffinic fractions; see Section III); bands characteristic of *n*-alkanes and branched–cyclics (Table IV) were detected.

In Table IV, it should be noted that the relative intensities of the 1370-

TABLE IV *Frequency of Infrared Vibrations Detected in Coal Tar Extracts*

Frequency (cm^{-1})			
Normal paraffins	Branched-chain and cyclic paraffins	Origin of band	Remarks
720	720	$-CH_2$ group in a chain	$>CH_2$ rocking
730–735			
770[a]	—	Terminal ethyl group	$-C_2H_5$
885	—	Normal alkyl chain	
—	965	Naphthenic ring	
—	1150–1170	$-CH(CH_3)_2$	Skeletal vibration of isopropyl group
1300	1305	$-(CH_2)_2$	
1340–1355	1355	Normal alkyl chain	
—	1370–1380[b]	$-CH(CH_3)_2$	C–H deformation vibration modes of isopropyl groups
1370–1380	—	$-\overset{\vert}{\underset{\vert}{C}}-$ or $-CH_3$	CH_3 symmetrical deformation
1465	1450–1470	$-(CH_2)$ and $-CH_3$	C–H deformation vibration
2830–2970	2850–2990	$-CH_2-$, $-CH_3$, and $-CH-$	$\rangle$C–H stretching vibration

[a] Not detected by Spence and Vahrman (1967).
[b] Peaks of equal intensity detected by Spence and Vahrman.

and 1380-cm^{-1} bands provide an index of the number of terminal isopropyl groups: two isopropyl groups in a molecule give approximately equal band intensities, whereas with a single group the intensity at 1370 cm^{-1} is reduced. Further, the strong band at 735 cm^{-1}, attributed to the $>CH_2$ rocking vibration of the $-(CH_2)_3-$ chain, is useful for the characterization of isoprenoid structure (Bendoraitis *et al.*, 1963); it is shifted progressively to a lower frequency limit of 720 cm^{-1} as the number of methylene groups in a chain segment increases.

Investigating analytical applications of spectroscopy in the near infrared to mixtures of hydrocarbons, Bernhard and Berthold (1975) charted the frequency ranges of the low-molecular-weight *n*-alkanes, isoalkanes, and cycloalkanes in the region 810–840 cm^{-1}. For a brief discussion of other spectroscopic techniques, see Chapter 22.

VI. THE DISTRIBUTION OF PARAFFINS IN COAL AND COAL PRODUCTS

The paraffinic hydrocarbons play such a vital role in the determination of the origins of coals and other sediments as to justify much detailed analytical investigation. This section discusses the results obtained from the separation, isolation, and identification methods described in the preceding sections. It may be useful to note that, in organic geochemistry, the term *organic diagenesis* usually refers to reactions up to 50–100°C; *thermal alteration* (cracking) occurs at higher temperatures, while the term *organic metamorphism* tends to be applied (Hunt, 1974) to reactions at 200°C or more, involving conversion to carbon, as for coals.

A. *n*-Alkanes

1. *n-Alkanes in Coals*

n-Alkanes are the most plentiful paraffinic constituents of coal and coal-derived materials, usually amounting to more than 60% of the total paraffins. The distributions observed are generally in the range C_{10}–C_{36}, but higher members of the series (up to C_{60}) have been detected in small quantities (Drake and Jones, 1977). The lower members may be lost in the extraction procedure. Thus C_1–C_7 *n*-alkanes identified in the gas from coal steamed at only 300°C were apparently originally present in the coal since only minute quantities of hydrogen were liberated simultaneously (Palmer and Vahrman, 1972).

2. *n-Alkanes in Coal Tars*

McNeil (1961) reported the analyses for paraffins by silica gel chromatography of distillate oils from 61 tars. These varied between ~3% of the dry tars for continuous vertical retort tars and <1% for coke oven tars. As part of a characterization index scheme, equations were derived relating the percentage of paraffins in the oils to the specific gravity of the tar, the phenols content of the 0–250°C distillate oil, and the benzene-insoluble content of the dry tar.

A number of more detailed analyses for individual *n*-alkanes have been made, especially for low temperature tars. Thus, Spence and Vahrman (1965) reported a range between C_{10} and C_{37} with a maximum at C_{17}–C_{21}; very similar results were reported by Bartle *et al.* (1970) for a tar produced by carbonization in a fluidized bed, where a Gaussian distribution (Fig. 3b) centered at C_{19} was noted. Vahrman (1970) has also

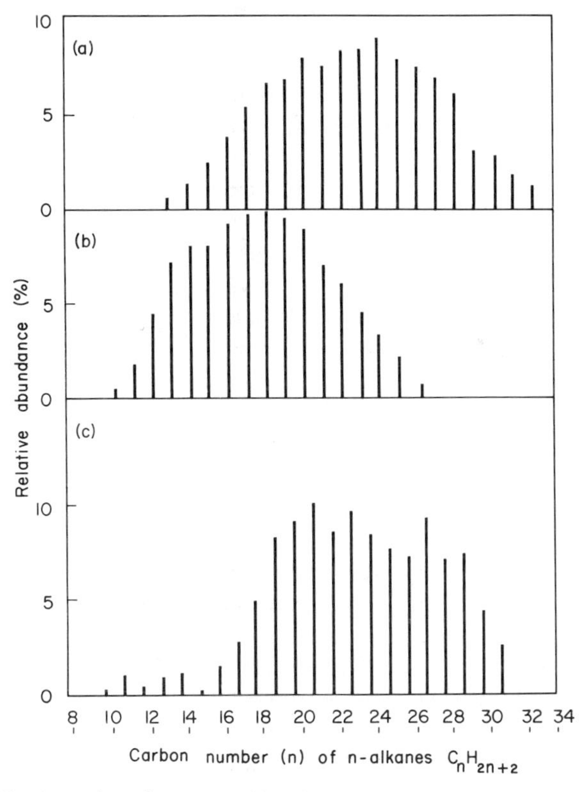

Fig. 3 Distribution of *n*-alkanes in (a) solvent extract of coal (Birkofer and Pauly, 1969); (b) neutral oil (Bartle *et al.*, 1970); and (c) gas extract of coal (Bartle *et al.*, 1975).

studied the composition of *n*-alkane mixtures from a variety of experimental tars.

3. *n-Alkanes in Coal Extracts and Low Temperature Pyrolysis Products*

Alkane extracts of immature or low rank terrestrial sediments generally exhibit a predominance of straight chains up to C_{31} (Bray and Evans, 1965), and a large proportion of the *n*-alkanes present in the tars may be obtained from the original coal by a long period of Soxhlet extraction (Spence and Vahrman, 1970) or by short autoclave (Rahman and Vahrman, 1971) or supercritical gas extraction (<400°C). Detailed studies have been made of the variation of *n*-alkane distributions in extracts (Brooks and Smith, 1969; Leythaeuser and Welte, 1969; Teichmüller, 1974) with coal rank. For 18 Australian and New Zealand coals, Brooks and Smith (1969) related differences in composition of the *n*-alkanes fraction in the C_{22}–C_{33} region to metamorphic changes. Leythaeuser and Welte (1969) studied the *n*-alkanes from 55 coals from the Saar district, ranging from subbituminous to medium volatile bituminous. As coalification increases up to rank of 30% volatiles, the concentration of high-molecular-weight *n*-alkanes increases relative to that of the lower homologs; with further coalification, the trend is reversed. Analytical errors and the variations in petrographic composition from one coal seam to another may also be partly responsible for the differing ranges of alkane distributions reported.

Marked preferences for odd-carbon-number *n*-alkanes (*vide infra*) were observed for many extracts in these studies, and in those by Birkofer and Pauly (1969) for a chloroform extract (Fig. 3a). Vahrman (1970) noted that, in his experiments, the preference for odd carbon numbers was maintained only in the small amounts of *n*-alkanes first extracted or liberated by heat; a smoother distribution was obtained for the remainder. Bartle *et al.* (1975) similarly extracted quite large quantities (i.e., ~0.4% of the coal) of *n*-alkanes with a supercritical gas at 350°C, but they reported an odd-over-even predominance superposed on a pronounced bimodal distribution (Fig. 3c), in contrast to the *n*-alkane distribution in a low temperature tar derived from the same coal (Fig. 3b) (Bartle *et al.*, 1970); thus, no alkenes were produced. Those alkene compounds present in the tar are thought to arise by degradation of the *n*-alkanes by demethanation and deethanation (Vahrman, 1970). Bacterial degradation, which could be responsible for the disappearance of *n*-alkanes from some paraffinic fractions extracted from oil sources, has been demonstrated by Rogers *et al.* (1972) and by Wehner (1974). Thermolysis of *n*-octacosane has been shown to lead to a complex mixture of alkenes (Henderson *et al.*, 1968a).

Allan and co-workers (1975) reported a strong predominance of the C_{27} and C_{29} homologs in an extract of a Nottinghamshire vitrinite. Pyrolysis of this extracted vitrinite under nitrogen at 275°C for 24 hr yielded further quantities of alkanes between C_{19} and C_{35}, with C_{31} as the major component. For a number of sporinites and vitrinites (Allan and Douglas, 1974) of varying rank pyrolyzed at 275°C, the distribution became smoother (with maximum at C_{21}) as rank increased. A similar trend to lower average molecular weight with increasing rank was observed for the *n*-alkanes produced by pyrolysis at 375°C, in agreement with the work of Leythaeuser and Welte (1969). Allan and Douglas (1974) found that, in the lower ranks, sporinite gave 20 times as much saturated hydrocarbons as vitrinite at 375°C; the ratio was much higher for pyrolysis at 275°C and decreased for higher ranks. For both macerals, *n*-alkanes ranged overall from C_{15} to C_{34} at 375°C and C_{17} to C_{33} at 275°C.

4. *n-Alkanes in Earth Waxes Associated with Coal*

Montan wax extracted from certain European brown coals contains a high proportion of aliphatic components (Wollrab and Streibl, 1969). Thus a Bohemian Montan wax contained (Wollrab *et al.*, 1963) C_{22}–C_{37} *n*-alkanes, among which odd carbon numbers predominated; C_{29} (39%), C_{31} (28%), and C_{27} (14%) were by far the largest contributors. In further work the range was shown to extend to C_{12}–C_{37}, in trace amounts (Jarolimek *et al.*, 1965). Similar distributions have been found in German Montan waxes (Presting and Kreutzer, 1965).

The naturally occurring waxy mineral hatchettite found in coal beds (Firth and Eglinton, 1972) consists almost entirely of *n*-alkanes with an even distribution and with maximum abundances between C_{24} and C_{29}. As one passes from high to medium volatile bituminous-associated coals, the maxima within the distribution curves shift toward lower homologs.

More recently, *n*-alkane distributions in some "fossil resins" associated with lignite have been studied by Douglas and Grantham (1974) by fingerprint gas chromatography.

5. *n-Alkane Distributions in Relation to Coal Origin*

As with a number of other sediments, the *n*-alkane fractions of coals and associated earth waxes exhibit experimentally a predominance of odd carbon numbers over even (Welte, 1967). A maturity (Phillippi,

1965) or carbon-preference index (CPI) may be defined (Maxwell *et al.*, 1971):

$$\mathrm{CPI} = \frac{1}{2}\left(\frac{\Sigma \text{ concentrations of odd } n\text{-alkanes } \mathrm{C}_{17}\text{–}\mathrm{C}_{31}}{\Sigma \text{ concentrations of even } n\text{-alkanes } \mathrm{C}_{16}\text{–}\mathrm{C}_{30}}\right. + \left.\frac{\Sigma \text{ concentrations of odd } n\text{-alkanes } \mathrm{C}_{17}\text{–}\mathrm{C}_{31}}{\Sigma \text{ concentrations of even } n\text{-alkanes } \mathrm{C}_{18}\text{–}\mathrm{C}_{32}}\right)$$

Other versions differ in the *n*-alkane range (e.g., Bray and Evans, 1961; Brooks and Smith, 1969; Kvenvolden, 1966).

The predominance of odd-numbered high-molecular-weight *n*-alkanes, as shown by the CPI, was first reported by Stevens *et al.* (1956); it occurs widely in the indigenous extracts of Recent sediments, as well as in many living organisms (Schenck, 1969). CPI values for the coals studied by Leythaeuser and Welte (1969) varied systematically on coalification from 1.59 to 1.00. A wider variation in CPI from 15.3 to approximately 1.0 was reported by Brooks *et al.* (1969) for Australian coals as the carbon percentage of the coal changed from 67% to over 90%. On the other hand, the CPI of *n*-alkanes from samples of hatchettite (Firth and Eglinton, 1972) varied between only 0.96 and 1.06.

Now the surface waxes of contemporary plants show a marked predominance of odd over even *n*-alkanes (Douglas and Eglinton, 1966), which is also found in Recent sediments—CPI = 2.4–5.5 (Bray and Evans, 1961; Kvenvolden, 1966). Indeed the conclusive (rather than spurious) identification of even-numbered members from natural sources came long after the detection of odd-numbered members (Eglinton and Hamilton, 1963). On the other hand, crude oils—CPI = 0.91–1.13 (Bray and Evans, 1961), CPI $\leqslant$ 1.3 (Schenck, 1968) and ancient sediments—CPI = 0.98–2.3 (Bray and Evans, 1961)—show a much smoother distribution (Eglinton *et al.*, 1964; Johns *et al.*, 1966; Oró and Nooner, 1967).

The extra contribution of the even-numbered *n*-alkanes in older sediments may be from the alcohols, acids, or other even-numbered oxidized species in the contributing biological materials obtained either by direct reduction or by hydrogenation (Douglas and Grantham, 1974). Thus an unusual predominance of *even*-numbered *n*-alkanes in geological materials (Welte and Ebhardt, 1968) has been shown in alginites and resinites (Allan, 1975); wurtzilites (Douglas and Grantham, 1974); and in pyrolysis products from vitrinite and sporinite (Allan and Douglas, 1974), kerogen (Dungworth, 1972), and recent sediments (Coates and Douglas, 1973). Highly reducing environments during diagenesis may account for these exceptions.

Didyk and McCarthy (1971) observed that the *n*-alkane distributions of crude oils either had a smooth distribution around one maximum at C_{16} (typified by crudes known to be of marine origin) or exhibited two maxima at C_{17} and C_{23} and thus contained a higher proportion of heavier hydrocarbons (typified by materials of nonmarine origin). These authors also took the view that the metamorphism process could remove any preponderance of odd-carbon-number alkanes present in the precursor lipid.

In studies of deep-sea cores of ages ranging from 1 to 100 × 10^6 years, Simoneit and Burlingame (1974) interpreted the predominantly odd-carbon-number distribution of the rather low concentration of the C_{22}–C_{33} *n*-alkanes found in a sample from 40 m below the seabed to minor terrigenous contributions. At this depth, low *n*-alkanes (peaked at C_{17}) were much more abundant. Simoneit and Burlingame attributed the absence of carbon number preference in these sediments to marine derivation and diagenetic maturation of the organic matter. From a consideration of earlier studies of thermal treatment of organic matter, Baker (1974) concluded that a sediment capable of producing alkanes without any preference for odd carbon numbers should be a good source of petroleum (Hellman, 1975).

The progressive variation of composition of the mixture of *n*-alkanes from coal with increasing diagenesis has been attributed either to thermal or catalytic alteration of the original straight chain alcohols, fatty acids, and alkanes (Henderson *et al.*, 1968a) or to progressive release of *n*-alkanes from the long chain wax components of brown and subbituminous coals which have no odd–even preference (Brooks and Smith, 1969). From experiments on fossil wood, Allan *et al.* (1975) concluded that the waxy material which is later capable of generating alkanes at a further stage of coalification is incorporated at the peat stage. Among earth waxes, Montan wax is mainly composed of little-changed higher plant *n*-alkanes (C_{27}, C_{29}, C_{31}, and C_{33} are ubiquitous major components of alkane extracts of plant matter) (Douglas and Eglinton, 1966), whereas hatchettites are considerably matured.

In support of these explanations, one may cite extractions of some *n*-alkanes with odd–even preference from Recent sediments, but generation of larger quantities with no such preference on thermal treatment (Baker, 1974). Schenck (1969) found that appreciable difference between the compositions of crude oil and rock extracts is reduced at greater depths, whereas *n*-alkanes extracted from similar sediments at increasing depths in a single formation differ in weights and in distribution (Maxwell *et al.*, 1971). In particular, from analysis of a long core of sediment, Albrecht and Ourisson (1969) found that large quantities of *n*-alkanes were generated from oxygen-containing or straight-chain compounds in the kerogen at depths between 1200 and 2200 m, just as heating Yallourn lignite caused an increase in *n*-alkane concentration, as detailed later (Connan, 1974; Brooks and Smith, 1969). The decrease in *n*-alkane concentration for depths greater than 2200 m was attributed to thermal and catalytic cracking at higher temperatures (Henderson *et al.*, 1968a). Under geological conditions, kerogen (which represents many kinds of compounds, chemically interlinked and cross-linked) is thermally cracked to lower *n*-alkanes, cyclics, and aromatics (Henderson *et al.*, 1968a). Geochemical processes that have been suggested as feasible for alkane generation include decarboxylation of fatty acids, deamination of proteins, and low temperature cracking (Baker, 1974).

By heating an Australian lignite (Yallourn mine, Gippsland Basin) in sealed tubes for several days at 300°C, Connan (1974), as mentioned in Section VI,B, found that isoprenoid alkanes (absent in the starting material) were generated; moreover, the *in vitro* neogenesis of *n*-alkanes yielded an odd–even predominance (C_{25}–C_{29} range) and increased concentrations by a factor of 100 over 3 days. High saturated hydrocarbon content is usually associated with a high degree of maturity (Poulet and Roucaché, 1970; Tissot *et al.*, 1971). In summary, one may say that the conventional attitude to the observed alkane carbon number distributions is to take them at their face value and interpret in terms of the coalification process.

Vahrman (1970) takes quite a different view of the observed odd–even distribution of *n*-alkanes, pointing out that many extracts examined may comprise only a part of the total extractable material. Thus, he suggests that the distribution actually observed may be, in consequence, a fortuitous result of the mode of extraction. Vahrman notes that the odd-carbon-number preference in the small amount of *n*-alkanes first extracted or liberated from coal when heated gives way to a smooth distribution of the rest, so that there may be two distinct origins. He proposes, in agreement with the foregoing theories, that the CPI values greater than 1 in the *n*-alkanes first extracted arise from the odd-carbon predominance of the hydrocarbons of plant cuticles, which are more easily extractable (like those in the coal). However, he then suggests that the smooth distribution of further *n*-alkanes is a consequence of their origin in the interior cells of plants. The *n*-alkanes of heartwoods (Grice *et al.*, 1968; Cocker and Shaw, 1963; Cocker *et al.*, 1965; Bennett and Cambie, 1967), seed oils (Kuksis, 1964), and the interiors of leaves (Herbin and Robins, 1970) show a smooth pattern extending over a wide range of carbon numbers.

B. Acyclic Isoprenoids in Coal

Acyclic isoprenoid alkanes are distributed throughout the geosphere in coals, crude oils, and ancient sediments. Their major precursor has generally been regarded as phytol (C_{20}) or the phytyl side chain (Bendoraitis *et al.*, 1962a), present at the time of deposition (Albrecht and Ourisson, 1971; Maxwell *et al.*, 1971; de Leeuw *et al.*, 1974); phytol is the alcohol which esterifies the acid group of the chlorin nucleus derived

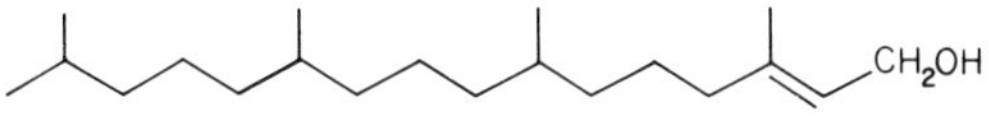

CH_2OH

from the chlorophyll pigment. Brooks *et al.* (1977) have recently suggested alternatives to phytol as the source of the acyclic isoprenoids, namely, naturally occurring higher terpenoids, such as squalene or lycopene, and sesquiterpenoids, such as farnesol. Although neither phytol nor dihydrophytol has been isolated from metamorphosed coal, the presence of six derived hydrocarbons has been demonstrated in a variety of coal extracts and tars (Table V); these are the C_{14}, C_{15}, C_{16}, C_{18}, C_{19}, and C_{20} isoprenoids.

Isoprenoids with chains larger than C_{20} have been found in the lithosphere—C_{21} in crude oils and sediments (Johns *et al.*, 1966; McCarthy *et al.*, 1967; Bendoraitis *et al.*, 1962b); C_{22}–C_{25} series in crude oil (Han and Calvin, 1969); C_{21}, C_{23}–C_{25} isoprenoids, and possibly C_{26}, C_{28}, and C_{30} in seep oil (Haug and Curry, 1974)—but not so far in coal; high-molecular-weight terpenoids such as the carotenoid lycopane and the triterpane solanesol have been proposed as precursors (McCarthy *et al.*, 1967). Nor has the C_{17} isoprenoid alkane, 2,6,10-trimethyltetradecane, been identified in coal; although isolated from Antrim shale (McCarthy and Calvin, 1967), it is present only in sediments in low abundance. Connan (1974) has probably detected in chromatograms the generation of C_{17} isoprenoid alkane from Yallourn (Australian) lignite heated *in vitro* for a day at 300°C, and has inferred that such isoprenoid alkanes arise by degradation of wax esters.

A systematic study of the phytane and pristane contents of Australian coals has been reported by Brooks *et al.* (1969). The pristane content of the saturated hydrocarbon fraction of chloroform–methanol extracts begins to increase in subbituminous coals of more than 76% carbon, but no significant general increase in concentration of phytane occurs until much later in the diagenetic process, corresponding to 83–85% carbon. At this stage (Fig. 4), the pristane/phytane ratio is a maximum. Vahrman (1970) has also commented on the greater concentration of pristane over

TABLE V *Isoprenoid Paraffinic Hydrocarbons Identified in Coal Extracts and Tars*

Compound and structure	Extracts					Tars	
	Cyclo-hexane	Chloroform–methanol or benzene–methanol	Ether	Supercritical gas (toluene)		Brown coal tar	Low temperature tar
				Bitu-minous	Turkish lignite		
C_{20} Phytane	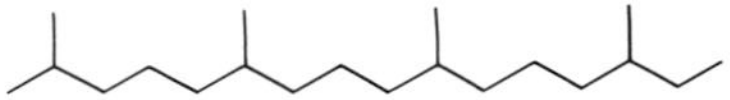 *a*	*b, l*		*d*	*m*		*g, h, i, j, k*
C_{19} Pristane	 *a*	*b, l*	*c*	*d*	*m*	*c*	*f, g, h, i, j, k*
C_{18} Norpristane	 *a*			*d*	*m*	*e*	*f, j, k, i*
C_{16} 2,6,10-Tri-methyltridecane	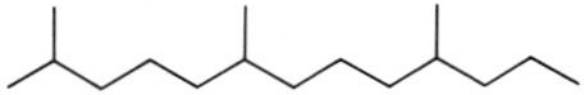 *a*			*d*	*m*		*f, j, k, i*
C_{15} Farnesane				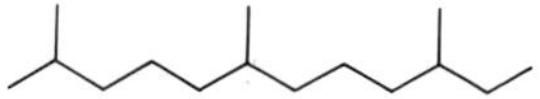 *d*	*m*	*e*	*f, j, k, i*
C_{14} 2,6,10-Tri-methylundecane				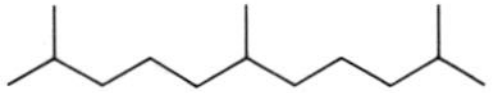 *d*	*m*		*f, h. i, j, k*

a. Birkofer and Pauly (1969)
b. Brooks *et al.* (1969)
c. Leythaeuser and Welte (1969)
d. Bartle *et al.* (1975)
e. Kochloefl *et al.* (1963)
f. Boyer and Payen (1964)
g. Spence and Vahrman (1967)
h. Vahrman (1970)
i. Pichler *et al.* (1970)
j. Bartle *et al.* (1977a)
k. Bartle *et al.* (1977c)
l. Douglas *et al.* (1970)
m. Bartle *et al.* (1977b)

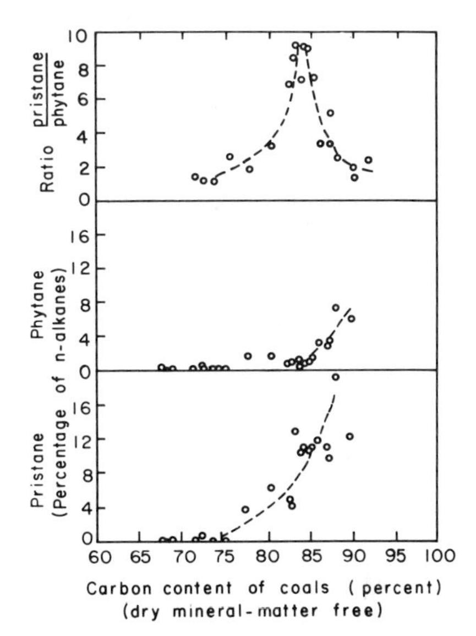

Fig. 4 Pristane and phytane contents of Australian coals. (After Brooks *et al.*, 1969.)

phytane in tars and extracts from British coals, while Bartle *et al.* (1977c) have illustrated the wider application of the curves of Fig. 4; the contents of pristane and phytane (whether measured as percentage of paraffins or as a pristane/phytane ratio, *r*) in coals of British (bituminous) and Turkish (lignite) origin also fitted the curves derived from measurements on Australian coals.

For syngenetic bitumen, Maximov *et al.* (1974) showed that the pristane/phytane ratio *r* is influenced by the organic matter (OM) composition in the original rocks, here from the upper protozoic; *r* increases with H/C which, in turn, increases with the proportion of lipid in the initial OM. Maximov's group concluded that the pristane may not originate exclusively by thermocatalytic disintegration of phytane, but it may be a constituent of plankton (Barghorn and Smith, 1969). Presumably the variation in *r* for oils from different sediments is a consequence of compositional differences among the accompanying OM. Differences in burial depths and in geological times may be expected to influence the resulting chemical composition, although Douglas *et al.* (1970) have suggested that the straightforward chemical reaction involved may not require geological times.

The mechanism of formation of the C_{14}–C_{20} isoprenoids from phytol has been the subject of much research. Maxwell *et al.* (1972) demonstrated by two methods that the stereochemistry of 80% of the pristane from an ancient sediment corresponded to the 6(R),10(S) configuration;

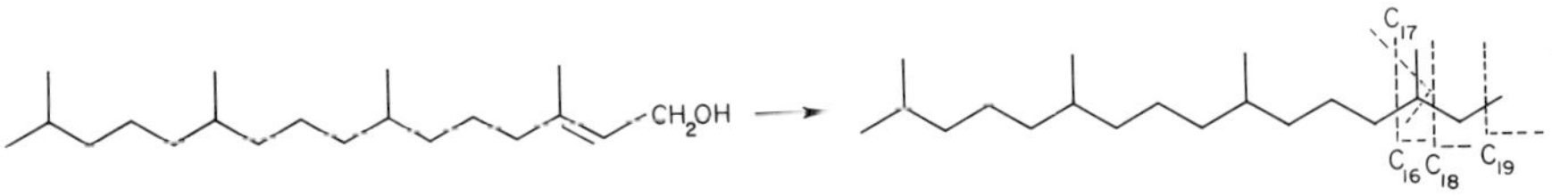

Fig. 5 Generation of isoprenoids by bond scission in phytol.

this is consistent with an origin from 3,7(R),11(R),15-tetramethyl-hexadec-*trans*-2-en-1-ol, the configuration of phytol from chlorophyll (Burrell *et al.*, 1959; Crabbé *et al.*, 1959; Burnett *et al.*, 1966; Maxwell *et al.*, 1971; McCarthy and Calvin, 1967). McCarthy and Calvin accounted for the formation of isoprenoids by the cracking of the C–C bonds of phytol (Fig. 5) and pointed out that double scission was necessary to form the rare C_{17} isoprenoids, although recombination reactions may also play a part (Eismar and Jurg, 1969). A scheme leading from phytol to the C_{16} and C_{18} isoprenoids has also been proposed by Johns *et al.* (1966). A rationalization of the diagenetic and maturation reactions to which phytol and its derivatives may have been subject in sediments has been attempted (Cox *et al.*, 1972). The scheme outlines principal pathways through which the alkanes, alkenes, alcohols, ketones, and acids may pass.

Brooks *et al.* (1969) attempted to rationalize their observations on the predominance of pristane and the increasing contribution of isoprenoids to alkanes in coals of increasing rank by decarboxylation to pristane of phytanic acid (present as esters in brown coals, and formed by oxidation of phytol in the early stages of chlorophyll decomposition). Both the clay mineral bentonite and the insoluble fraction of coal were shown to be effective decarboxylation catalysts.

A number of other laboratory alteration experiments have been reported. Thus Borgohain (1971) also showed that bentonite effected thermal catalytic cracking of phytol with hydrogen transfer to yield all the C_{14}–C_{20} isoprenoids except C_{17}. The conversion of ^{14}C-labeled phytol into phytanic and pristanic acids in a sediment (Brooks and Maxwell, 1974), and a biochemical pathway for converting phytol into pristane but not phytane (Avigan and Blumer, 1968), have been demonstrated.

C. Steranes and Terpanes in Coal

Steranes and terpanes are common constituents of ancient sediments and crude oils (Eglinton and Calvin, 1967; Hills *et al.*, 1970; Albrecht and Ourisson, 1971; Maxwell *et al.*, 1971). They probably derive over geological time from steroids and terpenoids which are biologically abundant and are constituents of bottom muds and Recent sediments

(Henderson *et al.*, 1971, 1972). This view is supported by three lines of evidence: (a) thermal alteration experiments with a radiolabeled steroid incorporated into a sediment (Rhead *et al.*, 1971) and other laboratory simulation processes (Steel *et al.*, 1972); (b) identification of individual steranes and terpanes (Erskine and Whitehead, 1975), some of which are exactly defined optically and by establishment of their complete structures and stereochemistries by x-ray crystallography (Hills *et al.*, 1968), ^{13}C-NMR (Balogh *et al.*, 1972, 1973) and ^{1}H-NMR (Mulheirn and Ryback, 1975), and optical rotatory dispersion (Mulheirn and Ryback, 1975); and (c) isolation of intact steroids and triterpenoids from geological sources (Ikan and Kashmann, 1963; Attaway and Parker, 1970; Henderson *et al.*, 1971) including brown coal (Ruheman and Raud, 1932; Münch, 1934; Jarolim *et al.*, 1961) and lignite (Ikan and McLean, 1960). Triterpanes are widespread enough in petroleum to allow fingerprinting (Pye, 1975; Gallegos, 1973).

Earlier, steroids were found in higher plants but, with their discovery in lower plants such as algae, steranes are now thought to derive from marine as well as terrestrial sources (Oró and Nooner, 1967). Henderson *et al.* (1969) illustrate saturated hydrocarbon (steranes and triterpanes) skeletons for some steroids and triterpenoids known to occur in nature, and suggest how they might derive from the squalene precursor of appropriate conformation by bacterial or thermal diagenesis in a geological environment. Following mass spectrometric and gas chromatographic studies of Texan crude oils, Bendoraitis *et al.* (1963) have proposed a maturation scheme for the conversion of a pentacyclic triterpene through a cracking sequence to bicyclic hydrocarbons.

Until recently, however, there had been few reports of steroid and triterpenoid paraffinic hydrocarbons in coal materials (Allan *et al.*, 1977), presumably because of degradation. Birkhofer and Pauly (1969), as mentioned in Section IV, assigned to perhydrophenanthrene derivatives the GC peaks in the region of steranes and terpanes from coal. While fichtelite **(B)** and a number of other paraffins derived from diterpenoids have often been identified in lignite (Simonsen and Barton, 1961; Briggs, 1937; Maxwell, 1967), triterpenoid paraffins were restricted to the unidentified triterpanes found in Czechoslovakian brown coal (Jarolim *et al.*, 1965; Maxwell, 1967). In Connan's (1974) *in vitro* diagenesis experiments (referred to in Section VI,B in connection with the abundance of *n*-alkanes generated), significant chromatographic peaks from steranes and triterpanes persisted when the Yallourn lignite was heated at 300°C for 3 days.

A number of reasons may account for these differences between coal and petroleum: first, the different conditions experienced by plant material in conversion to coal and its precursors (e.g., peat and lignite) initially in swamp environments (Cooper and Murchison, 1969; Given, 1972); second, the apparent thermal instability of steranes and terpanes in the coal matrix. Although these compounds may be liberated (Allan *et al.*, 1975) from coal by heating vitrinites of bituminous rank at 275°C for 24 hr, they do not contribute significantly to alkane extracts obtained at higher temperatures (e.g., 375°C for 24 hr). Allan and Douglas (1974) and Allan *et al.* (1977) studied vitrinites and sporinites extracted from bituminous coals and, from GC/MS, they established the presence of C_{27}–C_{31} triterpanes including, in particular, adiantane. Further, Tissot *et al.* (1971) have shown that the processes of maturation of sediments of the Paris Basin led to progressive degradation of these compounds with concomitant increases in content of derived aromatic structures. A third possible reason is the apparent aromatization of terpenoids without yielding a derived paraffin. Thus, while the diterpenoid abietic acid **(A)**, a common constituent of tree resins, eventually yields (Maxwell *et al.*, 1971) retene **(C)** by hydrogenation–dehydrogenation reactions, involving the paraffin fichtelite **(B)** which is, like **C**, a well-known geolipid, triterpenoids undergo aromatization in coal without previous removal of oxygen functional groups.

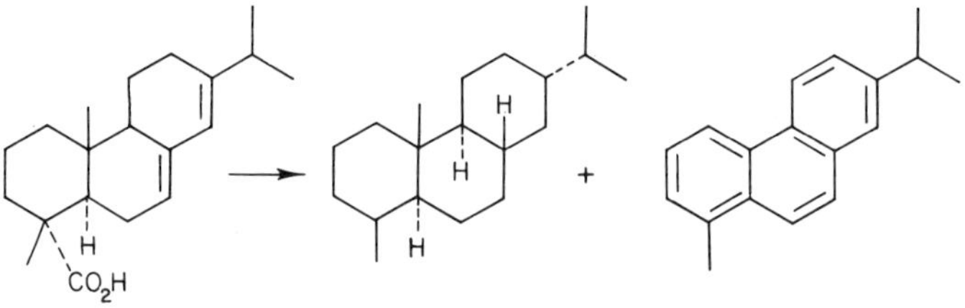

A B C

Thus, Jarolim *et al.* (1961, 1965) have identified numerous triterpenoids and their keto and alcohol derivatives, principally of the hopane and friedelane series, with corresponding aromatic hydrocarbons based on picene. The hydrocarbon 1,2,3,4-tetrahydro-2,2,9-trimethylpicene **(H)** is of particular interest, having been found to be identical with a compound isolated (Carruthers and Watkins, 1964) from crude oil, and is also found in a Yubari coal extract (Imuta and Ouchi, 1967, 1971). Compound **H** is thought to arise through dehydrogenation of the oleanane series of triterpenoids, in particular, allobetul-2-ene **(D)**. The predominance of the oxygenated species found by Jarolim provides evidence that the lignite has suffered only mild alteration during coalification. Laboratory studies showed that the following stepwise sequence (**E** → **H**) is possible with consecutive aromatization of rings and elimination of angular methyl groups (Streibl and Herout, 1969). Similar sequences have been suggested by Hills *et al.* (1968).

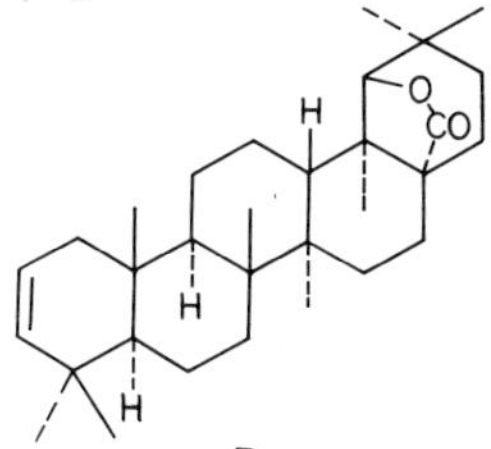

D

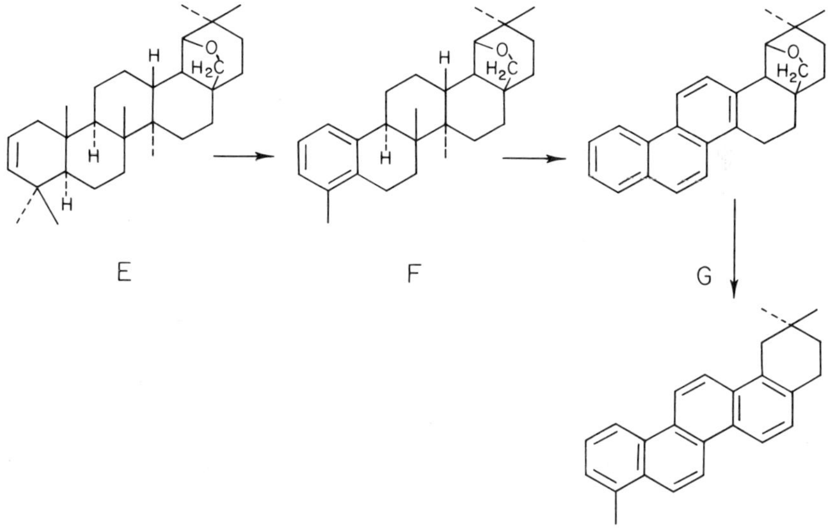

E F G

H

More recent work on the chemotaxonomy of a variety of sediments has shown (Ensminger *et al.*, 1974; Van Dorsselaer *et al.*, 1974; Kimble *et al.*, 1974) that the most widespread geolipids derived from terpenoid biolipids comprise the hopane series. These compounds are providing vital information—because of the structural specificity of the polycyclic skeleton, which maintains its stereochemistry—in particular, the role of microbial activity in sedimentation (Van Dorsselaer *et al.*, 1974; Bird *et al.*, 1971). The series of extended and degraded hopanes listed in Table VI has been identified (Van Dorsselaer *et al.*, 1974) in a chloroform extract of a Lorraine bituminous coal (275×10^6 years old) and in a Yallourn lignite (25×10^6 years old) by comparison with authentic samples by GC and GC/MS (coinjection on capillary columns—labeled [b] in Table

TABLE VI *Hopane Derivatives Identified in Lorraine Coal*[a]

Structure	Substituents
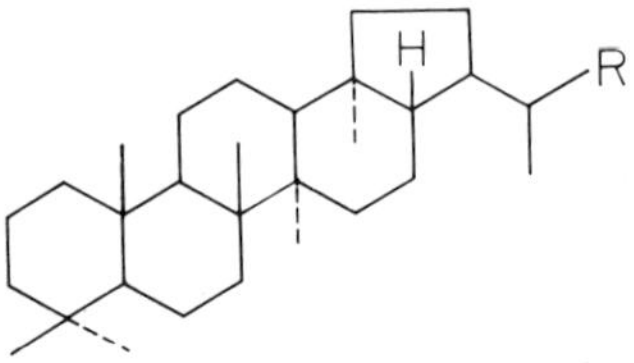 Hopane skeleton with R side chain	R = H[b]; R = CH_3[b]; R = C_2H_5[b]
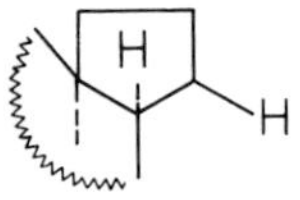 Ring E with H	[b]
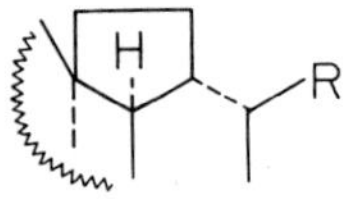 Ring E with R side chain	R = H,[b] CH_3,[b] C_2H_5,[b,d] n-C_3H_7,[b,d] C_4H_9,[c] C_5H_{11},[c] C_6H_{13}[b]

[a] Van Dorsselaer *et al.* (1974).
[b,c] See text.
[d] Also identified in a Yallourn lignite, present as sole major component.

VI) and more tentatively by inferences from GC/MS (labeled [c] in Table VI). Ensminger *ei al.* (1974) identified C_{27}, C_{29}, C_{30}, and C_{31} triterpanes as major components of the branched and cyclic alkane fraction (40–80 ppm of dry coal) for each compound (see also Section VI,D).

Ensminger *et al.* (1974) have suggested routes whereby these pentacyclic triterpanes may arise as major components in geological sources such as coals, as well as in a wide variety of sediments and crude oils. Microbiological degradation and alkylation of a C_{30} precursor, or degradation of a C_{35} precursor could generate the less stable 17-β(H),21-β(H)-hopane series (detected in living organisms), from which the more stable 17-β(H),21-α(H) or 17-α(H),21-β(H) (not actually detected in biological material) could form by epimerization. Lower organisms such as prokaryotes yield diploptene **(I)** from which hopane-type hydrocarbons can be obtained biochemically or abiotically.

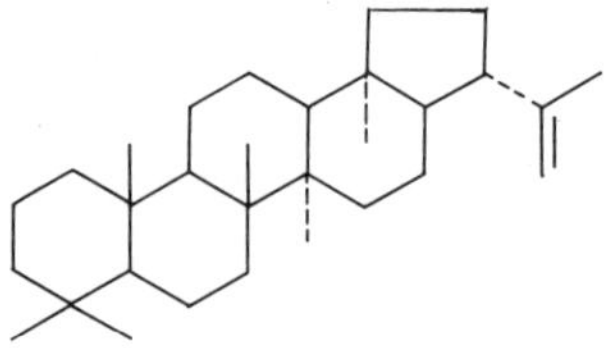

I

D. Singly Branched Alkanes

Iso-(2-methyl)- and ante-iso-(3-methyl)alkanes are present in small quantities in a number of sediments (Johns *et al.*, 1966; Van Hoeven *et al.*, 1966) and are thought to be derived from plant waxes (Eglinton and Hamilton, 1963; Douglas and Eglinton, 1966) in which they also occur. As mentioned in Section IV,B, Birkofer and Pauly (1969) identified small concentrations of every member of the iso-C_{15}–C_{31} series in a chloroform extract of coal, as well as the ante-iso C_{23} hydrocarbon; 5-methylnonacosane was also present. Trace amounts of the C_{16}–C_{35} iso and ante-iso series were also found in Montan wax (Jarolimek *et al.*, 1965). In a low temperature tar, Pichler *et al.* (1970) separated a series of singly branched alkanes at each carbon number from C_{12} to C_{18}—generally the 2, 3, 4, 5, 6, and 7 isomers, with the 2-methyl compound mostly predominant.

VII. CONCLUSION

Even though they are present only in comparatively low concentration, the alkanes of coal can serve to characterize different coals. Moreover, information about alkane distributions, especially when quantitative, can provide crucial evidence about the origin of the world's major fossil hydrocarbon and its relation to other carbonaceous sediments. For these purposes as "biological markers," the steranes and pentacyclic triterpanes, with their high structural and stereochemical specificity and chemical stability, are likely to be at least as valuable as the acyclic polyisoprenoid "molecular fossils" or the distributions of *n*-alkanes.

There may be close similarities (Welte, 1974) between the structures of kerogen (insoluble organic matter in rock sediments) and coal in that both have been suggested to consist of micelles or aggregates of chemically inert aromatic complexes immersed in a more reactive paraffinic matrix. Extension of our knowledge about the range of paraffin hydrocarbons, including the linear ones, should contribute materially to solving the enigma of coal structure: for example, the validity of the notion of a molecular sieve-like structure in coal and the controversy about "polyamantane" versus aliphatic-linked aromatic concepts of coal constitution (Chakrabartty and Kretschmer, 1972). For a detailed discussion of the latter, see Chapter 22, Section III.

Not surprisingly, challenging problems are presented by the separation, identification, and structural determination of very many closely similar molecules, themselves representing in total only a small fraction of the total carbonaceous materials. Capillary-column GLC is probably the preeminent technique here and its combination with mass spectrometry is powerful indeed, although ultimately destructive. Now that sensitivity problems are being surmounted, infrared spectroscopy is being overtaken by high field ^{1}H-NMR and ^{13}C-NMR for characterization of individual molecules. Fourier transform ^{13}C-NMR is the spectroscopic technique likely to make a large contribution in the near future to the study of paraffin hydrocarbons in coal and coal products.

REFERENCES

Adler, N. (1974). *Goriva Maziva* **13**(4), 15.

Ahmed, M. S. A., and Vahrman, M. (1966). *J. Appl. Chem.* **16,** 105.

Albrecht, P., and Ourisson, G. (1969). *Geochim. Cosmochim. Acta* **33,** 138.

Albrecht, P., and Ourisson, G. (1971). *Angew. Chem., Int. Ed. Engl.* **10,** 209.

Allan, J. (1975). Ph.D. Thesis, Univ. of Newcastle.

Allan, J., Bjorøy, M., and Douglas, A. G. (1977). *In* "Advances in Organic Geochemistry, 1975" (R. Campos and J. Goni, eds.), p. 633. Enadimsa, Madrid.

Allan, J., and Douglas, A. G. (1974). *In* "Advances in Organic Geochemistry, 1973" (B. Tissot and F. Bienner, eds.), p. 203. Technip, Paris.

Allan, J., Murchison, D., Scott, E., and Watson, S. (1975). *Fuel* **54,** 283.

Anderson, N. A., and Falcone, M. S. (1969). *J. Chromatogr.* **44,** 52.

Andrzejak, A., and Ignasiak, B. (1967). *Zesz. Nauk.* **211,** 193.

Andrzejak, A., and Ignasiak, B. (1968). *Freiberg. Forschungsh. A* **429,** 7.

Angla, B. (1947). *C. R. Acad. Sci.* **224,** 402.

Angla, B. (1949). *Justus Liebigs Ann. Chem.* **4**(12), 639.

Attaway, D., and Parker, P. L. (1970). *Science* **169,** 674.

Avigan, J., and Blumer, M. (1968). *J. Lipid Res.* **9,** 350.

Baker, B. L. (1974). *In* "Advances in Organic Geochemistry, 1973" (B. Tissot and F. Bienner, eds.), p. 137. Technip, Paris.

Baldwin, R. M., Golden, J. D., Gary, J. H., Bain, R. L., and Long, R. J. (1975). *Chem. Eng. Prog.* **71,** 70.

Balogh, B., Wilson, D. M., and Burlingame, A. L. (1972). *In* "Advances in Organic Geochemistry, 1971" (H. R. V. Gaertner and H. Wehner, eds.), p. 163. Pergamon, Oxford.

Balogh, B., Wilson, D. M., Christiansen, P., and Burlingame, A. L. (1973). *Nature (London)* **242,** 603.

Barghorn, E. S., and Smith, V. J. (1969). Cited in Maximov *et al.* (1974).

Barrall, E. M., and Baumann, F. (1964). *J. Gas Chromatogr.* **2,** 256.

Bartle, K. D. (1972). *Rev. Pure Appl. Chem.* **22,** 79.

Bartle, K. D., and Jones, D. W. (1969). *Fuel* **48,** 21.

Bartle, K. D., and Jones, D. W. (1972). *Adv. Org. Chem.* **8,** 317.

Bartle, K. D., and Smith J. A. S. (1965). *Fuel* **44,** 109.

Bartle, K. D., and Smith, J. A. S. (1967). *Fuel* **46,** 29.

Bartle, K. D., Smith, J. A. S., and Wilman, W. G. (1969). *J. Appl. Chem.* **19,** 283.

Bartle, K. D., Jones, D. W., Martin, T. G., and Wise, W. S. (1970). *J. Appl. Chem.* **20,** 197.

Bartle, K. D., Martin, T. G., and Williams, D. F. (1975). *Fuel* **53,** 226.

Bartle, K. D., Jones, D. W., and Pakdel, H. (1977a). *In* "Molecular Spectroscopy" (A. R. West, ed.), p. 127. Heyden, London.

Bartle, K. D., Jones, D. W., and Pakdel, H. (1977b). Unpublished results.

Bartle, K. D., Calimli, A., Pakdel, H., and Jones, D. W. (1977c). Unpublished measurements.

Bartz, K. W., and Chamberlain, N. F. (1964). *Anal. Chem.* **36,** 2151.
Bellamy, L. J. (1968). "Advances in Infrared Group Frequencies." Methuen, London.
Bellas, T. E. (1975). *Chromatographia* **8,** 38.
Bendoraitis, J. G., Brown, B. L., and Hepner, L. S. (1962a). *Anal. Chem.* **34,** 49.
Bendoraitis, J. G., Brown, B. L., and Hepner, L. S. (1962b). *World Pet. Congr., Frankfurt.*
Bendoraitis, J. G., Brown, B. L., and Hepner, L. S. (1963). *World Pet. Congr., Proc., 6th, Frankfurt, 1962* Pap. 15, Sect. 5.
Benger, F., and Schlenk, W. (1949). *Experientia* **5,** 200.
Bennett, C. R., and Cambie, R. C. (1967). *Phytochemistry* **6,** 883.
Bernhard, U., and Berthold, P. H. (1975). *J. Prakt. Chem.* **317,** 1.
Bieser, H. (1977). U.S. Pat. No. 4,006,197.
Binns, E. H. (1962). Rep. 0286. Coal Tar Res. Assoc., (now part of B.C.R.A., Chesterfield, England).
Bird, C. W., Lynch, J. M., Pirt, S. J., and Reid, W. W. (1971). *Tetrahedron Lett.* p. 3189.
Birkofer, L., and Pauly, W. (1969). *Brennst.-Chem.* **50,** 376.
Blunt, G. V., and Vahrman, M. (1960). *J. Inst. Fuel* **33,** 522.
Blytas, G. C., and Peterson, D. L. (1967). *Anal. Chem.* **39,** 1434.
Borgohain, M. (1971). Ph.D. Thesis, Univ. of Bristol.
Boyer, A., and Payen, P. (1964). *Chim. Ind. (Paris)* **92,** 367.
Bray, E. E., and Evans, E. D. (1961). *Geochim. Cosmochim. Acta* **22,** 2.
Bray, E. E., and Evans, E. D. (1965). *Am. Assoc. Pet. Geol., Bull.* **49,** 248.
Briggs, L. H. (1937). *J. Chem. Soc.* p. 1035.
Brooks, J. D., and Smith, J. W. (1967). *Geochim. Cosmochim. Acta* **31,** 2389.
Brooks, J. D., and Smith, J. W. (1969). *Geochim. Cosmochim. Acta* **33,** 1183.
Brooks, J. D., and Stevens, J. R. (1964). *Fuel* **43,** 87.
Brooks, J. D., Gould, K., and Smith, J. (1969). *Nature (London)* **222,** 257.
Brooks, P. W., and Maxwell, J. R. (1974). *In* "Advances in Organic Geochemistry, 1973" (B. Tissot and F. Bienner, eds.), p. 977. Technip, Paris.
Brooks, P. W., Maxwell, J. R., Cornforth, J. W., Butline, A. G., and Milne, C. B. (1977). *In* "Advances in Organic Geochemistry, 1975." Technip, Paris.
Brunnock, J. V. (1966). *Anal. Chem.* **38**(12), 1648.
Burlingame, A. L., and Schnoes, H. K. (1969). *In* "Organic Geochemistry" (G. Eglinton and M. T. J. Murphy, eds.), p. 89. Longman, London.
Burrell, J. W. K., Garwood, R. F., Jackman, L. M., Oskay, E., and Weedon, B. C. L. (1966). *J. Chem. Soc. C* p. 2144.
Burrell, J. W. K., Jackman, L. M., and Weedon, B. C. L. (1959). *Proc. Chem. Soc., London* p. 263.
Carman, C. J., Tropley, A. R., and Goldstein, J. H. (1973). *Macromolecules* **6,** 719.
Carruthers, W., and Watkins, D. A. M. (1964). *J. Chem. Soc.* p. 724.
Chakrabartty, S. K., and Kretschmer, H. O. (1972). *Fuel* **51,** 160.
Chen, N. Y., and Lucki, S. J. (1970). *Anal. Chem.* **42,** 508.
Clutter, D. R., Petrakis, L., Stenger, R. L., and Jensen, R. K. (1972). *Anal. Chem.* **44,** 1395.
Coates, R. C., and Douglas, A. G. (1973). Cited in Allan and Douglas (1974).
Cocker, W., and Shaw, S. J. (1963). *J. Chem. Soc.* p. 677.
Cocker, W., McMurray, T. B. N., and Ntamila, M. S. (1965). *J. Chem. Soc.* p. 1962.
Cole, R. D. (1968). *J. Inst. Pet., London* **54,** 288.
Cole, R. D. (1971). *Nature (London)* **233,** 545.
Coleman, W. M., Wooton, D. L., Dorn, M. C., and Taylor, L. T. (1976). *J. Chromatogr.* **123,** 419.
Connan, J. (1972). *Bull. Cent. Rech. Pau* **6,** 1,195.
Connan, J. (1974). *In* "Advances in Geochemistry, 1973" (B. Tissot and F. Bienner, eds.), p. 73. Technip, Paris.

Cooper, B. S., and Murchison, D. G. (1969). *In* "Organic Geochemistry" (G. Eglinton and M. T. J. Murphy, eds.), p. 699. Longman, London.

Cox, R. E., Maxwell, J. R., Ackman, R. G., and Hooper, S. N. (1972). *In* "Advances in Organic Geochemistry, 1971" (H. R. Van Gaertner and E. Wehner, eds.), p. 263. Pergamon, Oxford.

Crabbé, P., Djerassi, C., Eisenbraun, E. J., and Liv, S. (1959). *Proc. Chem. Soc., London* p. 264.

CSIRO (1964). *Coal Res. CSIRO* **23,** 17.

CSIRO (1966). *Coal Res. CSIRO* **29,** 6.

Cummins, J. J., and Robinson, W. E. (1964). *J. Chem. Eng. Data* **9,** 304.

de Leeuw, J. W., Correia, V. A., and Schenck, P. A. (1974). *In* "Advances in Organic Geochemistry, 1973" (B. Tissot and F. Bienner, eds.), p. 993. Technip, Paris.

Denisenko, A. N., Sidorov, R. I., Petrova, V. I., and Gamayunova, P. B. (1976). *Khim. Tekhnol. Topl. Masel* No. 8, 56.

Didyk, B. M., and McCarthy, E. D. (1971). *Nature (London) Phys. Sci.* **232,** 103.

Douglas, A. G. (1969). *In* "Organic Geochemistry" (G. Eglinton and M. T. J. Murphy, eds.), p. 161. Longman, London.

Douglas, A. G., and Eglinton, G. (1966). *In* "Comparative Phytochemistry" (T. Swain, ed.), p. 57. Academic, New York.

Douglas, A. G., and Grantham, P. J. (1974). *In* "Advances in Organic Geochemistry, 1973" (B. Tissot and F. Bienner, eds.), p. 261. Technip, Paris.

Douglas, A. G., Eglinton, G., and Henderson, W. (1970). *In* "Advances in Organic Geochemistry, 1966" (G. D. Hobson and G. C. Speers, eds.), p. 369. Pergamon, Oxford.

Drake, J. A. G., and Jones, D. W. (1977). *Fuel,* in press.

Drake, J. A. G., Jones, D. W., and Games, G. E. (1977). Unpublished measurements.

Dungworth, G. (1972). Ph.D. Thesis, Univ. of Newcastle-upon-Tyne.

Edstrom, T., and Petro, B. A. (1968). *J. Polym. Sci. Part C* **21,** 171.

Eglinton, G., and Calvin, M. (1967). *Sci. Am.* **216,** 32.

Eglinton, G., and Hamilton, R. J. (1963). *In* "Chemical Plant Taxonomy" (T. Swain, ed.), p. 187. Academic, New York.

Eglinton, G., and Murphy, M. T. J. (1969). *In* "Organic Geochemistry" (G. Eglinton and M. T. J. Murphy, eds.), p. 576. Longman, London.

Eglinton, G., Scott, P. M., Belsky, T., Burlingame, A. L., Richter, W., and Calvin, M. (1964). *Science* **145,** 263.

Eglinton, G., Scott, P. M., Belsky, T., Burlingame, A. L., Richter, W., and Calvin, M. (1966). *In* "Advances in Organic Geochemistry" (G. D. Hobson and M. C. Louis, eds.), p. 41. Pergamon, Oxford.

Eismar, E., and Jurg, J. W. (1969). *In* "Organic Geochemistry" (G. Eglinton and M. T. J. Murphy, eds.), p. 676. Longman, London.

Ensminger, A., Van Dorsselaer, A., Spyckerelle, C., Albrecht, P., and Ourisson, G. (1974). *In* "Advances in Organic Geochemistry, 1973" (B. Tissot and F. Bienner, eds.), p. 245. Technip, Paris.

Erskine, R. L., and Whitehead, E. V. (1975). *Iran. J. Sci. Technol.* **3,** 221.

Ettre, L. S. (1964). *Anal. Chem.* **36,** 31A.

Fetterly, L. C. (1950). *U.S. Pat.* No. 2,520,715 and 716.

Fetterly, L. C. (1951). *U.S. Pat.* No. 2,569,984 and 986.

Fetterly, L. C. (1952). *U.S. Pat.* No. 2,499,820.

Fetterly, L. C. (1957). *Hydrocarbon Process. Pet. Refiner.* **36,** 145.

Fetterly, L. C. (1964). *In* "Non-stoichiometric Compounds" (L. Mandelcorn, ed.), p. 491. Academic Press, New York.

Firth, J. N. M., and Eglinton, G. (1972). *In* "Advances in Organic Geochemistry, 1971" (H. R. Von Gaertner and H. Wehner, eds.), p. 613. Pergamon, Oxford.

Friedel, R. A. (1970). *In* "Spectrometry of Fuels" (R. A. Friedel, ed.). Plenum, New York.
Gallegos, E. J. (1970). *Am. Chem. Soc., Div. Pet. Chem., Prepr.* **15,** 1439.
Gallegos, E. J. (1973). *Anal. Chem.* **45,** 1399.
Giddings, J. C. (1963). *Anal. Chem.* **35,** 1999.
Given, P. H. (1972). *In* "Advances in Organic Geochemistry, 1971" (H. R. Von Gaertner and H. Wehner, eds.), p. 69. Pergamon, Oxford.
Goodman, R. A., Oldfield, E., and Allerhand, A. (1973). *J. Am. Chem. Soc.* **95,** 7553.
Gouw, T. H. (1973). *Anal. Chem.* **45,** 987.
Gouw, T. H., Whittlemore, I. M., and Jentoft, R. E. (1970). *Anal. Chem.* **42,** 1394.
Grant, D. W. (1971). "Gas Chromatography," p. 161. Van Nostrand-Reinhold, New York.
Grant, D. M., and Paul, E. G. (1964). *J. Am. Chem. Soc.* **86,** 2984.
Grantham, P. J. (1973). Ph.D. Thesis, Univ. of Newcastle-upon-Tyne.
Green, L. S., Schmauch, L. J., and Worman, J. C. (1964). *Anal. Chem.* **36,** 1512.
Grice, R. E., Locksley, H. D., and Scheinmann, F. (1968). *Nature (London)* **218,** 892.
Groenendijk, H., and Van Kemenade, A. W. C. (1968). *Chromatographia* **1,** 472.
Grudzien, J. (1967). *Koks, Smola, Gaz.* **12,** 91.
Guardino, X., Albaigés, J., Firpo, G., Rodriguez-Vinals, R., and Gassiot, M. (1976). *J. Chromatogr.* **118,** 13.
Habgood, H. W., and Harris, W. B. (1960). *Anal. Chem.* **36,** 663.
Han, J., and Calvin, M. (1969). *Geochim. Cosmochim. Acta* **33,** 733.
Haug, P., and Curry, D. G. (1974). *Geochim. Cosmochim. Acta* **38,** 601.
Hedfjall, B., and Ryhage, R. (1975). *Anal. Chem.* **47,** 666.
Hellman, H. (1975). *Erdoel Kohle, Compendium 1974/1975 (24th Ger. Miner. Coal Chem. Conf., Hamburg, 1974)* p. 931.
Henderson, W., Eglinton, G., Simmonds, P. G., and Lovelock, J. E. (1968). *Nature (London)* **219,** 1012.
Henderson, W., Wollrab, V., and Eglinton, G. (1969). *In* "Advances in Organic Geochemistry" (P. A. Schenck and I. Havenaar, eds.), p. 181. Pergamon, Oxford.
Henderson, W., Reed, W. E., Steel, G., and Calvin, M. (1971). *Nature (London)* **231,** 308.
Henderson, W., Reed, W. E., and Steel, G. (1972). *In* "Advances in Organic Geochemistry, 1971" (H. R. Von Gaertner and H. Wehner, eds.), p. 335. Pergamon, Oxford.
Hendrickson, J. G., and Moore, J. C. (1966). *J. Polym. Sci. Part A-1* **4,** 167.
Herbin, G. A., and Robins, P. A. (1970). Personal communication, cited in Vahrman (1970).
Hills, I. R., Smith, G. W., and Whitehead, E. V. (1968). *Nature (London)* **219,** 243.
Hills, I. R., Smith, G. W., and Whitehead, E. V. (1970). *J. Inst. Pet., (London)* **56,** 127.
Hirsch, D. E., Hopkins, R. L., Coleman, H. J., Cotton, R. O., and Thompson, C. J. (1972). *Am. Chem. Soc., Div. Pet. Chem., Prepr.* A **65,** 9–14.
Hsieh, B. C. B., Wood, R. E., Anderson, L. L., and Hill, G. R. (1969). *Anal. Chem.* **41,** 1066.
Hughes, R. H., and Martin, R. J. (1957). *Symp. Composition Pet. Oils, Determ. Eval. Am. Soc. Test. Mater., Philadelphia* p. 127.
Hunt, J. M. (1974). *In* "Advances in Organic Geochemistry, 1973" (B. Tissot and F. Bienner, eds.), p. 593. Technip, Paris.
Ignasiak, T., Ignasiak, B. S., and Montgomery, D. S. (1975). *Fuel* **54,** 133.
Ikan, R., and Kashmann, J. (1963). *Isr. J. Chem.* **1,** 502.
Ikan, R., and McLean, J. (1960). *J. Chem. Soc.* p. 813.
Imuta, K., and Ouchi, K. (1967). *Nenryo Kyokai-Shi* **46,** 889.
Imuta, K., and Ouchi, K. (1971). *Nenryo Kyokai-Shi* **50,** 88.
Jarolim, V., Hejno, K., Streibl, M., Horak, M., and Sorm, F. (1961). *Collect. Czech. Chem. Commun.* **26,** 451, 459.
Jarolim, V., Hejno, K., Hemmert, F., and Sorm, F. (1965). *Collect. Czech. Chem. Commun.* **30,** 873.

Jarolimek, P., Wollrab, V., Streibl, M., and Sorm, F. (1965). *Collect. Czech. Chem. Commun.* **30,** 880.
Johns, R. B., Belsky, T., McCarthy, E. D., Burlingame, A. L., Haung, P., Schnoes, H. K., Richter, W., and Calvin, M. (1966). *Geochim. Cosmochim. Acta* **30,** 1191.
Johnson, R. L., and Jones, L. A. (1968). *Anal. Chem.* **40,** 1728.
Jurriens, G. (1965). *Ital. Sostanze Grusse* **42,** 116.
Karr, C., and Chang, T. (1959). *Anal. Chim. Acta* **21,** 474.
Karr, C., and Chang, T. (1961). *Anal. Chim. Acta* **24,** 343.
Karr, C., and Chang, T. (1962). *Anal. Chim. Acta* **26,** 410.
Kimble, B. J., Maxwell, J. R., Philp, R. J., Eglinton, G., Albrecht, P., Ensminger, A., Arpino, P., and Ourisson, G. (1974). *Geochim. Cosmochim. Acta* **38,** 1165.
Kochloefl, K., Schneider, P., Reritha, R., and Horak, M. (1963). *Chem. Ind. (London)* 692.
Kovats, E. Z. (1958). *Helv. Chim. Acta* **41,** 1915.
Kuksis, A. (1964). *Biochemistry* **3,** 1086.
Kvenvolden, K. A. (1966). *Nature (London)* **208,** 573.
Levy, E. J., and Paul, D. G. (1967). *J. Gas Chromatogr.* **5,** 136.
Leythaeuser, D., and Welte, D. H. (1969). *In* "Advances in Organic Geochemistry, 1968" (P. A. Schenck and I. Havenaar, eds.), p. 429. Pergamon, Oxford.
Lindeman, L. P., and Adams, J. Q. (1971). *Anal. Chem.* **43,** 1245.
McCarthy, E. D., and Calvin, M. (1967). *Tetrahedron* **23,** 2609.
McCarthy, E. D., Van Hoeven, W., and Calvin, M. (1967). *Tetrahedron Lett.* **44,** 37.
McNeil, D. (1961). *J. Appl. Chem.* **11,** 90.
McTaggart, N. G., Luke, L. A., and Wood, D. (1971). *In* "Gas-chromatography 1970" (R. Stock, ed.), p. 35. Inst. Pet., London.
Maher, T. P. (1963). *J. Chromatogr.* **10,** 324.
Maher, T. P. (1966). *J. Gas Chromatogr.* **4,** 355.
Mair, B. J. (1964). *Oil Gas J.* **62,** 130.
Marin-Mudrovcić, S., Mühl, J., and Šateva, M. (1972). *Nafta Broz* **12,** 593.
Maximov, S. P., Botneva, T. A., Rodionova, K. P., Larskaya, E. S., and Safonova, G. I. (1974). *In* "Advances in Geochemistry, 1973" (B. Tissot and F. Bienner, eds.), p. 349. Technip, Paris.
Maxwell, J. R. (1967). Ph.D. Thesis, Glasgow Univ.
Maxwell, J. R., Pillinger, C. T., and Eglington, G. (1971). *Q. Rev., Chem. Soc.* **25,** 571.
Maxwell, J. R., Cox, R. E., Ackman, R. G., and Hooper, S. N. (1972). *In* "Advances in Organic Geochemistry, 1971" (H. R. Von Gaertner and H. Wehner, eds.), p. 277. Pergamon, Oxford.
Merritt, C., Jr. (1969). *In* "Ancillary Techniques of Gas Chromatography" (L. S. Ettre and W. H. McFadden, eds.), p. 325. Wiley (Interscience), New York.
Mold, J. D., Means, R. E., and Ruth, J. M. (1966). *Phytochemistry* **5,** 59.
Münch, W. (1934). *Oel Kohle* **2,** 564.
Mulheirn, L. J., and Ryback, G. (1975). *Nature (London)* **256,** 301.
Murphy, M. T. J., McCormick, A., and Eglinton, G. (1967). *Science* **157,** 1040.
Myers, M. K., Stallsteimer, J., and Wims, A. M. (1975). *Anal. Chem.* **47,** 2010.
Novotny, M., Segura, R., and Zlatkis, A. (1972). *Anal. Chem.* **44,** 9.
O'Connor, J. G., Burrow, F. H., Norris, M. S., and Mathew, S. N. (1962). *Anal. Chem.* **34,** 82.
Oelert, H. H., and Hemmer, E. A. (1970). *Erdoel Kohle* **23,** 87.
Oelert, H. H., and Köser, H. J. K. (1977). *In* "Molecular Spectroscopy" (A. R. West, ed.), p. 915. Inst. Pet., London.
Oró, J., and Nooner, D. W. (1967). *Nature (London)* **213,** 1082.
Palmer, T. J., and Vahrman, M. (1972). *Fuel* **51,** 14.

Peterson, R. M., and Rodgers, J. (1972). *Chromatographia* **5,** 13.
Phillippi, G. T. (1965). *Geochim. Cosmochim. Acta* **29,** 1021.
Pichler, H., Ripperger, W., and Schwarz, G. (1970). *Erdoel Kohle* **23,** 91.
Poulet, M., and Roucaché, J. (1970). *Rev. Inst. Fr. Pet.* **25**(2), 127.
Presting, W., and Kreutzer, T. (1965). *Fette, Seifen, Anstrichm.* **67,** 334.
Pye, J. C. (1975). *Anal. Chem.* **47,** 1017.
Quader, S. A., and Vaidegaswaran, R. (1966). *Indian J. Technol.* **4,** 128.
Rahman, M., and Vahrman, M. (1971). *Fuel* **50,** 318.
Rhead, M. M., Eglinton, G., Draffen, G. H., and England, P. J. (1971). *Nature (London)* **232,** 327.
Richter, H. K., and Calvin, M. (1966). *Geochim. Cosmochim. Acta* **30,** 1191.
Rogers, M. A., Bailey, N. J. L., Evans, C. R., and McAlary, J. D. (1972). *Int. Geol. Congr., 24th, Montreal* Sect. 5, p. 48.
Rowan, R. J. (1961). *Anal. Chem.* **33,** 658.
Ruheman, S., and Raud, H. (1932). *Brennst. Chem.* **13,** 341.
Sawatzky, H., George, A. E., Smiley, G. T., and Montgomery, D. S. (1975). Divisional Rep. 74/74-RBS. Energy Res. Cent., London.
Sawatzky, H., George, A. E., Smiley, G. T., and Montgomery, D. S. (1976). *Fuel* **55,** 329.
Schenck, P. A. (1969). *In* "Advances in Organic Geochemistry" (P. A. Schenck and I. Havenaar, eds.), p. 261. Pergamon, Oxford.
Schlenk, W., Jr. (1951). *Fortschr. Chem. Forsch.* **2,** 92.
Scott, C. G. (1966). *J. Gas Chromatogr.* **4,** 4.
Scott, C. G., and Phillips, C. S. G. (1965). *In* "Gas Chromatography" (A. Goldup, ed.) p. 266. Inst. Pet., London.
Scott, C. G., and Rowell, D. A. (1960). *Nature (London)* **187,** 143.
Shlyakhov, A. F., Koreshkova, R. I., and Telkova, M. S. (1975). *J. Chromatogr.* **104,** 337.
Simoneit, B. R. T., and Burlingame, A. L. (1974). *In* "Advances in Organic Geochemistry, 1973" (B. Tissot and F. Bienner, eds.), p. 629. Technip, Paris.
Simonsen, J. L., and Barton, D. H. R. (1961). "The Terpenes," Vol. 3, p. 337. Cambridge Univ. Press, London and New York.
Sista, V. R., and Srivastava, G. C. (1976). *Anal. Chem.* **48,** 1582.
Smith, A. E. (1950). *J. Chem. Phys.* **18,** 150.
Smith, J. W. (1966). *Fuel* **45,** 233.
Smith, P. V. (1954). *Am. Soc. Pet. Geol., Bull.* **33,** 377.
Snyder, L. R. (1968). "Principles of Adsorption Chromatography," p. 160. Dekker, New York.
Snyder, L. R., and Fett, R. (1965). *J. Chromatogr.* **18,** 461.
Sodhi, J. S., and Vesely, V. (1977). *Erdoel Kohle, Erdgas, Petrochem. Brennst.-Chem.* **30,** 42.
Sojak, L., Hrivnak, J., Majer, P., and Janak, J. (1973). *Anal. Chem.* **45,** 293.
Spence, J. A., and Vahrman, M. (1965). *Chem. Ind. (London)* p. 1522.
Spence, J. A., and Vahrman, M. (1966). *Analyst* **91,** 324.
Spence, J. A., and Vahrman, M. (1967). *J. Appl. Chem.* **17,** 143.
Spence, J. A., and Vahrman, M. (1970). *Fuel* **49,** 395.
Steel, G., Reed, W. E., and Henderson, W. (1972). *In* "Advances in Organic Geochemistry 1971" (H. R. Von Gaertner and H. Wehner, eds.), p. 353. Pergamon, Oxford.
Stevens, N. P., Bray, E. E., and Evans, E. D. (1956). *Am. Assoc. Pet. Geol., Bull.* **40,** 975–983.
Streibl, M., and Herout, V. (1969). *In* "Organic Geochemistry" (G. Eglinton and M. T. J. Murphy, eds.), p. 44 Longman, London.
Streibl, M., Jarolimek, P., and Wollrab, V. (1964). *Collect. Czech. Chem. Commun.* **29,** 2522.
Swern, D. (1955). *Ind. Eng. Chem.* **47,** 216.
Teichmüller, M. (1974). *In* "Advances in Organic Geochemistry, 1973" (B. Tissot and F. Bienner, eds.), p. 379. Technip, Paris.

Tissot, B., Califet-Debyser, Y., Deroo, G., and Ouchi, J. L. (1971). *Am. Assoc. Pet. Geol., Bull.* **55,** 2177.
Vahrman, M. (1970). *Fuel* **49,** 5.
Van den Dool, H., and Kratz, P. D. (1963). *J. Chromatogr.* **11,** 463.
Van Dorsselaer, A., Ensminger, A., Spyckerelle, C., Dastillung, M., Sieskind, O., Arpino, P., Albrecht, P., Ourisson, G., Brooks, P. W., Gaskell, S. J., Kimble, B. J., Philp, R. J., Maxwell, J. R., and Eglinton, G. (1974). *Tetrahedron Lett.* p. 1349.
Van Hoeven, W., Haug, P., Burlingame, A. L., and Calvin, M. (1966). *Nature* (*London*) **211,** 1361.
Vcelák, V. (1959). *Chem. Technol. Montanwachses, Prague, Vydavatelstvi ČSL Akad. Věd.* pp. 142–161.
Vidal-Madjar, C., and Guiochon, G. (1974). *In* "Separation and Purification Methods" (S. P. Edmond and J. V. Eligruska, eds.), Vol. 2, p. 1. Dekker, New York.
Vigdergauz, M. S., and Martynov, A. A. (1971). *Chromatographia* **4,** 463.
Wachowska, M., Nandi, B. M., and Montgomery, D. S. (1972). Divisional Rep. FRC 72/94-RBS. Fuels Res. Cent., DEMR.
Waddington, W. (1961). *Coal Tar Sci.* **10,** 18.
Watkins, P. V. (1967). "Standard Methods for Testing Tar and Its Products," p. 494. Standardization of Tar Products Tests Committee (now part of B.C.R.A., Chesterfield, England).
Wehner, H. (1974). *In* "Advances in Organic Geochemistry, 1973" (B. Tissot and F. Bienner, eds.), p. 409. Technip, Paris.
Wehrli, F. W., and Wirthlin, T. (1976). "Interpretation of Carbon-13 NMR Spectra," p. 41. Heyden, London.
Welte, D. H. (1967). *Erdoel Kohle* **20,** 65.
Welte, D. H. (1970). *In* "Advances in Organic Geochemistry, 1966" (G. D. Hobson and G. C. Speers, eds.), p. 111. Pergamon, Oxford.
Welte, D. H. (1974). *In* "Advances in Organic Geochemistry, 1973" (B. Tissot and F. Bienner, eds.), p. 3. Technip, Paris.
Welte, D. H., and Ebhardt, G. (1968). *Geochim. Cosmochim. Acta* **32,** 465.
Whitehead, E. V. (1974). *In* "Advances in Organic Geochemistry, 1973" (B. Tissot and F. Bienner, eds.), p. 225. Technip, Paris.
Whitehead, J. C., and Williams, D. F. (1975). *J. Inst. Fuel* **48,** 182.
Wiener, H. (1947). *J. Am. Chem. Soc.* **69,** 17.
Williams, R. B., and Chamberlain, N. F. (1963). *World Pet. Congr. Proc., 6th, Frankfurt, 1962* Sect. V, p. 17.
Wilman, W. G. (1966). "Chemical Examination of Road Tars from the Hercies Road Experiment," Rep. 0359. Coal Tar Res. Assoc. (now part of B.C.R.A., Chesterfield, England).
Wise, W. S. (1970). *Chem. Ind.* (*London*) p. 950.
Wollrab, V., and Streibl, M. (1969). *In* "Organic Geochemistry" (G. Eglinton and M. T. J. Murphy, eds.), p. 577. Longman, London.
Wollrab, V., Streibl, M., and Sŏrm, F. (1963). *Collect. Czech. Chem. Commun.* **28,** 1316.
Wollrab, V., Streibl, M., and Sŏrm, F. (1965). *Collect. Czech. Chem. Commun.* **30,** 1654.
Wooton, D. L., Dorn, H. C., Taylor, L. T., and Coleman, W. M. (1976). *Fuel* **55,** 224.
Wooton, D. L., Dorn, H. C., Taylor, L. T., and Coleman, W. M. (1978). *Fuel* **57,** 17.
Wszolek, P. C., Emilio, G., and Burlingame, A. L. (1972). *In* "Advances in Organic Geochemistry, 1971" (H. R. von Gaertner and H. Wehner, eds.), p. 229. Pergamon, Oxford.
Zimmerschied, W. J., Dinerstein, R. A., Weitkamp, A. W., and Marschner, R. E. (1950). *Ind. Eng. Chem.* **42,** 1300.

Part VI

MINERALS IN COAL

Chapter 26

Analysis of Mineral Matter in Coal

R. G. Jenkins *P. L. Walker, Jr.*
DEPARTMENT OF MATERIALS SCIENCE AND ENGINEERING
THE PENNSYLVANIA STATE UNIVERSITY
UNIVERSITY PARK, PENNSYLVANIA

I. INTRODUCTION

Coals are complex mixtures of organic and inorganic species. The inorganic fraction† is primarily composed of mineral species and, to a lesser extent, associations of inorganic elements with organic material, e.g., organometallic compounds and exchangeable cations. The phrase *mineral matter content* generally refers to the "noncoal" material contained within a coal. In this chapter we discuss only methods for the analysis of the mineral phases. Elsewhere in these volumes there are sections on the identification of minerals in lignites, including determination of the exchangeable cations (Chapter 27) and trace element analyses (Volume I, Chapters 11–15).

† The term *inorganic fraction* is used somewhat loosely because elements such as carbon, oxygen, and sulfur are contained in many mineral species (e.g., carbonates, oxides, and sulfides).

ISBN 0-12-399902-2

A. Nature and Occurrences of Minerals Associated with Coals

There are several sources of mineral matter found in coals. First, there is inorganic material generated from the plants which form the coal swamp. In addition, inorganic compounds were introduced from outside sources by mechanisms such as erosion either into the decaying vegetation or, at a later stage, into the coal seam by percolation through cracks or fissures (e.g., cleat filling). Mackowsky (1956) used the term syngenetic for minerals that were introduced during coal formation and the term epigenetic for those introduced after formation. There are other ways of classifying mineral matter in coals. One of the most widely used classifications (Edgecombe and Manning, 1952) is to subdivide mineral matter into either (a) inherent mineral matter, that which generally arises from plant material in the coal swamp and is so intimately associated with the organic fraction that it cannot be easily removed by physical methods, or (b) adventitious mineral matter, which is material that can be separated from the coal substance. It can be seen that inherent mineral matter is generally of syngenetic and epigenetic materials.

Another widely used classification is to apply terminology which has been utilized in describing sediments and sedimentary rocks. In this classification minerals which were transported by erosion (wind or water) are called detrital and those minerals which were formed within the coal swamp (or at a later stage in coalification) by precipitation of ions from solution are termed authigenic. Examples of detrital minerals are quartz and clays, and the most important examples for authigenic minerals are pyrite and carbonates.

As Given and Yarzab (1975) point out, the precise distribution of minerals in a coal will depend on its geological environment. Such factors as the surrounding geological setting and the nature of the groundwater are of prime importance. It has to be remembered that some of the mineral matter can be derived from seams of minerals which intersect with a coal seam. Such minerals can be included with the coal during the mining process.

From this discussion it can be seen that there are many potential problems in obtaining a representative sample of the mineral matter associated with a specific coal. As extreme examples, a sample taken from the mining process almost certainly will contain large amounts of mineral matter which will be removed during cleaning. On the other hand, specially selected samples of macerals (often used for research purposes) can contain very small amounts. O'Gorman and Walker (1972), not unexpectedly, found significant differences in the amount and composition of mineral matter among samples of whole coals and constituent lithotypes.

Many mineral species have been reported as being associated with coals (Nelson, 1953; Watt, 1968; O'Gorman and Walker, 1972; Gluskoter, 1975). A partial list of these minerals is given in Table I. In general, it is found that most of these minerals appear only in small concentrations in coals. The most commonly occurring types are quartz, clays, carbonates, and sulfides. Watt (1968) suggests that, for practical purposes, it is sufficient to consider that mineral matter is generally composed of six minerals, at most. These minerals would be one or two clays, one or two carbonates, pyrite, and quartz.

As a group of minerals, clays are the most frequently occurring inorganic constituents of coals. They are aluminosilicates which can contain a wide range of other cations, Ca^{2+}, Na^{+}, and K^{+}, for example. The most commonly reported clay minerals found in coals are montmorillonite,

TABLE I *Minerals Found in Coals*

Silica minerals:	Quartz (trigonal), SiO_2
Chlorite:	$(Mg, Al, Fe)_{12}[(Si, Al)_8O_{20}](OH)_{16}$
Serpentine:	$Mg_3[Si_2O_5](OH)_4$
Clay minerals:	Kaolinite group, $Al_4[Si_4O_{10}](OH)_8$
	Illite, $K_{1-1.5}Al_4[Si_{7-6.5}Al_{1-1.5}O_{20}](OH)_4$
	Montmorillonite group $(\frac{1}{2}Ca, Na)_{0.7}(Al, Mg, Fe)_4[(Si, Al)_8O_{20}](OH)_4 \cdot nH_2O$
Feldspar group:	Alkali feldspars, $(K, Na)[AlSi_3O_8]$
	Plagioclase, $Na[AlSi_3O_8]$–$Ca[Al_2Si_2O_8]$
Sulfates:	Gypsum, $CaSO_4 \cdot 2H_2O$
	Anhydrite, $CaSO_4$
	Hemihydrate, $CaSO_4 \cdot \frac{1}{2}H_2O$ (bassanite)
	Barytes, $BaSO_4$
Sulfides:	Pyrite (cubic), FeS_2
	Marcasite (orthorhombic), FeS_2
	Pyrrhotite, $Fe_{1-x}S$
	Chalcopyrite, $CuFeS_2$
	Sphalerite, ZnS
Carbonates:	Ankerite, $Ca(Mg, Fe^{2+}, Mn)(CO_3)_2$
	Calcite (trigonal), $CaCO_3$
	Aragonite (orthorhombic), $CaCO_3$
	Magnesite, $MgCO_3$
	Rhodochrosite, $MnCO_3$
	Siderite, $FeCO_3$
	Dolomite, $CaMg(CO_3)_2$
	Strontianite, $SrCO_3$
	Witherite, $BaCO_3$
Rutile:	TiO_2

illite, and kaolinite. Mixed layer clays are also present, the most abundant being mixed layer illite–montmorillonite.

The carbonates generally listed as occurring in coals are calcite, dolomite, ankerite, and siderite. However, in many cases the composition of the carbonates is somewhat complex because these compounds form solid solutions (Watt, 1968; Gluskoter, 1975).

By far the most frequently occurring inorganic sulfur compound found in coals is pyrite. Marcasite has also been reported; marcasite and pyrite have the same chemical composition but differ in their crystallographic structures. The reduced sulfide pyrrhotite ($Fe_{1-x}S$) is absent or in very low concentrations in coals. Occasionally, other sulfide minerals, e.g., galena and sphalerite, have been found in small concentrations. Sulfates, in general, are uncommon, although they can appear in weathered coals (Watt, 1968; Gluskoter, 1975).

Perhaps the most commonly detected mineral phase is quartz. It is readily identified by optical microscopy, and its occurrence has been reported for coals from many locations.

B. Importance of Minerals in Coal Utilization

In recent years, with the increased interest in using large quantities of coal to fulfill the world's energy requirements, it has become increasingly obvious that there is a need to understand the nature and behavior of mineral matter during coal utilization. Mineral matter content, generally, represents a significant proportion of a coal's composition. The amount varies from seam to seam; values of up to 32 wt % have been reported (O'Gorman and Walker, 1972). Gluskoter (1975) has estimated an "average" amount of 15 wt % for North American coals. For seams with high mineral matter content the economic viability of mining such a seam comes into question.

In general, mineral matter is considered to be undesirable and detrimental in coal utilization. Its presence affects almost every aspect of mining, preparation, and utilization. In coal mining and transportation mineral matter is a diluent and, therefore, undesirable. Coal preparation and beneficiation are aimed at reducing the quantity of these diluents and improving the quality of coal feedstocks. The efficient use of these techniques is very much dependent on the concentration and composition of mineral matter, because the methods utilized are related to a number of its properties, e.g., specific gravity, friability, size, and shape. From a knowledge of mineral composition a coal preparation engineer can adjust the preparation techniques accordingly. The mineralogical composition of coal refuse is also important because it is a potential source of water pollution.

No matter how effective the coal preparation technique, there is always a significant amount of residual mineral matter. This residual material is of considerable importance in coal utilization. The quality of coke is related to its ash and sulfur content, which are both dependent on the mineral composition of the feed coal. It is thought that the inorganic constituents of coking coals can have a marked effect on yields of carbonization products, structure, strength, and reactivity of the resulting coke. The presence of inorganic species in coke can be advantageous since some of them can act as catalysts and, thus, increase reactivity (see Chapter 32, Section III,B,5).

When coal is burned in a combustion unit, mineral matter undergoes major changes which lead to problems of clinker formation, fly ash, slagging, and boiler tube corrosion (Ely and Barnhart, 1963; Borio *et al.*, 1968; Watt, 1969). Recently Mitchell and Gluskoter (1976) have determined some of the mineral transformations which take place under oxidizing atmospheres at high temperatures. As a result of environmental considerations, one of the most important aspects of coal analysis has become that of the forms of sulfur. The concentration of pyrite (FeS_2) is of special significance in the production of oxides of sulfur. The efficiency of a combustion unit is related to the amount of ash produced, since it is a diluent. Disposal of ash can result in large capital expenditures. On the positive side, ash has been utilized as a construction material and is a possible source of refractories.

The renewed interest in coal gasification and liquefaction has also produced a need for a better understanding of the behavior of minerals in these processes. The role of minerals as catalysts for gasification reactions is well covered in Chapter 32. However, consideration must also be given to possible poisoning of methanation catalysts by products of reactions involving minerals. The behavior of coal minerals in liquefaction processes is still not understood but is the subject of widespread research (e.g., Given, 1974; Walker *et al.*, 1975, 1977; Tarrer *et al.*, 1977; Whitehurst *et al.*, 1977). One of the major problems encountered in coal liquefaction is the removal of insolubles (minerals, unreacted coal, and insoluble products) from the product stream. Filtration rates are generally found to be slow, partially because of the very fine size of the minerals. It is also known that the use of certain coals in liquefaction units leads to enhanced abrasion of valves and pumps. Such a phenomenon can be related to the mineralogical composition of the feed coal. Walker *et al.* (1977) have shown that during liquefaction most of the major mineralogical components of bituminous coals undergo only minor changes, e.g., dehydration. However, these workers and many others have found that pyrite is reduced to a pyrrhotite ($Fe_{1-x}S$). It has been suggested that iron sulfides play a catalytic role in coal liquefaction. Pyrrhotites have widely differing magnetic properties (Ward, 1970; Walker *et al.*, 1977), which is of major interest in the proposed use of magnetic separation techniques for mineral removal from liquefaction product streams (Maxwell *et al.*, 1976).

Analyses of the minerals in coals are important for accurate determination of the ultimate analyses of coals. This topic is fully discussed in this volume (Chapter 20, Section V,A) and elsewhere (Given and Yarzab, 1975; Given, 1976).

II. SEPARATION OF MINERALS FROM COAL

Although the concentration of minerals in a coal is appreciable, for analytical purposes it is desirable that they be separated from the coal in an unaltered form. In early studies, techniques were developed which were based on density separation. These techniques are considered to be unsatisfactory because they tend to produce mineral-enriched materials rather than fractions of isolated minerals. The solution to these problems has been found in methods which oxidize the carbonaceous constituents of coals and leave the minerals essentially unchanged. It should be noted that high temperature oxidation of coal produces an ash, the composition of which is very different from the constituent minerals. For example, under ashing conditions prescribed by ASTM (1973) standards (750°C), pyrite is oxidized to ferric oxide and sulfur dioxide, carbonates form oxides, and clays lose all water (both adsorbed and interlayer). About the only mineral which remains unaltered is quartz. Methods were developed in which coal was oxidized in a stream of molecular oxygen at atmospheric pressure, at comparatively low temperatures (300–400°C) (Watt, 1968). However, even under these conditions pyrite was totally oxidized to iron oxides and sulfur dioxide; the sulfur dioxide evolved reacted with carbonates to produce sulfates.

Many of the aforementioned problems were overcome with the advent of radio-frequency low temperature ashing, developed by Gleit (1963). In this method low pressure oxygen (1–3 torr) is activated by a radio-frequency discharge. The excited oxygen-containing atoms and free radicals oxidize the carbonaceous constituents at comparatively low temperatures (~150°C) (O'Gorman and Walker, 1972). Gluskoter (1965) found this method to be suitable for the liberation of coal minerals in a comparatively unaltered form. Since then the method has been widely used for the quantitative analyses of minerals associated with coals (Gluskoter, 1967; Estep *et al.*, 1968; O'Gorman and Walker, 1971, 1972; Rao and Gluskoter, 1973; Walker *et al.*, 1975, 1977; Painter *et al.*, 1978). Low temperature ashing (LTA) units are commercially available. In these instruments coal is spread out on dishes and operating conditions, such as radio-frequency (rf) power and oxygen flow rate, can be altered.

Frazer and Belcher (1973) studied the applicability of LTA as a method of determining mineral matter content for a series of Australian coals. They found that some pyrite can be oxidized and that, to some extent, organic sulfur can be fixed as sulfates. However, their samples were of comparatively low sulfur content. The rates of these reactions are functions of operating conditions, e.g., rf power level and oxygen flow rate. As an example, these investigators showed that at a high power level (167 W) and a high oxygen flow rate (1000 cm^3/min), siderite remained unchanged, whereas pyrite was oxidized to hematite. However, at lower settings of these variables (50 W, 300 cm^3/min) pyrite was stable and there was no irreversible water loss from clays. Gluskoter (1965) and Frazer and Belcher (1973) report that during LTA gypsum ($CaSO_4 \cdot 2H_2O$) is dehydrated to hemihydrate ($CaSO_4 \cdot \frac{1}{2}H_2O$).

Miller (1977) has made an extensive study of the changes in mineralogical composition which occur during LTA. From experiments with powdered specimens of a "pure" pyrite (from Rico, Colorado) mixed with a graphite, he found that the fraction of pyrite oxidized in 24 hr was proportional to the rf power level, e.g., after 24 hr at 50 W about 7% of the pyrite was oxidized. However, in a series of experiments in which several pyrites from coals and mineral sources were exposed alone to the same ashing conditions (50 W of rf power), somewhat variable results were noted. In some samples oxidation to hematite was rapid, whereas others remained unchanged after 48 hr, which suggests that the presence of defects or impurities in pyrite can catalyze or inhibit its oxidation rate. However, in experiments with several coals, Miller (1977) found that pyrite oxidation in these samples was comparatively slow; it would only become a problem if ashing lasted for a period of about 5 days. Oxidation of pyrite in coal was minimized by ashing for 48 hr, with frequent stirring.

The fixation of organic sulfur as inorganic sulfates has also been examined by Miller (1977). He found that with decreasing rf power there is an increase in sulfate formation. Below about 50 W of rf power the effect is most marked. During ashing, organic sulfur is oxidized to SO_3 which forms sulfuric acid or a hydrate of SO_3 in the presence of moisture. These acidic compounds can then react with carbonate minerals or, in the case of low rank coals (subbituminous and lignites), with exchangeable cations associated with carboxylic acid groups. It is suggested that the reason for the observed increase in fixation with decreasing rf power is that at lower power levels, the chamber temperature is lower, which enhances the condensation of acidic compounds onto the partially ashed sample. Ashing of low rank coals is always difficult; rates are much slower than those for bituminous coals and sulfate fixation by the exchangeable cations is a serious drawback. Removal of these cations by dilute acid washing does speed up ashing rates and removes sulfate fixation, but it must be remembered that this treatment will leach out acid-soluble minerals. It is presumed that the exchangeable cations bring about deactivation of active oxygen species which results in a retardation of oxidation rate.

Other factors affect rates of LTA, e.g., particle size and depth of sample bed. Oxidation initially takes place at the surface of the coal particles, and as a layer of mineral matter is liberated it inhibits the diffusion of active species into the center of the particle. Thus, small particles are desirable. However, if extremely fine material is used, care must be taken during evacuation and repressurization to prevent these particles from being blown out of the sample boat. Coal should also be spread out in a thin layer in the sample dish so that diffusion through the bed does not retard oxidation. Frazer and Belcher (1973) suggested a sample layer density of 70 mg/cm^2 of sample boat, but their ashing times (4–5 days) are now considered to be too long. Miller (1977) found that with a sample layer density of about 25 mg/cm^2 of sample dish and with stirring every 2–3 hr during the first day, between 95 and 99% of the combustible material is removed in 48 hr. It should be remembered that these observations are for ashing bituminous coals.

In light of the preceding discussions, Miller (1977) defined a routine method for the determination of mineral matter content for bituminous coals and anthracites. This method, which is now summarized, has been found to be very useful for the preparation of LTA for analysis.

In all his experiments Miller used a single-chamber oxygen plasma unit produced by International Plasma Corporation California (Model 1001B). It can produce 150 W of rf power at an efficiency (discharge power/reflected power) in excess of 95%. Samples are contained in 9-mm-diameter Pyrex Petri dishes. Vacuum is produced by a rotary pump. The ashing method outlined by Miller is as follows:

(i) A known weight (~1.5 g) of vacuum-dried coal (<80 mesh, Tyler) is spread evenly in a preweighed sample dish so that the sample layer density does not exceed 30 mg/cm^2 of the sample dish.

(ii) Ashing should take place at 50 W net rf power with an oxygen flow rate of 100 cm^3/min. Chamber pressure should be maintained at about 2 torr.

(iii) Samples are stirred three times within an initial 8-hr period and then oxidation is allowed to proceed for a further 16 hr.

(iv) Following the 24-hr period, the ashing chamber should be purged with dry air. Then the dish and its contents are removed, and the sample is stirred in a desiccator prior to weighing.

(v) Samples are reinserted into the asher. After 2 hr samples are removed, stirred, weighed [as in step (iv)], and replaced into the asher. This procedure should be repeated every 2 hr until weight loss is less than 2 mg in a 2-hr period. Ashing may continue for a further 24 hr but, in general, a total of 30–36 hr should be sufficient.

(vi) A representative sample of the final LTA should be analyzed for residual carbon content after carbonates have been removed by washing with 3 *N* HCl.

Thus, by the application of this procedure samples of LTA can be prepared for subsequent analyses by methods described in the following section. Enough coal should be ashed to provide a minimum of 500 mg of LTA.

III. ANALYTICAL PROCEDURES

Generally, it can be said that no single method will yield a complete analysis of the mineral matter in coal; it is often desirable to employ a combination of methods. In the following descriptions of analytical procedures it is assumed that samples are fully representative of the coal under investigation. It is imperative that samples be stored in dry, inert atmospheres, thus precluding the effects of oxidation.

A. X-Ray Diffraction

All analyses using x-ray diffraction should be performed on LTA.

1. *Qualitative and Semiquantitative Analyses*

X-ray diffraction is considered to be the best method for mineral identification. However, its application can be limited because of orientation effects. Therefore, it is necessary that a reliable method of sample preparation be used. Rao and Gluskoter (1973) investigated several methods of sample preparation for semiquantitative and quantitative analyses. They found that water smears and alcohol water smears can produce chemical reactions and orientation effects. It was noted that small amounts of water can react with sulfates, producing sulfuric acid which then reacts with carbonates. They suggested that a cavity-mount technique be used in all x-ray analyses. This technique consists of placing finely divided LTA (ground in an agate mortar and pestle) into a thin aluminum holder, covered by a clean glass slide. The powder is then gently tapped into the cavity, leveled, and covered by another clean glass slide. Finally, the sample holder is inverted and the slide covering the top surface removed carefully prior to placing the holder in the x-ray diffraction unit.

In the Mineral Constitution Laboratories, Pennsylvania State University, it has been found that such stringent precautions need not be applied for qualitative analysis (Suhr and Gong, 1977). Samples of −200 mesh (Tyler) LTA are ground in absolute alcohol for 3–5 min and then dispersed as a slurry onto a glass slide. The sample is then dried before it is placed in a diffractometer.

X-ray diffraction profiles are determined by use of a conventional diffractometer system,† e.g., a Phillips Norelco x-ray diffraction unit. X-radiation must be monochromatic; the most commonly used is Cu$K\bar{\alpha}$. For qualitative analysis the specimen is scanned over a wide angular range [2–50° (2θ) for Cu$K\bar{\alpha}$] to ensure that all of the major diffraction peaks of the component minerals are recorded. Diffraction spacings are then calculated from the peak positions. Qualitative analysis is carried out by use of standard tables (or card indices) of diffraction spacings. Table II lists the principal x-ray diffraction spacings of commonly occurring coal minerals.

It is also possible to identify the presence of expandable layer clays (e.g., montmorillonite) by subjecting samples to treatment with ethylene glycol vapor at 60°C for 1 hr (O'Gorman and Walker, 1972). This treatment causes a characteristic swelling of montmorillonite layers, resulting in an increase of basal spacing to about 17 Å.

Semiquantitative x-ray diffraction data can be obtained by measuring the relative intensities of characteristic diffraction peaks from selected minerals (Rao and Gluskoter, 1973). For this type of determination, the scanning speed of the diffractometer should be comparatively slow (0.5°/min); the sample must be in a cavity-mount and it is recommended that the sample be rotated by use of a spinner. The data obtained by these measurements can only be considered to be approximate estimates of mineral concentration because the method does not take into account differences in absorption coefficients of the minerals.

† For detailed descriptions of procedures and apparatus the reader is referred to Klug and Alexander (1974).

TABLE II *Principal x-Ray Diffraction Spacings of Commonly Occurring Coal Minerals*[a]

Mineral	Diffraction spacing (Å)
Kaolinite	7.15(100), 3.57(80), 2.38(25)
Illite	10.1(100), 4.98(60), 3.32(100)
Montmorillonite	12.0–15.0(100)
Chlorite	14.3(100), 7.18(40), 4.79(60), 3.53(60)
Mixed layer illite–montmorillonite	10.0–14.0(100)
Calcite	3.04(100), 2.29(18), 2.10(18)
Dolomite	2.88(100), 2.19(30)
Siderite	3.59(60), 2.79(100), 2.35(50), 2.13(60)
Aragonite	3.40(100), 3.27(52), 1.98(65)
Pyrite	3.13(35), 2.71(85), 2.42(65), 2.21(50)
Marcasite	3.44(40), 2.71(100), 2.41(25), 2.32(25)
Quartz	4.26(35), 3.34(100), 1.82(17)
Gypsum	7.56(100), 4.27(50), 3.06(55)
Rutile	3.26(100), 2.49(41)
Feldspars	3.18–3.24(100)

[a] Relative intensities are shown in parentheses.

2. *Quantitative Analyses*

X-ray diffraction procedures are used for quantitative analysis of pyrite, calcite, and quartz in LTA (Rao and Gluskoter, 1973; Walker *et al.*, 1975; Suhr and Gong, 1977). The first step of the technique is the preparation of mineral mixtures of known compositions, which are used for calibration purposes. The calibration mixtures are made up of known proportions of calcite, pyrite, quartz, and clay mixture (equal amounts of illite, montmorillonite, and kaolinite). The proportions of these minerals are varied to give a wide range of relative compositions. Calcium fluoride (CaF_2) is the most widely used internal standard. A 1-g sample of a calibration mixture is thoroughly mixed with 0.2 mg of internal standard. It is recommended that mixing take place in a Spex or Wig-L-Bug mixer mill for about 0.5 hr. Good mixing and grinding is critical because it improves the uniformity of particle size and composition. About 0.5 g of the calcium fluoride mixture is placed in a cavity-mount, by the method described earlier. The loaded sample holder is then positioned in the x-ray beam in a rotating sample mount (a "spinner"). Use of a spinner improves reproducibility because it reduces orientation and sampling errors (Walker *et al.*, 1975). When the sample holder is large enough, a relatively divergent primary beam should be used (2–4°)

in order that a larger specimen volume be irradiated (Klug and Alexander, 1974). The specimen is then scanned at a comparatively slow speed [0.25–0.50° (2θ)/min] over the required angular range. As in any quantitative x-ray procedure, it is important that the x-ray generator and counter electronics be stabilized. The counting device (proportional or scintillation counters) should have a short resolving time (1–2 μsec) and a large range of linear response. Other instrumental parameters, such as chart speed, should be set so that accurate peak height determinations can be made. It is suggested that duplicate diffraction profiles be recorded.

After the profiles are recorded, peak height measurements are made on the following reflections: (001) quartz, (104) calcite, (311) pyrite, and (111) calcium fluoride. The ratio of the intensity of a mineral peak to that of the internal standard is determined for each of the calibration mixtures. For each mineral, it is then possible to draw a calibration curve of peak height ratio versus weight fraction. An example is given in Fig. 1, which is for quartz. It can be seen that this plot is linear over a wide range of compositions.

For the analysis of LTA, about 0.5 g of ash is mixed with 0.2 g of calcium fluoride, in the manner just described. Sample preparation and

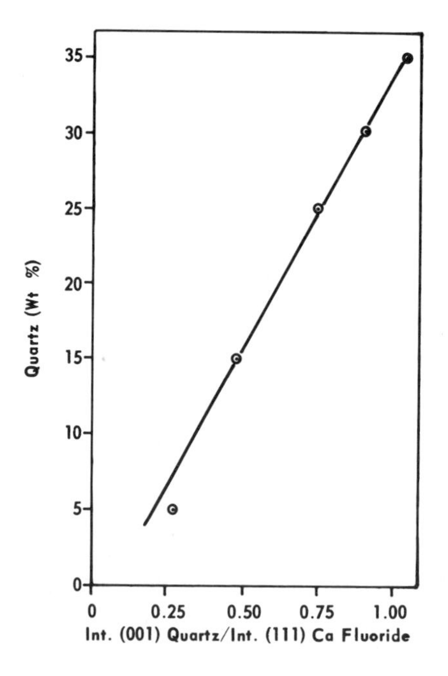

Fig. 1 X-ray calibration plot for quartz.

instrumental conditions should be exactly the same as used in the calibration runs. The peak height ratios (mineral to standard) are calculated and then converted into absolute weight fractions by reading off the appropriate calibration curve. Profiles should be run in duplicate and where possible two peaks for each mineral should be used. These precautions act as checks on sampling, intensity counting, and possible peak interferences.

Although the use of quantitative x-ray procedures has met with some success for mineralogical analyses of coal LTA (Rao and Gluskoter, 1973; Walker *et al.*, 1975), it does have some drawbacks. It cannot be considered as a high precision method for the following reasons:

(i) It is difficult to achieve good, uniform mixing.

(ii) Calibration curves are not always linear (care must be taken in selection of "pure" minerals for calibration standards).

(iii) Minerals that show preferred orientation, e.g., platelike or needle-shaped, are extremely difficult to measure quantitatively. Thus, it is not recommended for routine analyses of clay mineral content. For details of clay mineral determinations the reader is referred to Rao and Gluskoter (1973) and Klug and Alexander (1974).

(iv) For samples with high illite and pyrite concentrations there are problems with overlapping peaks.

Partially because of these reasons the best precision of this type of analysis is about ±10%. However, when one considers the problems associated with obtaining representative samples of coal, and hence coal minerals by LTA, this degree of precision is not unreasonable.

In conclusion, x-ray diffraction provides us with, perhaps, the best tool for mineral identification. Its application to quantitative analysis has met with some success for pyrite, quartz, and calcite, but for clay minerals it is somewhat limited. As with all techniques for the mineralogical analyses of coals, it is prudent to supplement these analyses with other techniques, e.g., infrared spectroscopy and chemical methods.

B. Infrared Spectroscopy

This section is divided into two subsections, the first deals with infrared analyses as determined by "classical," dispersive instruments, and the second gives details of the exciting potential of Fourier transform infrared (FTIR) spectroscopy to the analysis of coal minerals.

TABLE III *Infrared Absorption Bands for Standard Minerals*

Mineral	Source	Absorption bands (cm^{-1})
Kaolinite	Georgia	3695, 3665, 3650, 3620, 1108, 1025 1000, 910, 782, 749, 690, 530 460, 422, 360, 340, 268
Illite	Fithian, Illinois	3620, 1640, 1070, 1015, 920, 820 750, 510, 460
Sodium montmorillonite	Wyoming	3625, 3400, 1640, 1110, 1025, 915 835, 790, 515, 460
Calcite	Chihuahua, Mexico	1782, 1420, 871, 842, 710, 310
Dolomite	Lee, Mass.	1435, 875, 730, 390, 355, 310
Pyrite	Rico, Colorado	411, 391, 340, 284
Quartz	Minas Gerais, Brazil	1160, 1065, 790, 770, 687, 500 450, 388, 362, 256
Gypsum	Washington County, Utah	3605, 3550, 1615, 1150, 1110, 1090 1010, 660, 595, 450

1. *Analysis by Use of Dispersive Instruments*

Prior to the advent of satisfactory LTA techniques, the direct mineralogical analysis of coals by infrared spectroscopy was severely limited because the broad bands of the organic phase overlapped those from the constituent minerals. However, with the development of LTA for the removal of the organic fraction of coal, Karr *et al.* (1967) and Estep *et al.* (1968) demonstrated that a number of minerals occurring in coal could be identified and analyzed by classical infrared techniques. Table III lists the infrared absorption bands for a series of minerals which commonly occur in coals (Walker *et al.*, 1975).

Samples for examination by infrared spectroscopy are prepared by taking a small, representative amount (1 mg) of hand-ground LTA and grinding it with 200–300 mg of dried spectroscopically pure potassium bromide (KBr). At this concentration level the requirements of the Beer–Lambert law are met, i.e., absorbances are a linear function of concentration. Estep *et al.* (1968) have discussed the importance of sample preparation, the grinding time of which is especially important. As an example, Fig. 2 is a plot of the absorbance for the 1085-cm^{-1} band of quartz versus grinding time in a Wig-L-Bug (Painter *et al.*, 1978a). It will be noted that about 30 min is required to achieve maximum absorbance. It is important that the particles of the sample be reduced to such a size that they are smaller than the wavelength of the light to be used in infrared spectroscopy (van der Maas, 1972). Otherwise the

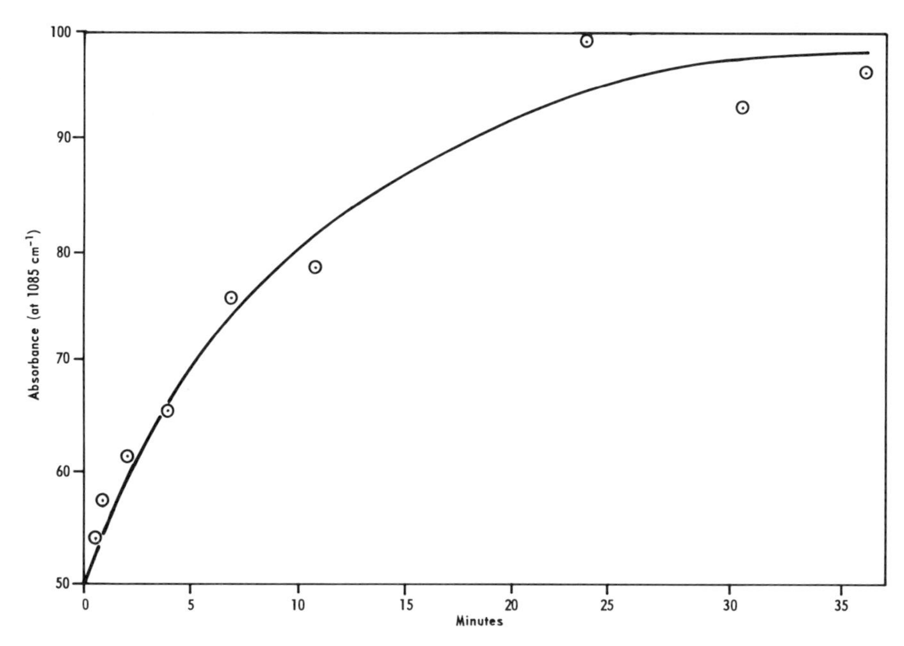

Fig. 2 Plot of peak height of 1085-cm^{-1} band of quartz versus grinding time. (From Painter *et al.*, 1978a.)

Christiansen effect, which is related to the difference between the refractive index of the sample and that of the matrix (KBr), becomes significant. This effect causes distortions in band shape.

The KBr mixture is then made into a pellet by use of a vacuum die and high pressures (40 tons/in.^{2}) so that a relatively optically clear sample is produced. An infrared scan is then made from 4000 to 200 cm^{-1} using a conventional dispersive instrument (e.g., a Perkin-Elmer 621 grating infrared spectrometer). For bands in the 650- to 200-cm^{-1} region, which includes pyrite bands, cesium iodide (CsI) should be used for the matrix powder instead of KBr because of its superior transparency at lower wave numbers.

Quantitative analyses of LTA using dispersive infrared instruments have met with only limited success because in such a complex mixture there is considerable band overlap. However, this method has been successful in the determination of kaolinite and gypsum concentrations (O'Gorman and Walker, 1972; Walker *et al.*, 1975; Suhr and Gong, 1977). The amount of kaolinite present in an LTA can be determined from the 910-cm^{-1} absorption band by use of the baseline method (O'Gorman and Walker, 1972). Figure 3 is used to illustrate this method; the

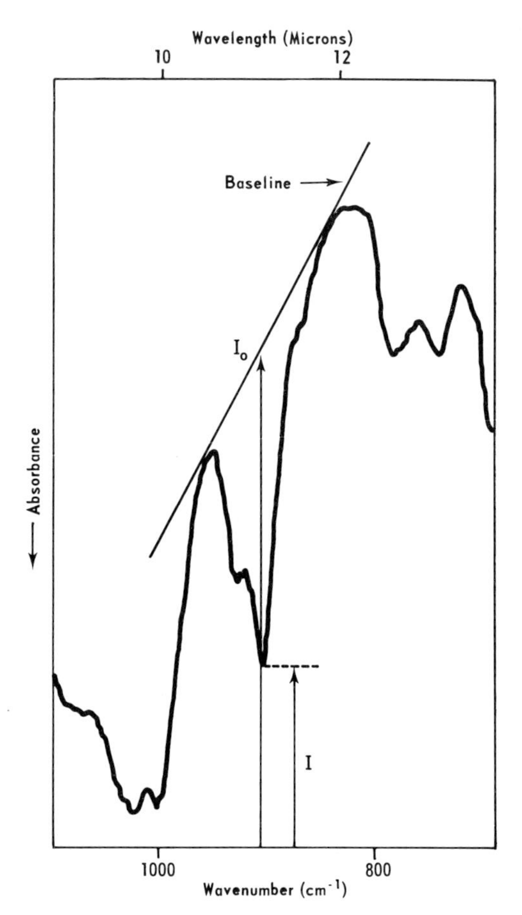

Fig. 3 Method of drawing baseline for plotting composition versus $\log(I_0/I)$. (From O'Gorman and Walker, 1972.)

transmitted radiation (I) is measured at the point of maximum absorption and the value of the incident radiation (I_0) is obtained by drawing a tangent to the shoulders of the band. The absorbance is calculated from $\log(I_0/I)$. A calibration curve is prepared from scans of several synthetic mineral mixtures which contain various proportions of a kaolinite mineral standard. A plot is made of concentration versus absorbance.

Gypsum is not determined directly because it will be recalled that during LTA some gypsum is dehydrated to hemihydrate. Thus, it is advisable to heat the ash to about 500°C to ensure complete dehydration to anhydrite ($CaSO_4$). The baseline method is then used in evaluating the 670-cm^{-1} anhydrite band. Calibration curves are determined in the same manner as for kaolinite. It then is a simple task to convert the concentration of anhydrite into that for gypsum.

In conclusion, conventional infrared spectroscopy is useful for identification purposes and for quantitative analysis for kaolinite and gypsum.

2. *Fourier Transform Infrared Techniques*

As described previously, the analysis of complex mixtures by infrared spectroscopy has been limited because many bands overlap and are superimposed. The introduction of FTIR spectroscopy has led to major improvements in the characterization of polymeric materials (Koenig, 1975). Recent studies in these laboratories (Painter *et al.* 1978a,b; Walker *et al.*, 1977) have shown that this technique can be successfully applied to the characterization of coals, coal-derived materials, and mineral matter.

Dispersive instruments are limited in their application because of energy-throughput considerations. The infrared beam is dispersed by prisms or gratings so that only a narrow frequency range falls on a set of slits. The prisms or gratings are rotated so that the intensity of the beam, modulated by the sample, is recorded as a function of frequency. In order to obtain high resolution spectra the slits must be narrow, severely limiting the amount of energy reaching the detector. For highly absorbing materials, e.g., coals, this problem can become acute. The major difference between FTIR and dispersive instruments is that the former do not contain dispersive elements or slits but employ an interferometer (for a full description, see Koenig, 1975). Utilization of an interferometer allows the detector to "see" all the energy at all times. In general, the energy throughput is between 80 and 100 times greater than for a dispersive instrument. Not all this advantage is achieved in practice, however, because there are inefficiencies in the detector.

The main drawback of using interferometers has been that an interferogram is produced rather than a spectrum. However, the spectrum can be obtained by performing a Fourier transform on the interferogram. The development of the fast Fourier transform and the advent of minicomputers have led to the production of commercial FTIR instruments. Using an interferometer, the total spectrum of a material is recorded at once; resolution depends on a movable mirror displacement. At a resolution of 2 cm^{-1} the complete spectrum is obtained in about 2 sec. The computer is used to coadd successive interferograms of the same sample. The signal-to-noise ratio of the final spectrum increases as the square root of such "scans." When weak bands are to be identified, as in the analysis of LTA, extremely high signal-to-noise ratios are of the utmost importance. Therefore, it is necessary to coadd about 400–500 scans for the analyses of these materials. For coals, an even higher number of scans may be necessary. Another advantage of the FTIR system is that spectra are stored in digital form on magnetic disks or tapes; thus, they can be recalled for purposes of comparison and mathematical manipulation at any time.

If there is no vibrational coupling between the constituents of a mixture, then its infrared spectrum can be represented as a sum of its constituents, assuming the Beer–Lambert law holds. Thus, the analysis of a multicomponent mixture by FTIR is based on the successive subtraction of the spectra of the individual components from the spectrum of the mixture. The spectrum of the component is subtracted by the computer until one of its characteristic bands just disappears from that of the mixture, i.e., the band is reduced to the baseline absorbance. If the weight of material in each sample (alkali halide pellet) and the fraction of the spectrum that is subtracted are known, the method is quantitative. The weight fraction x_1 of component 1 in a mixture is given by (Walker *et al.*, 1977; Painter *et al.*, 1978a)

$$x_1 = (K_B/K_A)(W_1/W_M) \tag{1}$$

where W_1 and W_M are the weights of component 1 and the mixture in the alkali halide pellets, and K_A and K_B are constants entered into the computer, which are adjusted on a trial-and-error basis until the band of component 1 is eliminated from the spectrum of the mixture. These constants (K_A and K_B) are multiplication factors which determine the fractional amount of the spectrum to be subtracted.

For the analysis of LTA, the spectra of individual component minerals are recorded on magnetic disks and stored in order to form a library of suitable reference spectra. In the FTIR studies in these laboratories, pellets were prepared by grinding samples in KBr (or CsI) in a Wig-L-Bug. Alkali halide disks were desiccated overnight prior to analysis. Spectra were recorded by a Digilab model FTS 15B FTIR spectrometer. Four hundred scans at a resolution of 2 cm^{-1} were used to obtain high signal-to-noise ratio spectra.

The power of this technique as an analytical tool is illustrated in Fig. 4. A 1 : 1 mixture (by weight) of kaolinite and illite, two commonly occurring clays in coals, was prepared and its spectrum recorded (Fig. 4A). If this spectrum is compared to that of kaolinite alone (Fig. 4B), it can be seen that both spectra are almost identical because kaolinite absorbs much more strongly than illite. From this observation it can be

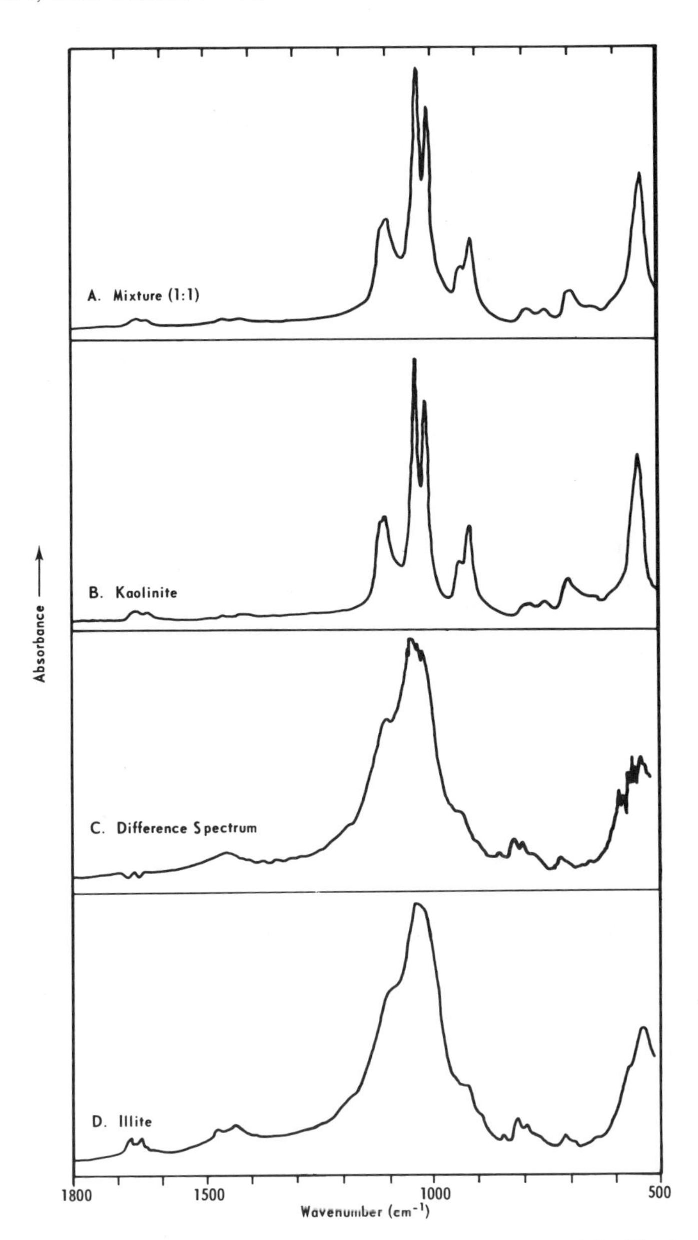

Fig. 4 Absorbance infrared spectra of (A) a 1:1 mixture (by weight) of kaolinite and illite, (B) kaolinite, (C) a difference spectrum obtained by subtracting spectrum B from A, (D) illite. (From Painter, *et al.*, 1978a).

stated that by conventional infrared methods the composition of the mixture could not be correctly identified. In addition, it reveals that in the analysis of an LTA, which contains significant amounts of these two clays, by conventional infrared methods values of kaolinite concentration can be in error. The spectrum of Fig. 4C was obtained by subtracting the spectrum of kaolinite from that of the mixture. As described previously this operation was performed by computer by the correct selection of the multiplication factors. If too much of the kaolinite spectrum had been subtracted, then the characteristic bands would appear negative, i.e., below the baseline. On the other hand, if not enough had been subtracted, then kaolinite bands would still appear, although the absolute absorbance intensities would be reduced. Generally, this trial-and-error approach takes only about 2 min. Finally, if the difference spectrum (Fig. 4C) is compared to that of "pure" illite, it can be seen that both spectra are very similar. It should be noted that all spectra shown in Fig. 4 were automatically scale-expanded by the instrument so that the strongest band in the spectrum is plotted to full scale. The use of automatic scale expansion is useful for comparative purposes when the absolute absorbance values of a difference spectrum are much lower than those for the mixture.

In its direct application to mineral matter analysis, this method proves to be successful partly because of the ability to "remove" kaolinite from the mixture. Its removal reveals the characteristic bands of such minerals as pyrite, illite, and montmorillonite. Although overlapping bands can, in certain cases, be accurately subtracted, it is usually advantageous to select specific bands which are characteristic of a particular mineral; thus, values of K_A and K_B can be determined with a minimum of error. To determine the reliability of this technique many mixtures of pure minerals were analyzed. Table IV lists some typical results for this type of analysis. It can be seen that the FTIR results compare extremely favorably with the known compositions of the mixtures. In general, analysis of these types of mixtures by conventional x-ray and infrared techniques tends to overestimate quartz and kaolinite for the previously described reasons.

Results of analyses of two different size fractions of LTA from Illinois #6 coal (Burning Star Mine) are shown in Table V. The conditions of ashing were those described by Miller (1977). Agreement between the two different approaches is quite good, although kaolinite was overestimated by the conventional infrared method and the quartz values obtained by x-ray methods do show significant variability. Inspection of Table V reveals that gypsum is reported, which may or may not be an

TABLE IV *Comparison of Amounts of Minerals in Mixtures Determined by FTIR with Known Values*

Mineral	Wt % as prepared	Wt % by FTIR
Analysis of Mixture I		
Kaolinite	6	5
Gypsum	17	15
Quartz	46	45
Calcite	10	14
Illite	11	11
Montmorillonite	11	10
Total	101	103
Analysis of Mixture II		
Kaolinite	9	9
Gypsum	16	15
Quartz	20	18
Calcite	14	18
Illite	29	32
Montmorillonite	12	14
Total	100	104
Analysis of Kaolinite and Pyrite		
Kaolinite	15	16
Pyrite	40	36
Kaolinite	25	26
Pyrite	30	27
Kaolinite	35	33
Pyrite	25	23

artifact of ashing (see Section II). For the determination of gypsum in LTA, the reference mineral has to be the dehydrated form.† If pure gypsum were used, then errors would be introduced into the subtraction routine because of spectral differences between the pure and dehydrated mineral. Pyrite determinations are only possible if the far infrared region is examined. In the FTIR system employed in this study the most commonly used beam splitter only permits the recording of spectra down to 450 cm^{-1}. However, the major bands of pyrite are found at lower frequencies; therefore, a different beam splitter which allows

† This material was prepared by placing a sample of ground gypsum in the LTA apparatus for 3 days.

TABLE V *Analysis of LTA by FTIR*

Mineral	Wt % by FTIR	Wt % by conventional ir and x-ray diffraction
Illinois #6, Burning Star (<45 μm)		
Kaolinite	12	16
Gypsum	6	ND[a]
Quartz	23	16
Calcite	6	6
Illite	13	ND
Montmorillonite	6	ND
Pyrite	32	27
Total	98	
Illinois #6, Burning Star (45–70 μm)		
Kaolinite	10	16
Gypsum	10	ND
Quartz	18	34
Calcite	9	10
Illite	18	ND
Montmorillonite	9	ND
Pyrite	29	30
Total	101	

[a] Not determined.

spectra to be recorded between 550 and 200 cm^{-1} was used. With conventional infrared methods it is very difficult to determine pyrite in the presence of appreciable quantities of kaolinite, but since the FTIR technique allows the subtraction of kaolinite the situation is much improved. It was found that the 415-cm^{-1} band of pyrite could be used for analytical purposes.

In addition to analyzing LTA, attempts have been made to analyze mineral matter by a more direct method (Painter *et al.*, 1978b). This method is also based on the ability of FTIR to produce difference spectra. The first step of the procedure is to obtain a very high signal-to-noise ratio spectrum of the whole coal. Then the coal is "demineralized" by acid washing (HCl/HF treatment under nitrogen) and its spectrum recorded. Subtraction of the spectrum of the demineralized coal from that of the parent coal should be characteristic of the acid-soluble minerals. Preliminary data (Fig. 5) obtained by this approach show much promise. In Figs. 5A and 5B are shown the spectra for the raw and demineralized coal (Illinois #6, Burning Star Mine) and the difference spectrum. When this difference spectrum is compared to the spectrum

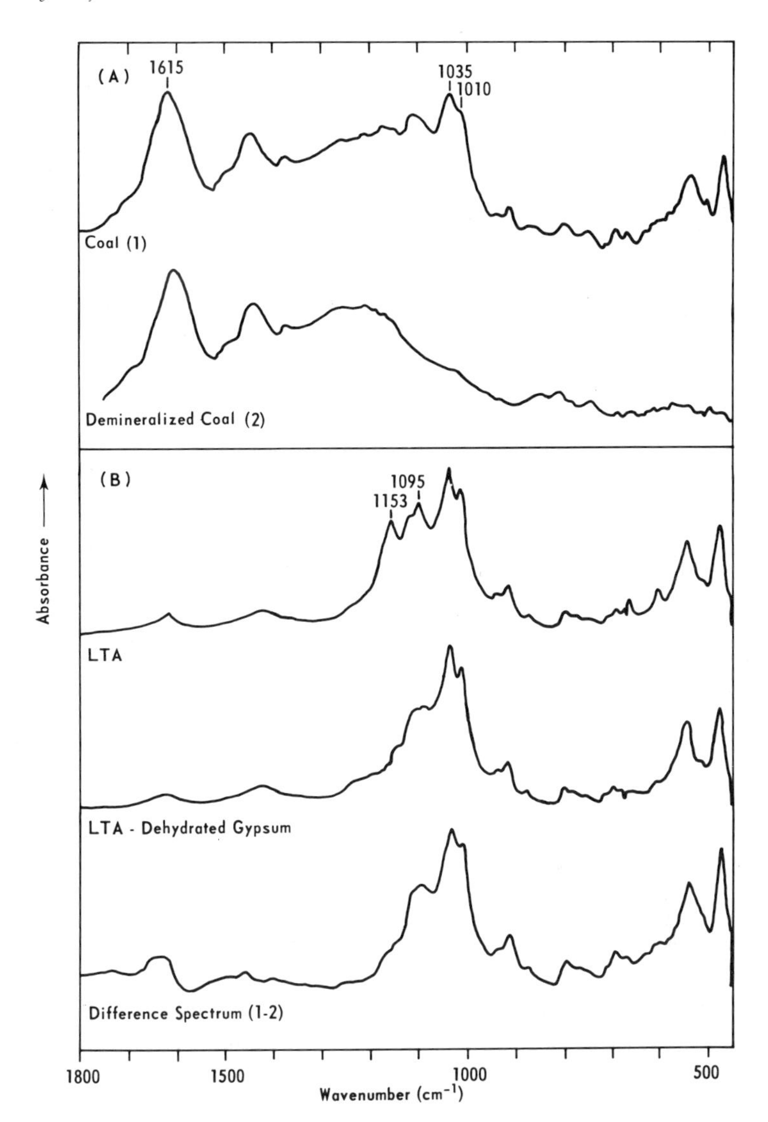

Fig. 5 Absorbance infrared spectra of raw and demineralized Illinois #6 (Burning Star Mine) coal (A), low temperature ash, and difference spectra (B). (From Painter *et al.*, 1978b.)

of an LTA produced from this coal (Fig. 5B), it can be seen that they are very similar. The most important difference is that the LTA spectrum contains bands at 1153 and 1095 cm^{-1} which are characteristic of dehydrated gypsum. If the spectrum of this latter mineral (hemihydrate) is subtracted from that of the LTA (Fig. 5B), then the resulting spectrum is very similar to the difference spectrum for the raw and demineralized coal. These results demonstrate directly that calcium sulfate was formed during the LTA process. The main advantage of the demineralized coal method would, of course, be that it does not require any form of ashing. The difference spectrum produced by this method can then be subjected to further subtractions, in the manner described previously, for quantitative analysis. The number of successive subtractions which can be performed is limited, because the signal-to-noise ratio decreases each time. At some stage the noise level will make further subtraction impractical. In general, some five or six subtractions can be employed; however, the actual number is dependent on the initial signal-to-noise ratio. Therefore, if the number of scans were increased to, say, 1000, more subtractions might be possible.

In conclusion, it is suggested that the application of FTIR spectroscopy will lead to significant advances in the analysis of mineral matter in coals. As with conventional infrared methods, it is somewhat limited for identification purposes. Therefore, the best approach would be to use x-ray diffraction for the identification of mineral species and then to use FTIR for quantitative analysis.

C. Chemical Analyses

If the chemical analysis of a mineral mixture is known, it is possible to estimate the various amounts of each mineral present by a "normative" or "rational" method (Watt, 1968; O'Gorman and Walker, 1972). In such a procedure the known concentrations of each analyzed element in the ash (usually expressed as an oxide) are ascribed to a mineral, or minerals, by stoichiometric relationships. As an example, aluminum is derived from clay minerals, silicon from quartz and clays, iron from pyrite, marcasite, siderite, and so on. For such an analysis it is important to know qualitatively the mineralogical composition of the mixture. Generally, such information is derived from x-ray diffraction of LTA.

In general, the first step of such an analysis is to determine the concentration of pyritic sulfur present (see Volume I, Chapter 9, Section V,B). This value is then used to estimate the amount of pyrite in the sample. The amount of iron associated with the pyrite is then subtracted from the total iron content of the mixture; the unaccounted iron is ascribed to other iron-containing minerals, usually siderite. In a similar fashion, if the concentrations of quartz and kaolinite in the mixture are known from x-ray or infrared methods, then equivalent amounts of silicon and aluminum can be accounted for. Remnant amounts of silica and alumina are then ascribed to various clay minerals. Sulfate sulfur can be determined and then calculated as gypsum, and the equivalent amount of calcium is subtracted from the total calcium content. Any remaining calcium is calculated as calcite. Other elements can be ascribed to various minerals in a similar way.

At best, such methods can be considered as yielding hypothetical mineralogical compositions. There are many drawbacks to this type of analysis because it is often difficult to ascribe entirely certain elements to specific minerals; e.g., although calcium can be equated to gypsum and calcite, it is also present as cations in certain aluminosilicates. Another objection to these methods is that one has to assume specific, "average" compositions of clays which can lead to significant errors.

D. Optical and Scanning Electron Microscopy

Many other methods have been used to characterize the mineralogical components of coals. In this section we briefly discuss two of these techniques: optical microscopy and scanning electron microscopy.

1. *Optical Microscopy*

Optical methods for the identification and distribution of minerals associated with coals are based on detailed microscopic examination of polished or thin sections of coal in transmitted or reflected light. Details of these techniques are described, at length, in Volume I, Chapters 1 and 2. In principle, identification of a mineral type is made by observing several of its optical properties, e.g., morphology, reflectance, refractive index, and anisotropy. These methods have been widely applied by petrographers and, hence, much information has been collected on the occurrences of mineral species in a wide range of coals (Watt, 1968).

Quantitative analyses can be performed for specific minerals by counting techniques, but these methods are very tedious and are limited by the optical resolution of the microscope. Even under the best circumstances, only minerals which are larger than 1 μm diameter can be observed. However, optical methods are very useful for describing types of occurrences of minerals in coals. Information can be gathered on various associations of mineral types with each other and with the maceral components.

Because pyrite has a much higher reflectance than coal it can be well characterized by optical microscopy. Several automated optical systems have been developed to determine the concentration of pyrite in coal and to define its particle size distribution. Automated reflectance systems and their applications to the characterization of pyrite are described in Volume I, Chapter 2, Section VII.

2. *Scanning Electron Microscopy*

The use of scanning electron microscopy (SEM) for the characterization of minerals in coal has grown considerably since the early 1970s because of the increased use of accessory x-ray fluorescence techniques. A full description of the application of x-ray fluorescence via the electron microprobe method to coal analysis is given in Chapter 27, Section III. By applying these analytical methods, it is possible to identify the elemental components of an SEM image and deduce its mineral type. Examination can be made on samples of LTA or whole coal. In general, SEM studies have been used to identify the composition and nature of minerals in coals and to determine the associations of various minerals with each other (Gluskoter, 1975; Russell, 1977).

White and co-workers (Thaulow and White, 1971; White *et al.*, 1972; Lebiedzik *et al.*, 1973; Troutman *et al.*, 1974) have developed a computerized system (CESEMI—computer evaluation of scanning electron microscopy images) which has been used to characterize minerals in coal mine dusts, coals, and coal liquefaction residues (Walker *et al.*, 1975, 1977). In this system, when mineral grains are located, they are sized and their elemental compositions are determined by monitoring seven x-ray channels (Al, Si, S, Ca, Fe, K, and Ti). From the various combinations of these elements, it is possible to characterize most of the commonly occurring minerals in coals. In a typical run, some 2000 particles are examined. After the data are recorded, the computer presents the results in a summary which lists average size, size distribution, shape factor, and volume and number percentages for each of the compositional types. The CESEMI system is generally applied to ground particles of coal mounted and polished in an epoxy resin. For quantitative analysis the method can be applied only to mineral grains larger than about 1 μm in diameter because it is limited by the lower limit of resolution of the x-ray system.

ACKNOWLEDGMENTS

We appreciate the financial support of the Office of Coal Research and the Energy Research and Development Administration (formerly), the Department of Energy (currently), and the Electric Power Research Institute of our studies in the characterization of mineral matter in coal. Their support made the writing of this chapter possible.

The authors acknowledge the great assistance of Dr. M. M. Coleman and Dr. P. C. Painter, Department of Material Sciences, The Pennsylvania State University, in the preparation of the section on Fourier transform infrared techniques.

REFERENCES

Astm (1973). "Annual Book of ASTM Standards," Part 19, ASTM Stand. D1374-73, pp. 438–439. Am. Soc. Test. Mater., Philadelphia, Pennsylvania.

Borio, R. W., Hensel, R. P., Ulmer, R. C., Wilson, E. B., and Leonard, J. W. (1968). *Combustion* **39**(8), 12–20.

Edgecombe, L. J., and Manning, A. B. (1952). *J. Inst. Fuel* **25,** 166–187.

Ely, F. G., and Barnhart, D. H. (1963). *In* "Chemistry of Coal Utilization" (H. H. Lowry, ed.), Suppl. Vol., pp. 820–891. Wiley, New York.

Estep, P. A., Kovach, J. J., and Karr, C., Jr. (1968). *Anal. Chem.* **40**(2), 358–363.

Frazer, F. W., and Belcher, C. B. (1973). *Fuel* **52,** 41–46.

Given, P. H. (1974). "Relation of Coal Characteristics to Liquefaction Behavior." Semi-Annu. Rep. to NSF (RANN) for Grant GI-38974.

Given, P. H. (1976). *Fuel* **55,** 256.

Given, P. H., and Yarzab, R. F. (1975). "Problems and Solutions in the Use of Coal Analyses," Tech. Rep. No. 1. Coal Res. Sect., Pennsylvania State Univ. to ERDA, Rep. FE-0390-1.

Gleit, C. E. (1963). *Am. J. Med. Electron.* **2,** 112–118.

Gluskoter, H. J. (1965). *Fuel* **44,** 285–291.

Gluskoter, H. J. (1967). *J. Sediment. Petrol.* **37,** 205–214.

Gluskoter, H. J. (1975). *Ad. Chem. Ser. No. 141, 1–22.*

Karr, C., Jr., Estep, P. A., and Kovach, J. J. (1967). *Chem. Ind. (London)* **9,** 356–357.

Klug, H. P., and Alexander, L. E. (1974). "X-ray Diffraction Procedures," 2nd Ed. Wiley, New York.

Koenig, J. L. (1975). *Appl. Spectrosc.* **29,** 293–308.

Lebiedzik, J., Burke, K. G., Troutman, S., Johnson, G. G., Jr., and White, E. W. (1973). *Scanning Electron Microsc., Proc. Annu. SEM Symp., 6th* pp. 121–128.

Mackowsky, M. T. (1956). *Proc. Int. Comm. Coal Petrol.* **2,** 31–34.

Maxwell, E., Kelland, D. R., and Akoto, I. Y. (1976). *IEEE Trans. Magn.* **12,** 507–510.

Miller, R. N. (1977). Ph.D. Thesis, Pennsylvania State Univ., University Park, Pennsylvania.

Mitchell, R. S., and Gluskoter, H. J. (1976). *Fuel* **55,** 90–96.

Nelson, J. B. (1953). *BCURA Mon. Bull.* **17,** 41–55.

O'Gorman, J. V., and Walker, P. L., Jr. (1971). *Fuel* **50,** 135–151.

O'Gorman, J. V., and Walker, P. L., Jr. (1972). "Mineral Matter and Trace Elements in U.S. Coals," Res. Dev. Rep. No. 61, Interim Rep. No. 2. Off. Coal Res., U.S. Dep. Inter., Washington, D.C.

Painter, P. C., Coleman, M. M., Jenkins, R. G., Whang, P. W., and Walker, P. L., Jr. (1978a). *Fuel* **57,** 337–344.

Painter, P. C., Coleman, M. M., Jenkins, R. G., and Walker, P. L., Jr. (1978b). *Fuel* **57,** 125–126.

Rao, C. P., and Gluskoter, H. J. (1973). *Ill. State Geol. Surv., Circ. No.* 476.

Russell, S. J. (1977). *Scanning Electron Microsc., Proc. Workshop Materials and Component Characterization, Chicago,* pp. 95–100.
Suhr, N. H., and Gong, H. (1977). Person communication, Pennsylvania State Univ., University Park, Pennsylvania.
Tarrer, A. R., Guin, J. A., Pitts, W. S., Healey, J. P., Prather, J. W., and Styles, G. A. (1977). *In* "Liquid Fuels from Coal" (R. T. Ellington, ed.), pp. 45–61. Academic Press, New York.
Thaulow, N., and White, E. W. (1971). *Powder Technol.* **5,** 377–379.
Troutman, S., Johnson, G. G., Jr., White, E. W., and Lebiedzik, J. (1974). *Am. Lab.* Feb., pp. 31–38.
van der Maas, J. H. (1972). "Basic Infrared Spectroscopy," 2nd Ed. Heyden, London.
Walker, P. L., Jr., Spackman, W., Given, P. H., White, E. W., Davis, A., and Jenkins, R. G. (1975). "Characterization of Mineral Matter in Coals and Coal Liquefaction Residues." 1st Annu. Rep. to EPRI, Project RP-366-1. Palo Alto, California.
Walker, P. L., Jr., Spackman, W., Given, P. H., White, E. W., Davis, A., and Jenkins, R. G. (1977). "Characterization of Mineral Matter in Coals and Coal Liquefaction Residues." EPRI AF-417, Project RP-366-1. Palo Alto, California.
Ward, J. C. (1970). *Rev. Pure Appl. Chem.* **20,** 175–206.
Watt, J. D. (1968). "The Physical and Chemical Behavior of the Mineral Matter in Coal Under Conditions Met in Combustion Plant," Part 1. BCURA (Br. Coal Util. Res. Assoc.) Lit. Surv., Leatherhead, England.
Watt, J. D. (1969). "The Physical and Chemical Behavior of the Mineral Matter in Coal Under Conditions Met in Combustion Plant," Part 2. BCURA (Br. Coal Util. Res. Assoc.) Lit. Surv., Leatherhead, England.
White, E. W., Mayberry, K., and Johnson, G. G., Jr. (1972). *Pattern Recognition* **4,** 173–179.
Whitehurst, D. D., Farcasiu, M., Mitchell, T. O., and Dickert, J. J., Jr. (1977). "The Nature and Origin of Asphaltenes in Processed Coals." Annu. Rep. to EPRI, Project RP-410. Palo Alto, California.

Chapter 27

Separation and Identification of Minerals from Lignites

Walter W. Fowkes†
GRAND FORKS ENERGY RESEARCH CENTER
U.S. DEPARTMENT OF ENERGY
GRAND FORKS, NORTH DAKOTA

I. BACKGROUND AND PRESENT STATUS OF INFORMATION

The nature and properties of the minerals in coal§ have and continue to receive much attention among coal researchers as their importance in respect to rational utilization of fuel values continues to grow. Initially, the primary objective of facilitating production of relatively "clean" fuel was adequately served by handpicking bulk impurities. As the scope and volume of coal usage grew, it became a major problem to extract this extraneous nonfuel fraction by such primitive methods. It was no great flash of genius to realize improvements could easily be accomplished in eliminating much of the debris by more careful mining, and by "breaking" the more friable coal substance for screening from

† *Present address:* 422 W. Farmer Avenue, Independence, Missouri.

§ In speaking of or writing about the mineralization of coal, it might be well to point out first that the term can have slightly different shades of meaning to different audiences. To some it will be confined to those entities associated with coal but strictly defined and characterized in the mineralogical sense: to others the term will embrace all components of a given coal mass which are not obviously organic in character. I prefer to consider the first as defining coal minerals and the second as the mineral matter of coal: a fine distinction, but *ad hominem*.

ISBN 0-12-399902-2

the more refractory portions of the bulk impurities, although the procedure of handpicking persists on 4-in. and larger sizes.

With the need for and advent of more sophisticated methods of coal cleaning, the character of the mineral content assumed ever greater importance in the selection and design of cleaning methods, aside from the fact that coal chemists consider the problem of its origin to be a part of the problem of coal's origin. It is accepted that the nature and distribution of the mineral matter in coal give some information regarding conditions of deposition of the parent material, and it is believed that some inorganic constituents of the mineral matter have reacted with organic compounds, so playing a direct part in the transformation of plant material into coal substance.

The mineral constituents of coal have been variously classified, a generalization of which might be (1) residues of mineral constituents of the plants from which the coal derived, (2) detrital materials, eolian or alluvial, which settled into the deposit, (3) salt deposits originating from the water with which the plant residues came into contact prior to and during coalification, (4) crystalline deposits from circulating groundwaters using the coal beds as aquifers, and (5) products of decomposition of these minerals and of the interactions between themselves and the coal substance. It is apparent, then, that the character and extent of mineralization should vary with the age and geologic environment of the coal, as well as the chemical and physical nature of the coal itself. This is ultimately confirmed by the fact that mineralization of the different petrographic entities of the coal will itself vary in character and composition (Lessing, 1925).

As a factor in determining the generic classification and, indirectly, the nature of mineralization of coal seams, age is per se not a limiting condition. Kreulen (1948, pp. 38–41), in his book, "Elements of Coal Chemistry," cites several examples of chronologically older coals which exhibit chemical and physical characteristics of younger coals; these would be expected also to exhibit a mineralization different from that which might be anticipated on the basis of age alone. In general, however, the older and more highly metamorphosed coals show an effect of these changes in that less inorganic matter is organically bound to the coal substance (inherent mineralization), and there is a greater proportion of extraneous mineralization. This is due to the chemical structure of these older coals, the general features of geologic environment, and the physical structure of the coal seam, all of which are factors relating to the metamorphic process, and hence to age. In these coals there are few, if any, sites for the ionic attachment of cations; in the younger coals, metallic cations find a ready "home" by chelation in the coal substance or by direct ionic bonding thereto. It is assumed in this case that the metallic cations which must have been bound to the younger coal were lost, along with the active sites for bonding, in the metamorphosis to higher rank. The factor of geologic environment influences these differences in mineralization in that the circumstances which resulted in metamorphic change of the coal usually accomplished similar changes in the nature of the surrounding strata, presenting a different source of detrital material for subsequent deposition in the coal seam by groundwaters as well as changing the pattern for circulation of subsurface water in most cases. The third interrelated factor influencing the mineralization of coal of different ranks, that of physical structure, acts to present a relatively low permeability over the general crosssection of the bed. This forces mineral-laden streams to follow essentially fixed paths along vertical cleats or horizontal partings resulting from contraction of the coal substance during metamorphosis, depositing their less soluble mineral burden as massive inclusions rather than the finely divided, frequently micron-sized particles found for the most part in the younger coals.

This hypothesis is not without exception, given the highly variable nature of coals and their associated environments. For the most part, however, it explains some of the basic differences in mineralization evidenced by coals of various ranks and from different localities. Disturbing exceptions do occur in the case of lignites and brown coals over relatively small areas, frequently within the same seam and mine. These irregularly occurring discrepancies are difficult to explain: We can only surmise that variations in the character and path of groundwaters over a long period of time might account for these vagaries—possibly associated also with the presence of some long-gone geologic formation, the victim of time and erosion. Whatever the cause, the effect is to produce distinct differences in the mineralization of coals, varying with locale, rank, and conditions of origin. There can be no standard, only an acceptable working mean, and each coal must be examined for its own peculiar characteristics. Fortunately for those involved in coal cleaning or preparation, these variations become less pronounced with increase in rank of the coal, owing largely to the greater proportion of extraneous debris as opposed to inherent inorganic fraction.

The fact that the largest part of the inorganic phase in lignite, and brown coals, is ionically bound or so finely divided as to be virtually inseparable has inhibited investigation regarding its nature (Durie, 1961). The more massive inclusions and partings are easily separable, but no greatly variable suite of minerals has been reported, although a large number of elements may be determined by meticulous examination of these low rank coals or their combustion residues. These elements are assumed to be a part of the organometallic component, by chelation or by ionic bonding. For example, very shallow lying seams of lignite have in the recent past been exploited as a source of uranium although little uraniferous mineralization has been reported as separate entities. It is mentioned that megascopic inclusions of uranium minerals have been found in certain lignites and carbonaceous sandstones of northwestern South Dakota (autunite, zeunerite, torbenite, and metatyuyamunite) but not in the rather extensive occurrences of uranium-bearing lignites in North Dakota, Montana, and Wyoming. Explanations regarding origin of the uranium generally favor the epigenetic hypothesis, i.e., introduction into the coal following coalification from groundwater which derived the element from hydrothermal sources or associated volcanic rocks. Plants which selectively concentrated the element in their living tissue and subsequently became a part of the coal mass undoubtedly made some contribution. Whatever means accomplished its fixation, the most economic process for recovery involved destruction of the organic components by combustion and treatment of the residual ash. Other minerals identified include gypsum, analcite, jarosite, limonite, and quartz (*U.S. Geol. Surv.*, 1959).

This same line of reasoning may be applied to numerous other metallic elements whose recovery from low rank coals has been proposed at one time or another. Semiquantitative spectrographic analyses of core samples show a wide range of elements in micro to semimicro amounts, but the responsible minerals are not identified. Some of these elements are found to remain as trace quantities in coals of higher rank, but in general the reduced amount indicates their loss due to elimination of active sites for ionic bonding or of chelation ability, as previously mentioned. Majumdar *et al.* (1959) report a finding of goethite in lignite of South Arcot (India) but do not specify other associated minerals. During the investigation of various lignite beds it was noted that uranium was usually concentrated in the upper portions of the thicker beds, gradually diminishing in quantity with depth. This would appear to support the theory of epigenetic origin, assuming mineral-laden groundwater percolating downward through the coal bed and exchanging the heavier elements preferentially during its progress. It was demonstrated in a limited investigation at the Grand Forks Energy Research Center that higher valency elements displaced those of lower valence from the coal when present in ionic form; i.e., Fe^{3+} or Al^{3+} displaces Ca^{2+} which displaces Na^{+} which displaces H^{+}—excess H^{+} will displace Na^{+}. By implication, this could be expected to occur also in the case of other heavy, higher valence cations, preferential exchange accounting for the observed vertical distribution of such elements in coal beds sufficiently thick as to effect appreciable separation.

Kemežys and Taylor (1964) have reported on a rather detailed study of mineral occurrences in Australian coals, using transmitted and reflected light microscopy, as well as electron microscopy and diffraction. The list of minerals identified is impressive but also serves to emphasize the variability of coal mineralization and its highly dispersed character, particularly in the case of low rank coals. Further, it demonstrates the significant factor of locale in determining the variations in coal mineral occurrences. The coals are identified according to geologic age but not otherwise classified.

A cursory examination of various lignites was carried out several years ago by the author to determine the proportions of soluble and insoluble sodium minerals in lignites of the North Dakota fields. Using multiple samples with variable sodium content, as determined by analyses of the ash, representing widely separated geographic locales, it was found that 10–20% of the sodium originally present could be extracted with distilled water; 92–98% was removed in 3% hydrochloric acid, the small residue remaining in the extracted coal. Recovery of soluble and exchangeable sodium varied with the original sodium content, appeared to be essentially independent of locale, and was quite reproducible at any given level of initial sodium content. The results match those of numerous investigations carried out on brown coals and lignites of Europe and Australia, in particular. However, contrary to many of these latter examinations, no more than traces of chlorine were found in the North Dakota samples.

In view of the importance of the sodium content as a factor in determining the fouling tendencies of a coal when fired in large boilers, this initial investigation was somewhat expanded in scope and scale to examine the feasibility of sodium removal by ion exchange (Paulson and Fowkes, 1968). Significant reduction in sodium content could be effected by treating relatively small particle sizes with water containing only small amounts of calcium ion; calcium and magnesium were removed by treatment with a solution of trivalent ions, i.e., iron or aluminum; and similar results were obtained by treating the coal with dilute acid. It was a little surprising to note a slight reduction in sulfur content along with the removal of sodium, calcium, or magnesium. This last observation prompted an examination of mineral separation by gravity methods although it was considered that only a very small portion of the inorganic constituents would be present as extraneous minerals (Paulson *et al.*, 1972).

The separation procedure was carried out in carbon tetrachloride, using lignites from six different mines in the Montana–Dakota area: crystalline structures were established from x-ray diffraction patterns, and elemental analyses were performed by electron microprobe examination. Separable fractions ranged from 2 to 15% of the total mineral matter as represented by the original ash content of the coal: one extreme case showed separation of about 40% of the total ash materials. It should be noted, however, that the "sink" or heavier fraction upon ignition indicated that it had carried along 40–60% combustible material. It is logical to assume also that an appreciable amount of finely divided mineral matter might have been occluded in the float material. This is corroborated by the fact that ash from the float material contained significant amounts of silica and alumina, indicating most probably the presence of very finely divided clay minerals. The principal minerals separated from coal samples of the six mines were identified as nacrite [$Al_2Si_2O_5(OH)_4$], barite ($BaSO_4$), pyrite (FeS_2), hematite (Fe_2O_3), quartz (SiO_2), gypsum ($CaSO_4 \cdot 2H_2O$), and calcite ($CaCO_3$), along with fragments of a mixture of α-quartz and nacrite. Observations based on diffraction patterns and microprobe examination indicated that the nacrite particles were not pure, but contained traces of magnesium, sodium, calcium, and iron in quantities too small to alter the crystal structure of the principal material. Calcite, gypsum, and α-quartz were uncontaminated when observed. No separable minerals were found in which sodium was a major constituent.

Two other small-scale investigations, carried out in part by the author, gave further indications of the minerals which may be found in the coals of the western United States. The first, an intensive search for soluble sodium compounds, examined 22 samples of lignite and subbituminous coal from the Montana–Wyoming–Dakota region (Fowkes, 1972). Exhaustive extraction of the samples with distilled water gave recoveries of solubilized material ranging from 0.13 to 1.43% (averaging ~0.5%) based on the weight of coal treated. The excellent diffraction patterns obtained on the water-soluble material indicated only sodium sulfate (thenardite, metathenardite, and form III) and calcium sulfate (anhydrite and hemihydrate); the hemihydrate was apparently formed in slightly overheating the evaporation residue:

$$CaSO_4 \cdot 2H_2O \xrightarrow{128^\circ} CaSO_4 \cdot \tfrac{1}{2}H_2O \xrightarrow{163^\circ} CaSO_4$$

The second investigation involved examination of x-ray fluorescence and electron microprobe techniques as applied to analysis of lignite, ash, and fireside boiler deposits (Beckering *et al.*, 1970). Careful scanning of small particles of lignites with the electron beam served primarily to confirm the thesis that a major part of the associated sodium, calcium, iron, aluminum, magnesium, and sulfur is an integral part of the organic substance in the coal. Silicon generally appeared as discrete particles (quartz) of mainly 5-μm diameter or less. Occasional particles of pyrite were observed, along with very finely divided complex silicates and sodium or calcium sulfates. Subbituminous and bituminous coals showed a somewhat different pattern, with more discrete mineral inclusions appearing in the scan, reflecting the differences in mineralization due to metamorphic progression, as alluded to earlier.

The sedimentology of the strata associated with the Dakota lignite beds is relatively fine-grained and usually poorly consolidated. The strata range from mudstone and siltstone to sandstone with pods and lenses of impure limestone. The lignite beds are usually within or bounded by gray clay and silt units. Both bentonitic and kaolinitic clays are found in the same area. Clay partings are common, but the mining methods serve to produce a relatively "clean" product. It is general practice to crush and screen the coal at the mine tipple, rejecting fines which would include most of the clay inadvertently included in mining: in most cases the coal is pulverized at the power plant before firing. Information is available on a single sample regarding mineralization of lignite from the other major producing and consuming area of the United States (Texas) wherein identified mineral species include kaolinite, illite, siderite, quartz, pyrite, gypsum, hematite, and rutile (Stefanko *et al.*, 1973). This listing essentially matches that for lignites of the Montana–Dakota area, which could be expected for most lignites of the Gulf Coast Province in view of the similarity of associated strata and apparently similar depositional conditions.

If we now step across that nebulous line of demarcation separating "mineralization" from "inorganic fraction," we find that numerous investigators have identified, qualitatively or semiquantitatively, a long list of elements other than carbon, hydrogen, nitrogen, oxygen, and sulfur associated with coals of all ranks. These range in amounts present from submicro to semimicro. In very few cases is there any attempt to specify the mineral species responsible for the given element, so we could assume on the basis of limited experimental evidence that most of the elements are ionically (perhaps covalently) bound to or chelated by the organic substance of the coal. However, as the tonnages of coal burned continue to increase significantly, these "trace" elements assume ever greater importance as potential toxic pollutants—they must be more precisely fixed, and means must be found to alleviate their detrimental effects. In proposed gasification or liquefaction procedures involving coal, some of these elements may prove to have beneficial effects, and it may also become both economic and essential to treat them as desirable by-products of coal utilization.

In any case, the list includes (in addition to major elements routinely determined in coal and its combustion products) barium, strontium, boron, manganese, copper, molybdenum, vanadium, chromium, nickel, lead, cobalt, scandium, zirconium, gallium, yttrium, tin, ytterbium, beryllium, silver, germanium, lanthanum, lithium, arsenic, cerium, zinc, neodymium, selenium, cadmium, and mercury—all of which have been detected in lignite and in higher rank coals at threshold limits or greater. Mercury, in particular, has received considerable attention (Joensuu, 1971), and a limited investigation of Dakota lignites with respect to this element and to boron was conducted. Results indicated mercury contents of lignites and subbituminous coals to be in the range 25–350 ppb, with most samples in the range 35–60 ppb. This experimental work was extended to soils closely associated with the lignite deposits, and the results followed closely the data obtained in analyzing the coals. Samples of bituminous coals obtained from mines in Utah and Washington, as well as subbituminous coals from Montana, Wyoming, Arizona, and New Mexico, indicated mercury contents in the range 25–60 ppb. Analyses for boron were performed on the ash and/or ash deposits from these same coals; no satisfactory method has been developed for determining boron directly on the coal. Boron contents from 300 to 950 ppm were indicated. Assuming an average ash content of 8%, these figures could represent a 12-fold enrichment in the ash since it is anticipated that no boron would be lost in the volatile combustion products. The analytical methods used here, while accurate and reproducible, are laborious and time-consuming; other procedures must be developed since it is obvious that additional information is needed.

Of equal importance will be the determination of breakdown products from both minerals and inherent inorganic fraction which might appear in volatile and solid products of combustion, along with material retained in the sludge and liquor discharge from wet scrubbers, which are rapidly coming into favor with power plants, old and new. Additional knowledge regarding the inorganic materials present in the fuel utilized is required in order to determine the variety of potentially harmful substances which may necessitate special handling of combustion by-products if emission standards are to be satisfied.

The hitherto limited and localized use of low rank western United States coals placed a low priority on study of these fuels. Although we may assume some characteristics of their nature from similar work on higher rank coals, it is becoming obvious that these low rank coals are unique in some respects, with significant variations appearing even within closely related deposits, applying in particular to the nature of their "mineralization."

II. MINERAL SEPARATION

Outside of the laboratory, mineral separation is carried out for the primary purpose of providing a clean, pollutant-free coal for burning. Most methods depend on differences in gravity between coal and mineral matter, although the procedure must be varied in many cases to adapt to the nature of the mineral to be removed. The cleaning of coal is both an art and a science, since many factors interrelate to determine the ease and efficiency with which mineral matter may be separated from coal, and significant differences will be noted between coals of the same rank taken from different geologic environments.

Up to this point no serious efforts have been made to separate minerals from lignite on a commercial scale, although consideration has been focused on "desalting" of the highly saline brown coals of East Germany and Poland. The mineralization of Australian brown coal and of U.S. lignites has also recently received much attention in respect to its influence on fouling and corrosion of heat exchanger surfaces. As would be expected, the reports indicate that mineral entities physically associated with the coal substances are very finely divided and generally of less importance in the problems of fouling and corrosion. In some coals of East Germany and Poland, high contents of sodium chloride are reported, but for the most part, alkali and alkaline earth elements are organically bound and, therefore, not physically separable. This is also true of the typical Australian brown coal and U.S. lignite.

Prior to the introduction of pulverized-fuel firing in large boilers, lignite was delivered in a range of lump sizes prepared at the mine tipple by crushing and screening, wherein fines that were high in mineral matter were usually rejected. Large inclusions of extraneous minerals were removed on picking tables, but there was no further attempt to clean the delivered coal. Lignite was considered a relatively "clean" coal as mined, and the methods of firing did not encounter the severe fouling problems which appeared with the introduction of pulverized-fuel firing. With these new problems came an intensified interest in the nature of lignite mineralization. Of more recent interest, environmental considerations dictate a comprehensive knowledge regarding the release of possibly noxious trace elements during combustion of the coal and, in particular, of the low sulfur western lignites and subbituminous coals which appcar to be the logical interim energy source prior to economic production of acceptable liquid and gaseous fuels, or use of nuclear power on a wider scale.

The application of classical gravity methods for separation of minerals proved relatively ineffective in the case of lignite. A rather small fraction of the extraneous minerals is present in sufficiently large particle size to be thus separated. In reaching the very small sizes usually present, the physical differences which ordinarily would make separation possible are diminished to the point that reasonably clean separation is impossible. Lignite is a fibrous, tough material not easily pulverized, and therefore difficult to separate from the finely divided mineral inclusions. Further, pulverization beyond about 325 U.S. Standard Mesh tends to produce a colloidal suspension in liquid media, and defies classification by air suspension. It is therefore necessary to adapt methods for mineral identification which do not require physical separation of the compound (or element). As implied, this often is accomplished only by conjecture based on elemental, usually semiquantitative, analyses of extremely small entities.

One other procedure which may be applied to mineral separation is that of recently developed low temperature ashing in a microwave-excited oxygen plasma. Properly conducted, the process may be carried out at temperatures sufficiently low as to recover most minerals unaltered. However, it has been shown that inorganic compounds not originally present may be formed (Fowkes, 1972), and that, with lignite in particular, the oxidation is quite slow, usually requiring 5–6 days to accomplish 90% oxidation of the organic carbon. And even at this relatively low temperature, given the extended time of exposure, it is not impossible that a small portion of the less refractory trace elements (e.g., mercury and gallium) might be lost. The procedure does, however, offer the best compromise between recovery or loss of unaltered minerals from coal. Incidentally, coals of progressively higher rank oxidize more readily in the oxygen plasma.

Chemical methods, as dissolution or ion exchange, have obvious limitations. In order to expose effectively the soluble, or exchangeable, material to solvent action the coal must be very finely pulverized, compounding the difficulties of complete recovery. Working with lignite of 150 μm particle size and smaller, exhaustive extraction with distilled water would yield an average of no more than 10% of the mineral matter originally present as determined by standard laboratory ashing procedures. This was subsequently identified as various forms of calcium and sodium sulfates by x-ray diffraction; no other patterns were discernible. The procedure was repeated for lignites from six different mines (several samples from each) and for subbituminous coals from three different mines; all results were essentially the same. Exchange reactions, of course, account for only the cation associated with the organic substance of the coal. Whether chelated or ionically bound, the associated cation may be exchanged for one of higher valence, as Fe^{3+} or Al^{3+} exchanges for Ca^{2+}, which exchanges for Na^+. The metallic cations may be removed by an excess of H^+, converting the humates back to the acid form.

As would be expected in the generalized treatment of any subject, there are exceptions to the description given for lignite mineralization. Silicified wood, sulfur balls of appreciable size, and even relatively pure crystalline deposits of sodium sulfate have appeared in lignite seams. Of course, clay and sand partings and lenses do occur and are cleaned away unless very minor in extent or depth. At the surface of one seam uncovered at an approximate depth of 50 ft, there appeared a sporatic but collectible layer of jarosite (a mixed sulfate of iron and magnesium or potassium). Siderite is reported occasionally, and rutile in rare instances (most lignite ashes will contain up to 0.1% TiO_2). Although spectrographic analyses of ashes from lignites or brown coals show a long list of other elements, few other mineral species have been identified, and these only as traces or as questionable occurrences. It is reasonable to assume, then, that the organic substance of lignite retains a large variety of metallic or semimetallic cations by some form of bonding, albeit in micro or submicro quantities.

However, these minute quantities become items of particular interest when considering the megatons of the coals in question which are certain to be required as fuel if we are to bridge the gap between environmentally acceptable energy sources and the growing demand of our energy-oriented economy. In addition to dealing with them at present as combustion products, we will be required to account for them as by-products of proposed coal conversion processes.

III. MINERAL IDENTIFICATION

Several classic methods are available for identification of minerals separable in quantity and in relatively pure form. Properties related to crystalline structure may be used for positive identification by ordinary visual means or by optical microscopy, depending on the size of the pure sample. The examination of dust dispersion slides or embedded thin sections by optical microscopy will reveal much information regarding the nature of the mineral matter based on observation of the refraction in air, color, and other optical properties dependent on the transmission of light. With opaque or nearly opaque substances, reflectance measurements, bifringence between particle and medium, observations in polarized light, or noting differences between light and dark field examination will provide much more information. Such data are generally reliable down to particle sizes of about 10 μm, and by using a combination of methods may be extended with some confidence to the range of perhaps 5 μm. Thaer's (1954) treatise gives a very thorough treatment on the application of optical microscopy to analysis of respirable dusts and the limitations of these methods. The work of Kemežys and Taylor (1964) on the Australian coals relies heavily on optical microscopy, supplemented by diffraction and electron microscopy, apparently on thin sections (CSIRO, 1961, 1965).

Thus, it is possible to characterize mineral inclusions of very small particle size, assuming the mineral is a separate and distinct entity within the coal matrix. In extremely fine division, difficulties may be introduced by the presence of isomorphic or polymorphic forms. This situation has been noted in attempts to characterize minerals from U.S. lignites by obtaining single-crystal or powder diffraction patterns in x-ray analysis, strong patterns obscuring weaker ones or the combined patterns producing a diffuse ambiguous pattern. Combining the information obtained from a clear diffraction pattern with the data furnished by x-ray emission (fluorescence) analysis of the same sample refines the identification process to confirm a particular mineral or to distinguish between different morphologic forms. X-ray emission spectroscopy is based on the fact that atoms excited by bombardment with a high energy electron beam, or high energy x-rays, emit at a characteristic frequency in the x-ray region. The emission spectrum is resolved by analyzing crystals, and a continuous count record for a range of elements may be recorded sequentially by varying the angle of the analyzing crystal. The recorded emission may be quantified, with appropriate corrections for interelement effects, to give an accurate elemental analysis. This procedure, however, becomes increasingly less reliable as the atomic weight of the element decreases, owing to lower emission energies with consequent increase in background interference. For practical purposes, sodium is the lowest weight element which may be determined with acceptable accuracy, and that with some qualifications. Also, the procedure requires about 3 g of sample for routine analysis, although smaller amounts may be handled with modifications in the preparation of the specimen and some loss of reliability in the data (Beckering *et al.*, 1970; CSIRO, 1963; Sweatman *et al.*, 1963). This method has been applied to the whole coal, making it unnecessary to ash the coal or separate the mineral matter for elemental analysis (Kuhn *et al.*, 1973).

The electron microprobe operates in a similar manner in that a high energy electron beam is focused on a spot about 1 μm square, resulting in the emission of characteristic radiation for all elements heavier than beryllium. Here, too, the emitted spectrum is analyzed by a crystal, the selected frequency picked up by a detector set at the appropriate angle, and the emission converted to an electrical signal which is recorded on a strip chart (Figs. 1–3). The sample is usually embedded in a resin, and a polished surface is prepared for examination. The sample may be moved relative to the electron beam, analyzing a narrow strip several millimeters long or an area approximately 320 μm square. Additionally, some of the impinging electrons are absorbed, producing a small sample current; and a certain fraction is reflected, producing a backscatter current. By monitoring the backscatter current with a beam scanner, a topographic view of the sample surface is presented via a cathode ray tube. The cathode ray image can be photographed, giving an "elemental map" of the sample surface to depths varying from 3 to 8 μm (Figs. 4–6). Examination of the figures indicates the relative accuracy of the microprobe in pinpointing mineral inclusions and in emphasizing the generalized distribution pattern of those elements bound to the coal substance. The linear recordings of Figs. 1–3 represent an analyzed strip approximately 1 μm wide and 500 μm long; the topographic views in Figs. 4–6 reproduce an area approximately 140 μm square, representing a magnification of 800. Quantification of the microprobe analysis, however, is at best semiquantitative. On the other hand, the ability to present elemental information on a spot source, combined with accurate elemental analysis by x-ray emission and possible definitive pattern from powder diffraction, simplifies the mineralogic characterization of coals (cf. CSIRO, 1965; Boateng and Phillips, 1976; Sutherland, 1975).

Recent exploratory work indicates that the scanning electron microscope with concurrent analyzing facility may improve on such mineralogic characterizations by furnishing a clear picture of particle contour and orientation in the micron and submicron range, along with an elemental analysis of the particle examined (Fisher *et al.*, 1976).

Several investigators have utilized diagnostic bands in the 650- to 200-cm^{-1} region of the infrared spectrum to characterize minerals from coal (Karr *et al.*, 1967, 1968; Barber, 1967; Angino, 1967; Oinuma and Hayashi, 1965; O'Gorman and Walker, 1971; Rekus and Haberkorn, 1966; Fripiat, 1960). (See also Chapter 26, Section III,B.) In this procedure as applied to coals, the minerals should be separated from the organic matrix. In most cases this is accomplished by low temperature ashing in an oxygen plasma produced by radiofrequency excitation. As previously mentioned, this sometimes leads to ambiguities in the resultant mineral mixture. However, minerals obtained in relatively pure form are quickly characterized, and the method has been used extensively in the examination of clay minerals in mixture.

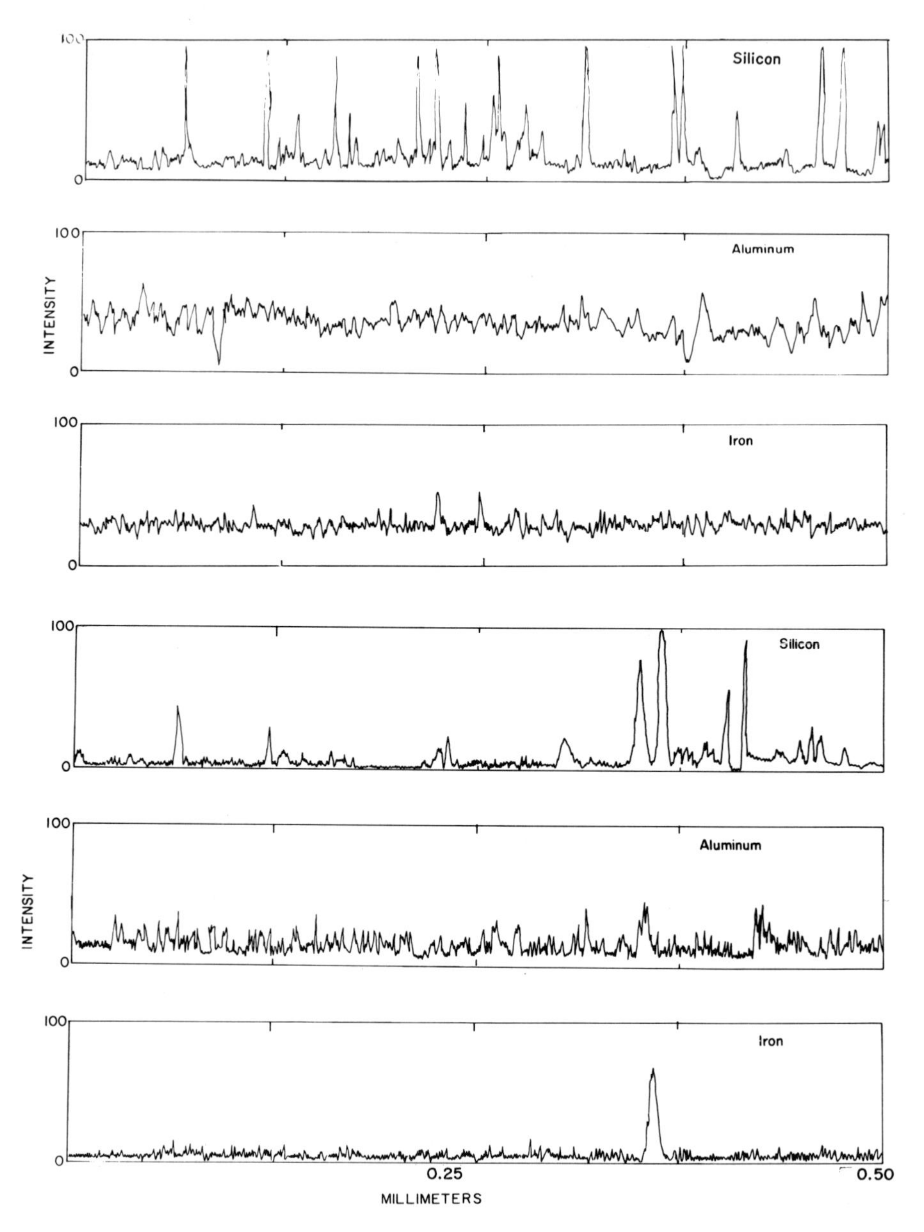

Fig. 1 Microprobe scan of a typical lignite (upper tracings) and a typical subbituminous coal (lower tracings).

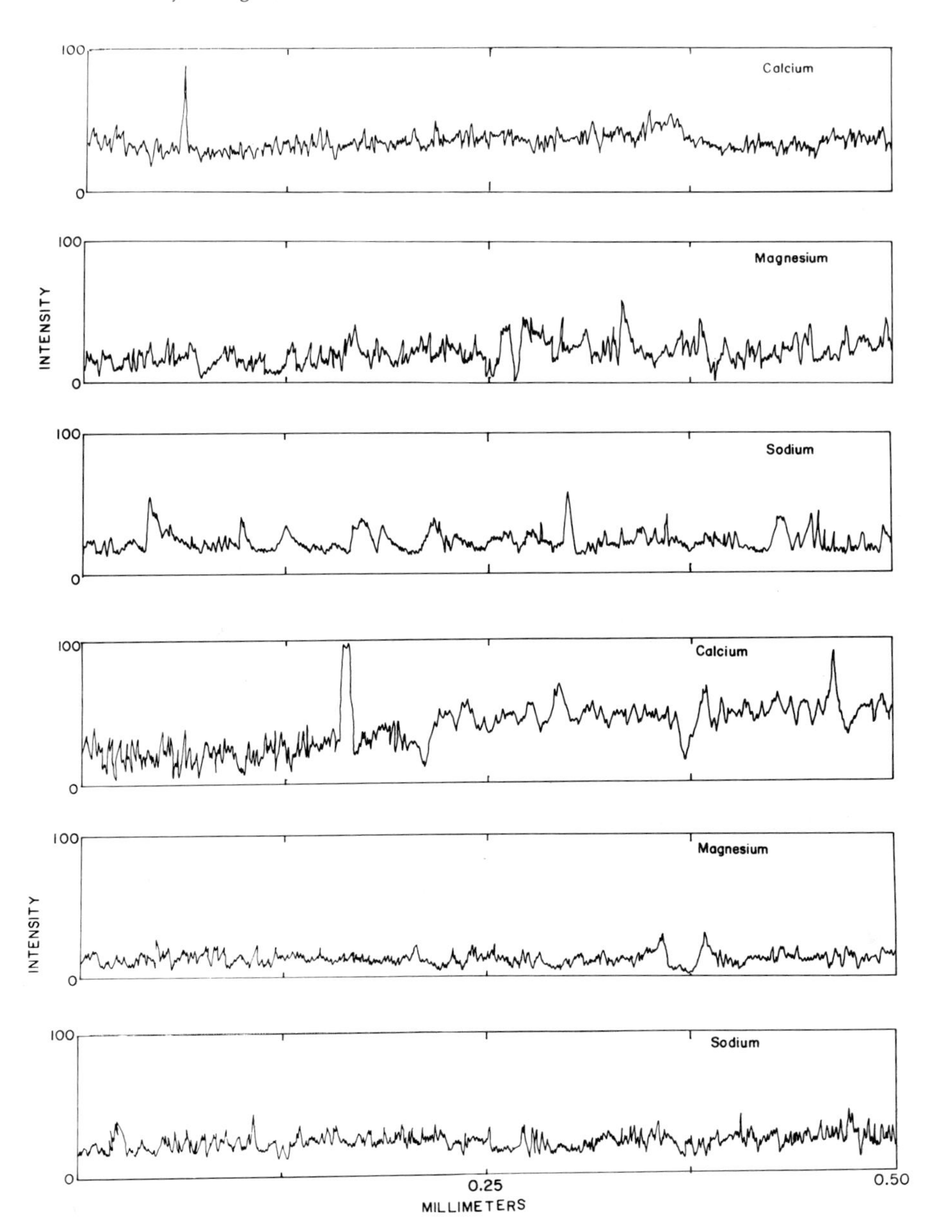

Fig. 2 Microprobe scan of a typical lignite (upper tracings) and a typical subbituminous coal (lower tracings).

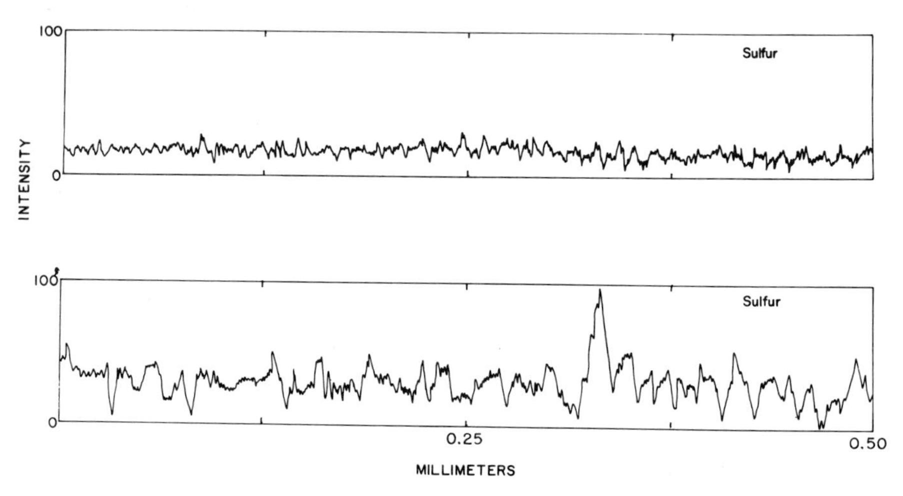

Fig. 3 Microprobe scan of a typical lignite (upper tracing) and a typical subbituminous coal (lower tracing).

Another procedure which has been applied to characterization of minerals in coal (Warne, 1965; Gordon and Campbell, 1955; Mukherjee *et al.*, 1972), as well as to pure mineral mixtures (Campbell *et al.*, 1959; Vaughan and Wiedemann, 1965; Bredahl, 1965), employs differential thermal analysis, with or without concurrent gravimetric measurement recorded during the temperature increase (TGA). This method has also been used to characterize (or rank) whole coals (Glass, 1955; Berkowitz, 1957; Kessler and Romováčková, 1961). (See also Chapter 37.) The method appears to be widely used, although the results have generated considerable controversy owing to the superimposition and masking of peaks in the recorded curves which are the visible indication of reactions or phase changes within the sample. It has further been noted that the response curves appear to vary with the thermal history of a given sample. On this basis, it would seem the method is a poor choice for mineral characterization in mixture, although further refinements in procedure and interpretation may enhance its utility.

IV. SUGGESTED PROCEDURES

As is the case in all investigative work, the validity of the sample is of prime importance, assuming it is desired to characterize in detail the mineralization of a given coal seam rather than collect random information regarding mineral content. The ideal sample would be a column, 6 in. square or slightly greater, comprising the entire thickness of the seam and obtained from a freshly cut face. The column should be carefully removed, the stratigraphic position of each section indicated, all natural partings noted, and the material comprising the separation interval collected. Since other tests may be needed to characterize this

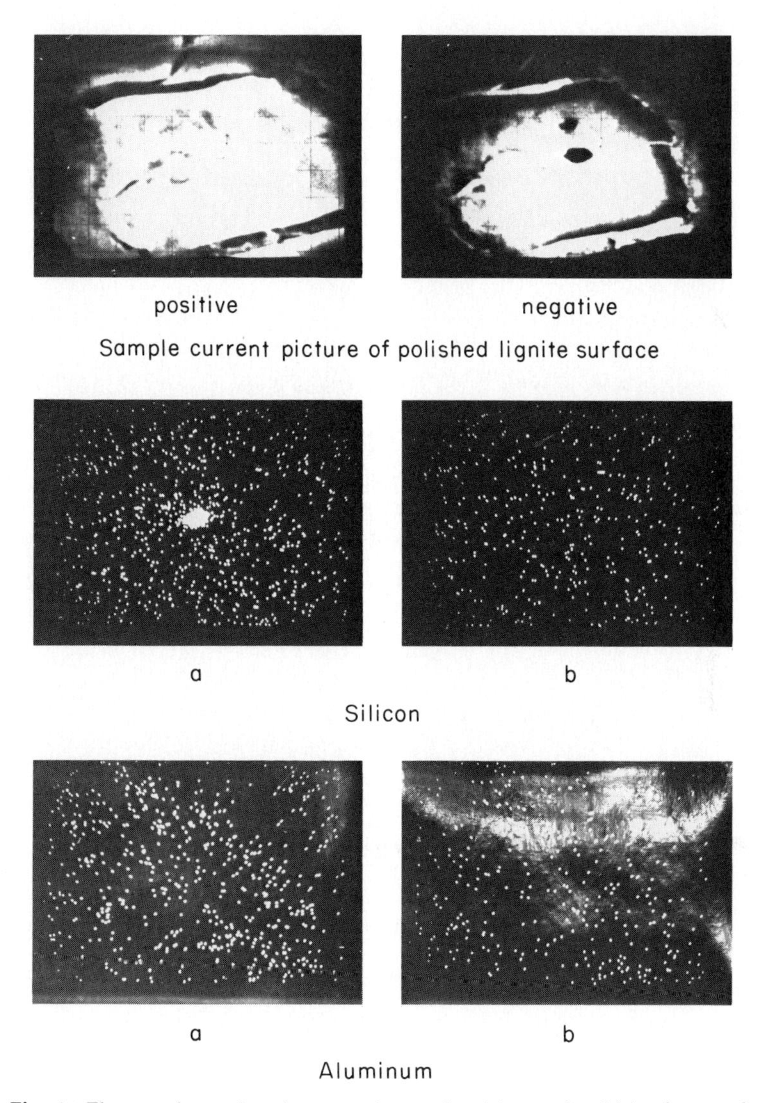

Fig. 4 Elemental map by electron microprobe: (a) sample, (b) background.

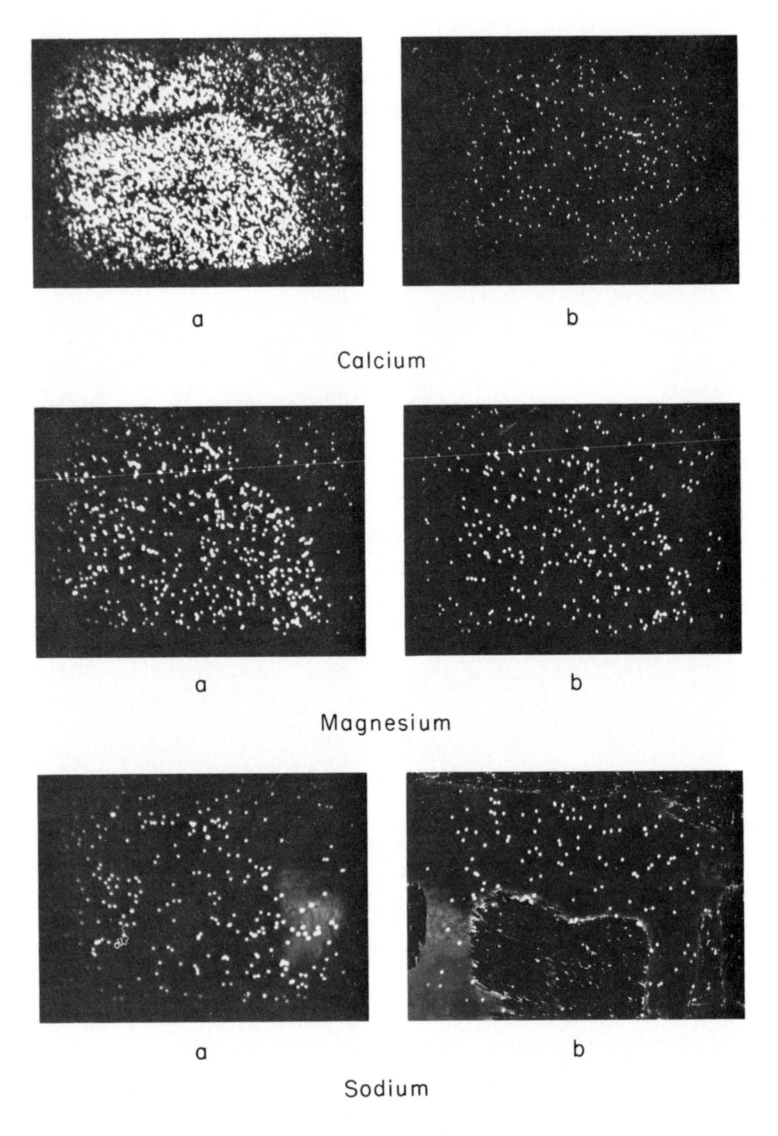

Fig. 5 Elemental map by electron microprobe: (a) sample, (b) background.

same sample, the column (or sections) should be wrapped in a moisture-proof covering for transport to the laboratory. Any evidence of unusual mineralization or other abnormalities should be noted at the collection site. Sampling carried out in exploratory mapping of reserves will generally be in the form of drill cores. These, of course, can be used

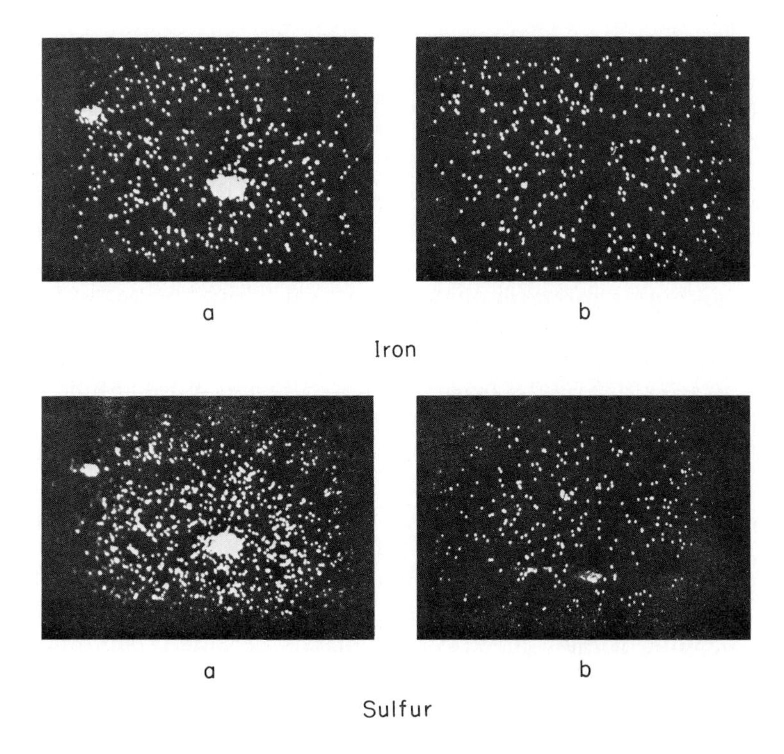

Fig. 6 Elemental map by electron microprobe: (a) sample, (b) background.

for mineral characterization and other tests, but are somewhat less desirable owing to the smaller cross-sectional area of the core and the possibility of contamination by chance inclusion of mineral particles from the overburden.

When the column (channel) sample arrives at the laboratory, it should first be carefully examined for macroscopic inclusions of minerals, including any material collected in or from natural partings in the coal seam. The top inch of the column should be sliced off and thin-section slides prepared from any areas of the fresh surface which visual examination indicates might be of interest. The remainder of this 1-in. slab should then be ground to a particle size of 16–20 mesh and dried *in vacuo* or at 105°C, preferably in an inert atmosphere. A preliminary separation should then be performed on the dry material, using carbon tetrachloride as the density-differentiating medium. (It has been observed that the use of several liquids of varying density accomplishes no noticeable improvement in the separations performed on lignites.) The float and the sink fractions, along with the prepared thin sections, may then be examined by optical and/or electron microscope, electron microprobe, and x-ray diffraction. At this point it will be observed that, with lignite, the gravity separation has, in general, been far from satisfactory, each fraction being more or less contaminated with inclusions of its opposite number. The two fractions should then be separately ground to 80–100 mesh and additional separations carried out on each in turn, again using carbon tetrachloride. Grinding to finer particle size will only serve to increase intolerably the time required for separation, finally reaching a point (about 325 mesh) at which the particles form a rather stable suspension. The separations attained after two or three suspensions of the 80- to 100-mesh material are as effective as may be expected, considering the extremely fine division of the mineral occurrences in lignites. Using these final fractions, several avenues are open for further characterization of the mineral constituents. Low temperature ashing in oxygen plasma may be used to remove the last traces of organic matter, bearing in mind the fact that artifacts not originally present may be formed from metallic cations and the sulfur or nitrogen oxides produced by the oxidation procedure. Diffraction patterns of this material must be carefully interpreted to reflect accurately the original mineralization of the coal. Information thus obtained, combined with that resulting from examination of the thin sections previously prepared, will give a relatively complete picture of the mineral content of the sample.

Elemental analyses may be carried out on the gravity separation fractions, before or after low temperature oxidation, or on the whole coal. This may be accomplished by x-ray fluorescence analysis, atomic absorption spectroscopy, neutron activation analysis, or emission spectroscopy. One recent innovation, inductively coupled argon–plasma spectroscopy applied to analysis for mineral matter in lignites, has yet to be validated. The scanning electron microscope has demonstrated some promise for this type of investigation, but here also data are insufficient for a positive judgment on its capability. Spectra recorded in the far infrared may also be used with samples free of organic matter, giving direct identification of mineral species but with some controversy concerning precise interpretation of certain bands, particularly in the region between 400 and 200 cm^{-1}. The choice of procedures will depend largely on the available instrumentation. The method of combined differential thermal and thermogravimetric analysis may be utilized with some reservations, particularly to mixtures of clay types.

Following the examination described for the top inch of the column sample, the investigator must decide on the degree of detail desired in characterizing the coal sampled. Columns have in the past been examined inch by inch, largely in regard to the nature of the parent organic material which contributed to the combustible fraction of the coal. Such minute detail might be irrelevant to a given objective, and this procedure may be modified as desired. It has been noted, however, that mineralization, and elemental analysis of the mineral matter, does very often change significantly with depth in the seam, and this fact may be of importance under some circumstances: This situation is particularly true of lignite deposits. The investigator must therefore weigh the validity of the data obtained against the effort required to produce it in any given case. On the positive side, however, interest in low rank coals of the United States is mounting rapidly, and coupled with significant refinements in analytical technology, will most certainly result in improved data regarding mineralization and elemental composition of the inorganic fraction of these coals. Such data are of prime importance in predicting combustion characteristics, environmental pollution hazards, and conversion potential of coals previously undeveloped or underdeveloped.

REFERENCES

Angino, E. E. (1967). *Am. Mineral.* **52,** 137.
Barber, T. L., Sr. (1967). *Proc. Annu. Mid-Am. Symp. Spectrosc., 18th* pp. 15–18.
Beckering, W., Haight, H. L., and Fowkes, W. W. (1970). *U.S. Bur. Mines, Inf. Circ.* **IC 8471,** 89–102.
Berkowitz, N. (1957). *Fuel* **36,** 355–373.
Boateng, D. A. D., and Phillips, C. R. (1976). *Fuel* **55,** 318–322.
Bredahl, R. G. (1965). M.S. Thesis, Dept. Chem. Eng., Univ. of North Dakota, Grand Forks.
Campbell, C., Gordon, S., and Smith, C. L. (1959). *Anal. Chem.* **31,** 1188–1191.
CSIRO (1961). *Coal Res. CSIRO* **14,** 2–3.
CSIRO (1963). *Coal Res. CSIRO* **20,** 6–10.
CSIRO (1965). *Coal Res. CSIRO* **25,** 7–12.
Durie, R. A. (1961). *Fuel* **40,** 407–422.
Fisher, G. L., Chang, D. P. Y., and Brummer, M. (1976). *Science* **192,** 553–555.
Fowkes, W. W. (1972). *Fuel* **51,** 165–166.
Fripiat, J. J. (1960). *Groupe Fr. Argiles, Bull.* **12,** 25–41.
Glass, H. D. (1955). *Fuel* **34,** 253–268.
Gordon, S., and Campbell, C. (1955). *Anal. Chem.* **27,** 1102–1109.
Joensuu, O. I. (1971). *Science* **172,** 1027–1028.
Karr, C., Jr., Estep, P. A., and Kovach, J. J. (1967). *Chem. Ind. (London)* **9,** 356–357.
Karr, C., Jr., Estep, P. A., and Kovach, J. J. (1968). *Am. Chem. Soc., Div. Fuel Chem., Prepr.* **12**(4), 1–12.
Kemežys, M., and Taylor, G. H. (1964). *J. Inst. Fuel* **37,** 389–97.
Kessler, M. F., and Romováčková, H. (1961). *Fuel* **40,** 161–170.
Kreulen, D. J. W. (1948). "Elements of Coal Chemistry." Nijgh & Van Ditmar, Rotterdam.

Kuhn, J. K., Harfst, W. F., and Shimp, N. F. (1973). *Am. Chem. Soc., Div. Fuel Chem., Prepr.* **18**(4), 72–77.
Lessing, R. (1925). *J. Soc. Chem. Ind., London* **44,** 277–278.
Majumdar, S. K., Banerjee, N. G., and Lahiri, A. (1959). *Brennst.-Chem.* **40,** 261–263.
Mukherjee, S. N., Nag, A. K., and Majumdar, S. K. (1972). *J. Mines, Met. Fuels* **20,** 363–373.
O'Gorman, J. V., and Walker, P. L., Jr. (1971). *Fuel* **50,** 135–151.
Oinuma, K., and Hayashi, H. (1965). *Am. Mineral.* **50,** 1213.
Paulson, L. E., and Fowkes, W. W. (1968). *U.S. Bur. Mines, Rep. Invest.* **RI-7176,** 18 pp.
Paulson, L. E., Beckering, W., and Fowkes, W. W. (1972). *Fuel* **51,** 224–227.
Rekus, A. F., and Haberkorn, A. R., III (1966). *J. Inst. Fuel* **39,** 474–477.
Stefanko, R., Ramani, R. V., and Ferko, M. R. (1973). Res. Dev. Rep. No. 61, Interim Rep. No. 7, 134 pp. Dep. Mineral Eng., Pennsylvania State Univ., University Park.
Sutherland, J. K. (1975). *Fuel* **54,** 132.
Sweatman, T. R., Norrish, K., and Durie, R. A. (1963). *CSIRO Misc. Rep.* **177,** 43 pp.
Thaer, A. (1954). *Staub* **38,** 555–570.
U.S. Geol. Surv. (1959). *Bull.* **1055,** 11–179.
Vaughan, H. P., and Wiedemann, H. G. (1965). *Vac. Microbalance Tech.* **4,** 1–19.
Warne, S. St. J. (1965). *J. Inst. Fuel* **38,** 207–217.

Chapter 28

Procedures for Analysis of Respirable Dust as Related to Coal Workers' Pneumoconiosis

Robert W. Freedman†
PITTSBURGH MINING AND SAFETY RESEARCH CENTER
BUREAU OF MINES
U.S. DEPARTMENT OF THE INTERIOR
PITTSBURGH, PENNSYLVANIA

I. INTRODUCTION

There is some diversity of opinion as to the cause of coal workers' pneumoconiosis (CWP). "Black lung" disease, as it is commonly referred to, often develops after accumulation of respirable coal dust deposits in the lungs. Cummins and Sladin (1930) believe that silica is involved, possibly synergistically. Morgan (1971) disputes this. The

† *Present address:* U.S. Bureau of Mines, 4800 Forbes Avenue, Pittsburgh, Pennsylvania.

ISBN 0-12-399902-2

presence of trace elements in miners' lungs may also be implicated in causing CWP. The chemical nature of organic constituents could conceivably affect the incidence of pneumoconiosis in coal mines (Warden, 1969; Freedman and Sharkey, 1972).

This chapter is devoted to analytical procedures involving respirable dust weight, free silica (principally α-quartz), trace elements, and organic components. The respirable dust is collected by size-selective samplers in accordance with Title 30 of the Bureau of Mines (1972). Owing to space limitations, the procedures covering these pollutants are described briefly in some cases, providing only selected specific details. Complete details can be obtained from the literature references.

II. MEASUREMENT OF RESPIRABLE COAL DUST WEIGHT

Mandatory dust standards for coal mines were established under the Federal Coal Mine Health and Safety Act of 1969. Sampling procedures are prescribed under Title 30, Code of Federal Regulations under Part 70 and Part 71, published in the Federal Register on April 3, 1970, and March 28, 1972, respectively. Under Section 202(a) of Title II, each coal mine operator is directed to take accurate samples of respirable coal mine dust, and detailed instructions are provided for this purpose. The more stringent 1972 standard permits no more than 2.0 mg of dust per cubic meter of air. Frequency, number, and location of sampling are stated in 30 CFR Part 70.

Mine operators collect size-selected samples with approved personal samplers on factory-preweighed membrane filters and send them to the central MESA (Mining Enforcement and Safety Administration) laboratory at the Pittsburgh Technical Support Center, Dust Group (MESA, 4800 Forbes Avenue, Pittsburgh, Pennsylvania 15213). The mine data card which accompanies the sample is filled out with items such as section identification, sampling time, miner's occupation, and method of mining.

A description of the laboratory operation is given by Jacobson and Parobeck (1971) and updated by Parobeck (1976). The average number of samples currently processed by MESA is 2500 per day. Specific data such as cassette number and initial weight are provided on the mine data card supplied with the sample. On receipt from the mine, each sample is weighed by MESA to ± 0.05 mg in a clean room environment carefully isolated from vibration. Static charges are removed with a radioactive ionizing unit.

A schematic of the sample handling system is shown in Fig. 1, which

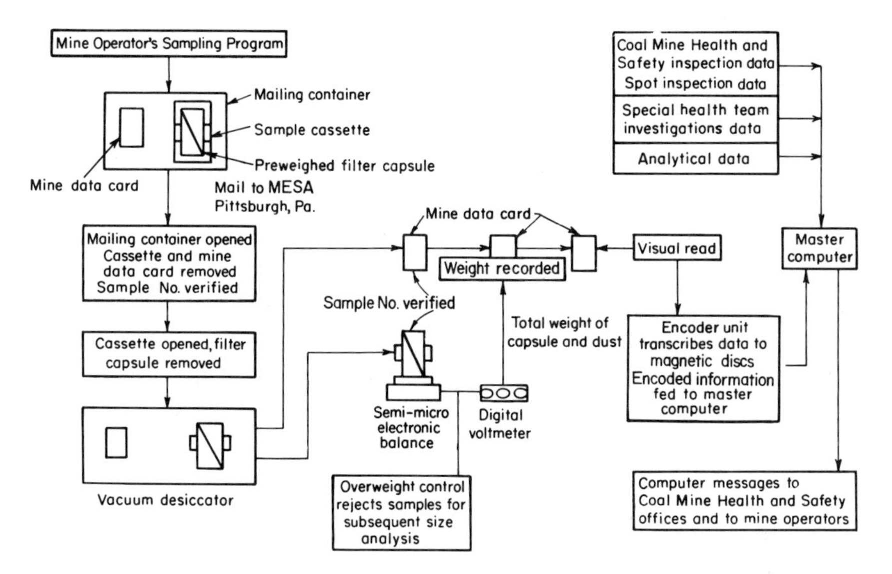

Fig. 1 Schematic of sample handling system. (From Parobeck, 1976, Fig. 2.)

is a copy of a figure in the report by Parobeck (1976). Data are recorded on flexible magnetic disk cartridges and transferred via a data phone to MESA's Division of ADP in Denver, Colorado. A Burroughs 5500 computer calculates dust concentration *D*, according to

$$D\ (\text{mg/m}^3) = \frac{\text{final weight (mg)} - \text{initial weight (mg)}}{\text{sampling time (min)} \times \text{rate of sample (m}^3\text{/min)}} \times 1.38 \qquad (1)$$

where 1.38 is the conversion factor to convert dust concentration obtained with a personal sampler, equipped with a pulsation damper, to an equivalent MRE concentration.

III. FREE SILICA

The polymorphs quartz, tridymite, crystobalite, and amorphous free silica are present in many minerals. However, the only form found in coal has been α-quartz. This simplifies the analysis to some extent. Good quantitative determination is best accomplished using some form of molecular spectroscopy as opposed to bulk chemical determination (Freedman, 1971). Anderson (1975), in a review of free silica analysis, covers both chemical and physical methods, and points out significant difficulties with all techniques. One problem with molecular spectroscopy is the variation of response with surface condition and particle size. In the respirable range, decreasing particle size yields an increase in infrared response (Tuddenham and Lyon, 1960) and a decrease in x-ray diffraction response (Bradley, 1972). This necessitates the use of chemically and physically similar standards for valid interlaboratory comparison. Nevertheless, use of either technique is more likely to produce accurate results than the phosphoric acid dissolution of silicates (Talvitie, 1951), the solution technique with the greatest acceptance. This approach requires exact control of digestion time and temperature, is very size dependent, and is also affected by the surface condition of the crystals. These effects necessitate considerable information about the sample to achieve any degree of confidence in the results.

A. Analysis of Quartz by Infrared Spectrometry

Three infrared procedures were developed by the Bureau for the analysis of quartz in respirable coal dust. A direct on-filter analysis of individual filter samples collected in the mine was developed by Mine Safety Applicances (MSA) (1974) on contract. In addition, a potassium bromide pellet procedure was described by Jacobson and Parobeck (1971) based on the author's in-house Method No. 19,† "Determination of Free Silica (Quartz) in Ashed Respirable Coal Dust by Infrared Spectroscopy." MESA updated Method 19 and issued it in 1971 as Standard Method No. A7. An on-filter method was developed at the Bureau (Freedman *et al.*, 1974) and is described in detail in the in-house procedure, Standard Method No. 46, "Determination of Free Silica in Respirable Coal Mine Dust on Membrane Filters by Infrared Spectroscopy."

1. *MSA Method*

Dust is collected on 35-mm preweighed MSA‡ FWS B membrane filters§ using size-selective samplers as prescribed in 30 CFR Part 70. The individual filters are placed in a rotating holder on a Perkin-Elmer Model 267 grating infrared spectrophotometer provided with a beam

† Several in-house methods are referred to in this publication and are obtainable from Dr. R. W. Freedman, U.S. Bureau of Mines, 4800 Forbes Avenue, Pittsburgh, Pennsylvania 15213.

‡ The use of commercial products does not imply endorsement by the Bureau.

§ Dust collection filters are of relatively coarse (5 μm) pore size for low pressure drop capability. Static charges which develop on polymer membranes serve to retain very fine particles in the dry state.

expander. The infrared beam covers a substantial portion of the filter area as compared to the small fraction covered without an expander. Readings are taken automatically on both sides of the 12.5-μm quartz doublet region. An additional correction is made for kaolinite by taking a reading at 10.95 μm and applying a factor for subtracting the 12.5-μm kaolinite peak from the quartz. Data processing is accomplished with a 12-bit analog-to-digital (A/D) converter, a 4K minicomputer, and an ASR teletype.

The method is very rapid (5-min analysis time), but the sensitivity of the system is marginal inasmuch as the current TLV is 100 μg. The best results contained in the MSA (1974) Final Report are shown in their figure 11 (page 61). In this plot silica is corrected for kaolinite. In the range 0–600 μg, the maximum permissible level of quartz (100 μg) produces an absorbance of only 0.0137. The standard deviation about the regression lines (standard error of estimate) is $\pm$39.3 μg. According to Bureau research chemists (Lang and Hay, 1975), modifications have been made on the system and precision improved to $\pm$20 μg. These workers obtained fairly good correlation with an x-ray technique, which indicated reasonable accuracy. The proposed TLV of 50 μg would rule out this technique. Consequently, attempts are underway at the Bureau to improve precision and accuracy.

2. *Potassium Bromide Pellet Procedure*

The procedure has been used by the Bureau, and currently by MESA, since 1969 with little change. The 12.5-μm (800-cm^{-1}) doublet is employed. Five to ten membrane filter samples are composited in order to provide sufficient sensitivity. A simplified version of the procedure (prepared from Methods 19 and A7) is as follows:

(i) Insert a sample of respirable coal mine dust, 1–4 mg, in the combustion pan. Determine weight of dust using a Cahn balance. Record as S.

(ii) Place the platinum weighing pan in a furnace, and ash the sample at 800°C. Reweigh the ashed sample and record the weight of ash.

The optimum weight of ash for an infrared determination is 0.40 $\pm$ 0.20 mg. If the quantity of ash resulting from step (ii) is less than 0.20 mg, rerun steps (i) and (ii) and use the combined ash for the infrared analysis. Should the quantity of ash resulting from step (ii) exceed 0.6 mg, rerun steps (i) and (ii), using a smaller dust sample.

(iii) Carefully remove combustion pan containing the ash. Experience has shown that the ash agglomerates and almost quantitative transfer occurs. If desired, the empty pan can be reweighed to establish any small correction.

(iv) Invert the combustion pan over a piece of tared glazed paper (4 × 4 in.) containing 360 ± 1 mg of potassium bromide, and gently crush and mix sample with a microspatula.

(v) Transfer to a stainless steel vial. Close vial and shake at least 15 times manually.

(vi) Return the sample to the glazed paper, and then transfer completely to an evacuable pellet disk die and insert in a pellet press.

(vii) Connect vacuum line to the die, and pump down for 3 min. Then apply 20,000 lb of total pressure for 3–4 min, while die remains under vacuum.

(viii) Release pressure, allow dry air to enter die, and remove die and pellet.

(ix) Weigh pellet to 0.1 mg and record as W.

(x) Place sample pellet and reference potassium bromide pellet in their respective holders of a double-beam infrared spectrophotometer.

(xi) Set gain at midrange, set to fast scan, and traverse a short region covering the 780/795-cm^{-1} doublet.

(xii) Determine absorbance A using baseline measurement of the 795-cm^{-1} peak.

(xiii) Construct a calibration curve using 8–50 mg of quartz mixed to form 360 ± 1 mg pellets to yield a plot of absorbance versus weight of quartz.

(xiv) Determine weight of quartz in the sample from absorbance using the calibration curve, and record as Q.

(xv) Calculate percent quartz based on the original coal dust weight S in milligrams, the original ash potassium bromide weight W_0 in milligrams, the pellet weight W in milligrams, and the weight of quartz Q in micrograms, by the following formula:

$$\% \text{ quartz} = QW_0/10SW$$

Currently MESA uses Method A7 exclusively. They claim a precision of ±10.2%. About 22,000 samples per year, submitted by federal mine inspectors, are analyzed.

3. *Standard Method No. 46*

This method bears some similarity to the MSA procedure but employs conventional infrared laboratory instruments in the scanning

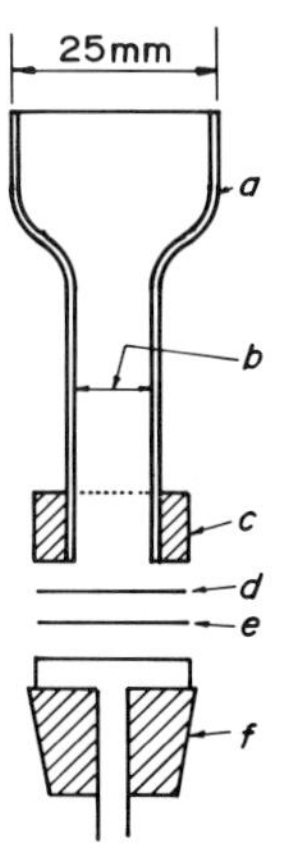

Fig. 2 Modified Millipore filtration apparatus. *a*, Upper part—flared glass funnel; *b*, accurately known internal diametec; *c*, Bakelite shoulder cast onto glass tubing; *d*, membrane filter; *e*, glass backup filter; *f*, lower part.

mode. Dust is transferred from a single-field filter to a membrane filter having excellent transmission in the 800-cm^{-1} region (Gelman DM 450).† Filters can be analyzed directly or ashed to eliminate clay mineral interference, followed by redeposition on a second DM 450 filter. Redeposited coatings are evenly distributed, greatly reducing absorption error. By using small-diameter filters considerable increase in sensitivity is available without beam concentration A 10-mm-diameter filter, for example, will be covered by most of the infrared beam and will yield about 10–15 times increase in infrared absorption.

A condensed version of the step-by-step procedure is as follows:

(i) Place collection filter into a 50-ml beaker in 15 ml of isopropanol and insert beaker in ultrasonic bath for 2 min.

(ii) Remove filter from beaker with tweezers, and rinse with isopropanol to remove all deposit from filter.

(iii) With slight vacuum applied, place a glass backup filter on filter base of modified Millipore xx102514 filtration apparatus (shown in Fig. 2). The upper part is a flared glass funnel with a stem of accurately known diameter *b*.

(iv) Select a 49-mm Gelman DM 450 filter having no pinholes, cut in half, and place the superimposed halves glossy side down on top of the glass fiber filter.

(v) Redisperse slurry in beaker for about a minute, and filter using small isopropanol washes.

(vi) Remove filter and, using a scriber, make four indentations just

† The 450 refers to the 0.45-μm pore size required for liquid filtration.

outside circumference of deposit, marking top and bottom overlaying halves of filter. Allow to air-dry.

(vii) Place top half in sample holder and bottom half in reference sampler holder. Holders are brass plates having holes of diameter b [step (iii)] and spring clips to hold the filter down.

(viii) Scan from 850 to 700 cm^{-1} and measure absorbance A by baseline measurement. If visually difficult (as may be the case with unashed dust), make a maximum-to-minimum measurement. This is done by measuring the vertical distance† between a tangent to the top of the 800-cm^{-1} peak and a tangent to the valley minimum of that peak. Use the same method of measurement for both the sample and calibration infrared traces.

(ix) Prepare a calibration plot using a series of filter deposits prepared from pipetted aliquots of a weighed batch of pure quartz in 50 ml of isopropanol. A convenient concentration range is 25–150 mg/cm^2 deposited on 10- to 18-mm filter areas. Plot absorbance A as ordinate versus concentration C (mg/cm^2) to give a straight line passing close to the origin and having a slope A/C equal to absorptivity a. A typical slope is 1.4×10^{-3}.

(x) From measured absorbance A of sample filter, calculate sample concentration C from $C = A/a$ in milligrams per square centimeter. Determine quartz sample weight, in micrograms, by multiplying C by calculated filter area in square centimeters. Then calculate percent quartz Q from original dust weight S in milligrams:

$$\%Q = 100Q/S$$

The precision of the method is ±5% expressed as standard deviation. The accuracy of this and other silica methods is discussed later in the section.

B. Analysis of Quartz by X-Ray Diffraction

X-ray diffraction has some advantages over infrared analysis for use in metal and nonmetal mines where, unlike coal mines, several polymorphs of free silica such as crystobalite, tridymite, and quartz occur and must be optically separated owing to differences in toxicity. However, the need of this specificity in coal dust does not exist. The principal advantage of having an x-ray method in addition to infrared

† Direct measurement of height is possible for absorbance measurements using instruments such as the Perkin-Elmer 180, which has an absorbance mode position. Otherwise, absorbance is calculated from transmittance measurements.

methods is to offer comparisons with an independent technique which can be helpful in determining accuracy as distinguished from precision.

The interferences of clay minerals such as muscovite, illite, and kaolinite are well known to x-ray crystallographers and where possible, these interferences should be removed by ashing the sample.† In infrared analysis, kaolinite does interfere (Hunt *et al.*, 1950), but allowance can be made for this interference by subtraction of calculated peak values as shown previously in the MSA procedure.

All the x-ray diffraction procedures described in this chapter are on-filter procedures. Sample collection is best accomplished on polymer membranes, and analysis can be conducted on either polymer or silver membranes after transferring collected dust. Collection rates of about 2 liters/min are used with personal samples in mines. Pressure drop requirements dictate that membrane pore size should be close to 5 μm. Polymer filters are much more retentive than silver using this pore size.

1. Silver Membrane Procedures

On-filter analysis of quartz by x-ray diffraction using silver membrane filters is in common use. This technique was first investigated by Leroux and Powers (1969). Correction for thickness of dust layers for mass absorption was shown to be small up to 200 mg/cm². Silver membrane filters have several advantages such as insolubility in solvents and low interference in the quartz regions. One disadvantage, according to Knight *et al.* (1972), is that the large pore size (5 μm) required for low resistance collection of dust can result in loss of respirable dust in the lowest size ranges. Results showed that the ratio of observed to calculated x-ray intensities decreases markedly below 200 μg/cm², presumably because fine quartz passes either through or into the relatively coarse silver pores where the analyte is screened from the x-rays. Thus it becomes difficult to obtain satisfactory results at low levels.

In spite of the difficulties just outlined, silver membrane filters have a good general utility, especially in noncoal mines where quartz concentrations are high and where collection practices do not present difficulties. To get around problems of sensitivity, at least in part, Leroux *et al.* (1973) recommended transfer from a larger organic membrane collection filter to a 13-mm silver membrane.

2. Organic Polymer Membrane Procedures

Polymeric membranes are of limited general utility. Most workers in the field, such as Leroux and Powers (1969), point to the wide diffraction

† For x-ray diffraction of low temperature ash see Chapter 26, Section III,A.

lines which lead to interference with quartz, as opposed to silver membranes which result in a clean background in the quartz angular region using molybdenum radiation. However, where it is possible to select membranes with specific low interference regions, polymers have several advantages. The static charge which develops tends to retain very fine particles, even with large pore sizes. Thus much lower pressure drop and more practical mine collection practices can be used. In addition, the collection filter can be ashed together with the dust, and the inorganic components redeposited on a suitable polymer membrane.

In the development of an on-filter x-ray procedure, it was desirable to select a redeposition membrane filter of the same material as is used in the infrared method. Thus, since both techniques are nondestructive, a direct comparison of both spectra can be made using a single filter. It was found that the DM series (polyvinyl chloride–acrylonitrile copolymer) supplied by Gelman has window regions for quartz for both infrared and x-ray diffraction. The Gelman DM 450 material was selected for redeposition filters. Other membrane types such as the MSA VM series (polyvinyl chloride) can be employed for collection.

An in-house procedure, Standard Method No. 47, "Determination of Quartz in Respirable Mine Dust on Membrane Filter by X-ray Diffraction," was developed at the Bureau. It served as the basis for the previously cited publication of Freedman *et al.* (1974) in which statistical comparisons of infrared and x-ray methods were made. A condensed version of the procedure is now given.

a. Calibration of Instrument. (i) Set proper conditions for x-ray equipment used. For example, with a GE-XRD6 diffractometer, typical settings are 45 *K*VP at 25 mA; a medium resolution soller slit and a 1° beam slit are employed. Detector voltage (adjusted daily) is typically 1.5 *K*V. The pulse height selector cutoff $EL = 2$ V and the window with $\Delta E = 6$ V are used.

(ii) For the calibration, select a 0.008-mm nickel filter to provide strong Cu $K\bar{\alpha}$ intensity.

(iii) Choose a 0.1° receiving slit, and adjust pulse height selector to provide the maximum signal-to-background ratio.

(iv) Set the goniometer close to 26.65° to maximize quartz response using a Novaculite† external standard.

(v) Determine the net peak height of the Novaculite (quartz) standard at its strongest peak ($2\theta = 26.65°$). In this and all other instances in which a net peak height is determined, three 90-sec counts are taken: a background count on each side of the peak and a count at the peak maximum. Net peak height is the peak count minus the average of the

† Commercially available pure silica standard.

two background counts. Compare the unit peak height obtained each day with that obtained during preparation of a quartz calibration curve, which is described in the following section. This will provide the external standard correction factor for sample analysis according to the equation

$$\text{external standard correction factor} = \frac{\text{daily novaculite peak height}}{\text{calibration novaculite peak height}} \quad (2)$$

b. Preparation of Calibration Curves Using Filter Standards. (i) Prepare a series of quartz standards by depositing known weights of pure quartz in the respirable range. Employ exactly the same procedure as previously used in the infrared method (Standard Method 46). A good working range is 25–150 mg/cm^2.

(ii) Determine the net peak height for the $2\theta = 26.65°$ peak in the Novaculite for use as an external standard. All other external standard counts will be set in proportion to this count during the calibration.

(iii) Measure the net peak height of the $2\theta = 26.65°$ peak in all quartz standard samples.

(iv) Plot the net quartz peak height (ordinate) versus concentration (abscissa).

c. Analysis of On-Filter Dust Samples. (i) Mount the membrane filter half containing an ashed respirable dust sample in the sample holder. The regular x-ray powder sample holder is modified with a device to hold a membrane filter against the holder. A thin piece of brass or aluminum (the same dimensions as holder) with a window slightly larger than the sample deposit area is hinged at both ends with transparent tape. Make sure that the deposit area is centered with respect to both holder and mask windows. A light behind the holder is helpful for this purpose.

(ii) Insert the mounted sample into position and scan over the desired range if qualitative analysis is required.

(iii) For quantitative measurements, obtain the net peak height for the strongest quartz peak ($2\theta = 26.65°$ with Cu $K\bar{\alpha}$ radiation).

(iv) Apply the daily external standard correction to the measured net peak height.

(v) Determine the quartz concentration (μg/cm^2) in the sample from the appropriate calibration curve.

(vi) Calculate the percentage of quartz in the sample using deposit area and original sample weight according to the equation

$$\%\ \text{quartz} = \frac{\text{concentration } (\mu\text{g/cm}^2) \times \text{deposit area } (\text{cm}^2) \times 100}{\text{sample weight } (\mu\text{g})} \quad (3)$$

C. Accuracy and Precision of Infrared and X-Ray Methods for Quartz

The bulk chemical method for quartz unfortunately is inferior to molecular spectroscopic procedures. It would have been possible, in a chemical procedure, to weigh out a given amount and assume that this was an *accurate* standard. With infrared and x-ray, instrumental response is highly dependent on particle size, physical condition of surface layers resulting from mechanical handling, thermal treatment, and so on. Thus while the ability to replicate the precision of individual batch analyses can be excellent, the comparison to a fixed standard (accuracy) is an unknown quantity simply because no single invariate standard exists. Interlaboratory comparison of "accuracy" is possible using the same samples nondestructively. Similarly we have compared infrared and x-ray samples within our own laboratory using the same on-filter samples. "Accuracy," as we determined it, is in reality a measure of methodic comparison.

We compared (Freedman *et al.*, 1974) 47 field samples of respirable coal mine dust redeposited from collection filters using three techniques (see Table I). These were infrared (Method 46) unashed, infrared (Method 46) ashed and redeposited, and x-ray (Method 47) ashed and redeposited. The true value of quartz concentration of each sample was completely unknown.

It can be observed from Table I that results for the second pair or for ashed samples, by the two independent methods, agree reasonably well. This is an indication that either one could be useful for interlaboratory measurement of accuracy if a single batch of silica could be agreed

TABLE I *Intercomparison of Quartz Methods*[a]

Comparison	N[b]	Correlation coefficient (r)	Slope	S y/x (standard error of estimate) ($\pm\mu g$)	y intercept (μg)
X ashed ir Y unashed ir	44	.966	1.07	20.5	1.63
X ashed ir Y ashed x-ray	47	.976	0.92	14.6	0.62

[a] From Freedman *et al.* (1974).
[b] Number of samples.

upon as a standard. A more detailed discussion of accuracy is given by Freedman *et al.* (1974).

The accuracy of any of the silica procedures is in considerable doubt, and this explains the philosophy of rule-making in which a single procedure, namely the KBr pellet method, has been selected by MESA and used exclusively. What this researcher takes issue with is the compositing of 5–10 samples for each analysis. This tends to mask the high values of individual samples which could be in violation of the law.

IV. TRACE ELEMENTS

A. Analysis of Lung Tissue

The supposition that certain trace elements play a role, either directly or synergistically, in the development of coal workers' pneumoconiosis has not been proven by epidemiological evidence. In 1964, the U.S. Public Health Service undertook a program involving chemical analysis of the minerals and metals in the lungs and pulmonary lymph node tissue of coal miners.

Crable *et al.* (1967, 1968) freeze-dried lungs or lung tissue, ground the specimens in a mullite ball mill, and digested aliquots of the ground tissue using the procedure of King and Gilchrist (1945). The mineral dust residue was ignited at 380°C and the coal content was determined by weight loss. The residual mineral fraction (ash) was analyzed for quartz and other crystalline materials by x-ray diffraction and for trace elements by emission spectrography. Levels of elements commonly occurring in coal and found in miners' lungs, were compared with levels found in normal lungs (see Table II).

The concentrations given in columns 2 and 3 of Table II are the mean values taken from the concentration ranges presented by Crable *et al.* (1967) in their Table II. The M/N ratios in the last column are high for certain elements such as Al, Ba, B, Ag, and V. This may or may not be significant.

B. Analysis of Respirable Dust for Trace Elements

The levels of trace elements found in coal are quite low as compared with those concentrated in miners' lungs. Two frequently used techniques employed for analysis are atomic absorption (AA) and spark source mass spectrometry (SSMS). AA spectrometry in the flame mode is rapid and accurate but lacks sensitivity for several elements which may be significant. About one to two orders of magnitude increase in

TABLE II *Comparison of Levels of Elements in Miners' Lungs with Normal Lungs*[a]

Element	Mean concentration (μg per dried lung)		
	Miners' lungs (M)[b]	Normal lungs (N)[b]	Ratio (M/N)
Al	9535	205	47
Ba	283	2.5	113
Be	4.0	—	—
B	7.7	0.28	27.5
Cr	11.6	3.02	3.8
GE	1.1	—	—
Fe	1595	2410	0.66
Pb	23.7	13.9	1.7
Mg	98.8	495	0.20
Mn	8.1	4.38	1.8
Ni	37.0	20.2	1.9
Ag	1.1	0.01	110
Sn	20.0	23.13	0.87
Ti	215	50.1	4.3
V	45.2	1.73	26

[a] From Freedman and Sharkey (1972).
[b] Adapted from data of Crable *et al.* (1967).

sensitivity can be obtained using nonflame AA. However, analysis is slower, and very careful sample preparation must be undertaken to avoid contamination. SSMS has adequate sensitivity for most elements but is semiquantitative. Nevertheless, it is highly specific and provides for wide coverage not readily obtainable with most other techniques. For additional details on AA and SSMS see Volume I, Chapter 14.

1. *Analysis by Flame Atomic Absorption*

Coal dust is brought into solution either by alkaline fusion (e.g., with lithium tetraborate, Hamilton *et al.*, 1967) or by acid digestion employing hydrofluoric acid and an oxidant. We found that the procedure of Bernas (1968) using hydrofluoric acid in a Teflon-lined bomb could not be used for coal even with concentrated perchloric or nitric acid as an oxidant. However, fuming nitric provided the needed oxidation potential to cause complete destruction of the organic matter. The procedure is described by Hartstein *et al.* (1973). As can be observed from the result shown in that publication, accuracy and precision are adequate. However, in some cases, such as beryllium and the transition metals, other techniques such as nonflame AA or emission spectrometry would be helpful in providing better sensitivity. For this purpose, the wet digestion procedure for dissolving the dust used by Hartstein is given in Bureau Standard Method No. 37, "Decomposition of Coal by the Use of an Acid Digestion Bomb." A condensed version of this procedure is as follows:

(i) Prepare a calibration curve containing 5, 10, 15, and 20 ppm of a given element by dilution of a commercial 1000-ppm AA standard solution. Aspirate in an air–acetylene or O_2–acetylene flame under conditions given in the instrument manufacturer's manual. Plot absorbance (ordinate) versus concentration in parts per million (abscissa).

(ii) Weigh 10–50 mg of coal dust (−350 mesh), transfer to a Teflon digestion cruciable, add 6 ml fuming nitric acid, assemble bomb, and digest at 150°C for 2–2.5 hr.

(iii) Add 4 ml of hydrofluoric acid to the bomb when cool, and digest at 150°C for an additional 15 min.

(iv) Transfer contents of bomb to a 50-ml glass volumetric flask with the aid of distilled, deionized water. Immediately add 2.8 g boric acid (to complex the HF and prevent attack on borosilicate glass) and make up to mark. Transfer to a polyethylene bottle within 2 hr.

(v) Conduct AA analysis along with analysis of standards and a reagent blank.

(vi) Obtain concentration C of each element from net absorbance and the calibration plot, and calculate the percentage of each element based on weight S of the original dust sample according to the equation

$$\%X = 10^{-4}VC/S \tag{4}$$

where X is the element sought and V is the dilution volume.†

2. *Analysis by Spark Source Mass Spectroscopy*

A program was conducted at the Bureau of Mines by the Pittsburgh Energy Research Center involving the analysis of respirable coal dust samples by SSMS in order to obtain correlation of major, minor, and trace elements with coal workers' pneumoconiosis. Sixty-four elements ranging in concentration from 0.01 to 41,000 ppm were determined in anthracite and lower ranking coal as reported by Kessler *et al.* (1973). The instrument used in this investigation was a commercial Mattauch–Herzog mass spectrometer equipped with photographic and electrical detection systems and an rf spark source. The resolution of the

† As described, 50 ml was used, but this could be reduced to 20 ml where added sensitivity is required.

instrument was 1 part in 5000. All trace elements were determined from mass spectra recorded on Ilford Q-2 photographic plates. Major elements were determined using the electrical detection system.

Electrodes were prepared by mixing the samples with equal parts of pure graphite. To ensure homogeneity of mixing and to determine the plate sensitivity, 50 ppm of indium was added as an internal standard to the sample graphite mixture. The mixtures were pressed into electrodes in polyethylene slugs in a commercial isostatic die. The final electrode size was $\frac{1}{16} \times \frac{3}{8}$ in.

For the analysis of trace elements, a series of graded photoplate exposures (E) ranging from 1×10^{-1} to 3×10^{-7} C was made. Using the same sparking parameters and series of graded exposures as used for the samples, mass spectra were obtained for the standard, U.S. Geological Survey sample BCR-1 (basalt rock).

Determinations of each element were made by direct visual comparison with spark source mass spectra of the BCR-1 standard obtained by Flanagan (1969) using the expression

$$\text{concentration of trace element} = \frac{E \text{ sample}}{E \text{ standard}} \times \text{concentration of standard} \tag{5}$$

Major elements were determined by magnetic scanning with recording of peak heights of elements and concentrations calculated according to

$$\text{concentration of trace element} = \frac{E \text{ sample}}{E \text{ standard}} = \text{concentration of standard} \tag{6}$$

It should be noted that the values reported are semiquantitative and should be employed as a guide for confirming specific elements of interest such as atomic absorption, neutron activation, and the like.

V. ORGANIC COMPONENTS

Saffioti *et al.* (1965) demonstrated that polynuclear aromatic hydrocarbons, such as those prevalent in coal dust, produce some of the most severe lung irritants when combined with inorganic material. Thus, in addition to the well-known incidence of lung cancer, fibrosis also can result.

Gas and liquid chromatography can provide quantitative results for well-specified and separable components. However, the multiplicity of organic compounds in coal requires an analytical tool of extreme specificity. Thus, high resolution mass spectrometry is a more suitable technique even considering its limitation to qualitative or, at best, semiquantitative analysis. A major advantage of high resolution mass spectrometry is that literally hundreds of organic components of coal dust can be analyzed without prior separation of the sample. For example, eight compounds having a mass close to 184 occur within 0.1 amu and can be separated.

A report was prepared by the Pittsburgh Energy Research Center, Department of Energy (Schultz *et al.*, 1975), under contract to the Bureau of Mines, describing the analysis of organic material originating from coal. High resolution mass spectrometry was employed to identify and classify the organic compounds in respirable mine dust collected in mining regions of high and low incidence of coal workers' pneumoconiosis. Coal of various ranks and from several geographic areas was examined. In most cases the collected respirable mine dust was compared with ground seam coal taken from the same location.

Experimentally, mass spectra of the mine dusts and coal dusts were obtained on a Dupont 21-110B high resolution mass spectrometer using direct probe sample introduction and recording data on photographic plates. The mass spectra represent that portion of the dust which was vaporized at 300°C and 10^{-6} torr. The mass spectrum of perfluorokerosine was used to provide standard masses for computer calculation of precise masses and empirical formulas for the unknown spectrum.

Mine dusts were obtained primarily from seams in which the incidence of coal workers' pneumoconiosis was documented in the recent study by Morgan *et al.* (1973). The respirable mine dusts were collected on personal samplers; dust was brushed from the filters to eliminate solvent contamination and reaction with filter material. The coal dust fractions were prepared using an Anderson sampler. A complete description of the preparation of coal dust samples was reported by Kessler *et al.* (1971).

Mass spectral data obtained from the photoplate spectra included precise masses, elemental composition, and intensities of the recorded ions. An average of 800 lines was found between mass 70 and mass 450, the upper limit of the photoplate; these lines corresponded in mass to elemental combinations consisting primarily of C_xH_y, C_xH_yO, and $C_xH_yO_2$ structures. This report is concerned primarily with the carbon–hydrogen components.

The various coal seams and coal ranks were differentiated by differences in molecular weight distribution and difference in C–H ratios (related to aromaticity and saturation). Differences in molecular weight were found between respirable coal dust and ground seam coal related to it. Although these differences may be useful for further research, no firm correlation can be observed between coal workers' pneumoconiosis and such factors as molecular weight distribution, number of rings (aromatic or alicyclic), bond saturation, chain branching, and presence of heteroatoms.

VI. OTHER METHODS

A considerable array of techniques are available for quantitative measurements of compounds or elements which may have a bearing upon the development of coal workers' pneumoconiosis. Several additional methods are considered worthy of brief mention.

A. Free Silica

1. *X-Ray Using an Internal Standard*

The internal standard method has been employed to correct for rapid changes in machine output and, partially, for differences in matrix absorption. An example involving the use of a silver membrane on-filter technique is that described by Bumstead (1973) in which fluorite is employed as an internal standard. One problem inherent in the use of an internal standard is that the standard itself must be very homogeneous to avoid response variations due to particle size and surface effects. An external standard can be relied upon to yield an invariant response with identical behavior of the instrument.

2. *Infrared with Beam Condensation*

A very interesting procedure, which unfortunately has not yet been published, was developed under contract to the Bureau by the Pittsburgh Energy Research Center. Micropellets only 4 mm in diameter were prepared using potassium bromide for the analysis of calcite in respirable coal dust. The objective was to differentiate between rock dust and coal dust. In the 1420-cm^{-1} band, a standard deviation (precision) of 2.5% was found. A comparison of size and potential for sensitivity of the micropellet with the conventional 13-mm pellet can be made. The area of the micropellet is only about 0.09 as great as that of the larger pellet. Thus, using the same thickness, a roughly 11 times increase in sensitivity is obtainable with the semipellet. By reducing thickness, even greater sensitivity for any infrared analyte can be obtained. In practice, 17 mg KBr was used compared to 360 mg for the micropellet with a 1% admixture of dust for both size pellets. The increase in sensitivity is simply the increase of the weight ratio or 21 times. This technique could, in principle, be applied to quartz using a procedure similar to Standard Method No. 19. It would then be possible to analyze single-filter samples, eliminating the need for compositing.

3. *Soft X-Rays*

Hurley and White (1973) employed soft x-rays for the analysis of quartz in coal mine dust. Total silicon is measured at the Si $K\bar{\alpha}$ peak. α-Quartz produces a peak shift from which its concentration can be calculated. Good linearity was obtained, but sensitivity was too low to be useful for mine analysis.

B. Trace Elements

Techniques other than atomic absorption emission spectrometry and spark source mass spectrometry can be used. Most of these have limitations such as variation in sensitivity or limited coverage of elements capable of analysis. A few techniques are briefly discussed.

1. *Stripping Voltammetry*

This method of analysis can provide a sensitivity of 10^{-9} *M* for several elements. However, the number of elements is limited, and only a few can be analyzed at a time owing to overlapping half-wave potentials.

2. *Ring Oven*

The ring oven technique of West and Mukherji (1959) is simple and easily implemented. However, sensitivities range from 1 to 100 ppm in solution. A fair number of elements in coal can be analyzed, and this method can serve for qualitative as well as quantitative confirmation.

3. *Ultraviolet-Visible Spectrophotometry*

The use of this type of analysis is well known and requires no documentation. Sensitivity varies from very high to inadequate. Use for confirmatory backup is indicated.

4. *X-Ray Fluorescence*

This type of analysis is often employed in metallurgical analysis. A review of x-ray absorption by Campbell *et al.* (1966) provides discussions of fundamental developments and summaries of applications. Several other reviews have been published in *Analytical Chemistry*, the latest of which is written by Birks and Gilfrich (1976). Berman and Ergun (1968) describe measurements of several major elements in coal using x-ray fluorescence, which is a nondestructive and relatively simple procedure. It is relatively insensitivie for many elements, requiring samples on the order of 500 mg. It is thus best suited for qualitative analyses of major elements.

5. *Neutron Activation*

This technique is well documented in the literature and has wide application such as in forensic investigation, for determination of impurities in commercial products, and for the behavior of elements in nuclear technology. However, this analytical procedure does not lend itself readily to analysis of coal dust on a routine basis.

Activation is provided either by nuclear reactors with fluxes on the order of 10^{12}–10^{13} neutrons/sec/cm^2 or with fast neutron or deuteron sources producing fluxes several orders of magnitude less. Sensitivities vary very widely, and analysis costs are prohibitively high for coal dust analysis, particularly since the upper limit fluxes provided by reactors are required for many elements. For details on neutron activation analysis see Volume I, Chapter 12.

VII. SUMMARY

Procedures for elements and compounds which might be implicated in the incidence of coal workers' pneumoconiosis are described. In some cases these are discussed briefly; in others, procedures are outlined in stepwise fashion. Attention is directed to the determination of dust weight, free silica (as α-quartz), trace elements, and aromatic components.

REFERENCES

Anderson, P. L. (1975). *Am. Ind. Hyg. Assoc., J.* **36,** 767–778.

Berman, M., and Ergun, S. (1968). *U.S. Bur. Mines, Rep. Invest.* **RI-7124.**

Bernas, B. (1968). *Anal. Chem.* **40,** 1682–1686.

Birks, L. S., and Gilfrich, J. V. (1976). *Anal. Chem.* **48,** 273R–281R.

Brindley, G. W. (1972). *In* "The X-ray Identification and Crystal Structures of Clay Minerals" (G. Brown, ed.), p. 507. Mineralogical Soc., London.

Bumstead, H. E. (1973). *Am. Ind. Hyg. Assoc., J.* **34,** 150–158.

Bureau of Mines, U.S. Department of the Interior (1972). Title 30, Part 71.100. *Fed. Regist.* **37**(60), March 28.
Campbell, W. J., Brown, J. D., and Thatcher, J. W. (1966). *Anal. Chem.* **38,** 416R–439R.
Crable, J. V., Keenan, R. G., Wolowicz, F. R., Knutt, M. J., Holtz, J. L., and Gurski, C. H. (1967). *Am. Ind. Hyg. Assoc., J.* **28,** 8–12.
Crable, J. V., Keenan, R. G., Kinser, R. E., Smallwood, I. W., and Maver, P. A. (1968). *Am. Ind. Hyg. Assoc., J.* **29,** 106–110.
Cummins, S. L., and Sladin, S. F. (1930). *J. Pathol. Bacteriol.* **33,** 1095–1132.
Flanagan, F. J. (1969). *Geochim. Comsochim. Acta* **33,** 81–120.
Freedman, R. W. (1971). *U.S. Bur. Mines, Inf. Circ.* **IC 8521.**
Freedman, R. W., and Sharkey, A. G. (1972). *Ann. N.Y. Acad. Sci.* **200,** 7–16.
Freedman, R. W., Toma, S. Z., and Lang, H. W. (1974). *Am. Ind. Hyg. Assoc., J.* **35,** 411–418.
Hamilton, E. I., Minski, M. J., and Clery, J. J. (1967). *Analyst* **92,** 257–259.
Hartstein, A. M., Freedman, R. W., and Platter, D. W. (1973). *Anal. Chem.* **45,** 611–614.
Hunt, J. M., Wisherd, M. D., and Lawrence, C. B. (1950). *Anal. Chem.* **22,** 1478–1497.
Hurley, R. C., and White, E. W. (1973). *Am. Ind. Hyg. Assoc., J.* **34,** 228–234.
Jacobson, M., and Parobeck, P. S. (1971). *U.S. Bur. Mines, Inf. Circ.* **IC 8520.**
Kessler, T., Sharkey, A. G., Jr., and Friedel, R. A. (1971). *U.S. Bur. Mines, Tech. Prog. Rep.* **42.**
Kessler, T., Sharkey, A. G., and Friedel, R. A. (1973). *U.S. Bur. Mines, Rep. Invest.* **RI-7714.**
King, E. J., and Gilchrist, M. (1945). *Med. Res. Counc. (G.B.), Spec. Rep.* **250,** 21.
Knight, G., Stefenich, W., and Ireland, G. (1972). *Am. Ind. Hyg. Assoc., J.* **33,** 469–475.
Lang, H. W., and Hay, J. E. (1975). *Am. Ind. Hyg. Assoc. Conf., Minneapolis-St. Paul, Minnesota* (unpublished).
Leroux, J., and Powers, C. A. (1969). *Staub—Reinhalt. Luft* **29**(5), 197–200.
Leroux, J., Davey, A. B. C., and Paillard, A. (1973). *Am. Ind. Hyg. Assoc., J.* **34,** 409–417.
Mine Safety Appliances (*MSA*) (1974). Final Rep., Bur. Mines, Pb 241 859/AS. NTIS, Washington, D.C.
Morgan, W. K. C. (1971). *Am. Ind. Hyg. Assoc., J.* **32,** 29–43.
Morgan, W. K. C., Burgess, D. B., Jacobson, G., O'Brian, R. J., Pendergrass, E. P., Ruger, R. B., and Shoub, E. P. (1973). *Arch. Environ. Health* **27,** 221–226.
Parobeck, P. S. (1976). MESA IR 1045. Min. Enforce. Saf. Admin., Pittsburgh Tech. Support Cent., Pittsburgh, Pennsylvania.
Saffiotti, V., Cefis, F., Kolb, L. H., and Shubik, P. (1965). *J. Air Pollut. Control Assoc.* **15,** 23–25.
Shultz, J. L., Friedel, P. A., and Sharkey, A. G. (1975). PERC/RI-75/4. Pittsburgh Energy Res. Cent., Pittsburgh, Pennsylvania.
Talvitie, N. A. (1951). *Anal. Chem.* **23,** 623–626.
Tuddenham, W. M., and Lyon, R. P. J. (1960). *Anal. Chem.* **32,** 1630–1634.
Warden, H. F., Jr. (1969). *Min. Congr. J.* **55**(9), 78–83.
West, P. W., and Mukherji, A. K. (1959). *Anal. Chem.* **31,** 947–950.

Part VII

COAL CARBONIZATION PRODUCTS: COKE, PITCH

Chapter 29

Analysis of Metallurgical Cokes

John W. Patrick *Herbert C. Wilkinson*
BRITISH CARBONIZATION RESEARCH ASSOCIATION,
CHESTERFIELD, DERBYSHIRE,
ENGLAND

I. INTRODUCTION

A. Analytical Requirements for Coke Used within the Metallurgical Industry

The principal industrial use of coke is in the blast furnace. The main constituents of coke which influence furnace performance and which

ISBN 0-12-399902-2

detract from its value as a source of carbon are moisture, sulfur, and ash. The former requires to be controlled to as low a value as possible since excess moisture requires additional heat for evaporation and results in higher fuel consumption. The amount of ash present affects slag volume and slag chemistry, influences coke consumption, and has a marked effect on both the technology and economics of iron production.

Most, if not all, of the sulfur which is retained in the metal originates from the coke. The distribution of the sulfur between slag and metal is affected by the sulfur load, and the corrective processes needed to reduce the amount of sulfur transferred to the metal are costly and reflect on productivity.

Consequently, industrial analysis of coke is frequently limited to what is generally termed "proximate analysis," involving the determination of moisture, ash, and volatile matter, the latter being merely an indication of the temperature to which coke has been heated during carbonization, and sulfur.

The determinations of moisture, ash, and volatile matter are largely empirical in nature and test conditions have therefore to be carefully specified. A reliable value of the volatile matter can only be obtained if a meaningful correction is made for the moisture simultaneously released during the volatile matter determination. Since the moisture content is frequently obtained by drying coke only at 105°C, it does not represent the total moisture released at the higher temperatures of the volatile matter test. Volatile matter results that are much too high are therefore frequently reported because of an inadequate moisture correction as shown by Wilkinson (1965).

Hydrogen content is a much more meaningful indication than volatile matter of the extent of carbonization of coke. Since it can be rapidly determined simultaneously with carbon, as described by Mott and Wilkinson (1957), direct determination of carbon and hydrogen is to be recommended in the interests of the accurate analysis of coke. It eliminates the need to make the erroneous assumption that the carbon content of coke ("the fixed carbon") can be accurately measured by subtracting the sum of moisture, ash, and volatile matter from 100. Sulfur is readily determined by various methods.

There is less awareness of the importance of minor constituents such as chlorine, nitrogen, and phosphorus and of the effects of substances present in even smaller amounts, such as alkalis. Methods are available for the determination of most of these constituents as well as for the determination of arsenic which is required when coke is used as a fuel in the brewing industry.

The physical properties of coke are probably of greater significance than the chemical properties in relation to the efficient operation of a blast furnace. Coke is the only solid material present in the regions of higher temperatures within the blast furnace, and in lump form, it confers the required degree of permeability to the burden in the melting zone. It must have sufficient physical strength not only to support the weight of the burden but also to be capable of withstanding severe degradation during its descent in the furnace shaft.

It is therefore of paramount importance to be able to estimate coke strength and many tests have been devised for this purpose. Such tests normally fall into two groups which measure the resistance of coke to breakage, either by a drop-type test, simulating impact breakage, or by a drum test, in which size degradation by breakage and abrasion processes are combined. The coke strength in each test is normally assessed from the proportion of the original sample remaining above a given size after testing.

However, although such tests provide general guidance in relation to the physical strength of coke and their results are of assistance in day-to-day control processes, they do not give an insight into the fundamental properties which control coke breakage and coke strength. For this reason new approaches are now being developed to obtain an insight into the properties of coke, analogous to those of other brittle materials, which affect its structure and strength. Such properties include porosity, tensile strength, and structural parameters, and a wide variety of instrumental techniques are now available for such studies.

Finally, since coke is used at high temperatures, methods are being developed to extend the information available from many types of tests by modifying them to operate at temperatures of 1000°C or above.

B. The Importance of Sampling

Coke is normally used under conditions of continuous operation and sampling systems are usually based on the collection of increments of material over a period of time, the gross sample obtained being regarded as representative of the coke used within that period. There are some exceptions to this generalization, notably the sampling of individual truck, train, or ship loads but in any case large quantities of material are involved.

The particle size of industrial coke, which may be within the range of 10–150 mm, must be taken into account in devising a sampling system.

The gross sample contains material of all particle sizes within a given range and the distribution of coke in various size fractions within this overall size range should be the same as the coke from which the sample was taken. Not only does the coke sample consist of material of widely different sizes, but it contains an in-built heterogeneity arising from slight differences in the heating rates of different coke ovens within a battery and from the effects due to the normal temperature difference between oven center and oven wall of any one oven. Such temperature effects can produce variation in both structural and chemical properties of coke.

It is also relevant to point out that the analysis sample is normally of about 100 g in mass, and because of the refractory nature of coke and the fact that many analytical procedures used involve a combustion process, the particle size of the analytical sample is generally about 200 μm. Preparation of the analysis sample from the gross sample therefore involves, in broad terms, the size reduction of several hundred kilograms of lump coke to about 100 g of <200-μm material.

II. SAMPLING AND SAMPLE PREPARATION

The importance of the need for careful sampling of the bulk coke and systematic preparation of subsidiary samples has been outlined. Two groups of samples are usually required to enable an assessment of coke quality to be made. These consist of the bulk samples of lump coke for physical and mechanical tests and the laboratory analysis samples. For special tests, such as the determination of reactivity, further samples may be required.

Once a representative gross sample has been obtained, preparation of the subsidiary samples is carried out on well-defined lines. The gross sample is dried and then subjected to size analysis. The bulk samples for physical and mechanical tests are compounded from aliquot proportions of the size fractions produced in this size analysis so that their size distribution is the same as that of the gross sample and therefore representative of the coke used during the period in which the sample was taken.

The sample for chemical analysis is produced from the gross sample by a succession of sample division and size reduction processes in which the mass and particle size of the coke are progressively reduced by mechanical treatment in crushers, crushing rolls, and various types of high speed mills. A sample division is carried out after each stage of size reduction, until a final analysis sample of about 100 g in mass and of a particle size less than 200 μm is obtained.

Suitable procedures are described in many standard specifications which give details of sampling and sample preparation methods appropriate to widely differing conditions, e.g., ISO 2309, "Coke Sampling," ASTM D346-75, "The Sampling of Coke," and British Standard BS 1017 Part II (1960), "The Sampling of Coke." These must be rigidly followed if subsequent analyses are to be meaningful. Care must be taken to avoid contamination of the sample, especially by ferruginous materials or limestone during the sample preparation procedure.

III. CHEMICAL ANALYSIS

A. Standard Methods

1. *Determination of the Major Constituents of Coke*

The details of the procedures for the determination of various constituents of coke have been fully described in Volume I, Chapter 6. For convenience, a summary is given in Table I of the principal methods used both as national and international standards for the determination of the more common chemical constituents of coke.

For certain purposes a knowledge of the arsenic, phosphorus, and oxygen content of coke may be required. Standard methods are available for the determination of arsenic and phosphorus, but in the case of the latter element, recent developments based on a combustion technique enable the determination to be carried out more rapidly.

2. *The Determination of Phosphorus*

Both international and national standards (e.g., ISO 926, B.S. 1016 Pt. 9, and DIN 51725) are available for this determination. All are basically similar, the phosphorus being extracted by acid digestion from the coke ash. It is then precipitated as ammonium phosphomolybdate and determined by a gravimetric or volumetric process. Alternatively, the phosphoric acid is converted to phosphomolybdic acid by treatment with ammonium molybdate and then reduced to the characteristic molybdenum blue complex which can be determined by suitable instrumental methods.

TABLE I *Standard Methods of Analysis for the Principal Constituents of Coke*

Constituent	Standard methods of test		
	International[a]	National	
Moisture (total)	ISO 579	AFNOR M03-028 ASTM D346-75	B.S. 1016 Pt. 2 1973 DIN 51718
Moisture (analysis sample)	ISO 687	ASTM D3173-73 AFNOR M03-029	B.S. 1016 Pt. 4 1973 DIN 51718
Ash	ISO 1171	ASTM D3174-73 AFNOR M03-030	B.S. 1016 Pt. 4 1973 DIN 51719
Volatile matter	ISO 562	ASTM D3175-73	B.S. 1016 Pt. 4 1973 DIN 51720
Carbon	ISO 609 ISO 625	AFNOR M03-032 ASTM D3178-73	B.S. 1016 Pt. 7 1977 DIN 51721
Hydrogen	ISO 609 ISO 625	AFNOR M03-032 ASTM D3178-73	B.S. 1016 Pt. 7 1977 DIN 51721
Sulfur	ISO 334 ISO 351	AFNOR M03-008 ASTM D3177-75	B.S. 1016 Pt. 7 1977 DIN 51724
Chlorine	ISO 352 ISO 587		B.S. 1016 Pt. 8 1977 DIN 51727
Nitrogen	—	AFNOR M3-018 ASTM D3179	B.S. 1016 Pt. 7 1977 DIN 51723

[a] Described in Volume I, Chapter 6.

Interference from a number of elements, particularly iron, titanium, and arsenic, is possible in this determination. The effects of iron are eliminated during the reduction of the phosphomolybdic acid, stannous chloride being a particularly effective reducing agent. Interference by titanium is counteracted by the strength of the nitric acid used in the initial digestion, and the amounts of arsenic normally present in coke are insufficient to affect the determination.

3. *The Determination of Arsenic*

International Standard Method ISO 601 is available for the determination of arsenic. It is applicable to samples containing not more than 0.0016% of arsenic but can be readily adapted to extend its range to cokes containing larger amounts of arsenic.

Coke is decomposed by wet oxidation, using a mixture of nitric and sulfuric acid, or by combustion at 800°C in the presence of Eschka mixture. In each case, arsenic is extracted into acid solution and reduced to the trivalent state. It is then evolved as arsine by reaction with zinc and sulfuric acid.

The arsine produced is absorbed in iodine solution where it is oxidized to arsenic acid. This solution is treated with ammonium molybdate in the presence of a reducing agent such as hydrazine sulfate, thereby producing a molybdenum blue coloration. An instrumental technique is used to measure the optical density of this solution which is compared with that produced from solutions containing known amounts of arsenic. The data are presented as a calibration curve, which enables the amount of arsenic present in the coke to be determined.

B. Other Conventional Laboratory Methods

1. *Direct Oxygen Determination*

The analytical method surveyed by Crawford *et al.* (1961), for the determination of oxygen in coke, is based on the Schütze–Unterzaucher method for the determination of oxygen in organic compounds. It has the disadvantage that pyrolysis, resulting in the conversion of oxygen to oxides of carbon, does not distinguish between organically combined oxygen and that derived from the decomposition of the mineral matter. However, this complication, although of significance when oxygen is to be determined in coal, is less important in the case of coke, where the oxygen combined in the coke ash is largely present as silicates and oxides which do not decompose under the conditions of pyrolysis employed in the analytical method.

Kirk and Wilkinson (1970) have described a method for the determination of oxygen in coke, in which it was not necessary to demineralize the coke. The method had an accuracy of about ±0.2%.

2. *Carbon, Hydrogen, and Nitrogen by Elemental Analyzers*

Although instrumental techniques are in widespread use for the determination of C, H, and N in organic compounds, the relatively refractory nature of coke presents some problems in the use of such instruments for these determinations. However, minor adjustments to the combustion or pyrolysis technique enable satisfactory results to be obtained. The accuracy of determination within the ranges of C, H, and N of 85–95, 0.2–1.5, and 0.8–1.5%, respectively, is approximately 0.3% C, 0.1% H, and 0.1% N, and this is adequate for routine purposes.

3. *Analysis with the Schöniger Flask*

Modified oxygen flask methods for the determination of sulfur, based on the Schöniger flask, are described in Volume I, Chapter 8, Section II.

The technique may also be used for the determination of phosphorus as described by Kirk and Wilkinson (1964). After combustion the phosphorus is converted into phosphomolybdate and reduced by a solution containing ascorbic acid and potassium antimonyl tartrate. The determination is completed colorimetrically and may be carried out in 45 min.

4. *Ash Fusion*

The fusion temperature of solid fuel ash is of considerable importance in many combustion processes and Standard methods for this determination are available, e.g., ISO Recommendation R540 (1967), British Standard 1016 Part 15 (1970), DIN 51730 and ASTM D 1857 (1974). ISO 540 and ASTM D 1857 are described in Volume I, Chapter 6, Section XI.

There are minor differences among the methods, but they are the same in principle. An artifact of ash is formed in a mold and inserted into a tube furnace through which a gas mixture passes at a controlled rate. The gas mixture may simulate a reducing or an oxidizing atmosphere. The furnace temperature is raised at a controlled rate and recordings are made of the three characteristic temperatures of deformation, softening, and flow of the ash. Since the maximum working temperature of most electrically heated furnaces is limited to about 1450°C, it is not always possible to attain the flow temperature of some coke ashes.

The British and German Standards permit the use of an alternative method employing a heating microscope.

C. Analysis for Minor Constituents and Trace Elements

1. *Conventional Ash Analysis*

Conventional analysis normally consists of the determination of the oxides of iron, aluminum, silicon, calcium, magnesium, sodium, potassium, titanium, phosphorus, sulfur, and occasionally manganese.

The British Standard B.S. 1016 Part 14 (1963) and the ASTM Standard D2795-69 are essentially similar. A weighed amount of ash is brought into solution by fusion with alkali and after suitable dilution a solution is obtained which is used for the determination of silica and alumina. A second solution is prepared by acid digestion of the ash and is used for the determination of the remaining components of the ash. A more rapid procedure, based on spectrochemical analysis, is offered as an alternative method in the British Standard. The German Standard DIN 51729 makes no provision for the determination of manganese, and the titanium content of the ash is not determined separately from the aluminum content. There is no International Standard for this method of analysis.

2. *Analysis for Trace Elements*

There is a growing awareness of the possible effects upon the environment of the accumulation in the atmosphere or terrain of many inorganic (and organic) materials produced by the processing and combustion of solid fuels.

Some 60 chemical elements have been detected in coals. On carbonization the majority of these elements are retained in the solid residue, but some (e.g., Sn, B, Cu, Co, Ni, Zn, Be, Cr, V, Mo, and Ge) exhibit some volatility within the temperature range of 300–700°C, whereas mercury, bromine, lead, and antimony have been shown to be released from coals at lower temperatures.

Since many of these elements are only present in coal in its natural state to the extent of a few parts per million, the amounts volatilized or retained in residues after processing are extremely small and present considerable difficulty in their determination. Great care is necessary in sampling and sample preparation; in particular, contamination by nickel, chromium, or copper can occur if the coal sample is crushed with certain types of equipment.

Considerable work of an exploratory nature has been undertaken in recent years to establish the best technique for determining the majority of the elements considered. The methods examined include neutron activation, optical emission, atomic absorption, x-ray fluorescence, and ion-selective electrode. Most suffer from some deficiencies, but a consensus of opinion favors atomic absorption spectrophotometry as the principal method of analysis, and optical emission spectroscopy as a second choice. For example, it is known that atomic absorption spectrophotometry may be used to determine Be, Cd, Cr, Cu, Mn, Ni, Pb, V, and Zn in ashes.

D. On-Line Methods of Analysis

1. *Moisture by the Neutron Gauge*

This method, in which the principle of neutron moderation can be applied, is particularly suitable for determining the moisture content of large quantities of lump coke. It is in fairly general use and typical reported results indicate that the 95% limits of the determination are within ±1–2%.

A crushed sample of coke is not required and the method is suitable for the examination of coke in bunkers. It is subject to the effects of bulk density and the results are affected by the hydrogen content of the coke.

2. *Moisture by Other Methods*

The moisture content of crushed coke may be determined using nuclear magnetic resonance, but the method is sensitive to the presence of ferromagnetic material and the degree of graphitization of the coke.

3. *Ash*

At the present time there is little evidence of the application of on-line methods to the determination of ash in coke, although such methods are available for the determination of ash in coal.

4. *Sulfur*

Feasibility studies have been made which suggest that an on-line method based on the use of x-ray fluorescence (XRF) techniques may be used for the determination of sulfur in coke. A comparison of results obtained by this method and by a classical method is given in Table II. The XRF data take into account the contribution made by anode radiation and the $K\bar{\alpha}$ value of silicon.

The results indicated that the mean difference between the two methods was 0.027% sulfur. The analysis may be made manually or by a computer-controlled automatic method, whereby the result is available

TABLE II *Comparison of the Sulfur Content of Cokes Determined by Chemical Analysis and by X-Ray Fluorescence*

Sample	Sulfur (%)		A − B
	Chemical method (A)	XRF method (B)	
A	0.70	0.710	−0.010
B	1.03	1.062	−0.032
C	1.41	1.357	+0.053
D	1.62	1.591	+0.029
E	1.85	1.860	−0.010

in 1 min. Powdered or pelletized samples may be used. Since the method is of the dispersive type there is little interference from the presence of other elements (e.g., Fe).

E. Tests of Reactivity to Oxygen and Carbon Dioxide

1. *Critical Air Blast Test*

This test is described in British Standard 1016 Pt. 13, "Tests Special to Coke." It is of particular relevance to the behavior of carbonized solid fuels in domestic heating appliances, the results giving guidance to the ignition properties of the fuel and its behavior under conditions of slow combustion.

The test measures the minimum rate of an air blast which will maintain combustion of a closely sized bed of coke after ignition under standard conditions. Typical values of the critical air blast (CAB) of different fuels are as follows:

Low temperature coke	0.014–0.020 ft^3/min
Anthracite	0.03–0.04 ft^3/min
High temperature oven coke	0.055–0.075 ft^3/min

2. *Ignition Test*

There is no standard ignition test for general use but many tests have been described, and the test developed by Blayden *et al.* (1943) is suitable for general use with a wide range of cokes. In this test approximately 0.2 g of <65-μm coke contained in an Alundum crucible is heated at 5°C/min in a current of dry air. Ignition is considered to have occurred when there is a sudden rise in the temperature of the coke. Duplicate determinations are made and the mean of the two determinations is taken as the ignition temperature. One test which has been standardized is described in ASTM D2677-71, "The Lightability of Barbecue Briquets."

3. *ECE Reactivity Test*

This test measures the rate of reaction of carbon in coke with carbon dioxide under standardized conditions of temperature, gas flow rate, and particle size of the coke.

Seven grams of coke, of particle size 1–3 mm, are introduced into the reaction tube which is purged with carbon dioxide gas for approximately 3 min. This assembly is then introduced into the constant-temperature zone of a furnace at 1000 ± 3°C and carbon dioxide is passed through the bed of coke at a rate of 7.2 liters/hr for a period of 15 min. The exit gas is then diverted to pass through a graduated potassium hydroxide absorber and the percentage of carbon dioxide remaining in the gas stream after reaction with the coke is determined. A velocity constant for the reaction, K_m, is calculated from the formula

$$K_m = -0.8689 \left[2 \ln \frac{2C_L/100}{1 + C_L/100} + \frac{1 - C_L/100}{1 + C_L/100}\right] \quad (1)$$

where C_L is the concentration of carbon dioxide in the gas at the moment of sampling. K_m is regarded as the reactivity of the coke.

Many variations have been made to this basic test. The consumption of carbon may be established by weight loss measurements, or the coke may be examined in the form of a cylinder instead of a particulate bed.

Typical values of K_m for coke oven cokes are as follows:

Domestic coke	0.40–0.55
Blast furnace coke	0.20–0.35
Foundry coke	0.10–0.16

The significance of the results in relation to coke consumption under industrial conditions is doubtful. Apart from the major differences between the size of industrial coke and that used in the reactivity test, together with differences in temperature, the reaction is highly susceptible to catalytic effects which are well known to occur under industrial conditions.

4. *Other Laboratory Tests of Burn-off in CO_2*

In recent years interest has been renewed in the reactivity of coke to CO_2 with regard to blast furnace performance with the result that several laboratory tests have been devised, most of which differ only in experimental details (see Chapter 32).

In our laboratories use has been made of a test carried out on cylindrical coke specimens (1.5 cm diameter and 1 cm thickness) drilled by means of a diamond-tipped core drill from coke pieces of suitable size, i.e., greater than ~3 cm. The use of such specimens enabled the results to be related directly to the coke tensile strength determined on such cylindrical specimens as described in Section V,C. The heterogeneity of coke necessitates the examination of many specimens in order to obtain a statistically reliable result and it has been found that about 50 results for each coke are required to achieve this. The apparatus consists of a multitube electrically heated furnace in which each tube is individually fed by a controlled flow of gas. One cylindrical coke specimen in the air-dried condition and contained in a small silica boat is placed in each tube, and the reaction temperature of each tube is monitored by a thermocouple inserted into the center of the tube. A nitrogen stream is maintained in the tube as the temperature is raised to 1000°C at which stage 10% CO_2 is added to the gas stream and maintained at a flow rate of 90 ml/min (equivalent to a complete change of the tube atmosphere every 2 min) for a period of 5 hr, after which the specimen is cooled in nitrogen.

The sample weight and dimensions are recorded before and after the gasification and from the weight loss, expressed as a percentage of the original weight of the coke specimen, a relative reactivity is obtained. From the change in specimen dimensions and hence in the apparent density an estimate can be obtained of the relative extent of internal and external burning which takes place during the reaction.

Variations of this test include the use of cubes of coke instead of cylinders, charges of several hundred grams of coke of graded particle size between 12 and 30 mm either in fixed beds or in revolving drums, measurement of CO in the outlet gas as in the ECE test (Section III,E,3) instead of weight loss of the coke, and the use of 100% CO_2, with the time taken to achieve a particular percentage of burn-off (usually 20%) being the measure of relative reactivity. Despite these variations the relative order of cokes, as measured by the CO_2 reactivity, remains substantially the same.

IV. PHYSICAL TESTS

A. Density Determinations

1. *Apparent Density*

A full description of the methods is given in ASTM D167-73, B.S. 1016, Part 13, and ISO Recommendation 1014. The method described in the British Standard has recently been revised and differs from the ASTM specification in that the coke used is between 40 and 60 mm in size.

2. *Real Density*

A comparison of the method described in ASTM D167-73 and of that described in ISO Recommendation 1014 is reported in Volume I, Chapter 6, Section IX. The British Standard Method B.S. 1016, Part 13 is identical with the ISO method except that 2 g of coke is used in the former and 5 g in the latter.

3. *Bulk Density*

Standard specifications for the determination of the bulk density are given in International Standards 567 and 1013 for small and large containers, respectively, B.S. 1016, Part 13, Section 5, and ASTM Standard D292-29 (reapproved 1972).

The principle of each test is similar in that the weight of a known volume of coke is obtained, but differences between the size of the container, the maximum allowable size of coke, and the method of leveling the filled container exist in the individual methods. These differences are outlined in Table III.

The struck method of leveling is effected by drawing a straight edge over the coke to give as level a surface as possible. The eye leveling method ensures that the projections of coke above the surface balance the depressions in the bed below the top plane. The accuracy of the determination depends on the mass, shape, and size of the particles and the wall effect, i.e., the relation between container size and shape and maximum coke particle size. To estimate the bulk density in large containers, e.g., large railway cars or blast furnaces, the results of standard tests can be processed according to the method described by Lee and Mott (1956).

TABLE III *Principal Differences in the Standard Methods Used for the Determination of the Bulk Density of Coke*

Standard method	Volume of container (m^3)	Maximum[a] coke particle size (mm)	Method of leveling
ISO 567	0.200	150	Struck
ISO 1013	Capacity 3000 kg	None stated	Struck
B.S. 1016 Pt. 13	0.057	120	Eye
ASTM D292-29	0.227	150	Eye
ASTM D292-29	0.028	30	Eye

[a] *Note.* The particle size of the coke is expressed in terms of round-aperture metric sieves. The conversion factor (inches square to millimeters round) is $R/S = 25.4 \times 1.17$.

B. Porosity

Coke is a spongelike material consisting of a network of pores of various dimensions and shapes, some of which are closed, but the majority of which are interconnected. The pore sizes range from ultrafine micropores of less than 1 nm involved in gas adsorption to macropores of several millimeters linear size which are visible to the naked eye. There is no single determination of the porosity which covers this entire range. It is, therefore, fortunate that for industrial purposes it is rare for a measure of the porosity over the entire pore size range to be required. Generally the porosity of industrial coke is determined indirectly from the ratio of the apparent and true relative densities or directly from penetration of a fluid, the usual one being mercury, or by microscopic assessment. Microporosity in the pore size range from 10 Å is measured by capillary condensation of nitrogen, a technique that is described in Volume I, Chapter 4, Section IV,A.

1. *From Density Measurements*

The volume porosity can be calculated from measurements of the true and apparent relative densities from the equation

$$\%\ \text{porosity} = 100\left(1 - \frac{\text{apparent density}}{\text{true density}}\right) \tag{2}$$

This method constitutes a British Standard (B.S. 1016, Part 13) and for lump coke of larger than 25 mm size, an ASTM Standard (D167-73).

2. *From Mercury Porosimetry*

The theory and technique of mercury porosimetry have been described in Volume I, Chapter 4, Section IV,A.

This method is based essentially on the forcing of mercury under pressure into the pores and measuring the volume of liquid penetrating the pores as a function of the applied pressure. By assuming pores of circular cross section and using the relationship between the applied pressure p (kg/cm^2) and the pore radius r (Å),

$$r = 75{,}000/p \tag{3}$$

a pore size distribution, in the range of pore radii from a lower limit of about 20 Å to a maximum of about 0.25 mm, can be derived.

Several commercial instruments are available and numerous descriptions of equipment for mercury porosimetry can be found in the literature. All of these are based on the same principle, but they differ considerably in the degree of sophistication, ease of operation, and the means of detecting the small changes in the volume of mercury entering the pores.

It should also be emphasized that in common with all porosity determinations involving fluid penetration or displacement, it is only the open porosity which is being measured. The use of Eq. (3) also assumes values for the angle of contact and the surface tension, as well as assuming, incorrectly for coke in general, that the pores are cylindrical or of circular cross section. Contamination of the mercury can significantly affect the surface tension and angle of contact; hence, it is essential that the mercury used is free of chemical impurity and to a lesser extent free of physical contamination. Even with this proviso, however, there is still some uncertainty about the real values of the surface tension and contact angle. Other possible errors arise from possible crushing of the sample and a consequent change in the pore structure although there is some evidence that such changes are relatively small in the case of industrial coke (Jüntgen and Schwuger, 1966).

Thus, despite the errors, uncertainties, and assumptions underlying mercury porosimetry, it nevertheless gives a useful relative measure of the porosity over a wide range of pore sizes and effectively bridges the gap between the microporosity determined from gas adsorption (which also assumes a circular cross section of pore entrances) and that derived from low power optical microscopy.

3. *From Optical Microscopy*

The determination of coke porosity from microscopic examination of polished coke sections involves either lineal analysis or point counting. In lineal analysis the intercept sizes, i.e., the width of the pores and pore walls along a straight line traverse, are measured using an appropriate measuring stage and the sum of the pore sizes as a fraction of the total gives the porosity. In the method described by Abramski and Mackowsky (1952) the linear values of five different pore sizes are determined and the average porosity P is then calculated according to the formula

$$P = (5a + 4b + 3c + 2d + e)/5 \tag{4}$$

where a, b, c, d, and e are the linear values in the pore sizes <0.1, 0.1–0.2, 0.2–0.5, 0.5–1.0, and >1.0 mm, respectively.

In point counting the traverse of the specimen is stepwise at a series of regular intervals (0.33 mm is generally used in determinations of coke porosity) with recording of the nature of the material, i.e., in this case pore or pore wall, under the eyepiece crosswires. The method is analogous to sampling of a population by cross-sectioning and for simple measurement of volume porosity it is far less tedious than lineal analysis. The analysis can be speeded up further by the use of an automatic point counter whereby the stage traverse is automatically triggered by the operation of counting switches. The accuracy of the determination is governed by the number of points counted and it is estimated that 1000 counts are required for a 1% accuracy in the determination of a component present to the extent of about 50% of the total sample. In practice a compromise has to be made between the accuracy and speed of the determination, and in our experience with coke samples, a 500-point count generally gives a suitable compromise.

This problem can be overcome by the use of a microscope incorporating an automatic image analysis based on a television-type scanner system by means of which the total porosity is derived from the number of detected points of pore space relative to the total number of points in the field of view under examination. A count of a very large number of points is then possible (Patrick *et al.*, 1976).

In all of the methods involving optical microscopy care has to be taken to obtain representative samples and to prepare the polished sections without causing undue damage to the porous structure of the coke. The procedures leading to polished sections are still to a large extent an art, with the result that each laboratory appears to have developed its own method and technique, but generally the methods are basically similar. The method now described has been successfully used in our laboratories to prepare polished sections of various types of industrial coke.

Assuming standard sampling procedures have been followed to obtain samples as representative as possible, the coke pieces are cleaned in a sonic cleaning bath of water to which a few drops of detergent have been added until any accumulated debris has been completely removed. The specimens dried in an air oven at 110°C are set in an epoxy resin (Araldite MY 753) which has relatively low viscosity necessary for filling of the pores, good wetting properties, and does not contract on hardening. Using successively finer grades of silicon carbide on a cast-iron lap, the coke is cut back to obtain a flat surface and any unfilled pores are impregnated with resin which is repeatedly smeared over the coke surface. Further impregnation steps are carried out as necessary after the initial polishing stages. When virtually all the open pores are filled with resin, the excess resin is removed by lightly grinding with fine grade silicon carbide before polishing with successively finer grades of alumina on a copper lap. The final polish is carried out with a foam fabric polishing cloth. Because of the heterogeneity of coke it is necessary to examine many fields of view from several specimens in order to get representative data, but it must be noted that the porosity values obtained are resolution dependent. This variation with different microscope magnifications arises from the detection at higher magnification of micropores which are unresolved at low magnification. The magnification to be used is thus governed by the requirement of the industrial application for which the results are required.

C. Analysis of Porous Structure by Optical Microscopy

There is frequently a need for more detailed information about the coke porosity than that given by the volume occupied by pores of different size. These additional data can be obtained by microscopic examination and analysis either manually or by use of a computerized automatic image analysis system. In the former the measurements are generally restricted to linear measurement of the pores and pore walls, whereas the automated systems have scope for a much more detailed characterization including the derivation of various shape factors.

1. *Manual Counting Method*

A polished block of coke prepared as described in the preceding subsection is examined under a reflected light microscope at a magnification chosen according to the degree of fineness of detail required. Using a graduated eyepiece alternate measurements of pore and pore-wall sizes are made along complete linear traverses of the coke section, enabling mean size and size distributions of the pores and the carbon matrix forming the pore walls to be derived. By making measurements at right angles some information can also be obtained about the pore shape and orientation. This method has been used to analyze the development of the porous structure during carbonization.

Abramski and Mackowsky (1952) measured pores and pore walls in cokes in five size categories with the aid of a measuring stage and calculated mean pore and pore-wall sizes from which they derived a compactness factor, which is a ratio of the amounts of wall material and pores.

The manual method can give good results but it is slow and tedious and as such does not lend itself to routine use. These limitations are overcome, however, by the use of automated image analyzers.

2. *Automatic Image Analysis*

Automatic image analysis is concerned with the quantitative assessment of images obtained microscopically. Although the degree of instrumentation varies widely, most of the commercially available instruments are based on television systems carrying out essentially a lineal analysis with processing of the data by a computer which produces the final output. In the author's laboratories use is made of a Quantimet 720 image analyzer based on a Vickers microscope for the analysis of those aspects of the porous structure of coke which influence its strength (Patrick *et al.*, 1976). The methods used are described herein to give the reader an insight into the operation which typifies the use for structural analysis of automatic image analysis systems in general.

The preparation of the polished coke blocks is carried out as described in Section IV,B,3. Since the human eye is better than an automatic image analyzer at ignoring artifacts such as scratches, dirt, or relief, rather more care is required, however, to avoid these defects in the polished surfaces. To ensure correct classification of the large porous features present in industrial coke (these are especially important in our studies), the microscopic examination must be carried out at low magnification (usually with a ×2.5 objective) since at higher magnification too many large features are incorrectly classified because they extend beyond the image area inside which measurements are taken. An adequate guard region must be employed to reduce edge effects and the automatic *XY* stage is set to give steps which prevent overlap of the fields measured. With these conditions the linear size of a picture point, i.e., the smallest feature which can be measured, is 5.5 μm with the largest measurable feature being about 4 mm.

The program is set to give for each field scanned, total porosity, total perimeter of the pores, number of pores, and sized intercepts of pores and pore walls using a series of size ranges and employing "greater than" criteria. Several other parameters, e.g., Feret diameters at various angles, can be included in this analysis provided the appropriate instrumentation is available. By use of simple or multiple quotients of these parameters, the data can be used to derive shape factors by means of which further classification is possible. The output data from the Quantimet are processed by a programmable calculator to present it in a convenient form through a printer/plotter.

In our studies wherein one of the objectives is to relate the porous structure to the coke strength, the coke specimens examined are of the same form as those used in the diametral-compression test of tensile strength, i.e., 1.0- or 1.5-cm-diameter cores drilled from coke lumps. Two polished blocks, each containing 25 of the 1-cm-diameter cores or 12–13 of the 1.5-cm-diameter cores, are prepared for each coke sample and a total of about 200 fields of view are analyzed, 100 on each block. The use of two blocks provides a check on the sampling procedures and if there is good agreement between the results for the two blocks, the results are combined to derive the average values of the parameters measured.

The values obtained are of course resolution dependent. Using the method just described, some typical mean values for a metallurgical coke and the standard errors of the determinations are as follows:

Porosity	53.4%	SE ±0.5%
Pore size	163 μm	±4 μm
Pore-wall size	136 μm	±2 μm

D. Determination of Optical Anisotropy

The optical anisotropy generally developed during carbonization of coking coals is readily discernible under a polarizing microscope and provides a means of classifying cokes into different categories. Recently methods have been developed to quantify the anisotropy in terms of either the proportions of the various identifiable anisotropic entities (Patrick *et al.*, 1973b) or the size of the anisotropic units (Goldring, 1973).

1. *Classification by Means of the Anisotropic Composition*

Suitable samples for the microscopic examination can be prepared by mixing approximately 0.5 g of crushed coke with a few drops of freshly prepared epoxy resin (e.g., Araldite MY 753) and pressing the mixture into the form of a thin pellet 20 mm in diameter. This pellet is then mounted in resin and prepared as a polished block by grinding with Carborundum and polishing with successively fine grades of alumina paste.

The polished coke surfaces are then examined on a polarizing microscope using a sensitive tint plate to enhance the color differences, thereby improving the detection of different anisotropic structures. For a description of these colors, see Chapter 30, Section I,A,1. With the aid of an automatic point counter, determinations are made of the different types of anisotropic structures present in the coke. The categories into which the microscopically identifiable components are divided range from isotropic material through mosaic-type anisotropy of various grain sizes to flow-type anisotropy, this latter term referring to the appearance with no direct relation to the plastic or fluid properties of coal during carbonization. Basic anisotropy is the term used to describe the relatively featureless anisotropy found in some high rank coals which undergoes little or no change during carbonization. The grain size of the various mosaics is estimated to be ~0.3 μm in diameter for fine grain,

0.7 μm for medium grain, and 1.3 μm for coarse grain mosaic. For the identification of fine structure it is necessary to work at high magnification and, using a X100 objective to give an overall magnification of ×1500, an area of 30 μm^2 is assessed at each point. With a 300-point count the maximum error in assessing each anisotropic type is of the order of ±5%, the actual error depending on the nature of the anisotropy.

2. *Classification by Size of Anisotropic Units*

The anisotropy of many cokes and of most metallurgical cokes is composed largely of mosaic units. In this method of characterizing coke the sizes of these mosaic units are determined by direct comparison with a calibrated scale superimposed on the microscope eyepiece as described by Goldring (1973). The small size of some of the units necessitates the use of a high resolution system at high magnification, usually with a ×100 objective. The observations are made in polarized light with the polars almost crossed; for identification of the units it is necessary to be able to rotate either the sample or the polars. A sensitive tint plate is not essential, although it is preferred in some instances.

The coke sample is prepared as a polished block as described in the preceding subsection, and the surface is systematically scanned using a mechanical stage with a step of sufficient distance (~200 μm) to enable each measurement to be carried out on an independent coke particle. For each field of view the area of a circle of 10-μm diameter about the crosswires is used. The larger complete mosaic units within this area are identified and measured by comparison with the eyepiece scale to obtain what is deemed to be a representative figure for the true diameter of the mosaic units in that area. A mean mosaic size and a size distribution for the coke are then calculated from 200 measurements. The mean sizes of the mosaic units generally range from 0 to >20 μm, the size being coal-rank dependent.

E. X-Ray Analysis of Structure

When x rays are diffracted by matter the resultant scattering intensity curves reflect the structural order of the matter. Crystals give sharp interference maxima of the coherent scattered radiation but for crystallites of small size the interference peaks become progressively broadened as the size decreases, and under suitable conditions the crystallite dimensions can be estimated from the line broadening. This is the basis of the x-ray powder diffraction method of analysis of the structure of coke, which as a nongraphitic carbon shows only diffuse maxima at scattering angles corresponding to the positions of the dominant graphite peaks.

As described by Blayden *et al.* (1944) the specimen for the analysis is prepared from a paste of finely powdered coke moistened with an adhesive such as tragacanth. The coke paste is formed into a thin cylindrical specimen ~0.5 mm in diameter and 1–1.5 cm in length by compaction in a suitable former, e.g., a glass capillary tube. The dried specimen is mounted in the specimen holder and accurately centered in the cylindrical x-ray diffraction camera which records the scattering curve. After exposure using monochromatic radiation, the intensity distribution of the developed x-ray film is determined and recorded by a suitable microdensitometer. From the width of the (002) and (100) lines the diffraction broadening β is determined and used to calculate the crystallite sizes; L_c, the crystallite height, and L_a, the crystallite diameter, respectively, from the Scherrer equation

$$\beta = \frac{K\lambda}{L \cos \theta} \tag{5}$$

where K is a constant of the order of unity, λ the wavelength of the radiation used, L a measure of the crystallite size perpendicular to the reflecting planes, and θ the Bragg angle.

Also by determination of the diffraction angles from the spacing of the diffraction peaks and use of the Bragg equation, the lattice parameters, $c/2$, the interlayer spacing, and a, the layer diameter, can be calculated from the (002) and (100) lines, respectively.

It must be emphasized that this is a very simplified procedure; the accurate determination of the crystallite dimensions involves many corrections and refinements, with the result that it is both difficult and time-consuming. On the other hand, in the industrial situation it is rarely necessary to determine the true values, and for most purposes the apparent values of L_c and L_a, determined as just described, are adequate and suitable for use as comparable relative figures.

V. MECHANICAL PROPERTIES

Conventional methods of measuring the mechanical properties of coke are concerned with an assessment of the resistance of coke to breakage by the combined effects of fracture and abrasion. The results determined by these tests are largely based on the measurement of the extent of size reduction of the coke after subjecting it to a known amount of work and expressing the results of these tests in the form of various indices. Such indices are empirical, and are influenced not only by the quality of the coke but by the size of the material forming the test sample as reported by Wilkinson (1972).

The tests consist of two types: drop tests, in which breakage is by impact alone, and drum tests, in which breakage takes place both by impact and by abrasion. The latter group of tests are the most numerous.

A. Shatter Test

This well-known test is described in the following specifications: ISO Standard 616, ASTM Standard D 3038-72, and British Standard 1016, Part 13.3.

Each specification describes a method of test whereby a sample of coke (25 kg in weight) is compounded from a gross sample such that the proportions of the test sample above a minimum size (2 in. or 50 mm) are the same as in the gross sample. This sample is then subjected to impact breakage by allowing it to drop four times from a height of 1.83 m (6 ft). The broken coke is then subjected to a size analysis and the shatter index reported as the percentage by weight of the original material remaining above a specified sieve size.

ASTM Standard D 3038-72 recognizes the effect of the initial size of coke and describes two alternative procedures, for cokes containing more or less than 50% of material above 4 in. in size. The British Standard test uses coke sized above 2 in. and the ISO Standard specifies a minimum coke size of 50 mm. All dimensions in the ISO Standard are metric and square aperture sieves are employed.

B. Drum Tests

1. *Tests at Ambient Temperature*

Drum tests subject the coke to degradation by two processes: breakage by impact, which takes place during the early stages of the test when coke breaks principally along the planes of the natural massive fissures, and size reduction by abrasion, which is most pronounced during the later stages of the test. This behavior leads to the production of a material having a pronounced bimodal size distribution. The initial size of the test sample is significant in relation to the degree of breakage of the coke and to the magnitude of the test results. Because of the

TABLE IV *Principal Features of Drum Tests for Measuring the Strength and Abrasion of Coke*

Test parameter	ASTM			MICUM				U.S.S.R.	
	D294-64	D2490-70	D3402-75	ISO / Micum Irsid	B.S. 1016.13	DIN 51717	Japanese	Large drum	Small drum
Drum dimensions									
Internal diameter (mm)	914	914	Combines	1000	1000	1000	1500	2000	1000
Internal radius (mm)	457	457	D294-64	500	500	500	750	1000	500
Internal length (mm)	457	457	(Procedure A)	1000	500	900	1500	800	1000
Lifting flights			with						
Number	2	2	D2490-70	4	4	4	6	Barred	4
Depth of face (mm)	51	51	(Procedure	100	100	100	250	drum	100
Depth of transverse support (mm)	51	51	B)	63	63	63	Small	—	63
Rotation									
Angular velocity (rpm)	24	24	—	25	25	25	15 15	10	25
Revolutions	1400	700	—	100 500	100	100	30 150	150	100
Test sample									
Mass (lb)	22	22							
(kg)	10	10	—	50	25	50	10 10	410	50
Lump size (in.)	2–3	1–2							
(mm)	—	—	—	>20 mm	>60 mm	—	>50 >50	>25	—
Sieve type	Square	Square	—	Round	Round	Round	Square Square	Square	—
Moisture (%)	1.0 max.	1.0 max.	—	3.0 max.	3.0 max.				
Strength indices	Stability factor % >1 in. (25 mm) Hardness factor % >¼ in.	Stability factor (modified) % >1 in. (25 mm) Hardness factor (modified)	— — — —	M40 I40 M30 I30 M20 I20 M10 I10	M40 = % >40 mm M10 = % <10 mm	— — M10 % <10 mm	Crushing strength = % >15 mm	% >25 mm % 10–25 mm % <10 mm	— — —

changing pattern of blast furnace operation in recent years, during which the use of smaller coke or coke screened within certain size limits has become more prevalent, the various National and International Standard tests have been updated to take account of such developments. The salient features of the most widely used tests are given in Table IV. The following should be noted:

The German Standard DIN 51717 classifies coke into various categories, e.g., foundry cokes, blast furnace cokes grades I, II, and III, and cokes of three size grades. The particle size of the test sample and the diameter of the sieve apertures used for establishing strength indices are varied according to the type of coke tested. The M10 index is always used as the abrasion index.

The small drum, used in the U.S.S.R., is identical with the ISO Micum drum and the test procedure is likely to follow that of the ISO method.

The number of the new ISO Standard, "Drum Test for Coke Greater Than 20 mm in Size," is not yet available. The indices M40, M30, I40, I30, etc., refer to the percentage of coke remaining on 40- or 30-mm-round-aperture sieves after the Micum and Irsid tests, respectively.

2. *Tests at Elevated Temperatures*

The principal use of coke is in blast furnace practice. Before it is converted into a gaseous reducing agent, it passes through a zone in the furnace where it supports the massive load of the overburden and is in contact with molten and partially molten iron, iron ore, and slag. Under these conditions, the coke must be resistant to breakdown and retain its size to an extent that the permeability of the furnace is not impaired.

The strength of coke at high temperatures is thus of considerable importance and a number of tests have been devised to study this topic. Because of their complex nature, no attempt has been made to derive a standardized form of test.

The collective evidence from these tests (Echterhoff, 1961; Bradshaw and Wilkinson, 1969; Wasa *et al.*, 1974) may be summarized as follows:

(i) Coke strength is reduced and coke abrasion increased when the testing temperature exceeds the carbonizing temperature.

(ii) The strength of coke measured at ambient temperature is not related to the high temperature strength.

(iii) The majority of workers were able to establish a satisfactory correlation at ambient temperature between the results given by their testing equipment and those given by conventional tumbler tests.

(iv) Evidence of annealing of the coke, resulting in an increase in strength, was observed between 600 and 1000°C.

C. Tensile Strength

The direct determination of the tensile strength of a heterogeneous porous brittle material such as coke is difficult largely because of the complications associated with the preparation of suitably shaped specimens. For this reason use has been made of indirect methods, and of these, the diametral-compression test (also termed the Brazilian or Carneiro test) has been found to be a method which lends itself to the testing of industrial coke (Patrick and Stacey, 1972a,b, 1975; Patrick *et al.*, 1972, 1973a).

The principle of the method involves the application of a compressive stress across the diameter of a thin disk of the material, breakage occurring due to the tensile stress developed at right angles to the line of the applied load. The applied load W and the tensile strength P are related by the equation

$$P = 2W/\pi dt \tag{6}$$

where d and t are the diameter and thickness of the specimen, respectively.

Theoretically the equation applies to an isotropic, homogeneous material which behaves elastically up to breakage and ideally requires line loading. Since these requirements are not completely fulfilled the results can only be regarded as comparative figures whose relationship to the true tensile strength as determined by the direct method is uncertain. Nevertheless it should be observed that values consistent within themselves are obtained; relationships between tensile strengths determined by the diametral-compression test and those determined by other methods have been deduced for several materials.

The heterogeneity of industrial coke means that it is essential to adopt a statistical approach. The method now described has been found to be suitable in that it provides reliable, representative values.

From a 25-kg sample of coke, every lump of adequate size, i.e., greater than 2–3 cm, is drilled to provide cores of 1 or 1.5 cm in diameter from which disks ~1 cm in thickness are sliced. From this bulk sample of coke disks, standard sample division techniques are used to separate the 50–100 specimens required for the tensile strength test.

The dimensions of the air-dried specimens are obtained by use of a micrometer and the specimens are placed edge-on between the platens of a testing machine. The load applied across the diameter of the specimen is progressively increased at a fixed rate of loading (in our laboratories we find it convenient to standardize on a machine crosshead speed of 0.5 mm/min) until breakage occurs as indicated by the abrupt fall in the load shown on the chart which automatically records the load applied. The tensile strength is calculated from the load at breakage using Eq. (6) and the average strength is calculated from a minimum of 50 tests. Median values of the tensile strength of industrial cokes range typically from 2 to 4.5 MN/m^2, the median value being used as the average because a skewed distribution of the values is frequently found.

The test can also be readily carried out at elevated temperatures by the use of an appropriate furnace.

D. Compressive Strength

By using a standardized procedure compressive strength results of a comparative nature may be obtained. Although this is adequate for most analyses it should be noted that as with the tensile strength tests, the values obtained are dependent on the sample shape and size, and on the experimental conditions used.

In the laboratories of the British Carbonization Research Association, the compressive strength of coke is determined on prismatic circular cylinders 1 or 1.5 cm in diameter and 1–2 cm in length with the proviso that the ratio of length to diameter is >1. The samples are obtained from coke lumps by means of a core drill and cut to the required length by a diamond-tipped cutting wheel which ensures flat and parallel end faces. These samples are centered between the steel platens of a test machine of sufficient capacity to ensure adequate stiffness of the machine. The loading rate is arbitrarily fixed at a machine crosshead speed of 0.5 mm/min and the load applied is automatically recorded on a chart; breakage is judged to have occurred when the chart records a sudden drop in the applied load of about 50%. In general some 50 tests are required for a reliable average value, assuming suitable precautions are taken at each stage of the sampling and sample division.

A similar type of test has been carried out at high temperatures (Holowaty and Squarcy, 1957; Zorena *et al.*, 1960).

E. Young's Modulus

Young's modulus of elasticity can be obtained from the stress–strain relationship derived from the so-called static tests in which the stress is applied relatively slowly or from dynamic tests in which the modulus is calculated from the resonance frequency of longitudinal or transverse vibration, in a specimen of given shape and size.

1. *Static Tests*

For an elastic material the stress–strain relationship is linear and the slope of the line is a measure of the Young's modulus. Industrial coke shows this type of behavior. As used by Patrick and Stacey (1972a) a relatively simple method of determining the stress–strain curve is to utilize strain gauges to measure the strain developed in a coke disk during a diametral-compression test (Section V,C).

Two important factors in the application of strain gauges for measurement of linear strain are the use of a suitable cement to ensure adhesion of the gauge to the test surface and the use of gauges having a small ratio of sensitivity across and along the length of the gauge. The cyanoacrylate type of adhesives (e.g., Kodak 910 or Permabond), which rapidly form strong bonds under relatively slight pressure at ambient temperature and 3 mm foil strain gauges, have been used satisfactorily and are recommended. The gauge is cemented across that diameter of the disk specimen along which tension is developed during the diametral-compression test, and included along with a similar unmounted gauge acting as a temperature compensator in an initially balanced bridge circuit whereby the subsequent unbalance voltage can be measured on a recording millivoltmeter. The records of stress and strain can then be used to plot a stress–strain curve from the slope of which the Young's modulus is determined.

Alternatively, the testing machine may be fitted with an extensometer system based on any one of a variety of transducer elements whereby either the strain may be recorded on a separate *XY* recorder or the stress–strain curve may be plotted directly on the machine chart recorder.

2. *Dynamic Tests*

There are several experimental advantages in measuring Young's modulus by dynamic methods, especially for measurements at elevated temperatures. The methods, the objective of which is to excite and detect resonance in the specimen and then to measure the frequency of this resonance, are well documented and attention is drawn to two methods which have been successfully used with industrial cokes (Blayden, 1966; Billyeald and Patrick, 1975).

The velocity V with which a stress wave is propagated through a homogeneous material is related to the effective elastic modulus E and the density ρ of the material by the equation

$$E = \rho V^2 \tag{7}$$

This forms the basis of a test in which measurements are made of the resonant frequencies of longitudinal stress waves generated in small cylindrical specimens. As described by Blayden (1966) a signal generator capable of providing sinusoidal oscillation, between 15 Hz and 30 MHz, is connected to an energizing coil surrounding a portion at one end of a ferrite rod transducer. A second coil near to the other end, to which the coke specimen is cemented, is connected to a voltage amplifier, the output from which is then fed to an oscilloscope and a digital frequency meter connected in parallel. The coils have a self-inductance of 15 mH and each is shielded by enclosure in a metal can, the second coil being energized by the stress changes in the transducer which is in the form of a rod 1 cm in diameter and 4 cm in length with a natural frequency of about 60 kHz. Torsional oscillations, which are produced simultaneously with the desired longitudinal oscillations, are suppressed by subjecting the transducer to an appropriate longitudinal magnetic field derived from a powerful permanent magnet.

The coke specimens in the form of 1-cm-diameter cores 1–5 cm long are obtained by core-drilling coke lumps and cutting to the desired size, the ends of the specimens being ground flat where necessary. The dried specimens are attached to the end of the transducer by means of a cyanoacrylate adhesive and the composite rod is laid, without any special support, within the two coils. Experimentally the frequency of the signal generator output is progressively increased and observations are made of the critical frequency at which the amplitude of the oscilloscope trace shows a sudden momentary increase. The fundamental frequency of the composite rod is determined from the usual simple numerical relationships between the fundamental and harmonic frequencies. Once the resonance frequency for the transducer alone has been determined, the natural frequency of the coke specimen is calculated from the equation

$$m_1 f_1 \tan(\pi f / f_1) + m_2 f_2 \tan(\pi f / f_2) = 0 \tag{8}$$

where m_1 and m_2 are the masses of the transducer and specimen, respectively, and f_1, f_2, and f are the natural frequencies of the transducer, specimen, and composite rod, respectively.

The resonance conditions in the rod are determined by the length of the rod l and the velocity V of the stress wave, and for the fundamental frequency

$$V = 2lf \tag{9}$$

Once the stress velocity V has thus been obtained, the Young's modulus is calculated by substitution in Eq. (7).

For successive determinations on different lengths of the same specimen there is little variation in stress velocities, and with different specimens the stress velocity can be obtained with a standard deviation of better than 1%. The Young's modulus determined by the dynamic method just described is generally $1\frac{1}{3}$ times greater than that given by the static method described earlier.

This method, however, is not suited to tests at elevated temperatures; for such tests a method involving flexural, or transverse, vibrations has been used (Billyeald and Patrick, 1975).

The apparatus consists of a signal generator giving a sinusoidal waveform, the frequency of which is continuously variable from 1 Hz to 30 kHz; this unit is coupled to a 10-W peak-to-peak power amplifier and then to a mechanical vibrator. The detector circuit comprises a high output crystal cartridge pickup connected to a voltage amplifier and a digital frequency meter which measures frequency to an accuracy of 1 Hz. A graphite resistance furnace is a suitable means of achieving the elevated temperatures. The coke specimens in the form of rectangular prisms on the order of $4.5 \times 1.0 \times 0.3$ cm are roughly cut and then accurately ground to the required dimensions and shape. The specimen is suspended in the center of the furnace by two spun asbestos threads of 0.2-mm thickness, tied close to the ends of the specimen, one thread being attached to the vibrator and the other to the crystal pickup. The furnace enclosure is evacuated and outgassed at 300°C until a pressure of less than 10^{-3} torr is achieved, after which the furnace temperature is reduced to ambient and the fundamental flexural frequency is determined as follows.

The signal generator frequency is progressively increased until the resonance condition is reached as shown by a large and sudden increase in the specimen vibrations detected by the crystal pickup, the amplified voltage output being fed to the *Y* plates of an oscilloscope. The *X* plates of the oscilloscope are fed by the signal generator and when both *X* and *Y* plates are receiving the same frequency at the same time, a Lissajou pattern is obtained, the maximum vertical amplitude of this pattern occurring at the resonance frequency. This value is read directly from the frequency meter.

The Young's modulus E is calculated using the equation

$$E = \frac{1.009ml^3f^2}{t^3b} \tag{10}$$

where m is the mass; l, t, and b are the length, thickness, and breadth of the specimen, respectively; and f is the fundamental frequency.

By this method the Young's modulus can be determined at any temperature up to the maximum attainable by the furnace.

VI. CONCLUDING REMARKS

There has been a marked increase since the late 1960s in the scope of analytical techniques available for the measurement of those coke properties considered to have a significant influence on its behavior in industrial processes. In particular there is greater awareness of the fundamental significance that the texture and microstructure of coke have upon its strength and behavior on combustion.

Coke is a heterogeneous material, and a study of its properties often demands a statistical approach. Only by the use of automated, computerized methods can this approach be realized. This has led to the application, to coke studies, of methods used for the structural analysis of metals, e.g., the automatic image analysis microscope. In addition, however, the general application of optical methods for the investigation of coke structure has enabled a much wider knowledge of structural parameters and of their effect on coke properties to be obtained.

There has also been growing recognition of the importance of measuring coke properties at high temperatures. Thus, much development work, both on micro- and macroscales, is in progress to measure strength characteristics at temperatures within the range of 1000–1500°C.

Conventional methods of coke analysis have been consolidated, rather than developed. The proven value of well-established methods has enabled them to be retained; the modifications that have taken place are largely concerned with improving the speed and accuracy of the methods. This is illustrated by the fact that no fewer than 19 methods for coke analysis have been published as ISO Recommendations since 1967 and recently 14 of them have been confirmed as ISO Standards. National Standards have developed in a similar way.

While it is unlikely that there will be significant further development of traditional methods of analysis, instrumental techniques are likely to supplant or replace conventional chemical methods.

REFERENCES

Abramski, C., and Mackowsky, M.-T. (1952). *In* "Handbuch der Mikroskopie in der Technik" (H. Freund, ed.), Vol. II, pp. 311–410. Umschau-Verlag, Frankfurt.

Billyeald, D., and Patrick, J. W. (1975). *Extended Abstr. Bienn. Conf. Carbon, 12th, Pittsburgh* pp. 147–148.

Blayden, H. E. (1966). *Year-Book Coke Oven Managers' Assoc.* pp. 197–216.

Blayden, H. E., Riley, H. L., and Shaw, F. H. (1943). *Fuel* **22,** 32–38, 64–71.

Blayden, H. E., Gibson, J., and Riley, H. L. (1944). *Proc. Conf. Ultrafine Struct. Coals Cokes, London, 1943* pp. 176–231.

Bradshaw, K., and Wilkinson, H. C. (1969). *J. Inst. Fuel* **42,** 112–117.

Crawford, A., Glover, M., and Wood, J. H. (1961). *Mikrochim. Acta* pp. 46–58.

Echterhoff, H. (1961). *Stahl Eisen* **81,** 992–1000.
Goldring, D. C. (1973). *Meet. Int. Comm. Coal Petrogr., 26th, Paris.*
Holowaty, M. O., and Squarcy, C. M. (1957). *J. Met.* **45,** 577–581.
Jüntgen, H., and Schwuger, M. (1966). *Chem.-Ing.-Tech.* **38,** 1271–1278.
Kirk, B. P., and Wilkinson, H. C. (1964). *Fuel* **43,** 105–109.
Kirk, B. P., and Wilkinson, H. C. (1970). *Talanta* **17,** 475–482.
Lee, G. W., and Mott, R. A. (1956). *In* "A Special Study of Domestic Heating in the United Kingdom—Present and Future," pp. 43–54. Inst. Fuel, London.
Mott, R. A., and Wilkinson, H. C. (1957). *Fuel* **36,** 39–42.
Patrick, J. W., and Stacey, A. E. (1972a). *Fuel* **51,** 81–87.
Patrick, J. W., and Stacey, A. E. (1972b). *Fuel* **51,** 206–210.
Patrick, J. W., and Stacey, A. E. (1975). *Fuel* **54,** 213–217.
Patrick, J. W., Stacey, A. E., and Wilkinson, H. C. (1972). *Fuel* **51,** 174–179.
Patrick, J. W., Stacey, A. E., and Wilkinson, H. C. (1973a). *Fuel* **52,** 27–31.
Patrick, J. W., Reynolds, M. J., and Shaw, F. H. (1973b). *Fuel* **52,** 198–204.
Patrick, J. W., Sims, M. J., and Stacey, A. E. (1976). *Dtsch. Keram. Ges., Carbon '76 Prepr., Baden-Baden* pp. 211–214.
Wasa, A., Nakai, T., and Sasaki, S. (1974). *Int. Organ. Stand.* **ISO TC27/SC 3,** No. 12.
Wilkinson, H. C. (1965). *Fuel* **44,** 191–198.
Wilkinson, H. C. (1972). *In* "Assessment of Physical Properties of Blast-Furnace Coke" (J. K. Wilkinson, ed.), pp. 96–108. Nat. Coal Board, Coal Res. Establ., Cheltenham, England.
Zorena, P. J., Holowaty, M. O., and Squarcy, C. M. (1960). *Blast Furn. Steel Plant* **48,** 443–451.

Chapter 30

The Formation and Properties of Anisotropic Cokes from Coals and Coal Derivatives Studied by Optical and Scanning Electron Microscopy

Harry Marsh *Janet Smith*
NORTHERN CARBON RESEARCH LABORATORIES, SCHOOL OF CHEMISTRY
UNIVERSITY OF NEWCASTLE UPON TYNE
NEWCASTLE UPON TYNE, ENGLAND

I. INTRODUCTION

Studies in the Northern Carbon Research Laboratories are currently concerned with the more fundamental aspects of mechanisms of conversion of coals and substances derived from coals into cokes, i.e., metallurgical cokes or the needle cokes more suitable for graphite production. There is interest in the structure of these cokes, and polarized light optical microscopy (with its relevance to optical texture) and scanning and transmission electron microscopy have proved to be very informative instruments. Most cokes undergo some degree of degradation following oxidation, e.g., metallurgical coke in the blast furnace is gasified by carbon dioxide and then by oxygen, and the graphitic arc electrode suffers severe gasification by air during melting operations. Scanning

ISBN 0-12-399902-2

electron microscopy has proved to be very effective in examinations of the topographical features of gasification processes. Transmission electron microscopy has been used to resolve the molecular structure within cokes. This chapter essentially describes the recent studies of the Laboratories as developed from earlier experiences. It is not intended as a thorough appraisal of the use of microscopy in studies of coals and cokes.

The studies of the Laboratories are concerned with an understanding of the mechanisms of the processes which lead from the isotropic fluid coal or pitch substance to the distinct but continuous anisotropic structures to be found in the cokes derived from these substances. These anisotropic structures vary appreciably in size, from submicrometer to hundreds of micrometers. The reasons for these large variations in carbonization behavior are yet to be found. A further development of this theme is to attempt to elucidate the importance of these anisotropic structures in cokes and to ask if they have any significant role and are not just incidental and unimportant to coke production. Importance of these anisotropic structures can be considered in three contexts, i.e., in terms of mechanical strength when cold, in terms of ability to withstand extremes of stress induced by temperature gradients and temperature, and to withstand oxidation treatments without breakage into unacceptable smaller fragments.

The studies of the Laboratories are sponsored by industry and an attempt is made, always, to relate the findings of the Laboratories to the industrial situation. Metallurgical coke for use in the blast furnace is a very difficult material to study. It is very heterogeneous in structure and its porosity requires a thorough statistical approach to analyses of data. The growing demand for metallurgical coke suitable for use in the new large blast furnaces (weekly output of pig iron greater than 50,000 metric tons) and the diminishing supply of suitable coking coals create a situation in which blending procedures for coals fed into coking ovens are a prerequisite. The technology of coke-making is currently refined in terms of blending procedures, with other procedures such as drying, preheating, and stamp charging, all contributing to an improved charge density and improved coke quality. Very recent developments in Japan (Nakamura *et al.*, 1977) incorporate briquets made of low rank coals and petroleum pitch into the charge being fed into the coke oven. This addition is claimed to be as effective as increased charge density associated with preheating prior to charging. All of these modifications or developments to coking procedure must influence the coking mechanism. For its industrial specifications, metallurgical coke must possess a high mechanical strength to withstand the pressures of the burden of the blast furnace and also must have a low reactivity to carbon dioxide so that solution losses in the middle regions of the stack of the blast furnace are minimized. These properties have to be understood, and again the question can be asked as to the relevance of anisotropic structures within the coke substance.

In order to have as wide an experience as possible in discussions of the origins of anisotropic structures in cokes, materials other than coals have been carbonized. These include pure organic compounds and pitches derived from coal and petroleum as well as the solvent-refined coals and Gilsonite pitch.

A. Use of Optical Microscopy

1. *Optical Texture*

Microscopic methods of examination of the solid products of carbonization represent the principal experimental approaches. Optical microscopy has, of course, a long established history in such examinations. The microscopes in our studies are Vickers M41 polarized light research microscopes. Polished sections of carbon or coke are examined, either mounted in a resin block or unmounted. The polished surface is usually monitored using parallel polars with a half-wave plate inserted between the specimen surface and the analyzer to create the interference colors, usually yellows, blues, and purples which can characterize the anisotropic carbon in terms of size of colored (isochromatic) areas and the actual color. This overall appearance is referred to as the optical texture. These colored areas can vary in size from several micrometers to submicrometer (called a mosaic structure) to hundreds of micrometers (called domains, if isometric, or flow type if elongated). Nomenclatures have been developed for their more precise description by Patrick *et al.* (1973), Sanada *et al.* (1973), and White and Price (1974).

The yellow, blue, and purple colors are used to assess the orientation of the constituent lamellar planes of the anisotropic carbon at the polished surface. Yellows and blues are indicative of prismatic edges exposed in the polished surface, yellows changing to blue and vice versa on rotation of the specimen stage by 180° or by reversal of the half-wave plate. The purples are indicative of basal planes, i.e., the surfaces of constituent lamellar planes of the anisotropic carbon, lying parallel to the polished surface. The purple color remains unchanged during rotation of the specimen stage of the microscope; it presents essentially an isotropic surface. It is to be noted that the nongraphitizing isotropic carbons when so examined in polished sections also exhibit a purple color which in practice is easily distinguished from the anisotropic purple, which is rather darker in shade. Thus, the optical microscope is used to assess the presence or absence of anisotropy, the shape and size of anisotropic constituent structures, and the orientation of these structures within the polished surface. (For classification by size of anisotropic units, see Chapter 29, Section IV,D,2.) The surface is photographed and colored micrographs illustrate reports, and the like. Sizes of mosaics and domains are tabulated. When a surface texture is not homogeneous, then large areas of sample, about 10 × 10 mm, are examined by a point counting, moving stage technique to measure the relative proportions of anisotropic structures of various sizes, these data being presented as histograms. This approach finds application in examinations of cokes from pitch blends. A major effort, in this area of mechanism of carbonization, of the Laboratories is the elucidation of the relationship of size and shape of mosaic and domain with the chemical composition of the parent material undergoing carbonization, including the conditions of carbonization.

2. *Quantitative Reflectance Microscopy*

The optical microscope can be used also as a quantitative instrument to measure the percentage of polarized light reflected from polished surfaces of anisotropic carbon. In this way, the microscope can be used in structural analyses. In rather basic terms, anisotropic carbons can be considered as highly imperfect crystals of graphite. Single-crystal graphite, found rarely, if ever, and approximated in stress-annealed pyrolytic graphites and natural crystals, e.g., as found in Ticonderoga, New York, is composed of parallel layers (lamellae) of carbon atoms that are arranged in hexagons as in benzene. There is in-plane σ bonding between the carbon atoms and out-of-plane π bonding in between the layers (but not across the layers). The amount of polarized light reflected from a polished surface of anisotropic carbon is a function of the interaction of the electric field of the incident electromagnetic radiation of polarized light and the electric fields within the carbonaceous lattice. The interaction is a maximum when the plane of polarized light is parallel with the layer planes (lamellae) and when basal planes are presented to the surface; it is a minimum when the plane of polarized light is at right angles to the lamellae. Thus, the percentage of light reflected from the surface is a function of the mode of presentation of the lattice to the light as well as the degree of perfection of lamellae. Curvature or buckling of the lamellae, the presence of heteroatoms, vacancies, or other defects, and hydrogenation of the lattice all contribute to diminish the percentage of the incident light reflected from the surface. The difference between the maximum and minimum value for reflectivity (i.e., bireflectance) from the same area of the specimen is indicative of the stacking perfection of the lamellae which constitute the anisotropic carbon.

Reflectivity measurements are used extensively to characterize coals (see Volume I, Chapters 1 and 2) and carbonized vitrinites, usually using oil objective lenses (Murchison, 1958; Goodarzi and Murchison, 1972, 1973a,b, 1976; Hook, 1973). As such, an indication is obtained of the degree of aromaticity within coals of different rank and this has proved to be probably more suitable for predicting coking behavior than any other single property of the coal (see also Gibson, 1972). The extension of this technique to examine the detail of coke structure is comparatively recent. Coals are easier to examine than the cokes. Large surfaces are available and computerized optical microscopy is applicable. For the results to be meaningful in examinations of cokes, however, the light spot of the microscope (micrometers in size) has to be focused precisely onto the selected areas of anisotropic carbon and to remain on these selected areas when the specimen stage is rotated to obtain the maximum and minimum values of reflectivity. This can be a very tedious operation. Consequently, it has proved easier to examine the reflectivities of the larger mosaics and domains.

An overall objective of such studies is to assess if reflectivity varies in a meaningful way with the size of the optical texture, mosaics, and domains. The optical texture is essentially a macroproperty whereas the reflectivity is measuring a property of the constituent lamellar molecules and their stacking arrangements. It can be asked, in this context, if it is the size and shape of the constituent lamellar molecules which govern or control the ultimate sizes of the optical texture. With this in mind, the Laboratories have accordingly examined anisotropic carbons prepared from known source materials. Alternatively, Murchison and co-workers have directed their studies toward cokes derived from coal macerals (see Chapter 31).

B. Use of Scanning Electron Microscopy

Scanning electron microscopy (SEM) (Thornton, 1967) is now established as a means of characterization of surface topography, mainly because of the large depth of focus available and the high magnifications currently possible (up to ×200,000). The Laboratories have used SEM directly in studies of morphological changes occurring during carbonization and gasification. Specimens of carbons are mounted on suitable stubs and examined in a "Stereoscan" Mark II (Cambridge Instrument Company), and more recently in the scanning reflection mode of JEOL 100C STEM. The surfaces of the carbon are usually coated with gold or gold–palladium alloy to prevent charging of the specimen. It must be stressed that SEM techniques, as used here, give topographical information only, including that of porosity, and cannot be used to provide structural information directly, as is the optical texture revealed by optical microscopy. Hence, in order to obtain specimens suitable for SEM examination in terms of shape and size of features and topography, the experimental carbonization procedures must be designed specifically for SEM examination. For example, Marsh *et al.* (1971a, 1973a) found that the use of hydraulic pressures (about 200 MPa) isolated the individual anisotropic units growing from the pitch and prevented their coalescence. This afforded, then, a direct method of SEM examination of the growth of anisotropic units.

In studies of gasification, a combination of optical microscopy and SEM has proved to be very informative. A polished surface of coke or carbon is prepared suitable for gasification experiments. It is a cube ~12 mm in size that is polished by hand, and is not mounted in resin. The optical texture is recorded and presented as a photomontage; the surface is often marked, for reference, with a micromanipulator. The polished surface is now photographed using SEM, recording the areas to be characterized by gasification. (With many of these anisotropic carbons, using SEM at magnifications of about 1000, no gold coating was necessary.) At this stage the SEM micrograph is not too informative, showing only the flat surface (a fracture surface can be interesting) and contained porosity. The carbon is then subjected to gasification conditions, e.g., by atomic oxygen and hydrogen, molecular oxygen, and/or carbon dioxide. On reexamination, it is found that mild hydrogenation (<1% burn-off) causes the disappearance of the reflectance interference colors of yellow, blue, and purple, even where the surface is otherwise not visually affected. Gasification with carbon dioxide results in the surface adopting an overall purple color. After weight losses of several percent by gasification, the optical microscope ceases to be of much value because the initial polished surface has been pitted and clear focusing is impossible. However, the SEM can now be used to advantage. Generally, the overall smooth polished surface of the carbon has disappeared. It is observed that gasification does not proceed by an even ablation of the surface, but rather by the selective, preferential gasification of parts of the surface, other parts remaining relatively unchanged. Areas of gasification develop initially as pits, but more prolonged gasification creates fissures running into the interior of the carbon specimen. By a direct comparison of optical texture by means of an optical microscope, with the topography created by gasification, using the SEM of the same area of surface, a correlation of gasification behavior with optical texture can be established.

Scanning electron microscopy can also be used to augment studies of coke strength and its variation with heat treatment. The procedure is similar to that just described except that gasification treatments are replaced by heat treatments. To simulate the conditions of the blast furnace, coke, as obtained from the ovens (but with polished surfaces), is heated in an inert atmosphere to about 2100 K. Generally, as revealed by rotating drum methods, the cokes lose strength, the various cokes behaving differently. A significant factor in assessments of strength is the development of shrinkage fissures, their size, shape, frequency, and relative orientations, as well as their origins in terms of optical textures of original surfaces. Scanning electron micrographs, when compared with the colored optical micrographs, reveal the positions of development of shrinkage fissures in terms of optical texture (perhaps only on cooling from 2100 K; the exact condition of the coke at 2100 K poses an interesting problem). Further, by using direct interfacing with computerized microscopy, e.g., "Quantimet," or using epidiascopic attachments, the size, shape, and frequency of fissures and pores can be statistically described.

The use of "stereopairs," obtained in the SEM by a change in tilt angle of the specimen to the electron collector, facilitates the assessment of heights or depths of topographical features using photogrammetry and stereological approaches (Boyde, 1967; Boyde and Ross, 1975; Underwood, 1970). Stereopair photography allows the eyes (and brain) to synthesize a three-dimensional image of the surface. Stereopairs can be analyzed in appropriate instrumentation to enable, e.g., the depths of pores or fissures to be measured as well as other relevant topographic information. Such an analysis can be carried out on the prepared surface and then followed by similar analyses of new surfaces created by progressive grinding and polishing. Thus, there can be built up, by this stereological approach, a description of porosity or fissure content within the bulk of the sample of coke, graphite, or carbon. The method is, however, rather time-consuming and expensive.

II. THE FORMATION AND PROPERTIES OF ANISOTROPIC COKES

A. Historical

Almost without exception, cokes or carbons prepared from solid parent substances which do not fuse or melt during carbonization, are isotropic in structure and will not graphitize. Common examples are charcoals prepared from wood and copra, and a useful laboratory porous carbon which is prepared from polyvinylidene chloride. Conversely, carbons prepared from parent substances which fuse or melt during carbonization are nearly always anisotropic in structure and will graphitize. An exception is the carbon prepared from saccharose. Common examples are carbons prepared from coal tar pitch, petroleum pitch, solvent-refined coal (particularly when hydrogenated), polyvinyl chloride, and medium to low volatile coals. That such anisotropic carbons will graphitize is indicative of crystallographic order existing within the carbons. Such carbons are seen to grow from the isotropic melt of pitch or coal and it is intriguing to know the exact mechanism of this growth process. Is it a form of crystallization or perhaps a precipitation? What factors control the shape, size, and degree of coalescence of the growth units which ultimately form the anisotropic carbon?

Optical microscopy has been used for many years in studies of cokes and coke formation. Ramdohr (1928) described the microscopic appearance and optical properties of graphite and graphitizing carbons and emphasized the high reflectance of graphitizing carbons. Marshall (1945) made considerable use of optical microscopy to characterize the anisotropic properties of vitrinites from bituminous coals. In such coals, metamorphosed by dikes and sills, Marshall (1945) observed a "new" granular material about 3 μm in size constituting a mosaic, and, like Ramdohr (1928), emphasized the striking optical properties of the coke formed in the affected coal seam. Marshall (1945) was not able to offer an explanation of its origin. Mackowsky (1949) stated that the fine anisotropic mosaic structure, observed in cokes from coals, originated by a continuous ordering of crystallites during the period of fluidity of the carbonization process. And, in 1952, Abramski and Mackowsky (1952) stressed the importance of the existence of a plastic zone in the carbonization process, pointing out that the size of the anisotropic units increased with the increased fluidity of coals of higher rank. This aspect of the relationship of size of anisotropic unit in cokes with rank of coal was developed by Alpern (1956). Kipling *et al.* (1964; Kipling and Shooter, 1966a,b) studied the anisotropic carbons prepared from polyvinyl chloride, a white solid which melts and decomposes to a black pitchlike material. However, not all the fluid phases studied by Kipling *et al.* (1964) formed anisotropic carbon.

Taylor (1961) published a study of vitrinites in the Wongawilli seam of New South Wales, Australia. The seam had been affected by an igneous dike which had penetrated the coal seam and established a positive thermal gradient through the coal seam toward the dike. The effect over some hundreds of meters from the dike was the establishment of a very gradual thermal gradient reaching toward the dike to maximum temperatures which passed through and exceeded the temperatures of formation of anisotropic coke from coal. It was the slow rates of heating which had occurred and the very narrow temperature zone of formation of anisotropic carbon which provided the first clues as to the nature of this unique property. Previous laboratory experiments had all used too rapid a heating rate and the detail of formation of anisotropic carbon had been missed. Approaching the dike, Taylor (1961) noted that the vitrinite had evidently become fluid or plastic and, on later cooling, was isotropic; i.e., the original bedding (inherent) anisotropy of the vitrinite had been destroyed. However, on further approaching the igneous dike, small spheres initially of micrometer size, were observed in the isotropic vitrinite, the spheres on further heating growing at the expense of the vitrinite and coalescing together to form the mosaic structures observed by earlier workers.

Taylor (1961) recognized that this anisotropic sphere provided the clue to the development of graphitizing carbons. The molecular arrangements within the spheres deduced from optical properties and confirmed by electron diffraction closely resembled those of the nematic phase of substances called *liquid crystals* (Gray, 1962; Brown *et al.*, 1971; de Gennes, 1974; Gray and Winsor, 1974; Priestley, 1975).

Brooks and Taylor (1968) subsequently developed this concept and thus provided the basis of the mechanism by which anisotropic carbon is formed from pitchlike substances. Recent years have seen the exploitation of this concept and its application to industrial systems (Marsh, 1973, 1976; Patrick *et al.*, 1973; Patrick, 1976; White, 1976; White and Zimmer, 1976; Sanada *et al.*, 1972, 1973; Yamada *et al.*, 1974; Marsh and Cornford, 1976).

B. Nematic Liquid Crystals, Mesophase, and Anisotropic Carbon

1. *Pitch to Semicoke*

Liquid crystals have been recognized since 1888 (Reinitzer, 1888) although before 1968, they were often just a curiosity in the laboratory (Brown and Doane, 1974). The term was originally used to describe systems obtained, e.g., by melting cholesteryl benzoate which had unusual fluid properties. These systems were anisotropic when viewed in thin sections between crossed polarizers of the optical microscope. They thus possessed more structural order than found in normal isotropic liquids, but were not genuinely crystalline. Friedel (1922) suggested that the term *mesophase* (intermediate state) would obviate the apparent inconsistency of "a crystalline liquid," but "mesophase" is rather too imprecise a term in itself. However, it is used currently and this use is discussed in the following.

Of the several categories into which liquid crystals may be placed, it is the nematic system which has relevance to carbonization. The arrangement of molecules in the nematic liquid crystal is usually envisaged as one in which the molecules lie parallel to one another, but with no order in the stacking sequence. This is the type of order Brooks and Taylor (1968) observed in their anisotropic spheres within the vitrinite matrix. These spheres, in polished sections, exhibited rather unusual optical properties which were consistent with anisotropic stacking of the constituent lamellar molecules parallel to an equatorial plane with a more radial arrangement of lamellae at the sphere surface. An alternative possibility of circumferential stacking, as in an onion, was discounted.

The formation of anisotropic carbon from within the fluid phase of pitch can be envisaged somewhat as follows. Initially, the fluid phase, if viewed in a suitable hot stage by optical microscopy, would appear to be quite isotropic and, with a half-wave plate and parallel polars, would exhibit a purple color. Between 630 and 700 K, depending on parent substance and heating rate, anisotropic units would become visible in the microscope, obviously growing from units submicrometer in size. With coal tar pitch and some petroleum pitches these units would appear as spheres and grow as spheres as illustrated in the optical micrograph of Fig. 1. However, it should be stressed that spherical growth, although common, is *not* an essential feature of growth of anisotropic units. In the pyrolyzing pitch system there is a continuous growth in the size and concentration of lamellar constituent molecules. At a critical concentration, dependent on temperature and chemical structure, these lamellar constituent molecules (of molecular weight greater than about 1000 amu) are able to "assemble together," by physical forces only, to create the nematic liquid crystal. This is considered to be of short duration. It is reversibly formed as shown in the careful experimentation of Lewis (1975). Pyrolysis chemistry still continues as the carbonization temperature increases such that polymerization within the nematic liquid crystals occurs and their dissociation into constituent molecules is no longer possible. The resultant polymer, with its structure of parallel stacked molecules based on the structure of nematic liquid crystals, still retains fluidity, plasticity, or measurable viscosity. This polymer is termed the *mesophase* in carbonization studies and is the plastic, grow-

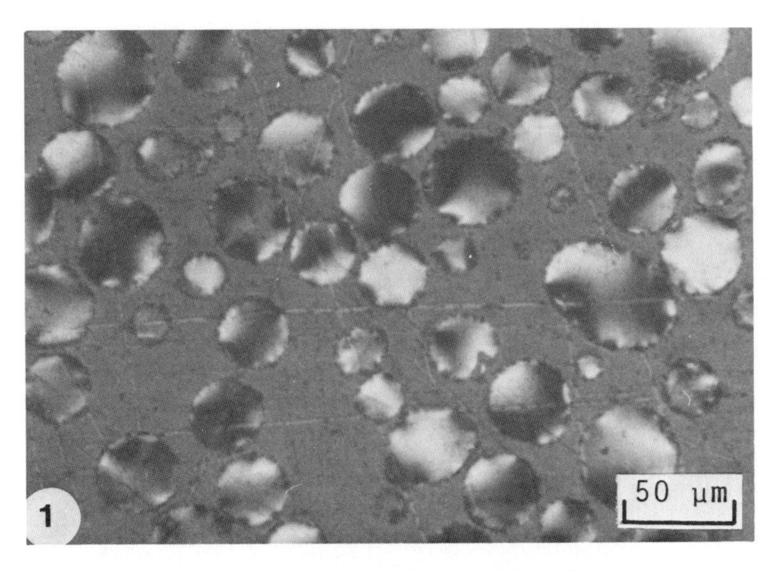

Fig. 1 Optical micrograph of polished surface of carbon, Orgreave lean coal tar pitch, heat treatment temperature (HTT) 670 K, 5 K/hr, 24 hr soak (9/2/77/13).

ing anisotropic phase seen in conjunction with parent pitch substances. That growth units of mesophase are often spherical is due solely to the fact that in such systems, usually aromatic and of relatively small molecular weight, the viscosity of the mesophase is sufficiently low to enable the requirement of minimum surface area to dictate a spherical shape to a deformable phase surrounded by a liquid.

In other carbonization systems, the degree of polymerization within the mesophase at comparable temperatures can vary significantly. This degree of polymerization influences the viscosity of the mesophase, and this property alone, more than any other, probably influences the resultant optical texture of the carbon or coke.

2. *Mosaics and Domains*

Marsh *et al.* (1971a) developed a carbonization procedure which enabled the growth of mesophase to be studied comparatively easily and quickly, in order to make a survey of controlling influences. The procedure was one of carbonization of material in sealed gold tubes to 875 K under hydraulic pressures of 200–300 MPa. These extremely high pressures prevented the ultimate coalescence of the mesophase. (This can vary under normal conditions to establish the large domains, several

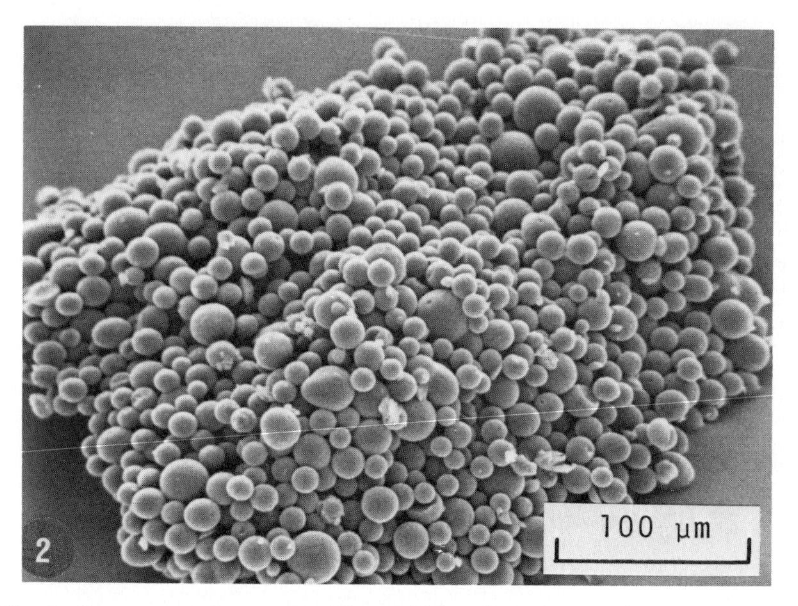

Fig. 2 Stereoscan micrograph of botryoidal, spherical mesophase, acenaphthylene, HTT 823 K, 5 K/min, 0.5 hr soak, 260 MPa (4/102).

hundreds of micrometers in size, seen in petroleum-derived cokes.) Thus, if viewed by SEM, the growth units of mesophase can be monitored for size and shape.

Under the conditions of carbonization under hydraulic pressures Marsh *et al.* (1973a) observed the growth of mesophase as spheres which together gave a grapelike (botryoidal) appearance. Figure 2 is a scanning electron micrograph of such spheres prepared from acenaphthylene (823 K; 260 MPa). Should the growth units of mesophase possess too high a viscosity, because of cross-linkage, then the shape of the growth units will not respond to surface tension effects and will not be spherical. The growth of elongated units of mesophase (spaghetti-like or tadpole-like) with the constituent lamellar molecules parallel to the short axis (as in a stack of pennies) is seen in Fig. 3 which is a scanning electron micrograph of mesophase from anthracene (763 K; 250 MPa). These elongated units can change to spherical units as the viscosity is lowered by increasing carbonization temperatures, passing through an intermediate ovoid shape. Figure 4, a scanning electron micrograph of mesophase from a mixture of anthracene with phenanthrene (3 : 7; 823 K; 300 MPa), contains elongated, ovoid, and spherical units of mesophase.

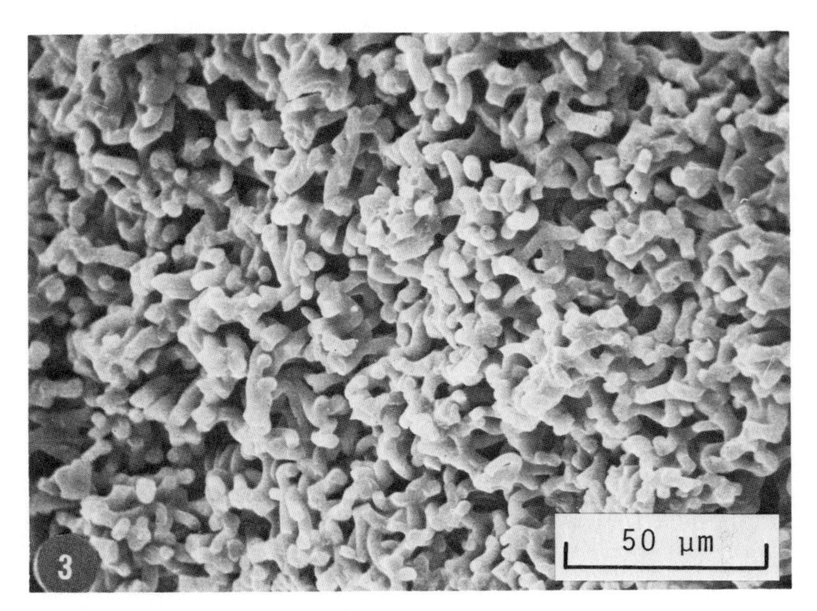

Fig. 3 Stereoscan micrograph of spaghetti mesophase, anthracene, HTT 763 K, 5 K/min, 0.5 hr soak, 250 MPa (3/73).

Fig. 4 Stereoscan micrograph of mesophase of mixed shape, anthracene–phenanthrene (3 : 7), HTT 823 K, 0.5 K/min, 300 MPa (4/171).

This transition from elongated to spherical units of mesophase is dependent on a lowering of viscosity accompanying increasing temperatures of carbonization. In many carbonization systems this cannot occur and the growth units of mesophase adopt and maintain irregular shapes. Such shapes are to be seen in mesophase from pyranthrone (a large dyestuff molecule) in Fig. 5 (773 K; 120 MPa) and in mesophase from Gilsonite pitch (Marsh *et al.*, 1971b) in Fig. 6 (683 K; 0.1 MPa). The relatively high initial viscosities, probably associated with chemical cross-linkage and irregular packing, prevent both the adoption of spherical shapes by the growth units of mesophase and ultimately the coalescence of the mesophase to form semicoke.

The mode of coalescence of the mesophase (or its apparent absence) is of major importance in determining the physical and chemical properties of the resultant semicokes, cokes, and graphites. Growth of the units of mesophase, in particular those of spherical shape, as in Fig. 1, continues from within the fluid phase of carbonization until the spheres make contact with each other. At this stage, such spheres will coalesce together to make a single larger sphere. But, in so doing, the very considerable internal movements within the sphere seriously reduce the stacking order of the constituent molecules within the initial spheres,

Fig. 5 Stereoscan micrograph of mesophase, pyranthrone, HTT 773 K, 5 K/min, 0.5 hr soak, 120 MPa (5/590).

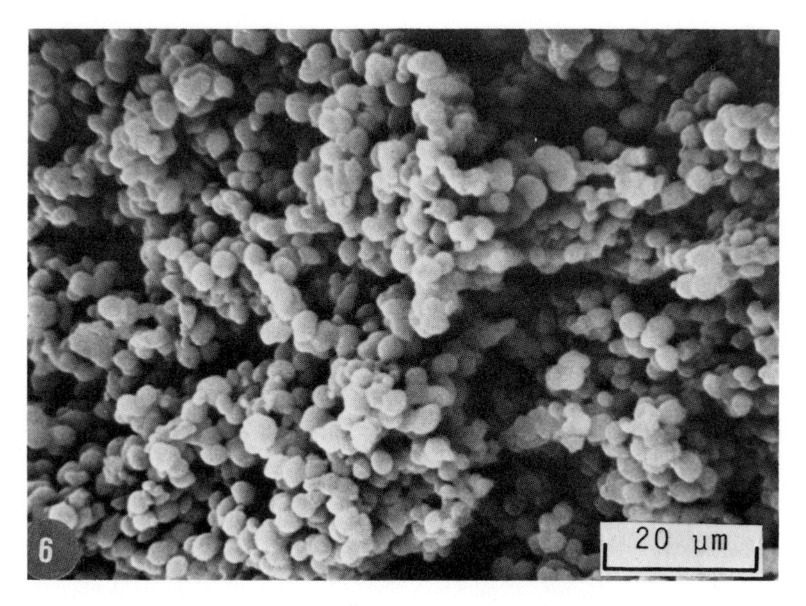

Fig. 6 Stereoscan micrograph of mesophase, Gilsonite pitch, HTT 683 K, 0.5 K/min, 4 hr soak, 0.1 MPa (3/422).

and so the defects and polycrystallinity of cokes and synthetic graphites are introduced. This process of movement and flow of mesophase (it can exhibit thixotropy) continues after the initial coalescence of spheres until the entire fluid phase is converted to anisotropic semicoke. The flow characteristics of the mesophase, however, are not entirely random processes. They follow the known flow characteristics of the classical nematic liquid crystals, and, as such, establish within the semicoke a sequence of macrodefects, known as disclinations, which are identifiable and characterizable. The elaborations of the origins, structure, properties, and industrial relevance of these disclinations have been made by White *et al.* (1967), Dubois *et al.* (1970), White and Price (1974), and White and Zimmer (1976) (see also Patrick, 1976).

C. Shrinkage Fissures

An optical micrograph of a polished surface of a carbon prepared from coalesced mesophase is shown in Fig. 7. This micrograph, of A200 Ashland petroleum pitch, shows large domains of isochromatic areas of optical texture, with shrinkage cracks running parallel to the plane of stacking of the constituent lamellar molecules. Figure 8 is an optical

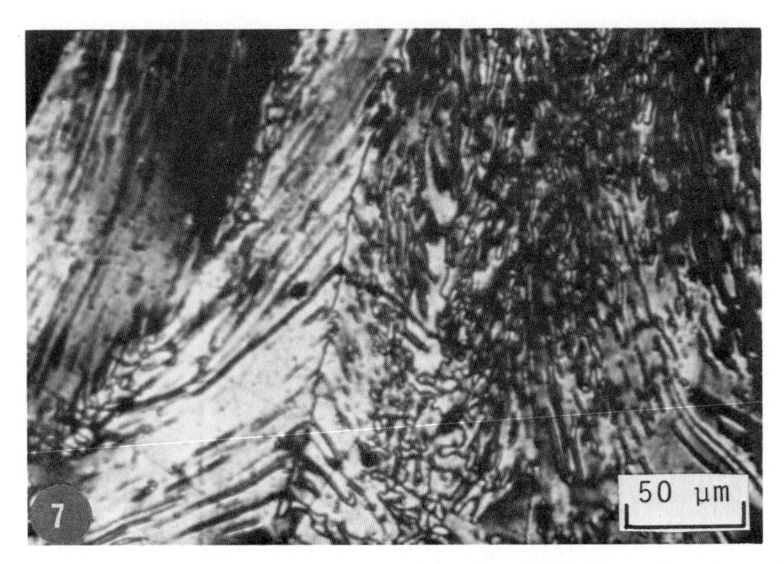

Fig. 7 Optical micrograph of polished surface of carbon, A200 Ashland petroleum pitch, HTT 800 K, 70 K/hr.

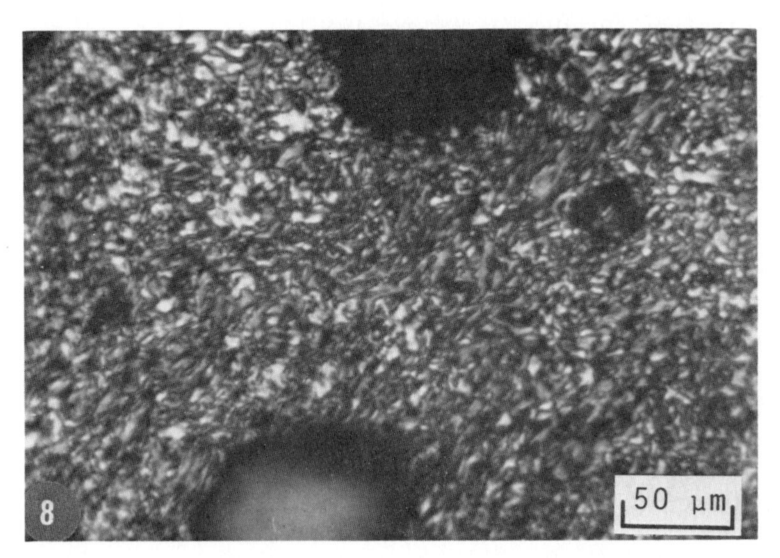

Fig. 8 Optical micrograph of polished surface of carbon, Gilsonite pitch, HTT 1100 K, 5 K/min.

micrograph of a polished surface of coke prepared from Gilsonite pitch. This micrograph shows the mosaics of isochromatic areas, only a few micrometers in size. During the carbonization of this pitch, the units of mesophase grow only to a few micrometers in size until with increasing numbers of such units, the fluid phase is converted entirely to semicoke. However, with this material, coalescence, as described for the spherical growth units, does not occur. These smaller, irregularly shaped growth units of mesophase merely "fuse" together without significant deformation. The exact nature of this fusion is not yet fully understood. When a coking coal or coking coal blend is carbonized, the growth units of mesophase are even smaller, with Patrick *et al.* (1973) classifying them from fine grain mosaic 0.3 μm in diameter to coarse grain mosaic 1.3 μm in diameter. Figure 9 is an optical micrograph of a polished surface of metallurgical coke. The smallness of the anisotropic growth units can be seen, in association with a larger grouping of these small mosaics, to establish a supermosaic structure. The exact nature of the bonding or binding between these mosaic units is not yet elucidated. There could be an interface of isotropic material which is adsorbed onto the growth units of mesophase during their development or, despite the disparity in crystallographic orientation between the growth units, there could be some form of chemical bonding. Whatever the exact nature of this interface condition, it is certainly not weak in view of the considerable strengths exhibited by the coke substance.

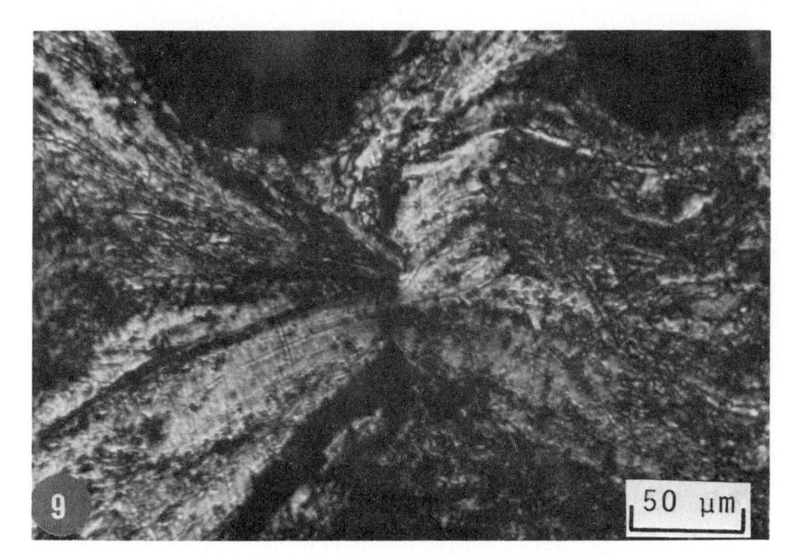

Fig. 9 Optical micrograph of polished surface of a metallurgical coke.

Petroleum needle cokes, made from petroleum pitch in a delayed coker (those pitch substances which usually give spherical mesophase and large isochromatic domains), are characterized by low values of coefficient of thermal expansion (CTE) and low mechanical strength. The shrinkage fissures as seen in Fig. 7 can thus accommodate the stresses set up as a result of the internal thermal expansions of the coke or graphite. But, since the fissures are large and easily interconnected, so we have a resultant relatively weak material. Metallurgical coke, on the other hand, is relatively hard and is resistant to thermal shock. The detail of the coke structure has to be examined to account for these properties. Despite the heterogeneous origin and growth processes in the carbonization of coal and coal blends, the final components which constitute the composite of a good metallurgical coke are essentially compatible and bonded to each other. This compatibility is indicated in scanning electron micrographs of surfaces of metallurgical coke subject to temperatures of about 2000 K. Figures 10–12 are scanning electron micrographs of the surface of coke prepared from a Durham coking coal (Sacriston Victoria seam). The larger shrinkage fissures, about 10 μm in diameter and 100 μm in length, are usually contained within the cell wall material of the coke. No fissuring is apparently associated with the

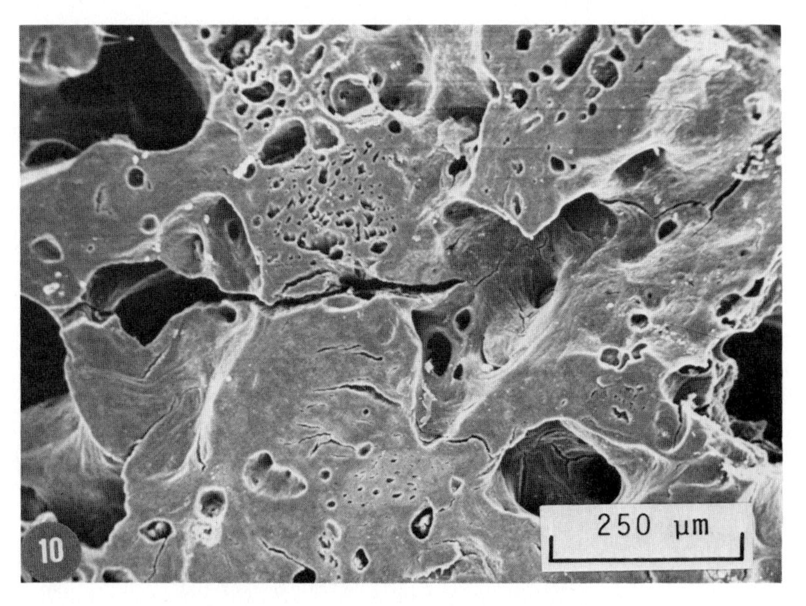

Fig. 10 Stereoscan micrograph of surface of metallurgical coke, Sacriston Victoria seam Durham, HTT 2000 K (10/400).

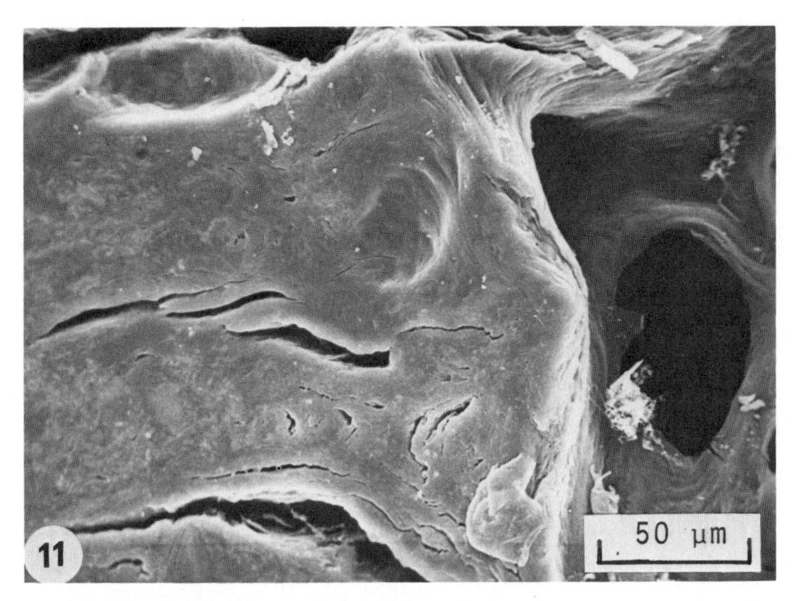

Fig. 11 Stereoscan micrograph of surface of metallurgical coke, Sacriston Victoria seam Durham, HTT 2000 K (10/399).

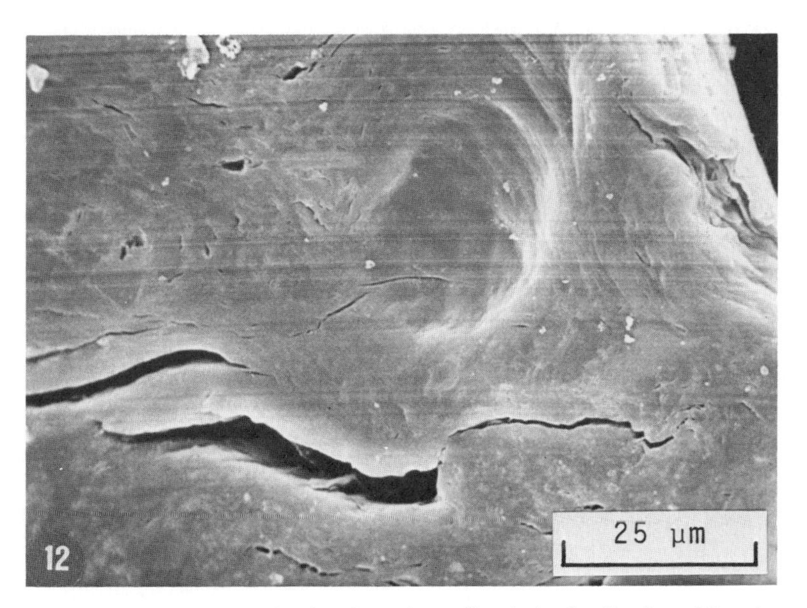

Fig. 12 Stereoscan micrograph of surface of metallurgical coke, Sacriston Victoria seam Durham, HTT 2000 K (10/398).

small mosaic structure of the coke (see Fig. 9). The larger fissuring (60–100 μm) of a petroleum coke (A200 Ashland), heat treatment temperature (HTT) 1223 K, is seen in Fig. 13. The smaller fissuring, less than about 10 μm in length of the Gilsonite coke (HTT 2300 K) seen in Fig. 14, is undoubtedly associated with the smallness of mosaic size of the optical texture.

It must be emphasized that such considerations as those of fissure development and strength form only part of a wider investigation into coke strength. Patrick and co-workers (Patrick *et al.*, 1972; Patrick and Stacey, 1972a,b) have investigated the tensile strengths of many coke substances in which variations in porosity and cell wall thickness must be a major controlling factor. The structure of cell wall coke substance must also be of relevance, particularly where there are variations in thermal and gasification resistance of those cokes which, when tested prior to loading onto the charging skip of the blast furnace, appear to be identical.

It can be argued that the anisotropic texture has a significant role to play. The nonporous isotropic carbons usually do not possess good resistance to thermal shock. It was because of this that Easton and Jenkins (1976) were unable to prepare stable glassy carbon shapes in other than

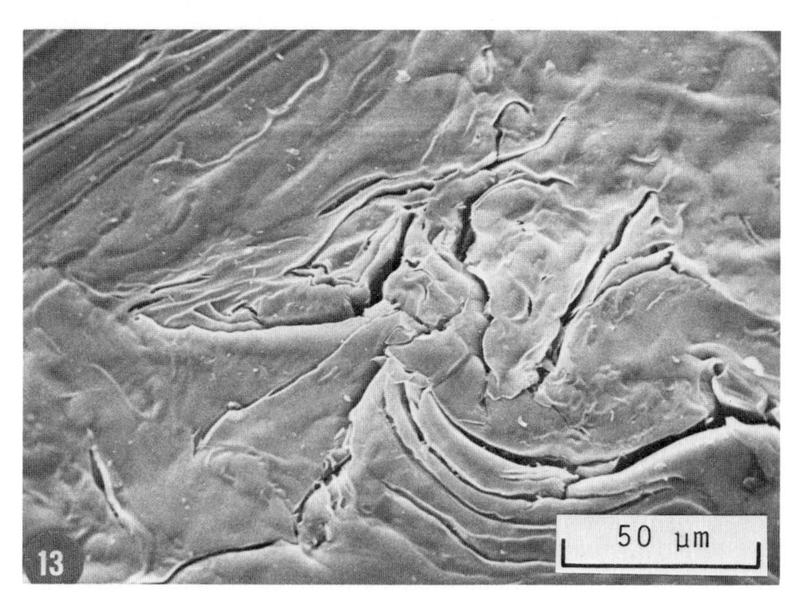

Fig. 13 Stereoscan micrograph of surface of coke, A200 Ashland petroleum pitch, HTT 1223 K, 5 K/min, 0.5 hr soak (9/217).

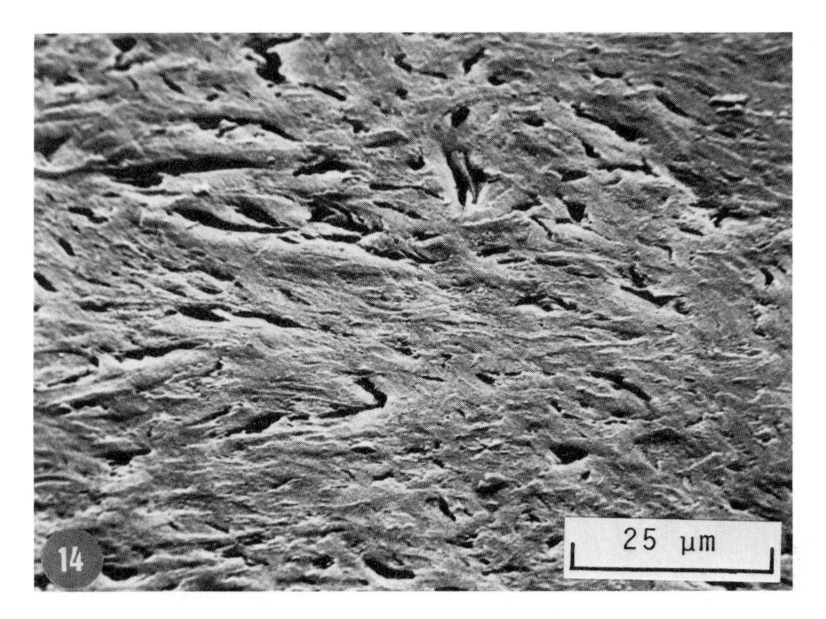

Fig. 14 Stereoscan micrograph of surface of coke, Gilsonite pitch, HTT 2300 K (11/30).

modest sizes. These carbons fractured extensively on cooling. The mosaics of anisotropic carbon may make at least two contributions to thermal stability. First, it has been observed in high resolution scanning electron microscopy (Marsh *et al.*, 1977a) that metallurgical cokes contain anisotropic units of size 0.05–0.1 μm, i.e., below the resolution of the optical microscope, as well as those of size 0.1–1.3 μm. The SEM analysis indicates that the smallest units have retained their identity and have fused on touching, leaving space or porosity between the units. Such space can accommodate expansions associated with high temperatures. The second contribution is associated with the independence of structural alignments of the mosaic units and the random alignment of the units. Shrinkage fissures occur parallel to the layer planes (the interlayer distance is sensitive to HTT). These fissures thus remain small and within the unit. Since the units are in random alignment, the fissure alignment will be also. Thus any expansion is accommodated by small "concertina" movements within the mosaic units, in the three dimensions of the coke substance, hence restricting rather than promoting the larger shrinkage fissures. It is this form of analysis that is being applied to coke substances of different properties in the Northern Carbon Research Laboratories.

D. Gasification Fissures

It is a general feature of the gasification of carbons and graphites that the surfaces of these carbons and graphites do not gasify smoothly or evenly (Thomas, 1965). Rather, there occurs preferential gasification—some areas of the surface remain apparently quite inert whereas other areas undergo severe gasification. Although some of these effects are undoubtedly due to the presence of catalytic impurities, there is no doubt that there is a relationship between anisotropic optical texture and the size, shape, and degree of reaction anisotropy or preferential gasification of areas of surface.

Examples of this preferential gasification can be found with all types of optical texture. Petroleum cokes, e.g., from A200 Ashland petroleum pitch, exhibit the flow-type anisotropy of the needle cokes [where the domains of mesophase have flowed and elongated while plastic, this being induced by convection currents within the fluid phase, or passage of gas bubbles through the fluid (Rester and Rowe, 1974)]. Gasification of flow-type anisotropic carbon induces gasification fissures running into the interior of the coke particle. Figures 15 and 16 are scanning electron micrographs that show this effect rather clearly for needle coke gasified in carbon dioxide to 40% burn-off (Marsh and Macefield, 1977). The

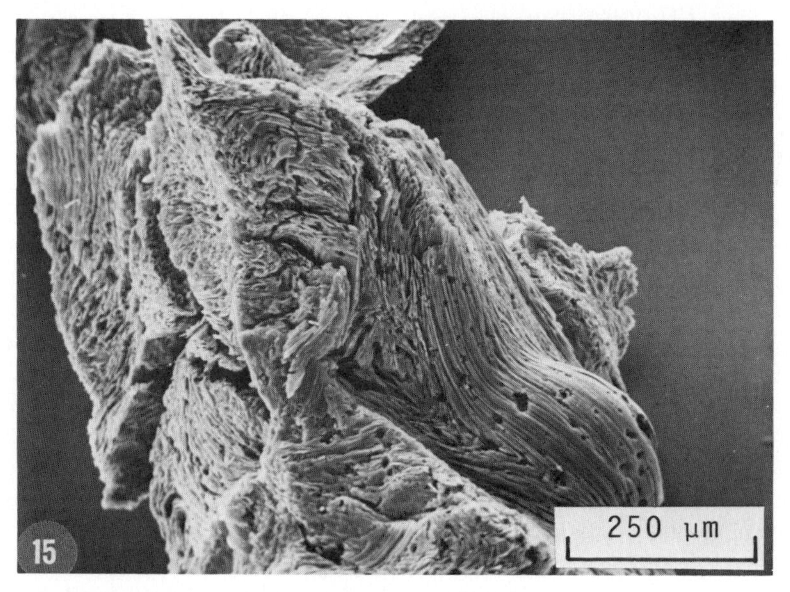

Fig. 15 Stereoscan micrograph of surface of coke, A200 Ashland petroleum pitch, HTT 1223 K, 5 K/hr, 40% burn-off (bo) CO_2, 1123 K (9/222).

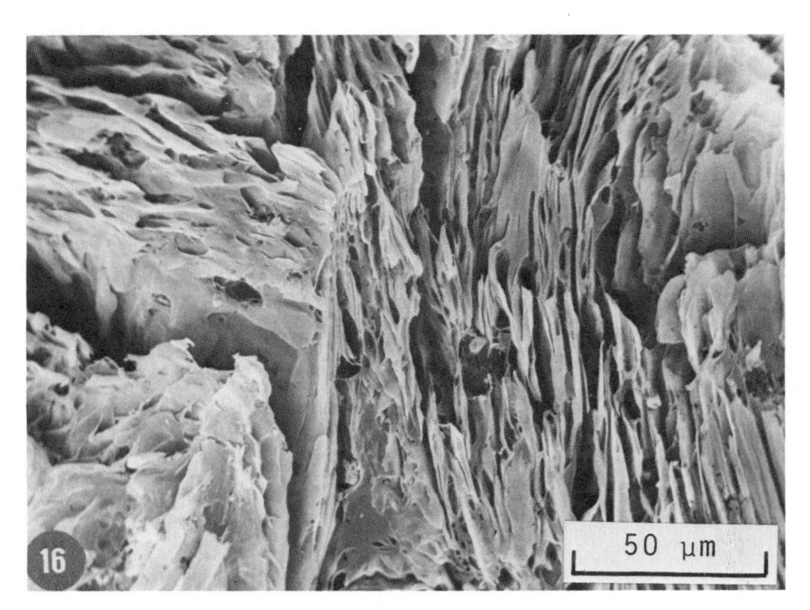

Fig. 16 Stereoscan micrograph of surface of coke, A200 Ashland petroleum pitch, HTT 1223 K, 5 K/hr, 40% bo CO_2, 1123 K (9/221).

gasification fissure can be 10 μm wide and run deeply into the interior. This fissuring must seriously influence the strength of the gasified coke. Gasification of cokes of smaller optical size is quite different. Figures 17 and 18 are scanning electron micrographs of surfaces of a coke prepared from a solvent-refined coal (HTT 1223 K; anthracene oil as solvent) (Kimber and Gray, 1976). The optical texture was about 5 μm in size as revealed by the fracture surface of Fig. 17. Gasification in carbon dioxide to 65% burn-off did not produce fissures but only a flaking of the surface (Marsh and Macefield, 1977).

Since metallurgical cokes are a complex assembly of optical textures, they can exhibit a variety of topographical features created by gasification in carbon dioxide. Figure 19 is a typical scanning electron micrograph of a surface of a blast furnace coke in which porosity 50–500 μm in size is a common feature. After gasification in carbon dioxide at 1193 K to 33% burn-off, extensive pitting of this surface is found as in Fig. 20, as well as pore widening and pitting (Fig. 21), together with fissuring and a high density of pitting as seen in Fig. 22 (Adair *et al.*, 1972).

Gasification of cokes of different optical texture by atomic oxygen at 300 K also produces topographic changes which are quite distinctive. The A200 Ashland petroleum coke (HTT 800 K) reacts with atomic oxy-

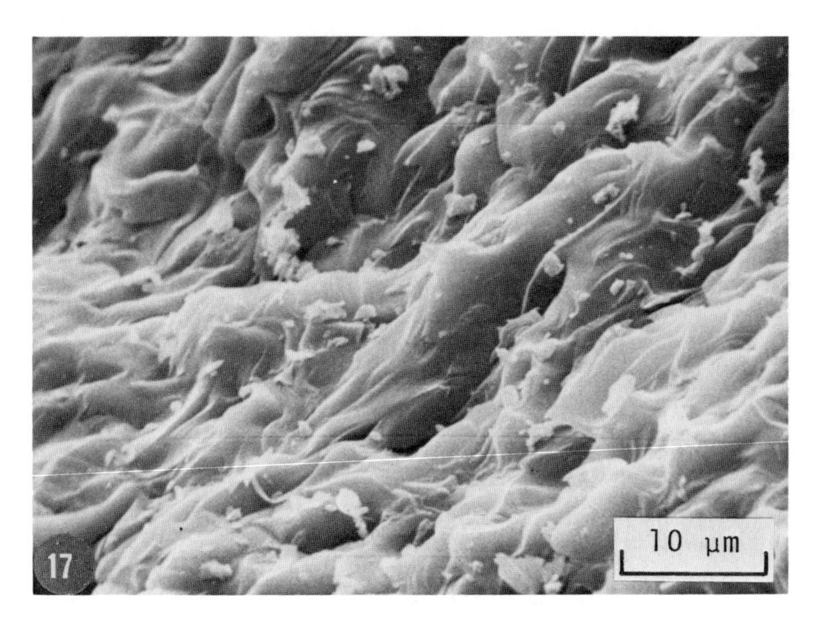

Fig. 17 Stereoscan micrograph of surface of coke, solvent-refined coal, HTT 1223 K, 5 K/hr, 0.5 hr soak (9/183).

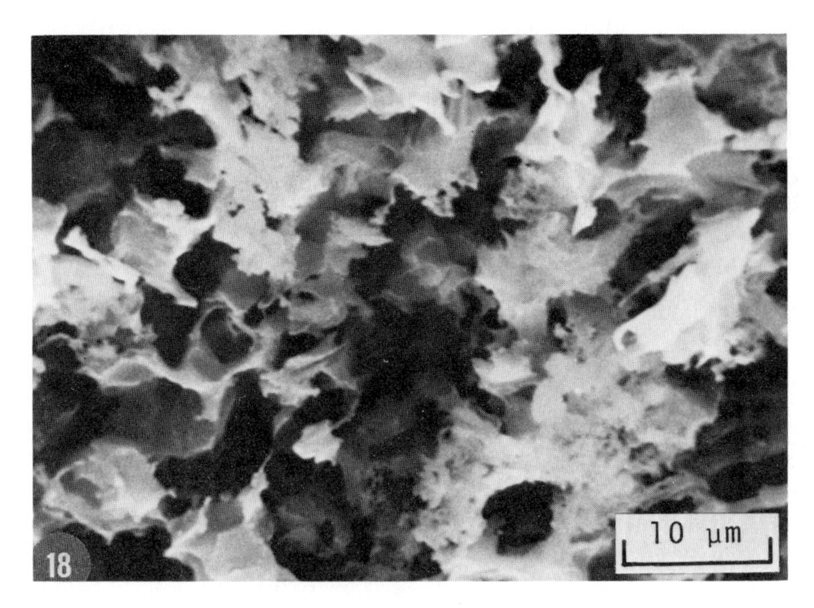

Fig. 18 Stereoscan micrograph of surface of coke, solvent-refined coal, HTT 1223 K, 5 K/hr, 0.5 hr soak, 65% bo CO_2, 1123 K (9/191).

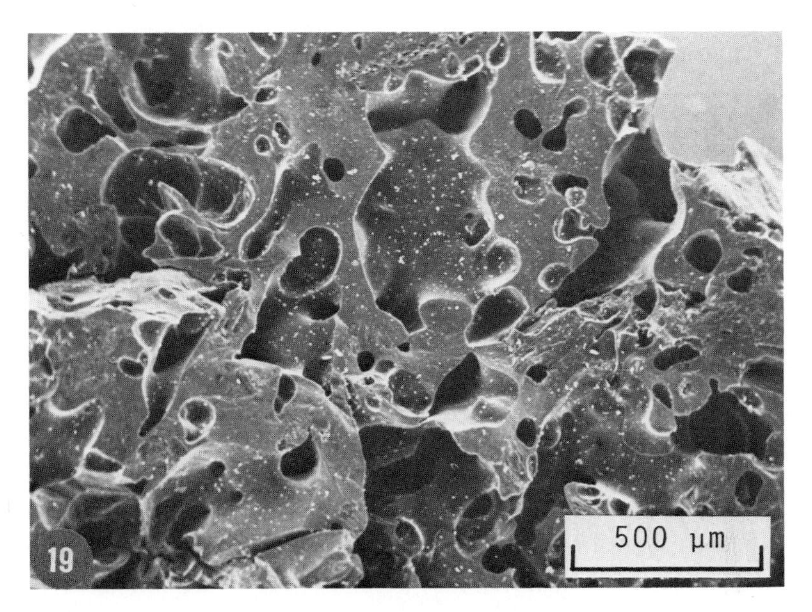

Fig. 19 Stereoscan micrograph of surface of a metallurgical coke (3/339).

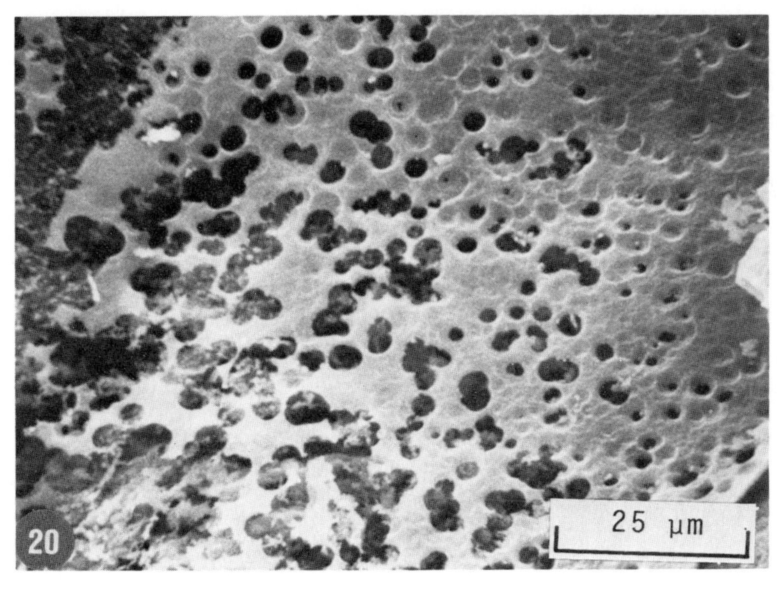

Fig. 20 Stereoscan micrograph of surface of a metallurgical coke, gasified at 1193 K in CO_2 to 33% bo in 22 hr (3/383).

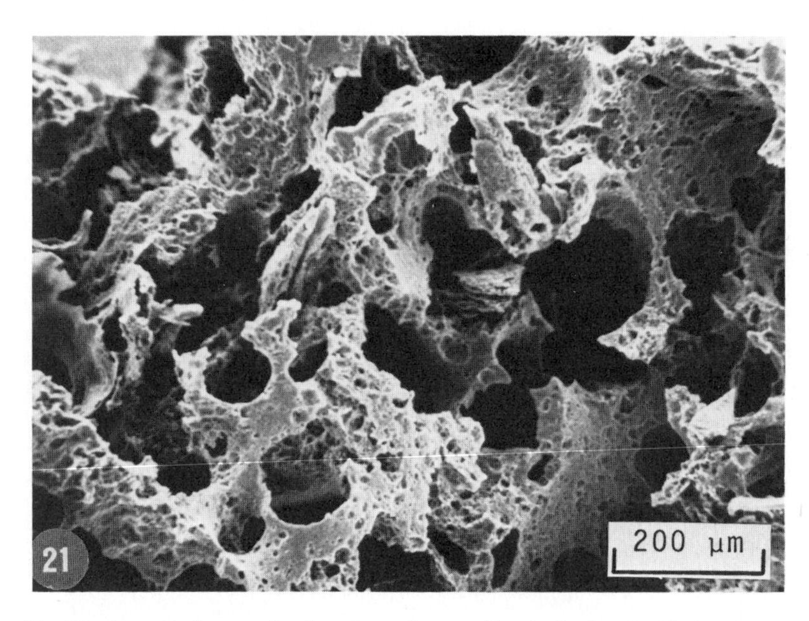

Fig. 21 Stereoscan micrograph of surface of a metallurgical coke, gasified at 1193 K in CO_2 to 71% bo in 77 hr (3/445).

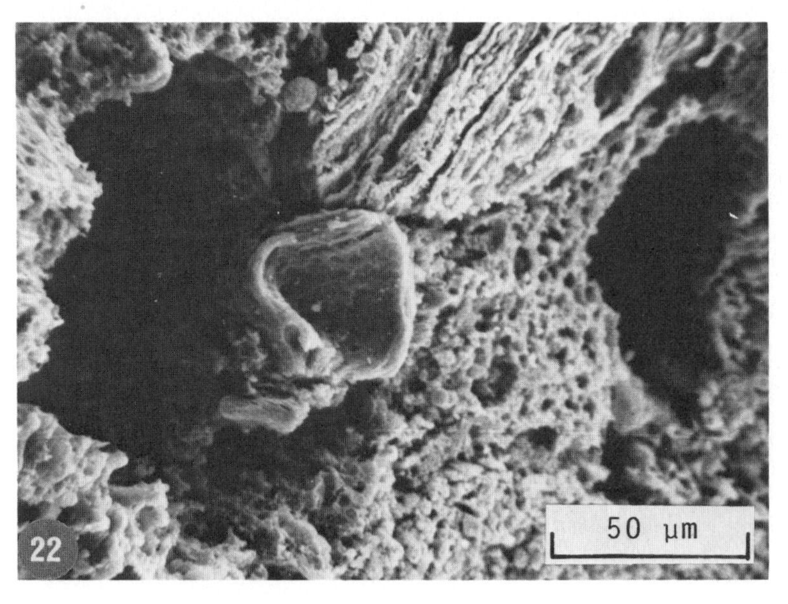

Fig. 22 Stereoscan micrograph of surface of a metallurgical coke, gasified at 1193 K in CO_2 to 33% bo in 29 hr (3/484).

gen to produce a series of gasification channels running parallel and across the optical texture (Figs. 23 and 24) (Marsh *et al.*, 1977a). The mechanism of this is not yet understood. With a coke from Gilsonite pitch (HTT 820 K) the effect is as shown in Fig. 25. It would appear as though each isochromatic anisotropic unit has gasified independently of its neighbors to produce a pitted (etched) surface, the size of the etchings being those of the optical texture. Gasification of metallurgical coke (Sacriston Victoria coal) by atomic oxygen for 3 min at 300 K creates an etch pattern which is closely related to the optical texture of the coke (Fig. 26). The etching of flow-type anisotropy is seen on the left of the micrograph and the pitting of mosaics is seen at the top, right-hand side (Marsh and French, 1977). In a similar study, Patrick *et al.* (1977) used the technique of argon ion etching to enhance the structural features of polished surfaces of cokes from seven handpicked vitrains of different rank. Using scanning electron microscopy to examine the etched surface they confirmed that the units (<2 μm in size) which constitute the mosaic anisotropy were formed from distorted spheres, that the size of the units is directly related to coal rank, and that the flow-type anisotropy results from the alignment and overlapping of the growth units.

Studies into gasification mechanisms of metallurgical cokes are of

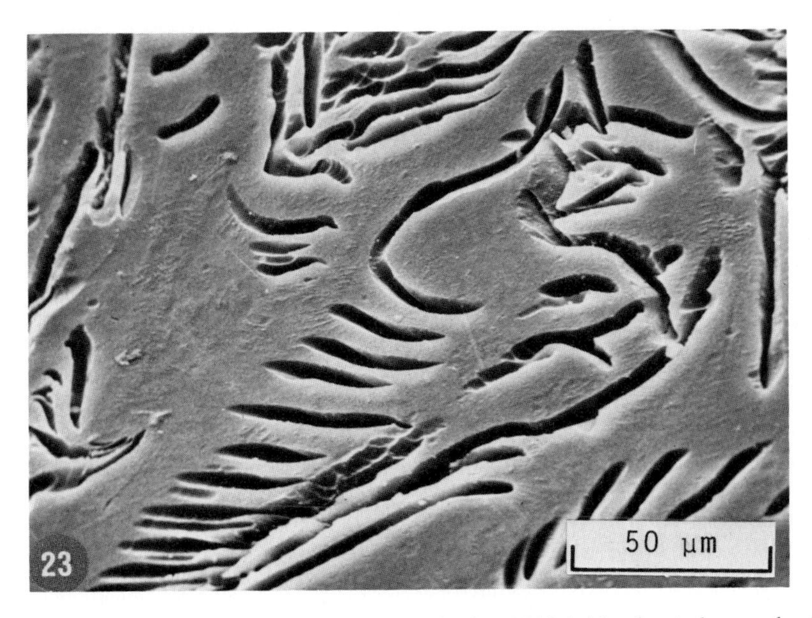

Fig. 23 Stereoscan micrograph of surface of coke, A200 Ashland petroleum coke, HTT 800 K, 5 K/hr, reacted with atomic oxygen for 3 min at 300 K (10/37).

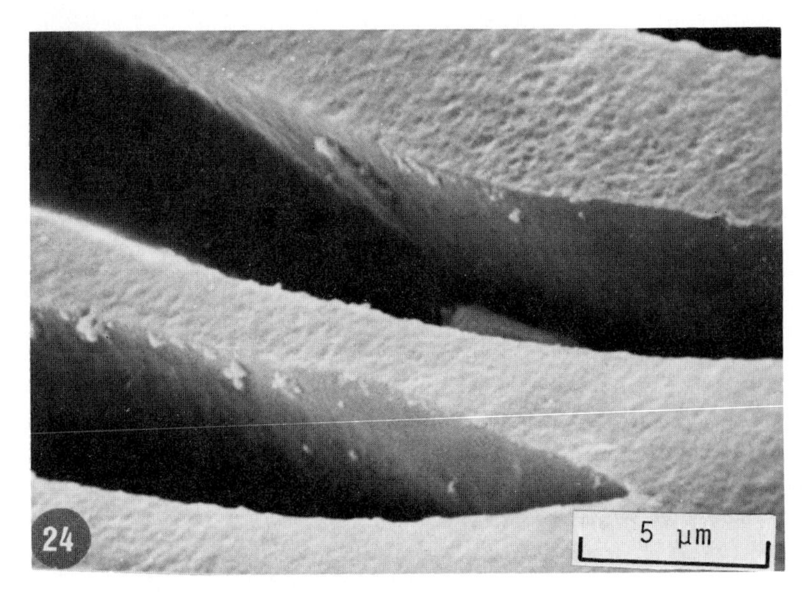

Fig. 24 Stereoscan micrograph of surface of coke, A200 Ashland petroleum coke, HTT 800 K, 5 K/hr, reacted with atomic oxygen for 3 min at 300 K (10/39).

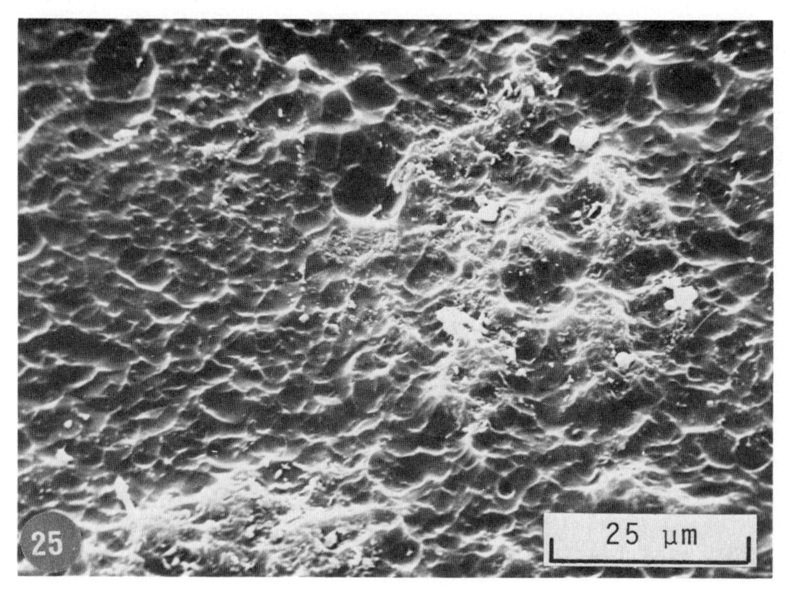

Fig. 25 Stereoscan micrograph of surface of coke, Gilsonite pitch, HTT 820 K, 5 K/min, 0.5 hr soak, reacted with atomic oxygen for 3 min at 300 K (11/458).

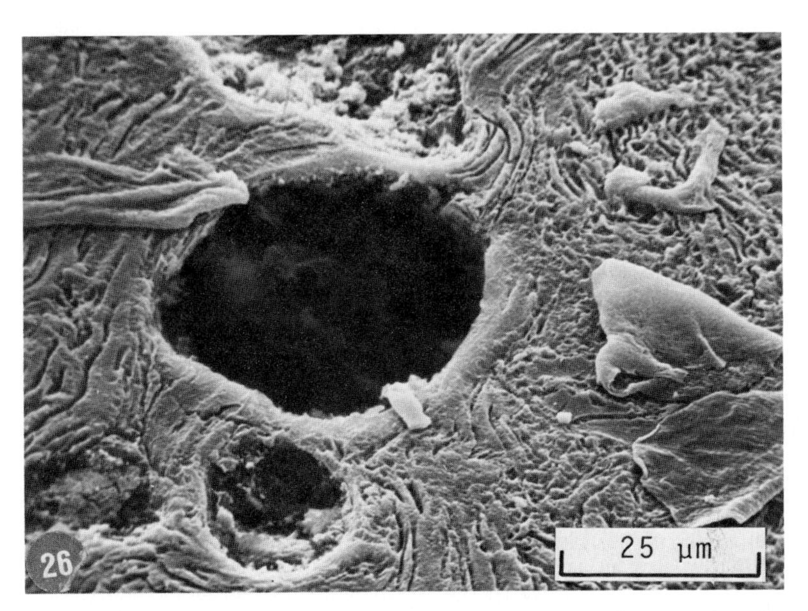

Fig. 26 Stereoscan micrograph of surface of metallurgical coke reacted with atomic oxygen for 3 min at 300 K (10/365).

growing industrial relevance. With the progressive increase in size of the blast furnace and the accompanying increase in pig iron output per cubic meter of capacity, the specification for acceptability of metallurgical cokes becomes more stringent. Good blast furnace operation requires a minimum in coke rate, a low blast pressure, and a minimum thermal load on the walls of the furnace. This can be achieved by using "good" coke which is able to withstand the loading by the burden and the gasification by carbon dioxide and oxygen without fragmentation into a range of smaller sizes. These smaller sizes tend to accumulate around the tuyere zone and depress the permeability of the blast furnace and hence its output of pig iron.

The iron-making industry has always had problems of specifically identifying the desirable properties of its cokes in terms of their structure. Often, cokes which have identical properties measured prior to charging, have quite different operational performances in the blast furnace. This could be due to differences in the detail of the topography of gasification (e.g., extent of fissuring) induced by both uncatalyzed and catalyzed reactions, and to differences in thermal stability of the cokes, gasification, and thermal stability being interdependent. Factors that may influence these two properties are the mode and the strength of attachment of the components of coke substance. The detail of this attachment or bonding has not been studied extensively.

Taylor and Marsh (1976) examined the structure of carbon composites made in the laboratory and the changes induced by gasification in carbon dioxide. Carbons (called first-generation carbons) were prepared initially from acenaphthylene (HTT 850, 1400, and 2330 K), polyfurfuryl alcohol (HTT 850 and 1400 K), and coal tar (HTT 850 and 1400 K). These carbons were then mixed with acenaphthylene and carbonized under hydraulic pressure to 850 and 1200 K. The carbon from this acenaphthylene is called second-generation carbon. Resultant composites were polished and examined by optical microscopy to establish the overall optical texture and, in particular, the orientation and bonding of the carbons at the interface positions (i.e., between the first- and second-generation carbons). The composites were also gasified in carbon dioxide at 850 and 1125 K, followed by similar examination.

It was observed that the first-generation carbons of low HTT (850 K), with a relatively high hydrogen content and high unpaired electron concentration (Jackson and Wynne-Jones, 1964), can chemically bond into a carbon composite system, in contrast to carbons of HTT 1400 K. This chemical bonding between the first- and second-generation carbon is indicated by the observation of layer plane alignment in the latter *perpendicular* to the boundary surface between the two particles. This compares with the more usual observation of alignment *parallel* to an inert surface of mesophase growth and semicoke as with the higher temperature first-generation carbons. On gasification of the well-bonded composites, severe disruptive oxidation of the second-generation carbon occurs, usually at prismatic edges to produce fissures. With composites made with the higher HTT first-generation carbons gasification occurred readily at the boundary between the two components of the composite, thus separating them. At boundaries at which perpendicular alignment of layer planes occurred (using lower HTT first-generation carbons) no gasification took place. This is considered to be due to the fact that the oxidizing gas could not gain access to reactive edge sites in these areas.

Thus, variation in the type of bonding between composites in metallurgical cokes, and arising as a result of differences in origin and prior heat treatment of components, could cause differences in coke strength subsequent to gasification and heat treatment within the blast furnace (see also Edington and Johnston, 1971; Grainger, 1972; Craig, 1973; Hyslop, 1974).

E. Blending of Pitchlike Materials

The industrially important parent materials, such as coals, coal tar pitch, solvent-refined coal (SRC), petroleum pitch, and asphaltenes, are complex mixtures of organic compounds. Despite the complexity of the chemistry of pyrolysis the materials do carbonize to give anisotropic cokes with appreciable differences in their optical textures. Among the petroleum pitches there are those which give large domains and those which give small mosaics in resultant cokes. However, blending means the cocarbonization of materials which, when carbonized singly, would produce different optical textures in their cokes. It is relevant to know the nature of the optical textures of cokes from blends. Evans and Marsh (1971) and Marsh *et al.* (1973b,c, 1974) carried out a survey type of study using model organic compounds to investigate the causes of small and large size of optical texture. (Evans and Marsh incidentally observed that even when starting from a single compound such as anthracene the resultant pitch of HTT 625 K contained about 100 chemical species, and it was from this pitch that growth of mesophase occurred.) The results are summarized in detail (Marsh, 1976), but some of the basic principles may be quoted here. It is possible to reduce the size of domains of a mesophase (e.g., from anthracene) by additions of a nongraphitizing (no mesophase formed during carbonization) compound, e.g., diphenyl- or dibenzothiophene. The anthracene can usually accommodate several percent of the additive without effect, but at about 10% of addition the size of the mesophase growth units begins to diminish until, with further addition, they disappear below the resolution of the optical microscope. Additions of small-sized inert material, e.g., carbon black, reduces the size of the mesophase because the carbon black, adsorbed onto the surfaces of the mesophase units, prevents coalescence and growth. Additions of molecules of relatively high chemical reactivity, e.g., aromatic hydroxyl and carboxylic derivatives, promote cross-linkage between the lamellar stacked units of mesophase, and reduce viscosity and hence growth and coalescence characteristics. If the parent molecule is relatively large (e.g., pyranthrone) compared with anthracene, then effects of diffusivity and growth in molecular size at too early a stage in the carbonization process restricts the development of a low viscosity and the resultant coke has a small optical texture. However, cocarbonization of anthracene, naphthalene, acenaphthylene, and phenanthrene, which, singly, give isochromatic domain via quite different intermediate chemical compounds (Evans and Marsh, 1971), resulted in large isochromatic domains indistinguishable from the parent compounds. Thus, in some systems in which there is homogeneity of type of chemical compound, the growth of mesophase units is not sensitive to the detail of chemical composition. But with severe heterogeneity of type of chemical compound, mesophase growth may be markedly affected.

The possibility of eutectic formation in carbonizing systems has been put forward by Marsh *et al.* (1974). It was observed that some organic compounds when carbonized singly produced an isotropic carbon, but when carbonized together (e.g., carbazole and pyromellitic dianhydride) produced anisotropic mosaics (5 μm in diameter). This is an effect to look for in industrial situations.

The increased cost of petroleum in recent years has increased the value of petroleum residues, i.e., pitches, and created an economic incentive to use these materials profitably (Scott and Conners, 1971). The shortage of metallurgical, medium volatile coals has created the need to use other sources of "fluidity" in carbonization systems. Hence there is a growing interest in the use of petroleum pitch or similar material in coal blends. Again, little is known of the nature of the interaction of pitch substances with coals in terms of formation and properties of mesophase. There are at least four ways of adding petroleum pitch to the coal blend: (1) Direct addition of pitch to the coal being carbonized. Technically this may not be easy because of sticking problems associated with pitch and problems with making a homogeneous blend. (2) Addition of petroleum coke to the coal blend. Hyslop (1974) considers that the effect of addition of petroleum coke to carbonization blends is similar to that of coke (coal) breeze, but has more compatibility with high volatile coals with less critical abrasion sensitivity, although a higher proportion is needed for equivalent improvement of coke size and strength. (3) Addition to the coal blend of briquets of low rank coal bonded with petroleum pitch. This approach, essentially developed by the Japanese, is considered to be very successful (Nakamura *et al.*, 1977). When the coal blend is carbonized with the briquets (up to 30%), the briquets expand with marked evolution of gas during softening and melting. As a result, the surrounding coal blend is compressed, and becomes more coherent and compact, resulting in increase in coke strength. (4) Addition to the coal blend of precarbonized briquets made from noncoking coals and petroleum pitch.

Marsh and French (1977) have some initial findings from studies of pitch–coal interactions. Additions of petroleum pitch, coal tar pitch, and solvent-refined coal have been made to a Sacriston Victoria coking coal (ratio of 1 : 3). Generally, on carbonization, the pitch dissolves the coal substance such that the identity of the original coal particles is lost. Also, the optical texture of the coke from the blend is close in size to that of the pitch additive, despite the fact that it is a minor component. The study of Marsh and French (1977) has been extended to coals of lower rank where similar effects are noted. Further, two pitches which produce cokes of different optical textures have been cocarbonized in various proportions. It is observed that minimum additions (15–25%) of the pitch which gives the larger optical texture are usually sufficient to produce a large optical texture in the coke from the cocarbonized blend. This effect, called the *dominant partner* effect by Marsh *et al.* (1977b), is relevant not only to aspects of carbonization of blends but also to considerations of the carbonization of extracts of coals (using pyridine, chloroform, etc.) and the carbonization together of the extracts (e.g., pyridine-soluble with -insoluble fractions). This suggestion may contribute to an explanation of such studies as made by Mochida *et al.* (1977) of soluble and insoluble fractions.

An observation of some relevance concerns the miscibility of the pitches. It was found that a simple mixing of two viscous, solvent-refined coals was not sufficient to ensure a homogeneous fluid phase despite the fact that between 375 and 475 K the fluid phase was of quite low viscosity. The resultant carbon showed two distinct phases originating from the separate pitches, with shrinkage fissures developing at the boundaries. However, extensive stirring at 375 K produced a more homogeneous optical texture in the resultant coke. This observation can be applied to the use of the delayed coker which collects pitch fractions from different sections of the oil refinery and cocarbonizes the mixture (see, e.g., Rose, 1971). The heterogeneous nature of petroleum coke (Whittaker and Grindstaff, 1969) is often attributed to the selective growth of mesophase from a homogeneous liquid phase; another explanation is that the fluid phase is not homogeneous, the fractions added separately not mixing as would be anticipated in a liquid system (Marsh and Smith, 1977).

F. Results Using Quantitative Reflectance Microscopy

In the Northern Carbon Research Laboratories three experimental projects have been carried out using reflectance microscopy to monitor, quantitatively, the structural changes taking place in the carbons.

1. Growth of Mesophase from Orgreave Lean Coal Tar Pitch

Marsh and Cornford (1975) carbonized Orgreave lean coal tar pitch (93.2% C, 6.4% H) under nitrogen at 5 K/min to between 600 and 1200 K, soak time 15 min. The solid products were mounted in resin and polished. Reflectivity measurements were made with polarized light, 546 nm, using a Leitz Orthoplan reflectance polarizing microscope (see Volume I, Chapter 2, Fig. 13) fitted with an EM1 6094 photomultiplier. Specimen rotation enabled maximum and minimum reflectivities to be obtained from the anisotropic mesophase units. A diamond surface, reflectivity of 17.3%, served as a standard (see Volume I, Chapter 1, Table V).

Mesophase spheres of size <0.5 μm first appeared in the carbonization system at an HTT of 685 K. At 750 K, three phases coexisted, pitch, mesophase spheres, and coalesced mesophase. The anisotropic carbon, HTT 900–1000 K, showed the development of shrinkage fissures. Reflectivities of the pitch (invariant on specimen rotation) and of mesophase (maxima and minima) are given in Fig. 27. For the anisotropic mesophase, mean reflectivities were calculated from maximum and minimum values.

Figure 27 shows that there is no marked change in mean reflectivities between the pitch and mesophase (C and D, Fig. 27) at the temperature when pitch is converted to mesophase. The observed small rise from C to D could be due to differences in densities between the pitch and mesophase, or to small differences in polishing characteristics. The development of conjugated bonding within the pyrolysate is indicated

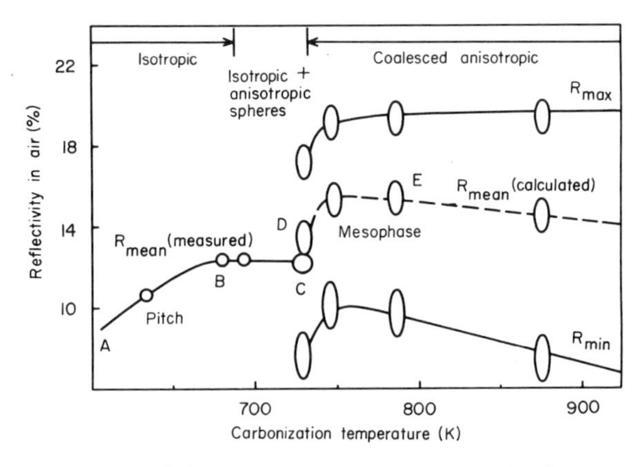

Fig. 27 The variation with heat treatment temperature of reflectivity in air of cokes from Orgreave lean coal tar pitch. Lettered parts are identified in text.

by the rise in mean reflectivity (A to B) of the pitch. The development of mesophase (B to C) does not appear to be associated with further development of conjugated sp_2 bonding. The flat portion of the curve, coincident with the origins of the mesophase, is associated with sp_3 rather than an sp_2 type of bonding of molecules. Once the mesophase is observable, then increasing carbonization temperature produces increases in mean reflectivities (D to E), as well as R_{max} and R_{min} associated with increased conjugated bonding and improved stacking of lamellae. Thus, Fig. 27 indicates that conjugated bonding occurs both prior and subsequent to mesophase formation and is not a characteristic of its formation. This is in agreement with the sequence of other stereospecific aromatic pyrolysis reactions; e.g., naphthalene initially dimerizes with single bond formation, to give dinaphthyl, which subsequently gives the five-ringed molecule of perylene (Evans and Marsh, 1971).

Reflectivity is a function of the refractive index and electronic absorption of the reflecting substance as given by Beer's equation:

$$R = \frac{(n - n_i)^2 + n^2K^2}{(n + n_i)^2 + n^2K^2} \tag{1}$$

where R is reflectivity, n the refractive index of carbonaceous material, n_i the refractive index of measuring medium, air or oil, and K the absorptive index of carbonaceous material.

It is generally true that an increase in size of a conjugated aromatic molecule shifts the electronic absorption maxima to longer wavelengths as measured by ultraviolet spectroscopy; the corresponding refractive index resonance feature is similarly shifted. An increase in molecule size by sp_3 bonding of carbon atoms has little or no effect on the positions of the absorption maxima. Thus, an increase in reflectivity, at a given wavelength, may be correlated with an increase in the average size of the aromatic molecules. At ultraviolet and visible wavelengths, the absorption by aromatic hydrocarbons (Yogev *et al.*, 1974) and graphite single layers (Ergun, 1968) is highly anisotropic, being of maximum value for light polarized in the basal plane, and of minimum value for light polarized perpendicular to the basal plane. The refractive index shows similar directional variations. The randomly oriented molecules in the pitch will show a mean reflectivity intermediate between the average molecular maximum and minimum.

2. *Reflectivities of Cokes and Carbons of Different Origins*

The purpose of this study is to examine the variation of reflectivity with heat treatment temperature of carbons prepared from different parent materials, the resultant carbons having different optical textures. Such a study could provide guidelines in discussions of whether or not there are significant differences in the size and orientation (stacking perfection) of the lamellar constituent molecules in anisotropic isochromatic areas of optical texture and varying in size from the domains and flow-type anisotropy to the small mosaics.

This study is an extension, using single organic compounds and pitch substances, of that reported by Goodarzi and Murchison (1973b) who measured the reflectivities of cokes from coals of different rank (82.5% C, 88.0% C, and an anthracite). Their data are reproduced in Fig. 28, for comparative purposes.

Marsh and Cornford (1977) carbonized acridine, 1,10-phenanthroline, and 2-(γ-hydroxyphenyl)benzothiazole (Augustyn *et al.*, 1976) under hydraulic pressure (Marsh *et al.*, 1971a) to 850 K at 5 K/min, 0.5 hr soak, 220 MPa pressure. The resultant carbons were mounted in resin, polished, and examined for optical texture and reflectivity. The optical textures were acridine carbon 60 μm, 1,10-phenanthroline carbon 15 μm, and benzothiazole carbon 1–2 μm. The maximum air reflectivities which could be obtained are plotted in Fig. 28. The benzothiazole-derived carbon has a reflectivity of 12.4% and is coincident with the reflectivities of cokes from the bituminous coals as studied by Goodarzi and Murchison (1973b). The acridine- and phenanthroline-derived carbons, with reflectivities of 15.5%, lie above the curve for cokes from

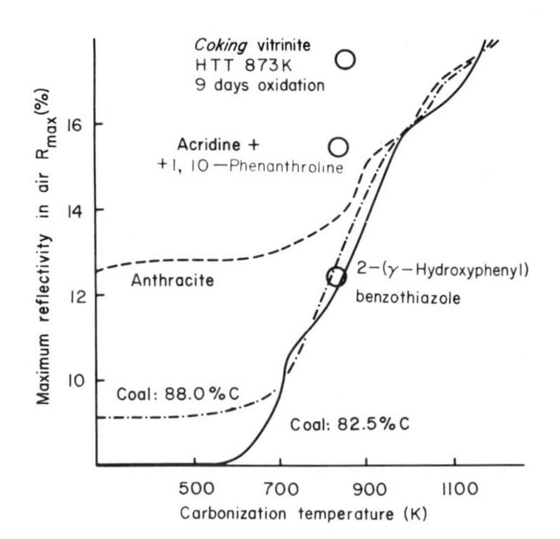

Fig. 28 The variation with heat treatment temperature of maximum reflectivity in air of carbons prepared from single organic compounds and three coals. (Coals after Goodarzi and Murchison, 1973b.)

anthracite. It must be mentioned that it is extremely difficult to know if one is measuring exactly a single mosaic or anisotropic isochromatic area when the mosaic sizes are about the same size as the aperture of the optical microscope, i.e., about 1–2 μm. There is always the doubt that the lower value of 12.4% could be attributed to the light spot (beam) overlapping more than one mosaic area, with one area not contributing to the maximum reflectivity. However, there appears to be no doubt that the acridine- and phenantholine-derived carbons have higher reflectivities, and possibly aromaticity, than the coke from the anthracite.

This initial study was extended to include other parent materials over a longer range of heat treatment temperatures. Thus carbons were prepared at 5 K/min, 0.5 hr soak, from polyvinyl chloride, HTT $\leq$3000 K; A200 Ashland petroleum pitch, HTT $\leq$2400 K; and Gilsonite pitch, HTT $\leq$2400 K. Three solvent-refined coals (SRC) were obtained from the National Coal Board (U.K.): SRC-A and SRC-B, which were carbonized in open boats, and SRC-C, which was carbonized under hydraulic pressure.

The carbons from polyvinyl chloride and A200 petroleum pitch had domains and large flow-type anisotropy in their optical texture (50–200 μm in size; see Fig. 7). The Gilsonite pitch and SRC-derived cokes had much smaller optical textures (5–15 μm in size; see Fig. 8).

The reflectivities in air of these carbons are plotted in Fig. 29. At the lowest HTT values, the reflectivities are somewhat higher than the reflectivities of the anthracite-derived coke (Fig. 28). The acridine- and phenanthroline-derived carbons have reflectivities which lie on this

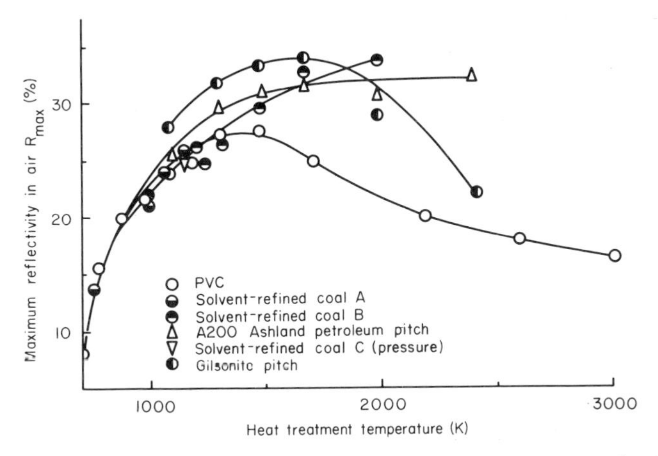

Fig. 29 The variation with heat treatment temperature of maximum reflectivity in air of carbons prepared from several pitch substances.

curve of Fig. 29. It would appear that up to HTT values of ~1100 K the reflectivities of the graphitizing carbons are indistinguishable. The reflectivities of carbons from coals and the benzothiazole could be lower, up to 1100 K. This would imply that the lamellar constituent molecules of the graphitizing carbons may also be indistinguishable by this method of measurement of reflectivity; i.e., in the temperature range of 800–1100 K the differences in chemical composition of parent materials cease to be important and are not apparent in the structure of the constituent molecules. However, beyond HTT values of 1100 K the carbons behave quite differently.

Figure 29 shows that the reflectivity values of cokes from A200 Ashland petroleum pitch reach a maximum of about 31% at 1600 K and are then almost invariant with increasing HTT. The reflectivity values of cokes from Gilsonite pitch rise to a maximum of 33% at an HTT of about 1600 K but subsequently decrease rapidly with increasing HTT. The reflectivity values of cokes from SRC-B continue to rise, reaching a maximum value of 33.5% at an HTT of 2000 K. The reflectivity values of cokes from polyvinyl chloride maximize at 27% (HTT 1400 K) and then decrease with progressive increase in HTT (Khurasani, 1974) to about 17% at an HTT of 3000 K.

Thus, at this stage of the investigation, the reflectivity values of cokes are changing in a way that does not appear to be related to the optical texture of the cokes. Forrest *et al.* (1977) discuss some possible reasons for this behavior. It could be due to the development of microfissures whereby the amount of carbon under the beam of light of the reflectance microscope diminishes as the microcracks widen. A second explanation is that, at a certain stage of the graphitization process, the lamellar constituent molecules become strained and deformed by bond-shortening mechanisms. This zigzag structure of molecules has been recognized in phase contrast fringe-imaging transmission electron microscopy carried out in the Northern Carbon Research Laboratories. A third explanation is that an absorption edge is moving into the cokes of increasing HTT and reflectivities will be modified by significant changes in electronic absorption processes within the lattice of the graphite. A fourth explanation is essentially an experimental artifact, namely that some form of deformation of surface occurs during the polishing of the carbons from polyvinyl chloride (PVC) and Gilsonite. Maximum reflectivities require basal plane alignment either parallel or perpendicular to the light beam. Polishing may misalign the layers so that true maximum reflectivities are not obtained.

Overall, the result of this study would imply that in the HTT range up to ~1100 K the maximum values of reflectivity (%) of polarized light from polished surfaces of carbon show some sensitivity to the chemical origins of the carbon and possibly its optical texture. Perhaps the reflectivity of anisotropic carbon in carbons (cokes) derived from coals is lower in this temperature range than the reflectivity of the graphitizing carbons, e.g., PVC carbons. The lower reflectivity may be associated with smaller mosaic size (as with the benzothiazole-derived carbon) which, in turn, may be associated with smaller, less planar, less aromatic constituent molecules of the carbons. In the HTT range 1100–1400 K the reflectivity is essentially insensitive to the size of the optical texture, i.e., the origins of the carbons. Above 1400 K reflectivities diverge in a way not yet understood.

A very promising experimental approach to examine directly the structure of cokes and other carbons derived from coal substances is the use of phase contrast, high resolution electron microscopy (see Ban *et al.*, 1975; Crawford and Marsh, 1977). This approach facilitates the direct imaging of fringes derived from the defective graphitelike lattice of the carbons and provides structural information in an elaboration of detail which cannot be obtained from any diffraction technique. A logical extension of studies is the correlation of reflectivity measurement with such structural information.

Although the detail of molecular structure at the temperature of formation of mesophase may reflect the different origins of the mesophase, the subsequent aromatization of the lamellar molecules, the removal of hydrogen and heteroatoms, and the growth in size of the layers all contribute to create similar molecular constituents of carbons. Defects or vacancies in the layers, which may ultimately affect the detail of the graphitizability of the carbons, appear not to be important in reflectivity considerations.

3. *Reflectivities of Chars from Preoxidized Vitrinites*

Vitrinites may possess an inherent anisotropy associated with the geological conditions of formation in the coal measures. The vitrinites of the medium and low volatile coals lose this inherent anisotropy in the fluid phase of carbonization prior to the onset of formation of anisotropic semicoke via the mechanism of growth of mesophase. However, if the vitrinites are oxidized prior to carbonization, then although the ability to fuse is lost, and hence the ability to form mesophase and anisotropic semicoke by that route is also lost, the effect of oxidation is to cross-link the molecules of the vitrinite and to create essentially a polymeric material with a structure very closely related to that of the inherent anisotropic structure of the vitrinite.

Goodarzi and Murchison (1972, 1973a,b, 1976) measured the reflectivities of chars prepared from vitrinites with and without preoxidation and noted a significant increase in the reflectivities of chars from preoxidized vitrinites. The aromaticity of the chars from preoxidized vitrinites was higher than that of chars from fresh vitrinite. Goodarzi and Marsh (1977) extended this study by carbonizing the fresh and preoxidized vitrinites under hydraulic pressures of about 200 MPa in sealed gold tubes (Marsh *et al.*, 1971). Under these conditions the volatile matter remains in the tube and contributes to the development of molecular structure in the char during the carbonization of the vitrinite.

A handpicked vitrinite of a Northumbrian coking coal (87.9% C), ground to a particle size of less than 210 μm, was oxidized at 378 K in air for periods of 1–20 days, and subsequently carbonized to 873 $\pm$ 5 K, 5 K/min, soak period of 0.5 hr, at 1 atm pressure under nitrogen and in the sealed gold tubes under hydraulic pressure. Reflectivities at 546 nm in oil were measured on relief-free polished surfaces of blocks of char from each level of preoxidation in the vitrinites (Goodarzi and Murchison, 1973b). Maximum and minimum reflectivities were measured. Figure 30 shows the variation of reflectivity in oil with extent of preoxidation of the vitrinite for chars (HTT 868 K). Figure 31 shows the variation of bireflectance ($R_{max} - R_{min}$, %) with extent of preoxidation of the vitrinites for chars (HTT 868 K).

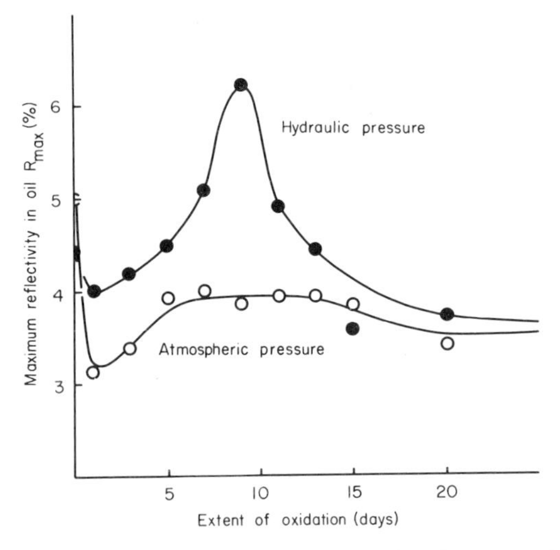

Fig. 30. The variation of maximum reflectivity in oil with extent of oxidation at 378 K of Northumbrian coking vitrinite carbonized at 1 atm and under hydraulic pressure to 868 K at 5 K/min.

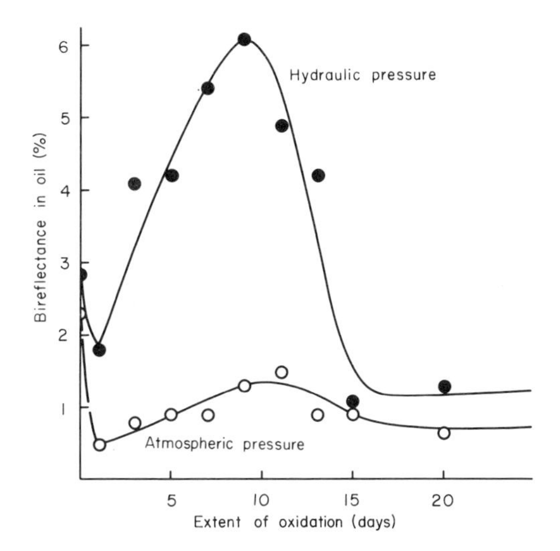

Fig. 31 The variation of bireflectance in oil with extent of oxidation at 373 K of Northumbrian coking vitrinite carbonized at 1 atm and under hydraulic pressure to 868 K at 5 K/min.

The variation of optical properties of the chars shows three different stages of oxidation identifiable with those describing the optical texture and morphology of samples from the vitrinites reported in an earlier paper (Goodarzi *et al.*, 1975). In the *first* stage (1 day) the optical properties of chars from the vitrinite prepared at atmospheric and hydraulic pressure show a distinct decrease during this stage of minimum oxidation. In the *second* stage (1–9 days) the chars prepared at atmospheric pressure show an initial increase in their reflectivities with extent of oxidation, subsequently to remain constant (Fig. 30). The chars prepared under hydraulic pressure have reflectivities which continuously increase during this stage (Fig. 30). The bireflectance values for the two series of chars increase during this stage (Fig. 31), but with a marked enhancement for the chars from the hydraulic pressure apparatus. In the *third* stage (9–20 days) the chars, prepared at 1 atm pressure, initially maintain the reflectivities of the second stage but then slowly decrease with extent of preoxidation. The oil reflectivities of chars carbonized under hydraulic pressure decrease sharply initially, and then remain constant with further extent of oxidation (Fig. 30). The bireflectance values of chars (1 atm) decrease initially and then remain constant. The bireflectance values of chars prepared under hydraulic pressure decrease sharply and become constant (Fig. 31).

The effect of preoxidation initially is to restrict progressively the formation of anisotropic carbon under both conditions of carbonization, via the formation of nematic liquid crystal and mesophase. As the size of the optical texture of the chars diminishes, so does the reflectivity. Eventually, after 3–4 days, this ability to form anisotropic carbon from a fluid phase of carbonization is lost, and is replaced by a mechanism of enhancement of the inherent anisotropy of the fresh vitrinite. A combination of preoxidation and hydraulic pressure during carbonization not only maintains this inherent anisotropy but enhances it very considerably such that both the reflectivities and bireflectances of the chars so prepared are very much higher than those of chars prepared at 1 atm pressure, where the loss of volatile matter must have disturbed the molecular stacking, which creates the anisotropy. The maximum reflectivity in oil (6.2%; in air 17%) of the char (HTT 873 K) prepared under hydraulic pressure is close to reflectivity values of carbons prepared from other sources (Fig. 29), indicating that the aromaticity of the chars from the vitrinite and possibly graphitizability are comparable to that of petroleum pitch cokes and cokes from solvent-refined coal. Thus anisotropic carbons, prepared via two mechanisms, enhancement of inherent anisotropy in the solid phase and creation of semicoke via the fluid phase and the mesophase, are comparable in optical properties.

ACKNOWLEDGMENTS

We acknowledge, with appreciation, financial support from the European Coal and Steel Community (Grant No. 7220-EB/8/807) and the National Coal Board which enables these studies of the Northern Carbon Research Laboratories to continue. Also we are grateful for the continuous encouragement of our Director and Head of Department, Professor D. H. Whiffen, F.R.S., and the support of Patricia Barbour, Maggie French, and Marion Poad.

REFERENCES

Abramski, C., and Mackowsky, M.-T. (1952). *In* "Handbuch der Mikroskopie in der Technik" (H. Freund, ed.), Vol. II, pp. 311–410. Umschau-Verlag, Frankfurt.

Adair, R. R., Boult, E. H., and Marsh, H. (1972). *Fuel* **51,** 57.

Alpern, B. (1956). *Brennst.-Chem.* **37,** 194.

Augustyn, D., Hermon, G., and Marsh, H. (1976). *Proc. Conf. Ind. Carbon and Graphite, 4th, Soc. Chem. Ind., London* p. 61.

Ban, L. L., Crawford, D., and Marsh, H. (1975). *J. Appl. Crystallogr.* **8,** 415.

Boyde, A. (1967). *J. R. Microsc. Soc.* **86,** 359.

Boyde, A., and Ross, H. F. (1975). *Photogramm. Rec.* **46,** 408.

Brooks, J. D., and Taylor, G. H. (1968). "Chemistry and Physics of Carbon" (P. L. Walker, Jr., ed.), Vol. 4, p. 243. Arnold, London.

Brown, G. H., and Doane, J. W. (1974). *Appl. Phys.* **4,** 1.

Brown, G. H., Doane, J. W., and Neff, V. D. (1971). "Review of Structure and Properties of Liquid Crystals." Butterworth, London.

Craig, J. (1973). *Year-Book Coke Oven Managers' Assoc.* pp. 114–125.
Crawford, D., and Marsh, H. (1977). *J. Microsc.* **109**(1), 145.
de Gennes, P. G. (1974). "The Physics of Liquid Crystals." Oxford Univ. Press (Clarendon), London and New York.
Dubois, J., Agace, C., and White, J. L. (1970). *Metallography* **3,** 337.
Easton, A., and Jenkins, G. M. (1976). *Proc. Ind. Conf. Carbon and Graphite, 4th, Soc. Chem. Ind., London* p. 304.
Edington, M. D., and Johnston, D. J. (1971). *Year-Book Coke Oven Managers' Assoc.* pp. 266–277.
Ergun, S. (1968). "Chemistry and Physics of Carbon," Vol. 3, p. 45. Dekker, New York.
Evans, S., and Marsh, H. (1971). *Carbon* **9,** 733, 747.
Forrest, R. A., French, M., Marsh, H., Griffiths, J. A., and White, J. L. (1977). *Abstr., Conf. Carbon, 13th, Am. Carbon Soc., Irvine, Calif.*
Friedel, G. (1922). *Ann. Phys. (Paris)* **18,** 273.
Gibson, J. (1972). *Year-Book Coke Oven Managers' Assoc.* pp. 182–205.
Goodarzi, F., and Marsh, H. (1977). Unpublished results.
Goodarzi, F., and Murchison, D. G. (1972). *Fuel* **51,** 322.
Goodarzi, F., and Murchison, D. G. (1973a). *Fuel* **52,** 90.
Goodarzi, F., and Murchison, D. G. (1973b). *Fuel* **52,** 164.
Goodarzi, F., and Murchison, D. G. (1976). *Fuel* **55,** 141.
Goodarzi, F., Hermon, G., Iley, M., and Marsh, H. (1975). *Fuel* **54,** 105.
Grainger, L. (1972). *Year-Book Coke Oven Managers' Assoc.* pp. 126–149.
Gray, G. W. (1962). "Molecular Structure and the Properties of Liquid Crystals." Academic Press, New York.
Gray, G. W., and Winsor, P. A. (1974). "Liquid Crystals and Plastic Crystals," Vols. 1 and 2. Ellis Horword, Chichester, England.
Hook, W. (1973). *Year-Book Coke Oven Managers' Assoc.* pp. 143–160.
Hyslop, W. (1974). *Year-Book Coke Oven Managers' Assoc.* pp. 285–298.
Jackson, C., and Wynne-Jones, W. F. K. (1964). *Carbon* **2,** 227.
Khurasani, G. K. (1974). Unpublished results.
Kimber, G. M., and Gray, M. D. (1976). "Petroleum Derived Carbons," ACS Symposium Series, No. 21, pp. 444–450. Am. Chem. Soc., Washington, D.C.
Kipling, J. J., and Shooter, P. V. (1966a). *Carbon* **4,** 1.
Kipling, J. J., and Shooter, P. V. (1966b). *Proc. Conf. Ind. Carbon and Graphite, 2nd, Soc. Chem. Ind. London* p. 15.
Kipling, J. J., Sherwood, J. N., Shooter, P. V., and Thompson, N. R. (1964). *Carbon* **1,** 315.
Lewis, R. T. (1975). *Abstr., Conf. Carbon, Am. Carbon Soc., 12th, Pittsburgh, Pa.* p. 215.
Mackowsky, M.-T. (1949). *Brennst.-Chem.* **30,** 44.
Marsh, H. (1973). *Fuel* **52,** 205.
Marsh, H. (1976). *Proc. Conf. Ind. Carbon and Graphite, 4th, Soc. Chem. Ind., London* p. 2.
Marsh, H., and Cornford, C. (1975). *Abstr., Conf. Carbon, 12th, Am. Carbon Soc., Pittsburgh, Pa.* p. 235.
Marsh, H., and Cornford, C. (1976). "Petroleum Derived Carbons," ACS Symposium Series, No. 21, pp. 266–281. Am. Chem. Soc., Washington, D.C.
Marsh, H., and Cornford, C. (1977). Unpublished results.
Marsh, H., and French, M. (1977). Unpublished results.
Marsh, H., and Macefield, I. (1977). Unpublished results.
Marsh, H., and Smith, J. (1977). Unpublished results.
Marsh, H., Dachille, F., Melvin, J., and Walker, P. L. (1971a). *Carbon* **9,** 159.
Marsh, H., Akitt, J. W., Hurley, J. M., Melvin, J., and Warburton, A. P. (1971b). *J. Appl. Chem.* **21,** 251.

Marsh, H., Dachille, F., Iley, M., Walker, P. L., and Whang, P. W. (1973a). *Fuel* **52,** 253.
Marsh, H., Foster, J. M., Hermon, G., and Iley, M. (1973b). *Fuel* **52,** 234.
Marsh, H., Foster, J. M., Hermon, G., Iley, M., and Melvin, J. N. (1973c). *Fuel* **52,** 243.
Marsh, H., Cornford, C., and Hermon, G. (1974). *Fuel* **53,** 168.
Marsh, H., French, M., Smith, J., and White, J. L. (1977a). *Abstr., Conf. Carbon, 13th, Am. Carbon Soc., Irvine, Calif.* p. 134.
Marsh, H., Macefield, I., and Smith, J. (1977b). *Abstr., Conf. Carbon, 13th, Am. Carbon Soc., Irvine, Calif.* p. 304.
Marshall, C. E. (1945). *Fuel* **24,** 120.
Mochida, I., Amamoto, K., Maeda, K., and Takeshita, K. (1977). *Fuel* **56,** 49.
Murchison, D. G. (1958). *Brennst.-Chem.* **39,** S47–S50.
Nakamura, N., Togina, Y., and Adachi, T. (1977). "Coal, Coke and the Blast Furnace." Conference, Met. Soc., London.
Patrick, J. W. (1976). *Year-Book Coke Oven Managers' Assoc.* pp. 201–218.
Patrick, J. W., and Stacey, A. E. (1972a). *Fuel* **51,** 81.
Patrick, J. W., and Stacey, A. E. (1972b). *Fuel* **51,** 206.
Patrick, J. W., Stacey, A. E., and Wilkinson, H. C. (1972). *Fuel* **51,** 174.
Patrick, J. W., Reynolds, M. J., and Shaw, F. H. (1973). *Fuel* **52,** 198.
Patrick, J. W., Shaw, F. H., and Willmers, R. R. (1977). *Fuel* **56,** 81.
Priestley, E. B., ed. (1975). "Introduction to Liquid Crystals." Plenum, New York.
Ramdohr, P. (1928). *Eisenhuttenwesen* **1,** 669.
Reinitzer, F. (1888). *Monatsh. Chem.* **9,** 421.
Rester, D. O., and Rowe, C. R. (1974). *Carbon* **12,** 218.
Rose, K. E. (1971). *Hydrocarbon Process.* July, p. L85.
Sanada, Y., Furuta, T., Kimura, H., and Honda, H. (1972). *Carbon* **10,** 644.
Sanada, Y., Furuta, T., Kimura, H., and Honda, H. (1973). *Fuel* **52,** 143.
Scott, C. B., and Conners, J. W. (1971). *J. Met.* July, p. 19.
Taylor, D. W., and Marsh, H. (1976). *Proc. Conf. Ind. Carbon and Graphite, 4th, Soc. Chem. Ind., London* p. 360.
Taylor, G. H. (1961). *Fuel* **40,** 465.
Thomas, J. M. (1965). "Chemistry and Physics of Carbon," Vol. 1, p. 122. Dekker, New York.
Thornton, P. R. (1967). "Scanning Electron Microscopy." Chapman & Hall, London.
Underwood, E. E. (1970). "Quantitative Stereology." Addison-Wesley, Reading, Massachusetts.
White, J. L. (1976). "Petroleum Derived Carbons," ACS Symposium Series, No. 21, pp. 282–314. Am. Chem. Soc., Washington, D.C.
White, J. L., and Price, R. J. (1974). *Carbon* **12,** 321.
White, J. L., and Zimmer, J. E. (1976). "Surface and Defect Properties of Solids," Vol. 5, pp. 16–35. Chem. Soc., London.
White, J. L., Guthrie, G. L., and Gardner, J. O. (1967). *Carbon* **5,** 517.
Whittaker, M. P., and Grindstaff, L. I. (1969). *Carbon* **7,** 615.
Yamada, Y., Imamura, T., Kakiyama, H., Honda, H., Oi, S., and Fukuda, K. (1974). *Carbon* **12,** 307.
Yogev, A., Margulies, L., Strasberger, B., and Mazur, Y. (1974). *J. Phys. Chem.* **78,** 1400.

Chapter 31

Optical Properties of Carbonized Vitrinites

Duncan G. Murchison
ORGANIC GEOCHEMISTRY UNIT, DEPARTMENT OF GEOLOGY
UNIVERSITY OF NEWCASTLE UPON TYNE
NEWCASTLE UPON TYNE, ENGLAND

I. INTRODUCTION

Modern coke microscopy spans more than 50 years and stems from papers by Beilby (1922) on the structure, origin, and development of coke and by Ramdohr (1928) on the microscopy of cokes and graphites. The principal difficulty in these early studies of cokes was that no truly satisfactory microscopical method had been developed for their investigation. With the high optical absorption of cokes, reflected light micros-

ISBN 0-12-399902-2

copy was a virtual necessity, but the results of such examination were so low in quality, that although Marshall (1936) used polished sections to study thermally metamorphosed coals, investigators preferred to face the challenge of preparing ultrathin sections (<0.4 μm) and then employing transmitted light microscopy (Ueji, 1932; Marshall, 1945).

The successful construction of oil immersion objectives for use in reflected light by the German firm Ernst Leitz (Wetzlar) some 40 years ago, and the continuing development of these optical systems since then, has revolutionized the microscopical study of highly absorbing substances. As a result, a substantial literature relating to natural, industrial, and laboratory-type cokes, which has a breadth far beyond the scope of this chapter, has been published since World War II. Much of the work has been qualitative and essentially descriptive. The need, however, to understand the factors underlying coke formation and, in the industrial environment, the way in which coke quality might be improved, has led to an increasing number of quantitative microscopical studies that attempt to define coke properties more precisely.

In the Organic Geochemistry Unit at Newcastle, there is interest in cokes in natural and industrial environments. Recent experimental work in this Unit has been aimed at establishing the relative importance of a number of influences that modify the behavior of the optical properties of coal macerals during carbonization and so improve present interpretations of the thermal histories of cokes and the constituents from which they are derived. The results of this work, although still incomplete, and their comparison with similar data from other investigations form the basis for this chapter. First, however, the carbonization process is described in sufficient detail to allow later understanding between the molecular structural changes and the principal optical phenomena that affect reactive coal macerals during carbonization. Since the quantitative microscopical methods employed to study cokes are similar to those applied to other forms of crustal organic matter, these are only considered briefly. More detailed consideration is given to the derivation and function of the optical parameters used to define the optical characters of coals and cokes and particularly to the precision and accuracy of these parameters when determined for cokes. The remainder of the chapter discusses the results of optical studies of cokes in different environments and finally the relationship of the observed optical variations to changes in molecular structure.

II. THE CARBONIZATION PROCESS

Carbonization is a complex process in which polymerization plays a dominant role. The process is virtually complete at a temperature of 1500°C when the product will approximate to elementary carbon. To reach this condition the original organic matter undergoes loss of its noncarbon elements as the temperature rises. These elements are evolved as gases, usually in combination with part of the carbon, and this process is associated with concomitant development of the residual carbon atoms into condensed aromatic systems.

Three overlapping phases (occasionally four) are usually identified in the carbonization process. The boundaries between the phases are defined by temperature levels whose absolute values vary with the character of the material undergoing carbonization. Unless the original organic matter is already low or high in carbon, all three phases will appear.

In the first phase of "precarbonization" (between ~150 and ~400°C), although some important molecular rearrangements, early condensation reactions, and "molecular stripping" take place, there is little change to the optical properties of the majority of organic materials. The significant optical changes occur in the two later but highly contrasting phases.

The second intermediate phase of "active decomposition" (from ~400 to ~650°C) sees extensive devolatilization of the organic matter and condensation reactions resulting in the formation of a residual carbon. More detailed descriptions of the nature of these reactions can be found in reviews of the carbonization process by Berkowitz (1967) and Blayden (1969). An observation arising from these reviews that has a considerable bearing on the optical behavior of carbonizing organic substances, is that the degree of ordering in the resulting carbonization product is principally related to whether or not the condensation reactions lead to three-dimensional bonding early in aromatic condensation. If they do, the carbon formed will be highly disordered. If, however, the aromatic condensation takes place without such bonding, then the carbon product will have much greater ordering due to the development and progressive alignment of aromatic layers or lamellae in groups which form "crystallites."

Although the first phase of the carbonization process may be partially involved, it is during the second phase that softening or formation of liquid from the original organic matter may accompany the pyrolysis, which leads to the greater molecular mobility that is necessary for the development of a high degree of ordering. A number of changes of state (Fig. 1) can be identified during this second carbonization phase and certain of these changes may also be reflected by variations in the pattern of the optical properties. It is while the organic matter is in this fluid condition that an optically anisotropic mesophase may form from the isotropic melt. Development of mesophase is most satisfactorily observed in pitch fractions, in some natural bitumens, in certain polymers, and other simple organic substances.

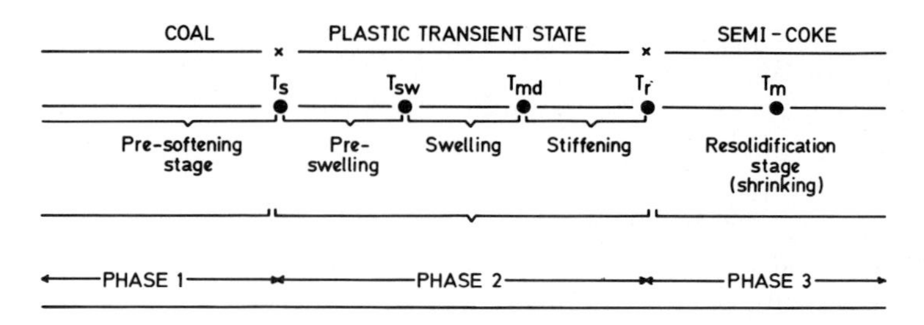

Fig. 1 Important temperature levels during the carbonization process at which changes in state occur. T_s is the softening temperature, T_{sw} the temperature of initial swelling, T_{md} the temperature of maximum devolatilization rate, T_r the resolidification temperature, and T_m the onset of molecular reorganization in the solid. (Modified after van Krevelen, 1961.)

Although many authors have discussed its formation, the first description of mesophase was given by Taylor (1961), followed by later comment by Brooks and Taylor (1965) and Taylor and Brooks (1965). Droplets, each with a similar type of lamellar development and alignment in relation to its geometry, develop in isotropic liquid. As the temperature continues to rise, the droplets enlarge, maintaining a simple but characteristic optical anisotropy that is related to their molecular structure. Eventually, the droplets coalesce and become deformed to produce a mosaic texture, which then displays a more complex optical anisotropy than the droplets. Although Kisch and Taylor (1966) have described the formation of mesophase droplets in thermally metamorphosed coals, for coal macerals, even those of the liptinite group, which have high fluidity, there is no unequivocal evidence for widespread initial development of anisotropic droplets in an isotropic matrix, either under natural or laboratory conditions. The mosaic texture (Fig. 2) that develops in many macerals which have been carbonized is, however, essentially similar in appearance to that formed from a pyrolized pitch or bitumen (see, e.g., Chapter 30, Figs. 7–9). It is possible that mesophase formation in coal macerals occurs at a scale below that resolvable with the polarizing light microscope. It should also be emphasized that development of mosaic texture in coal macerals is dependent on both rank and type. Macerals of coals lower than bituminous rank, severely oxidized macerals, and inertinite macerals do not display mosaic textures and form chars; vitrinites and exinites of bituminous rank show mosaic texture and form cokes; and anthracites yield graphitic carbon with no coke stage. For a more detailed discussion of mesophase formation and mosaics, see Chapter 30, Section II,B.

In the third phase of carbonization (~650 to ~1500°C), after resolidifi-

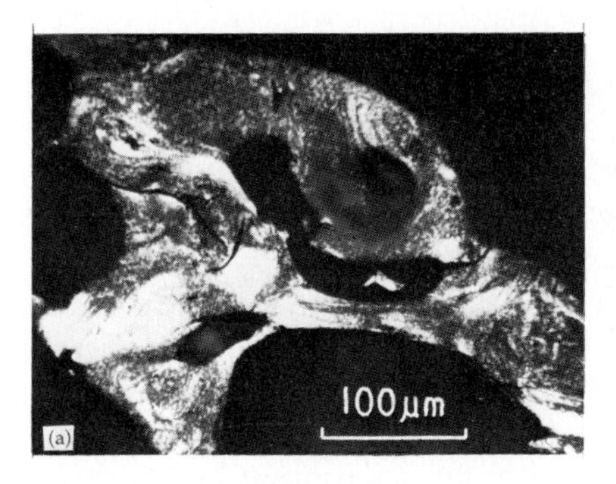

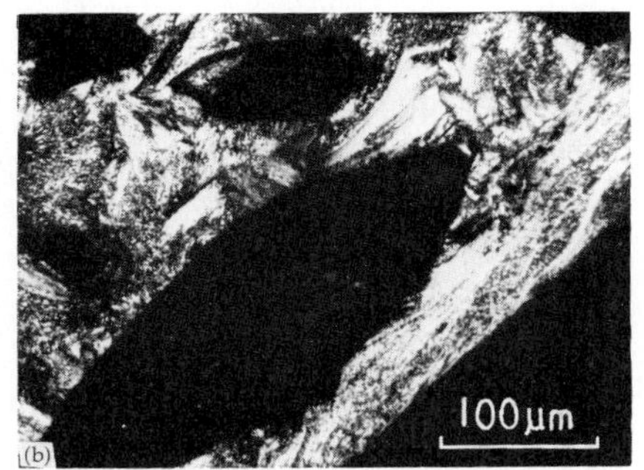

Fig. 2 Mosaic textures developed in the vitrinite [$\%R_{oil}$(max) = 1.24 (546 nm)] of a medium volatile bituminous coal carbonized at 700°C using two different heating rates with a residence time of 1 hr: reflected light, oil immersion, slightly uncrossed polars. (a) Fine to medium granular mosaic produced at a heating rate of 10°C/min. (b) Medium to coarse granular mosaic with incipient development of flow-type mosaic produced at a heating rate of 60°C/min.

cation, reactions take place principally in the solid state. Although the second phase is characterized by the loss of a range of complex volatiles, at temperatures >650°C, volatile products are chemically simple substances, with hydrogen the predominant product of the pyrolysis. Within this phase, however, although the microscopical appearance of the coke changes little, its properties, including its optical properties, alter radically. The changes all point to progressive lateral growth of aromatic systems, although there is strong evidence for some structural disorganization and reorganization occurring in the earlier part of the phase. The molecular structural changes that take place as coal macerals pyrolyze are discussed in more detail later in relation to the observed variations in optical properties.

III. EQUIPMENT AND METHODS

A. Carbonization Equipment

Carbonization experiments (<1000°C) in this Organic Geochemistry Unit have almost exclusively been carried out in simple silica tube furnaces linked to programmable controllers. Samples have been ground to pass BSS 72 mesh (<210 μm), then carbonized in silica boats at rectilinear heating rates varying between 1 and 60°C/min employing temperature intervals of either 25 or 50°C. Except for experiments involving extended residence times, "soak periods" of 1 hr, controlled to

±2°C, have always been used. Furnaces have been flushed from the beginning of each carbonization experiment through to final cooling with oxygen-free nitrogen.

The carbonization equipment and procedures employed by other investigators cited in this chapter vary to some degree from those used in this Unit. Chandra and Bond (1956), for example, employed a double-tube electric furnace, while de Vries *et al.* (1968) contained their samples in dilatometer tubes which were then inserted in a heating block before exposure to differing heating rates. Ghosh (1968) gives no information on the type of equipment used for carbonization. It is difficult to predict how much, if any, influence the variations in equipment specification play in the optical differences observed between individual investigations, but probably any variations of the optical properties introduced by variation in equipment design will be submerged by the greater influence of other factors. Even substantial differences in the particle size of the initial charge for carbonization seem to have little effect on the reflectance of the resulting coke (de Vries *et al.*, 1968).

Most carbonization studies have been carried out at atmospheric pressure. Understandably, there is much greater potential variation in experimental conditions where elevated pressures have been introduced. Table I shows the time–pressure–temperature combinations that have been used in recent studies, the majority of which utilized hydrothermal bombs as the pressure apparatus.

B. Samples

Table II lists the properties of samples which have been employed in the carbonization experiments carried out in the Organic Geochemistry

TABLE I *Time–Pressure–Temperature Combinations Employed in Recent Optical Studies of Coals*

Reference	Approximate time (hr)	Approximate maximum pressure (atm)	Approximate maximum temperature (°C)
Huck and Patteisky (1964)	24–48	8,000	350
Chandra (1965)	≤600	6,000	350
Hryckowian *et al.* (1967)	≤62	20,500	900
Bostick (1971)	≤720	1,000	450

TABLE II *Properties of Some Vitrinite Concentrates Employed in Recent Carbonization Studies, Organic Geochemistry Unit, University of Newcastle upon Tyne*

Seam	Colliery/ location	Ultimate analysis (%daf)[a] C	H	S	N + O (by difference)	$\% \overline{R}_{oil}$(max) (546 nm)
Clown Bright[b]	Shireoak, Yorkshire	80.0	5.3	1.2	13.5	0.67
Victoria[c]	Netherton, Northumbria	82.5	5.2	0.6	11.6	0.74
Bottom Gray[d]	Acorn Bank, Northumbria	83.1	5.2	1.3	10.4	0.62
Parkgate[b]	Houghton Main, Yorkshire	85.4	5.3	1.0	8.3	1.08
Unnamed[b]	Whitonstall, Northumbria	87.9	5.3	1.1	5.7	1.24
Garu[c]	Cymtillery, South Wales	88.0	5.0	0.8	6.2	1.25
Nine Feet[d]	Windsor, South Wales	88.6	5.0	1.8	4.6	1.24
Betteshanger[b]	Kent No. 6, Kent	90.0	4.8	1.5	3.7	1.55
Big Vein[c]	Cynheidre, South Wales	93.2	3.2	0.8	2.8	3.08
Rock[d]	Wernos, South Wales	93.3	3.3	0.8	2.6	3.32
Red Vein[b] (dmmf)[e]	Abernant, South Wales	93.5	3.4	0.7	2.4	3.11
Pumpquart[b] (dmmf)[e]	Pentramawr, South Wales	94.2	3.0	0.9	1.9	4.10

[a] daf-dry, ash-free.
[b] Goodarzi and Murchison (1979).
[c] Goodarzi and Murchison (1972).
[d] Marshall and Murchison (1971).
[e] dmmf-dry, mineral-matter free.

Unit and referred to in this chapter. Appropriate data for samples carbonized by other investigators are quoted in the text–figure captions when this information was available. When reference is made in the text to a stage of organic metamorphic development within the coalification series, U.S. terminology is employed in relation to the chemical or physical properties quoted (Table III).

TABLE III *Stages of Coalification Based on the North American (ASTM) Classification and Defined by Reflectance, Volatile Matter Yield, and Carbon Content*[a]

	% Vitrite		
Rank stage (ASTM)	Reflectance R_{oil}(max)	Volatile matter (daf)[b]	Carbon (daf)[b]
Peat	0.2		
		68	
		64	~60
Lignite	0.3	60	
		56	
		52	
Sub-bitum. C	0.4		
Sub-bitum. B		48	~71
Sub-bitum. A	0.5		
High volatile bituminous C	0.6	44	~77
High volatile bituminous B	0.7	40	
High volatile bituminous A	0.8	36	
	1.0	32	
Medium volatile bituminous	1.2	28	~87
	1.4	24	
Low volatile bituminous		20	
	1.6	16	
	1.8		
Semi-anthracite	2.0	12	
		8	~91
Anthracite	3.0	4	
	4.0		
Meta-anthracite			

[a] Modified after Stach *et al.* (1975).
[b] daf-dry, ash-free.

C. Sample Preparation for Photometry

Because of the relatively small amounts available, after grinding and mixing with liquid Bakelite, carbonized samples were mounted in small holes (10 mm in diameter and 3 mm deep) drilled in larger cylindrical Bakelite blocks. The blocks were prepared for reflectance measurements by a widely used and standard procedure, namely grinding on a diamond lap to produce a flat, then grinding on two grades (200 and 400) of wet silicon carbide paper until the surface was matt, but scratch free, followed by successive polishing stages with three grades of alumina (5/20, 3/50, and "gamma") dispersed on wet, Selvyt-covered, brass laps.

D. Microscope Photometry

Reflectances in this Unit are either measured with Leitz or Zeiss microscope–photometer systems (see Volume I, Chapter 2, Fig. 13) using air and stable immersion oil ($n_{546\,\mathrm{nm}} \sim 1.520$) as the measuring media. Both instruments contain secondary magnification systems which allow reliable measurements on field areas with diameters as small as 5 μm. Photomultipliers (type EMI 9592B), with a spectral response extending from 200 to 800 nm and peaking at 500 nm, are used as detectors, with a pen recorder (see also later) possessing 10-mV full-scale deflection to record measurements. The majority of reflectances have been measured at 546 nm, but where reference is made to reflectance dispersion curves in the visible region, seven wavelengths (403, 448, 490, 502, 546, 652, and 709 nm) have been isolated with Schott IL interference filters with half-band widths of 12 nm and peak transmissions of ~30%.

Other investigators have, in general, used similar measuring systems. Chandra and Bond (1956), Chandra (1958, 1962, 1965), and Ghosh (1968), however, measured reflectances with Berek visual microphotometers. This instrument involves visual matchings of light intensities by the observer. Because of its construction, measurements on substantially larger field areas than with modern microscope photometers are almost inevitable.

E. Optical Parameters and Methods of Measurements

Descriptions of quantitative optical methods in coal and coke microscopy can be found in Volume I, Chapters 1 and 2; a more extended treatment is given by Stach *et al.* (1975). Four optical properties are of particular interest in the study of carbonized vitrinites, namely reflectance, bireflectance, refractive index, and absorption index.

The reflectance of an unknown sample is determined by comparison against a standard of known optical properties. Calculation of reflectance is through the equation

$$R_u = R_s \cdot D_u / D_s \quad (1)$$

where R_u is the percentage reflectance of the unknown sample, R_s the percentage reflectance of the standard, D_u the reading for the unknown from whatever recording system is employed, and D_s the reading for the standard from the recording system.

Although practice varies between laboratories, it would be usual to measure the reflectance of at least 50 particles on each polished block to obtain the mean reflectance of the sample. Oil immersion objectives would generally be used, but objectives computed for use in other media are necessary if refractive and absorption indices are to be derived from reflectance measurements. At least two standards of different reflectance should be employed, usually stable glasses for reflectance measurements in the range of fresh and little altered vitrinites, and diamond or silicon carbide against which the reflectances of chars or cokes would be determined. Measurements nowadays are most frequently carried out at a wavelength of 546 nm, but in the past, measurements have frequently been referred to a wavelength of 530 nm.

Bireflectance (the difference between the mean maximum and minimum reflectances, $\overline{R}_{max} - \overline{R}_{min}$) is a variable of considerable significance in studies of thermally metamorphosed and laboratory carbonized coals which estimates the anisotropy of the samples. In many investigations only the maximum reflectance is required and then the value can be read directly from a digital voltmeter. When bireflectances are needed, it is most satisfactory to use a pen recorder which will give a continuous trace of the signal from the specimen as the microscope stage is rotated through 360°. The record for measurements on 10 particles of an anisotropic vitrinite of coking rank, bracketed by readings on two isotropic standards, is shown in Fig. 3. Two maxima and two minima are recorded for the principal directions of each vitrinite section on one revolution of the microscope stage. Each pair of appropriate values would then be averaged to give the maximum and minimum reflectances for the particle. Vitrinites of bituminous rank usually behave analogously to optically negative uniaxial materials, while anthracitic vitrinites probably all show departure to a biaxial condition (Cook *et al.*, 1972), although the shift is not severe. Consequently, in any group of randomly oriented vitrinite particles, values of true or close to true maximum reflectance will be recorded. Any minimum value, ranging from the true minimum reflectance (yielding maximum bireflec-

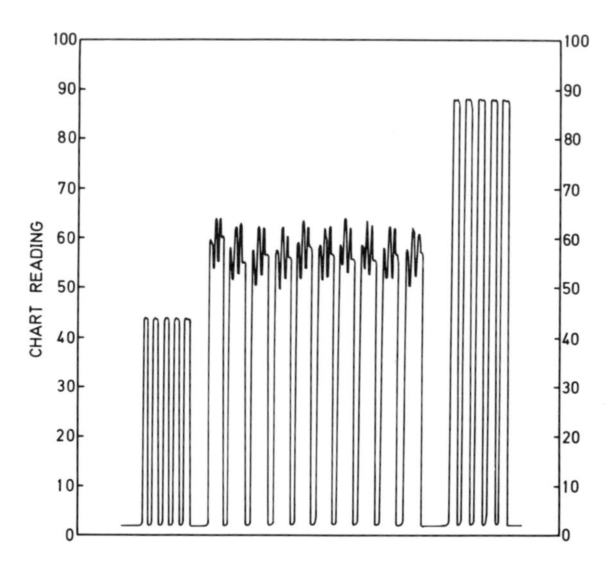

Fig. 3 Recorder chart presentation of measurements on 10 different points of an optically negative uniaxial material which shows some inherent variation compared against measurements on two isotropic glass standards [R_{oil}(546 nm) = 0.695 and 1.429%].

tance) to the maximum reflectance (giving zero bireflectance) may, however, be recorded.

Although reflectance is the most used, directly measured, experimental variable in microscopical studies of coals and cokes, reflectance is not a fundamental property of organic matter, although like other optical properties it responds to thermal and other influences. The reflectance of an optically absorbing material is dependent on the refractive index (n_s) and the absorption index (κ_s) of the material, as well as on the refractive index (n_m) of the medium in which the reflectance measurement is made. These quantities are related by Beer's equation:

$$R_s = \frac{(n_s - n_m)^2 + n_s^2\kappa_s^2}{(n_s + n_m)^2 + n_s^2\kappa_s^2} \tag{2}$$

The absorption or extinction coefficient k is sometimes calculated instead of the absorption index κ:

$$k = n\kappa \tag{3}$$

Reflectance is primarily employed as a diagnostic and "fingerprinting" variable. Although bireflectance can serve a similar purpose, it also reflects the degree of ordering of the molecular structure of a maceral or coke. Refractive index and some property of absorption are, however, the principal variables used to estimate the molecular structural development of organic matter. They are used for that purpose in this chapter. To derive n and κ, measurements of reflectance must be made in at least two media of differing refractive index. Normally air and immersion oil ($n_{546\text{nm}} \sim 1.520$) are used for this purpose.

Equation (2) may be conveniently transposed into the forms

$$n_s = \frac{(n_0^2 - n_a^2)/2}{n_0[(1 + R_0)/(1 - R_0)] - n_a[(1 + R_a)/(1 - R_a)]} \tag{4}$$

and

$$\kappa_s^2 = \frac{R_a(n_s + n_a)^2 - (n_s - n_a)^2}{n_s^2(1 - R_a)} \tag{5}$$

or

$$\kappa_s^2 = \frac{R_0(n_s + n_0)^2 - (n_s - n_0)^2}{n_s^2(1 - R_0)} \tag{6}$$

where R_a and R_0 are, respectively, the reflectances of the substance in air ($n_m = n_a$) and oil ($n_m = n_0$).

F. Precision and Accuracy of Optical Parameters

Detailed consideration in this section is given only to the accuracy and precision of optical parameters required in carbonization studies. Where proximate or ultimate chemical analyses are quoted, the parameters are subject to the particular specifications of the appropriate national standard. For vitrinites concentrated for use in pyrolysis experiments, some indication of the likely accuracy of purity estimation by standard point counting techniques can be gained from recent considerations on the standardization of petrographic analysis (Juckes and Pitt, 1977). The large standard deviations between laboratories recorded by these authors will be considerably reduced within single laboratories. In later discussion in this chapter, the numerical values quoted for certain x-ray parameters in comparisons of the scale of development of molecular structures must be considered in a relative rather than an absolute sense.

The following remarks bear generally on optical measurements on thermally affected macerals, but are based principally on observations made in the Newcastle Unit. Precision and accuracy of reflectance measurements in coal petrology are discussed at length in the "International Handbook for Coal Petrography" (ICCP, 1971) and by Stach *et al.* (1975). These publications tacitly assume in any statement of a standard deviation for either the maximum or average reflectance that the reflectances are measured on uniaxial negative materials and that the reflectance distribution for the sample is approximately normal. To extend rigorous statistical treatment to the refractive and absorption indices derived from reflectance measurements on coke mosaics would be unrealistic, if only because of the character of many of the mosaic surfaces that must be employed for measurements (see, e.g., Fig. 2). Further, the form of Eq. (2) indicates that a complex relationship exists between the different optical parameters and that both reflectance measurements will be subject to errors from a variety of experimental causes. Because of these constraints, a more empirical approach to the accuracies of the fundamental parameters is desirable.

Recently Cook and Murchison (1977) discussed the accuracy of refractive and absorption indices derived from reflectance measurements on low reflecting materials. Individual sources of error affecting the reflectances were not considered at length, but precautions to reduce the size of certain of the principal errors were suggested. The surfaces were assumed to be homogeneous in appearance. Instead of employing standard deviations of reflectances, the effect of different sizes of percentage relative errors of reflectance on the refractive and absorption indices was examined. The principal conclusion of this study, which is of relevance to this chapter, is that in the central part of the visible spectrum, i.e., in the green region, both the precision and accuracy of reflectance measurements are sufficiently high to yield acceptable values of n and κ derived through Beer's relationship.

Within individual laboratories, precision of reflectance is probably ~2% relative at the 95% confidence limits. Accuracy is more difficult to quantify. It seems, however, that although standards and unknown may be individually affected, experimental errors will generally either be cancelled out or have the same sense for the measurements in the two media. Only when an incorrect refractive index value for the immersion oil is used, will there be an error in the measured oil reflectance which will not be compensated by an error similar in sign in the air reflectance.

Because of the complex relationship between the parameters in Beer's equation, the size of relative errors in n and κ varies depending on the level of the measured reflectances. The relative errors in n for materials with air and oil reflectances of the general level shown by cokes are likely to be higher than for the lower reflecting, original vitrinites of any bituminous rank. In contrast, the errors in κ for the cokes will be lower than those affecting the absorption indices for vitrinites of bituminous rank. The worst situation that can be identified is a positive error in air reflectance and a negative error in oil reflectance. Table IV shows the errors in n and κ for different levels of error in reflectances of a typical coke under these circumstances.

With modern microscope photometers, relative errors in reflectance of 2% can usually be anticipated. A photometer system operating only to give relative reflectance errors of ~5% would be unacceptable to most laboratories. Consequently, for homogeneous surfaces, upper limits of ~2.5% for Δn and ~5% for $\Delta\kappa$ can probably be set for the relative errors in n and κ, which, considering the indirect method of obtaining these variables, are not unreasonable levels. Many of the errors obtained would almost certainly be substantially below these worst estimates, perhaps one-half or even one-third of the levels quoted in Table IV. Precision and accuracy will, however, fall moving away from the green region of the spectrum toward either the red or blue regions (Cook and Murchison, 1977).

There is a further factor that complicates any reflectance measurement on a char or coke (Goodarzi and Murchison, 1972) and which must reduce the suggested accuracies attainable in estimates of n and κ. The accuracies quoted in Table IV refer strictly to values of n and κ derived from reflectance measurements on homogeneous surfaces. Figure 2 shows that a polished coke surface may comprise a mosaic of small units. While the size of the units of any mosaic is governed by the relative influence of a number of factors (discussed later), in the majority of cokes, individual units rarely attain diameters that are sufficiently large to allow discrete reflectance measurements upon single units, even with the small field size restrictions possible in modern microscope–photometer systems. A diameter of 5 μm is probably the smallest realis-

TABLE IV *Percentage Errors in Calculated Values of n and κ*[a]

Percentage relative error in reflectance (R)	Percentage relative error in n and κ for $+R_{air} - R_{oil}$
$\Delta R_0 = 1\%$	$\%\ \Delta n = +2.0$
$\Delta R_a = 1\%$	$\%\ \Delta\kappa = -4.0$
$\Delta R_0 = 5\%$	$\%\ \Delta n = +11.0$
$\Delta R_a = 5\%$	$\%\ \Delta\kappa = -24.0$
$\Delta R_0 = 5\%$	$\%\ \Delta n = +4.8$
$\Delta R_a = 1\%$	$\%\ \Delta\kappa = -11.0$

[a] Resulting from errors in reflectance measurement at 546 nm in air and in oil of refractive index 1.516 of a substance with $R_{air} = 16.00\%$, $R_{oil} = 6.35\%$, $n = 1.91$, and $\kappa = 0.42$, optical properties comparable with those of a 750°C coke. (After Cook and Murchison, 1977.)

tic field size whose reflectance can be safely measured. Consequently, apart from exceptional circumstances, any reflectance measurement upon a carbonized maceral with a mosaic texture will be on a field comprised of a number of units. It is this factor that poses the question of how to measure and to calculate the reflectances of such fields, not only to determine how reflectance varies and how the fundamental properties n and κ respond during pyrolysis, but also to assess any development in the degree of ordering of the coke that may be established through its bireflectance.

A field occupied by a number of mosaic units of a carbonized maceral will often appear to be essentially structureless in plane-polarized light. Only when the analyzer is placed in the optical train will the true textural complexity of the field become apparent. There are fields of the more strongly coking macerals which, even in plane-polarized light, indicate the presence of mosaic texture. Charts recording measurements on successive fields of this general character do not show the typical regular features of records of particulate preparations of optically negative uniaxial materials (Fig. 3), but are much more irregular in character (Fig. 4), resembling at least superficially, measurements on particles of

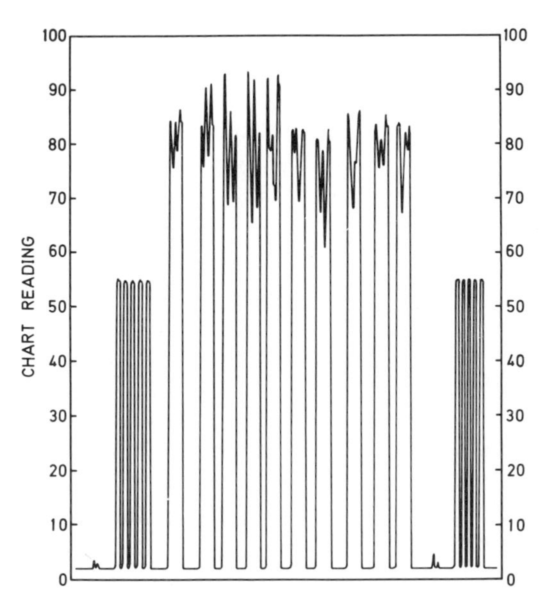

Fig. 4 Recorder chart presentation of measurements on 10 different points of an optically biaxial material compared against measurements on diamond as a standard [R_{oil}(546 nm) = 5.307%].

an optically biaxial substance. The principal modification to the uniaxial record is that there are considerable fluctuations in the maximum values, as well as in the minimum readings. The irregularities of the record are a consequence of two factors: (a) the degree of ordering of the mosaic units and (b) the degree of ordering developed within each individual unit. Recognition of these two levels of ordering should not be allowed to conceal the complexity of the optical structures involved in the deformed mosaic, but as the degree of ordering improves, the fluctuations occurring in the maximum reflectances appear to be reduced.

With such materials the most reliable method of calculating and presenting reflectance data and related optical parameters will only be decided once measurements have been analyzed from a wide range of samples. In this Unit an empirical approach has been employed. The customary method of averaging 50 individual estimates of maximum reflectance to give $\bar{R}_{max}$ for each sample has been used and 50 such measurements in air and oil have been used to derive n and κ. With the variability of the maximum reflectances, some fall in the accuracy of n and κ must be anticipated. The assessment of anisotropy by bireflectance poses another problem and two approaches have been used. First, the mean bireflectances have been calculated from the averages of the highest 20% of maximum readings and the lowest 20% of minimum values. The second method of approach has been to use the 20% of individual measurements showing the largest differences between maximum and minimum reflectance; these have been meaned to give bireflectances. No significant differences between the bireflectance trends resulting from the two methods have yet been observed. There is some variation in absolute values, however, but at this stage of development of the application of optical methods to cokes, there could be no claim or expectation that individual bireflectance values would be highly specific. Chandra (1972) believes that both these methods of determining bireflectance will "emphasize the anisotropy," but the case against this view is argued by Marshall and Murchison (1972).

The comments on lack of specificity of determined values must also apply to the derived refractive and absorption indices, but much less so to the maximum reflectances. What this discussion thus emphasizes is that the errors generated in individual values for the optical parameters are likely to be greater for mosaic surfaces than for structureless surfaces. Many carbonized macerals, however, do not develop mosaic textures, and consequently measurements on these untextured surfaces should not suffer this loss of accuracy and precision.

IV. THE RESPONSE OF THE OPTICAL PROPERTIES OF VITRINITES TO CARBONIZATION

A. Reflectance

The earlier description of the carbonization process summarized some of the principal chemical and physical changes caused by pyrolysis, particularly those with relevance to variations in optical properties. The rate and amount of change induced in all the properties of a developing coke can be substantially modified by the experimental or environmental conditions. In this section is described the response of the two directly measurable optical properties during carbonization, namely, reflectance and bireflectance, to some of the important experimental factors: temperature, rank, residence time, heating rate, oxidation prior to carbonization, and pressure. The relative importance of such influences is considered in later discussion.

1. *Variation with Temperature and Rank*

The general pattern of reflectance variation with temperature is illustrated in Fig. 5 (refer also to Fig. 1) and, as might be expected, the patterns for both the air and oil reflectances are similar. Vitrinites of bituminous rank show little change in reflectance throughout phase 1 of the carbonization process, although the reflectance of the lowest rank vitrinite begins to rise at approximately 300°C. With increasing rank, the initial reflectance rise is gradually delayed to higher temperatures until with the anthracitic vitrinite, there is little sign of change until a temperature of ~500°C is reached. Once the reflectance does begin to rise, however, it does so quite rapidly and at 950°C the reflectance is still increasing without any sign of a fall in the rate. The rates of reflectance rise appear comparable for all three vitrinites. However, at a temperature of 950°C, the air and oil reflectances of the anthracitic vitrinite lie at levels slightly below those for the two vitrinites of bituminous rank.

Comparisons of reflectance–temperature trends established by Goodarzi and Murchison (1972, 1979) with similar data produced by other authors (Chandra and Bond, 1956; Ghosh, 1968; de Vries *et al.*, 1968) are illustrated in Figs. 6 and 7. Where possible, carbonized vitrinites or coals of equivalent starting rank and pyrolysis treatment are compared, but such direct comparison has not proved possible for some of the earliest data produced (Chandra and Bond, 1956) because of a lack of equivalence between starting ranks. Chandra and Bond (1956) and

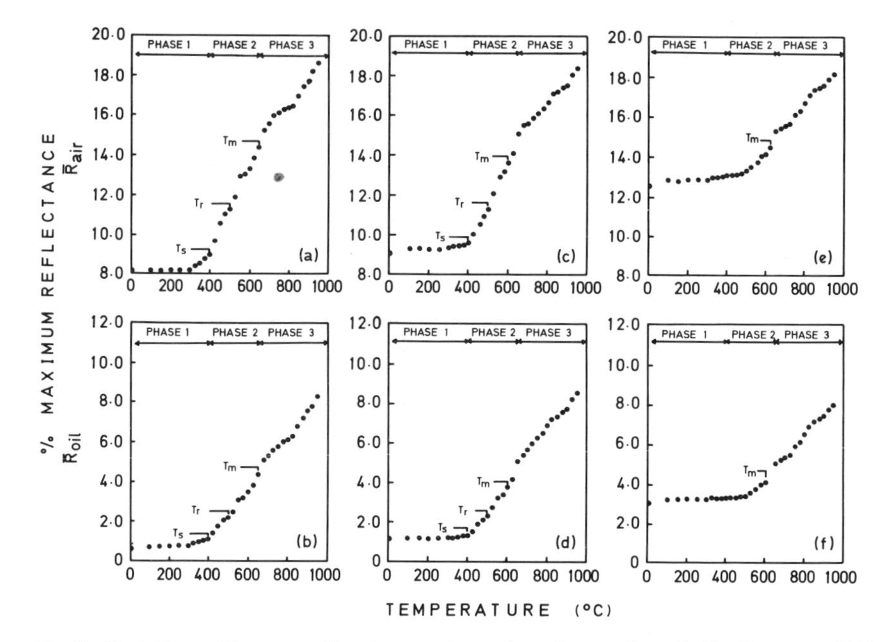

Fig. 5 Variation with preparation temperature of maximum air and oil reflectances (546 nm) of vitrinites from three coals carbonized over the temperature range 20–950°C. (a,b) Vitrinite from high volatile bituminous coal: $\%R_{oil}$(max) = 0.74; (c,d) vitrinite from low volatile bituminous coal: $\%R_{oil}$(max) = 1.25; (e,f) vitrinite from anthracite: $\%R_{oil}$(max) = 3.08. T_s is the onset of plasticity, T_r the onset of resolidification, and T_m the onset of molecular reorganization in the solid. (After Goodarzi and Murchison, 1972.)

Ghosh (1968) employed Berek visual microphotometers for their measurements.

2. *Effect of Heating Rate*

Although a number of authors have commented on the qualitative influence that different heating rates have upon the optical properties of carbonized macerals, principally in relation to the development of anisotropy, only two systematic studies have considered the quantitative aspects of heating rate upon reflectance. Figures 8a and 8b compare data produced by Ghosh (1968) and Goodarzi and Murchison (1979) for coals varying in starting rank from subbituminous A to anthracitic carbonized at a heating rate of 1°C/min. The two sets of results show similarity to the extent that the general pattern of reflectance–temperature change is similar. There, however, the similarity ends. The results produced by Ghosh (1968) for higher heating rates of 3 and

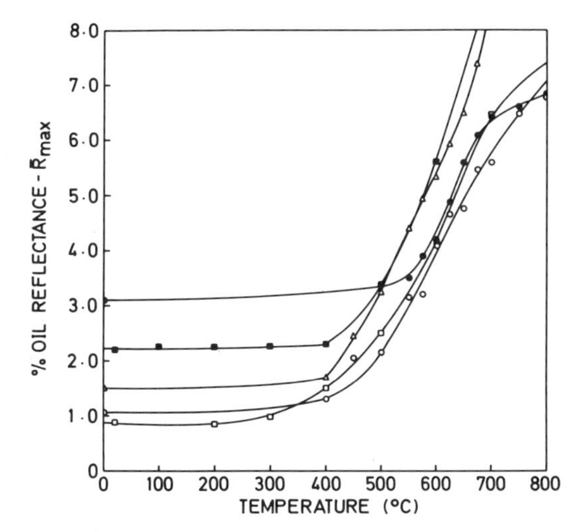

Fig. 6 Variation with preparation temperature of maximum oil reflectances (546 nm) of vitrinites from coals of different rank carbonized within the temperature range 20–800°C. $\%R_{oil}$(max) = (○) 1.08, (△) 1.55, and (●) 3.11. Heating rate 1°C/min, residence time 1 hr. (After Goodarzi and Murchison, 1979.) $\%R_{oil}$(max) = (□) 0.87 and (■) 2.20. Heating rate 1.24°C/min, residence time 1 hr. (After Chandra and Bond, 1956.)

5°C/min show a progressive fall in reflectance at the same temperature for a rising heating rate. This observation directly conflicts with recent data of Goodarzi and Murchison (1979) for a larger number of vitrinites, who show that when heating rate has an effect, then the rate of rise of reflectance increases with heating rate and does not fall (Figs. 8b–8d).

3. *Effect of Residence Time*

A factor likely to have an influence on optical properties is the length of time the organic matter is exposed to the maximum temperature attained (the "residence time" or "soak period"). Investigations reported under this heading fall into two groups: studies of reflectance behavior above or below the decomposition points of the thermally affected constituents.

In an investigation in which the coals had passed through their decomposition points, Ghosh (1968) concluded that differing but constant levels of reflectance were reached after residence times of 5 hr or less: The time taken to attain equilibrium depended on the heating rate and temperature employed. The data of de Vries *et al.* (1968), however, show reflectance continuing to rise after heating for residence times of 72 hr

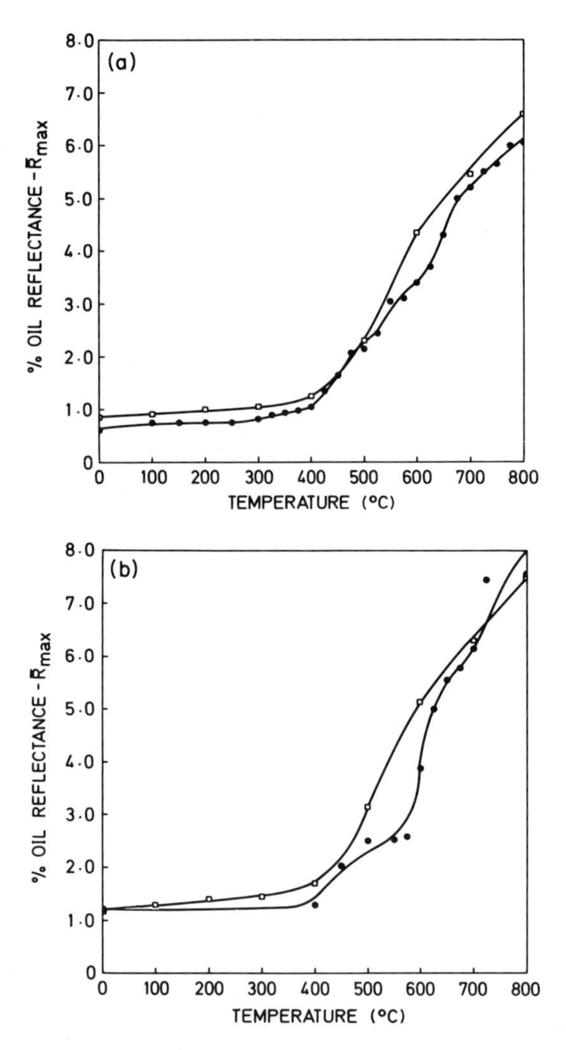

Fig. 7 A comparison between the results of different investigators of the variation with preparation temperature of maximum oil reflectances (546 nm) of vitrinites from coals of different rank carbonized within the temperature range 20–800°C. (a) ●, After Goodarzi and Murchison (1972): heating rate 2.45°C/min, vitrinite $\%R_{oil}$(max) = 0.74. □, After Ghosh (1968): heating rate 3°C/min, vitrinite $\%R_{oil}$(max) = 0.85. (b) ●, After Goodarzi and Murchison (1979): heating rate 1°C/min, vitrinite $\%R_{oil}$(max) = 1.24. □, After Ghosh

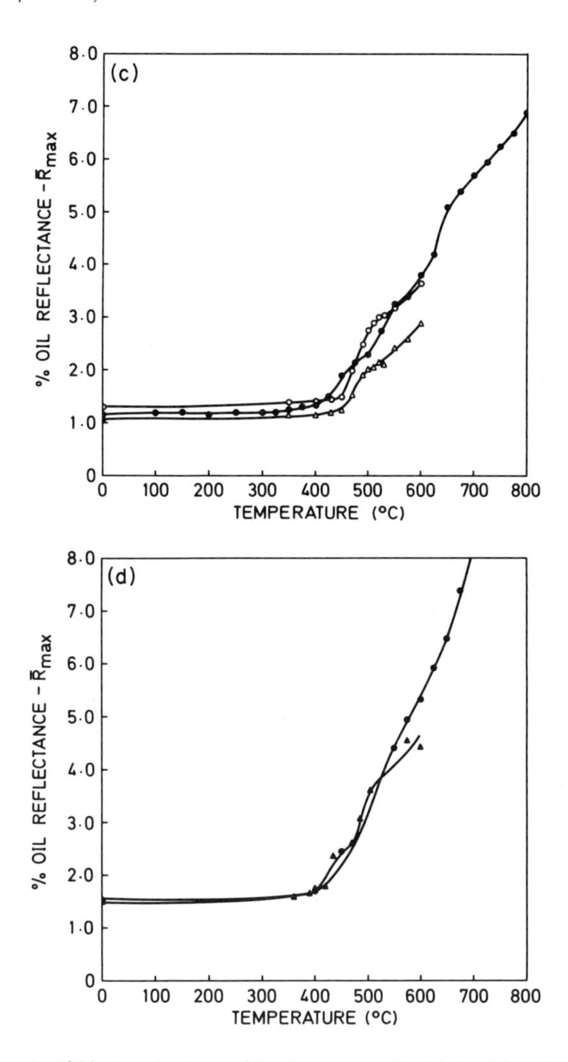

(1968): heating rate 3°C/min, vitrinite $\%R_{oil}$(max) = 1.25. (c) ●, After Goodarzi and Murchison (1972); heating rate 2.45°C/min, vitrinite $\%R_{oil}$(max) = 1.25. △, After de Vries *et al.* (1968): heating rate 3°C/min, vitrinite $\%R_{oil}$(max) = 1.17. ○, After de Vries *et al.* (1968): heating rate 3°C/min, vitrinite $\%R_{oil}$(max) = 1.55. (d) ●, After Goodarzi and Murchison (1979): heating rate 1°C/min, vitrinite $\%R_{oil}$(max) = 1.55. ▲, After de Vries *et al.* (1968): heating rate 3°C/min, vitrinite $\%R_{oil}$(max) = 1.63. Residence times: Goodarzi and Murchison 1 hr, Ghosh 1 hr, de Vries *et al.* unknown.

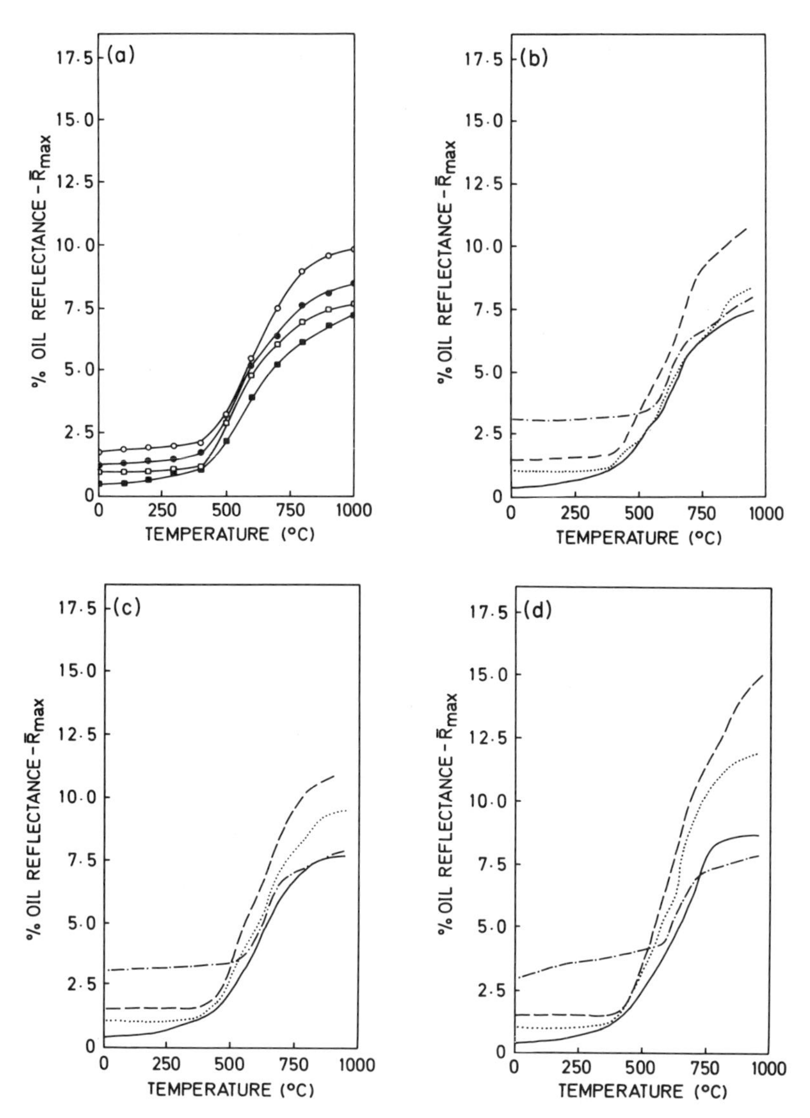

Fig. 8 Effect of different heating rates on the maximum oil reflectances (530 nm) of vitrinites carbonized within the temperature range 20–1000°C. (a) After Ghosh (1968) (530 nm): $\%R_{oil}$(max) = (■) 0.50, (□) 0.85, (●) 1.25, and (○) 1.70. Heating rate 1°C/min, residence time 1 hr. (b,c,d) After Goodarzi and Murchison (1979) (546 nm): $\%R_{oil}$(max) = (——) 0.67, (· · ·) 1.08, (– – –) 1.55, and (–·–) 3.11. Residence time = 1 hr, heating rate (b) 1°C/min, (c) 10°C/min, (d) 60°C/min.

(Fig. 9). The plots of reflectance versus time show that the rate of rise is dependent not only on the heating rate and temperature level reached but also on the rank of the vitrinite carbonized.

Data from similar investigations carried out at lower temperatures support the results of de Vries and co-workers. Mackowsky (1961) heated a low rank coal over periods of 24–48 hr at temperatures below those which would normally cause decomposition during laboratory carbonization and observed an increase in reflectance. With no "soak period," however, there was no increase of an initial vitrinite reflectance of ~0.8%, even after heating at 380°C. Goodarzi and Murchison (1977) used a temperature (350°C) similar to that used by Mackowsky, as well as a much lower temperature (150°C), but heated the vitrinites, which ranged in rank from high volatile bituminous to anthracitic, over periods up to 32 weeks. The reflectances of all the vitrinites were affected, but much less at 150°C than at 350°C (Fig. 10).

4. *Effect of Oxidation Prior to Carbonization*

In view of the serious effects which oxidation has on the properties of coals intended for use in a variety of technological processes, it is surprising that such limited information is available on the effect that oxidation prior to carbonization has on the optical properties of the resulting cokes. Reflectance data are, however, available for oxidized and weathered coals (Chandra, 1958, 1962; Benedict and Berry, 1964). Possibly the recognition that the loss of swelling and plastic properties through oxidation of fresh coals can be detected more readily and at

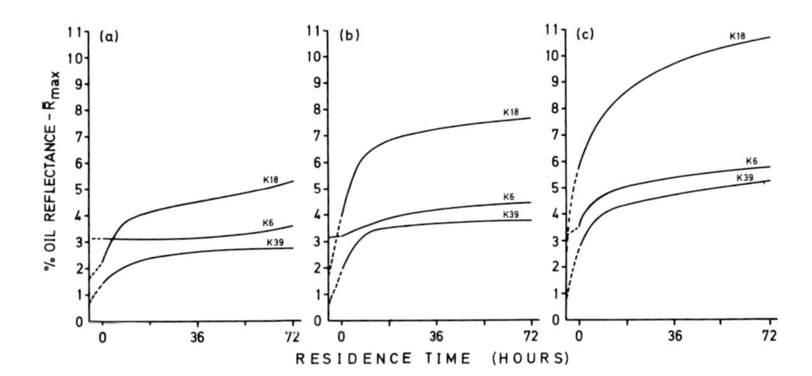

Fig. 9 Effect on the maximum oil reflectances (546 nm) of cokes from vitrinites of different rank heated for residence times up to 72 hr at different temperatures: (a) 450°C, (b) 520°C, (c) 600°C. Vitrinites: K6, $\%R_{oil}(max) = 0.70$; K18, $\%R_{oil}(max) = 1.63$; K39, $\%R_{oil}(max) = 3.16$. (After de Vries *et al.*, 1968.)

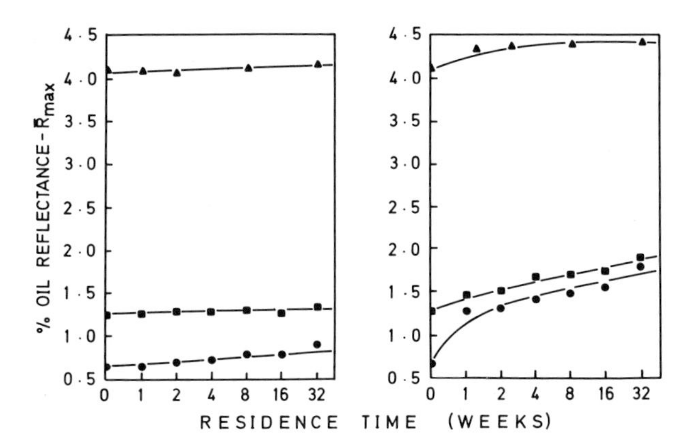

Fig. 10 Effect on the maximum oil reflectances (546 nm) of vitrinites heated for residence times of up to 32 weeks at different temperatures: (a) 150°C, (b) 350°C. $\%R_{oil}$(max) = (●) 0.66, (■) 1.24, and (▲) 4.10. (After Goodarzi and Murchison, 1977.)

much lower levels of oxidation by methods other than microscopical has diverted investigators from this particular area of the carbonization field. Only the results of one optical study are available.

Goodarzi and Murchison (1973a) reported on the behavior of the reflectances of three vitrinites (high and low volatile bituminous and anthracitic) that were deliberately oxidized severely for 14 days at 150°C prior to carbonization over the temperature range 300–800°C. Figure 11 compares the reflectance–temperature tracks of carbonized preoxidized and carbonized fresh vitrinites. There is relatively little effect on the reflectance–temperature track of the anthracitic chars. For the vitrinites

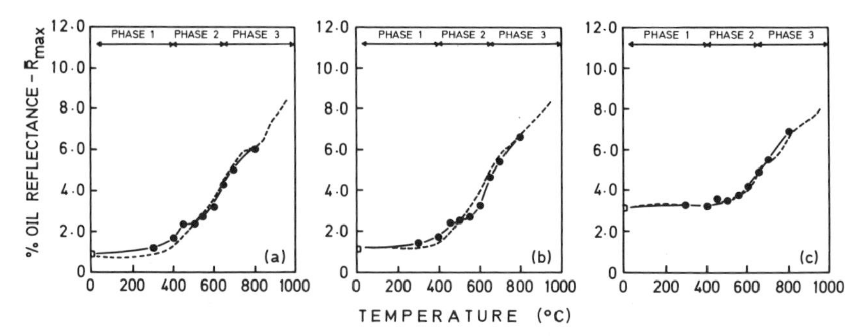

Fig. 11 Comparison between the results for the variation with preparation temperature of the maximum oil reflectances (546 nm) of vitrinites of different rank when carbonized (– – –) and preoxidized and carbonized (———). $\%R_{oil}$(max) (a) 0.74, (b) 1.25, and (c) 3.08. Heating rate 2.45°C/min, residence time 1 hr. (After Goodarzi and Murchison, 1973a.)

of lower rank, the reflectance–temperature tracks become crosscutting. The reflectances of the preoxidized vitrinites are higher at the approximate resolidification point (~500°C), while the reverse relationship holds above this temperature.

5. *Effect of Pressure*

Literature on the effect of pressure on the optical properties of heat-treated macerals is understandably more sparse, principally because of the greater experimental difficulties and expense involved in such investigations. Huck and Patteisky (1964) produced limited data on the behavior of vitrinite reflectance in response to pressure variation at an elevated but constant temperature of 350°C. From an initial value of 0.91% for the fresh vitrinite, the reflectance rose to 1.51% under a vacuum of 0.5 torr, was reduced to 1.21% at 1 atm, and was lower (1.01%), but still above the reflectance of the fresh vitrinite, at 8000 atm.

The results of a more detailed study were published by Chandra (1965). At temperatures of 325 and 350°C, reflectance increased with rising pressure over periods of 5 days. The rate of reflectance increase was greater for the higher temperature. Again, at different pressure levels, but using the same two temperatures, reflectance increased with time over a period of 25 days, the highest reflectance over the period being associated with the highest pressure. While Chandra states that certain anomalies exist in his results, he observed a correlation of the highest maximum reflectance with the highest pressure at both temperatures employed.

Hryckowian *et al.* (1967), experimenting with coals of anthracitic rank, found that the effects of pressure at constant temperature on the reflectances of the anthracites were almost negligible. However, at any given pressure, the reflectances of all samples employed rose with increasing temperature, as they also did as the time of the experiments was extended. These findings are partially supported by the results of Bostick (1971) who also observed a reflectance increase with rising temperature and time of exposure. Since pressure was allowed to rise concomitantly in the hydrothermal bombs, the effect of pressure cannot be isolated.

B. Bireflectance

The development of anisotropy in organic materials as pyrolysis progressed was discussed in Section II, in a summary of the carbonization process. There is now a substantial literature on the more qualitative and morphological aspects of the textures that form during the early stages of carbonization. This literature also includes considerable discussion on the mode of origin of these textures. Although the many contributions in this field cannot be reviewed here, the principal conclusions that have been reached and which have a bearing on the quantitative determination of bireflectance are now summarized. A considerable understanding of this field can be gleaned from recent publications (Dubois *et al.*, 1970; Honda *et al.*, 1971; Sanada *et al.*, 1973; Patrick *et al.*, 1973).

The two most recent of these papers are concerned with the development during carbonization of (a) anisotropic mesophase in carbonaceous materials and (b) optical anisotropy in vitrains. Although varying degrees of anisotropic texture related to the factors of rank, temperature level, and residence time, have been recognized for many years (see, e.g., Abramski and Mackowsky, 1952), Sanada *et al.* (1973) and Patrick *et al.* (1973) attempt to quantify the proportion of different types of mosaic structures that can be identified in cokes. The textural progression is from a finely granular mosaic through medium and coarsely granular forms of mosaic to a "flow type" (Patrick *et al.*, 1973) or its approximate equivalent "fibrous" and "leaflet" types (Sanada *et al.*, 1973). As rank increases, or the temperature rises, or the residence time is extended, the mosaic gradually coarsens and, under appropriate conditions, an increasing proportion of flow-type mosaic is produced.

Other factors, besides those just noted, are involved in the degree to which a mosaic develops. Goodarzi and Murchison (1976a) showed that when vitrinites of bituminous rank are severely oxidized prior to carbonization, losing all swelling and plastic properties, not only are the original shape and petrography of the particles retained on carbonization to high temperatures, but no mosaic forms, although high levels of anisotropy under crossed polars can be observed in individual particles. A similar reduction in swelling and plastic properties, but on a less striking scale, was recorded in carbonized telinites, which were probably pseudovitrinites (Benedict *et al.*, 1968), in which finely granular mosaics developed in the solid state, there being no obvious change in the shape of carbonized particles (Goodarzi and Murchison, 1976b). In contrast, Goodarzi and Murchison (1978) have more recently illustrated the pronounced coarsening of mosaic structures and the formation of flow-type mosaics as the heating rate applied to pyrolyzing vitrinites is increased. All these textural variations show some correlation with the quantitative changes that can be expressed by bireflectance.

1. *Variation with Temperature and Rank*

Three investigations have considered the systematic variation of bireflectance with temperature and rank. Chandra (1965), using data from earlier work (Chandra and Bond, 1956; Chandra, 1963), produced tracks of maximum oil reflectance plotted versus minimum oil reflectance for cokes from coals varying in rank through the bituminous and anthracitic ranges (Fig. 12). Ghosh (1968) also referred his results to this diagram and found that his plots of maximum versus minimum reflectance for cokes formed from coals with similar ranks to those carbonized by Chandra generally corresponded with tracks of Fig. 12.

More recently, Goodarzi and Murchison (1972) published bireflectances for carbonized vitrinites using much smaller temperature intervals between measurements than had been used in earlier studies. The detailed bireflectance variations that are characteristic for vitrinites of bituminous and anthracitic rank are shown in Fig. 13. The bireflectance of the bituminous rank vitrinite shows a fall to a minimum in the region of 400°C before beginning to rise again. Although the bireflectance increases quite rapidly from this temperature, the rate becomes much

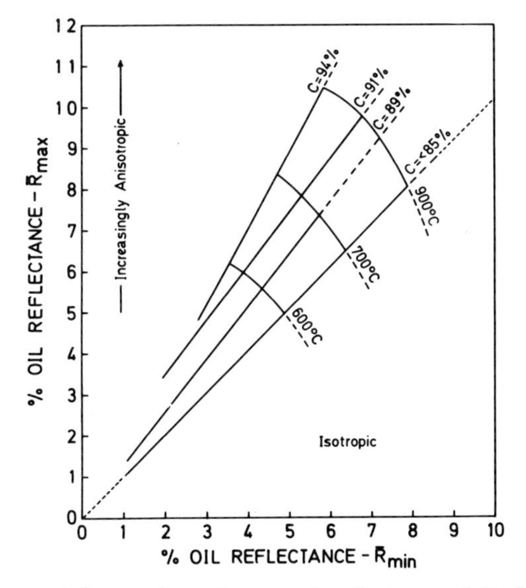

Fig. 12 Development lines of maximum oil reflectance plotted as a function of minimum oil reflectance (530 nm) for carbonized coals of different rank based on data from Chandra and Bond (1956) and Chandra (1963, 1965). A heating rate of 1.25°C/min with a residence time of 1 hr is assumed for all measurements. Lines of constant temperature have been interpreted from Chandra's original diagrams.

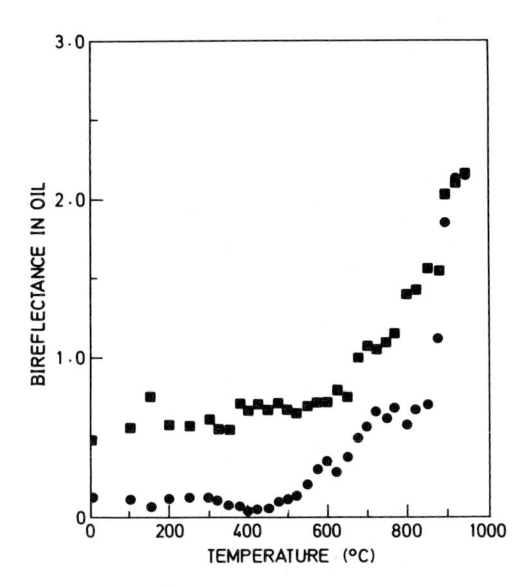

Fig. 13 Variation with preparation temperature of oil bireflectances at 546 nm of vitrinites from two coals carbonized over the temperature range 20–950°C. ●, Vitrinite from high volatile bituminous coal, $\%R_{oil}(max) = 0.74$; ■, vitrinite from anthracite, $\%R_{oil}(max) = 3.08$. Heating rate 2.45°C/min, residence time 1 hr. (After Goodarzi and Murchison, 1972.)

greater at ~500°C. A similar sharp increase affects the anthracitic vitrinite at this temperature, but there is no earlier fall in bireflectance to a minimum value.

2. *Effect of Heating Rate*

Figure 14 illustrates what appear to be the only systematic quantitative results available showing the effect of varying heating rates on the bireflectance of carbonized vitrinites of different rank (Goodarzi and Murchison, 1978). In the first tenfold increase of heating rate (Fig. 8b), it is primarily the bireflectances of the cokes from the vitrinite of medium volatile rank that are affected. The further rise of the rate to 60°C/min shows all the cokes displaying increased bireflectances at temperatures higher than 500°C, but again those of the cokes of the medium volatile rank vitrinite are most affected.

3. *Effect of Residence Time*

No quantitative data are available for any variation of bireflectance that might take place with residence time.

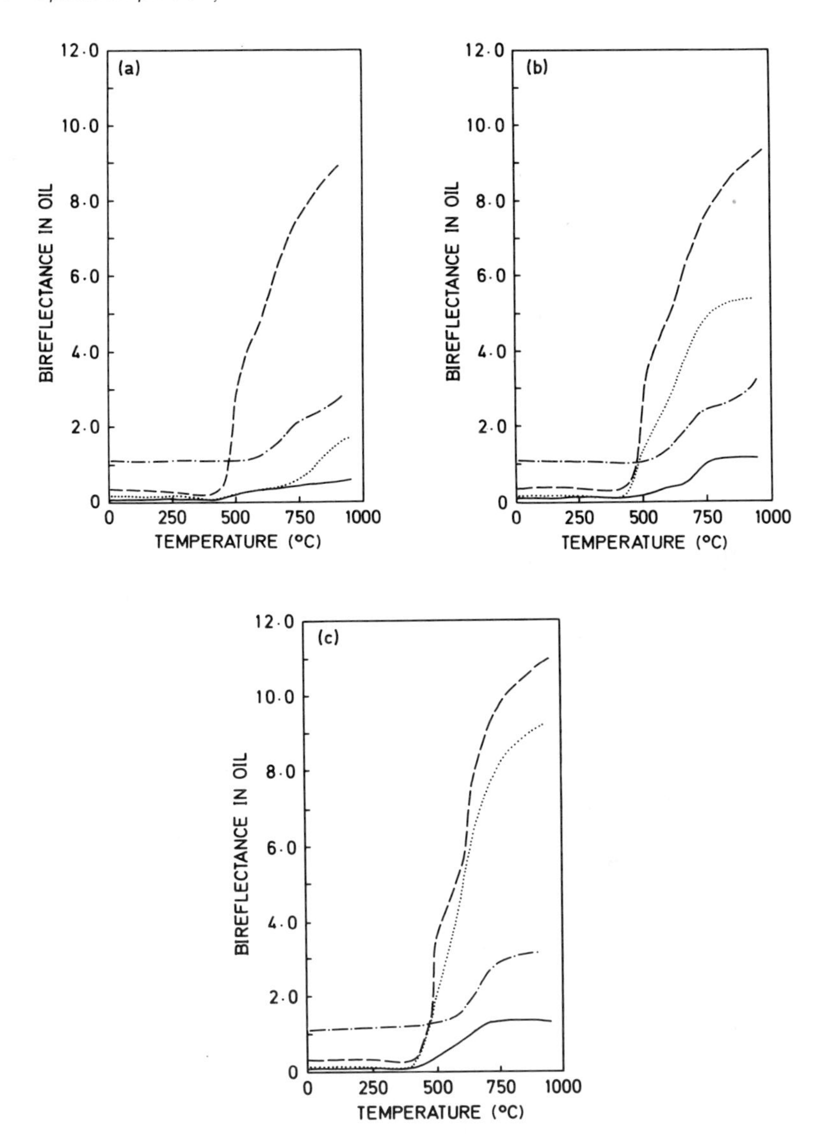

Fig. 14 Variation with preparation temperature of oil bireflectance (546 nm) of vitrinites from coals of different rank carbonized over the temperature range 20–950°C. $\%R_{oil}$(max) = (—) 0.67, (· · ·) 1.08, (– –) 1.55, and (–·–) 3.11. Residence time = 1 hr, heating rate (a) 1°C/min, (b) 10°C/min, (c) 60°C/min. (After Goodarzi and Murchison, 1978.)

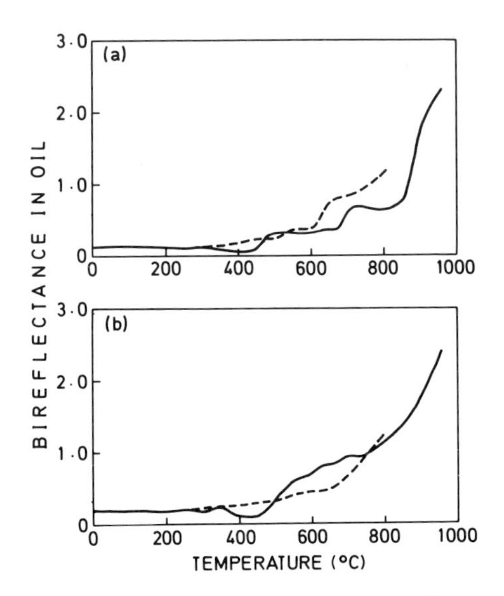

Fig. 15 Variation with preparation temperature of oil bireflectances (546 nm) of vitrinites from two coals, carbonized (———) and preoxidized carbonized (– – –). $\%R_{oil}$(max) = (a) 0.74 and (b) 1.25. Heating rate 2.45°C/min, residence time 1 hr. (After Goodarzi and Murchison, 1976a.)

4. *Effect of Oxidation Prior to Carbonization*

Again few results exist for the behavior of bireflectance under these conditions. Goodarzi and Murchison (1976a), in what was an essentially qualitative study of the petrography and anisotropy of vitrinites that had been heavily oxidized prior to carbonization, compared the bireflectances of the carbonized oxidized and carbonized fresh vitrinites (Fig. 15).

5. *Effect of Pressure*

Influence of pressure on the bireflectances of carbonized vitrinites is reported by Chandra (1965) for a single high volatile bituminous coal and by Hryckowian *et al.* (1967) for coals of anthracitic rank (Fig. 16). The same high volatile bituminous coal carbonized by Chandra at 325°C at 6000 atm (not included in Fig. 16) showed an increase of bireflectance to 0.36. Apart from the experiment at 850°C by Hryckowian and co-workers, in which bireflectance falls with rising pressure, there is a general trend of rising bireflectance with increasing pressure at constant temperature for the anthracites.

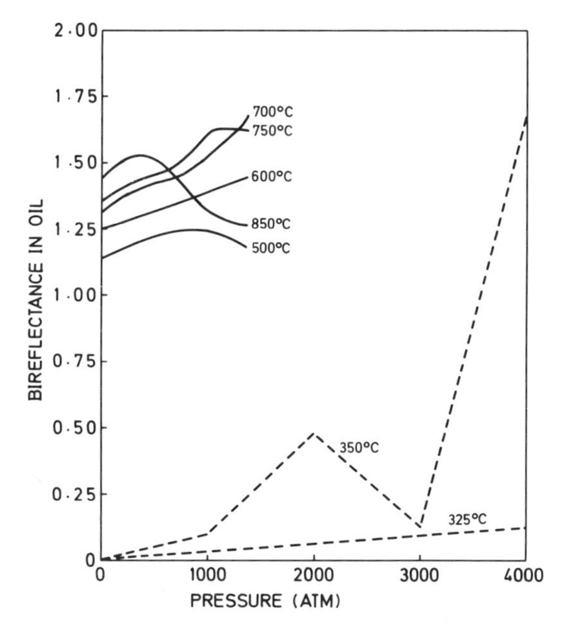

Fig. 16 Variation with pressure and preparation temperature of the oil bireflectances of a high volatile bituminous coal and an anthracite. (---) Vitrinite from high volatile bituminous coal, $\%R_{oil}(\max) = 1.00$ (530 nm). (After chandra, 1965.) (—) Vitrinite from anthracite, $\%R_{oil}(\max) = 2.63$ (wavelength unstated, but probably between 520 and 550 nm). (After Hryckowian *et al.*, 1967.)

C. Refractive Index and Absorption Index

There are relatively few results available for the variation of the fundamental optical parameters n and κ of carbonized vitrinites with temperature. Since these variables are derived from experimental determinations of reflectance, they are less likely to be used for diagnostic purposes in the same way as reflectance or bireflectance. Indeed, their principal function, as the later discussion shows, is to give an insight into the molecular structural changes that affect vitrinites during carbonization.

1. *Refractive Index*

Bond *et al.* (1958) produced data for the variation with carbonization temperature of the refractive indices of two high volatile bituminous coals and an anthracite. The refractive indices of the low rank coals rose to a maximum at ~700°C and then fell again, although no such trend was observed for the refractive indices of the anthracite. These early results were partially confirmed by Goodarzi and Murchison (1972), who also examined the behavior of the refractive indices of a low volatile bituminous vitrinite. The general pattern of change of refractive indices of vitrinites with temperature throughout the rank range appears to be similar (Fig. 17), namely a rise to a maximum at ~700°C, followed by a fall in the index which continues up to a temperature of 950°. The indices for the cokes from the low volatile bituminous vitrinite peak at a lower level of refractive index and at a lower temperature than do the indices of vitrinites of both lower and higher rank.

Although the experimental results have not yet been fully analyzed, variations of heating rate also produce alterations to the refractive index–temperature track (Fig. 18). For the reactive vitrinites in the middle of the bituminous range, it appears that as the heating rate increases, the peak for the refractive index becomes sharper, moving at the same time to a lower temperature (550–600°C at 60°C/min). The fall from the maximum at this high heating rate is dramatic and the refrac-

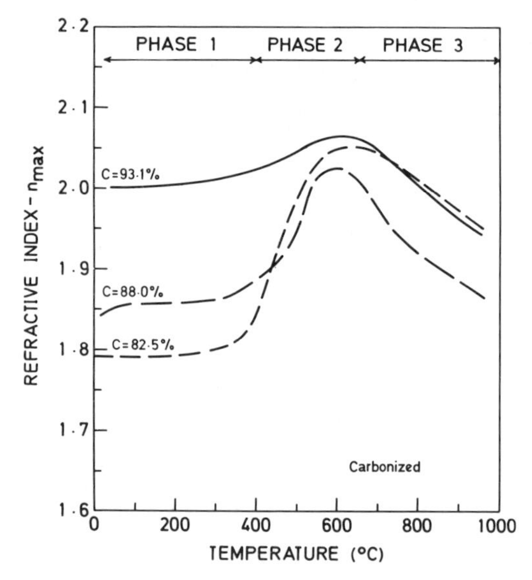

Fig. 17 Variation with preparation temperature of the maximum refractive indices (546 nm) of vitrinites from three coals carbonized over the temperature range 20–950°C. (– – –) Vitrinite from high volatile bituminous coal, $\%\bar{R}_{oil}(max) = 0.74$. (——–) Vitrinite from low volatile bituminous coal, $\%R_{oil}(max) = 1.25$. (———) Vitrinite from anthracite, $\%R_{oil}(max) = 3.08$. Heating rate 2.45°C/min, residence time 1 hr. (After Goodarzi and Murchison, 1972.)

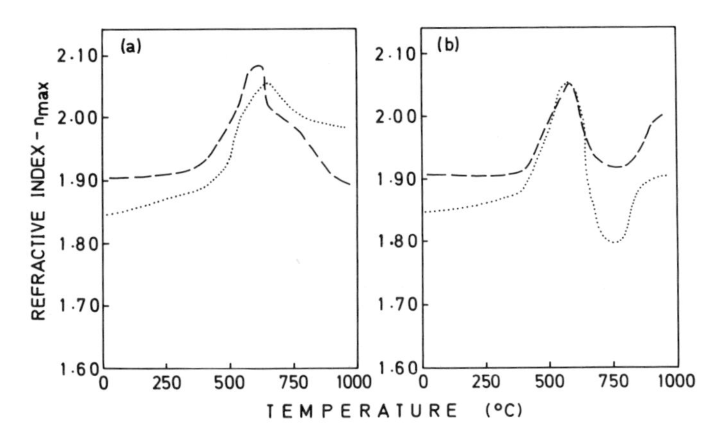

Fig. 18 Variation with preparation temperature of the maximum refractive indices (546 nm) of vitrinites from two coals carbonized over the temperature range 20–950°C at different heating rates. $\% R_{\text{oil}}(\text{max}) = 1.08$ (··········) and 1.55 (———). Residence time 1 hr, heating rate (a) 1°C/min, (b) 60°C/min. (After Goodarzi and Murchison, 1979.)

tive indices have begun to rise again at ~750°C, whereas at the lower heating rate of 1°C/min, the indices are still falling, as they were at a heating rate of 2.45°C/min (Fig. 17).

Increasing residence time produces a slow but gradual rise in refractive index with time, the index following a course that is similar to that for the reflectances under the same experimental conditions (Fig. 10) (Goodarzi and Murchison, 1977). The same authors (Goodarzi and Murchison, 1973a) have also given data for the changes in the refractive indices for vitrinites that were severely oxidized prior to carbonization (Fig. 19). The refractive index–temperature tracks for the three oxidized vitrinites are much closer together than if the vitrinites were carbonized fresh and they now peak at approximately the same temperature.

No data have been published that illustrate the variation of refractive index of carbonized vitrinites under pressure.

2. *Absorption Index*

Bond *et al.* (1958) also published results for the variation of absorption index with temperature for carbonized low rank and anthracitic coals. Many of the calculations of absorption index using Beer's relationship, however, yielded unreal values for κ^2, a situation that indicates experimental errors of considerable scale in the reflectances (Cook and Murchison, 1977). The positive values of absorption index obtained by Bond's group do not show any clear trend with temperature, although,

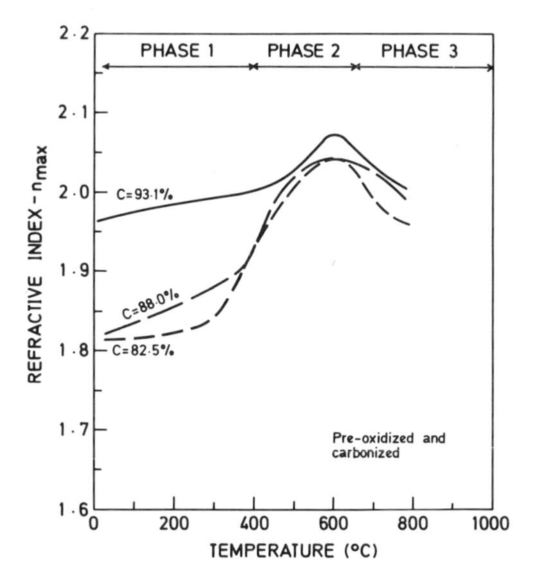

Fig. 19 Variation with preparation temperature of the maximum refractive indices (546 nm) of vitrinite from three coals carbonized over the temperature range 20–950°C after preoxidation for 14 days at 150°C. (– – –) Vitrinite from high volatile bituminous coal, $\%R_{oil}(max)$ = 0.74. (— — —) Vitrinite from low volatile bituminous coal, $\%R_{oil}(max)$ = 1.25. (———) Vitrinite from anthracite, $\%R_{oil}(max)$ = 3.08. Heating rate 2.45°C/min, residence time 1 hr. (After Goodarzi and Murchison, 1973a.)

in general, values for the carbonized anthracite are higher than those for the low rank vitrinites.

Results published by Goodarzi and Murchison (1972) show absorption index following a closely similar trend to reflectance as temperature rises, although the track for the low volatile bituminous vitrinite is clearly separated from the other tracks for lower and higher rank vitrinites above a temperature of 600°C (Fig. 20). Increasing the rate of heating causes the absorption indices of the more reactive bituminous rank vitrinites to rise much more rapidly (Fig. 21). Oxidation prior to carbonization causes the same crosscutting trends for the absorption index–temperature tracks to develop as they did for the reflectance–temperature tracks, but the relationships are not so clear (Fig. 22). With increasing residence time, absorption index, in the same way as reflectance and refractive index, slowly rises (Goodarzi and Murchison, 1977).

There are again no data available for the influence of pressure on the absorption indices of carbonized vitrinites.

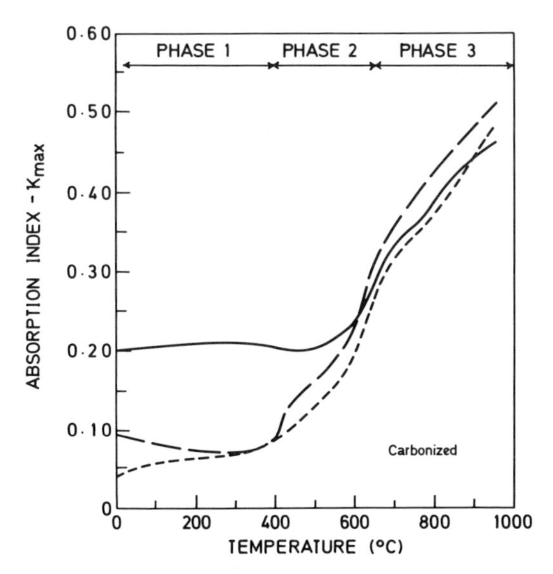

Fig. 20 Variation with preparation temperature of the maximum absorption indices (546 nm) of vitrinites from three coals carbonized over the temperature range 20–950°C. (– – –) Vitrinite from high volatile bituminous coal, $\%R_{oil}(max) = 0.74$. (——–) Vitrinite from low volatile bituminous coal, $\%R_{oil}(max) = 1.25$. (———) Vitrinite from anthracite, $\%R_{oil}(max) = 3.08$. Heating rate 2.45°C/min, residence time 1 hr. (After Goodarzi and Murchison, 1972.)

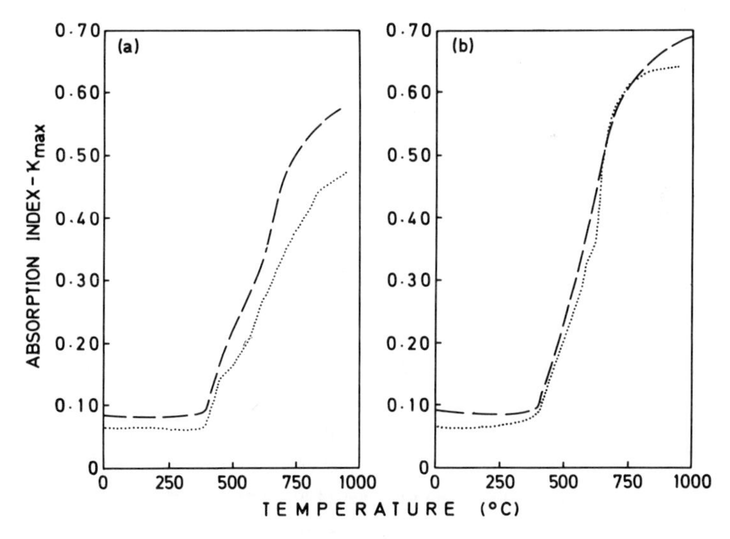

Fig. 21 Variation with preparation temperature of the maximum absorption indices (546 nm) of vitrinites from two coals carbonized over the temperature range 20–950°C at different heating rates. $\%R_{oil}(max) = 1.08$ (· · ·) and 1.55 (– – –). Residence time 1 hr, heating rate (a) 1°C/min, (b) 60°C/min. (After Goodarzi and Murchison, 1979.)

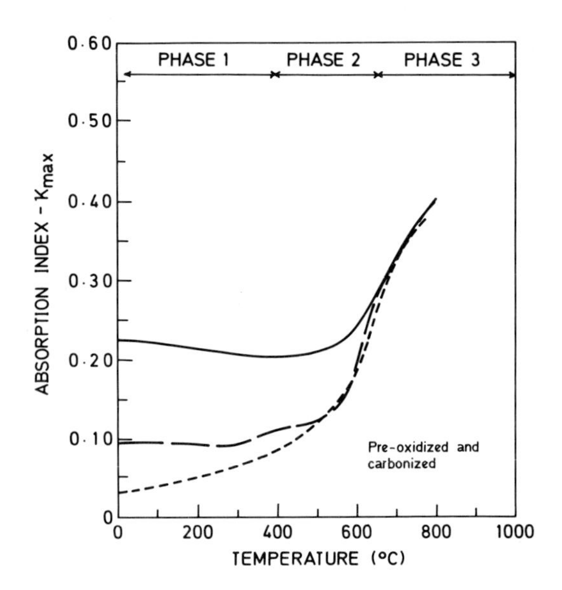

Fig. 22 Variation with preparation temperature of the maximum absorption indices (546 nm) of vitrinites from three coals carbonized over the temperature range 20–950°C after preoxidation for 14 days at 150°C. (– – –) Vitrinite from high volatile bituminous coal, $\%R_{oil}(max) = 0.74$. (——–) Vitrinite from low volatile bituminous coal, $\%R_{oil}(max) = 1.25$. (———) Vitrinite from anthracite, $\%R_{oil}(max) = 3.08$. Heating rate 2.45° C/min, residence time 1 hr. (After Goodarzi and Murchison, 1973a.)

V. DISCUSSION

A. Reflectance

There is agreement between the trends found by different workers for the relationship between reflectance and temperature as carbonization proceeds, but the agreement is only general and in no way specific. There has never been any deliberate attempt to standardize experimental conditions between laboratories and this may certainly be a cause for the lack of detailed correlation. Some of the discrepancies between the sets of data from different laboratories can certainly be attributed to known differences in the initial ranks of coals used, the heating rates, residence times, and other experimental factors.

One factor that certainly causes modification to the detailed course of the reflectance–temperature curves is the use of a smaller temperature interval (25°C) by Goodarzi and Murchison (1972) compared with larger intervals, often 100°C, by other investigators (Chandra and Bond, 1956; Ghosh, 1968; de Vries *et al.*, 1968). With the smaller interval, it becomes possible to relate, to some degree at least, the known changes of state that take place as carbonization proceeds in phase 2 (Fig. 1), to variations in the rate of change of reflectance with temperature (Fig. 5). The positions of the changes have been marked by reference to the reflectance curves, wherever there is a change of rate in reflectance close to the appropriate temperature.

The temperature interval alone will not explain all the differences between the reported data. For example, the correct relative levels between the reflectance–temperature tracks are maintained in Fig. 7a where Ghosh (1968) used a vitrinite of slightly higher starting rank and a higher heating rate than did Goodarzi and Murchison (1972). The data of Fig. 7b (Ghosh, 1968; Goodarzi and Murchison, 1979) are, however, conflicting and on other recorded experimental evidence, the relationship of the two reflectance–temperature curves in Fig. 7b should be reversed. There is good agreement, however, between the results of Goodarzi and Murchison (1972, 1979) and de Vries *et al.* (1968), despite the use of different heating rates to produce the data of Fig. 7d. It is possible, however, that the reflectance–temperature tracks of de Vries and co-workers may follow a lower trend at temperatures above 600°C. Figure 6 suggests that the reflectances measured by Chandra and Bond (1956) follow a substantially higher trend for equivalent temperatures than do the reflectances of Goodarzi and Murchison (1979), and consequently the discrepancy between the results of Chandra and Bond and those of de Vries and associates may be substantial. It should also be noted that the highest rank vitrinite carbonized by Chandra and Bond (1956) is quoted as having a carbon content of 92.0% (dry, ash-free), i.e., it is anthracitic in rank. The vitrinite, however, has an initial reflectance of a semianthracitic vitrinite and behaves optically on carbonization in a manner most similar to a low volatile bituminous vitrinite.

On the basis of experimental evidence, despite the broad agreement that exists in trends, it would be difficult at present to define sets of specific experimental conditions and then to predict the precise reflectance levels these would generate. The lack of specificity and some of the discrepancies between sets of data from different laboratories can certainly be attributed to variations induced by the different experimental factors and to the failure to take into account the effects of initial rank, heating rate, residence time, and other influences. For example, heating rate clearly can have a pronounced effect on the rate of reflectance rise of certain ranks of vitrinites that are carbonized. It has relatively little influence on the reflectances of high volatile bituminous and anthracitic vitrinites, but the reflectances of low volatile bituminous vitrinites are affected to some degree, while the medium volatile bituminous vitrinites show a marked reflectance change as heating rate rises. The data of Ghosh (1968), which show a decrease in the rate of rise of reflectance with increasing heating rate, are at the very least surprising; most other evidence suggests that the reverse relationship holds.

Results from a number of investigations are available for the effect of residence time during thermal treatment. A strong body of thought favors time as a secondary, but still important factor in the increase of coalification. Apart from the results of Ghosh (1968), experimental evidence supports this opinion. Even in the laboratory, it is possible to induce reflectance increases in vitrinites at temperatures well below the temperatures of the normally accepted decomposition points for vitrinites in short-term carbonizations. At atmospheric pressure, for example, after heating at 350°C for periods extending over many weeks, there is considerable evidence of the lowering of the decomposition point of vitrinites of true coking rank (medium to low volatile bituminous), since devolatilization vacuoles and mosaics develop (Goodarzi and Murchison, 1978). At any temperature, the influence of residence time on reflectance is the more noticeable the lower the rank of the vitrinite. The increase of reflectance is particularly striking over the first 2 weeks of treatment (Fig. 10). It should be emphasized, however, that the total reflectance increase, even after a residence time of 32 weeks, does not approach the order of the reflectance rise recorded by de Vries *et al.* (1968) for cokes held at constant temperatures above the vitrinite decomposition points over relatively short periods of 72 hr.

The limited response of vitrinite reflectance to oxidation effects, even when these are severe, is confirmed by the work of Goodarzi and Murchison (1973a). The modification of the reflectance–temperature track for fresh carbonized vitrinites that is caused by severe oxidation of the vitrinite prior to carbonization is, however, quite measurable, particularly in phases 1 and 2 of the carbonization process. Fresh vitrinites, when carbonized, display little if any change of reflectance up to a temperature of ~400°C, despite the occurrence of some important chemical changes at lower temperatures. The reflectance–temperature tracks of the carbonized preoxidized vitrinites of bituminous rank lie at a distinctly higher level up to a temperature of ~500°C. After the crossover point at this temperature, it is difficult to assess how much separation of the curves there is at higher carbonization temperatures (Fig. 11).

The data produced in the few investigations that introduce variations of pressure as well as temperature and time are the most conflicting. Huck and Patteisky (1964) and Chandra (1965) agree that varying the pressure at a constant temperature causes modification of reflectance, but Huck and Pattiesky maintain that reflectance falls with rising pressure, whereas Chandra observed reflectance rising with increasing pressure, the more surprising result. Hryckowian *et al.* (1967), in contrast, found that varying pressure has virtually no influence on reflectance when the temperature is held constant.

A possible explanation for the results of Hryckowian and co-workers is that they used anthracites for their studies, while Huck and Patteisky and Chandra experimented with more reactive coals of lower rank. Indeed, the known high resistance of anthracites to change is well illustrated in Fig. 23 in which the reflectance of the Middle Mammoth anthracite (initial oil reflectance 2.63%) is plotted against temperature for all pressures used in the experiments (i.e., up to ~1350 atm) by Hryckowian's group. The authors attribute the sharp reflectance rise in the region of 600°C to an observed increased volatile evolution that may be related to a change in the pore system of the coal or to the development of a plastic state. The two other development lines on the graph refer to the anthracitic vitrinite carbonized by Goodarzi and Murchison (1979) at atmospheric pressure and two heating rates of 1 and 60°C/min. The relatively close coincidence of all the tracks is added support for the opinion of Hryckowian's group that pressure has little influence on the reflectance, at least on the reflectance of anthracites. There is greater agreement between all the investigators that at constant temperature and pressure, increase of time will produce a reflectance rise. Final resolution of the differing opinions on the effect pressure has on the reflectance of macerals of lower rank at elevated temperatures will only come after more extensive systematic studies.

B. Bireflectance

Bireflectance is a property that gives an indirect estimate of the degree of ordering of the molecular structure of organic constituents. Thus, the increasing bireflectance of vitrinites with rising coalification reflects the

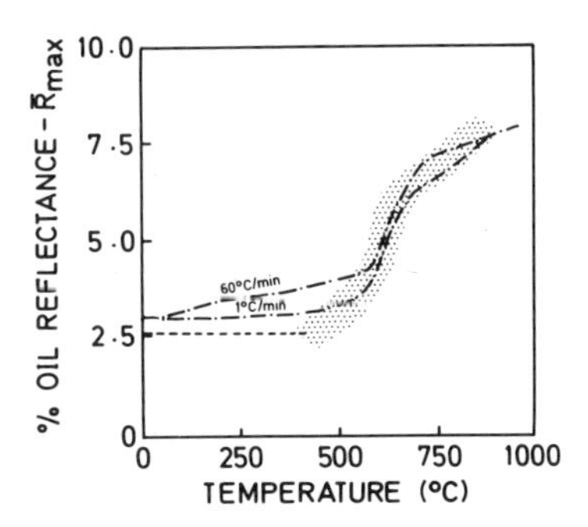

Fig. 23 Comparison between the variation with preparation temperature of the maximum oil reflectances of two anthracitic vitrinites, one carbonized at atmospheric pressure at two different heating rates, the other at pressures varying between 135 and 1350 atm plotted to show total spread of results. (– ·) $\%R_{oil}(max) = 3.11$ (546 nm). (After Goodarzi and Murchison, 1979.) Dotted area: $\%R_{oil}(max) = 2.63$ (wavelength unstated, but probably between 520 and 550 nm). (After Hryckowian *et al.*, 1967.)

progressive alignment of the "crystallites" (which are formed of groups of aromatic layers), which takes place as aromatic condensation proceeds. Figure 13 shows the bireflectance of a high volatile bituminous vitrinite falling to a minimum at ~400°C. This reduction in bireflectance indicates that the original inherent natural anisotropy is reduced, or even entirely destroyed, as the vitrinite passes through this temperature level. Most vitrinites of bituminous rank display this minimum bireflectance in this temperature region as the molecular structure breaks down when the vitrinite softens in the plastic stage of phase 2 of the carbonization process. Indeed, the vitrinite need not be what is regarded as a truly softening vitrinite for the bireflectance to decrease. The bireflectances of anthracitic vitrinites do not pass through a minimum since they do not soften. The original degree of ordering of the molecular structure of anthracitic vitrinites remains intact until rising temperature begins to improve the ordering at a temperature of ~500°C.

Beyond the region of disorder displayed by the bituminous rank vitrinites, the molecular structure begins to reorder itself. The crystallites soon attain a higher degree of ordering than that possessed by the original vitrinite although Fig. 13 shows that only when the coke of the high volatile bituminous vitrinite has reached a temperature of ~800°C does the level of ordering approximate to that of fresh anthracitic vitrinite. Between 800 and 850°C, both vitrinites display a marked rise in the rate of bireflectance increase. Molecular reorganizations during phase 3 of the carbonization process are often interrupted in this temperature region and it seems that the rate of ordering of the molecular structure is related to this interruption to the process.

Figure 12 contains the most comprehensive early data based on the bireflectances of carbonized coals, but these are in an inappropriate form for direct comparison with bireflectances plotted versus temperature. These data (Chandra, 1965) have therefore been transposed to the form of Fig. 24a, where they are compared with bireflectances measured on similar ranks of coal by Goodarzi and Murchison (1978) (Fig. 24b). It is assumed that all the transposed measurements were made on coals carbonized at ~1.25°C/min, the heating rate reported by Chandra and Bond (1956) for their earlier measurements. The general form of the resulting curves in the two plots is similar, but the relationships between the curves from cokes arising from similar ranks of coal are not. For example, the bireflectance of the low volatile bituminous vitrinite of Goodarzi and Murchison shows an extremely rapid rise, but there is no indication of a similar rapid bireflectance increase in the cokes from the vitrinite of similar rank cited by Chandra.

It must be stressed that the methods of measuring bireflectance in the

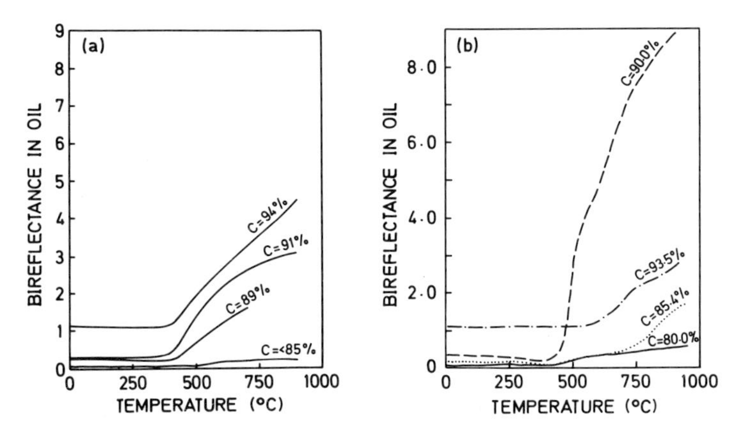

Fig. 24 Comparison of the variation with preparation temperature of oil bireflectances of vitrinites from similar ranks of coal carbonized under approximately the same conditions over the temperature range 20–950°C. (a) Modified after Chandra and Bond (1956) and Chandra (1963, 1965). Heating rate 1.25°C/min, residence time 1 hr, 530 nm. (b) After Goodarzi and Murchison (1978). Heating rate 1°C/min, residence time 1 hr, 546 nm.

two investigations were not the same. Chandra (1965) averaged all maximum and all minimum reflectance values before subtracting these to obtain bireflectances; Goodarzi and Murchison (1972) used the method described in Section III,F of this chapter. Comparison of the data of Fig. 24 do not support the contention by Chandra (1972) that using this method leads to exaggeration of the anisotropy. As employed here, the method seems to give a more realistic appraisal of how anisotropy develops and of the relationship between bireflectance tracks of cokes from vitrinites of different rank.

The differences between the two plots of Fig. 24 suggest that the original diagram published by Chandra (1965), and on which Fig. 12 is based, is an oversimplification. Used initially by Chandra to estimate the temperature to which thermally metamorphosed coals have been exposed, provided minimum pressure was involved, Chandra extended the use of the diagram to include predictions of when thermally metamorphosed coals had been subjected to high pressure. For example, the thermally metamorphosed coals described by Brown and Taylor (1961) were interpreted as having been subjected to high pressure because of their extremely high anisotropies. Such an interpretation, although it may be correct for these coals, does not preclude other factors that may also generate high bireflectances.

A factor that can clearly produce high bireflectances during carbonization at atmospheric pressure is rank; e.g., the low volatile bituminous vitrinite, illustrated in Fig. 14, is little affected by increase of heating rate, since even at 1°C/min the vitrinite develops very high bireflectance by the time it has been carbonized to 950°C. Equally true is that vitrinites of slightly lower rank can respond to increasing heating rates by developing high bireflectances, e.g., the high to medium volatile bituminous vitrinite, also illustrated in Fig. 14. Such high heating rates admittedly would not be encountered under natural conditions, but the possible influences of rate of heating should be taken into account when using bireflectance to estimate temperature levels or pressure effects during organic metamorphism.

Of other factors that may affect bireflectance, nothing is known of the influence of residence time. Bireflectance almost certainly does rise as the residence time is extended, particularly as mosaic structures form in the vitrinite particles, as reported by Goodarzi and Murchison (1977), after samples are heated for 32 weeks at 350°C. Likewise, little is known about the quantitative effect of oxidation on bireflectance, prior to carbonization. The few results available (Fig. 15) do not suggest that distinct and separate tracks are followed by the two differently treated vitrinites once they are carbonized. One distinct feature, however, stands out in each diagram. In the tracks of the vitrinites oxidized before carbonization, no minimum bireflectance occurs in the region of 400°C—instead, the bireflectance rises slowly. Further study of the influence of preoxidation would seem to be justified. Qualitative observations (Goodarzi and Murchison, 1976a) indicate high levels of anisotropy developing in preoxidized carbonized vitrinite particles, and similar observations were made by Goodarzi *et al.* (1975) when preoxidized vitrinites were carbonized under pressure.

With so few data available, the influence of pressure on bireflectance is difficult to evaluate. Increasing anisotropy in the natural coalification series is not unreasonably attributed, in part at least, to rising pressure. Although the experimental evidence is rather conflicting, even the scanty data available up to now generally confirm this trend (Fig. 16).

C. Variations in Refractive Index and Absorption Index in Relation to Molecular Structural Changes

The description of the trends of optical properties of vitrinites with rising temperature in Section IV has shown a close similarity between the rising patterns of air and oil reflectance and absorption index. Bireflectance also follows approximately the same general course with modification in detail. The refractive index curves at all levels of rank and under differing experimental conditions, however, show a pronounced contrast to the curves for other properties, displaying a reversal in the 600–700°C region, perhaps the most interesting single feature displayed by any of the curves.

Goodarzi and Murchison (1972, 1973a,b, 1976a,b, 1977, 1978, 1979) have discussed the relationship between the trends of different optical parameters and modifications to molecular structure under different experimental conditions. Although rise of reflectance can be correlated with increasing aromaticity, and changes of state in the carbonized product can be identified by variations in the rate of change of reflectance with rising temperature, fundamental molecular structural changes are most satisfactorily interpreted through the patterns of refractive and absorption indices. Some consideration of past x-ray investigations of coals is first necessary for this interpretation. In Section III only the general nature of the carbonization process was considered without any reference to the detailed molecular structural changes that take place during coalification or carbonization.

The structure of the relatively large molecules of coal macerals and their carbonized products is most satisfactorily considered in terms of condensed aromatic layers that may occur singly or in groups of two, three, or more layers which are stacked parallel to one another. Hirsch (1954), on the basis of his x-ray studies, concluded that the molecular structure of vitrinites becomes progressively more ordered as rank rises. The increased ordering is accompanied by a pronounced increase in aromaticity and a corresponding loss of aliphatic material. These changes in molecular structure influence the optical properties of vitrinites and other macerals substantially. For vitrinites of the normal coalification series, the molecular changes are reflected in two distinct optical patterns for refractive index and absorption index.

(i) A rise of the refractive indices of vitrinites takes place up to a carbon content of approximately 89%, when the index stabilizes or may even fall (see, e.g., McCartney and Ergun, 1958, 1967). The rise is attributable in part to improvement in stacking order.

(ii) A slow progressive rise of absorption index to approximately 89% carbon is followed by a much more rapid increase in this index to higher rank levels due to electronic absorption in the aromatic lamellas as they increase in diameter.

Coalification is then essentially a process of condensation of aromatic layers, ordering of these layers, and then flattening of this structure. The dispersion curves of refractive index and absorption index within the visible spectrum confirm this general development of aromatic condensation. As rank increases, the refractive index curve which rises from the red to the blue region for high volatile bituminous vitrinites, gradually changes until it is almost flat for low volatile bituminous vitrinites, and falls from the red to the blue region for anthracitic vitrinites. This pattern is consistent with materials that display a gradually increasing degree of aromatization and/or condensation with concomitant development of absorption bands in the ultraviolet region (Gilbert, 1960; Ergun and McCartney, 1960; McCartney and Ergun, 1967). Marshall and Murchison (1971) showed that this same pattern of change in the dispersion of refractive index held for vitrinites carbonized up to 750°C, while Goodarzi and Murchison (1973b) demonstrated the same shift in the refractive index curves of progressively oxidized vitrinites which show rising aromaticity with increasing oxidation (van Krevelen, 1961).

The molecular changes established by Hirsch through x-ray investigation and reflected in the behavior of the fundamental optical properties of vitrinites of the normal coalification series are brought about principally through temperature rise. Similar changes would therefore be expected in the optical properties of macerals of laboratory carbonized coals and industrial cokes. Parallel studies on carbonized and thermally metamorphosed coals were undertaken by Franklin (1951) and Diamond (1960) at approximately the same time as the work of Hirsch (1954) on fresh coals. The investigation yielding results that afford easiest comparison with the data of this chapter, however, was carried out earlier by Blayden *et al.* (1944) who examined the molecular structural changes that occurred in coals carbonized up to a temperature of 1750°C. Two principal parameters were employed to illustrate the molecular changes:

(i) L_c, which estimates the height of the stacks of aromatic layers, but also reflects such features as buckling of the aromatic layers due to input of thermal energy, and
(ii) L_a, which estimates the diameter of the aromatic layers.

Figure 25 illustrates the variations that take place in the stack heights and in the aromatic layer diameters for coals of bituminous and anthracitic rank (Blayden *et al.*, 1944). The L_c curves can be interpreted in the following way. An initial flattening and layering of the lamellas cause an increase in the heights of the aromatic layer groups in vitrinites of bituminous rank up to ~600°C. There is no modification of the molecular structure of the vitrinite of anthracitic rank. Over the next 350–400°C, there is disordering and/or disruption of some of the layer planes, probably due mainly to gas evolution, which leads to a reduction in the L_c dimensions of all the samples, including the anthracitic

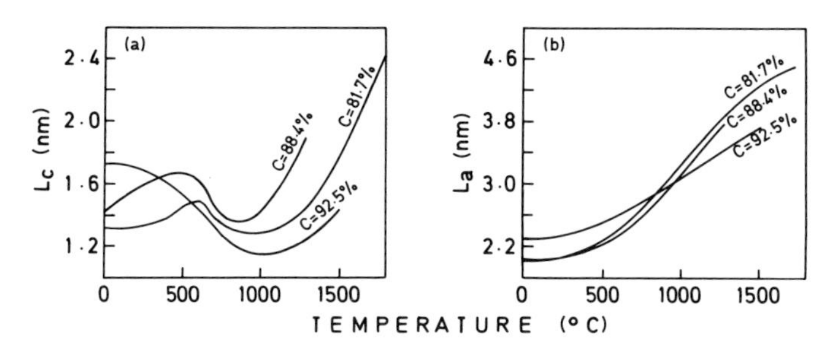

Fig. 25 Variation with temperature of the diameters of the aromatic layer groups (L_a) and of the heights of the aromatic layer groups (L_c) in high volatile bituminous coal (daf carbon content 81.7%), low volatile bituminous coal (daf carbon content 88.4%), and anthracite (daf carbon content 92.5%). (After Blayden *et al.*, 1944.)

vitrinite. Beyond a temperature of 1000°C, renewed stacking of the lamellae causes a further rise in the L_c curve. While these structural changes are occurring, there is a steady increase in the L_a curve, indicating gradual growth of the aromatic layer diameters of all the coals.

The similarity of the L_c and L_a curves of Fig. 25 to the curves in earlier figures of this chapter in which refractive index and absorption index are plotted as a function of temperature is quite clear. The behavior of the refractive index curves is obviously closely related to the changes that take place in the height of the crystallites and also to the number of aromatic lamellas present in the crystallites. The satisfactory correlation between the trends of the L_c and the refractive index curves suggests that refractive indices derived from reflectance determinations in two media are probably an adequate substitute for tracing gross molecular changes on carbonization. The same consideration applies to the substitution of absorption index for L_a. The relationships have recently been shown to be even more sensitive by Khavari-Khorasani *et al.* (1978) in bitumen studies.

The pattern of variation for L_c is largely reproduced in the refractive index curves of Fig. 17. The curve for the vitrinite of low volatile bituminous rank follows a different course, however, from the curves of the vitrinites of higher and lower rank. Goodarzi and Murchison (1972) interpreted this difference in relation to the differing behavior of vitrinites that would be regarded as essentially nongraphitizing (high volatile bituminous vitrinites) or nongraphitizing until a high temperature (>2000°C) was reached (anthracitic vitrinites) and the potentially graphitizing vitrinites (those of low volatile bituminous rank). The relationship between nongraphitizing and graphitizing carbons is considered by Franklin (1951) and van Krevelen (1961). There are pronounced differences between these two general types of carbon which account for the contrast between the trends of the refractive index curves in the 0–1000°C temperature region.

Modification of the experimental conditions causes the rate and amount of change affecting the molecular structure to alter. Goodarzi and Murchison (1973a), for example, showed that if a graphitizing vitrinite were oxidized, then the refractive index curve for its carbonization products would correspond much more closely with those of the non-graphitizing vitrinites (Fig. 19). It is also clear from the refractive index curves of Fig. 18 that increasing heating rate causes quite marked alterations to the way in which the molecular structures of vitrinites respond to the input of thermal energy. At a low heating rate (1°C/min) the decrease in stack height and the associated buckling goes on over a temperature range that extends for several hundreds of degrees. At a high heating rate, not only is the increase in stack height achieved rapidly, it also reaches its peak at a lower temperature. The disordering process is equally dramatic, but again, at ~800°C, the aromatic layer groups have begun to rebuild, a feature not seen below a temperature of 1000°C at lower heating rates.

The variations that are produced in the absorption index curves (Figs. 20–22) are much less striking. The more rapid increase in aromatic layer diameter with increasing heating rate is, however, quite clear for the vitrinites of medium and low volatile rank. On the indirect basis of absorption index, both vitrinites (Fig. 21b) have larger aromatic layer diameters at 1000°C when heated rapidly than when heated slowly. Oxidation prior to carbonization produces three absorption index curves that are virtually coincident from ~600°C, which can only indicate close similarity in the rate of growth of the layer diameters of the three vitrinites.

VI. CONCLUSIONS

1. The optical properties of the principal coal constituent, vitrinite, show wide-ranging changes when carbonized over the temperature range 0–1000°C. Although the optical properties of graphite are not reached, at 950°C vitrinites of low volatile bituminous rank and vitrinites of medium volatile bituminous rank when subjected to a high heating rate (~60°C/min) can attain reflectances in air and oil at 546 nm of ~27% and ~15%, respectively [cf. the 0001 plane of a natural surface of graphite: R_{air}(max) = 31.40% and R_{oil}(max) = 17.85% at 546 nm (Kwiecinska *et al.*, 1977)].

2. There are disagreements between the results produced by different laboratories for the variation of the optical properties of coals and vitrinites with temperature. None of the optical properties, either measured or derived, can yet be regarded as sufficiently certain to allow a precise relationship between specific optical properties and particular sets of experimental conditions. Nevertheless, the data obtained so far are useful in both industrial and natural contexts.

3. The lack of specificity of the optical properties in relation to experimental conditions can be largely related to insufficient data being available on the effect that different factors play in modifying the course of optical properties of organic matter with temperature during carbonization and thermal metamorphism. The principal influences are rank, temperature level, heating rate, residence time, degree of oxidation prior to carbonization, and pressure. Rank and temperature level have an importance that is apparent throughout the coalification series, while heating rate has a more restricted influence, primarily affecting the medium to low volatile bituminous vitrinites. Residence time apparently has a less dramatic effect, but one that may be quite general within the coalification series. The further experimentation that is required to understand this influence more clearly will have greater relevance to rank studies in the natural environment. Oxidation, prior to carbonization, despite its profound effect on other properties of coals, has relatively little influence on the course of the quantitative optical parameters during carbonization, although the shift in refractive index of preoxidized carbonized vitrinites is marked. The response of severely oxidized vitrinites that would normally react strongly to heating changes when fresh, has not, however, been established, but the effect under these circumstances might be pronounced. There are generally too few results from experiments involving pressure to be sure about its precise influence.

4. Because of these constraints, the directly measurable optical properties, reflectance and bireflectance, are perhaps more satisfactory when applied to thermal studies in a natural rather than in an industrial environment. The coking industry seeks highly specific data to promote an overriding aim, namely the beneficiation of cokes. In the natural environment, where relative trends are often acceptable, a less rigorous approach can be tolerated. The use of reflectance and bireflectance in rank studies of regions and of borehole successions situated where parts of the crust have been subjected to high heat flows has been successful (see, e.g., Ridd *et al.*, 1970; Jones and Creaney, 1977). The combination of reflectance–bireflectance data, however, with quantitative information on the textural characteristics (mosaic size, proportion of mosaic

types, etc.) may yield information that is sufficiently specific for many purposes of the coke technologist, even at this stage of development.

5. The two indirectly determined variables, refractive index and absorption index, have their most important role in revealing the changes that take place in molecular structure during carbonization. Recent studies on bitumens (Khavari-Khorasani *et al.*, 1978) have shown that a sensitive correlation exists between refractive index and the heights of the aromatic layer groups (L_c) and also between the interlayer spacing between the aromatic groups (d-spacing), confirming the generality of the relationship between refractive index and the stack heights of the aromatic layer groups referred to earlier in this chapter. The less well-defined relationship between the structural and optical parameters for vitrinites is perhaps due to the greater heterogeneity and more complex composition of vitrinites than bitumens which cause changes of state during the carbonization of vitrinites to take place gradually over a range of temperature rather than at a specific temperature. Absorption index, which reflects the increasing diameter of the aromatic layer groups, can often be replaced by the reflectance for this purpose, since reflectance follows a similar trend to absorption index.

6. Finally, for the future, the influence of temperature on the optical properties of other macerals, particularly those of the liptinite group, will require examination. Furthermore, while the influence of certain factors in carbonization and during thermal metamorphism is becoming increasingly understood, the role of pressure is obviously complex and unclear in both industrial and natural situations. Below the decomposition point, when coals are undergoing so-called normal coalification, progressive alignment of the aromatic lamellae of the macerals takes place under rising overburden pressure giving increased bireflectance within a uniaxial symmetry. That pattern can easily be distorted by lateral pressures during orogenic movements, producing the lower biaxial optical symmetry (Cook and Murchison, 1972; Jones *et al.*, 1973). Overburden or lateral pressures (or indeed pressures within a coke oven) that operate when temperatures are sufficiently high to bring the organic matter close to or through the decomposition point probably give one of the most complex situations that can exist in the development of optical properties of organic matter. More extensive and controlled experimentation than has so far been undertaken will be needed to gain complete understanding of how the course of optical properties is modified when these different combinations of variables are at play.

ACKNOWLEDGMENTS

The author would like to thank the Natural Environment Research Council for grants (No. GR/3/1924—"The Content and Distribution of Organic Residues in the Rocks of

Northern England," and No. GR/3/1783—"Organic Geochemistry and Petrology of Recent and Fossil Sediments Including Coals"). He is also most grateful to Professor A. C. Cook for critically reading the manuscript and to Miss Y. Rogerson and Mrs. A. Summerbell for their considerable assistance in the typing of the manuscript and in the preparation of diagrams.

REFERENCES

Abramski, C., and Mackowsky, M.-T. (1952). *In* "Handbuch der Mikroskopie in der Technik (H. Freund, ed.), Vol. II, pp. 311–410. Umschau-Verlag, Frankfurt.

Beilby, G. (1922). *Fuel* **1,** 225–238.

Benedict, L. G., and Berry, W. F. (1964). "Recognition and Measurement of Coal Oxidation," 41 pp. Bituminous Coal Res., Monroeville, Pennsylvania.

Benedict, L. G., Thompson, R. R., Shigo, J. J., and Aikman, R. P. (1968). *Fuel* **47,** 125–143.

Berkowitz, N. (1967). *Symp. Sci. Technol. Coal, Dep. Energy, Mines Resour., Ottawa* pp. 149–155.

Blayden, H. E. (1969). *J. Chim. Phys.* Spec. No., pp. 15–20.

Blayden, H. E., Gibson, J., and Riley, H. C. (1944). *Proc. Conf. Ultra-fine Struct. Coals Cokes, 1943, BCURA, London* pp. 176–231.

Bond, R. L., Chandra, D., and Dryden, I. G. C. (1958). *Rev. Ind. Miner.* Spec. No., pp. 171–181.

Bostick, N. H. (1971). *Geosci. Man* **3,** 83–92.

Brooks, J. D., and Taylor, G. H. (1965). *Nature* (*London*) **206,** 697–699.

Brown, H. R., and Taylor, G. H. (1961). *Fuel* **40,** 211–224.

Chandra, D. (1958). *Econ. Geol.* **53,** 102–108.

Chandra, D. (1962). *Fuel* **41,** 185–193.

Chandra, D. (1963). *Fuel* **42,** 69–74.

Chandra, D. (1965). *Econ. Geol.* **60,** 621–629.

Chandra, D. (1972). *Fuel* **51,** 88.

Chandra, D., and Bond, R. L. (1956). *Proc. Int. Comm. Coal Petrol.* No. 2, pp. 47–51.

Cook, A. C., and Murchison, D. G. (1972). *Fuel* **51,** 180–184.

Cook, A. C., and Murchison, D. G. (1977). *J. Microsc.* (Oxford) **109,** 29–40.

Cook, A. C., Murchison, D. G., and Scott, E. (1972). *Fuel* **51,** 180–184.

de Vries, H. A. W., Habets, P. J., and Bokhoven, C. (1968). *Brennst.-Chem.* **49,** 105–110.

Diamond, R. (1960). *Phil. Trans. R. Soc., Ser. A* **252,** 193–223.

Dubois, J., Agache, C., and White, J. L. (1970). *Metallography* **3,** 337–369.

Ergun, S., and McCartney, J. T. (1960). *Fuel* **39,** 449–454.

Franklin, R. E. (1951). *Proc. R. Soc., Ser. A* **209,** 196–218.

Ghosh, T. K. (1968). *Econ. Geol.* **63,** 182–197.

Gilbert, L. A. (1960). *Fuel* **39,** 393–400.

Goodarzi, F., and Murchison, D. G. (1972). *Fuel* **51,** 322–328.

Goodarzi, F., and Murchison, D. G. (1973a). *Fuel* **52,** 164–167.

Goodarzi, F., and Murchison, D. G. (1973b). *Fuel* **52,** 90–92.

Goodarzi, F., and Murchison, D. G. (1976a). *Fuel* **55,** 141–147.

Goodarzi, F., and Murchison, D. G. (1976b). *J. Microsc. (Oxford)* **106,** 49–58.

Goodarzi, F., and Murchison, D. G. (1977). *Fuel* **56,** 89–96.

Goodarzi, F., and Murchison, D. G. (1978), *Fuel* **57,** 273–284.

Goodarzi, F., and Murchison, D. G. (1979). *Fuel* to be published.

Goodarzi, F., Hermon, G., Iley, M., and Marsh, H. (1975). *Fuel* **54,** 105–112.

Hirsch, P. B. (1954). *Proc. R. Soc., Ser. A* **226,** 143–169.

Honda, H., Kimura, H., and Sanada, Y. (1971). *Carbon* **9,** 695–697.

Hryckowian, E., Dutcher, R. R., and Dachille, F. (1967). *Econ. Geol.* **62,** 517–539.
Huck, G., and Patteisky, K. (1964). *Fortschr. Geol. Rheinl. Westfalen* **12,** 551–558.
International Committee for Coal Petrology (ICCP) (1971). "International Handbook for Coal Petrography," Supplement to the 2nd Edition. CNRS, Paris.
Jones, J. M., and Creaney, S. (1977). *J. Microsc. (Oxford)* **109,** 105–118.
Jones, J. M., Murchison, D. G., and Saleh, S. A. (1973). *Proc. Yorks. Geol. Soc.* **39,** 515–525.
Juckes, L. M., and Pitt, G. J. (1977). *J. Microsc. (Oxford)* **109,** 13–21.
Khavari-Khorasani, G., Blayden, H. E., and Murchison, D. G. (1978). *J. Microsc. (Oxford)* to be published.
Kisch, H. J., and Taylor, G. H. (1966). *Econ. Geol.* **61,** 343–361.
Kwiecinska, B., Murchison, D. G., and Scott, E. (1977). *J. Microsc. (Oxford)* **109,** 289–302.
McCartney, J. T., and Ergun, S. (1958). *Fuel* **37,** 272–282.
McCartney, J. T., and Ergun, S. (1967). *U.S. Bur. Mines, Bull.* No. 641, 49 pp.
Mackowsky, M.-T. (1961). *4th Int. Conf. Coal Sci., Le Touquet, Fr.* Pap. No. 91, 16 pp.
Marshall, C. E. (1936). *Trans. Inst. Min. Eng.* **91,** 235–260.
Marshall, C. E. (1945). *Fuel* **24,** 120–126.
Marshall, R. J., and Murchison, D. G. (1971). *Fuel* **50,** 4–22.
Marshall, R. J., and Murchison, D. G. (1972). *Fuel* **51,** 252.
Patrick, J. W., Reynolds, M. J., and Shaw, F. H. (1973). *Fuel* **52,** 198–204.
Ramdohr, P. (1928). *Arch. Eisenhuettenwes.* **1,** 669–672.
Ridd, M. F., Walker, D. B., and Jones, J. M. (1970). *Proc. Yorks. Geol. Soc.* **38,** 75–103.
Sanada, Y., Furuta, T., Kimura, H., and Honda, H. (1973). *Fuel* **52,** 143–148.
Stach, E., Mackowsky, M.-T., Teichmüller, M., Taylor, G. H., Chandra, D., and Teichmüller, R. (1975). "Stach's Textbook of Coal Petrology." Borntraeger, Berlin.
Taylor, G. H. (1961). *Fuel* **40,** 465–472.
Taylor, G. H., and Brooks, J. D. (1965). *Carbon* **3,** 185–193.
Ueji, T. (1932). *Suiyokai-Shi* **7,** 632–637.
van Krevelen, D. W. (1961). "Coal." Elsevier, Amsterdam.

Chapter 32

Reactivity of Heat-Treated Coals

O. P. Mahajan *P. L. Walker, Jr.*
DEPARTMENT OF MATERIALS SCIENCE AND ENGINEERING
THE PENNSYLVANIA STATE UNIVERSITY
UNIVERSITY PARK, PENNSYLVANIA

I. INTRODUCTION

The importance of the conversion of coal to gaseous fuels has had a curious history in the United States. In the first half-century, the production of low and medium Btu gas via reactions with steam, air, and CO_2 was of considerable importance as a source of energy for industry and the home. The coals primarily used were obtained from the bituminous and anthracite deposits. But with the abundance of cheap natural gas in the 1950s and 1960s, coal gasification plants became dormant.

Interest in coal gasification has suddenly been revived, initially because of interest in the production of synthetic natural gas and lately also because of the realization that industry will again need low and medium Btu gas produced from coal. The interest not only encompasses the coals of the eastern United States but also, in a most significant way, focuses on the western subbituminous and lignite deposits. Interest in

ISBN 0-12-399902-2

the western coals is stimulated by their high reactivity as well as their abundance.

Ironically, perhaps, even though there was little interest in coal gasification processes and research in the 1950–1970 period, during this time we probably obtained our greatest understanding of the fundamentals of the gasification of carbon. The impetus for this understanding was concern about the gasification of moderator graphite used in high temperature gas-cooled nuclear reactors. These studies have given us the scientific base to better understand the gasification of more complex coal-derived chars. In fact, we now understand the major factors which affect gasification rates. In this chapter we consider approaches to measure gasification rates and then discuss how rates which are found to be widely different for coal-derived chars depend on measurable properties of the chars. This chapter reproduces much of the material found in a Department of Energy report (Mahajan and Walker, 1978).

II. EXPERIMENTAL MEASUREMENT OF REACTIVITY

It was discussed in Volume I, Chapter 4, that when microporous coals are thermally heated, they lose volatile matter, planar regions grow in size, and cross-links are broken. The heat treatment of coals invariably results in the production of some kind of coke or char.[†] The structure of a char and, hence, its reactivity are influenced in a complex manner by a number of variables, such as heating rate, maximum heat treatment temperature (HTT), soak time at maximum temperature, atmosphere under which heating occurs, particle size, inorganic impurities present and their distribution in the coal precursor, maceral composition, and whether the coal behaves as a thermoplastic or thermosetting material. In order to have a meaningful comparison of char reactivities in different gasification atmospheres, we have used in this laboratory the same set of chars derived from the same size fraction (40 × 100 mesh) of coals of different rank. Noncaking coals were carbonized in a fluidized-bed reactor, while the caking coals were carbonized in a horizontal tube reactor. In each case, coals were heated at a rate of 10°C/min to 1000°C. An inert atmosphere of N_2 was used during heat treatment. Soak time at 1000°C was 2 hr. Chars prepared from noncaking and weakly caking coals were used as such, whereas chars prepared from strongly caking coals were reground and 40 × 100 mesh fractions used for reactivity measurements.

[†] If, upon heating, the coal precursor goes through a liquid state, the resulting solid is termed a coke. If, however, negligible softening occurs, the resulting solid is termed a char. For convenience, we have used the term char throughout this chapter.

When a thermobalance is used for reactivity measurements, the chars can be prepared from the coal precursors in the thermobalance itself. Jenkins *et al.* (1973) found that the reactivity of a char to air (1 atm) at 500°C was essentially the same whether the char was prepared in the thermobalance or in a fluidized-bed reactor under identical experimental conditions. As discussed in Volume I, Chapter 4, chemisorption of oxygen on freshly prepared chars upon exposure to air, even at ambient temperature, can result in significant alteration of char pore structure, which in turn might affect its subsequent reactivity. Therefore, in order to desorb the chemisorbed oxygen, the chars prepared in the fluidized and horizontal tube reactors must again be heated, prior to reactivity measurements, in an inert atmosphere up to the maximum HTT used during the char preparation.

Reactivities of chars can be measured by the methods (techniques) now discussed.

A. By Weighing

There are a number of commercial thermogravimetric analyzer (TGA)† units available to measure weight changes during chemical reactions. Each unit comprises basically two components—an electrobalance having a sensitivity in the microgram range and a linear temperature programmer. In this laboratory we have used two different TGA units for reactivity measurements: a Fisher TGA unit, Series 300, and a DuPont 951 TGA unit. The two units can handle maximum loads of 2.5 and 1.0 g, respectively. However, the starting weight of a char used for reactivity measurements is usually a few milligrams (<10 mg). This is because under the chosen experimental conditions char reactivity should be independent of bed depth (weight). That is, resistance for diffusion of reactant gas molecules down through the bed should be minimal.

For most of the reactivity studies in this laboratory we have used a Fisher TGA system. Therefore, we describe in detail the experimental procedure followed for this particular unit. However, with minor modifications the procedure should be valid for other TGA units. For reactivity measurements, the char sample is contained in a platinum pan which in turn is supported by a nickel hangdown wire connected to one end of an electrobalance. The thermocouple can be adjusted manually; it is maintained at a fixed position, ~1 mm below the sample pan. To allow for a controlled atmosphere during the experimental run, a

† Thermogravimetric analysis of coals and coal ashes is described in Chapter 37.

quartz tube is positioned around the hangdown wire and sample. For heating, a sample furnace is positioned around the quartz tube.

In the TGA unit, gases can be introduced through the conventional inlet, which is located at the rear of the balance chamber (the chamber has a volume of 4 liters), or through the conventional outlet, which is located in the thermocouple assembly; this assembly can be connected to the quartz reactor through a ground glass joint. When the former flow system (referred to as downward flow in the text) is used during reactivity measurement, the gasification rates obtained may be erroneous for the following reason. During a reactivity run when N_2 (used during preheat treatment of the sample) is replaced by the reactant gas at the reaction temperature, a volume of about 4 liters of N_2 needs to be displaced by the reactant gas. This means that until the outlet gas concentration equals the inlet reactant gas concentration, the char will be gasified at progressively increasing partial pressures of reactant gas. Since gasification rates are dependent on partial pressure of reactant gas, as is shown later in this chapter, the char reactivity measured in varying partial pressures of reactant gas will be different from that measured in the "pure" reactant gas, i.e., in the absence of dilution of reactant gas by N_2.

For these reasons, Soledade (1976a) used the conventional gas outlet in the TGA unit as the inlet for introducing gases. Using this flow system (referred to as upward flow in the text), the volume of N_2 to be displaced by reactant gas, before the char sample "sees" the inlet pressure of reactant gas, is less than 50 cm^3. In our reactivity work we have used a gas flow rate of 300 cm^3/min. Thus, when N_2 is replaced by reactant gas, it will take less than 1 min to displace N_2. The gases were vented off through a port located below the weighing chamber. During reactivity runs, N_2 is continuously passed through the weighing chamber to avoid any possible damage to the weighing mechanism from the hot and/or corrosive product gases; N_2 is vented off to atmosphere through the same port as that used for the reactant gas.

Soledade (1976a) used the following procedure to measure char reactivity. Prior to making reactivity runs, the system was flushed with N_2 (300 cm^3/min) for 30 min to displace the air present. The char sample was then heated (at a rate of 20°C/min) from ambient to the maximum HTT seen by the char during its preparation and held at this temperature until sample weight became essentially constant. During heat treatment, the char lost weight due to desorption of chemisorbed oxygen as CO and CO_2† Following heat treatment, the char sample was cooled to

† During heat treatment, the char also showed an apparent weight loss due to buoyancy effects. Sample weights prior to onset of gasification at the chosen reaction temperature were corrected for buoyancy effects.

reaction temperature at a rate of 20°C/min. The char was held at reaction temperature for 20 min to allow for temperature stabilization. Nitrogen was then replaced by the reactant gas at the same flow rate. Changes in sample weight due to gasification were then recorded on a strip chart recorder as a function of time.

The effect of the gas flow system used during reactivity measurements is illustrated by burn-off versus time plots (Fig. 1) obtained for a Texas

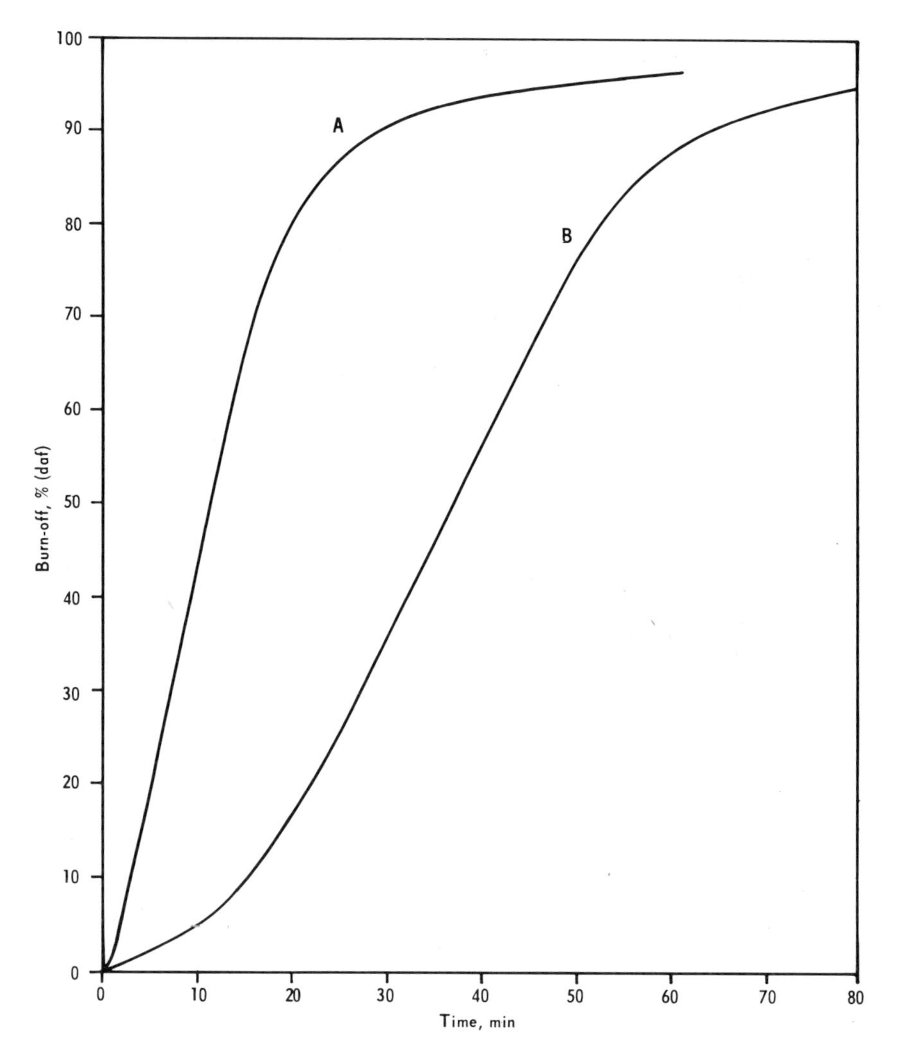

Fig. 1 Effect of gas flow system on reactivity of a lignite char in air at 500°C as measured in a TGA unit. (A) upward flow; (B) downward flow. Effect is related to amount of N_2 to be displaced before obtaining an air environment. (From Soledade, 1976b.)

lignite char when it is reacted with air (1 atm) at 500°C. It is seen that at any stage gasification rate is appreciably greater when the upward gas flow system is used instead of the conventional downward flow system. These results bring out clearly the importance of using the proper gas flow system to obtain unambiguous gasification rates.

Because of the large effect of temperature on rates of char gasification, it is critical always to duplicate closely positioning of the thermocouple relative to the sample pan. Soledade (1976b) has suggested that the TGA system be calibrated by determining the temperatures corresponding to the decomposition of hydrated calcium oxalate ($CaC_2O_4 \cdot 2H_2O$) to CaC_2O_4, $CaCO_3$, and CaO.

The carbon–hydrogen reaction is thermodynamically the least favorable gasification reaction (Walker *et al.*, 1959). Conversion to methane is enhanced at high pressures. In this laboratory, the char–hydrogen reaction was studied at 980°C and a H_2 pressure of 400 psi (Tomita *et al.*, 1977). It was ascertained that under these conditions the reaction was not limited by equilibrium.

Tomita *et al.* (1977) used a DuPont 951 TGA balance, in conjunction with a 990 Thermal Analyzer, to monitor weight changes during the hydrogasification reaction. A schematic diagram of the balance and the gas flow system used is shown in Fig. 2. A platinum pan (6 × 7 × 5 mm) containing ~10 mg of char was suspended from the quartz beam of the balance. The flexible end of the Chromel–Alumel thermocouple, which is attached to the balance housing, was placed in close proximity to the sample. The reactor was made of a quartz tube (25 mm, od) with a reduced end on one side and an aluminum retainer ring on the other side which fitted into the balance housing. The H_2 flow was divided into two parts. The gas entered through inlets A and C and was vented through outlet B; this outlet is a hole in the reactor tube near the metal ring. Outlet D was always kept closed. The volume of the quartz reactor tube was reduced to a minimum to ensure a rapid replacement of the

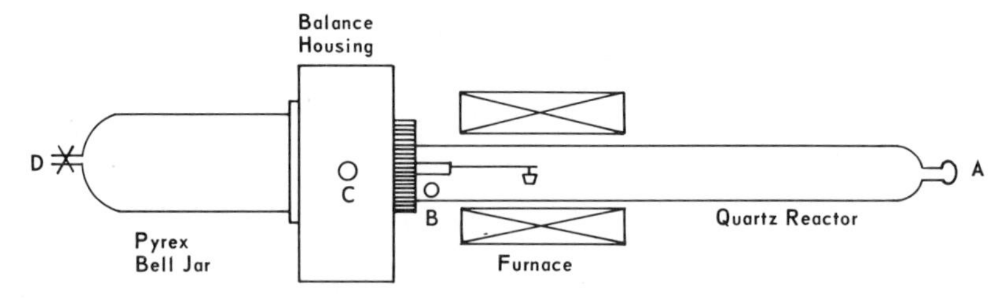

Fig. 2 Schematic diagram of a horizontal TGA balance used to measure gasification rates at elevated pressures in an autoclave. A and C are gas inlets, B and D are gas outlets.

inert atmosphere by H_2. The reactor tube was surrounded by a tube furnace† (wound with molybdenum wire) of inner diameter only slightly greater than the outer diameter of the reactor tube. The balance assembly along with the reactor and furnace was mounted in a carbon steel pressure vessel (built by Autoclave Engineers, Erie, Pennsylvania) which was provided with a quick opening door with Conax pressure-seal outlets. After introducing the sample, the pressure vessel was closed and evacuated, following which it was pressurized with N_2 to 400 psi. Helium at 400 psi was introduced into the reactor at a flow rate of 1.2 liters/min (STP). After a period of 15 min, the furnace was activated to raise the sample temperature to 980°C at a heating rate of 20°C/min. The temperature was kept constant at 980°C for 15 min to ensure thermal stability. At this stage, helium in the reactor tube was replaced by H_2 at 400 psi at the same flow rate. Sample weight was then continuously recorded. The volume of the reactor tube was ~120 cm^3. Therefore, using a flow rate of 1.2 liters/min (STP) or 41 cm^3/min at 400 psi, it took ~3 min to displace helium by H_2 over the char sample. In all reactivity measurements, starting time for the reaction was considered to be when helium was displaced by H_2.

Tomita *et al.* (1977) ascertained in a few preliminary runs that the use of a quartz bucket for a given char gave essentially the same weight losses at different reaction times as the use of a platinum bucket, indicating that the platinum bucket catalyzed the hydrogasification reaction to a negligible extent.

Johnson (1974) used a thermobalance to study the kinetics of bituminous coal char gasification with gases containing steam and hydrogen at pressures between 1 and 70 atm and reaction temperatures in the range 815–1093°C. The apparatus is shown schematically in Fig. 3. The coal sample (20 × 40 mesh) was contained in the annular space of a wire mesh basket bound on the inside by a hollow stainless steel tube and on the outside by a wire mesh screen. In order to facilitate mass and heat transfer between the carbon bed and the gaseous environment, the bed thickness was only two to three particle diameters. Sufficiently large gas flow rates were used in the reactor so that gas conversion during the reaction was minimal.

For determining reactivity, the following procedure was used. The wire mesh basket containing the coal/char particles was initially held in an upper cooled portion of the reactor in which a downward flow of an

† The DuPont furnace (which is an integral part of the TGA system) is a low capacity furnace. It was found to be ineffective to raise the sample temperature to the desired reaction temperature, i.e., 980°C, at a H_2 pressure of 400 psi.

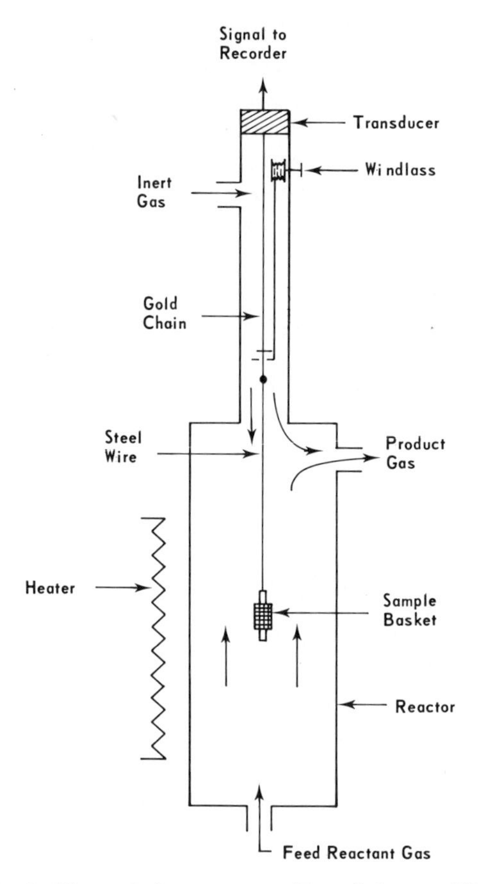

Fig. 3 Thermobalance reactor. (From Johnson, 1974.)

inert gas was maintained. When the desired temperature and pressure conditions were established in the lower portion of the reactor in the presence of the reactive gas, the basket was lowered into the heated reaction zone; this procedure took 5–7 sec. The sample took ~2 min to attain reaction temperature. The sample was kept in the heated zone for different intervals of time while its weight was recorded continuously. The experimental run was terminated by raising the basket back to the upper cooled portion of the reactor. During an experimental run, Johnson measured the dry feed gas flow rates by an orifice meter and the dry product gas flow rates by a wet test meter. Samples of product gas were analyzed periodically by a mass spectrometer. Feed and product steam flow rates were measured gravimetrically.

B. By Gas Analysis

Perusal of the literature shows that a number of experimental approaches have been used to measure char reactivities by gas analysis. Among these approaches were those used by Ergun (1956), Daly and Budge (1974), Fuchs and Yavorsky (1975), Kayembe and Pulsifer (1976), and Hippo (1977). Basically in each of these approaches product gases are analyzed by suitable techniques such as gas chromatography, mass spectrometry, and infrared. From a material balance, the amount of carbon gasified at different time intervals can be calculated. A small fixed-bed or fluidized-bed reactor is used under isothermal conditions; the fluid bed offers the advantage of more efficient heat and mass transfer between reactant and product gases and char particles.

Recently, Hippo (1977) used a fluidized-bed reactor to measure reactivities in steam of chars derived from raw and calcium-exchanged lignite. A schematic diagram of the apparatus is shown in Fig. 4. The gas inlet section consisted of a water feed system and a preheater. Water was contained in a steel feed tank (B). Helium was bubbled through water to remove dissolved O_2. Steam was vaporized in a boiler (D) which was equipped with an automatic feed pump (C). The desired steam pressure was obtained by controlling the electrical power input. Steam left the generator through $\frac{1}{4}$-in. copper tubing; excess steam was discharged

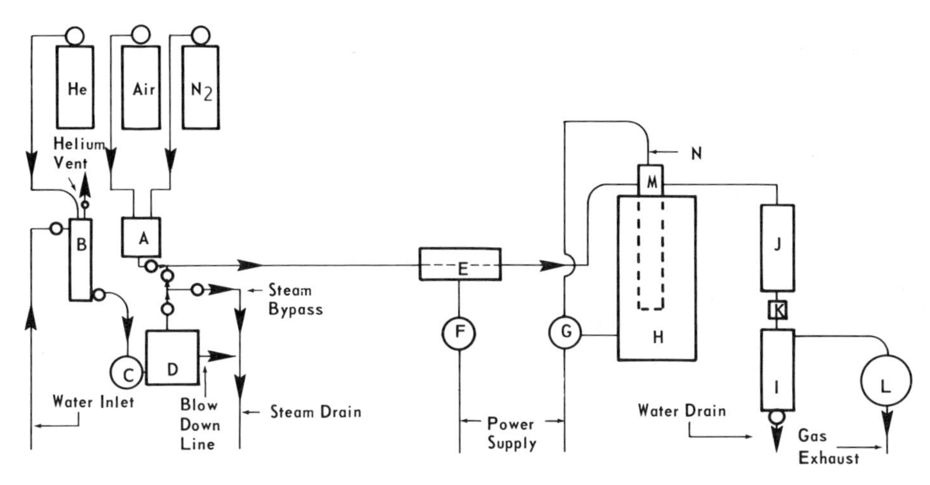

Fig. 4 Schematic diagram of char–steam reaction system. A, Rotameters; B, feed water storage tank; C, feed pump; D, steam generator; E, preheater; F and G, temperature controllers; H, furnace; I, water tap; J, condenser; K, sampling port; L, wet test meter; M, reactor; N, thermocouple. (From Hippo, 1977.)

into a drainpipe. The gas line from the steam generator to preheater E was wrapped with heating tapes. The preheater, filled with reduced copper turnings, was maintained at 600°C. This arrangement facilitated the removal of traces of oxygen present in the gas steam. From the preheater, the gases were passed through Pyrex tubing (wrapped with heating tape) to the Vycor fluidized-bed reactor. The reactor was fitted with a Vycor fritted disk to support the char sample. A 10-mm-i.d. Vycor inlet tube wrapped around the body of the reactor allowed preheated gas to enter the reactor through an opening in the bottom. A graded seal was used to connect the main body of the reactor to the top of the reactor, which in turn consisted of a ground glass joint which fitted inside the reactor cap. A 7-mm-i.d. Vycor thermocouple well was connected to the reactor cap through a Pyrex–Vycor graded seal.

About 20 g oven-dried (110°C) char was taken in the reactor. The reactor was placed inside the furnace (H). The temperature of the sample in the center of the bed was read by a Chromel–Alumel thermocouple (N). Prior to reactivity measurements, the reactor was flushed with N_2 to displace air. The furnace was then activated and the reactor brought to the desired reaction temperature. After thermal equilibrium was attained, N_2 was replaced by steam. Product gases coming out of the reactor at M were passed through a condenser (J). The condensate was collected in separating funnels. The gases were then passed through a wet test meter and then vented off into the atmosphere. After different reaction times, samples of product gases were taken and analyzed for CO, CO_2, CH_4, and H_2. A Fisher gas partitioner (FGP) was used to analyze CO, CO_2, and CH_4, whereas a Hewlett-Packard chromatograph (HPC) with a Carbosieve B column was used to measure the concentration of H_2. Helium was used as a carrier gas in the FGP and Ar in the HPC. After the desired level of carbon burn-off was attained, the run was terminated. The residual sample was cooled in N_2 and finally weighed.

Fuchs and Yavorsky (1975) used a fluidized-bed reactor to measure reactivities of chars in CO_2 and steam at high pressures ($\leq$70 atm). The reactor was made of stainless steel. Two disks of porous stainless steel confined the char sample within the reaction zone. The reactor was heated by direct resistance heating in which the reactor itself served as the heating element. At the bottom of the reactor a preheater (in the form of a ceramic rod containing a heating wire) raised the feed gas temperature. For the steam experiments, the preheater was insufficient to vaporize the water completely. In such cases, heating tapes were wrapped around the water feed line to maintain the temperature above the boiling point of water at the experimental pressure. For high pressure steam experiments, a water reservoir was pressurized with helium. In order to circumvent the possibility of backreactions, care was taken to prevent a buildup of product gas concentration near the particle surface.

C. By Differential Scanning Calorimetry

Differential scanning calorimetry (DSC) is a thermal analytical technique in which the ordinate value of an output curve at any given temperature is directly proportional to the differential flow of heat between a sample and reference material and in which the area under the measured curve is directly proportional to the total differential caloric input. The term "scanning" implies that the temperature of both the sample and reference material is increased during the experiment at a known heating rate.

Recently, Mahajan *et al.* (1976, 1977) have used the DSC technique for measuring the heats involved during pyrolysis and hydrogenation of a wide spectrum of coals. Although this technique has not yet been used for measuring reactivities of chars, it holds great potential as a fast analytical tool in this area. Therefore, it is discussed in some detail in conjunction with studies being conducted on characterization of active surface area, which is an important consideration in carbon gasification. One of the prime factors governing gasification rates of carbons is the active surface area (ASA), or the concentration of carbon atoms located at the edges of layer planes. In coals, according to Cartz and Hirsch (1960), the average layer diameter and number of atoms per layer increases with increasing coal rank, i.e., ASA decreases with increasing coal rank. These characteristics of the coal precursors are thought to be carried over to the resultant chars. In this context, Diamond (1960) has reported that when chars are prepared from different coals at a given temperature, the average layer diameter increases with increase in the carbon content of the parent coal. It will be discussed shortly that the reactivity of a char in different gasification atmospheres decreases, in general, with increase in the rank of the parent coal. Taken in the context of the work of Cartz and Hirsch (1960) and Diamond (1960), this means that char reactivity decreases with decrease in ASA of the coal precursor and, hence, of the char derived from it.

It has not yet been established experimentally whether the ASA of a char, prior to gasification, is in any way a measure of or related to char reactivity during gasification. It is shown later that ASA of relatively pure carbons can be determined by oxygen chemisorption. However, it is uncertain if ASA of "impure" carbons such as chars, which are invariably associated with inorganic impurities, can be measured accurately. It is well known (Walker *et al.*, 1968) that the first step in the overall gasification process is the dissociative chemisorption of the reactant gas at the active sites. Therefore, the heat released during chemisorption of oxygen on a unit weight of char may be related to its ASA and, hence, to its reactivity during gasification.

With the DSC technique, it is possible to measure quantitatively the magnitude of exothermic heat involved during chemisorption of oxygen on chars under controlled experimental conditions. In this laboratory, Ismail (1977) has measured the heat of chemisorption of oxygen on Saran char at 100°C [Saran is a copolymer of polyvinylidene chloride (PVDC) and polyvinyl chloride (PVC) in the ratio of 9 : 1]. Saran char is highly microporous and has a structure similar to that of the coal-derived chars. For determining the heats, Ismail used the following procedure. About 10 mg char (150 × 250 mesh) was taken in a gold pan. It was ascertained that under the chosen experimental conditions no diffusional effects were involved. That is, the heat released was independent of bed height, char particle size, and O_2 flow rate. The sample was heated in the DSC cell under a N_2 atmosphere at a rate of 20°C/min up to 700°C, which is the maximum attainable temperature. Soak time at 700°C was 30 min. The sample was then cooled under N_2 to 100°C. After temperature stabilization, N_2 was replaced by O_2 at the same flow rate (45 cm^3/min). The DSC output curve, i.e., the exotherm, was followed for 30 min. Heat released during the reaction was calculated from the equation

$$\Delta H = \frac{60ABE\zeta}{m} \tag{1}$$

where ΔH is heat released (mcal/mg), A the area of the exotherm (in.^{2}), m the sample mass (mg), B the time-base setting (min/in.), E the cell calibration coefficient (dimensionless), and ζ the Y-axis sensitivity [mcal/(sec in.)]. If, as is usually the case, the heat released is expressed per unit weight of gas chemisorbed, it is imperative also to determine the amount of oxygen chemisorbed during chemisorption. Ismail determined weight increase occurring during chemisorption using a Cahn RG electrobalance under conditions simulating those in the DSC runs.

Mahajan *et al.* (1977) found that the heat released during the hydrogenation of coals of different rank, in the temperature range 200–570°C, could be used as a relative measure of reactivity of coals during their hydrogasification. In the light of this observation, it is probable that the heat released during the reaction of chars with gases, under conditions simulating those in the reactivity runs, may well be a measure of char reactivity during gasification. Furthermore, in designing gasification reactors, information is badly needed on heats involved during gasification. During gasification the char loses weight continuously. Mahajan *et al.* (1976) have argued that in such cases the DTA and DSC output curves are displaced (relative to the reference baseline) due to both weight and thermal changes involved during the reaction. These authors have emphasized that unless the output curves are corrected for displacement due to weight changes it is not possible to estimate quantitatively the thermal effects involved during the reaction. These corrections cannot be made in the DTA technique but can be made in the DSC technique (Mahajan *et al.*, 1976). Unfortunately, there appears to be no commercial equipment available at present to permit the DSC technique to be used above 700°C. Nevertheless, the work of Mahajan and coworkers shows that the DSC technique holds potential for measuring quantitatively thermal changes, and possibly reactivities as well, during char gasification.

III. SELECTED EXPERIMENTAL RESULTS

A. Burn-off versus Time Curves

Typical burn-off versus time plots obtained for the gasification in air at 405°C of four 1000°C chars derived from 40 × 100 mesh fractions of (A) lignite, (B) Sbb-A, (C) HVC, and (D) HVB coals are shown in Fig. 5. These plots are more or less typical of the plots found for all the chars reacted in other gasification media (CO_2, steam, and H_2). Inspection of Fig. 5 shows that all the plots are essentially of the same shape and that each curve has three distinct regions. The first region involves the so-called induction period; i.e., the reaction rate increases slowly with time. The plot then goes through a maximum in slope, followed by a lengthy region of decreasing slope as burn-off approaches 100%.

Qualitatively, at this time, the shape of the burn-off curves in Fig. 5 can be explained on the basis of what is known about the development of porosity and surface area in microporous chars as they undergo gasification. As discussed in Volume I, Chapter 4, chars before gasification contain closed porosity, i.e., porosity inaccessible even to helium. With the onset of gasification, two important phenomena occur: (i) enlarging of pores that were open in the unreacted char, and (ii) opening up of closed pores. Since the total number of pores is increased as well as their average radius, specific pore volume and specific surface area increase with increasing carbon burn-off. The specific surface area in-

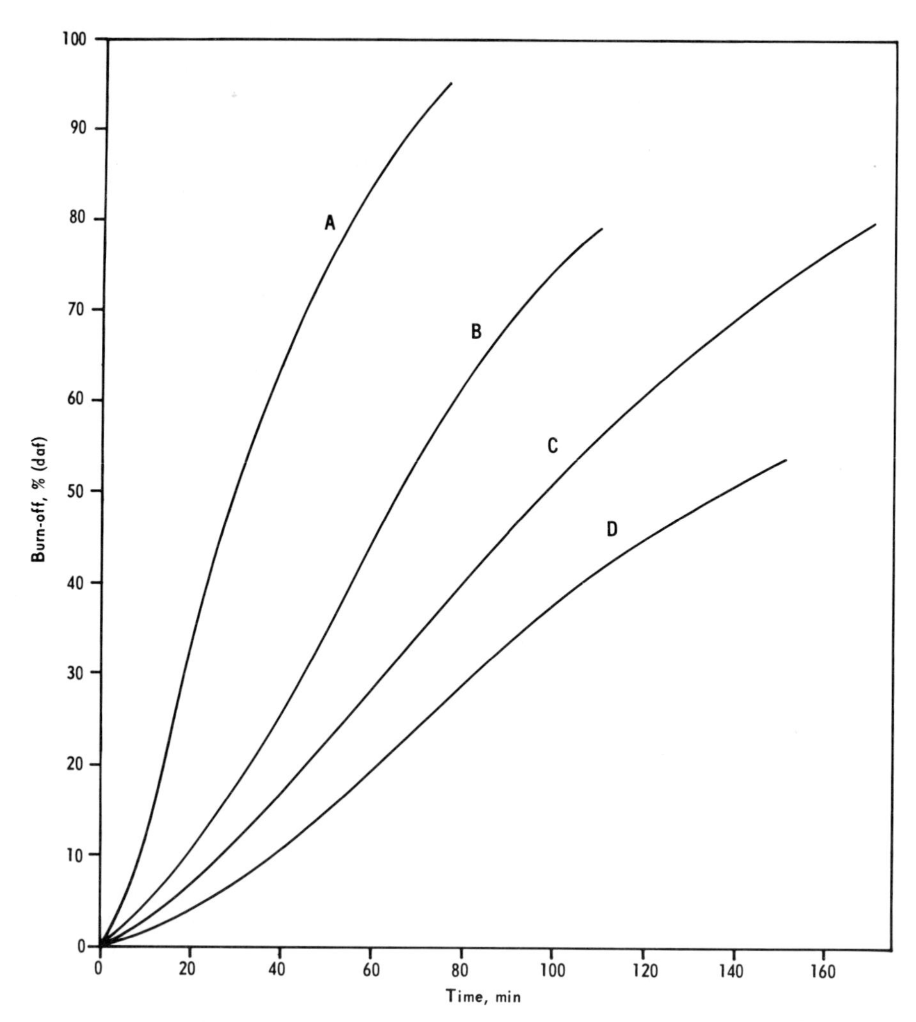

Fig. 5 Typical burn-off versus time curves in air at 405°C. (A) lignite; (B) Sbb-A; (C) HVC; (D) HVB. (From Soledade, 1976b.)

creases sufficiently rapidly as reaction time increases so that the product of specific area and char weight remaining, i.e., total area remaining, increases. The result is an increasing slope of the burn-off versus time plot. At some point, depending on the pore structure of the individual char, walls between existing pores are gasified away, and the total number of open pores commences to decrease. This leads to specific surface area ultimately going through a maximum as burn-off proceeds.

This results in the slope of the burn-off versus time plot going through a maximum value and continuing to decrease out to complete burn-off.

Perusal of the literature shows that several parameters have been used to express reactivities of chars during gasification.† In this laboratory, Jenkins *et al.* (1973) expressed reactivity as

$$R_T = \frac{1}{W} \cdot \frac{dW}{dt} \tag{2}$$

where R_T is reactivity [g/(hr g)], W the starting weight of char on a daf basis (g), and dW/dt the maximum rate of reaction (g/hr). Tomita *et al.* (1977) expressed reactivity as

$$R = \frac{1}{W_u} \cdot \frac{dW}{dt} \tag{3}$$

where W_u is the weight (daf) of char unreacted at time t and dW/dt is the slope of burn-off curve at the corresponding time t. Tomita's group found that over a certain burn-off range the kinetics followed a first-order rate expression with respect to the unreacted char.

B. Effect of Different Variables on Reactivity

1. *General*

The major factors which control the reactivity of carbonaceous solids to O_2, CO_2, H_2O, and H_2 are known (Walker *et al.*, 1959, 1968). They are (i) concentration of active sites, i.e., carbon atoms located at the edges of crystallites or building blocks; (ii) diffusional limitations on how rapidly the reactive gas molecules can reach active sites; and (iii) presence of catalytic inorganic impurities. Char properties undergo dynamic changes during the gasification process. As gasification proceeds, ASA can change as a result of opening of closed porosity and enlarging of existing porosity. The extent of change in ASA during gasification depends in complex ways upon such variables as the starting structure of the char, gasification temperature, reactive gas, nature and dispersion of inorganic impurities present, and the level of burn-off itself.

The majority of ASA in chars is located in micropores. Since most of the micropores in coal chars (prior to their gasification) are very small or of a size comparable to that of the reactant molecules, diffusion into these pores is activated (Walker *et al.*, 1966). Diffusion rates increase very sharply with small amounts of gasification (Patel *et al.*, 1972). Therefore, the utilization factor, η, i.e., the extent of utilization of active

† For reactivity of coke to CO_2, see Chapter 29, Section III,E,3.

sites, increases. The factor η can vary from 0 to 1, where a value of 1 indicates complete utilization of active sites, i.e., all active sites are "bathed" in the same concentration of reactive gas as that which exists in the main gas stream outside of the particles. During gasification, it is desirable for η to approach 1 as closely as is possible (and economically feasible).

It is well known that most inorganic impurities catalyze carbon gasification (Walker *et al.*, 1959, 1968). The extent to which an impurity is an active catalyst during gasification depends on the amount present, chemical form (i.e., a metal, an oxide, etc.), and degree of dispersion (particle size). For a given amount of catalyst, the greater the degree of its dispersion the greater its specific catalytic activity is expected to be. In coal chars there can be a wide range in the dispersion of impurities. Most of the inorganic impurities present as discrete minerals have a rather low dispersion, and their degree of dispersion is expected to change little during gasification. Conversely, most of the inorganic impurities present within the carbon matrix or associated with functional groups at the edges of the matrix are highly dispersed. This degree of dispersion would be expected to decrease markedly during gasification. The extent of change will depend on the impurity, its original degree of dispersion, the reactant atmoshere, and reaction temperature. These parameters determine the degree of mobility of impurities and, therefore, the possibility of their coalescence into larger particles (Ruckenstein and Pulvermarcher, 1973; Flynn and Wanke, 1974).

2. *Active Surface Area*

Laine *et al.* (1963) have discussed the importance of ASA on the carbon–oxygen gasification reaction. They oxidized Graphon, a highly graphitized nonporous carbon black devoid essentially of all impurities, to seven levels of burn-off varying between 0 and 35%. The total surface area (TSA) of each sample was measured by the BET (Brunauer–Emmett–Teller) method using N_2 at $-196°C$. The ASA was determined by chemisorption of oxygen at 300°C.

Reaction rate runs were then made at temperatures of 575, 625, and 675°C and an initial O_2 pressure of $\sim$40 mtorr. It was ascertained that the reaction rates were solely controlled by the intrinsic chemical reactivity of the carbon samples. The concentrations of reactant and product gases during the reaction were followed continuously using a mass spectrometer. The amount of stable oxygen complex building up on the surface during reaction was followed by a continuous material balance, and at the end of a run by outgassing. From these data, the unoccupied active surface areas (UASA) were calculated.

From the reactivity results, Laine *et al.* (1963) calculated rate constants for the disappearance of O_2 and formation of CO and CO_2. On the basis of TSA, the rate constants were calculated by an equation of the form

$$-\frac{dP_{O_2}}{dt} = k'_{O_2}(P_{O_2})(\mathrm{TSA}) \tag{4}$$

where $-dP_{O_2}/dt$ is the rate of decrease of O_2 pressure with time and TSA is the BET N_2 area for the particular Graphon sample. On the basis of ASA, the rate constants were calculated using an equation of the form

$$-\frac{dP_{O_2}}{dt} = k_{O_2}(P_{O_2})(\mathrm{ASA})(1-\theta) \tag{5}$$

where ASA is the active surface area for the particular Graphon sample and $1-\theta$ is the fraction of the ASA which is unoccupied with the complex under particular conditions of reaction time and temperature.

Laine's group found that for a particular Graphon sample reacted with O_2 at 625°C, the rate constants based on TSA decreased sharply with time, only beginning to level off as the amount of stable oxygen complex formed tended to saturation. In contrast, the rate constants based on UASA were found to change relatively little with time. The rate constants as a function of prior Graphon burn-off for reaction at 625°C are listed in Table I. The rate constants based on TSA increased monotonically and sharply with prior burn-off given to the graphon. In contrast, the rate constants based on UASA were essentially constant for Graphon samples with burn-offs between 3.3 and 34.9%. Laine *et al.*

TABLE I *Variation of Rate Constants with Amount of Prior Graphon Burn-off for Reaction with Oxygen at 625°C*[a]

Burn-off (%)	×10⁶ (sec⁻¹ m⁻², BET)				×10³ (sec⁻¹ m⁻², UASA)			
	k'_{O_2}	k'_{CO_2}	k'_{CO}	k'_{C}	k_{O_2}	k_{CO_2}	k_{CO}	k_{C}
0	5.2	2.4	2.7	5.1	9.8	4.8	5.4	10.2
3.3	11.0	3.2	9.0	12.2	5.1	1.4	4.3	5.7
6.4	16.3	3.7	20.1	23.8	4.5	1.1	5.7	6.8
8.5	21.3	4.5	26.5	31.0	5.0	1.0	6.0	7.0
14.4	27.4	6.0	33.1	39.1	4.5	1.0	5.9	6.9
20.8	34.4	7.1	38.0	45.1	4.9	0.9	5.4	6.3
25.8	39.8	9.2	48.8	58.0	4.8	0.9	5.9	6.8
34.9	49.7	11.0	59.1	70.1	4.9	1.0	5.8	6.8

[a] From Laine *et al.* (1963).

(1963) have offered a plausible explanation for the higher rate constants observed for the unactivated, i.e., 0% burn-off, Graphon sample.

It is instructive to emphasize that although the pioneering work of Laine and co-worker is of fundamental importance in understanding unambiguously the role of ASA in determining the kinetics of carbon gasification, their approach for expressing rate constants on the basis of UASA probably cannot be used in the case of chars. This is so because, as discussed earlier, during char gasification not only ASA but also diffusional effects and, hence, η, as well as the chemical form and degree of dispersion of the catalytic impurity can change significantly.

3. *Rank of Parent Coal*

Walker and his school have measured the reactivities of a number of chars (prepared at 1000°C from 40 × 100 mesh coals varying in rank from anthracite to lignite) in air (1 atm) at 405°C, CO_2 (1 atm) at 900°C, steam (0.022 atm)† at 910°C, and H_2 (27.2 atm) at 980°C. Reactivity parameters R_T, as defined by Eq. (2), for various chars in different gasification atmospheres are listed in Table II. It is seen that reactivities in air, CO_2, and steam decrease, in general, with increase in rank of the parent coal. However, reactivities in H_2 show a random variation with rank. Furthermore, the spread in char reactivities in the oxidizing atmospheres is far greater than in H_2; the reactivities of various chars in air, CO_2, steam, and H_2 show variations of about 170-, 160-, 260-, and 30-fold, respectively.

For chars derived from subbituminous and bituminous coals, Daly and Budge (1974) found a much wider variation (15,000-fold) in reactivities to CO_2 than that listed in Table II. These workers followed essentially the same experimental approach as that of Okstad and Hoy (1966) for measuring char reactivities. The reactant gas (CO_2) was passed at a known flow rate over the bed at temperatures between 750 and 1000°C. The reaction temperature was adjusted to give a conversion of ~10% under steady state conditions. The results were then recalculated to a standard temperature of 950°C assuming the activation energy for the $C–CO_2$ reaction to be 85 kcal/mole. The significance of a much wider variation in char reactivity observed by Daly and Budge should be considered in the context of two questionable points. First, these workers assumed, *a priori*, that the activation energy (E_a) for the $C–CO_2$ reaction for all cokes and chars was 85 kcal/mole. This is not necessarily true (Walker *et al.*, 1959, 1968). Daly and Budge themselves emphasize that any errors resulting from incorrect values for E_a have a greater effect

† This steam pressure was generated by bubbling prepurified N_2 through deaerated distilled water maintained at 20°C.

TABLE II *Reactivities of Chars in Different Atmospheres*

PSOC sample number	Parent coal ASTM rank	Parent coal C, daf (%)	Char ash, dry (%)	Reactivity, R_T [g/(hr g)] Air[a]	CO_2[b]	Steam[c]	H_2[d]
91	Lignite	70.7	11	2.7	6.3	2.9	0.91
87	Lignite	71.2	13	1.1	—	2.8	2.2
140	Lignite	71.7	12	1.3	3.4	1.5	1.0
138	Lignite	74.3	16	0.53	1.9	1.2	1.1
98	Sbb-A	74.3	12	0.60	1.3	1.0	0.73
101	Sbb-C	74.8	8	1.8	4.6	2.5	1.1
26	HVB	77.3	20	0.30	0.18	0.25	0.63
22	HVC	78.8	23	0.34	0.59	0.45	1.3
24	HVB	80.1	14	0.46	1.3	0.52	1.2
67	HVB	80.4	5	0.27	0.15	0.22	1.1
171	HVA	82.3	11	0.08	0.08	0.15	0.43
4	HVA	83.8	2	0.26	0.20	0.30	0.86
137	MV	86.9	19	0.11	0.09	0.10	0.26
114	LV	88.2	12	0.13	0.07	0.07	0.37
127	LV	89.6	7	0.016	0.07	0.011	0.07
81	Anthracite	91.9	6	0.11	0.13	0.13	0.81
177	Anthracite	93.5	5	0.08	0.04	0.11	0.38

[a] Soledade (1976a).
[b] Tomita (1976).
[c] Linares *et al.* (1977).
[d] Tomita *et al.* (1977).

on reactivity the farther the reaction temperature is from the standard reaction temperature (950°C). Second, the chars were not prepared at the same temperature; the bituminous chars were prepared at 1050°C, whereas the more reactive subbituminous chars were prepared at 920°C. As will be discussed shortly, the temperature at which a char is prepared has a profound effect on subsequent char reactivity. Since the chars listed in Table II were all prepared at 1000°C, comparison of their reactivities is thought to be more meaningful.

4. *Heat Treatment Conditions*

a. Carbonization Temperature. Jenkins *et al.* (1973) have studied the effect of HTT of chars derived from PSOC-138 (lignite), PSOC-24 (HVB), and PSOC-171 (HVA) on subsequent reactivity in air at 500°C. The precursor coals (40 × 100 mesh) were heated under N_2 at temperatures between 600 and 1000°C. Results summarized in Fig. 6 show that char reactivity decreases as HTT is increased. This was attributed to a combination of several factors, namely decrease of accessibility of active

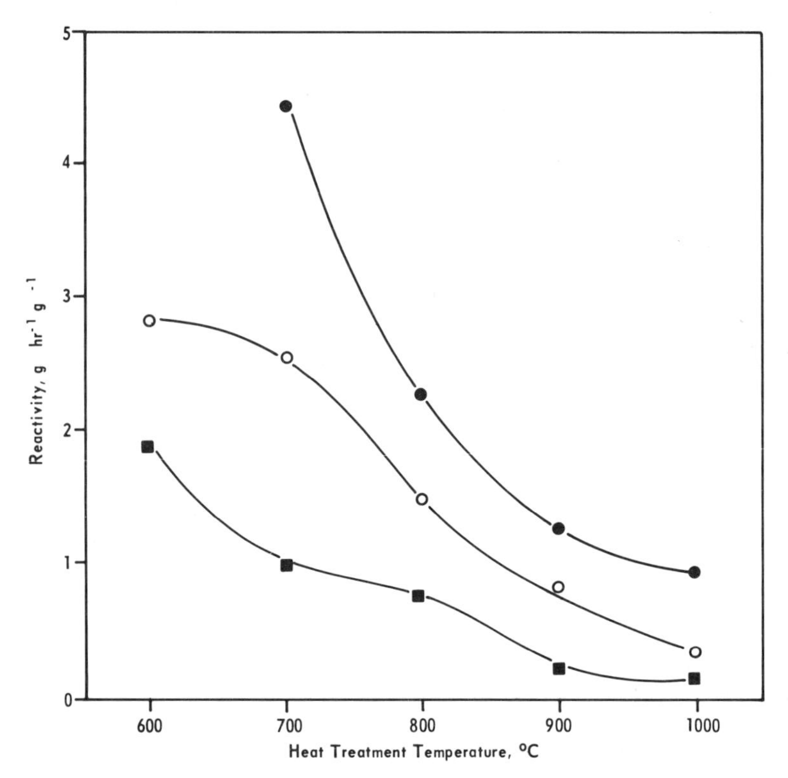

Fig. 6 Effect of temperature of heat treatment of chars on their reactivity in air at 500°C. ●, PSOC-138; ○, PSOC-24, ■, PSOC-171. (From Jenkins *et al.*, 1973.)

sites, decrease in volatile matter content, and decrease in catalytic activity of inorganic impurities as a result of their sintering as HTT is increased. Thus, it is important that chars not be taken to a temperature any higher than necessary if their reactivity (at some fixed temperature) is to be maximized.

b. Heating Rate during Carbonization. Recently, Ashu (1976) has shown that the use of very high heating rates up to pyrolysis temperature can lead to substantial increases in subsequent char reactivity. He prepared chars from a North Dakota lignite (70 × 100 mesh) by heating under N_2 to maximum temperature either in a fluidized bed at 10°C/min or in a laminar flow unit at ~(8 × 10^3)°C/sec. In both cases, soak time at maximum temperature was under 1 sec. The chars were subsequently reacted with air (1 atm) at 500°C. Reactivity parameters (R_T) for various samples are listed in Table III. It is seen that rapid heating to 800°C yields a char having a subsequent reactivity almost twice that of the

TABLE III *Char Reactivity Following Different Heat Treatments*[a]

Heat treatments	Weight loss on heat treatment, dry basis (%)	R_T [g/(hr g)]
10°C/min to 500°C	37.3	3.85
10°C/min to 800°C	43.8	2.79
8×10^3 °C/sec to 800°C	13.0 (22.8)[b]	5.17
10°C/min to 800°C followed by 8×10^3 °C/sec to 800°C	~43.8	3.10
10°C/min to 500°C followed by 8×10^3 °C/sec to 800°C	~37.3	4.89

[a] From Ashu (1976).
[b] 22.8 is weight loss during heating up to reaction temperature under N_2 in the TGA unit at a rate of 10°C/min.

char produced by slow heating to 800°C. In the context of the work of Jenkins *et al.* (1973), referred to earlier, that char reactivity decreases with increase in HTT of the char, it is noteworthy that rapid heating to 800°C more than counterbalances the use of slow heating to only 500°C insofar as subsequent reactivity is concerned. Rapid heating to 800°C of a char which was previously slowly heated to 500°C also produced a substantial increase in reactivity. Even rapid heating to 800°C of a char which was previously slowly heated to 800°C results in some enhancement of reactivity. These results show that in order to maximize char reactivity, it is desirable to maximize the heating rate during the carbonization of the precursor coal.

c. Concurrent Devolatilization and Gasification. When coals or low temperature chars are heated at higher temperatures, they lose volatile matter. The loss of volatile matter leaves behind nascent carbon sites which are highly reactive toward gasification (Walker *et al.*, 1977b). In the absence of a reactant gas, the nascent sites are deactivated by rehybridization. Therefore, it is desirable to devolatilize coals in the presence of a reactant gas. The advantage of such an approach is illustrated by the hydrogasification results of Johnson (1974) for an HVA bituminous char (obtained from an air-pretreated caking coal) containing, on a dry basis, 28.4% volatile matter. Johnson treated the char with H_2 in two different ways: (i) the char was first devolatilized under N_2 (500 psi) at 927°C followed by reaction with H_2 (500 psi) at 927°C, and (ii) the char was treated directly with H_2 (500 psi) at 927°C. From Fig. 7, it is seen that both the extent and the rate of hydrogasification are far greater when the char is hydrogenated directly. It is noteworthy that the weight loss,

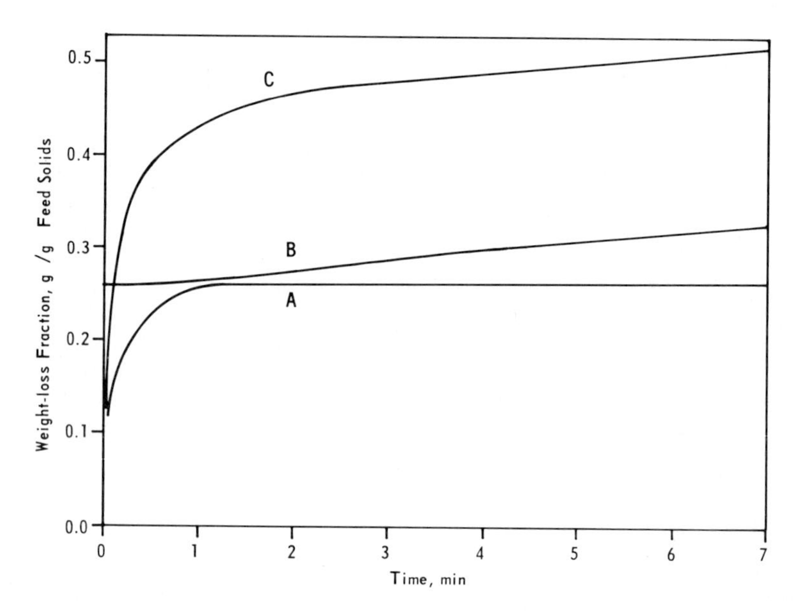

Fig. 7 Weight loss–time curves obtained for a char in N_2 and H_2 at 500 psi and 927°C. (A) Char–N_2; (B) char–N_2 followed by char–H_2; (C) char–H_2. (From Johnson, 1974.)

when hydrogasification is conducted following the devolatilization step, is only slightly higher than that involved during the devolatilization step itself.

5. *Inorganic Impurities*

Removal of inorganic impurities (mineral matter),† either from precursors prior to their carbonization or from chars prepared from raw coals, has a marked effect on char reactivity in air (Jenkins *et al.*, 1973), CO_2 (Hippo and Walker, 1975), steam (Linares *et al.*, 1977), and H_2 (Tomita *et al.*, 1977). In the case of chars derived from lower rank coals, removal of mineral matter decreases, in general, subsequent char reactivity. However, reactivity of chars prepared from higher rank coals increases upon removal of inorganic impurities. Removal of mineral matter from coals prior to their carbonization brings about profound changes in surface area and porosity of chars produced (Tomita *et al.*, 1977). The changes in char reactivity and surface area are much less pronounced when the raw chars rather than the coal precursors are acid-washed or demineralized (Linares *et al.*, 1977).

† By acid-washing with HCl or by demineralizing with a HCl–HF mixture.

Recently, Hippo (1977) has shown that reactivity in steam of chars produced from a Texas lignite could be further enhanced by the addition of exchangeable potassium, sodium, and calcium ions. At equivalent loading, the catalytic effect was most pronounced for potassium ions and about equal for sodium and calcium. Hippo prepared a number of chars containing different amounts of calcium (up to a maximum of 12.9%, by weight) by heating calcium-exchanged samples under N_2 at 800°C. Figure 8 summarizes results for reactivity of the various chars under 1 atm of steam at 650°C. It is seen that char reactivity increases linearly with increase in calcium content in the char and that the maximum reactivity achieved is about 12 times greater than that of the char prepared from the raw lignite.

Johnson (1975) also found a linear relationship between amounts of calcium and sodium added by ion exchange to North Dakota and Montana lignites and reactivity of product chars in H_2 at 35 atm and 927°C. For the same loading, sodium ions were found to be catalytically more active than calcium ions. Johnson also found that exchangeable sodium and calcium ions significantly enhanced char gasification in steam–H_2 mixtures, even more so than for gasification in H_2 alone. However, contrary to the behavior in pure H_2, the effect of calcium concentration on reactivity in steam–H_2 mixtures was the same as that of sodium at corresponding loading.

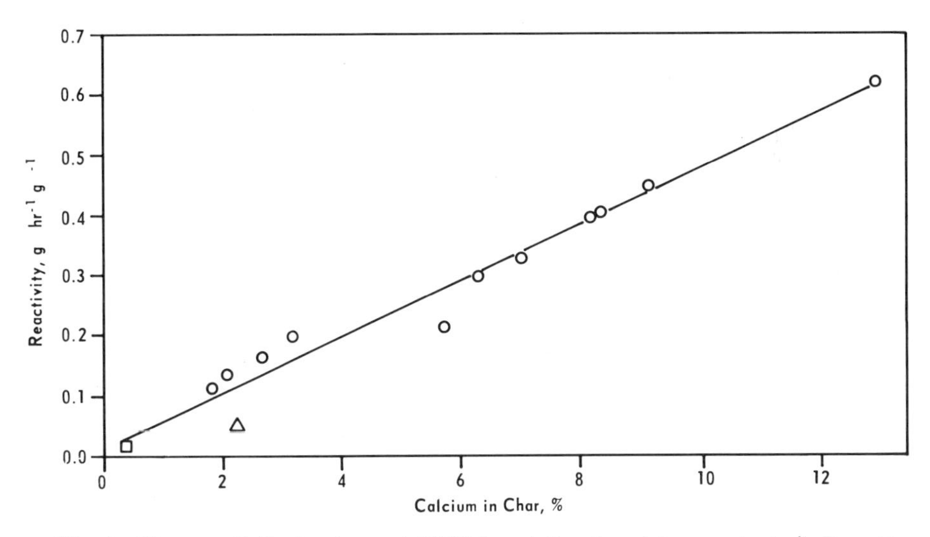

Fig. 8 Char reactivity in steam at 650°C in relation to calcium content. △, Raw; □, demineralized; ○, Ca-exchanged. (From Hippo, 1977.)

6. *Particle Size*

Hippo and Walker (1975) studied the effect of particle size on reactivity under CO_2 (1 atm) at 900°C of a relatively unreactive low volatile bituminous char (PSOC-127) and a highly reactive lignite char (PSOC-87). Particle size fractions studied were 40 × 100, 100 × 150, and 200 × 325 mesh. It was found that a decrease in particle size invariably increased char reactivity, indicating that the gasification reaction was partly diffusion-controlled. Whereas a reduction in particle size of PSOC-87 from 40 × 100 to 200 × 325 mesh resulted in a reactivity increase of only 2.7-fold, a similar particle size reduction of PSOC-127 resulted in a reactivity increase of 35-fold. These results show that the limitations which mass transport can impose on gasification rates of less reactive chars can be minimized by using increasingly smaller particle sizes in the reactor.

7. *Reaction Temperature*

Activation energies involved in the chemical step of char gasification are substantial (Walker *et al.*, 1959). Therefore, an increase in reaction temperature enhances gasification rates so long as the reaction is not limited primarily by mass transport or equilibrium. However, even for a chemically controlled reaction a given increase in reaction temperature may bring about different increases in reaction rates for different carbons. The reasons for this behavior have been discussed by Walker *et al.* (1959, 1968).

Tomita *et al.* (1977) determined reactivities of selected 40 × 100 mesh chars in H_2 (28.2 atm) at 875, 925, and 980°C. The reactivity of each char increased with temperature, as expected. The pseudoactivation energy for char gasification was found to increase with carbon burn-off and tended to level off toward an asymptotic value at higher burn-offs. This behavior, which was also observed by Zielke and Gorin (1955), was attributed by Tomita *et al.* (1977) to a reduction in diffusion control of gasification rates and/or a decrease in the extent to which gasification was catalyzed by impurities.

Linares *et al.* (1977) studied the reactivity of a Montana lignite char in steam (0.022 atm) in the temperature range 750–930°C. Below 890°C, the reaction was found to be chemically controlled, whereas above 890°C it was diffusion controlled. Kayembe and Pulsifer (1976) also found that gasification rates of two chars with steam (1 atm) in the temperature range 600–850°C were controlled by chemical reaction at the carbon surface.

8. *Pressure of Reactant Gas*

Walker and his school have studied the effect of pressure of the reactant gas on the reactivity of a 1000°C Montana lignite char in different gasification atmospheres. Figure 9 shows burn-off versus time plots for the char reacted at 405°C in various partial pressures of O_2 at a total O_2–N_2 pressure of 1 atm. As expected, the gasification rate decreases sharply with decreasing O_2 pressure. Soledade (1976a) calculated that the reactivity parameter (R_T) for the lignite char was proportional to $P_{O_2}{}^{0.71}$. R_T values for the lignite char under different partial pressures of CO_2 (900°C) and of steam (910°C) at a total reactant gas–N_2 pressure of 1

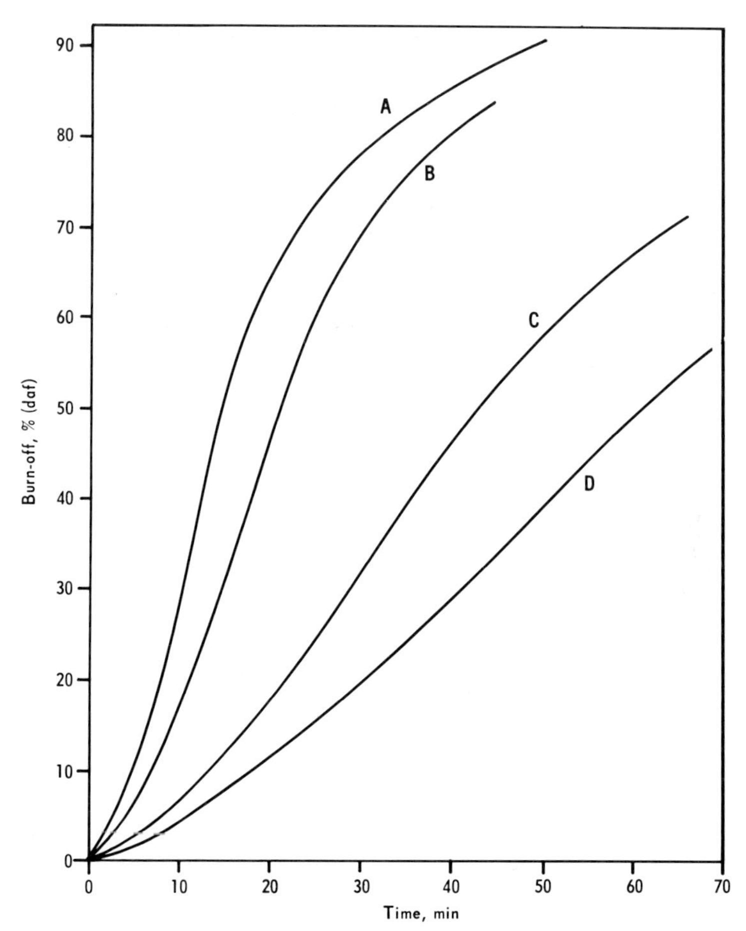

Fig. 9 Influence of oxygen concentration on reactivity of PSOC-91 char at 405°C. (A) 21% O_2; (B) 10% O_2; (C) 4% O_2; (D) 2% O_2. (From Soledade, 1976a.)

atm were found to be proportional to the 0.55 and 0.60 powers of CO_2 (Linares, 1976) and steam (Linares *et al.*, 1977) pressures, respectively. Tomita *et al.* (1977) found that in the hydrogen pressure range of 6.8–27.2 atm, the hydrogasification rate, when the kinetics followed a first-order rate expression with respect to the unreacted char, was proportional to the first power of H_2 pressure.

9. *Carbon Deposition*

Recently Kamishita *et al.* (1977) have studied the effect of carbon deposition (CD), resulting from the cracking of methane, on a lignite char on subsequent reactivity to air (1 atm) at 375°C. Different amounts of carbon were deposited at 855°C on raw and acid-washed chars prepared by the heat treatment in N_2 of a North Dakota lignite (40 × 100 mesh) at 855 and 1000°C. It was found that the reactivity decreased progressively with increasing amounts of CD. This was attributed to a decrease in ASA and deactivation of catalytic inorganic impurities due to coating with carbon. The results indicated that the deposited carbon was much less reactive to air than the lignite char. It was also found that surface area development in the char as a result of gasification was sharply reduced by prior CD. It was concluded that in order to maximize surface area development and, hence, char reactivity to oxidizing gases, carbon deposition from volatiles during the conversion of coal to char should be kept to a minimum.

C. Unification of Coal Char Gasification Reactions

In calculating char reactivities in different gasification atmospheres, we have so far used the simple expressions in Eqs. (2) and (3). As discussed earlier in this chapter, even though there are major differences in char reactivity as the rank of the parent coal from which chars are derived is changed (Table II), the shapes of the burn-off versus reaction time plots are quite similar (cf. Fig. 5). If this is so, all reactivity plots should be able to be normalized using an adjustable time parameter, τ, which can be conveniently used as a measure of differences in reactivity for a wide spectrum of chars. Walker *et al.* (1977a) have examined the feasibility of such a normalizing procedure. For this study, they considered burn-off plots for various chars in air (1 atm) at 405°C, CO_2 (1 atm) at 900°C, steam (0.022 atm) at 910°C, and H_2 (27.2 atm) at 980°C.

The principle of normalizing reactivity plots is shown in Figs. 9 and 10. Figure 10 shows that individual reactivity plots corresponding to different partial pressures of O_2 (Fig. 9) can be normalized using a di-

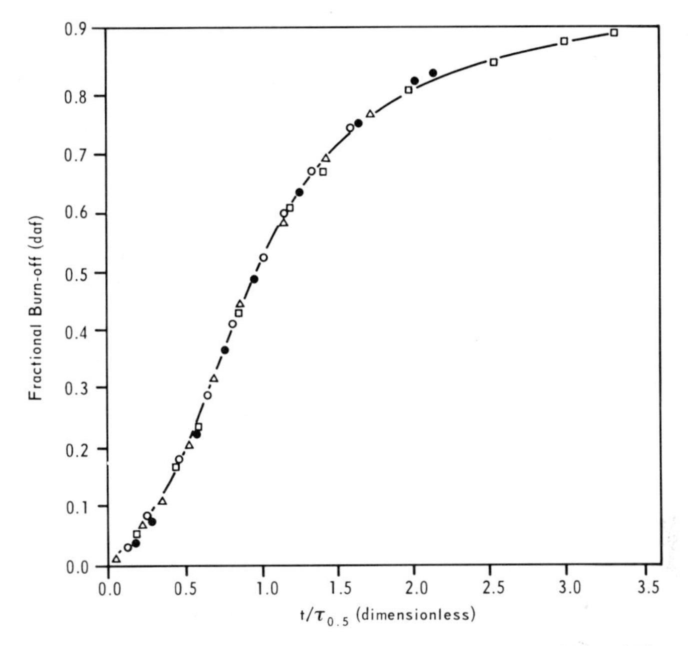

Fig. 10 Normalized plot for reactivity of PSOC-91 char at 405°C in different concentrations of O_2:

	O_2 (%)	$\tau_{0.5}$ (min)
○	2	61.0
△	4	42.6
●	10	20.4
□	21	14.8

(From Walker *et al.*, 1977a.)

mensionless time scale such that $t/\tau_{0.5} = 1$ at a fractional burn-off of 0.5. Values of $\tau_{0.5}$, or the times to reach a fractional burn-off of 0.5, decrease from 61.0 min to 14.8 min as the percentage of O_2 in the reactant mixture is increased from 2 to 21%.

As discussed in Section III,A of this chapter, Fig. 10 is more or less typical of the shape of burn-off versus $t/\tau_{0.5}$ plots found for all chars reacted in all gases. The explanation that was offered earlier to account for the observed shape of the burn-off curve ignores the effect which catalysis by inorganic impurities can have on the shape of the burn-off curve. For example, if a catalyst is initially very active but, as burn-off proceeds, becomes less active because of sintering or change in chemical state, the $t/\tau_{0.5}$ region over which the gasification rate is increasing can be shortened or indeed removed completely. That is, the maximum rate can be observed immediately as gasification commences. In this case, the catalytic effect on gasification is obviously overshadowing the effect of increase in specific pore volume and specific surface area.

In order to find out if one equation, with $\tau_{0.5}$ being the only adjustable parameter, can unify all the char reactivity data, Walker *et al.* (1977a) conducted a computer correlation of data for each gasification medium, as well as a computer correlation of data for all gasification media. Burn-off versus time data for a fractional burn-off up to 0.7 were used in all cases. The suitability of the following equations to correlate the data was tested: a linear equation between burn-off and $t/\tau_{0.5}$, an equation involving first- and second-power terms in $t/\tau_{0.5}$, an equation involving first- and third-power terms in $t/\tau_{0.5}$, and an equation involving first-, second-, and third-power terms in $t/\tau_{0.5}$. The last equation was found to give the best correlation of the data.

Tables IV and V summarize the results. Table IV shows the wide variation of $\tau_{0.5}$ values for the chars in each reaction medium as the rank of coal from which the chars were produced changes. Generally, $\tau_{0.5}$

TABLE IV $\tau_{0.5}$ *Values for Gasification Runs*[a]

PSOC coal number	Parent coal			$\tau_{0.5}$ for different reacting gases (min)			
	ASTM rank	Ash, dry (%)	C, daf (%)	Air	Steam	CO_2	H_2
89	Lignite	11.6	63.3	—	—	5.5	43.5
91	Lignite	7.7	70.7	14.8	10.6	5.0	36.5
87	Lignite	8.2	71.2	30.0	11.4	—	24.0
140	Lignite	9.4	71.7	29.6	19.6	10.3	34.0
138	Lignite	10.3	74.3	69.5	28.0	17.0	32.0
98	Sbb-A	8.4	74.3	66.4	—	26.0	50.0
101	Sbb-C	6.1	74.8	21.6	13.6	7.0	37.5
26	HVB	10.8	77.3	121	138	200	59.0
22	HVC	10.1	78.8	99.0	64.0	54.0	33.5
24	HVB	11.8	80.1	78.5	51.0	30.0	32.0
67	HVB	4.8	80.4	134	152	220	34.5
171	HVA	7.6	82.3	—	260	—	96.0
4	HVA	2.1	83.8	—	114	—	49.0
114	LV	9.8	88.2	—	—	—	126
81	Anthracite	7.8	91.9	—	255	270	51.5
177	Anthracite	4.3	93.5	—	330	—	110

[a] From Walker *et al.* (1977a).

TABLE V *Unification of Coal Char Gasification Reactions*[a]

Reactant	No. of reactant–char combinations	Cubic model				First order	
		a	b	c	$\mathcal{R}^2$(%)	Slope	$\mathcal{R}^2$(%)
Air	10	0.317	0.367	−0.182	96.0	0.756	94.1
CO_2	11	0.436	0.189	−0.122	99.2	0.728	78.4
H_2O	13	0.375	0.276	−0.148	99.1	0.761	87.4
H_2	16	0.349	0.283	−0.144	96.6	0.693	88.6
All	50	0.368	0.277	−0.147	98.2	0.727	87.5

[a] From Walker *et al.* (1977a).

values for each reactant gas fall in the same order, but there are exceptions, as expected, which reflect the uniqueness of each reaction.

Table V presents the best values for coefficients in the cubic equations between fractional burn-off and $t/\tau_{0.5}$ for each reactant, as well as for all reactants. $\mathcal{R}^2$ values give how much of the sum of variance, assuming no correlation between burn-off and $t/\tau_{0.5}$, can be removed by the particular cubic equation. It is obvious that for each individual reactant and also for all reactants the equations given yield a high correlation of results for burn-off versus $t/\tau_{0.5}$.

Since each char–reactant gas mixture shows some region of $t/\tau_{0.5}$ over which the reaction rate is first order in amount of char remaining, Walker *et al.* (1977a) made computer correlations with the first-order model, $\ln[1/(1 - \text{BO})] = k(t/\tau)$, where BO is fractional burn-off. These results are also summarized in Table V. If the data perfectly obeyed the first-order model, k must equal 0.69 since $t/\tau_{0.5} = 1.0$ at BO = 0.5. As seen in Table V, different reactions are seen to be more or less closely described, over a fractional burn-off range up to 0.7, by a first-order equation. A reasonably good correlation is found for the C–air reaction, and a low correlation is found for the C–CO_2 reaction.

From these studies, Walker *et al.* (1977a) concluded that a good parameter to use to correlate char reactivity data is the time required to reach a fractional burn-off of 0.5. Furthermore, they suggested that since char reactivity runs also exhibit a $t/\tau_{0.5}$ region in which the rate constant is first order in weight of char remaining, it probably is desirable to give first-order rate constants for each run over this region in order to adhere to the more conventional treatment of kinetic data.

ACKNOWLEDGMENTS

We appreciate the financial support of the Office of Coal Research and the Energy Research and Development Administration (formerly) and the Department of Energy (currently) of our studies on reactivity of coal-derived chars. Their support made the writing of this chapter possible.

REFERENCES

Ashu, J. T. (1976). M.S. Thesis, Pennsylvania State Univ., University Park.
Cartz, L., and Hirsch, P. B. (1960). *Phil. Trans. R. Soc. London, Ser. A* **252,** 557–604.
Daly, T. A., and Budge, C. F. (1974). *Fuel* **53,** 8–11.
Diamond, R. (1960). *Phil. Trans. R. Soc. London, Ser. A* **252,** 193–223.
Ergun, S. (1956). *J. Phys. Chem.* **60,** 480–485.
Flynn, P. C., and Wanke, S. E. (1974). *J. Catal.* **34,** 390–399.
Fuchs, W., and Yavorsky, P. M. (1975). *Am. Chem. Soc., Div. Fuel Chem., Prepr.* **20**(3), 115–133.
Hippo, E. J. (1977). Ph.D. Thesis, Pennsylvania State Univ., University Park.
Hippo, E., and Walker, P. L., Jr. (1975). *Fuel* **54,** 245–248.
Ismail, I. M. K. (1977). Unpublished results, Pennsylvania State Univ., University Park.
Jenkins, R. G., Nandi, S. P., and Walker, P. L., Jr. (1973). *Fuel* **52,** 288–293.
Johnson, J. L. (1974). *Adv. Chem. Ser.* No. 131, pp. 145–178.
Johnson, J. L. (1975). *Am. Chem. Soc., Div. Fuel Chem., Prepr.* **20**(4), 85–102.
Kamishita, M., Mahajan, O. P., and Walker, P. L., Jr. (1977). *Fuel* **56,** 444–450.
Kayembe, N., and Pulsifer, A. H. (1976). *Fuel* **55,** 211–216.
Laine, N. R., Vastola, F. J., and Walker, P. L., Jr. (1963). *J. Phys. Chem.* **67,** 2030–2034.
Linares, A. (1976). Unpublished results, Pennsylvania State Univ., University Park.
Linares, A., Mahajan, O. P., and Walker, P. L., Jr. (1977). *Am. Chem. Soc., Fuel Chem., Prepr.* **22**(1), 1–3.
Mahajan, O. P., and Walker, P. L., Jr., (1978). DOE Tech. Rep. FE-2030-TR8.
Mahajan, O. P., Tomita, A., and Walker, P. L., Jr. (1976). *Fuel* **55,** 63–69.
Mahajan, O. P., Tomita, A., Nelson, J. R., and Walker, P. L., Jr. (1977). *Fuel* **56,** 33–39.
Okstad, S., and Hoy, A. (1966). *Conf. Ind. Carbon Graphite, 2nd, Soc. Chem. Ind., London* pp. 100–106.
Patel, R. L., Nandi, S. P., and Walker, P. L., Jr. (1972). *Fuel* **51,** 47–51.
Ruckenstein, E., and Pulvermarcher, B. (1973). *J. Catal.* **29,** 224–245.
Soledade, L. E. B. (1976a). M.S. Thesis, Pennsylvania State Univ., University Park.
Soledade, L. E. B. (1976b). Unpublished results, Pennsylvania State Univ., University Park.
Tomita, A. (1976). Unpublished results, Pennsylvania State Univ., University Park.
Tomita, A., Mahajan, O. P., and Walker, P. L., Jr. (1977). *Fuel* **56,** 137–144.
Walker, P. L., Jr., Rusinko, F., Jr., and Austin, L. G. (1959). *Adv. Catal.* **11,** 134–221.
Walker, P. L., Jr., Austin, L. G., and Nandi, S. P. (1966). *In* "Chemistry and Physics of Carbon" (P. L. Walker, Jr., ed.), Vol. 2, pp. 257–371. Dekker, New York.
Walker, P. L., Jr., Shelef, M., and Anderson, R. A. (1968). *In* "Chemistry and Physics of Carbon" (P. L. Walker, Jr., ed.), Vol. 4, pp. 287–380. Dekker, New York.
Walker, P. L., Jr., Mahajan, O. P., and Yarzab, R. (1977a). *Am. Chem. Soc., Div. Fuel Chem., Prepr.* **20**(1), 7–11.
Walker, P. L., Jr., Pentz, L., Biederman, D. L., and Vastola, F. J. (1977b). *Carbon* **15,** 165–168.
Zielke, C. W., and Gorin, E. (1955). *Ind. Eng. Chem.* **47,** 820–825.

Chapter 33

Analysis of Coal Tar Pitches

*C. S. B. Nair**
CENTRAL FUEL RESEARCH INSTITUTE
DHANBAD, BIHAR, INDIA

* Present address: Research and Development Division, Fact Engineering and Design Organization, The Fertilisers and Chemicals Travancore Ltd., Udyogamandal 683501, Cochin, Kerala State, India.

ISBN 0-12-399902-2

I. INTRODUCTION

This chapter deals mainly with the methods currently available for the analysis and testing of coal tar pitches. Sections II–IV make essential but brief references to the methods of production of coal tar pitches, types of pitches, their fields of application, and characterization. These are areas which have been dealt with in more detail in other chapters, e.g., Chapter 34.

In view of the necessity to rely strictly on established standard procedures for testing of pitches, the treatment of the main subject in this chapter has been based mainly on the standard methods of testing and specifications laid down by various countries, as follows: (1) Canada; (2) Czechoslovakia; (3) Germany; (4) East Germany; (5) India; (6) Japan; (7) Poland; (8) Spain; (9) Turkey; (10) United Kingdom, (a) British Standard Specifications, (b) Standard Methods for Testing Tar and Its Products, published by the Standardisation of Tar Products Tests Committee, Gomersal; (11) United States, (a) American National Standards, (b) Federal Supply Service Standards, (c) American Association of State Highway Officials, (d) American Society for Testing and Materials, (e) International Conference of Building Officials, (f) National Roofing Contractors Association, (g) Underwriters Laboratories; (12) U.S.S.R. Specifications.

Efforts have been made to bring together the representative and reliable methods available for each individual test.

II. METHODS OF PITCH PRODUCTION

The residue left from the distillation of coal tar is called pitch. This usually has a boiling point above 360°C and forms 30–60% of the tar, depending on the conditions under which the tar was produced.

The chemical composition and properties of the pitches depend to a great extent on the following:

(i) On the carbonizing process by which the parent crude tars are obtained. There is much variation in the nature and composition of high, medium, and low temperature tars, water gas tars, producer gas tars, etc., which results in marked differences in the quality of pitches produced from them.

(ii) On the nature of the distillation process to which these tars were subjected, whether by batch distillation or by continuous distillation.

(iii) On the extent of distillation effected, which decides the quantity of lower boiling constituents left in the pitch.

(iv) On whether the pitches are straight run or blended.

(v) On whether the pitches were subjected to further treatment such as weathering, air blowing, or thermal treatment (Hoiberg, 1966; McNeil, 1961).

Coal tar pitch is a very complex material which is estimated to contain about 5000 chemicals. Distillation of pitches derived from coke oven tar under vacuum, or with steam, or under coking conditions gives a distillate which has been intensively studied by fractionation and further separation by chromatography by German workers. A total of 127 compounds have been identified or isolated from this fraction and have been well classified (Hoiberg, 1966; Schafer, 1956). Two hydrocarbons of formula $C_{36}H_{18}$ have been isolated from the acetone extract of pitch and another hydrocarbon of composition $C_{34}H_{22}$ was isolated from the benzene extract (Wieland and Muller, 1949).

Of the compounds positively identified, about two-thirds are condensed polynuclear aromatic hydrocarbons; the rest are heterocyclic compounds containing three to six rings. About 50% of the compounds are substituted, the substituents being methyl and hydroxyl groups. The molecular weights of these compounds range from 170 to 380.

Pitches derived from vertical and horizontal retorts were shown to contain more paraffinic compounds, hydrogenated ring compounds, and ring compounds substituted by one or more methyl or hydroxyl groups (Harris *et al.*, 1953, 1956). Pitch crystalloids (petroleum ether-soluble fractions) were studied by various workers (Wood, 1961; Greenhow and Smith, 1960a; Katona *et al.*, 1954; Wood and Phillips, 1954; Fair and Volkman, 1943) and the findings show that the pitch crystalloids are similar in composition to the distillate oils. The portion of pitch apart from distillate oils and pitch crystalloids consists of high-molecular-weight carbonaceous materials. Even though a great deal of work has been done to elucidate their structure, their chemical constitution has not yet been clarified (Janik, 1965; Greenhow and Smith, 1961; Volkman, 1959; Badenhorst and Perold, 1957; Von Krevelen and Chemin, 1954, 1957; O'Brochta, 1956; Franck, 1955; Wood and Phillips, 1955; Raich, 1952; Green and Olden, 1951; Sack, 1951; Mallison, 1950; Pfeiffer, 1950; Krenkler and Wagner, 1948; Fair and Volkman, 1943; Wieland and Muller, 1949; Berl and Schildwächter, 1928; Nellensteyn, 1928).

III. TYPES OF PITCHES

Crude tar on distillation gives various distillate fractions, the distillation being programmed to give the types of products desired. The extent to which the distillation is carried out decides the nature of the residue or pitch. The pitches are generally classified by their softening point as follows (see Section VI,B,2,a for a discussion of R&B):

(i) <40°C (R&B) Refined tar
(ii) 40–60°C (R&B) Soft pitch
(iii) 60–75°C (R&B) Medium soft pitch
(iv) 75–110°C (R&B) Medium hard pitch
(v) >110°C (R&B) Hard pitch

Coal tar pitches of low softening point are prepared either by stopping the distillation process at a lower temperature so that some of the higher boiling oils are retained in the residue or, alternatively, by fluxing the medium soft pitch with the required proportion of a suitable tar fraction. The softening point of a pitch may be as low as 30°C.

Hard pitches are prepared by continuing the distillation process preferably with passage of steam. Alternatively, such pitches could be produced by air-blowing medium soft pitch at temperatures in the range 300–370°C. The softening point may be raised to as high as 250°C. Such materials could be granulated by spraying the molten pitch into water or by other methods.

IV. FIELDS OF APPLICATION OF COAL TAR PITCHES

A. Electrodes

Carbon electrodes are used for a wide range of applications. The electrochemical industry uses large electrodes for the production of aluminum, alkalies, chlorine, and magnesium. Electrodes are used in electrothermal processes for the production of calcium carbide, silicon carbide, graphite, phosphorus, and ferrous and nonferrous alloys. Smaller electrodes are used for arc carbons and in primary batteries.

In all these cases, the electrodes are made out of an aggregate of carbonaceous material and a binder which usually forms 10–30% by weight of the mixture. These are heated and thoroughly mixed and formed into electrodes by tamping, pressing, or extrusion, or continuously added to self-baking electrodes. These green electrodes are baked at gradually rising temperatures to 750–1250°C over a period of 6–9 weeks to obtain various types of electrodes. For special applications, the electrodes are graphitized by heating them at temperatures in the range 2400–3000°C for 3–5 weeks.

1. Binder

The bulk of the binder used for electrode manufacture is derived from high temperature coal tar.† These pitches are characterized by

† For complete details on the analysis of coal tar binders for electrodes see Chapter 34.

(i) the highest carbon/hydrogen ratio for a specified softening point,
(ii) the highest yield of coke on baking than given by most other materials, and
(iii) desirable operational characteristics.

Pitches derived from low temperature tar are generally considered unsuitable for use as binders for carbon electrodes (Greenhow and Smith, 1960b; Pollack and Alexander, 1959; Darney, 1956).

2. *Carbon-to-Hydrogen Ratio*

The C/H ratio indicates the degree of aromaticity of a pitch binder and is a useful factor for predicting the quality of pitch binders. Highly aromatic binders result in pastes with reduced shrinkage and swelling characteristics (Thomas, 1960; Darney, 1958; Dell, 1958; Martin and Nelson, 1958; O'Brochta, 1956; Charette and Bischofberger, 1955).

3. *Coke-Forming Characteristics*

Coal tar pitches used as binder, particularly for anodes, usually give a yield of over 50% of coke residue. A high coke yield and the ability of the coke to bond the aggregates together determine the suitability of a pitch binder for electrodes (O'Brochta, 1956; Shea and Juel, 1950).

The other characteristics of pitches, such as softening point, viscosity, specific gravity, and insolubility in various solvents, also influence the behavior of a pitch as binder and have been well described by Branscomb (1966) and Olschenka (1974).

B. Refractories

Several U.S. patents (1962, 1963, 1964a,b,c,d,e, 1965) relate to the use of coal tar pitches as an auxiliary carbon bond for dolomite or magnesite refractory lining for converters using oxygen steel-making processes such as the Kaldo process, the LD process, and the rotor process. The incorporation of pitch inhibits the tendency of dolomite or magnesite to hydrate in contact with air. Usually 3–8% of pitch (of softening point ~65°C) is used. Such refractories are reported to be very resistant to slag attack. Direct impregnation of refractory bricks with molten pitch is also reported.

For the manufacture of pitch-bonded basic refractory bricks, burned dolomite or magnesite or their mixture is intimately mixed with a pitch having low softening point in a Sigma blade mixer. The green mix is molded into the desired shape in a brick press. The quantity of pitch used varies with the composition, porosity, and size of granules used for the blend.

C. Fiber Pipes

Pitch fiber pipes are being widely used for drainage, sewage, and irrigation. Pitch fiber pipes are also used as underground conduit for electric and telephone cables. These are made by forming tubes from fibrous materials such as wood pulp, asbestos, or wastepaper pulp which are felted under pressure, dried to a low moisture content, and then saturated by immersion in liquid pitch using a combination of pressure and vacuum saturation techniques. The final material generally has a pitch content of 70–75% (Caplin, 1964; Pitch Fibre Pipes, 1959, 1960).

An excellent study of the application of coal tar pitches for impregnating purposes has been made by Wall (1966).

The pitches used usually have a softening point (Ring and Ball) of 65–70°C; they should also be low in quinoline-insoluble content. It is essential that a pitch for fiber pipe applications contain a minimum of material boiling below 355°C, usually not more than 6% by weight. Excess of low boilers causes foaming of the pitch and fouling of the vacuum lines during the impregnation step.

Pitch obtained from light water gas tar has been found suitable for impregnation of fiber conduits. Continuous vertical retort pitch is not suitable because of its tendency to coke in the impregnation tank and also because of the development of cracks on the outer surface of pipes on exposure to sunlight.

Blending of pitches with high boiling tar distillates such as anthracene oil and creosote oil has been recommended (Dolkart, 1961; Thomas, 1960; Calaque, 1953; Herzog, 1944). There is controversy regarding the role of phenols: some workers claim increased bonding effect (Wagner, 1915) whereas others report deleterious effects (Massinon, 1956; Rosengren, 1954).

Viscosity, softening point, and penetration are other important characteristics of a fiber pitch. High viscosity and softening point cause difficulties in handling during manufacture while low values for these characteristics cause too much pitch to drain from the pipe.

Quinoline-insoluble matter consisting of minute particles clogs the pores of the fiber pipe during the impregnation step and so the QI content should not normally exceed 5% by weight.

Light tar recovered from the primary coolers has been found to be suitable for pitch fiber pipe impregnation. These contain up to 1.2% of

QI and on distillation give a pitch containing less than 5% QI. Centrifugation of heavy tars is also practiced to obtain low QI pitches.

D. Protective Coatings

The two drawbacks of straight coal tar pitch enamels are poor temperature susceptibility and high cold flow.

Many fillers, organic and inorganic, have been tried as pipe coating enamel extenders (Streiter, 1938). Powdered silica and cement are very abrasive and cause damage to pumps used for handling. Asbestos fines, though effective, reduce the working properties of the enamels. Mica and powdered schist have been shown to be good fillers but they are comparatively expensive and tend to settle down.

Pulverized hydrous magnesium and aluminum silicates have been found by experience to be by far the best fillers. These are comparatively inexpensive, nonabrasive, improve the mechanical properties of pitches, and have low settling characteristics.

Enamels are generally prepared by blending talc or mixtures of talc and mica with suitable coal tar pitch (cube-in-air softening point 85°C) in reaction kettles at 200°C. It is necessary to add aromatic tar to adjust the softening point within specification limits.

Primers suitable for this type of enamel are made by blending coal tar solvent naphtha with pitch [softening point (cube-in-air) method, 65°C] at 100°C and by thinning suitably with solvent.

Several patents relate to the use of coal digestion for improving the characteristics of coal tar pitches. Pulverized bituminous coal is added to coal tar with or without the addition of coal tar oil and digested by heating at 300–350°C. The coal disperses in the tar to form a colloidal suspension. By proper adjustment of the proportions of coal, oil, and tar, pitches of any softening point could be made.

In further modifications of the process, the coal-digested pitch is blended with heavy coal tar oil and ~30% of a mineral filler such as slate powder or talc to give a final product of softening point 105–125°C.

E. Roofing, Damp-Proofing, and Waterproofing

Coal tar pitch finds wide applications in the roofing industry for the weatherproofing of flat or nearly flat roofs of buildings. Its resistance to sunlight, low water absorption rate, resistance to oxidation, self-sealing properties, and adhesion to aggregate and felt make it an excellent roofing and waterproofing material.

Alternate layers of felt and coal tar pitch are used to form a built-up roof. Usually the assembly consists of three to five plies of felt saturated with pitch bonded with alternate layers of pitch. The roofs may be finished with coal tar pitch or with a layer of gravel or slag chippings.

Generally, roofing felts consist of hessian, asbestos felt, or woven cotton fabrics or heavy kraft paper saturated with pitch, and are made by passing the material through a bath of the appropriate pitch heated to ~80°C. The material is so treated to absorb 150% by weight of the pitch.

The American Society for Testing Materials (ASTM, 1955) designates three types of pitches for use in various applications: type A, a soft, adhesive, self-healing asphalt suitable for use below ground level; types B and C, a somewhat less susceptible asphalt for use above ground level. Type B is recommended for applications in which the material is exposed to temperatures not exceeding 50°C and type C for temperatures above 50°C.

There are several standards relating to coal tar products and materials used for damp-proofing and waterproofing (ASTM, 1970, 1972, 1973, 1974; Federal Specifications, 1974a, 1974b).

Coal-digested pitch-based enamels are used for wide-range enamel and primers and are made in special mechanically agitated vessels (Yeager, 1941). By proper control of the temperature and time of digestion, a blend of the desired softening point and penetration may be obtained without blending the pitch with tar, creosote, or other materials.

Primers suitable for such enamels are made by mixing the coal-digested pitch with suitable solvents. In coal-digested pitch primers, the primer is partially a solution and partially a dispersion of the pitch in the solvent.

Cold applied coatings are prepared by fluxing a straight pitch or coal-digested pitch enamels with suitable solvents to bring them to a consistency suitable for spraying or brushing. The inclusion of about 5% of chlorinated rubber is reported to be beneficial. They may also be prepared as water dispersions by dispersing the unplasticized base pitch in water to produce homogeneous emulsions. Inorganic fillers and pigments are also added to the formulation. Such dispersions are used in conjunction with a suitable primer.

The use of coal tar pitch modified by incorporation of polymers such as epoxies and polyurethane resins have proved satisfactory for the protection of steel structures, and for coating the inside of pipes carrying potable water.

F. Road Tars

Considerable quantities of coal tar pitches are converted to road tars to be used as paving material. Road tars are usually prepared by blending coal tar pitch with selected coal tar fractions to bring the pitch up to the appropriate specifications.

The specifications of the ASTM regarding road tars cover a wide range and represent the usual types of road tars used all over the world (ASTM, 1968). Other specifications are those of the American Association of State Highway Officials (AASHO, 1942) and the federal government (1942). Standards for road tars have been laid down by various other countries but the range of road tars mentioned herein could be considered as covering the entire range. These specifications represent various grades of road tars designated from RT-1 to RT-12 and RTCB-5 and RTCB-6 and are meant for certain specific uses. RT-1 is very thin whereas RT-12 is a very viscous material. RTCB denotes tars that have been cut back with a low boiling solvent.

The bulk of road tar is manufactured from high temperature coal tar. Water gas tar is also used to a limited extent for the manufacture of road tar. The process of production of road tar should ensure that components harmful to the performance of the tar are removed from it.

Phenols have been shown to be particularly susceptible to attack by atmospheric oxygen (Dickinson and Nicholas, 1949; Lee and Dickinson, 1954). German and Swiss road tar specifications state a maximum tar acid content of 3%.

Lee and Dickinson (1954) showed that naphthalene-type hydrocarbons tend to cause tar to age-harden in service. Again German and Swiss specifications have put limits to the naphthalene content in road tars, the German specifications tolerating up to 4% whereas Swiss specifications only up to 2% of naphthalene as the upper limit.

The presence of excessive amounts of anthracene is reported to have an adverse effect on the adhesiveness of road tar (Duriez, 1939). Skopnik (1931) found that anthracene oil of boiling range 315–355°C, freed of crystalline anthracene, has good lubricity which enables the surface of the coated aggregate to have the minimum of voids. Lee and Dickinson (1954) found that aromatic hydrocarbons of the naphthalene–anthracene type crystallize in the body of the tar below 15°C and thus cause an increase in viscosity.

G. Coal Tar Fuels

Use of coal tar pitch and tar oils as substitutes for petroleum fuel oil developed during World War II. These fuels have low fulfur content and the higher emissivity of their flame promotes rapid heat transfer.

The most important requirement of coal tar fuels (CTF) is their satisfactory atomization into a furnace chamber. Various grades of CTF are currently manufactured, e.g., CTF-50, CTF-100, CTF-200, CTF-250, CTF-300, and CTF-400. The numbers indicate the temperature in degrees Fahrenheit to which the CTF should be heated to achieve satisfactory atomization.

H. Pitch Coke

Pitch coke finds wide application as the solid grist in Soderberg anodes, graphite manufacture, certain types of ceramics, and speciality foundry fuels.

The coking of pitch is carried out in modified coke ovens which are faced with special silica bricks and have more taper than standard coke ovens.

The pitch to be charged into the coking ovens should have a softening point of 140–145°C (Ring and Ball). The pitch is melted and charged into the oven at a controlled rate according to a regular program. The charge is usually coked for 20 hr at a flue temperature of 1300°C. The coke yield is ~70% of the pitch charged.

In Germany, medium soft pitch is steam-distilled to produce a hard coke to be charged into the ovens. Russian practice is to blow air through a mixture of medium-soft coke oven pitch and recycle pitch coking oils at 370–380°C to raise its softening point (McNeil, 1966a).

The coke oven process causes a great deal of atmospheric pollution and so adoption of delayed coking processes has been suggested (Stepanenko *et al.*, 1974).

I. Briquetting, Core Binders, and Target Pitch

Coal briquets are made of blending coal fines with 8–10% of powdered medium-soft pitch at ~90°C. The soft mass so obtained is shaped into rectangular briquets in a ram press or formed into ovoids in a continuous roll press. The briquets are then heated to drive off the volatile matter.

Pitches of high softening point and high free carbon content find application in foundries for binding sand cores used for casting. Pitch for this use is generally sold in lump or flake form, and is ground to proper size before use.

The market for this type of application has reduced considerably with changes in foundry practices and the availability of other competitive casting techniques.

The application of pitch for targets makes use of the brittleness of pitch when hard. Coal tar pitch is blended with more than an equal volume of inert fillers and used for making clay pigeons for target shooting.

V. SOLVENT ANALYSIS OF PITCHES

Many workers concerned with the utilization of tar and pitches have attempted to characterize their products by means of solvent fractionation in an effort to find correlations between the components so isolated and the useful properties of the pitch or tar.

Solvent analysis methods may be divided into two types, the preparative and purely analytical. In the preparative type of analysis, a moderately sized sample is successively treated with different solvents to obtain distinct fractions which are then subjected to more detailed studies, e.g., elementary analysis, average molecular weight, melting point, and other characteristics.

In the purely analytical approach, different portions of the sample are treated with each solvent to estimate the matter insoluble in that solvent. The quantity of matter soluble in one solvent and insoluble in the next is obtained by difference.

An excellent comparative study of the characterization of pitches by solvent analysis has been made by McNeil (1966b). This study covers the work of Demann; Broche and Nedelmann; Adam, Shannan, and Sach; Dickinson, Mallison, Berl, and Schildwachter; Hubbard and Reeve; Duriez; Krenkler and Volkmann; and Rhodes and co-workers.

Of those methods, the one proposed by Mallison (1950) has been widely used, particularly in Europe, as a useful guide in defining the properties of pitch binders. In this system, the pitch is split into six fractions using anthracene oil, pyridine, benzene, methanol, and methanol–water mixture as follows.

(i) H-resins: Insoluble in anthracene oil and pyridine.
(ii) M_1-resins: Insoluble in pyridine but soluble in anthracene oil.
(iii) M_2-resins: Insoluble in benzene but soluble in pyridine.
(iv) N-resins†: Insoluble in methanol but soluble in benzene.

† Current terminology includes "asphaltenes" for coal-derived as well as petroleum-derived fractions soluble in benzene but insoluble in aliphatic solvents.

(v) m-oils: Insoluble in methanol–water mixture but soluble in methanol.
(vi) n-oils: Soluble in methanol–water mixture.

H; M, m; and N, n denote *Hoch Molekulare, Mittel Molekulare,* and *Niedrig Molekulare,* respectively.

Table I shows the relation between the fractions obtained by solvent analysis and their contribution to the functions of pitch in industrial practice, particularly as relating to the manufacture of electrodes.

McNeil (1961) made a detailed study of pitches obtained from a wide range of British tars and found that the solvent analysis methods are not fully satisfactory. He is of the opinion that many specifications for pitch for use in briquetting and electrode manufacture are empirical in nature.

Recent studies on the elucidation of the nature of pitches and correlation of pitch properties with their performance have generally been along the following lines:

(i) the composition and structure of pitches (Shaposhnikova and Gorpinenko, 1974; Sanada *et al.*, 1972; Stepanenko *et al.*, 1970; Kekin *et al.*, 1968; Babenko *et al.*, 1967; Schafer, 1967a,b; Weiler, 1967a,b);
(ii) the relation between coke-forming properties and fractions obtained by solvent fractionation (Izd., 1970; Kotlik *et al.*, 1966; Smith, 1966; Takaki and Miyasaka, 1965; Domitrovic *et al.*, 1962);
(iii) development of newer solvent fractionation techniques and modifications on existing techniques (Kekin *et al.*, 1975).

VI. ANALYSIS AND TESTING OF COAL TAR PITCHES

Quite a number of standards exist in various countries regarding coal tar pitches. The scope and field of coverage of these standards reflect the fields of application of coal tar pitches in these countries. The standards described in ASTM are very comprehensive and cover almost the entire range of standards described elsewhere. The general principles involved in testing pitches for various characteristics, and also the basic design of equipment employed for testing as described in the various standards, have close similarities.

In the description that follows the important tests used for defining various aspects of coal tar pitches are briefly described. The principles underlying the tests are highlighted as well as the manner of performing such tests.

TABLE I *Relationship between the Solvent Fractions of Coal Tar Pitch and the Performance of the Pitch*

Resin type	Fraction of pitch	Molecular weight	Known functions
High-molecular-weight resins	Insoluble in quinoline, anthracene oil, and pyridine (H-resins)	>2000	Increase the compressive strength of baked electrodes. Higher contents of these resins decrease the flow of binder through the voids between grist particles. When this type of resin is less than 3%, the flow between particles becomes too easy and the binder tends to segregate and drain out
Medium-molecular-weight resins	Insoluble in toluene, less insoluble in quinoline (β resins or C_2 resins); matter insoluble in benzene, less H-resins = M-resins	~1000	Confers binding properties on pitch
Low molecular weight resins	N-resins	~400	They assist the binder to wet the grist and have adhesive properties
Oils	m-oils n-oils	~350 ~250	These give fluidity to the binder and assist in wetting the grist during mixing. In the case of Soderberg-type anodes, these oils increase fluidity at the top of anode and ensure even distribution of electrode paste

A. Specific Gravity

Specific gravity is the ratio of the weight of a given volume of a material to the weight of an equal volume of water at the same temperature. There are two basic methods generally used for the measurement of specific gravity: (a) the pycnometer and (b) the displacement methods.

1. *The Pycnometer Method*

In this method, about 5 g of the prepared sample is placed in a tared pycnometer and weighed to obtain the amount of pitch. The pycnometer is then filled with freshly boiled distilled water, taking all precautions to ensure that no air is entrapped, and weighed again. Addition of a few drops of wetting agent and use of suction are recommended to ensure complete removal of entrapped air. The weight of the pycnometer filled with freshly boiled distilled water is also determined at the same temperature. The specific gravity is calculated from these data.

The STPTC recommends use of the Hubbard specific gravity bottle also.

Standards applicable

ASTM D 2320-66 (reapproved 1971)
ANS A 149.8-1969
STPTC S. No. GP. 1-62

2. *The Displacement Method*

In this technique a fragment of the sample is suspended by a thin wire (Nichrome is recommended) from an analytical balance and its weight both in air and in water at a particular temperature is determined. The specific gravity is calculated from the relationship

$$\text{specific gravity} = a/(a - b) \qquad (1)$$

where a is weight of the specimen in air and b is weight of the specimen in water. The test specimen in this case could be fragments of bulk sample weighing between 5 and 20 g. The specimens could also be prepared by casting in molds.

Standards applicable

ASTM D 71-72a
ANS A 37.72
ASTM D 70-72
ANS A 37.71-1974

B. Softening Point

1. *Scope and Significance*

Since pitch is a supercooled liquid, the transition from solid to liquid is not very distinct, and thus pitch does not have a true melting point. As its temperature is raised, pitch gradually softens and becomes less viscous. The process is continuous and there is no definite temperature at which softening point can be said to begin. It is therefore most important that the determination of softening point be made by a closely defined test method to ensure reproducible results.

Small amounts of moisture produce significant lowering of softening point and so adequate care in the preparation of the sample is of the utmost importance.

2. *Summary of Methods*

a. Ring and Ball Method. In this method, a disk of the sample is held within a brass ring and made to support a steel ball 0.953 cm in diameter and weighing ~3.5 g. The assembly is heated at a prescribed rate in a water or glycerin bath. The temperature at which the sample softens sufficiently to be forced down a distance of 2.54 cm is taken as the R & B softening point and is reported to the nearest 0.5°C or 1.0°F.

Standards applicable

ASTM D 36-70
ANS A 37.10-1974
STPTC S. No. PT 3-62

b. The Kraemer and Sarnow Method. In this method the pitch melted under specified conditions is poured into brass or stainless steel pitch containers. The containers are attached to glass tubes in the apparatus by small lengths of rubber tubing. The tubes are charged with 5 ± 0.05 g of mercury. The assembly is positioned in water in a beaker which in turn is surrounded by water in a larger beaker. The temperature of the water is raised at a specified rate and the temperature at which the mercury bursts through the envelope of viscous pitch is noted. The average of at least two determinations is reported as the softening point.

The test is not applicable to pitches which give a result below 30°C by this method. For hard pitches which give a softening point of 80°C by this method, the water in the baths is replaced by glycerin.

Standard applicable

STPTC S. No. PT 2-62

c. Cube-in-Water Method. In this method, the sample is formed into well-shaped cubes with sharp edges of 1.27 cm and a 2-mm hole in the center. The sample is suspended from an L-shaped copper hook, supported in water in a beaker, and heated at a definite rate. The temperature reported to the nearest 0.5°C (or 1.0°F) at which the cube sags and flows downward a distance of 2.5 cm is taken as the softening point.

Pitches having softening points above 80°C by this method should be tested by the cube-in-air method.

Standards applicable

ASTM D 61-75
ANS A 37.36-1974

d. Cube-in-Air Method. The sample is prepared exactly as described for the cube-in-water method. The sample suspended from the wire hook is assembled in a standardized air oven and heated at a definite rate. The temperature at which the cube sags and flows downward a distance of 6.0 mm, reported to the nearest 0.5°C (or 1.0°F), is defined as the softening point.

Standards applicable

ASTM D 2319-66
ANS A 1497-1969

e. Mettler Softening Point Method. This method gives results comparable to those of the cube-in-air method. The sample is placed in a chromium-plated brass cup having a 6.35-mm-diameter hole at the bottom. The cartridge assembly containing the sample is positioned in a standard furnace and heated at the rate of 2°C/min. At the softening point, the sample flows downward a distance of 19 mm to interrupt a beam of light; the temperature is displayed on the digital layout and reported to the nearest 0.1°C (or 0.2°F).

Standard applicable

ASTM D 3104-75

C. Viscosity

Coal tar pitches have been found to behave almost like a Newtonian liquid over a wide range of temperatures, particularly at temperatures above their R&B softening points. Some reduction of viscosity with increasing shearing stress has been noticed at lower temperatures and for pitches with a high content of toluene-insoluble material (McNeil, 1966a,b).

Since the degree of flow of a pitch under a set of conditions is dependent on the applied force, determination of both softening point and penetration is equivalent to viscosity measurement. Ductility and brittle point determinations which involve breaking a thread or film of the sample could also be considered as a measure of the equiviscous temperature.

The determination of viscosity of coal tar pitches and fluxed coal tar pitches is usually made using tar viscometers, an Engler viscometer, or a Saybolt furol viscometer. The methods measure the time for flow of a specified quantity of the sample under strictly controlled conditions through a standard orifice.

1. *Viscosity by Standard Tar Viscometer*

a. Scope and Significance. The determination of viscosity involves the measurement of the time of efflux of 50 ml of the sample through a standard orifice under strictly specified conditions. Two types of cup are specified for the instrument, one having a 10-mm-diameter orifice suitable for road tars, coal tar fuels, and similar fluxed pitches, and the other having a 4-mm-diameter orifice suitable for more fluid materials.

The working range of the instrument is limited. The efflux time in the case of the 10-mm cup must be between 10 and 140 sec. This is achieved by making determinations at temperatures which are suitable multiples of 5°C, not lower than 20°C.

b. Summary of the Method. The standard viscometer is assembled, and the water bath is filled to within 10 mm of the top. The contents of the water bath are brought to a temperature 0.05°C higher than the test temperature and maintained to within ±0.05°C throughout the test.

After proper preliminary treatment, the sample is warmed to about 0.5°C above the test temperature and poured into the cup to the level specified. When the sample has attained the proper temperature, the ball valve is lifted and the sample is allowed to flow into the receiver. The time for flow of 50 ml is noted and reported in seconds, noting the test temperature and the size of the orifice in the cup used.

Standard applicable

STPTC RT2-62

2. *Engler Specific Viscosity of Tar Products*

a. Scope and Significance. This method covers the determination of specific viscosity of tars and their fluid products. It is an empirical flow test and the requirements of the method should be closely adhered to in order to obtain reproducible results. The Engler specific viscosity is the ratio of the time required for the flow of 50 ml of the material using an Engler viscometer at a selected temperature to the time of flow of an equal volume of water at 25°C. The specific viscosities are usually determined at 25, 40, 50, and 100°C (77, 104, 122, and 212°F, respectively). Generally the temperature is so selected that the specific viscosity is not more than 45.

b. Summary of the Method. The Engler viscometer is first calibrated with distilled water under strictly specified conditions and the time for efflux of 200 ml determined as the mean of the averages of at least two series of determinations agreeing within 0.2 sec. The time for an acceptable viscometer is between 50.0 and 52.0 sec. The factor representing the efflux time for 50 ml water at 25°C has been found to be equivalent to the efflux time for 200 ml of distilled water at 20°C multiplied by 0.224. The determination of efflux for 50 ml of the sample is made at the desired temperature under strictly specified conditions. The temperature of the bath should not vary more than 1°C for test at 25, 40, or 50°C and not more than 2 or 3°C for test at 100°C.

The specific viscosity is calculated from the formula

$$\text{Engler specific viscosity at } t°\text{C} = \frac{\text{time in seconds for the flow of 50 ml of sample at } t°\text{C}}{\text{factor}} \qquad (2)$$

Standards applicable

ASTM D 1665-61 (reapproved 1973)
ANS A 37.112-1964 (reapproved 1969)

3. *Saybolt Furol Viscosity of Bituminous Materials at High Temperatures*

a. Scope and Significance. This method covers the determination of Saybolt furol viscosity of bituminous materials at temperatures of 250, 275, 300, 325, 350, 400, and 450°F (121, 135, 149, 163, 177, 204, and 232°C, respectively). The Saybolt furol viscosity is reported as the time in seconds for 60 ml of the sample to flow under strictly specified conditions through the Saybolt viscosity tube having a calibrated furol orifice at the appropriate temperature. The furol viscosity is approximately one-tenth of the universal viscosity and is recommended for those petroleum products having viscosities greater than 1000 sec (Saybolt universal).

b. Summary of the Method. The sample prepared as specified is heated to 20–25°F (11–14°C) above the test temperature with proper stirring. The sample is poured through the heated No. 20 (850 μm) sieve into the viscosity tube until the sample overflows into the gallery. The equipment is assembled as specified. The bath temperature is adjusted so that the temperature of the sample remains constant. The receiving flask (capacity 60 ± 0.005 ml) is placed in position and the sample is allowed to flow into the flask; the flow time is measured in seconds. The results are reported to the nearest 0.5 sec for viscosities below 200 sec and to the nearest whole second for viscosities 200 sec or above.

Standard applicable

ASTM E 102-62 (reapproved 1973)

D. Equiviscous Temperature

The equiviscous temperature (EVT) of a sample of tar or cutback coal tar pitch is defined as the temperature in degrees centigrade at which the viscosity of the sample is 50 sec when determined in a standard tar viscometer employing a 10-mm cup. The equiviscous temperature can be determined directly by means of an EVT viscometer. It could also be derived by calculation from determinations made in a standard tar viscometer.

1. *Direct Determination of EVT*

a. Scope and Significance. The EVT is determined by the observation of the viscous drag on a rotating cylinder partly immersed in the sample. The instrument has to be periodically calibrated by reference to a tar whose EVT has been determined by STPTC RT 3, Part I. The dimensions of the instrument are not critical since it is regularly calibrated against a sample of known EVT. This instrument is not recognized for reference purposes.

b. Summary of the Method. The instrument consists essentially of a cylindrical cup made of stainless steel or chromium-plated brass having an internal diameter of 22 mm and a depth of 38 mm in which is suspended a cylinder made of stainless steel or chromium-plated brass 16 mm in diameter and 34 mm in length. The cylinder is centered in the cup with a clearance of 3 mm between the bottom of the cylinder and the inner surface of the cup and hangs from a straight beryllium–copper alloy torsion wire 200 mm in length and 0.65 mm in diameter. The wire also carries a flywheel which is graduated in 45° divisions and rotates beneath a fixed pointer. The upper end of the torsion wire is attached to a torsion head which can be turned through 180° between two stops.

The tar cup and cylinder are completely immersed in a water bath which is equipped with a manually operated paddle stirrer. There is an arrangement for raising the temperature of the water bath at a controlled rate.

During a determination the specified quantity of the sample is taken in the cup. It is then positioned on the viscometer and water is poured into the bath to the indicator line so that the cup and cylinder are immersed. The pointer is set to coincide with one of the graduation lines on the flywheel.

The temperature of the bath is adjusted to not less than 10°C below the expected EVT of the sample and then heating is adjusted to raise the temperature at a rate of 1°C/min. The bath is agitated continuously. At periodic intervals, the torsion head is quickly turned through 180° and the extent of deflection of the flywheel is noted. The temperature at which an overswing of one 45° division is first attained is recorded as the EVT of the sample.

2. *EVT by Calculation from Determination Made in a Standard Tar Viscometer*

a. Scope and Significance. In connection with the sale and purchase of road tars, refined tars, etc., the EVT is generally understood to be that obtained by the method discussed in this subsection.

The viscosity of the sample is determined in a Standard Tar Viscometer. The 10-mm cup is used for tars or fluxed coal tar pitches of EVT at or above 17.5°C and the determination is made at the temperature which is a multiple of 5°C nearest the expected EVT or to 2.5°C above it. The EVT is obtained by correcting the temperature of the test for the viscosity in seconds by reference to given tables.

When the viscosity is outside the range 33–75 sec in the 10-mm cup, the approximate viscosity is ascertained from the accompanying table and the determination made at the correct temperature. The EVT is obtained by reference to the same table.

Standard applicable

STPTC RT 3-62

E. Viscosity Conversion Charts

There are some extremely useful charts relating to viscosity and temperature, for the conversion of kinematic viscosity to equivalents in other viscometer units.

1. *Viscosity Temperature Chart for Asphalts*

ASTM Standard D 2493-68 (reapproved 1973) provides a viscosity temperature chart which is a convenient means of plotting data for estimating the absolute viscosity of asphalts at any temperature within a limited range. Conversely the chart may be used to ascertain the temperature at which a desired viscosity is attained.

The chart is suitable for original asphalt cements, for asphalts recovered from laboratory aging tests or extracted from pavements, and for roofing asphalts.

2. *Temperature–Viscosity Relationships for Coal Tar and Pitches*

This nomogram is described in the appendix of STPTC. The nomogram provides information relating to viscosity of coal tars and pitches over a wide range of materials and temperature conditions. This is an extremely useful nomogram as it enables fairly close relationships to be obtained relating to a range of viscosities defined on one side by the tar viscometer and extending to the higher viscosity ranges defined by penetrometer measurements, EVT, and temperature.

3. *Conversion of Viscosities*

STPTC provides a viscosity conversion chart for the conversion of viscosities obtained by one instrument to corresponding values by other viscometers or to kinematic viscosities at the same temperature. The conversion of kinematic viscosity to Saybolt universal viscosity or to Saybolt furol viscosity is covered by ASTM Standard D 2161-74 (corresponding to ANS-Z-11.129-1975; Method 9101, Federal Test Method Standard No. 791b).

4. *Nomogram for Fluxing Coal Tar or Pitch with Tar Oils*

STPTC provides a nomogram which enables the proportion of tar oils (such as light creosote, heavy creosote, or still heavier oil) to be added to a base tar or pitch to produce a product of required equiviscous temperature. Alternatively, this nomogram could be used to find the EVT of a base pitch or tar which has been fluxed with a known quantity of a tar fraction.

5. *Nomogram for Raising the Viscosity of Tar or Soft Pitch by the Addition of Medium-Soft Pitch*

This nomogram, described in STPTC, is based on the use of medium-soft pitch of R&B 80°C which is equivalent to an EVT of ~100. When tar or soft pitch is blended with this material, the EVT of the base is raised, and can be obtained from the nomogram.

F. Penetration Number

1. *Scope and Significance*

Penetration is a measure of the consistency of a bituminous material and is expressed as the distance in tenths of a millimeter that a standard needle penetrates a sample vertically under known conditions of loading time and temperature.

2. *Methods*

The sample is melted, transferred to a sample container, and cooled under specified conditions. The sample container is placed on the stand of a penetrometer. The sample container must be completely covered with water at the specified temperature. A standard needle is positioned so that its tip is in contact with the surface of the pitch. The needle is released under a specified load to penetrate the pitch and the distance of penetration is measured in tenths of a millimeter.

The conditions of test are normally 25°C, 100 g load, and 5 sec time. Special tests call for other conditions, such as the following:

Temperature		Load	Time
°C	°F	(g)	(sec)
0	32	200	60
4	39.2	200	60
46.1	115	50	5

Standards applicable

ASTM D 5-73
ANS A 37.1
IP 49/71

G. Distillation Test

1. *Scope and Significance*

This test is a measure of the quantity of lower boiling constituents in a pitch. A 100-g sample of the pitch is distilled under carefully controlled conditions from an electrically heated 300-ml distillation flask. Unless otherwise specified, the fractions usually collected are (1) ≤270°C, (2) 270–300°C, and (3) 300–360°C. The weights of the fractions collected are reported as percentages to the nearest 0.1% based on the weight of water-free sample.

2. *Procedure*

A total of 100 ± 0.1 g of the prepared sample is transferred to a standard distillation flask and weighed. The flask assembly is positioned in the flask shield and connected to the condenser tube. Heating is started and controlled at the specified rate. The distillate fractions are collected in tared receivers and weighed.

Standards applicable

ASTM D 2569-75
ANS A 149.10
ASTM D 20-72
ANS A 37.9-1974

This is an adaptation of an earlier method applicable to road tars.

H. Volatile Matter

The quantity of lower boiling constituents expelled from a pitch sample when heated to temperatures in the range 925–970°C is measured by these methods.

1. *Platinum Crucible Method*

a. Scope and Significance. This method measures the quantity of volatile matter which is expelled from the sample of pitch at around the fusion point of potassium chromate (968.3°C).

b. Summary of the Method. The finely divided sample (1 ± 0.001 g) is placed in a tared platinum crucible of specified dimensions. It is closed with a lid and supported on a three-arm crucible support with silica points so arranged that when in position the bottom of the crucible is 10 mm above the burner, the crucible being completely enveloped by the flame. The gas pressure to the burner is adjusted to provide sufficient height and heat input by preliminary trials. The crucible is heated for 3 min, allowed to cool quickly, and then weighed. The loss in weight represents the volatile matter in the sample and is reported as percentage by weight.

Standard applicable

STPTC No. PT 5-62

2. *Muffle Furnace Method*

a. Scope and Significance. This method is very similar to the method used for determination of volatile matter in coal and uses the apparatus specified in BS 1016. The results obtained by this method are, however, a few percent lower than those obtained by the platinum crucible method.

b. Summary of the Method. The finely divided sample (1 ± 0.001 g) is weighed into a tared standard silica crucible having a lid. It is then placed on a crucible support in a muffle furnace which is maintained at 925 ± 5°C. The crucible is left in the muffle for exactly 7 min, removed, and cooled quickly by placing on a cold metal plate. It is then cooled to room temperature in a desiccator and weighed. The loss in weight is reported as percentage wt/wt of volatile matter.

Standard applicable

STPTC PT 6-62

I. Solubility Tests

1. Benzene Insolubles

a. Scope and Significance. This is an empirical method and covers the determination of benzene-insoluble matter. A weighed quantity of the prepared sample (adjusted to give between 150 and 250 mg of matter insoluble in toluene) is digested with hot toluene, then extracted with hot benzene in an Alundum thimble. The insoluble matter is dried and weighed.

b. Summary of the Method. The required amount of sample is taken in a tared 150 ml beaker and weighed to the nearest 0.5 mg. This is mixed thoroughly with 60 ml of toluene and heated to 95 ± 5°C, at which temperature it is maintained for 25 min.

Portions of the hot mixture are decanted into an Alundum thimble and allowed to filter through. The insoluble matter is finally transferred to the thimble with small quantities of toluene. The thimble and contents are worked once with benzene. While still wet, the thimble is positioned in the extraction apparatus and extracted with boiling benzene for 18 hr. The thimble is taken out and the benzene allowed to evaporate off at room temperature and then dried in an air oven at 105 ± 5°C for 60 min. It is cooled in a desiccator and weighed to the nearest 0.5 mg. The benzene-insoluble content is reported to the nearest 0.1%.

Standards applicable

ASTM D 2317-66
ANS A 149.5-1969

2. *Toluene Insolubles*

a. Scope and Significance. In this method, the sample of pitch is extracted with toluene and the insoluble matter is weighed in a filtering crucible. The weight of ash of the residue is also determined. The method could be used for determining insolubles in any other solvent with appropriate adjustment of the working temperature and the quantity of the solvent.

b. Summary of the Method. The sample (~1 g) is weighed into a beaker with an accuracy of 0.002 g. Toluene (100 ml) at 90–100°C is added and the sample is mixed thoroughly with a glass rod and allowed to settle while maintained hot. The supernatant solution is filtered through a filtering crucible under suction. The residue is treated with hot toluene (100 ml lots) three or four times and decanted through the crucible, the sediment is also transferred to the crucible, and washed with a further 200 ml of toluene. The crucible is dried at not more than 100°C, cooled in a desiccator, and weighed. The percentage by weight of insoluble matter in toluene is reported. A procedure using filter paper in place of the filtering crucible is also recommended. The residue in the crucible/filter paper is incinerated at a temperature of 650°C to constant weight and the weight of ash determined. The percentage weight of ash and ash-free insoluble matter in the sample is reported.

Standard applicable

STPTC S. No. RT 8-62

3. *Dimethylformamide Insolubles*

a. Scope and Significance. This is an empirical method and gives a measure of the dimethylformamide-insoluble (DMF-I) matter in tar and pitch. The sample (0.5 g) is digested in hot DMF and filtered. The insoluble material is washed, dried, and weighed.

b. Experimental Procedure. The sample (0.5 g) and Celite (0.45–0.55 g) are transferred to a 100-ml beaker. Dry formamide (25 ml) is added to the mixture with stirring and heated at 95–100°C in a water bath for about 30 min. When the digestion is completed, the mixture in the beaker is transferred gradually to a tared filtering crucible containing a mat of Celite (0.45–0.55 g). The beaker is washed with successive lots of formamide (2–3 ml) each at 100°C and transferred to the filter; the washings are repeated until the filtrate is of the same color.

Standard applicable

ASTM D 2764-71

4. *Quinoline Insolubles*

a. Scope and Significance. This is an empirical method and covers the determination of the quinoline-insoluble matter (QI) in tar and pitch. The sample is digested in hot quinoline and filtered. The insoluble material is washed, dried, and weighed.

b. Experimental Procedure. The sample (a suitable quantity that would yield between 75 and 150 mg of matter insoluble in quinoline) is transferred to a tared 100-ml beaker. Quinoline (25 ml) is added with stirring and heated in a water bath maintained at 75 ± 5°C for ~20 min. When the digestion is complete the hot quinoline–pitch mixture is poured into a tared crucible containing a prepared mat of Celite (0.45–0.55 g). The beaker is washed with successive portions of quinoline (2–3 ml each) at 75 ± 5°C and filtered through the filter. The residue on the filter is washed first with benzene (three times) and then with acetone (twice). The crucible is dried in an air oven at 105–110°C, to constant weight (±1 mg). The quinoline insoluble (QI) is reported to the nearest 0.1%.

Standards applicable

ASTM D 2318-66 (reapproved 1971)
ANS A 149.6-1969

5. *Anthracene Oil Insolubles*

Anthracene oil insolubles are sometimes determined in place of quinoline insolubles. This method gives results comparable to those of the quinoline-insoluble method. However, it is necessary to specify the properties of the anthracene oil used, such as its boiling range, and content of tar acids and bases, and other constituents. The method cannot be said to be fully standardized, but in the hands of experienced analysts consistent values for anthracene oil insolubles can be obtained.

J. Ash

1. *Scope and Significance*

This method relates to the determination of the ash of tars and pitches and is a measure of the mineral matter content. A weighed quantity of the sample is carefully volatilized and burnt and the carbonaceous residue is completely oxidized at 900°C in a muffle furnace.

2. *Experimental Procedure*

A representative portion of the sample (10 g) is transferred to a tared dish or crucible and placed in a cold muffle furnace. The temperature of the furnace is gradually raised at such a rate as to avoid mechanical loss from the sample by boilover or spattering (the sample may be heated on a hot plate or over a gas flame). The residue left after expulsion of volatiles is ignited in a muffle furnace at 900 ± 10°C until all the carbon has been burned off. The dish/crucible is cooled to ~100°C, then cooled further to room temperature in a desiccator, and weighed to the nearest 0.1 mg to constant weight. The ash percentage is reported to the nearest 0.01%.

Standards applicable

ASTM D 2415-66
ANS A 149.9-1969
STPTC S. No. PT 8-62

K. Coking Value† (Modified Conradson)

1. *Scope and Significance*

In this case a sample of the pitch is vaporized and pyrolyzed under specified conditions in standardized equipment and the residue obtained is reported as coking value. The coking value is a measure of the coke-forming propensities of the pitch under test.

2. *Experimental Procedure*

A weighed quantity of the representative sample (3.0 g) is transferred to a tared porcelain or silica crucible and placed in the center of a Skidmore crucible which in turn is positioned in the center of the metal crucible over a layer of sand. The Skidmore and metal crucibles are covered; the assembly is placed on a Nichrome heating triangle, positioned inside the insulator, and covered with the sheet iron hood and chimney.

The assembly is placed over a furnace (crucible furnace Hoskins No. FP 10+ or equivalent) set to give 900 ± 10°C. The sample is heated in the assembly for exactly 30 min and then allowed to cool for about 10 min. The porcelain crucible is placed in a desiccator, cooled, and weighed. The coking value is reported to the nearest 0.1%.

† For additional information on coking value see Chapter 34, Section III,I.

Standards applicable

ASTM D 2416-73
ANS A 149.12

L. Tests for Pitches Used for Pipeline Coatings

There are a number of tests for pitches particularly designed to measure their performance for various specific uses. Thus pitches used for coatings are generally tested for characteristics such as abrasion resistance, bendability, disbonding, impact resistance, chemical resistance, effect of weathering, water penetration, and measurement of film thickness. Pitches used for roofing, damp-proofing, and waterproofing are especially tested for flash point, ductility, and bitumen content.

1. *Abrasion Resistance*

a. Scope and Significance. This procedure defines a method for comparing the relative resistances of pipeline coatings to abrasion. Abrasion resistance may be used to specify the optimum coating thickness of candidate materials both in development and research work to study new coating systems or methods, and in quality control.

In this method the relative resistance of steel pipeline coatings to abrasion by a slurry of coarse abrasive and water is determined. The method is applicable to the testing of all types of electrical insulating pipeline coatings and tapes including thermoplastics, thermoset, and bituminous materials.

b. Summary of Test Method. A test specimen 406 mm (16 in.) long is cut from a representative length of production-coated 19.1 mm (nominal $\frac{3}{4}$ in.) steel pipe; the specimens should be holidayfree and used in duplicate. The coating thickness in each specimen is measured by means of an appropriate nondestructive type of thickness gauge. The specimens are positioned in the horizontal revolving drum of the test equipment such that they are electrically insulated from contact with the test apparatus. The apparatus is loaded with 13.6 kg (30 lb) of abrasive (aluminum oxide grit) and 5.68 liters (1.5 gal) of water. The drum is revolved at the rate of 30.5 m (100 linear ft)/min. The electrical resistance of the specimen is determined at 25-hr intervals for a test period of 200 hrs by means of a volt-ohmmeter. The thickness of the coating is also measured.

Precision data are limited to two adjacent specimens taken from the same production-coated pipe. The appearance of the coating and the thickness of the coating before and after testing are noted and reported, as are the electrical resistance measurements at 25-hr intervals over a 200-hr period.

Standard applicable

ASTM G 6-75

2. *Bendability of Pipeline Coatings*

a. Scope and Significance. This test provides information on the ability of coatings applied to pipe to resist cracking, disbonding, or other mechanical damage as a result of bending. Since the test is applied to coated pipes from commercial production, the results can be used directly in the selection of similar materials for service. The test also has application as a quality control method when variations in coating applications or material formulation will affect bending performance. The method covers the determination of the effect of short-radius bends on coatings applied to small-diameter pipe.

b. Summary of the Method. The test specimen is 2.59 m (100 in.) in length and should be representative of the production-coated pipe and be free of obvious coating flaws or defects. Coating specimens are applied to either 19 mm (0.75 in.) or 25.4 mm (1 in.) nominal diameter pipe and the coating not exceed 32 mm (1.25 in.). The conditioned specimen is placed in the V notch of the equipment and clamped in place.

The method consists of bending a small-diameter specimen of coated pipe around a mandrel to produce a range of short radius bends. Coating failure in the form of cracking or loss of adhesion is detected through visual and electrical inspection of the bent specimen.

Standard applicable

ASTM G 10-72

3. *Impact Resistance*

The ability of a pipe coating to resist mechanical damage during shipping, handling, and installation depends on its resistance to impact. There are two well-established testing methods. Both aim to measure the energy required to rupture the pipe coating under specified conditions of impact from a falling weight.

a. Limestone Drop Test. This test is intended to simulate the effects of backfilling after the pipe has been positioned in a trench. The impact resistance is determined by dropping weighed amounts of a specified type of limestone through a chute on the coated specimen. Results are reported as the number of drops required to pierce through the coating to bare metal as determined visually or electrically.

Standard applicable

ASTM G 13-72

b. Falling Weight Test. In this method a tup of fixed weight is dropped through varying heights to produce a point of impact on the surface of the pipe specimen. The resultant breaks in the coating are detected electrically.

Standard applicable

ASTM G 14-72

4. Chemical Resistance of Pipeline Coatings

a. Scope and Significance. The method is intended for evaluating the resistance of pipe coating materials to various concentrations of reagents or suspected soil contaminants and serves as a guide to compare the relative merits of pipe coating materials in specific environments.

The choice of reagents, concentrations, duration of exposure, temperature of the test, and properties to be reported are arbitrary and should be chosen to reflect conditions known to exist along the pipeline right-of-way.

b. Summary of the Method. The coated specimens are placed in long-term contact with both the liquid and vapor phase of the test reagent in a closed container. The specimens are inspected for visible signs of chemical attack.

The specimens may subsequently be tested for cathodic disbonding and penetration under load under specified standards in order to determine whether the specimens have undergone any loss of mechanical or bonding characteristics.

Standard applicable

ASTM G 20-72

5. Penetration Resistance of Pipeline Coatings (Blunt Rod)

a. Scope and Significance. This accelerated method covers determination of the relative resistance of steel pipeline coatings to penetration or deformation by a blunt rod under specified load. The method is intended to apply to the testing of all types of nonmetallic pipeline coatings and tapes including thermoplastics, thermosets, and bituminous materials.

b. Summary of the Method. The depth or rate of penetration or deformation that is caused by a weighted blunt rod to a coating system applied to steel pipe is measured over a period of time with a micrometer depth gauge. Three consecutive identical readings taken at specified intervals conclude the test.

Standard applicable

ASTM G 17-72

6. *Disbonding Characteristics of Pipeline Coatings by Direct Soil Burial*

a. Scope and Significance. Coated pipe is seldom, if ever, buried without some damage to the coating. Hence an actual soil burial test can contribute significant data, provided the method of testing is controlled, the test specimen monitored, and the relationship between the area disbonded, the current demand, and the mode of failure fully understood.

The method covers the determination of the relative disbonding characteristics of damaged coatings on steel pipe by cathodic protection potentials in direct soil burial. The method is intended to apply to the testing of all types of nonmetallic pipeline coatings and tapes including thermoplastics, thermoset, and bituminous materials.

The test is limited to nonconducting or nonmetallic pipe coatings and is not applicable to conducting materials such as zinc coatings on steel pipe.

b. Summary of the Method. Specimens in intentionally damaged areas are buried in soil at an outdoor site and electrically connected to a magnesium anode in a specified apparatus. After testing, the disbonded coating is removed, the exposed area measured, and comparisons are made to other specimens similarly exposed.

Standard applicable

ASTM G 19-72

7. *Cathodic Disbonding of Pipeline Coatings*

a. Scope and Significance. These methods cover accelerated procedures for determining comparative characteristics of insulating coating systems applied to steel pipe exterior for the purpose of preventing or mitigating corrosion that may occur in underground service where the pipe will be in contact with natural soils and may or may not receive cathodic protection. These are intended for use with samples of coated pipe taken from commercial production and are applicable to such samples when the coating is characterized by function as an electrical barrier.

b. Summary of the Method. Two methods are described in which the coating of the test specimen is subjected to electrical stress in a highly conductive electrolyte. The electrical stress is produced by connecting the test specimen to a magnesium anode. The coating is artificially perforated before starting the test. In method A, no electrical instruments are used. Results are determined by physical examination. In method B electrical instruments are used for measuring the current flowing in the cell. The specimen is physically examined on conclusion of the test.

Standard applicable

ASTM G 8-72

8. Effects of Outdoor Weathering on Pipeline Coatings

a. Scope and Significance. Since coated pipe may be stored outdoors for long periods before burial, weathering tests of the type described in this method are needed to evaluate the stability of these coatings stored outdoors. The results obtained should be treated only as indicating the general effect of weathering. Exposure conditions vary greatly from year to year, from one part of a year to another, and from locality to locality. The results of short-term exposure tests in the North are more meaningful if exposure is started in the summar followed by a winter season. In the South, where climatic conditions are more uniform throughout the year, the time of year when short-term exposure is started is less critical. In all localities, the longer the exposure period, the more reliable are the results obtained.

b. Summary of the Method. The effects of outdoor weathering on pipeline coatings after 6, 12, and 24 months' exposure are determined visually and by electrical means by comparing exposed samples of coated pipe with unexposed samples of coated pipe before and after impact and bending tests.

Standard applicable

ASTM G 11-72

9. *Atmospheric Environmental Exposure Testing of Nonmetallic Materials*

The recommended practice is a guide to those engaged in atmospheric environmental exposure testing in obtaining uniform results by indicating the variables that should be considered and specified. The methods of preparation of test specimens and particular exposure requirements of various materials are specified and are to be selected from appropriate specifications.

The samples are positioned on test racks made of any suitable material such as wood or aluminum with arrangements to adjust the angle of elevation and azimuth. Instruments are to be provided for recording temperature, relative humidity, solar radiation, wetness, rainfall, and contaminants such as NO_2, SO_2, and O_3. Information as to the exact environmental, exposure, mounting, and examination requirements should be provided for all specimens to be exposed.

Standard applicable

ASTM G 7-69 T

This method also refers to various ASTM standards relevant to various test procedures.

10. *Water Penetration into Pipeline Coatings*

a. Scope and Significance. This method covers the determination of the apparent rate of depth of water penetration into insulating coatings applied to pipe.

b. Summary of the Method. The method consists of an immersion-type test in which pipe specimens are suspended in an aqueous electrolyte for the duration of the test period. Electrical measurements of coating capacitance and dissipation are used to follow the water absorption rate of the test materials.

Standard applicable

ASTM G 9-72

11. *Nondestructive Measurement of Film Thickness of Pipeline Coatings on Steel*

a. Scope and Significance. This method covers the nondestructive measurement of the thickness of a dry, nonmagnetic coating applied to the external surface of a steel pipe. The method is recommended for coating thickness up to 3.0 mm (0.120 in.) and for any diameter pipe, but not smaller than 12.7 mm (0.5 in.).

Measurement of film thickness is an essential part of most test methods related to coatings on steel pipe. Adequate thickness of coating is essential to fulfill its function of preventing or mitigating corrosion of steel pipelines.

The accuracy of the measurements may be influenced by the deformability of the coating. The method is not applicable to coatings that are readily deformable under the force exerted by a probe of the measuring instrument.

b. Summary of the Method. The coating thickness is determined by a thickness gauge capable of being standardized over its range of intended use. It is designed so that variations in magnetic flux or magnetic attraction between its detection unit and the steel base can be calibrated to indicate the thickness of the coating material. It should be suitable for measuring thickness of dry, nonmagnetic coatings on either a flat or a circular base.

Standard applicable

ASTM G 12-72

12. *Joints, Fittings, and Patches in Coated Pipelines*

These methods describe the determination of the comparative corrosion preventive characteristics of materials used for application to joints, couplings, irregular fittings, and patched areas in coated pipelines. They are applicable to materials whose principal function is to act as barriers between the surface of the pipe and the surrounding soil environment. The test methods described employ measurements of leakage current, capacitance, and dissipation factor to indicate changes in insulating effectiveness of joint and patching materials.

a. Significance. The exposed metal surfaces at joints, fittings, and damaged areas in an otherwise coated pipeline will be subjected to corrosion if allowed to come into contact with soil environment. The performance of joints and patching materials designed to function as protective coverings will depend on such factors as the ability of the material to bind to both the pipe coating and exposed metal surfaces, the integrity of the moisture seal at lapped joints, and the water absorption characteristics of the joint material.

The significance of substantial leakage current through the coating joint, patch, or fitting is reliable evidence that the material has suffered a significant decrease in its performance as a protective barrier. In a similar manner, measured changes in the joint capacitance and dissipation factor are useful because they are related to the water absorption rate of the joint material. Water permeating an insulating barrier increases its capacitance and its progress can be measured through the use of a suitable impedance bridge.

b. Summary of Methods. Method A: This is an immersion test whereby the coated pipe specimens, each containing a simulated joint, tee, or patched area, are suspended in an electrolyte and placed under cathodic protection by connecting the specimen to the negative (−) terminal of a 6-V dc power supply. An anode, also immersed in the electrolyte and connected to the positive (+) terminal of the power supply, completes the test circuit. Joint or batch performance is followed through periodic determinations of leakage current measured as voltage drop across a calibrated resistor in the anode to cathode circuit.

Method B: Test circuit and specimen are identical to those in method A. However, capacitance and dissipation factor measurements are used to supplement the periodic leakage current determinations. These additional electrical measurements will furnish early indications of joint performance in cases in which water absorption leads to premature joint failure.

Standard applicable

ASTM G 18-72

M. Tests for Pitches Used for Roofing, Damp-Proofing, and Waterproofing

1. *Flash Point*

a. Scope and Significance. Coal tar pitches meant for use in roofing, damp-proofing, and waterproofing are usually tested for flash point, the usual equipment specified being the Cleveland open cup. Even though ASTM D 92-72, corresponding to ANS Z 11.6-1966 (reapproved 1972), does not specifically cover determination of flash point of pitches, this method is recommended in ASTM Specifications D 450-71 (corresponding to ANS A 109.6-1974).

b. Summary of the Method. The standard test cup is filled to the specified level with the sample. The temperature of the sample is increased at specified rates, rapidly at first, and then at a slow constant rate as the flash point is approached. At periodic intervals, a small test flame is passed across the cup. The lowest temperature at which application of the test flame causes the oil to ignite is taken as the flash point and is reported to the nearest 5°F or 2.5°C.

Standards applicable

ASTM D 92-72, corresponding to ANS Z 11.6-1966 (reapproved 1972)
American Association State Highway Officials Standard, AASHO No. T 48

2. *Ductility*

a. Scope and Significance. The ductility of a bituminous material is the distance to which it will elongate before breaking when a standard test specimen is pulled apart at a specified speed and temperature.

b. Summary of the Method. The sample is melted and strained through a wire cloth sieve under specified conditions and then poured into a mold. The sample is allowed to cool down to room temperature and then placed in the water bath maintained at the appropriate temperature. The specimen is positioned in the testing machine and pulled apart at a constant rate which is usually 5 cm/min ± 5.0%. The distance in centimeters through which the sample has been pulled to produce rupture is noted. The usual temperature at which the test is made is 77 ± 0.9°F (25 ± 0.5°C). The average of three normal tests is reported as the ductility of the sample.

Standards applicable

ASTM D 113-69
ANS A 37.11-1970
AASHO T 51

ACKNOWLEDGMENTS

The author is grateful to the Director of the Central Fuel Research Institute for permission to take on this assignment and for continued encouragement. He is also grateful to the help received from all members of the Tar and By-Products Section. He is particularly indebted to Mihir Baran Roy for helping in the compilation of the manuscript.

REFERENCES

American Association of State Highway Officials (AASHO) (1942). "Standard Specifications for Tar Use in Road Construction," M 52-42. Washington, D.C.

ASTM (1955). "Specifications for Coal-Tar Pitch for Roofing, Dampproofing, and Waterproofing," D450-41. Am. Soc. Test. Mater., Philadelphia, Pennsylvania.

ASTM (1968). "Standard Specification for Tar," D 490-47. Am. Soc. Test. Mater., Philadelphia, Pennsylvania. (Reapproved, 1968.)

ASTM (1970). "Coal Tar Saturated Roofing Felt for Use in Waterproofing and in Constructing Built-up Roofs," D227-56. Am. Soc. Test. Mater., Philadelphia, Pennsylvania. (Reapproved, 1970.)

ASTM (1972). "Woven Burlap Fabrics Saturated with Bituminous Substances for Use in

Waterproofing," D 1327-59. Am. Soc. Test. Mater., Philadelphia, Pennsylvania. (Reapproved, 1972.)
ASTM (1973). "Woven Glass Fabrics Treated for Use in Waterproofing and Roofing," D 1668-73. Am. Soc. Test. Mater., Philadelphia, Pennsylvania.
ASTM (1974). "Woven Cotton Fabrics Saturated with Bituminous Substances for Use in Waterproofing," D 173-74. Am. Soc. Test. Mater., Philadelphia, Pennsylvania.
Babenko, E. M., Sukhorukov, I. F., Oshchepkova, N. V., and Chalykh, E. F. (1967). *Coke Chem. USSR* No. 4, pp. 24–28.
Badenhorst, R. P., and Perold, G. W. (1957). *J. Appl. Chem.* **7,** 32.
Berl, E., and Schildwächter, H. (1928). *Brennstoff.-Chem.* **9,** 137.
Branscomb, J. A. (1966). *In* "Bituminous Materials, Asphalts, Tars and Pitches" (A. J. Hoiberg, ed.), Vol. 3, Part 12, pp. 372–380. Wiley (Interscience), New York.
Calaque, R. (1953). *Stahl Eisen* **73,** 1339.
Caplin, P. B. (1964). *Chem. Ind. (London)* pp. 44–49.
Charette, L. P., and Bischofberger, G. T. (1955). *Ind. Eng. Chem.* **47,** 1412–1415.
Darney, A. (1956). *Coal Tar Res. Assoc. Conf., Leeds.*
Darney, A. (1958). *Proc. Conf. Ind. Carbon Graphite, Soc. Chem. Ind., London* pp. 152–161.
Dell, M. B. (1958). *Am. Chem. Soc., Div. Gas Fuel Chem., Prepr.* pp. 1–8.
Dickinson, E. J., and Nicholas, J. H. (1949). *Road Res. (U.K.), Tech. Pap.* No. 16.
Dolkart, F. Z. (1961). *Chem. Abstr.* **55,** 27825 g.
Domitrovic, R. W., Stickel, R. M., and Smith, F. A. (1962). *Am. Chem. Soc., Div. Fuel Chem., Prepr.* pp. 54–64.
Duriez, M. (1939). *Int. Road Tar Conf., Brussels.*
Fair, W. F., and Volkman, E. W. (1943). *Ind. Eng. Chem., Anal. Ed.* **15,** 235.
Federal Government Specification (1942). RT-143. Washington, D.C.
Federal Specification (1974a). "Cotton Fabric, Woven Coal Tar Saturated," HH-C-591. Washington, D.C.
Federal Specification (1974b). "Pitch Coal Tar for Mineral Surfaced Built up Roofing, Water-proofing and Damp-proofing," RP-381. Washington, D.C.
Franck, H.-G. (1955). *Brennst.-Chem.* **36,** 12.
Franck, H.-G., and Wegener, O. (1958). *Brennst.-Chem.* **39,** 195.
Green, S. J., and Olden, M. J. F. (1951). *J. Appl. Chem.* **1,** 433.
Greenhow, E. J., and Smith, J. W. (1960a). *Aust. J. Appl. Sci.* **11,** 1, 169.
Greenhow, E. J., and Smith, J. W. (1960b). *CSIRO, Div. Coal Res., Tech. Comun.* No. 37.
Greenhow, E. J., and Smith, J. W. (1961). *Fuel* **40,** 69.
Harris, A. S., White, E. N., and McNeil, D. (1953). *J. Appl. Chem.* **3,** 433.
Harris, A. S., White, E. N., and McNeil, D. (1956). *J. Appl. Chem.* **6,** 293.
Herzog, E. (1944). *Vertraul Ber.* **82,** 1.
Hoiberg, A. J. (1966). "Bituminous Materials, Asphalts, Tars and Pitches," Vol. 3. Wiley (Interscience), New York.
Izd. "Nauka" Moscow (1970). *Chem. Abstr.* **72,** 14510q (1970).
Janik, M. (1965). *Brennst.-Chem.* **46,** 72–75.
Katona, A., Erdstrom, T., and Wash, T. J. (1954). *Am. Chem. Soc., 126th Meet.*
Kekin, N. A., Stepanenko, M. A., and Matusyak, N. I. (1968). *Coke Chem., USSR* No. 7, pp. 45–50.
Kekin, N. A., Belkina, T. V., Palaguta, T. S., Ikonomopulo, V. P., Vodolazhchenko, V. V., and Nikitina, T. E. (1975). *Khim. Tverd. Top.* No. 2, pp. 106–116. (*Chem. Abstr.* **83,** 100576z.)
Kotlik, B. E., Kolesnikova, R. Y., and Cherkasov, N. K. (1966). *Coke Chem., USSR* No. 5, pp. 52–54.
Krenkler, K., and Wagner, R. (1948). *Erdoel Kohle* **1,** 280.

Lee, A. R., and Dickinson, E. J. (1954). *Road Res. (U.K.), Tech. Pap.* No. 31.
McNeil, D. (1961). *J. Appl. Chem.* **11,** 90.
McNeil, D. (1966a). "Coal Carbonisation Products." Pergamon, Oxford.
McNeil, D. (1966b). *In* "Bituminous Materials, Asphalts, Tars and Pitches" (A. J. Hoiberg, ed.), Vol. 3, Part 5, pp. 154–158. Wiley (Interscience), New York.
Mallison, H. (1950). *Bitumen, Teere, Asphalte, Peche* **1,** 313.
Martin, S. W., and Nelson, H. W. (1958). *Ind. Eng. Chem.* **50,** 33–40.
Massinon, J. (1956). *Stahl Eisen* **76,** 331–333.
Nellensteyn, F. J. (1928). *J. Inst. Pet., London* **14,** 134.
O'Brochta, J. (1956). *Am. Chem. Soc., Div. Gas Fuel Chem., Prepr.*
Olschenka, P. (1974). *Freiberg. Forsch. A* **534,** 29–47. [*Chem. Abstr.* 23144W (1975).]
Pfeiffer, J. P. (1950). "The Properties of Asphaltic Bitumens," pp. 49–76. Elsevier, Amsterdam.
Pitch Fibre Pipes (1959). *Pipes Pipelines* **4**(3), 36–37.
Pitch Fibre Pipes (1960). *Pipes Pipelines* **5**(6), 85–89.
Pollack, S. S., and Alexander, L. E. (1959). *Am. Chem. Soc., Div. Gas Fuel Chem., Prepr.* pp. 135–150.
Raich, H. (1952). "Special Report on Colloid Structure of Bituminous Residuals." Mellon Inst. Ind. Res., Pittsburgh, Pennsylvania.
Rosengren, A. (1954). *Arch. Eisenhuettenw.* **25,** 11–18.
Sack, H. A. J. (1951). *J. Rech. CNRS* **16,** 21.
Sanada, Y., Furuta, T., Kumai, J., and Kimura, H. (1972). *Tanso,* **71,** 133–134. [*Chem. Abstr.* **78,** 87012n (1973).]
Schafer, H. G. (1956). *Freiberg. Forsch. A* **51,** 35.
Schafer, H. G. (1967a). *Erdoel Kohle* **20,** 416–419.
Schafer, H. G. (1967b). *Coke Rev.* (*Br. Coke Res. Assoc.*) **6,** 3.
Shaposhnikova, V. A., and Gorpinenko, M. S. (1974). *Zh. Metall. Abstr.* No. 2G 123. (*Chem. Abstr.* **81,** 155560d.)
Shea, F. L., Jr., and Juel, L. H. (1950). U.S. Pat. No. 2,500,208.
Smith, J. W. (1966). *Fuel* **45,** 233–244.
Stepanenko, M. A., Matusyak, N. I., and Kekin, N. A. (1970). *Coke Chem., USSR* No. 2.
Stepanenko, M. A., Pityulin, I. N., and Krysin, V. P. (1974). *Coke Chem., USSR* No. 7, pp. 33–39.
Streiter, O. G. (1938). *J. Res. Natl. Bur. Stand.* 20, 163.
Takaki, M., and Miyasaka, S. (1965). *Koru Taru* **17**(2), 33–35. [*Chem. Abstr.* **63,** 393b. (1966).]
Thomas, B. E. A. (1960). *Gas World—Coking* April 2.
U.S. Patent (1962). No. 3,070,449.
U.S. Patent (1963). No. 3,106,475.
U.S. Patent, (1964a). No. 3,141,783.
U.S. Patent, (1964b). No. 3,141,785.
U.S. Patent (1964c). No. 3,141,790.
U.S. Patent (1964d). No. 3,141,917.
U.S. Patent (1964e). No. 3,148,238.
U.S. Patent (1965). No. 3,168,602.
Van Krevelen, D. W., and Chermin, H. A. G. (1954). *Fuel* **33,** 79.
Van Krevelen, D. W., and Chermin, H. A. G. (1957). *Fuel* **36,** 313.
Volkmann, E. W. (1959). *Fuel* **38,** 445.
Von Skopnik, A. (1931). *Teer Bitumen* **29,** 26–31.
Wagner, J. (1915). *Stahl Eisen* **35,** 1289–1296.

Wall, E. J. (1966). *In* "Bituminous Materials, Asphalts, Tars and Pitches" (A. J. Hoiberg, ed.), Vol. 3, Part 14. Wiley (Interscience), New York.
Weiler, J. F. (1967a). *Blast Furn. Steel Plant* **55,** 238.
Weiler, J. F. (1967b). *Coke Rev. (Br. Coke Res. Assoc.)* **6,** 2.
Wieland, H., and Muller, W. (1949). *Justus Liebigs Ann. Chem.* **524,** 199.
Wood, L. J. (1961). *J. Appl. Chem.* **11,** 130.
Wood, L. J., and Phillips, G. (1954). *Nature (London)* **174,** 801.
Wood, L. J., and Phillips, G. (1955). *J. Appl. Chem.* **5,** 326.
Yeager, F. W. (1941). U.S. Pat. No. 3,96,904.

Chapter 34

Analysis of Coal Tar Binders for Electrodes

Laurence F. King†
RESEARCH DEPARTMENT
IMPERIAL OIL ENTERPRISES LTD
SARNIA, ONTARIO, CANADA

I. CARBON ELECTRODE MANUFACTURE

Large carbon electrodes are used in the electrochemical industry in the production of aluminum, magnesium, alkalies, and chlorine; in electrothermal processes for graphite, calcium, and silicon carbides; and in manufacturing steel and some nonferrous alloys. Small electrodes are

† Present address: 843 St. Clair Parkway, Mooretown, Ontario, Canada, NON 1MO.

ISBN 0-12-399902-2

employed in electric arc lighting and primary batteries (Cameron, 1950; Pearson, 1955; Bacon, 1964; Branscomb, 1966). The term "anode" refers to use in an electrolytic process, as in the manufacture of aluminum or sodium and chlorine; the term "electrode" signifies applications in which the function is primarily conduction of current.

The aggregate or filler which forms the bulk of the electrode is a form of carbon with a variable range of particle sizes. Many materials are utilized, sometimes in combination: calcined petroleum coke, anthracite, graphite, carbon black, metallurgical coke, and pitch coke. Typical properties of petroleum coke (Liggett, 1964) and a flow sheet for carbon electrode manufacture (Cameron, 1950; Mantell, 1975) have been reported. Because of its onion skin structure fluid coke is not as suitable as delayed coke, and both are amorphous (Stokes, 1975); the more crystalline needle coke is often preferred for graphite electrodes. The largest uses of petroleum coke are anodes for aluminum production and graphitized electrodes for electric arc steel furnaces (Shobert, 1975). Product quality control of fillers is by analysis for moisture, volatile matter (ASTM D 3176-74 and D 3180-74), sulfur (ASTM D 3177-75), water-soluble chlorides (ASTM D 1411-69), and real density (ASTM C 135-66) or electrical resistivity (Liggett, 1964).

In electrode manufacture addition of a binder, usually a pitch of coal tar origin but occasionally one from petroleum or other sources, is required to provide plasticity for pressing, tamping, and extrusion of the carbon paste. The coke produced by carbonization of the binder cements the aggregate particles together to form an electrode having the following properties: high apparent density and compressive strength, low porosity, uniformity, and resistance to thermal shock. These properties generally assure low electrode consumption, even distribution of electric current, and adequate thermal conductivity in service. The physical properties of amorphous carbon and graphite electrodes have been tabulated (Liggett, 1964; Branscomb, 1966).

The pitch binder, preferably in liquid form (Miller, 1963), is blended with the filler in a sigma-bladed mixer or a high shear kneader of the Baker–Perkins or Werner–Pfleiderer type at a temperature well above the softening point of the pitch. For larger electrodes typical mixes consist of calcined petroleum coke, varying in size from 1 to 2 cm or more to finer than No. 200 ASTM mesh, and 15–35 wt % of coal tar pitch (Hader *et al.*, 1954). The carbon paste or green mix may be compression molded or extruded to give products which are then baked at a gradually increasing temperature to a maximum of 750–1250°C, depending on the end use. The green strength of carbon paste, both hot and cold, is important since on carbonization a strong dense structure free of cracks must result. Green strength depends on many factors—the nature and particle size distribution of the filler; mixing conditions, extrusion/molding temperature, and pressure; and the amount and properties of the binder. Characteristics of the carbonized article are, in addition, a function of the heat-up rate during baking, the maximum temperature, and the cooling rate.

In aluminum production a common alternative to prebaking is a continuous, self-baking process in which the solidified paste in block form is suspended above a Soderberg anode. The paste slowly moves down and is fully baked when it arrives at the working face. To ensure adequate paste flow more binder is required than in the prebaked electrode process, since the temperature at the top of a Soderberg anode rarely exceeds 125°C. Controlled paste flow is of paramount importance, so that carbonization occurs with minimal shrinkage of the electrode and consumption is uniform at the electrolyte surface.

Performance data on prebaked and Soderberg electrodes in aluminum manufacture and consumption of carbon and graphite electrodes used in other industries are available (Mantell, 1975). About 1000 lb of carbon is consumed in the production of a ton of aluminum.

When graphitization is desired, prebaked electrodes are subsequently heated to 2500–3000°C over a long period.

II. PRODUCTION OF COAL TAR PITCH

A. Commercial Processes

The chemical and physical properties of a coal tar pitch binder depend on the process used in its manufacture. Pitches are normally classified as being derived from the corresponding coal tar. MacLeod *et al.* (1926) described various commercial coal carbonization processes having average temperatures ranging from 650°C for low temperature tar to 1300°C for ultranarrow coke oven tar, with vertical, inclined, and horizontal gas retorts, and ordinary coke ovens operating at intermediate temperatures. Low temperature tar contains paraffin wax and essentially no unalkylated condensed aromatics; in high temperature tar paraffins and aromatics with side chains are cracked leaving anthracene and naphthalene analogs. Analysis of the distillates from tars for naphthalene has been used to establish the probable source.

Products of high temperature treatment, especially coke oven pitch, as electrode binders gave less shrinkage on baking and a coke of lower porosity than a vertical retort pitch (Darney, 1958). McNeil and Wood

(1958) tabulated the properties of pitches from various sources; the specific gravity and aromaticity of coke oven products were generally the highest. Analytical data by infrared spectroscopy, solvent fractionation, and chemical methods were correlated with performance as an electrode binder. Pitches were preferred when prepared in batch stills, now practically obsolete, rather than in continuous stills. The longer residence time in a batch still maximized chemical reactions such as polymerization, cracking, cross-linking, and condensation which led to more aromatics in the pitch and higher binder coke production in an electrode. More high-molecular-weight resinous material was also produced in the pitch. Often referred to as free carbon—more accurately as benzene or toluene insolubles—these resins are universal constituents of coal tar pitch and distinguish it from aromatic pitches derived from other sources.

B. Modifications

The literature on electrode binders is replete with references to techniques for thermally and chemically modifying coal tar pitch to improve its performance, but conclusive data to substantiate improvement are rare. Thermal treatment yields products of increased softening point and resin content (McNeil and Wood, 1958; Mason, 1970), as does mild oxidation, e.g., by air blowing (McNeil and Wood, 1958; Tanaka *et al.*, 1965; Mitra *et al.*, 1974). Pitch from a low temperature coal tar was heated with sulfur to dehydrogenate the binder and increase the coking value (Liggett, 1964; Rao, 1964). Heating pitch with an inorganic nitrate increased the softening point (Mitra *et al.*, 1974). High coking binder compositions were prepared by adding mixtures of organonitrogen and chlorinated aromatic compounds with and without thermal carbon blacks (Shea and Juel, 1950). Various organic and inorganic chlorides when heated with pitch or mixed with carbon paste promoted chemical reactions that increased resin content, coking value, and apparent viscosity (Thrune, 1942; Brückner and Huber, 1950; Darney, 1958; McNeil and Wood, 1958). A disadvantage of the use of inorganic salts is the resulting increase in ash content. Although benzene-insoluble material was formed by treatment of Soderberg paste with organic compounds such as carbon tetrachloride and benzotrichloride, the slight gain in performance—lower electrode consumption—did not justify the cost of the additives (Darney, 1958).

Addition of lampblack to a carbon paste resulted in improved graphitizability (Darney, 1958; Fialkov, 1958). Since coal tar pitch is not merely a mechanical mixture of oil and various resinous and carbonaceous compounds but a complex colloidal system, addition of carbon black in general should be regarded only as increasing the content of filler. As much as 25 vol % based on the filler has been recommended (Swallen and Nelson, 1950). Small particle size, high surface area, reinforcing or semireinforcing furnace blacks in the nonpelletized form may behave differently. Relatively minute amounts, less than 1 wt % of the paste, were useful in upgrading coal tar binders of borderline quality, test electrodes of higher density and compressive strength being produced (King *et al.*, 1967; King and Robertson, 1973a,b).

It should be pointed out that none of the foregoing modification techniques has gained wide acceptance commercially, probably because of the history of general availability of coal tar pitch of adequate, if not outstanding, quality and the difficulty of justifying economically the use of additives.

III. PITCH BINDER PROPERTIES

When electrode components are mixed the molten binder enters the interstitial voids and the micropores of the solid particles. There is some evidence that binders which wet the filler less efficiently give stronger electrodes (Mason, 1970); the binder coke as it is formed modifies the properties of the solid particles (Mrozowski, 1958). The softening point of the pitch must be such as to provide a paste of suitable plasticity at the mixing temperature. The thermal coefficient of viscosity of the pitch should be low (McNeil and Wood, 1958). In commercial practice it is usual for the binder to be kept in hot storage (Miller, 1963), during which quinoline-insoluble and some benzene-insoluble components settle out. Some means, agitation or circulation, must be supplied to redisperse these (King and Robertson, 1973b). General binder requirements are reasonable cost, availability, and a low level of impurities such as metals and sulfur.

The particle size distribution of the filler and the binder content are both optimized for the process used and the end product. A compromise is often called for between two binder contents: one that is optimal for paste flow, the other for electrode properties (Mason, 1950; King and Robertson, 1973b). Particularly in the case of Soderberg electrodes, more pitch is added than may be indicated by requirements for maximum electrode density and strength. Desirable properties of electrodes are functions of the coked binder (coke residue × binder content) (King and Robertson, 1973a).

A pitch must be resistant to thermal cracking and production of excessive gas at low temperatures, which lead to porous electrodes (Bowitz *et al.*, 1958; Darney, 1958). Polymerization and similar reactions should predominate during coking, a requirement that is generally met by coal tar pitch with its condensed aromatic ring structures and only a few side chains (King and Robertson, 1968).

Properties of soft, medium, and hard pitches (80, 100, and 150°C softening point, respectively) have been tabulated (Liggett, 1964), as well as specifications or typical analyses pertinent to U.S. and foreign industry (Branscomb, 1966). No major differences are indicated between the two sets of specifications, though test procedures differ somewhat in detail. Table I lists typical properties of coal tar pitch for electrode binder use and fractions obtained by solvent extraction (McNeil and Wood, 1958; Liggett, 1964; Branscomb, 1966; King and Robertson, 1968).

It has long been realized that no single characteristic of a binder, as determined by laboratory tests, suffices as a criterion of electrode properties. Charette and Bischofberger (1955, 1961) and Charette and Girolami (1958, 1961) concluded that coking value and carbon/hydrogen atomic ratio, an index of aromaticity, are the best indices of pitch quality and designated the product of these two numbers as "correlation factor No. 1." They also suggested another: wt % benzene insolubles × C/H ratio of benzene insolubles × C/H ratio of benzene-soluble portion. Other investigators have listed specific gravity, softening point, aromaticity, coking value, resin content, and front ends of the pitch by distillation as best accounting for variations in electrode properties (Dell, 1959; Branscomb *et al.*, 1960; Mason, 1970). These pitch characteristics are interrelated. Current specifications in the main select those binders giving baked electrodes of highest density and strength and of lowest porosity.

A. Specific Gravity

The normal range of specific gravity, as determined pycnometrically by ASTM D 70-72 (semisolid pitch), D 2320-66 (hard pitch), or D 71-72, a displacement method, is 1.25–1.32. (For details on the pycnometer and displacement methods see Chapter 33, Section VI,A.) This property is a direct function of aromaticity, which in turn depends on pitch source. The most desirable binders, those from coke oven tar, have the highest specific gravity and normally contain more resinous material contributing selectively to coke formation (McNeil and Wood, 1958).

B. Softening Point

In purchase specifications softening point is one of the most important properties, inasmuch as it is an index of the suitability of a pitch for

TABLE I *Typical Analyses of Coal Tar Electrode Binders*[a]

	Whole pitch	Benzene insoluble ($C_1 + C_2$ resins)[b]	Benzene soluble (C_3 resins + oils)
Specific gravity	1.25–1.32	—	—
Softening point, C/A (°C)	90–105[c]	—	53
Apparent viscosity			
EVT_{15} (°C)	150	—	—
EVT_{1015} (°C)	110	—	—
T_c (P/°C)	25	—	—
Distillation (wt %)			
0–360°C	<6	—	(<9)
Resin content (wt %)			
C_1 (quinoline insoluble)	5–15 } 15–35	20–60	—
C_2 (quinoline soluble, benzene insoluble)	10–20 }	80–40	—
C_3 (benzene soluble, petroleum ether insoluble)	(35)	—	(50)
Carbon (wt %)	92–95	~95	93.5
Hydrogen (wt %)	4.3–4.5	3.0–3.5	5.2
C/H atomic ratio	1.6–1.8	2.2–2.6	1.5
Nonhydrocarbon components (wt %)			
S	0.3–0.8	—	—
N	0.5–1.5	—	—
O	1–2	—	—
Ash	0.2–0.5		
Coking value[d] (wt %)	50–60	92	37
Coking value × C/H atomic ratio[e]	85–110	180–230	56

[a] Values in parentheses are estimates.
[b] "Free carbon."
[c] Medium pitch, for vertical stub Soderberg anodes. May be 110–120°C for horizontal stub Soderberg or prebaked anodes. Hard pitch ~150°C.
[d] Direct heating 2.5 hr at 550°C, Alcan procedure.
[e] Correlation factor No. 1 (Charette and Bischofberger, 1955).

specific uses (Morrissey *et al.*, 1960). From the standpoint of coke yield, it is preferred to employ a binder of high softening point. Limiting this are the storage and handling systems available, e.g., the maximum temperature obtainable in mixing equipment. The optimum depends on whether the anode is prebaked or continuous, and, if continuous, whether it has vertical or horizontal stubs. In general, it is desirable to use a binder with a softening point as high as is compatible with adequate flow properties of the carbon paste. For similar types of pitch the coking value increases with softening point, as do apparent density and crushing strength of resulting electrodes. Electrical resistivity and consumption rate of anodes (Hodurek, 1920) tend to vary inversely as the softening point of the binder.

In the laboratory softening point is determined by procedures such as the Kraemer and Sarnow (K&S), the ring and ball (R&B), ASTM D 36-70, or the cube-in-air (C/A) method, ASTM D 2319-66. Values obtained by the R&B and C/A procedures are generally similar and 10–15°C higher than the K&S softening point (Darney, 1958; Branscomb, 1966).

There are three other ASTM methods: the Mettler, D 3104-75; the Mettler cup and ball, D 3461-75 (which gives results comparable to ASTM D 2319); and the cube-in-water test, D 61-75, for pitches having softening points below 80°C. Details on methods are presented in Chapter 33, Section VI,B.

Specifications for binders to be used in electrothermal and electrolytic anodes call for a C/A softening point of ~100°C, as do those for a Soderberg anode with vertical stubs. Prebaked anodes and Soderberg anodes with side pins utilize a pitch of 110–120°C C/A softening point (Fialkov, 1958; Branscomb, 1966). Experience has shown that binders having C/A softening points above the stipulated maxima permit too little flow of carbon paste, while those having softening points that are too low tend to thin out at mixing temperature and separate from the aggregate. Binder softening point (K&S) requirements and mixing temperatures for various other uses (e.g., pressed anodes, extruded electrodes, and rammed lining mixtures) are given in the literature (Darney, 1958).

McNeil and Wood (1958) tabulated data showing correlations among various viscosity functions, including softening point, penetration, ductility, brittle point, and glass transition temperature. At the K&S softening point the viscosity of a coal tar pitch is given as 55,000 P, at the R&B softening point, 8000 P. An equation was also shown from which an approximation to the viscosity at any temperature from 30°C below the R&B softening point to 150°C above it can be calculated if the viscosity at one temperature and the softening point are known.

The effect on softening point of air-blowing (McNeil and Wood, 1958) or heating pitch with a chemical condensing agent such as aluminum chloride is considerable (Thrune, 1942; Brückner and Huber, 1950). It is not normally feasible to effect a substantial improvement in a binder pitch by thermochemical means without unduly raising its softening point.

C. Apparent Viscosity

Coal tar pitches exhibit negligible departure from Newtonian behavior, except near the softening point, and thus differ from aromatic petroleum residua (King and Robertson, 1973b). The susceptibility of pitch viscosity to temperature is calculated from values determined at two or more temperatures (Evans and Pickard, 1931; Bowitz *et al.*, 1963). High- or medium-molecular-weight constituents, C_1, C_2, or C_3 resins, whether in true solution or colloidally dispersed, finely divided carbon, coke fines, and coal dust are all said to decrease the temperature susceptibility (Evans and Pickard, 1931; Volkmann *et al.*, 1936; Badenhorst and Perold, 1957; Bowitz *et al.*, 1963). Some investigators concluded that the viscous properties of a coal tar pitch are determined largely by its C_2 resin (quinoline-soluble, benzene-insoluble) content rather than by the quinoline-insoluble C_1 (Hodurek, 1920; Evans and Pickard, 1931).

In mixing fillers with a binder there is a critical range of viscosity for maximum adhesion corresponding to a temperature well above the R&B softening point (Darney, 1958). The equiviscous temperature, at which the viscosity is 250 P (McNeil and Wood, 1958), usually lies in that range. Alternatively, two equiviscous temperatures (EVT_{15} and EVT_{1015}), at which the pitch has a viscosity of 15 and 1015 P, respectively, may be determined graphically (King and Robertson, 1968). From these the temperature coefficient of viscosity, T_c, is calculated as follows:

$$T_c = 1000/(EVT_{15} - EVT_{1015}) \text{ P/°C}$$

Typical data for a 92°C C/A softening point coal tar pitch binder of the Soderberg type are EVT_{15}, °C = 150, EVT_{1015}, °C = 110, T_c (P/°C) = 25. The viscosity of electrode binders is often measured in the laboratory with a Brookfield viscometer at various temperatures and spindle rotation speeds.

Koppers vacuum capillary viscometers have also been used (King and Robertson, 1973b). Details on viscosity and equiviscous temperature determinations are presented in Chapter 33, Sections VI,C and VI,D, respectively.

D. Molecular Weight Range

The nominal boiling range of coal tar pitch is high because during production by distillation of a tar, to an extent depending on fractionation efficiency, lower boiling constituents are volatilized. Electrode binder specifications usually call for a maximum of 6% distilling below 360°C (Branscomb, 1966). ASTM D 2569-75 is a suitable test procedure. Light ends in the binder during baking of electrodes can lead to excessive gas, cracked, deformed, high porosity products, and objectionable fuming. Low and intermediate boiling components impart flow properties to carbon paste.

On the average, electrode binders consist of seven or eight linearly condensed aromatic rings, as determined by proton magnetic resonance spectrometry of carbon disulfide extracts (Dewalt and Morgan, 1962). Information on molecular weight of coal tar pitches from various sources was obtained via solvent fractionation: between 55 and 75%, depending on the source, consisted of material of less than 400 molecular weight (Wood and Phillips, 1955; McNeil and Wood, 1958). In a micropot-type still at 10^{-5} torr pressure, data on 20 pitches were obtained which when statistically analyzed suggested a possible correlation with the compressive strength of test electrodes (Charette and Bischofberger, 1961). King and Robertson (1968) distilled an electrode binder *in vacuo* and collected nine fractions, 41 wt % total, at 540°C maximum vapor temperature, atmospheric equivalent. The number average molecular weight was determined in a vapor pressure osmometer using toluene as solvent. The mean molecular weight of the distillate was 250, a range of three to six aromatic rings per mole. The higher boiling overhead fractions produced coke in a standard test. The excellent performance of the starting pitch as a binder was ascribed to the presence of distillable coke precursors and a distribution of aromaticity over the molecular weight range. Mason (1970) concluded that the consumption rate of a Soderberg electrode depends to some degree on the coking value of lower molecular weight components of the binder.

E. Resin Content by Solvent Fractionation

Many solvents have been employed to extract components of coal tar pitch selectively on the basis of chemical composition and molecular weight. Resinous and carbonaceous constituents, some of which are naturally present in the tar and others of which are formed by heat during production of pitch, vary with the source (McNeil and Wood, 1958). Selective solvents commonly used are, in order of increasing power, petroleum ether, benzene or toluene, pyridine, anthracene oil, and quinoline (Darney, 1958; Huth, 1970), and less often, acetone, carbon disulfide, dimethylformamide, and nitrobenzene.

Light petroleum fractions precipitate, from a benzene or toluene solution of pitch, material variously known as C_3 resins, γ resins, or resinoids which are akin to the asphaltenes of petroleum, but are of only one-third the molecular weight, have shorter alkyl side chains attached to aromatic rings and are of higher carbon content (Lapik and Kruzhinova, 1974). The C_3 resins of coal tar contribute significantly to the wetting and adhesion characteristics of an electrode binder (Denman, 1933), somewhat less to its viscous and coking properties. Like the oily constituents (crystalloids) also found in the benzene-soluble portion, the C_3 resins comprise a considerable proportion of a typical pitch.

Higher molecular weight materials, insoluble in benzene or toluene, exist in a pitch as nuclei, some of colloidal dimensions, surrounded by layers of low-molecular-weight compounds (Dickinson, 1945; Kreulen, 1946). Analysis of these high-molecular-weight resins is by ASTM procedure D 2317-66 or a more rapid version thereof (Domitrovic *et al.*, 1962; Takaki and Miyasaka, 1965). ASTM D 2764-71, utilizing dimethylformamide, gives similar results. As noted earlier, many of these components tend to precipitate from a pitch during hot storage. Benzene insolubles have for a long time been referred to as free carbon, a misleading term since they are not carbon but polycondensed aromatic and heterocyclic compounds.

Solvents more polar than benzene, e.g., pyridine, acetone, quinoline, nitrobenzene, cinnamaldehyde (Croy, 1962), or the condensed aromatics of anthracene oil, often dissolve 50–90% of the benzene-insoluble material. The soluble fraction is designated as C_2 or β resins, the remaining insolubles as C_1 or α resins (Table I).

Of all the components of coal tar pitch the C_2 resin fraction may have the greatest single effect on electrode binding properties (Hodurek, 1920; Nelson, 1963). European manufacturers often specify a minimum content of 20% C_2 resins. These compounds combine the ability to fuse and penetrate the pores and voids of an aggregate with the ability to form coke of high binding strength. Analysis of a typical coke oven pitch indicated this fraction to have a molecular weight of 1500–1800, whereas the C_3 resins had a mean molecular weight of only 560 (McNeil and Wood, 1958).

The fraction, generally 5–15%, of the pitch which is insoluble in quinoline or anthracene oil, as determined by ASTM D 2318-66 or a more rapid procedure (Domitrovic *et al.*, 1962), consists of finely divided, suspended particles varying in size from less than 1 to 10 μm or more (Charette and Girolami, 1958). In a series of pitches studied, the particle size of C_1 resins increased as the content increased from 2 to 35% of the pitch (Morgan *et al.*, 1960). Of high but indeterminate molecular weight, C_1 resins do not fuse, and hence behave more or less as a filler having a high C/H atomic ratio and specific gravity (Wood and Phillips, 1955). The amount of quinoline insolubles present often serves to identify the commercial source of pitches.

There is some uncertainty about the role played by C_1 resins in electrode manufacture. Of a series of binders investigated, those containing ~14% C_1 resins gave electrodes with the highest compressive strength (Charette and Girolami, 1961). On the other hand, a German patent disclosed that reducing the pyridine-insoluble content from 12 to 2% resulted in only slightly diminished electrode strength and a significantly lower electrode weight loss under identical oxidation conditions (Lahr, 1957). In other studies the content of quinoline insolubles was found to correlate with the aromatic hydrogen content of a carbon disulfide-soluble fraction, and both of these parameters were related to the binding properties of the pitch (Kini and Murthy, 1974). The surface of dispersed C_1 resins may serve as reactive sites for coke formation and there is an interfacial area between the surface of the resins and the surrounding pitch for which electrode properties are optimized (Morgan *et al.*, 1960).

Combinations of other than the usual solvents have been employed to fractionate electrode binders. The use of acetone and quinoline as separation media is the basis of an Alcan procedure (Charette and Girolami, 1961; Mason, 1970) and that of Dell (1959). Mallison (1950) described in detail a procedure involving anthracene oil–pyridine, benzene, and pyridine–methanol as solvents; for pitches there was good agreement with conventional techniques, but not for crude coal tars (see Chapter 33, Table I).

F. Carbon-to-Hydrogen Atomic Ratio

The carbon-to-hydrogen atomic ratio, an index of aromaticity and a criterion of electrode performance, is calculated from C, H values determined by ASTM D 3178-73, a semimicro method AMI 114.073 T, or a modification such as that proposed by Panicker and Banerjee (1957). Typical data are given in Table I: the C/H ratio is 1.6–1.8 (King and Robertson, 1968; Mason, 1970). When a binder was distilled *in vacuo* to maximum overhead the residue had a C/H ratio of 2.0 (King and Robertson, 1968).

A useful criterion of pitch quality is the aromaticity of the quinoline-insoluble fraction, for which a C/H ratio of 4.3 has been reported (Charette and Girolami, 1958; Martin and Nelson, 1958), and Branscomb (1966) confirmed it as being generally above 3. For the combined C_1–C_2 (benzene-insoluble) fraction a C/H ratio of 2.45 was determined (King and Robertson, 1968), from which it was calculated that the C_2 resins had a C/H ratio of 2.0. Other investigators reported a value of 1.8–1.9 for the C_2 resin fraction of coke oven tars (McNeil and Wood, 1958).

G. Aromaticity

Both the disperse and the continuous phases of coal tar pitch consist of condensed polyaromatic compounds and minor amounts of heterocyclics containing nitrogen, sulfur, and oxygen. The number of alkyl groups, primarily methyl, attached to an aromatic nucleus decreases with increasing molecular weight and, according to one authority (Volkmann, 1959), is zero for components boiling above 410°C. In the case of electrode binder fractions obtained by solvent extraction, the average number of aromatic rings per mole, as determined by infrared examination in the 3-μm region of the spectrum, led to the conclusion that the most probable pitch structure is a ring–chain configuration in which relatively small groups of aromatic rings are linked by single bonds or methylene bridges (McNeil and Wood, 1958). Some methyl substituents were indicated in the ortho position. Data of other workers who used proton magnetic resonance spectrometry to analyze carbon disulfide extracts supported a linearly condensed aromatic structure, with only 2–6% of the ring carbon atoms being substituted, mostly by methyl groups (Dewalt and Morgan, 1962; also see Chapter 23, Section IV,F). It was concluded by Dell (1959) that the apparent density and reactivity of carbon anodes were functions of the aromaticity of the binder as determined by the C/H atomic ratio, the infrared index of carbon disulfide extracts, and the sulfonation index (ASTM D 872) and refractive index of pitch distillates.

Based on the specific gravity, carbon and hydrogen contents, number average molecular weight, and sulfur and nitrogen contents of vacuum distillate cuts, the fraction of aromatic carbons, condensation index, and average number of aromatic rings per mole of the distillate were calculated as 0.92, 0.35 and ~4, respectively (King and Robertson, 1968). Optical densities of the adsorption bands have been determined from the aromatic and aliphatic stretching vibrations at 3046 and 2925 cm^{-1} (Bowitz *et al.*, 1963). It was observed that aromaticity increases during thermal treatment and mild oxidation of a binder. Pitches have been examined by electron spin resonance (ESR); the ESR concentration increased nonlinearly with quinoline-insoluble content (Kini, 1973; Kini and Murthy, 1974). Recently, structural analysis of a pitch was accomplished by combining data from conventional analysis with nuclear magnetic resonance spectrometry for computerization (Katayama *et al.*, 1975).

H. Nonhydrocarbon Components

Minor constituents of coal tar pitch are organic compounds of nitrogen, sulfur, and oxygen, and inorganics contributing to ash. Nitrogen, usually determined by ASTM D 3179-73, may amount to as much as 1.5%; sulfur, by ASTM D 3177-75 procedure, 0.3–0.8%; and oxygen, by difference, 1–2%. Ash, normally 0.2–0.5 wt % of the pitch as determined by ASTM D 2415-66 or equivalent procedure, contains iron (500 ppm) and traces of vanadium and boron (Liggett, 1964).

Organic nitrogen in a binder has not been reported to affect electrode manufacture or performance. Since sulfur content of the pitch is low as compared to that of fillers, this element is normally of little concern. When the sulfur content of the coke, for example, is excessive, certain adverse effects have been observed—corrosion of the anode casing, the iron or steel contact pins, or the fume collection equipment (Pearson, 1955). Sulfur compounds in electrodes may dissolve in the electrolyte, thereby reducing the current efficiency. Oxygen compounds, though rarely analyzed for in pitch, may account for as much as or more than nitrogen and sulfur together. Three analytical methods were proposed for determining hydroxyl groups in pitch: potentiometric titration in nonaqueous solutions, acetylation with acetic anhydride, and reaction with organomagnesium compounds to produce active hydrogen (Shaposhuikova *et al.*, 1972). The last method was found to be the most reliable.

Ash-producing constituents of electrodes also may contaminate certain products via dissolution in the electrolyte. Traces of metals such as iron, vanadium, and alkalies are said to catalyze the reaction of electrodes with the gases oxygen and carbon dioxide, thereby accelerating surface disintegration (Bowitz *et al.*, 1958). The elements Ca, Si, Fe, Al, Na, and Mg were reported to be present in trace amounts in carbon electrodes (Ehrlich *et al.*, 1961); with the exception of iron these would not have come from the binder.

I. Coking Value

Carbonization of a pitch binder within the matrix of an aggregate is the end result of a number of complex organic reactions involving loss of hydrogen and condensation of the molecular structure. Properties of the binder coke, which virtually determine those of an electrode, are governed primarily by the aromaticity and molecular weight range of the pitch. Other binder characteristics, high specific gravity, Newtonian viscosity behavior, and content of dispersed carbonaceous resins, are all manifestations of high aromaticity.

Although lower molecular weight components, those distillable in a high vacuum, contribute only about 6% of the total binder coke (King and Robertson, 1968), their effect may not be negligible in producing high quality electrodes (Mason, 1970), especially since they largely determine the degree of wetting and permeation of the filler particles. Four aromatic rings per mole were found to be a prerequisite for coke production, and increasing the number from four to six raised the coking value of a distillate fraction from 5 to 35 (King and Robertson, 1968). Data points for a higher molecular weight petroleum pitch from catalytic cracking fell on a continuation of the same correlation curve: Seven and eight aromatic rings per mole gave 43 and 48 coking value, respectively. These results were obtained using an Alcan laboratory coking procedure which involved heating the pitch in a reducing atmosphere for 2.5 hr at 550°C (Charette and Girolami, 1958), a modification of the Norske method previously used.

Many other laboratory techniques have been employed to estimate the coke-forming tendencies of various residua, and values differ according to the procedure used. The old ASTM D 271 procedure for volatile matter, now superseded, in which a 1-g sample is heated to 950°C for 7 min in a platinum crucible, was used by Dell (1959), but it is not generally considered to be meaningful because of the short heating period. Others have employed the German standard test method DIN, DVN 3725 (Brückner and Huber, 1950), and the Japanese (J.I.S.) method which consists of a 30-min preheat at 430°C, followed by 30 min in a muffle furnace at 800°C. The residue obtained in the J.I.S. test was said to be much less than that laid down by the binder in electrodes (Watanabe, 1963). An Elektrokemisk procedure and a revised version thereof are referred to in the literature but not described (Bowitz *et al.*, 1963; Branscomb, 1966). A long established coking procedure is that for Conradson carbon residue, ASTM D 189, which used an open flame for heating. This method gave a reasonably reliable indication of the coking potential of heavy residua: wt % coke $\approx 1.2 \times$ Conradson carbon residue. The modified version of this test, ASTM D 2416-73, employs an electric furnace and yields more consistent results. In the case of many petroleum stocks coke formation is closely related to asphaltene content (King, 1973); an aromatic petroleum pitch containing 44 wt % of material insoluble in petroleum ether produced the following weight percent coke by different methods: 42 by destructive distillation, 37 as Conradson carbon, and 38 by the Alcan procedure noted previously. The benzene-soluble fraction of a coal tar electrode binder, containing an estimated 50% of C_3 resins, which are superficially similar to asphaltenes (Lapik and Kruzhinova, 1974), had a coking value of 37 by the Alcan test (Table I). Part of this coke was attributable to heavy aromatic components soluble in petroleum ether.

The effects of coking on the carbon and hydrogen contents of (a) a coal tar pitch, (b) a benzene-insoluble fraction, and (c) a benzene-soluble fraction were reported (King and Robertson, 1968). By the Alcan test procedure the whole pitch lost 45% in weight and 40% of its hydrogen,

TABLE II *Coking Properties of Resin Fractions of a Typical Coal Tar Electrode Binder*[a]

Resin	Coking value (wt %)	Amount in pitch (wt %)	Contribution to coke (wt %)
C_1, quinoline insoluble	98[b]	10	9.5
C_2, quinoline soluble, benzene insoluble	~87[c]	20	17.5
C_3, benzene soluble, petroleum ether insoluble + oil (mostly heavy aromatics)	37	70	26
Total			53

[a] Conditions: 2.5 hr at 550°C (Alcan method).
[b] 95–97 Conradson carbon (Branscomb, 1966).
[c] Calculated from footnote *b* and data on ($C_1 + C_2$), Table I.

and the C/H ratio increased by 80%. In effect, the laboratory coking procedure reduced the entire pitch to something resembling the benzene-insoluble ($C_1 + C_2$ resins) fraction. The individual contribution to coking of each of the fractions obtained by solvent extraction of a typical electrode binder is summarized in Table II. The binder coke produced was split about evenly between the free carbon fraction (30% of the pitch) and the remaining 70%, which was benzene soluble.

Whether this distribution would apply strictly to coke produced during electrode manufacture is debatable. It has long been suspected that laboratory tests, mainly because of their short duration, do not correlate well with practice. Martin and Nelson (1958) observed that the amount of coke produced by some coal tar pitches varied inversely as the rate of heating and suggested a modified laboratory method with the object of better simulating conditions in large-scale commercial coking operations. It was found by Watanabe (1963) that the coking values of some pitch fractions separated by solvent treatment were higher when a slow heating method was used than by the conventional J.I.S. procedure. Some electrode manufacturers employ thermogravimetric analysis (TGA) to establish an optimum temperature cycle for baking electrodes so that the rate of evolution of hydrocarbon vapors can be controlled. According to the usual laboratory procedure, a weighed sample of pitch is placed on a balance pan enclosed under nitrogen in a furnace, the temperature of which is increased to 650°C or more over a period of 10 hr, and the loss in weight is determined at several temperatures. Data were reported using both this procedure and a modified version at a much lower heating rate—22 hr at each of seven temperatures from

250 to 800°C, total 154 hr (King and Robertson, 1973a). The latter gave a better correlation with prebaked test electrodes than either the Alcan test or the conventional, fast heating TGA procedure.

J. Miscellaneous

A few less familiar test methods, some of which are useful for specification purposes, should be mentioned. The thermal expansion or contraction of coal tar pitch may be determined by ASTM D 2962-71. Pechiney assesses the resolidification temperature of pitch in a Cerchar-type plastometer (Mason, 1970). A method, modified after Kattwinkel, for determining the agglutinating index of a pitch, has been described in detail (McNeil and Wood, 1958), and another procedure measures the proportion of coke abraded under standard conditions, the Roga index (Mason, 1970). A test for volatile matter based on a British standard test method has been carried out (Darney, 1958) but, like ASTM D 3175-73, gives higher weight losses than laboratory coking procedures because of a very fast heating rate. McNeil and Wood (1958) reported values of 57–90 wt % volatile matter for a series of coal tar pitches from various sources, for which normal coke residues by laboratory tests would be in the range 45–55 wt %. The Coal Tar Research Association of Great Britain measures the surface tension of electrode binders over the temperature range 120–220°C and the angle of contact of pitches on carbon surfaces at 155 and 200°C. These are indices of wetting characteristics (Mason, 1970).

IV. FLOW PROPERTIES OF CARBON PASTE

The aggregate grading affects the properties of both the green mix and the baked electrodes: The particle size distribution having the maximum bulk density usually yields the best electrodes (Pearson, 1955). Aggregates of petroleum coke, because of their somewhat elongated particles (Liggett, 1964), did not closely approximate the Weymouth maximum density gradation for spheres (King and Robertson, 1973a).

The mixing procedure is important. The temperature used for preparing Soderberg paste usually falls in the range 150–165°C, 5–20 P binder viscosity (King and Robertson, 1968). Prebaked electrode paste is mixed at a temperature which is dependent on the softening point of the binder and limitations of the equipment. Sometimes the coarse filler components are contacted for a shorter time to minimize attrition (King, 1973). The binder content is optimized for a given aggregate. When coal tar pitch is used exclusively this is possible by visual observation during the mixing cycle, but when a change is made to petroleum pitch, which wets the filler better and so gives a paste of drier appearance, the visual approach may lead to incorporation of too much binder (King, 1973; Pendley and Bullough, 1975). The optimum binder content for prebaked electrodes is generally less than that for self-baking electrodes: 15–25 wt % versus 25–30 wt % for Soderberg with horizontal pins and 27–35 wt % for Soderberg with vertical pins (Okada and Takeuchi, 1958; Okada, 1961; King and Robertson, 1973a; Pendley and Bullough, 1975). However, pastes for prebaked products often incorporate, in addition, up to 3% of a paraffinic mineral oil to facilitate extrusion, pressing, and molding. Okada (1961) calculated the amount of binder that would just fill the pores and voids of the aggregate and that required for extrusion. In general, less than the optimum binder content results in electrodes of inferior properties; more than the optimum leads to partial separation of the binder during baking, which has several adverse effects (Branscomb, 1966). A laboratory method of determining the specific surface area of small carbon particles has given information useful in assessing the amount of pitch required (Mizera, 1965). Reinforcing-type carbon blacks, which have very much larger surface area than quinoline insolubles of coal tar, had a pronounced effect on the flow rate of Soderberg pastes made with coal tar pitch at a concentration of less than 1 wt % of the paste (King and Robertson, 1973b).

Other parameters on which carbon paste flow depends are the shape and porosity of dispersed particles, including resins associated with the pitch, and viscosity of the binder. For a given type of filler and a standard mixing procedure particle shape is not normally a variable, but porosity may be, and a test for this property was developed which was applicable to both the filler and a coked binder (Bowitz *et al.*, 1963). Mason (1970) referred to this as an Elektrokemisk procedure. Mixtures of a coal tar pitch, itself Newtonian, with coke fines only were pseudoplastic and some commercial electrode pastes made with graded fillers were said to be dilatant (Bowitz *et al.*, 1963).

The sensitivity of paste flow to temperature should obviously be as low as possible, especially for a self-baking electrode. Bowitz *et al.* (1963) observed that small differences in viscosity and temperature coefficient of the binder had a surprisingly large effect on paste flow at 140 and 180°C. They also noted that the relative viscosity, $\eta_{\text{paste}}/\eta_{\text{binder}}$, increased more or less exponentially with the effective volume concentration of coke fines, i.e., the ratio of the actual volume concentration to the volume concentration at maximum density of the mixtures. In an elongation or flowability test, fine calcined coke is mixed with an equal weight of binder at 160°C and then molded into small test cylinders which are placed on an inclined test plate in an oven at 140 or 180°C for 60 min. The percent increase in length over the original is recorded. The validity of the data is subject to the qualifications that a graded filler is not used and more pitch is added than the normal complement of 25–35%.

Other investigators employed a similar technique but with a graded aggregate and less pitch, heating for 15 min at 255°C, an Alcan method, and at lower temperatures (King and Robertson, 1973b). The Alcan test is used for quality control of commercial Soderberg pastes, the amount of coal tar pitch needed to give elongations of 60 and 150% being determined.

A carbon paste spreading test at 150°C under extremely low load or no applied load has been suggested for simulating flow at the low shear rate and temperature prevailing at the top of a Soderberg anode. Correlations among the variables—spreading of paste, elongation or flowability of paste, and apparent viscosity of the pitch binder—were found to be complex and temperature dependent (King and Robertson, 1973b). At the highest test temperature, 255°C, binder content of the paste apparently exerted the major effect on paste fluidity; at lower temperatures viscosity of the binder played an increasingly important role.

A capillary rheometer was used by Bhatia (1973) to measure the flow properties of green carbon mixes consisting of petroleum coke, carbon black filler, and a pitch. At low shear rates the mixes behaved as Bingham bodies with a measurable yield value.

V. EVALUATION OF TEST ELECTRODES

A. Procedures

According to the Elektrokemisk method test electrodes are prepared in the laboratory from a 35/65 mixture of pitch and coke fines (Mason, 1970). Jones *et al.* (1960) described a procedure employing a graded aggregate, and their experimental data indicated that the apparent density, compressive strength, and electrical resistivity of baked electrodes correlated with (a) the binder quality index (coking value × C/H atomic ratio) (Charette and Bischofberger, 1955) and (b) the baking cycle. Adopting Jones' procedure with some modifications, others evaluated electrodes made with coal tar and petroleum pitch binders in a Soderberg formulation. They used petroleum coke in a commercial grading except for the coarsest component, which amounted to 7 wt % of the aggregate (King and Robertson, 1968, 1973b).

Girolami (1963) reported that coal tar pitch has a higher effective coking value when preheated with calcined coke fines and then coked, than when not preheated. This observation is the basis of the Alcan test for "saturation value," the pitch concentration at which the coking value shows a sharp increase. In a later study test electrodes were prepared with a graded petroleum coke by molding at 130°C and 5000 psig pressure, then baking in a controlled 48-hr cycle to 1000°C (King and Robertson, 1973a). It was observed that none of the conventional (rapid heating) laboratory coking or thermogravimetric (TGA) procedures on the pitch alone or slow heating of a mixture of pitch and graded coke (without prior compression) produced as much coke as was obtained by preparing test electrodes. Slow heating of the pitch in a modified TGA procedure gave the best correlation. Properties of the prebaked test electrodes were a function of the coked binder present: Apparent density and compressive strength passed through maxima and electrical resistivity through a minimum at a specific coked binder content which was optimal for a given pitch–coke system.

During the preparation of test electrodes gas evolution attained a maximum rate at 350–450°C and was essentially complete at 500°C (Smirnova *et al.*, 1965). These data were confirmed during slow heating of a binder in the modified TGA test (King and Robertson, 1973a). Preoxidation at 100–230°C of a mixture of petroleum coke and a pitch is said to reduce the yield of volatile compounds and the maximum gas evolution rate during coking (Rozenman *et al.*, 1974). Methods for evaluating shrinkage of Soderberg electrodes, which should be minimal to prevent crack formation around the stubs, have been reported (Sandberg *et al.*, 1957; Bowitz *et al.*, 1963). Baking of small carbon electrodes in a nitrogen atmosphere containing traces of oxygen gave stronger electrodes—twice that at optimum concentration—than baking in pure nitrogen, and a very close correlation was observed between electrode strength and binder coke yield (Lauer and Bonstedt, 1964).

B. Anode Consumption

There is general agreement that surface disintegration of carbon anodes during electrolysis of molten salts is influenced by preferential oxidation of the binder coke. This so-called dusting tendency is manifested as excessive consumption, expressed as a percentage of that corresponding to formation of carbon dioxide at 100% Faraday efficiency. It is particularly important to control anode consumption in the electrolytic production of aluminum, not only because loss of carbon is stoichiometrically high, but also because eroded carbon particles become mixed with the cryolite bath.

Prebaked test electrodes, made with binder and coke fines, were evaluated in the laboratory for oxidative reactivity by heating to 950°C in a vertical tube furnace in a slow stream of carbon dioxide (Bowitz *et al.*, 1958, 1963). The reactivity was related to (a) the microporosity of the electrode, i.e., the percentage of pores smaller than 6 μm, which may be determined pycnometrically using mercury or by displacement with steam (Darney, 1958); and (b) the difference between the specific surface of the binder coke and that of the dry aggregate. The reactivity of a coked mixture was reported to be higher by a factor of 10 than that of coked binder or dry aggregate determined singly (Bowitz *et al.*, 1958). Increasing the baking rate of an electrode led to higher porosity, greater differences in specific surface, and higher reactivity. All of these properties were related to the softening point, specific gravity, coking value, and aromaticity of the binder.

Watanabe (1963) carried out electrolytic tests on Soderberg-type anodes, made with binders from distillation of high temperature coke oven tar, in both aqueous sodium hydroxide solution and molten alumina–cryolite. In aqueous alkali, determination of crystalline structure of the electrode by x-ray diffraction indicated amorphous regions to be preferentially oxidized. In both media a reasonable correlation was observed between anode consumption and the degree of selective oxidation of the binder coke and dry aggregate. It was recommended that a pitch be used with as high a coking value as possible.

The possibility that consumption of carbon in excess of the stoichiometric in industrial cells is also due to reactivity with carbon dioxide was investigated. Scalliet (1963) observed a rough correlation with prebaked anodes but not with Soderberg anodes. Hollingshead and Braunwarth (1963), using a cryolite base electrolyte in a small externally heated cell, found that only a small part of the excess consumption of carbon anode specimens was attributable to formation of carbon monoxide; most was due to erosion of carbon particles via selective oxidation of binder coke. An Alcan test is based on this work.

Mason (1970) computed correlation coefficients between various pitch properties and anode reactivity. High values of the correlation coefficient were observed for carbon-to-hydrogen atomic ratio or coking value (Alcan method) of the following lower molecular weight pitch fractions: C_2 resins, C_3 resins, and acetone solubles, vis-à-vis the rate of anode consumption in the Alcan test. Mason concluded that current binder specifications, which are directed to production of anodes with adequate strength and porosity, do not necessarily ensure low anode consumption.

A laboratory apparatus was described recently which had been used for several years to assess the quality of coal tar and petroleum binders by the performance of carbon electrodes in molten salts (Pendley and Bullough, 1975). Two coal tar binders employed as standards had typical properties. Nineteen petroleum pitches having generally higher softening points, containing much less benzene-soluble material and very little quinoline insolubles, and of lower coking value and aromaticity than the coal tar pitches, produced electrodes of lower apparent density and higher electrical resistivity but of generally equivalent electrolytic reactivity in molten salts. These data appear to support Mason's contention that anode consumption is not predictable on the basis of current specifications of the binder alone.

C. Graphitization

To graphitize the carbon, baked electrodes are slowly heated to high temperatures, thus transforming amorphous into crystalline material by rearrangement of the carbon atoms. A small-scale procedure has been described for the preparation of electrodes from various mixtures of petroleum coke, pitch coke, lampblack, anthracite, and coal tar pitch by baking at 1300°C followed by graphitization at 2500°C (Fialkov, 1958). Graphitized test specimens were evaluated for specific gravity, electrical resistivity by the ammeter-voltmeter method, hardness by Shore's apparatus, and tensile strength.

Delayed petroleum coke and the carbon residue of coal tar pitch are the most graphitizable of raw materials. Conditions used for calcination of the filler affect the properties of graphite but baking of the carbon paste is probably the most critical step. The porosity of both baked and graphitized products is 25–30% and the apparent density is 1.6–1.7. Porosity, determined from the difference in volume calculated from the bulk density and the real density, may be reduced when desired by impregnation with a coal tar pitch having a lower softening point and less benzene-insoluble and quinoline-insoluble material than most electrode binders. Up to one-third of the total porosity may be closed pore volume, i.e., between the crystallites in the filler–binder matrix, the remainder consisting of voids between particles. During graphitization, crystallite growth, as indicated by x-ray diffraction patterns, commences at ~2200°C.

Fabrication of experimental graphites has been described using coal tar pitch as well as two unsaturated organic compounds as binders (Overholser, 1973). The surface tension coefficient and surface energy were calculated during synthetic graphite production (Drovetskaya *et al.*, 1974). By the use of ESR techniques data have been obtained on the pyrolysis of aromatic hydrocarbons, confirming the presence of stabilized free radicals during the formation of graphite (Csermak, 1974).

VI. SUMMARY AND CONCLUSIONS

There is no unanimity in the views advanced for the properties of the ideal electrode binder pitch and the various criteria that have been suggested are by no means of universal application. The specifications adopted by various companies, based on practical experience, have had the effect of selecting the most aromatic pitches. These give a high coke residue, which favors high electrode strength, low porosity, and possibly low electrical resistivity, but does not necessarily ensure adequate paste flow properties or low anode consumption. The optimum binder content adopted in practice is usually a compromise between two optima: one for paste flow, the other for electrode properties.

It is generally agreed that to minimize anode consumption during electrolysis the reactivities of the binder coke and the dry aggregate should not be too dissimilar, but precisely how this is to be accomplished is debatable. Many of the binder specification tests applicable to a coal tar pitch system cannot be transposed directly to another: For example, the consumption characteristics of anodes made with a series of petroleum pitches were equivalent to those of anodes made with coal tar pitch, though test data on the binders differed widely. Subcommittee XVII of the American Society for Testing Materials' D-8 Committee has been investigating current analytical methods and may propose further modifications or additional tests to establish the suitability of pitch for electrode manufacture.

REFERENCES

Bacon, L. E. (1964). *In* "Kirk-Othmer Encyclopedia of Chemical Technology" (A. Standen, ed.) 2nd Ed., Vol. 4, pp. 202–243. Wiley (Interscience), New York.

Badenhorst, R. P., and Perold, G. W. (1957). *J. Appl. Chem.* **7,** 32.

Bhatia, G. (1973). *Carbon* **11**(5), 437–440.

Bowitz, O., Bockman, O. C., Jahr, J., and Sandberg, O. (1958). *Proc. Conf. Ind. Carbon Graphite, Soc. Chem. Ind., London* pp. 373–377.

Bowitz, O., Eftestol, T., and Selvik, R. A. (1963). *Extr. Met. Alum., Proc. AIME Symp. Alum. Prod., New York. 1962* **2,** 331–349.

Branscomb, J. A. (1966). *In* "Coal Tars and Pitches" (J. A. Hoiberg, ed.), Vol. 8, pp. 359–384. Wiley (Interscience), New York.

Brückner, H., and Huber, G. (1950). *Gas- Wasserfach* **91,** 104.

Cameron, H. K. (1950). *Chem. Ind., (London) pp*. 399–403.

Charette, L. P., and Bischofberger, G. P. (1955). *Ind. Eng. Chem.* **47,** 1412–1415.

Charette, L. P., and Bischofberger, G. P. (1961). *Fuel* **40,** 99–107.

Charette, L. P., and Girolami, L. (1958). *Fuel* **37,** 382–392.

Charette, L. P., and Girolami, L. (1961). *Fuel* **40,** 89–98.

Croy, F. A. (1962). *Bitumen, Teere, Asphalte, Peche* **13**(2), 67–72.

Csermak, K. (1974). *Banyasz. Kohasz. Lapok, Ontode* **25**(6), 284–286.

Darney, A. (1958). *Proc. Conf. Ind. Carbon Graphite, Soc. Chem. Ind., London* pp. 152–161.

Dell, M. B. (1959). *Fuel* **38,** 183–187.

Denman, W. (1933). *Brenst.-Chem.* **14,** 121.
Dewalt, C. W., and Morgan, M. S. (1962). *Am. Chem. Soc., Div. Fuel Chem., Prepr.* pp. 33–45.
Dickinson, E. J. (1945). *J. Soc. Chem. Ind., London* **64,** 121.
Domitrovic, R. W., Stickel, R. M., and Smith, F. A. (1962). *Am. Chem. Soc. Div. Fuel Chem., Prepr.* pp. 54–64.
Drovetskaya, L. A., Nagornyi, V. G., Syskov. K. I., and Tsarev, V. Y. (1974). *Konstr. Mater. Osn. Grafita* **8,** 32–37.
Ehrlich, G., Gerbatsch, R., Jaetsch, K., and Scholze, H. (1961). *Reinstoffe Wiss. Tech. Int. Symp., Dresden* **1,** 421–428.
Evans, E. V., and Pickard, H. (1931). *Fuel* **10,** 352.
Fialkov, A. S. (1958). *Proc. Conf. Ind. Carbon Graphite, Soc. Chem. Ind., London* pp. 101–110.
Girolami, L. (1963). *Fuel* **42,** 229–232.
Hader, R. N., Gamson, B. W., and Bailey, B. L. (1954). *Ind. Eng. Chem.* **46,** 2.
Hodurek, R. (1920). *J. Soc. Chem. Ind., London* **39,** 622A.
Hollingshead, E. A., and Braunwarth, V. A. (1963). *Extr. Met. Alum., Proc. Int. Symp., 1st, New York, 1962* **2,** 31–49.
Huth, W., (1970). *Freiberg. Forschungsh. A* **472,** 45–59.
Jones, H. L., Simon, A. W., and Wilt, M. H. (1960). *J. Chem. Eng. Data* **5,** 84–87.
Katayama, Y., Hosoi, T., and Takeya, G. (1975). *Nippon Kagaku Kaishi* No. 1, pp. 127–134.
King, L. F. (1973). Unpublished data.
King, L. F., and Robertson, W. D. (1968). *Fuel* **47,** 197–212.
King, L. F., and Robertson, W. D. (1973a). *Am. Chem. Soc., Div. Fuel Chem., Prepr.* pp. 138–144.
King, L. F., and Robertson, W. D. (1973b). *Am. Chem. Soc., Div. Fuel Chem., Prepr.* pp. 145–154.
King, L. F., Robertson, W. D., and Steele, C. T. (1967). U.S. Pat. No. 3,316,183.
Kini, K. A. (1973). *J. Indian Chem. Soc.* **50**(9), 616–618.
Kini, K. A., and Murthy, G. S. (1974). *Fuel* **53**(3), 204–205.
Kreulen, D. J. W. (1946). *Fuel* **25,** 99.
Lahr, G. (1957). Ger. Pat. No. 965,207.
Lapik, V. V., and Kruzhinova, L. V. (1974). *Khim. Tverd. Topl.* No. 4, pp. 54–62.
Lauer, G. G., and Bonstedt, K. P. (1964). *Carbon* **1**(2), 165–169.
Liggett, L. M. (1964). *In* "Kirk-Othmer Encyclopedia of Chemical Technology" (A. Standen, ed.) 2nd Ed., Vol. 4, pp. 158–202. Wiley (Interscience), New York.
MacLeod, J., Chapman, C., and Wilson, T. A. (1926). *J. Soc. Chem. Ind., London* **45,** 401–406.
McNeil, D., and Wood, L. J. (1958). *Proc. Conf. Ind. Carbon Graphite, Soc. Chem. Ind., London* pp. 162–172.
Mallison, H. (1950). *Bitumen, Teere, Asphalte, Peche* **1,** 313.
Mantell, C. L. (1975). *Am. Chem. Soc., Div. Pet. Chem., Prepr.* **20**(2), 312–320.
Martin, S. W., and Nelson, H. W. (1958). *Ind. Eng. Chem.* **50,** 33–40.
Mason, C. R. (1970). *Fuel* **49**(2), 165–174.
Miller, J. A. (1963). *Extr. Met. Alum., Proc. Int. Symp., 1st, New York, 1962* **2,** 315–320.
Mitra, D. C., Banerjee, T., Banerjee, S. N., Raja, K., and Banerjee, N. G. (1974). *Indian Chem. J.* **8**(11), 31–37.
Mizera, J. (1965). *Chemik* **18,** 269–270.
Morgan, M. S., Schlag, W. H., and Wilt, M. H. (1960). *J. Chem. Eng. Data* **5,** 81–84.
Morrissey, H. A., Bullough, V. L., and Branscomb, J. A. (1960). *Am. Chem. Soc., Div. Gas Fuel Chem., Prepr.* **2,** 115–123.

Mrozowski, S. (1958). *Proc. Conf. Ind. Carbon Graphite, Soc. Chem. Ind., London* pp. 7–18.
Nelson, H. W. (1963). *Extr. Met. Alum., Proc. AIME Symp. Alum. Prod., New York, 1962* **2,** 321–330.
Okada, J. (1961). *Kogyo Kagaku Zasshi* **64,** 840–844.
Okada, J., and Takeuchi, Y. (1958). *Tokai Denkyoku Giho* **19**(1), 6–20.
Overholser, L. G. (1973). Oak Ridge Rep. Y-1857. *Nucl. Sci. Abstr.* **28**(3), 6386.
Panicker, A. R., and Banerjee, N. G. (1957). *J. Proc. Inst. Chem., Calcutta* **29,** 145–149.
Pearson, T. G. (1955). "The Chemical Background of the Aluminum Industry," Lect. Monogr. No. 3. R. Inst. Chem., London.
Pendley, J. W., and Bullough, V. L. (1975). *Am. Chem. Soc., Div. Pet. Chem., Prepr.* **20**(2), 379–382.
Rao, B. S. N. (1964). *Low-Temp. Carbon. Non-Caking Coals Lignites Briquettes Coal Fines, Symp. Hyderabad, India, 1961* **2,** 211–216.
Rozenman, I. M., Shein, L. N., and Kondrat'ev, I.A. (1974). *Khim. Tverd. Topl.* No. 2, pp. 126–132.
Sandberg, O., Olsen, L., and Eftestol, T. (1957). *J. Met.* **9,** 261–266.
Scalliet, R. (1963). *Extr. Met. Alum., Proc. Int. Symp., 1st, New York, 1962* **2,** 373–384.
Shaposhuikova, V. A., Gorpinenko, M. S., and Sukhorukov, I. F. (1972). *Sb. Nauchn. Tr., Nauchno-Issled. Inst. Electrod. Prom.* No. 4, pp. 77–82.
Shea, F. L., and Juel, L. H. (1950). U.S. Pat. No. 2,500,208; U.S. Pat. No. 2,500,209; U.S. Pat. No. 2,527,596.
Shobert, E. I. (1975). *Am. Chem. Soc., Div. Pet. Chem., Prepr.* **20**(2), 340–357.
Smirnova, A. S., Ryss, M. A., Dmitrieva, G. V., and Bazhenov, N. A. (1965). *Tsvetn. Metall.* **38**(11), 90–93.
Stokes, C. A. (1975). *Am. Chem. Soc., Div. Pet. Chem., Prepr.* **20**(2), 387.
Swallen, L. C., and Nelson, H. W. (1950). U.S. Pat. No. 2,527,595.
Takaki, M., and Miyasaka, S. (1965). *Koru Taru* **17**(2), 33–35.
Tanaka, T., Otanigawa, T., and Yamamoto, M. (1965). *Koru Taru* **17**(7), 324–328.
Thrune, R. I. (1942), U.S. Pat. No. 2,270,199.
Volkmann, E. W. (1959). *Fuel* **38,** 445–468.
Volkmann, E. W., Rhodes, E. O., and Work, L. T. (1936). *Ind. Eng. Chem.* **28,** 733.
Watanabe, T. (1963). *Extr. Met. Alum. Proc. Int. Symp., 1st, New York, 1962* **2,** 351–372.
Wood, L. J., and Phillips, G. (1955). *J. Appl. Chem.* **5,** 326–338.

Part VIII

COAL COMBUSTION PRODUCTS

Chapter 35

Sampling and Analysis of Emissions from Fluidized-Bed Combustion Processes—Part 1

Harvey I. Abelson *John S. Gordon*
THE MITRE CORPORATION
METREK DIVISION
MCLEAN, VIRGINIA

William A. Löwenbach
LÖWENBACH AND
SCHLESINGER ASSOCIATES, INC.
MCLEAN, VIRGINIA

I. INTRODUCTION

Fluidized-bed combustion (FBC) is an available technology which, when implemented on a commercial scale, can greatly increase the utility of coal as an environmentally acceptable source of energy. The technology involves the combustion of coal (sized at $\frac{1}{4}$ in. or less) in a bed of inert ash and limestone or dolomite that has been fluidized (held in suspension) by the uniform injection of air through the bottom of the bed at controlled rates.

Fluidized-bed combustion offers several advantages in comparison with conventional modes of firing coal (mechanical stokers, pulverized coal-fired units, and cyclone boilers), and potentially can be used in several major energy-user sectors including electric utilities and indus-

ISBN 0-12-399902-2

trial and institutional applications. The advantages envisioned for FBC include increased system efficiencies, relatively early commercial availability, a projected competitive cost relative to other near-term technologies, reduced emissions of SO_2 and NO_x, the ability to burn char, the ability to burn high ash and high sulfur coal and refuse, and the ability to burn all ranks of coal from lignite to anthracite. In addition, high heat releases and heat transfer directly from bed material to heat transfer surfaces in the combustor result in high steaming capacities from an exceptionally small boiler.

A National FBC Research, Development and Demonstration Program has in recent years been instituted in the U.S. Energy Research and Development Administration (ERDA). The intent of this multiagency government and industry cosponsored program is to advance FBC technology to a level of commercial acceptability such that combustion equipment vendors can satisfy marketplace demand with equipment that is capable of meeting user requirements and, at the same time, function in an environmentally acceptable manner.

A major task in the National FBC Program involves the environmental characterization of this developing energy technology through the acquisition of pertinent environmental data for all FBC process configurations regarded as candidates for commercialization, over a full range of feed parameters and operating/design variables. It is intended that the data will be acquired on a suitable experimental scale such that the environmental impact of future commercial units can be assessed, and on a time schedule compatible with the hardware development schedule published in the National FBC Plan. The need for a cost- and information-effective sampling and analysis program for environmental data acquisition has led to an EPA-sponsored study† (Abelson and Löwenbach, 1977) on which this and the following chapter are based.

Two variations of the technology, atmospheric FBC and pressurized (combined cycle) FBC, each a candidate for commercialization, are addressed herein. In each of these categories attention is focused on a proposed generic process configuration which is intended to be representative of a future commercialized unit (subject to our ability to forecast technological developments).

The intent of this chapter is twofold: to provide the reader with background relevant to FBC technology and to describe the sampling and analytical strategy recommended for its environmental assessment. In Chapter 36, procedures for both sampling and analysis of FBC systems are delineated.

† Study was conducted under the aegis of the Process Measurements Branch, Industrial and Environmental Research Laboratories, Research Triangle Park, North Carolina.

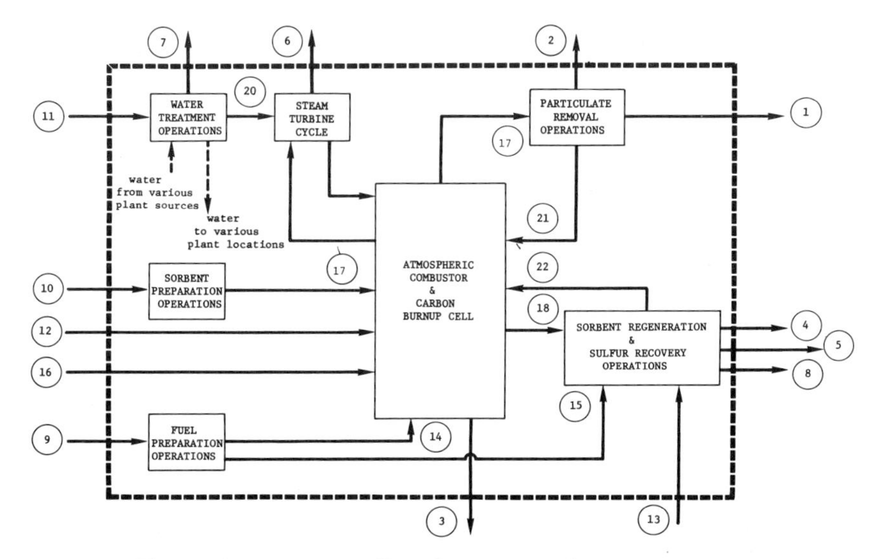

Fig. 1 Generic process flow sheet—atmospheric FBC of coal.

II. GENERIC FLUIDIZED-BED COMBUSTION PROCESSES

A. Atmospheric FBC Processes

A schematic flow diagram representing a generic atmospheric unit is presented in Fig. 1. The diagram (as well as subsequent flow diagrams) depicts only those streams and operations that are relevant to an environmental assessment or basic to an overall process description. A legend describing each numbered stream is provided in Table I. Detailed process descriptions, developmental history, and ongoing as well as anticipated work in atmospheric fluidized-bed combustion are discussed by Mesko *et al.* (1974) and Gordon *et al.* (1972).

Briefly, atmospheric fluidized-bed combustion occurs in the temperature range 788–843°C (1450–1550°F) with excess air values of 15–25%, at normal atmospheric pressure. Steam produced in tube bundles and/or water walls located within the combustor is converted to electrical energy in a conventional steam turbine cycle. Flue gas emissions of SO_2 are substantially reduced by having limestone† (which acts as a sorbent) serve as the noncombustible bed material. NO_x emissions are lower than EPA standards due to the relatively low combustion temperatures. Most of the ash present in the coal feed is normally elutriated from the bed and must be removed (along with attrited limestone and other particulates) prior to the release of flue gas to the atmosphere. Since the

† Use of calcitic limestone is projected for commercialized atmospheric FBC systems.

TABLE I *Stream Designations for Generic FBC Processes*

Stream No.	Designation
1	Stack gas
2	Particulate removal discard
3	Bed solids discard
4	Particulare removal discard—regeneration operations
5	Ash discard from regenerator
6	Blowdowns from steam turbine cycle
7	Blowdown from water treatment operations
8	Product from sulfur recovery (sulfur or sulfuric acid)
9	Raw coal to preparation
10	Raw sorbent to preparation
11	Intake water to treatment
12	Air to combustor
13	Air/steam to regenerator
14	Prepared coal feed to combustor
15	Prepared coal feed to regenerator
16	Start-up fuel feed
17	Prepared sorbent feed to combustor
18	Bed solids to regenerator
19	Flue gas to particulate removal
20	Makeup water to steam turbine cycle
21	Recycle from particulate removal
22	Recycle from regeneration and sulfur recovery
23	Gas to gas turbine inlet (pressurized units only)

ash may be high in carbon content, its direct disposal would result in a lowered combustion efficiency. Consequently, it is anticipated that atmospheric units will employ a carbon burn-up cell (CBC), a separate high temperature, high-excess-air bed to which the collected ash is fed and combusted. Gaseous products from the CBC would then undergo an additional particulate removal operation prior to release to the atmosphere. Alternatively, reinjection of collected ash into the combustor may be adopted to improve combustion efficiency. It should be noted that the "particulate removal operations" indicated in Fig. 1 (and subsequent figures) may include combinations of cyclones, filters, baghouses, precipitators, and other devices. Normally a cyclone or series of cyclones is employed initially to remove coarse particles from the flue gas while final cleanup is accomplished by other methods.

A dry solid sulfate material is formed upon reaction of the bed sorbent with the SO_2 released during coal combustion. For both atmospheric and pressurized FBC systems, problems associated with disposal of spent sorbent, as well as raw sorbent requirements, can be largely reduced by employing a regeneration process in which sulfated sorbent is withdrawn from the combustion bed, regenerated, and returned to the bed for reuse. Sorbent regeneration processes are discussed by Gordon *et al.* (1972) and Keairns *et al.* (1973). For commercialized systems, it is anticipated that the SO_2 (or H_2S)-rich gas produced in the regeneration process will be fed to a conventional sulfur recovery operation (located on-site) and converted to either liquid sulfur or sulfuric acid.

Single-stage regeneration of limestone with subsequent recovery of sulfuric acid is projected for atmospheric FBC systems. A flow sheet of these operations, again addressing mainly source assessment requirements, is shown in Fig. 2. Descriptions and flow diagrams of the contact process for sulfuric acid production are provided by Shreve (1967).

In brief, partially sulfated sorbent undergoes decomposition in reducing gases at high temperature and atmospheric pressure:

$$CaSO_4 + CO + H_2 \xrightarrow[\text{(1038–1093°C)}]{\text{1900–2000°F}} CaO + CO_2 + H_2O + SO_2$$

Reducing gases are produced by combustion of coal in the fluidized-bed regenerator. Tail gas from the acid production process is recycled to the boiler combustion air supply.

Generic coal and sorbent preparation operations relevant to both atmospheric and pressurized FBC systems are shown schematically in Fig. 3. From an environmental point of view, the coal preparation operations are similar to those associated with conventional coal combustion systems. Furthermore, sorbent preparation poses no new environmental problems. For start-up of both atmospheric and pressurized FBC sys-

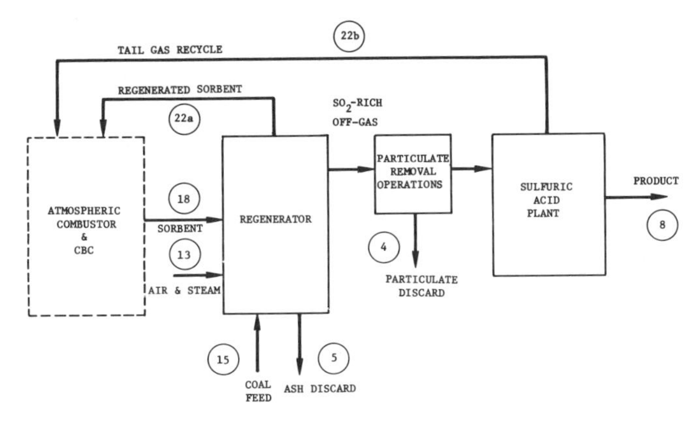

Fig. 2 Generic sorbent regeneration and sulfur recovery operations for atmospheric FBC systems.

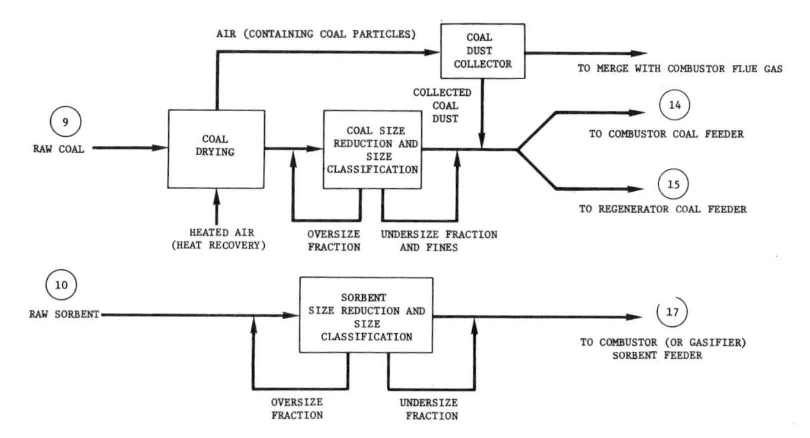

Fig. 3 Generic coal preparation and sorbent preparation operations (atmospheric and pressurized FBC processes).

tems, it is anticipated that natural gas, propane, and distillate fuels will be employed. No preparation is associated with these fuels.

B. Pressurized FBC Processes

A schematic flow diagram representing a pressurized, combined-cycle FBC unit is presented in Fig. 4. The generic operations and streams illustrated correspond identically with those depicted for the atmospheric FBC unit (Fig. 1), except for the inclusion of a gas turbine and one additional stream (No. 23). Detailed process descriptions, developmental history, and ongoing as well as anticipated work in pressurized fluidized-bed combustion are discussed by Keairns *et al.* (1973) and in a position paper by EPA/OR&M, Advanced Process Section (EPA, 1972).

Combustion again occurs in a fluidized bed of sorbent,† with excess air ranges similar to those found in the atmospheric boiler and at temperatures ~93°C (200°F) higher. Pressure within the combustor, however, is maintained at a design value of 4–10 atm, resulting in a dramatic reduction in combustor size requirements.

In addition to energy conversion achieved through a conventional

† Use of dolomite (i.e., dolomitic limestone) is projected for commercialized pressurized FBC systems. Dolomite undergoes attrition more rapidly than calcitic limestone but calcines readily to MgO–$CaCO_3$ at temperatures commonly found in pressurized FBC. The use of limestone in pressurized systems would require higher temperatures to achieve calcination and, in addition, limestone dust may erode gas turbine blades more rapidly.

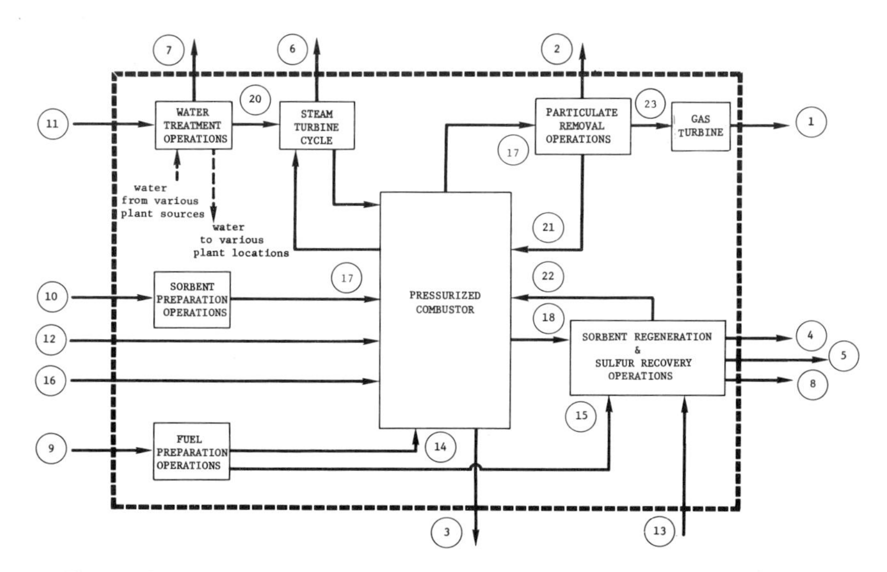

Fig. 4 Generic process flow sheet—pressurized, combined-cycle FBC of coal.

steam turbine cycle driven by steam produced in combustor tubes, the hot, high pressure gases exiting the combustor, after being cleansed of particulates, are expanded through a gas turbine to generate additional power and increase system efficiency. The power output of a typical combined-cycle unit is projected to be 80% steam turbine and 20% gas turbine generated. To maximize combustion efficiency, the larger ash particles removed by the first-stage cyclones are recycled to the combustor.

Employment of the Westinghouse two-stage dolomite regeneration process (Keairns *et al.*, 1973) with subsequent sulfur recovery in a Claus plant is projected for commercialized pressurized FBC systems. A flow sheet of these operations is shown in Fig. 5. Shreve (1967) provides descriptions and flow diagrams of the Claus sulfur production process.

Briefly, sulfated dolomite reacts with reducing gases at elevated pressure (~8 atm) in the first-stage fluidized-bed regenerator to produce calcium sulfide:

$$CaSO_4 + 4H_2 + 4CO \xrightarrow[\substack{1500°F \\ (876°C)}]{} CaS + 4H_2O + 4CO_2$$

Reducing gases are produced by conventional gasification of coal with air and steam. The reduced solids are then fed to the second-stage fluidized-bed regenerator where they are reacted at high pressure (12 atm) with steam and CO_2 (obtained by stripping a flue gas side stream) to form calcium carbonate:

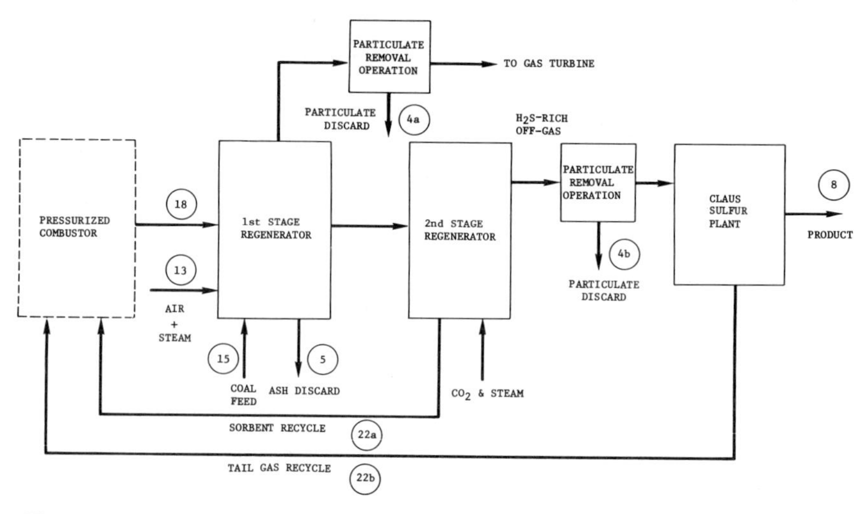

Fig. 5 Generic sorbent regeneration and sulfur recovery operations for pressurized FBC systems.

$$CaS + H_2O + CO_2 \xrightarrow[\substack{1100°F \\ (593°C)}]{} CaCO_3 + H_2S$$

Tail gas from the sulfur recovery operation is recycled to the combustor air supply.

Fuel and sorbent preparation, the steam turbine cycle, and water treatment operations for pressurized units are similar to the corresponding operations for atmospheric units. Since fuel and sorbent are injected into a pressurized combustion chamber (and regenerator), feed operations are necessarily more complex.

III. A PHASED APPROACH TO SAMPLING AND ANALYSIS

Because of constraints on resource allocations, a sampling and analysis program to support environmental assessment must be both cost-effective and information-effective. A strategy in which each sample is directly analyzed for an entire spectrum of possible components, individually, is obviously information-effective, but prohibitively expensive. On the other hand, direct analysis of a sample for a limited list of components known to be hazardous is relatively inexpensive but not information-effective since the risk of bypassing unsuspected hazardous materials is greatly increased.

A program strategy representing a compromise between these extremes employs a phased approach (Hamersma and Reynolds, 1976; Vlahakis and Abelson, 1976) in which two separate and distinct levels of

TABLE II *Characteristics of Level 1 and Level 2 Sampling and Analysis*

Issue	Level 1	Level 2
General		
Goal	Characterization of the pollutant potential of influent and effluent streams of a process. Planning basis for level 2 effort	Accurate quantitative identification of selected components in selected streams and determination of associated emission rates
Streams considered	All process influents and effluents including airborne and waterborne fugitive emissions	Selected influents and effluents based on level 1 output
Pollutant classes/species considered	Broad screening for elements, organics, and selected anions	Selected pollutant classes/species based on level 1 output
Process operating condition	Steady state representative condition	Steady state representative condition
Sampling		
Sampling techniques	Particulates	Particulates
	Single point, isokinetic at cross section point of average velocity	Full traverse of cross section, isokinetic at each point
	Convenient location, not in region of irregular flow	Specified distance from flow disturbances, as per EPA, Method 1
	Gases	Gases
	Convenient sampling site in unstratified location	Same sampling location as for particulates
	Single point, grab (e.g., displacement bomb)	Single point unless stratification exists (then, traverse required)
		Individual trains for specific components (primarily Federal Register methods)
	Liquids	Liquids
	Homogeneous—single point grab or tap	Homogeneous—single point grab or tap
	Heterogeneous—full stream cut grab at discharge or tap in well-mixed region of line	Heterogeneous—automatic sampler (full steam cut operation)
	Solids	Solids
	Pile or open holding container—depth integrated grab	Pneumatic transport line—automatic sampler (e.g., Vezin type)
	Stream discharge point—full stream cut grab	Belt conveyor—stopped belt (full stream cut)
	Belt conveyor—stopped belt (full stream cut)	

TABLE II *(Continued)*

Issue	Level 1	Level 2
Replications	None	Particulates: three minimum, separate analysis for each sampling Gas: compliance with Federal Register specifications Solids and liquids: three minimum, separate analysis for each sampling
Analytical		
Sensitivity	Sufficient to ensure detection of all class/species at trace levels	Sensitivity requirements will be determined by level 1 output
Accuracy	Target accuracy factor of ±2 on concentration	Target accuracy of ±10%
Specificity	Broad screening for inorganic gaseous species	Selected inorganic gases based on level 1 output
	Separation of organics into eight fractions on the basis of polarity	Identification of specific organic compounds from selected fractions
	Elemental analysis	Selected species based on level 1 output
	Selected anions from standard water analysis	Selected anions from level 1 elemental and standard water analyses
Physical characterization of solids and particulates	Morphology, sizing	More refined morphological and sizing determinations
Bioassay		
Analyses performed	Cytotoxicity, mutagenicity	Cytotoxicity, mutagenicity, carcinogenicity
Sample	Whole	Fractionated; specific components

sampling and analysis are utilized. A screening phase (level 1) characterizes the pollutant potential of all influent and effluent streams of a process and has a broad range of applicability. In this phase, sampling and analysis is designed to determine the presence or absence, approximate concentrations, and associated emission rates of inorganic and organic stream components. Perhaps the most important feature of the level 1 screening effort is the fact that no assumptions need be made, *a priori,* regarding the nature of the constituents in any given stream.

Biotesting,† an integral part of environmental assessment, is limited to a determination of the cytotoxic and mutagenic potential of a sample.

Using the output of the screening phase, priorities for additional testing can be established among streams within a given process and components within given streams, on the basis of pollution potential. In this manner, an optimized follow-up phase (level 2) can be planned and the appropriate resources allocated. The goal of the level 2 effort is the accurate, quantitative identification of specific components in selected streams (as dictated by the level 1 output) and the determination of associated emission rates. General characteristics of level 1 and level 2 sampling and analysis are summarized in Table II.

IV. SELECTION OF FBC PROCESS STREAMS FOR SAMPLING AND ANALYSIS

The screening phase (i.e., level 1) strategy requires that all process influents and effluents, including fugitive emissions, be sampled. In some cases, however, certain influent streams can be eliminated. As an example, for the generic processes addressed here, all intake water and air streams as well as the start-up fuel stream (No. 16) have been dropped from consideration.

Sampling of internal process streams is, in general, not proposed as part of an environmental assessment effort. Interest in these streams relates primarily to process and control device evaluations. (Accidental discharge from within-process streams, while posing a potential hazard to the environment, is considered beyond the scope of the present study.)

Sampling of airborne and waterborne fugitive emissions, while an integral part of process environmental assessment, is not addressed here. A discussion of this subject (including methods for assessing the leaching potential of FBC solid residues and feeds) is provided by Abelson and Löwenbach (1977).

Figure 6 shows all influent and effluent streams associated with the FBC systems under consideration and indicates those streams selected for sampling and analysis in the screening phase. Also noted are streams selected for biological testing as well as streams having associated fugitive emission potential.

Selection of streams for the level 2 phase, as noted earlier, is dependent on the screening phase output. Certain streams (and stream components) may be eliminated from further consideration, based on data

† Procedures for biological testing are not addressed here.

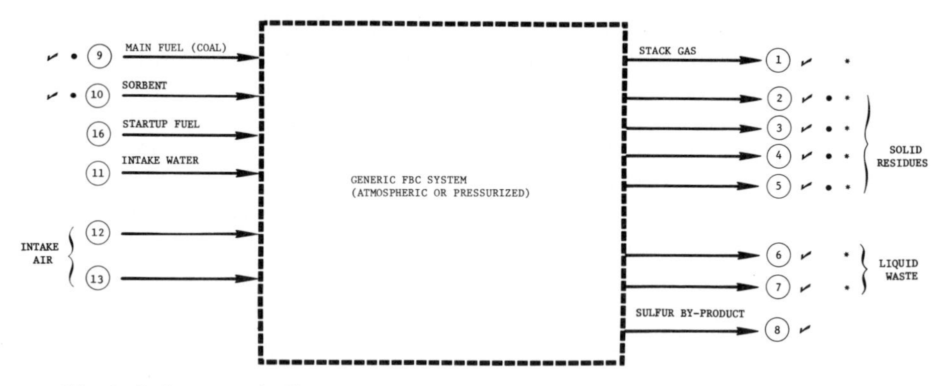

Fig. 6 Influents and effluents—generic FBC processes. Checks denote sampling at level 1; dots denote airborne and waterborne fugitive potential; and asterisks denote biological testing.

obtained at level 1. For the present effort, not having the benefit of level 1 results, nothing can be assumed regarding elimination of streams or stream components. Therefore, in Chapter 36, level 2 procedures are delineated for all streams that were considered in the screening phase.

Assessment of the physical conditions present in streams selected for

TABLE III *Physical Characteristics of Sampled Streams*

Stream No.	Stream designation	Physical composition	Temperature (°F)[a]	
			Atmospheric unit	Pressurized unit
1	Stack gas	Gas, particulates	240–300	275
2	Particulate removal discard	Solid	200–300	200–300
3	Bed solids discard	Solid	200–300	200–300
4	Particulate removal discard—regeneration operations	Solid	200–300	200–300
5	Ash discard from regenerator	Solid	200–300	200–300
6	Blowdowns from steam turbine cycle	Liquid	100	100
7	Blowdown from water treatment operations	Liquid	Ambient	Ambient
8	Product from sulfur recovery (sulfur or sulfuric acid)	Liquid	100	230
9	Raw fuel to preparation	Coal	Ambient	Ambient
10	Raw sorbent to preparation	Solid	Ambient	Ambient

[a] All measurements are made at atmospheric pressure.

sampling is prerequisite to the specification of appropriate sampling procedures. Two important parameters influencing technique selection are stream static pressure and temperature. For the range of system operating conditions under which sampling is planned, typical values of these parameters at proposed sampling sites are displayed in Table III for generic atmospheric and pressurized FBC units. Sampling of all streams will take place at approximately atmospheric pressure for both the atmospheric and pressurized FBC systems. Values for flow rate, an additional parameter of importance, could not be specified in a generic context.

V. AREAS OF ANALYTICAL INTEREST

A. Introduction

Although the level 1 screening/detection phase of environmental assessment eliminates the need for forecasting a list of pollutants which may be released to the environment, it remains of interest to consider briefly the major areas of analytical concern in terms of existing data from FBC processes as well as by comparison with conventional coal-fired systems.

The discussions presented in the following sections are organized around the following areas of analytical interest: organic compounds (gaseous and condensed), inorganic gaseous components, trace elements, and anions. Analytical procedures pertinent to these areas are considered in Chapter 36.

B. Organic Compounds

Data on specific organic compounds which are emitted during fluidized-bed combustion are extremely rare. Total hydrocarbons (THC) have been measured only on occasion, using a flame ionization technique. Although THC values may well be of use in measuring overall combustion efficiency, they are of relatively little help in predicting species concentrations. Typical values for stack hydrocarbon emissions (reported as parts per million of methane) are on the order of ~1000 ppm from atmospheric FBC (Gordon *et al.*, 1972) and ~100 ppm from pressurized FBC (Vogel *et al.*, 1975). Clearly, these values represent only a portion of organic emissions; of equal importance are organic emissions in the condensed phase. Relatively little work has been done to identify and quantify condensed organic compounds which occur with particulates.

Potentially, an extremely wide variety of organic compounds could form. Detailed kinetic or thermodynamic calculations are of limited value; both the identity of the true reacting species and the assumption of equilibrium between reacting species are often speculative. While a detailed discussion of the combustion of fuels in fluidized-bed processes is beyond the scope of this chapter, some basic considerations may provide a general understanding of chemical reactions and species formation.

Combustion is a combination of two competitive pathways: oxidative and pyrolytic. The oxidative path leads to oxidized products such as CO. The pyrolytic pathway leads to reduced species such as olefins, aromatics, and, of greater environmental concern, polycyclic aromatics such as benzo(*a*)pyrene. In addition, the products of pyrolytic reactions may play an important role in the formation and growth of particulate matter.

The nature of molecular reactions during combustion is determined by many factors, including the temperature and chemical nature of the environment. Locally elevated temperatures and less oxidizing environments, conditions common to shallow-bed atmospheric FBC (due to the nonuniformity of coal injection and airflow through the bed), favor bond rupture and dehydrogenation reactions which lead to less complex (i.e., lower molecular weight) reduced species. At lower temperatures, pyrolytic reactions involve polymerization to produce more complex species, the fate of which depends on the local conditions at which they mix with postflame gases.

C. Inorganic Gases

Data on inorganic gaseous species are incomplete. Continuous monitoring of stack gas is typically limited to sulfur dioxide, nitrogen oxides, carbon monoxide, carbon dioxide, and oxygen. Sulfur trioxide has also been measured on occasion for both atmospheric and pressurized units. There are, aside from the previously mentioned species, a number of additional gaseous compounds which may be synthesized during the combustion process. Such species might include, for example, ammonia, carbonyl sulfide, halogens, hydrogen halides, hydrogen cyanide, and hydrogen sulfide.

Although detailed calculations may be made to estimate the composition of gaseous effluents (generally the most common method is to minimize the chemical system's free energy), the values generated by such methods probably represent only trends. Necessarily, these methods assume chemical equilibrium which may or may not be at-

TABLE IV *Typical Stack Gas Composition Data from Fluidized-Bed Combustion*[a]

Species	Atmospheric FBC	Pressurized FBC
SO_2	800	140
NO (NO_x)	221	200
CO_2	20.79%	19.0%
O_2	3.46%	3.8%
Hydrocarbons (as methane)	675	—
CO	5275	35
Conditions		
Coal	Sewickley Seam 4.88% S	Arkwright 2.82% S
Bed temperature (°F)	1550	1550
Ca/S	2.8	2.9
Fluidizing velocity (ft/sec)	8.0	3.0
Additive	Grove limestone	Tymochtee dolomite

[a] All values in parts per million by weight unless otherwise noted.

tained. Calculations of this sort have been done primarily for coal gasification studies and do not seem to have been applied extensively to coal combustion studies.

Typical stack gas analysis data are shown in Table IV for atmospheric (Mesko *et al.*, 1974) and pressurized FBC (Keairns *et al.*, 1973).

D. Trace Elements

Virtually all elements below atomic number 92 are contained, in at least trace amounts (100 ppm), in coal. Occurrence frequency and typical ranges of concentrations of trace elements are shown in Figs. 7 and 8. Coal combustion releases trace elements to the environment via three pathways. One route is through particulate emissions. While a substantial fraction of trace elements present in the coal is retained with the fly ash removed by control devices, significant quantities of trace elements may still be emitted as submicron-sized particles because of collection inefficiencies inherent in these devices. (For fluidized-bed combustion systems, particulate includes attrited sorbent in addition to ash.) A second pathway of trace element release during coal combustion in-

H ND																	
Li 100	Be 100											B 100	C ND	N ND	O ND	F 100	
Na 100	Mg 100											Al 100	Si 100	P 100	S 100	Cl 100	
K 100	Ca 100	Sc 100	Ti 100	V 100	Cr 100	Mn 100	Fe 100	Co 100	Ni 100	Cu 100	Zn 100	Ga 100	Ge 100	As 100	Se 100	Br 100	
Rb 100	Sr 100	Y 100	Zr 100	Nb 100	Mo 100	Tc Nd	Ru 0	Rh 0	Pd 0	Ag 92	Cd 92	In Standard	Sn 100	Sb 92	Te 85	I 85	
Cs 100	Ba 100	La 100	Hf 46	Ta 62	W 69	Re 0	Os 0	Ir 0	Pt 0	Au 0	Hg 38	Tl 31	Pb 100	Bi 31	Po ND	At ND	
Fr ND	Ra ND	Ac ND	Ce 100	Pr 100	Nd 100	Pm ND	Sm 100	Eu 100	Gd 85	Tb 85	Dy 85	Ho 77	Er 77	Tm 0	Yb 62	Lu 38	
			Th 92	Pu ND	U 92												

Fig. 7 Occurrence frequency of elements in 13 raw coals as determined by spark-source mass spectrometry. All quantities in percent. ND, Not determined; O, checked but not detected.

volves the volatilization of certain elements (e.g., mercury) and subsequent emission in the gaseous phase. Finally, trace elements may be leached from solid residues and feeds and adversely impact groundwater and surface water quality. This is of particular concern in FBC systems.

Recent investigations have demonstrated that several trace elements (e.g., lead, cadmium, arsenic, and nickel) are preferentially concentrated in the smallest particles emitted from conventional coal-fired power plants (Natusch *et al.*, 1974; Kaakinen *et al.*, 1975; Klein *et al.*, 1975). The mechanism of trace element enrichment in submicron-sized particles is suggested to be volatilization of elements or compounds in

H ND																	
Li 4-163	Be 0.4-3											B 1-230	C ND	N ND	O ND	F 1-110	
Na 100-1000	Mg 500-3500											Al 3000-23.000	Si 5000-41,000	P 6-310	S 700-10,000	Cl 10-1500	
K 300-6500	Ca 800-6100	Sc 3-30	Ti 200-1800	V 2-77	Cr 26-400	Mn 5-240	Fe 1400-12,000	Co 1-90	Ni 3-60	Cu 3-180	Zn 3-80	Ga 0.03-10	Ge 0.03-1	As 1-10	Se 0.04-03	Br 1-23	
Rb 1-150	Sr 17-1000	Y 3-25	Zr 28-300	Nb 5-41	Mo 1-5	Tc ND	Ru <0.1	Rh <0.1	Pd <0.1	Ag <0.01-3	Cd <0.01-0.7	In Standard	Sn 1-47	Sb <0.1-2	Te <0.1-0.4	I <0.1-4	
Cs 0.2-9	Ba 20-1600	La 0.3-29	Hf <0.3-4	Ta <0.1-8	W <0.1-0.4	Re <0.2	Os <0.2	Ir <0.2	Pt <0.3	Au <0.1	Hg <0.3-0.5	Tl <0.1-0.4	Pb 1-36	Bi <0.1-0.2	Po ND	At ND	
Fr ND	Ra ND	Ac ND	Ce 1-30	Pr 1-8	Nd 4-36	Pm ND	Sm 1-6	Eu <0.1-0.4	Gd <0.1-3	Tb <0.1-2	Dy <0.1-5	Ho <0.1 0.4	Er <0.1	Tm <0.1	Yb <0.1-0.5	Lu <0.1-0.3	
			Th <0.1-5	Pu ND	U <0.1-1												

Fig. 8 Concentration range of elements in 13 raw coals analyzed by spark-source mass spectrometry. All quantities in parts per million by weight. ND, Not determined.

the combustion zone of the furnace and subsequent condensation or adsorption onto particulates. Because of higher surface-to-mass ratios for fine particulates, trace elements become concentrated on them more readily.

Since fluidized-bed combustion is carried out at temperatures well below those of conventional coal combustion systems (1700 versus 2800°F), trace element release by volatilization is reduced, in turn resulting in a minimization of the preferential concentration of trace elements in finer particulates. This decrease, however, is at the expense of higher concentrations in the sorbent bed and the larger collected particulates, which must ultimately be disposed of.

Representative trace element concentration data are presented in Table V for a pressurized fluidized-bed combustion experiment (Swift *et al.*, 1976).

TABLE V *Typical Trace Element Concentration Data from Pressurized FBC*[a]

Element	Starting bed	Coal	Final bed	Primary cyclone	Secondary cyclone	Flue gas
As		5	3.5	24		
Ba		[b]		350	860	
Be	0.8	0.7	0.8	2.6	6.0	
Br	6	13			3	
Ce		[b]		19	13	
Co		18		11	19	
Cr		100	12	180	300	
Dy		0.2		1.8	2.9	
F	100	25		20	10	8.1
Fe	240	1×10^4	5.5×10^3	5.9×10^4	3.6×10^4	
Ga	110					
Hf	1.7		97	2.9	6	
Hg	0.005	0.15	0.005	0.46	0.46	
K		580	660	3.7×10^3	5×10^4	
La		4.2	3.7	31	52	
Mn	*b*	26	39	110	140	0.32
Na	1.7×10^3	690	1.3×10^3	4.1×10^3	7.2×10^3	*b*
Pb	1.1	29	51	95	260	
Sb		0.3	0.7	3	6.2	
Sc	0.3	1.7	1.8	9	19	
Sm		0.8	0.1	*b*		
Yb			5.2	4	7.5	
Zn	*b*					
Zr			340			

[a] All values in parts per million by weight. Conditions: combustion temperature, 1550°F; pressure, 10 atm; gas velocity, 2.4 ft/sec; coal, Arkwright; bed material, alumina.

[b] Identification only.

E. Anions

A variety of anions are expected to be found in solid residues, solid feeds, and particulates from the FBC process as well as in leachate from these materials. Additionally, certain anions are associated with water treatment and steam cycle blowdowns, common to both conventional and fluidized-bed combustion processes. Analysis of solid residues, i.e., specific inorganic compound identification, has been limited primarily to determinations of the chemical state of sulfur in particulates (Craig *et al.*, 1974), spent FBC sorbent (Hubble *et al.*, 1976), and coal.

The problems associated with leaching are common to both conventional and fluidized-bed combustion systems. Leachate is liquid which has contacted solid material and has extracted and/or suspended constituents from it. (Whenever water comes into direct contact with solid materials, the potential for leaching exists.) Many species existing in solid materials may be readily soluble in water; still others may be solubilized by the action of leachate upon them. Leachate from coal (known as acid mine drainage) can be generated as a result of incident precipitation on storage piles. Solid residue (i.e., slagged ash) as well as collected fly ash from conventional combustion may be subject to leaching before and after disposal (e.g., piles, landfills). A similar problem exists for solid residue and collected particulate from FBC systems. The release of anionic species here, however, is of greater concern because of

TABLE VI *Leachate Characterization from Atmospheric and Pressurized Combustion Processes*[a]

	Atmospheric FBC		Pressurized FBC	
Constituent	Fly ash	Spent sorbent	Fly ash	Spent sorbent
pH	12.2	12.2	9.3	12.2
TDS	4212	4064	2124	4188
COD	105	58	45.6	51.4
Arsenic	2.5	5.0	*b*	5.0
Chloride	6.2[c]	22.8[c]	5.8	8.4
Nitrate	2	28	27	26
Phosphate	2	*b*	4	*b*
Sulfate	1600	1640	1190	1900

[a] All values in parts per million by weight. All leachates are generated by extraction of residues with distilled, deionized water unless otherwise noted.

[b] Not detected.

[c] Extracted with 1 *N* NaOH.

higher concentrations and the possibility of additional species in the sorbent. Table VI compares preliminary results of leaching experiments on collected particulate and spent sorbent from both atmospheric and pressurized FBC systems.

REFERENCES

Abelson, H. I., Löwenbach, W. (1977). "Procedures Manual for Environmental Assessment of Fluidized Bed Combustion Processes," EPA Rep. 600/7-77-009. Environ. Prot. Agency, Washington, D.C.

Craig, N. L., Harker, A. B., and Novakov, T. (1974). *Atmos. Environ.* **8,** 15–21.

Environmental Protection Agency (EPA) (1972). "Pressurized Fluidized-Bed Combustion of Coal." Advanced Process Section, OR&M, Durham, North Carolina.

Gordon, J. S., Glenn, R. D., Ehrlich, S., Ederer, R., Bishop, J. W., and Scott, A. K. (1972). "Study of the Characterization and Control of Air Pollutants from a Fluidized-Bed Boiler—the SO_2 Acceptor Process," EPA Rep. R2-72-021, PB-229242. Environ. Prot. Agency, Washington, D.C.

Hamersma, J. W., and Reynolds, S. L. (1976). "Field Test Sampling/Analytical Strategies and Implementation Cost Estimates: Coal Gasification and Flue Gas Desulfurization," EPA Rep. 600/2-76-0936. Environ. Prot. Agency, Washington, D.C.

Hubble, B. R., Siegel, S., Fuchs, L. H., and Cunningham, P. T. (1976). *Proc. Int. FBC Conf., 4th* pp. 367–391. MITRE Corporation, McLean, Virginia.

Kaakinen, J. W., Jordan, R. M., Lawasami, M. H., and West, R. E. (1975). *Environ. Sci. Technol.* **9,** 862.

Keairns, D. L., Archer, D. H., Hamm, J. R., Newby, R. A., O'Neill, E. P., Smith, J. R., and Yang, W. C. (1973). "Evaluation of the FBC Process. Vol. 1: PFBC Process Development and Evaluation," U.S. National Technical Information Service, PB Rep. 231 162, Springfield, Virginia.

Klein, D. H., Andrew, A. W., Carter, J. A., Emery, J. F., Feldman, C., Fulkerson, W., Lyon, W. S., Ogle, J. C., Talmi, Y., Van Hook, R. I., and Bolton, N. (1975). *Environ. Sci. Technol.* **9,** 973.

Mesko, J. E., Erlich, S., and Gamble, R. A. (1974). "Multicell Fluidized-Bed Boiler Design, Construction and Test Program," Rep. No. 90, PB-236254. Off. Coal Res., U.S. Bur. Mines, Washington, D.C.

Natusch, D. F. S., Wallace, J. R., and Evans, C. A. (1974). *Science* **183,** 202.

Shreve, R. N. (1967). "Chemical Process Industries." McGraw-Hill, New York.

Swift, W. M., Vogel, G. J., Panek, A. F., and Jonke, A. A. (1976). *Proc. Int. FBC Conf., 4th* pp. 525–543. MITRE Corporation, McLean, Virginia.

Vlahakis, J., and Abelson, H. I. (1976). "Environmental Assessment Sampling and Analytical Strategy Program," Rep. No. M76-35. MITRE Corporation, McLean, Virginia.

Vogel, G. J., Swift, W. M., Montagne, J. C., Lene, J. F., and Jonke, A. A. (1975). *Inst. Fuel Symp. Ser. No. 1, Proc.* **1,** D3/1—D3/11.

Chapter 36

Sampling and Analysis of Emissions from Fluidized-Bed Combustion Processes—Part 2

Harvey I. Abelson *John S. Gordon*
THE MITRE CORPORATION
METREK DIVISION
MCLEAN, VIRGINIA

William A. Löwenbach
LÖWENBACH AND
SCHLESINGER ASSOCIATES, INC.
MCLEAN, VIRGINIA

I. INTRODUCTION

In this chapter, recommended procedures for multimedia sampling and analysis of fluidized-bed combustion systems are presented. These procedures apply to the generic processes described in Chapter 35 and are consistent with the phased environmental assessment strategy discussed therein. The intent here is to summarize and highlight techniques as well as to provide key references wherein detailed instructions, for use by sampling and analytical personnel, can be found. Primary attention, however, is focused on the analytical rather than the sampling aspects of the assessment program.

ISBN 0-12-399902-2

II. SAMPLING PROGRAM

A. General

Sampling characteristics associated with the screening (level 1) and follow-up (level 2) phases of assessment have already been summarized in Table IV of Chapter 35. In accordance with these characteristics, recommended sampling procedures have been selected for both assessment levels and are summarized in Table I. Where a recommended technique is common to several streams, these streams are treated as a group (e.g., solid residues). Stream numbers and designations refer to the process flow diagrams and Table I of Chapter 35. In the following sections, highlights of the sampling program are discussed. The reader is referred to Abelson and Löwenbach (1977) and the references contained therein for a more in-depth treatment of the various procedures, including rationale for selection, site preparation requirements, and discussions of pretest procedures to be performed by the source testing contractor.

B. Particulate Sampling

The sampling of stack particulates represents one of the most costly and complex aspects of an assessment sampling program. The source assessment sampling system (SASS train), developed by EPA and the Aerotherm Corporation, is recommended for particulate sampling at both levels 1 and 2. The SASS train, depicted schematically in Fig. 1, has the following advantages:

(i) Collected particulate is fractionated into four size fractions, enabling adequate biological evaluation.
(ii) Sampling rates are high enough to permit an adequate amount of material to be collected in a reasonable amount of time from streams with lower mass loadings.
(iii) The train has the capability of collecting volatile trace elements and organic species in addition to particulates.

Briefly, the sampling train consists of three sections: a stainless steel probe; an oven module containing three cyclones, a filter, and a sorbent trap; and four impingers for collection of volatile inorganic species. Particulate matter is fractionated into four ranges: (1) >10 μm, (2) 3–10 μm, (3) 1–3 μm, and (4) <1 μm. Volatile organic material is collected in a trap using a porous polymer resin (XAD-2). Additionally, small amounts of volatile inorganic species will probably be trapped as a result of simple impaction. Volatile inorganic elements are collected in a series of impingers using the reagents specified in Table II. The pumping capacity is supplied by a 10-cfm high volume vacuum pump, while required pressure, temperature, power, and flow conditions are maintained by the main controller. A further and more detailed description of the train, operating procedures, and sample handling and transfer requirements is provided by Hamersma and Reynolds (1976).

For the level 1 effort, a convenient sampling site in the stack or breaching, removed from regions of irregular flow, should be selected. Single-point isokinetic sampling is performed at the cross section point of average velocity (as determined by a velocity traverse). Isokineticity is established by selecting an appropriate probe nozzle.

Level 2 particulate sampling entails a full cross-sectional traverse with isokinetic conditions at each traverse point. EPA Method 1 specifications for port placement and traverse point locations must be followed. Sampling sites are normally located on the stack, at least eight equivalent diameters downstream of any flow disturbance. The SASS train is designed to operate at a predetermined constant flow rate which, if changed, would alter the cut size of particulate captured by the train's cyclones. To maintain isokinetic conditions for a full traverse would entail a change of nozzles at each traverse point (a rather impractical procedure). An alternative involves locating the sampling site at a cross section where the traverse point velocities do not vary more than 10% from a mean value. The SASS train would then provide conditions that are isokinetic within level 2 accuracy limits, using a single nozzle. Another option is applicable to sampling locations where some traverse points (i.e., outlyer points) have a velocity variation greater than 10% of a mean value. If the number of outlyer points is small compared to the total number to be sampled, then nozzles may be changed to maintain isokinetic conditions.

C. Gas Sampling

For the screening phase (i.e., level 1), a simple displacement bomb technique is recommended for the sampling of most gaseous components from the stack. Gas displacement, liquid displacement, or evacuated bomb methods may be employed. Samples are extracted at a single point using a 3-liter bomb with a Teflon line as a probe and a glass wool plug as a prefilter for particulates. A convenient, readily accessible sampling site should be selected in a region in which there is little or no stratification. Bomb samples should be analyzed as soon as possible and stored away from direct sunlight and heat. For level 1 sampling of SO_3, HCl, HF, Hg, and nonvolatile organic compounds, use

TABLE I *Summary of Recommended Sampling Procedures*

Stream or group [stream number(s)][a]	Level 1		Level 2	
	Recommended technique(s)	Reference	Recommended technique(s)	Reference
Solid residues [2,3,4,5]	1. Simple grab (while discharging): For sampling site at discharge point of line or hopper	Toggert (1945)	1. Vezin-type automatic sampler: Used when material is conveyed by pneumatic transport. Installed in a vertical section of transport line	ASTM (1972) (D 2234)
	2. Pipeborer or auger: For sampling site at pile or open holding container (e.g., bin, hopper). If material is too dense for pipeborer, auger is used	Toggert (1945)	2. Pipeborer or auger: See level 1	
Solid feeds [9,10]	1. Stopped belt sampling: Used if raw feeds are transported by accessible belt conveyor. Full stream cut sampling	No reference	1. Stopped belt sampling: See level 1	
	2. Auger: See "solid residues" group, level 1	Toggert (1945)	2. Vezin-type automatic sampler: See "solid residues" group, level 2	
	3. Simple grab: For sampling site at pile	ASTM (1972) (D 2234)		
Liquid blowdowns [6,7]	1. Dipper (full stream cut): For sampling site at discharge of blowdown line (*Note:* These streams may have high solids content)	ASTM (1972) (D 510)	1. Automatic hi-volume sampler: Installed in well-mixed region of line	No reference
	2. Tap: Installed in well-mixed region of blowdown line	ASTM (1972) (D 860)		
Liquid sulfur/sulfuric acid product [8]	Tap: Installed in line or holding tank (*Note:* This stream is homogeneous in nature)	ASTM (1972) (D 510)	Tap: see level 1	

Stack gas—gaseous components [1]				
Acid gases				
SO_2	Simple displacement bomb: See text discussion	ASTM (1972) (D 1605)	1. EPA Method 6 (manual)	Federal Register (1971)
			2. EPA Performance Spec. No. 2 (continuous)	Federal Register (1975)
SO_3	SASS[b]: See text discussion	Acurex Corp. (1976)	Controlled condensation (manual)	West and Chiang (1974)
NO_x	Simple displacement bomb	ASTM (1972) (D 1605)	1. EPA Method 7 (manual)	Federal Register (1971)
			2. EPA Performance Spec. No. 2 (continuous)	Federal Register (1975)
H_2S	Simple displacement bomb	ASTM (1972) (D 1605)	EPA Method 3 (manual)	Federal Register (1971)
CO_2	Simple displacement bomb	ASTM (1972) (D 1605)	1. EPA Method 3 (manual)	Federal Register (1971)
			2. EPA Performance Spec. No. 3 (continuous)	Federal Register (1975)
HCl	SASS	Acurex Corp. (1976)	Impinger train: Similar to EPA Method 6 for SO_2	ASTM (1972) (D 2036)
HCN	Simple displacement bomb	ASTM (1972) (D 1605)	Impinger train	Ruch (1970)
HF and F	SASS	Acurex Corp. (1976)	Impinger train	Ruch (1970)
Other inorganic gases				
CO	Simple displacement bomb	ASTM (1972) (D 1605)	1. EPA Method 10 (manual): Essentially a Method 3 sampling system	Federal Register (1974)
			2. EPA Method 10 (continuous)	
O_2	Simple displacement bomb	ASTM (1972) (D 1605)	1. EPA Method 3 (manual)	Federal Register (1971)
			2. EPA Performance Spec. No. 3 (continuous)	Federal Register (1975)

TABLE I *(Continued)*

Stream or group [stream number(s)][a]	Level 1		Level 2	
	Recommended technique(s)	Reference	Recommended technique(s)	Reference
NH_3	Simple displacement bomb	ASTM (1972) (D 1605)	Kjeldahl Nesslerization (manual): This is an EPA method for NH_3 in wastewater and is suggested as an interim method for stack gas. Impingers contain 0.1 *N* H_2SO_4	Sittig (1974)
H_2O	EPA Method 4 (manual)	Federal Register (1971)	EPA Method 4 (manual): See level 1. For continuous monitoring for low concentration moisture, NDIR[c] may be used	
COS	Simple displacement bomb	ASTM (1972) (D 1605)	EPA Method 3 (manual)	Federal Register (1971)
CS_2	Simple displacement bomb	ASTM (1972) (D 1605)	EPA Method 3 (manual)	Federal Register (1971)
Hg	SASS	Acurex Corp. (1976)	Amalgamation method (manual): Adaptation of EPA Method 5 to collect Hg, As and particulate	Kalb and Baldeck (1972)
Organic components				
Volatile organics (C_1–C_6)	Simple displacement bomb	ASTM (1972) (D 1605)	EPA Method 3 (manual)	Federal Register (1971)
Nonvolatile hydrocarbons (C_6)	SASS	Monsanto (1976)	SASS (see Level 1)	
Stack gas—particulates [1]	SASS: see text discussion	Acurex Corp. (1976)	SASS (see Level 1)	

[a] See Chapter 35, Table I.
[b] Source assessment sampling system.
[c] Nondispersive infrared.

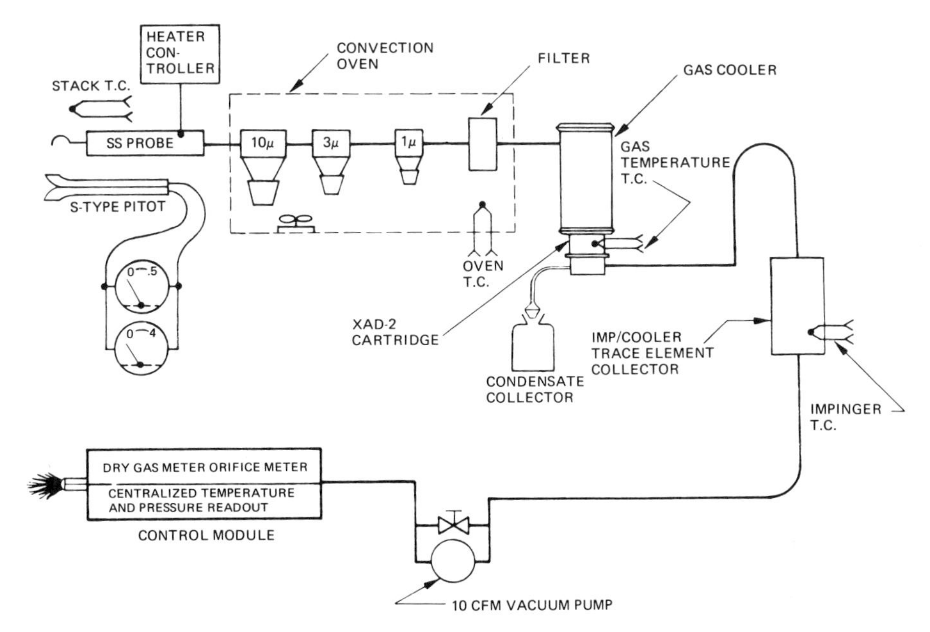

Fig. 1 Source assessment sampling system schematic.

of the SASS train (discussed earlier) is suggested. As noted in the preceding chapter, no sampling replications are called for in the screening phase.

As indicated in Table I, for the level 2 assessment phase, EPA Federal Register Methods and Performance Specifications are recommended for the sampling (and subsequent analysis) of most gaseous components. For SO_2, NO_x, CO_2, and O_2, a continuous as well as a manual option has been presented. It is desirable, from a cost-effectiveness standpoint, that the level 2 particulate sampling site be employed for gas sampling. Single-point sampling is satisfactory provided stratification does not exist at this location. If an investigation does indicate the presence of

TABLE II *SASS Train Impinger System Reagents*

Impinger	Reagent/quantity	Function
1	6 *M* H_2O_2/750 ml	Scrubber for reducing gases (e.g., SO_2) to prevent depletion of impingers 2 and 3
2 and 3	0.2 *M* $(NH_4)_2$ S_2O_8, 0.02 *M* $AgNO_3$/750 ml	Collection of volatile trace elements by oxidative dissolution
4	$CaSO_4$ (color indicating)/ 750 g	Moisture trap

stratification, a sampling routine involving multipoint traversing must be designed. The number of level 2 sampling replications should comply with Federal Register specifications.

For level 1 gas flow measurement, use of an "S"-type Pitot tube is suggested. For level 2 applications, however, the ellipsoidal-nose standard Pitot tube is recommended. Traverse points for velocity determinations at both levels are selected in accordance with EPA Method 1. Pitot tubes are inserted into ports employed for sample acquisition.

D. Liquid Sampling

For level 1 assessment, the liquid blowdowns from water treatment operations and the steam turbine cycle can be sampled using a dipper (full stream cut technique) if the discharge points are accessible. Alternatively, a tap may be installed in a well-mixed region of the blowdown line. For the level 2 phase, use of an automatic high volume sampler, also installed in a well-mixed region, is recommended. This device should be operated such that the intake velocity matches or exceeds the stream velocity at the sampling point. Because of the homogeneous nature of the liquid sulfur/sulfuric acid product stream, a tap installed in a holding tank or line should provide a representative sample for both phases of assessment. The maximum tap sampling rate recommended by ASTM is 500 ml/min.

E. Solids Sampling

Sampling alternatives for solid residue and solid feed (i.e., raw coal and sorbent) streams are presented in Table I for several potential sampling site locations. While the taking of a single sample is sufficient for the level 1 effort, each of the level 2 techniques involves collecting a number of increments which are then subject to a size reduction operation such as riffling to produce a representative sample. ASTM procedures for determining the weight and number of increments should be followed. All solid samples should be placed in airtight polyethylene containers or bags for transfer to the laboratory.

F. Other Considerations

Minimum sample sizes for level 1 assessment are listed as follows according to sample type (note that bioassay sample size requirements have not been considered here): particulate, 30 m^3; gas, 3 liters; liquid, 10 liters; and solid, 1 kg. Minimum sample sizes for the level 2 effort cannot be specified since these values depend in part on the results of the screening phase.

Since environmental assessment is concerned with the detection and quantification of pollutants which may be present in trace amounts, and since bioassay is an integral part of environmental assessment, it becomes extremely important that contamination and degradation of the sample does not occur. To minimize this possibility, the following precautions should be taken:

(i) Sampling apparatus and sample containers must be constructed of biologically inert materials and must not appreciably alter the concentration of sample constituents by erosion, adsorption, or any other mechanism.

(ii) Adequate sample preservation techniques should be employed.

(iii) Sampling equipment must be thoroughly cleaned before each use. Transfer of samples from sampling apparatus to containers must be performed in a consistent and careful manner in a clean environment. Sampling containers should be airtight.

III. ANALYTICAL PROGRAM

A. Introduction

In the following sections, recommended procedures are presented for phased, multimedia chemical characterization of the FBC process streams previously selected for sampling. An overview of the level 1 analytical scheme is provided in Figs. 2 and 3, while the level 2 overview is given in Figs. 4 and 5. (Some general characteristics of phased analysis have already been summarized in Table II of Chapter 35.) Procedures for a given analysis area (e.g., elemental analysis) are, for the most part, independent of sample type (e.g., solid, liquid, particulate) and, in addition, are insensitive to process configuration (i.e., atmospheric or pressurized FBC). Sample preparation procedures, however, may vary with the type of sample.

The level 1 scheme has been designed for broad screening capability, to ensure that potential pollutants, present even in trace quantities, do not go undetected. An accuracy factor of ± 2 is regarded as sufficient for level 1 detection. Chemical characterization in this screening phase encompasses the following areas:

(i) organic species in all influent and effluent streams,
(ii) inorganic elements in all influent and effluent streams,

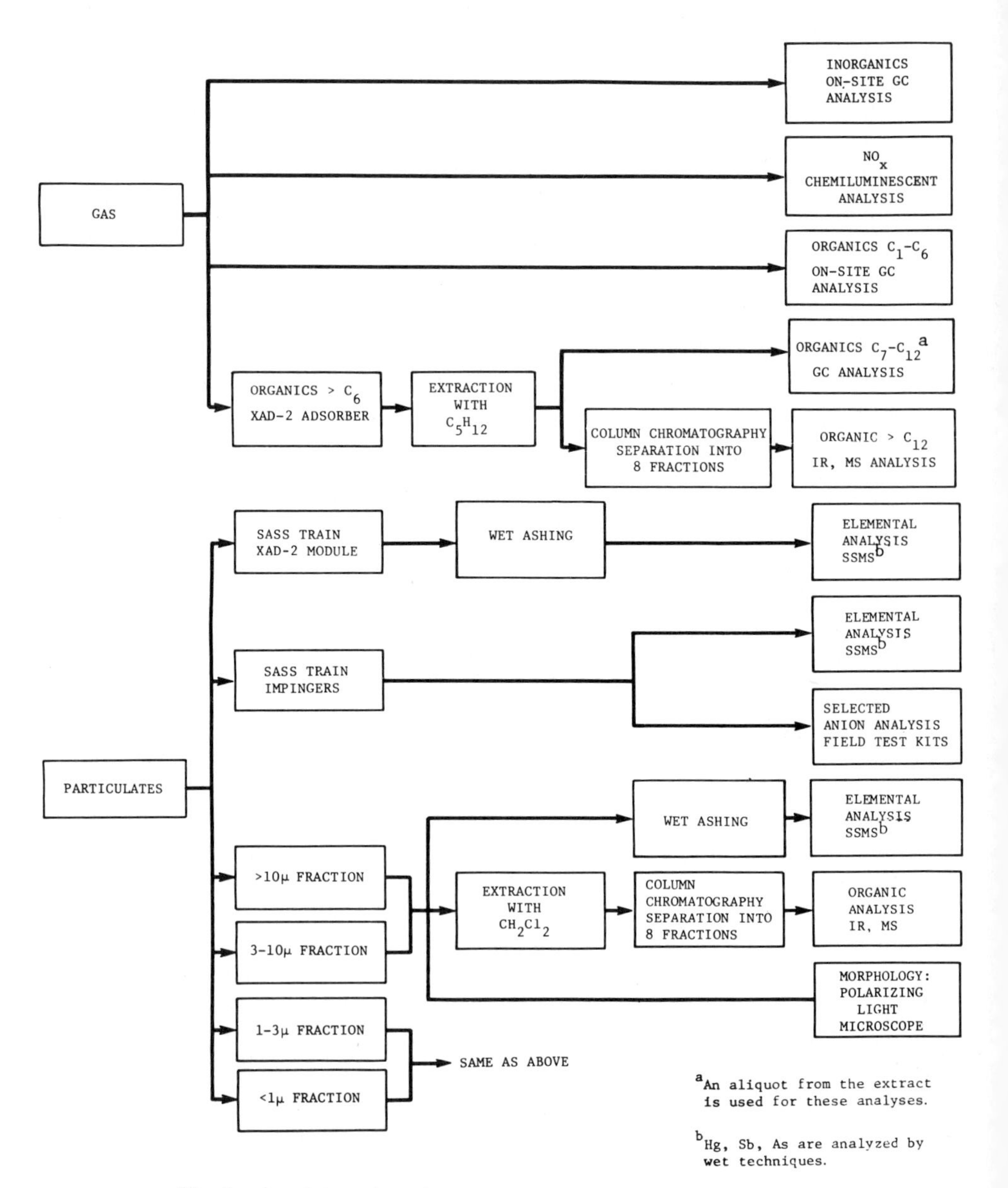

Fig. 2. Level 1 analytical scheme—gas and particulate samples.

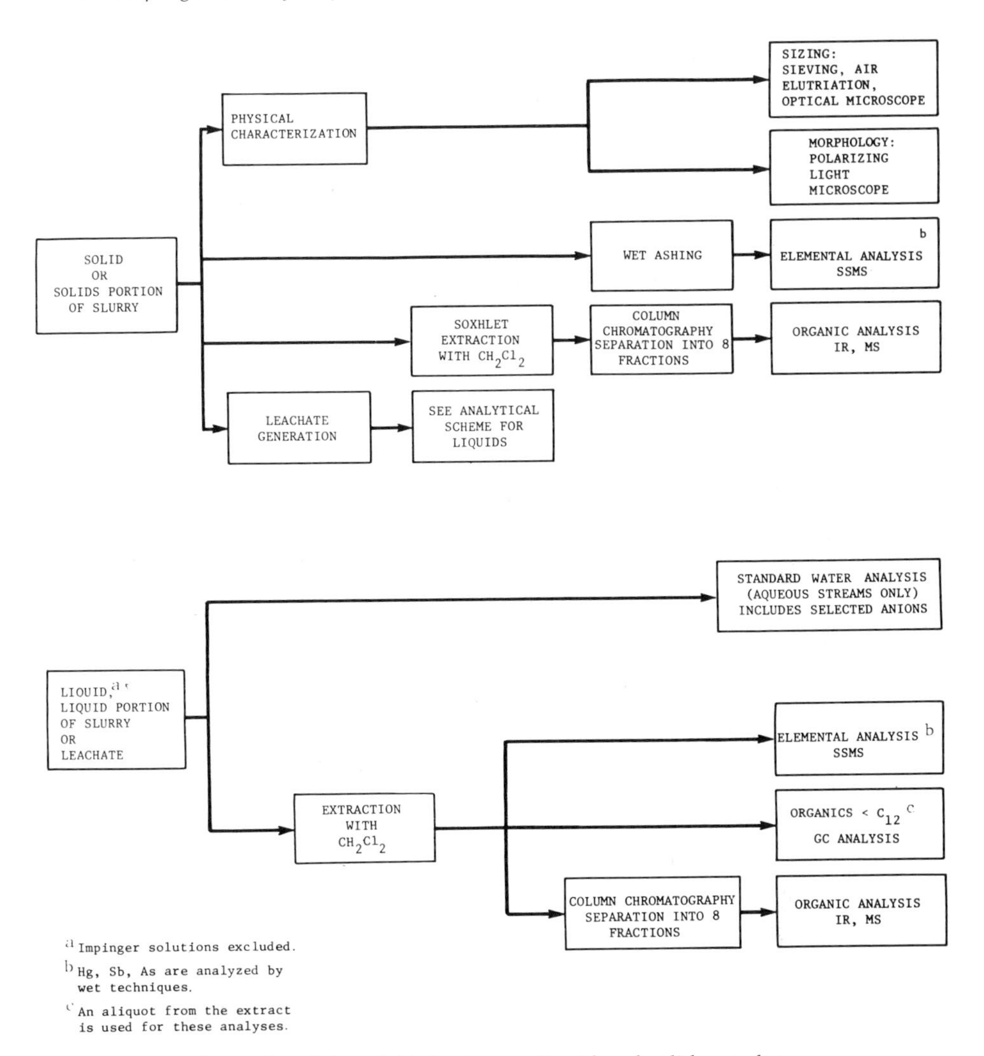

Fig. 3 Level 1 analytical scheme—liquid and solid samples.

(iii) gaseous species, including but not limited to SO_2, NO_x, CO, O_2, CO_2, N_2, H_2S, COS, NH_3, HCN, and $(CN)_2$,

(iv) pH, acidity, alkalinity, conductivity, BOD, COD, dissolved oxygen, dissolved solids, and suspended solids in aqueous streams, and

(v) leachable cations and anions from solid feeds and effluents.

Once the environmentally significant components (and streams) have been identified through screening, priorities can be established and a

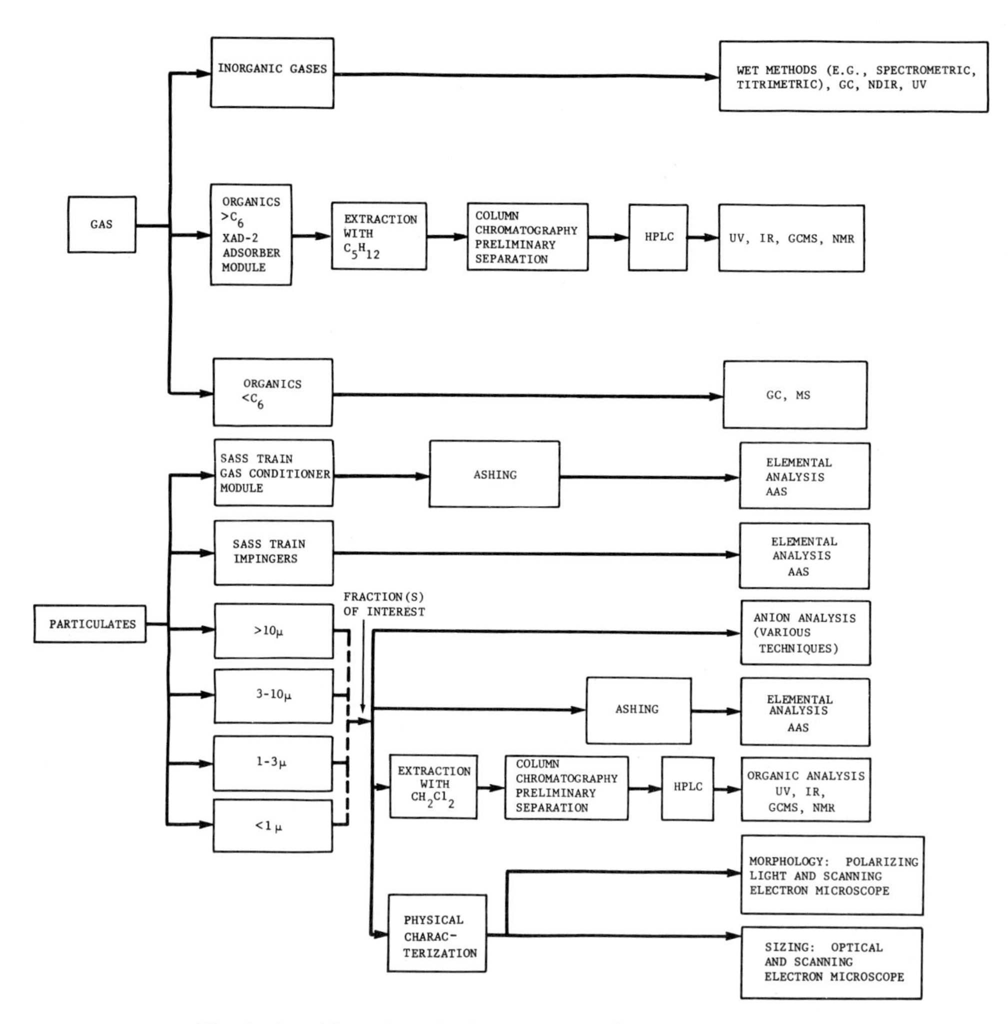

Fig. 4 Level 2 analytical scheme—gas and particulate samples.

plan devised for sampling and analysis in the follow-up (i.e., level 2) assessment phase.

The intent of level 2 analysis is to provide a more accurate, quantitative identification of specific components in selected streams, based on level 1 output. Many of the recommended methods have been chosen on an individual species basis. In addition, many of the selected techniques are standard EPA, ASTM, API, etc., methods. An accuracy and/or precision of ±10% or better has been specified for level 2 analysis.

B. Inorganic Gas Analysis

1. *Level 1*

For the level 1 effort, gas chromatography is recommended for the analysis of a variety of inorganic gaseous components of environmental interest. Specific analytical procedures are summarized in Table III. Sample concentrations are expected to range from parts per billion (ppb) levels for sulfur compounds to several percent for carbon dioxide, oxygen, and water vapor in the stack gas. Because the stability of many samples, particularly those containing hydrogen sulfide and other sulfur

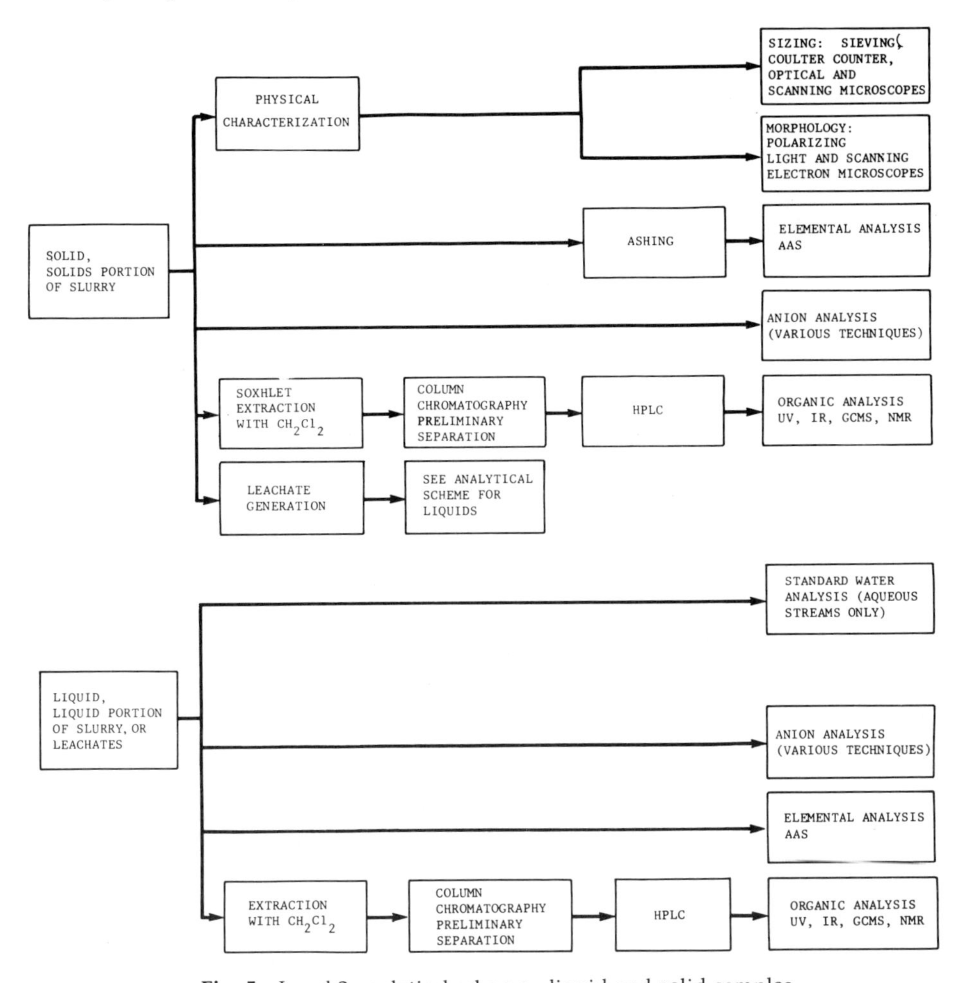

Fig. 5 Level 2 analytical scheme—liquid and solid samples.

TABLE III *Recommended Gas Chromatographic Procedures for Inorganic Gas Analysis*

Gaseous component(s)	Column/conditions	Detection limit	Reference
NH_3, HCN, $(CH)_2$	6 ft × 0.125 in. ss, Poropak Q (80/100) 40°C isothermal, thermal conductivity detector	25 ppm	ASTM (1972) (D 1426, D 2036)
H_2S, SO_2, and miscellaneous sulfur species (COS, CS_2, etc.)	18 in. × 0.125 in. Teflon, acetone-washed Poropak QS (80/100), 30°C for 1 min, program at 40°C/min to 210°C flame photometric detector	10 ppb	deSouza *et al.* (1975)
CO_2, CO, N_2, O_2, H_2O, NO_x	6 ft × 0.125 in. ss, molecular sieve 5A (80/100), isothermal at 40°C, thermal conductivity detector	25 ppm	Driscoll (1974)

species, is unknown (due to wall adsorption or possible chemical reaction), gas chromatographic (GC) analysis should be performed as rapidly as possible, in an on-site mobile laboratory. The displacement bomb containing the level 1 stack gas sample is attached to the gas chromatograph via an automatic gas sampling valve.

Although a specific list of inorganic species and proposed conditions is shown in Table III, this list does not exclude additional species which may be found. Indeed, it is not expected that the proposed set of conditions will be suitable for all mixtures and concentration ranges. When unique situations arise, appropriate analytical conditions must be selected.

Several gaseous species of environmental interest (NO, NO_2, halogens, hydrogen halides, mercury, and sulfur trioxide) are not measured routinely on a gas chromatograph. The level 1 analysis of NO/NO_2 concentrations is performed using the chemiluminescence method. Because a major portion of the analysis time associated with this technique involves instrument calibration, it is recommended that calibration sufficient to maintain a minimum accuracy factor of ± 2 $(+100\%, -50\%)$ be performed. Halogens, hydrogen halides, mercury, and sulfur trioxide are absorbed in the impingers of the SASS train and are analyzed by spark source mass spectrometery (SSMS) or flameless atomic absorbance spectroscopy (for mercury) at level 1.

2. *Level 2*

Recommended procedures for level 2 inorganic gas analysis are summarized in Table IV. In general, level 2 techniques are designed for

TABLE IV *Recommended Procedures for Level 2 Inorganic Gas Analysis*

Gaseous component(s)	Analytical method	Reference
NH_3	Nesslerization (0.05–1 ppm) or titration (>1 ppm) with 0.01 *N* H_2SO_4. NH_3 is absorbed in impingers containing 0.1 *N* H_2SO_4. The sample is buffered at pH 9.5 and the ammonia distilled into solution of boric acid	No reference
CO_2	Gas chromatography (GC) (see Table III)	Driscoll (1974)
	Nondispersive infrared spectroscopy (NDIR) is recommended for continuous measurement (>10 ppm)	
CO	GC (see Table III)	Driscoll (1974)
	NDIR is recommended for continuous measurement (>10 ppm)	Federal Register (1974)
HCl	Titration (>1 ppm) with $Hg(NO_3)_2$ and diphenyl carbazone/bromphenol blue indicator. HCl is absorbed in impingers containing distilled water	ASTM (1972) (D 512)
$HCN/(CN)_2$	GC (see Table III)	
	Spectrometric determination with chloramine-T (0.01–1 ppm) or titration with $AgNO_3$ (>1 ppm). HCN is absorbed in impingers containing 1 *N* KOH; sample is acidified and distilled into 1.25 *N* NaOH solution	ASTM (1972) (D 2036, D 1253)
HF	Specific ion electrode (SIE) determination. HF is absorbed in impingers containing distilled water and after buffering determined with fluoride SIE	No reference
H_2S	GC (see Table III)	deSouza *et al.* (1975)
Hg	Gold amalgamation technique. Hg is absorbed in impingers containing 10 g of gold chips. The chips are transferred to a furnace and Hg is determined by standard flameless atomic absorption spectroscopy (FAAS) techniques	Kalb and Baldeck (1972)
NO_x	Phenol disulfonic acid (PDS) method (EPA Method 7). NO_x is sampled with a 2-liter bomb containing 0.1 *N* H_2SO_4 and 0.1% H_2O_2 and determined spectrometrically	ASTM (1976) (D 1608)
	Chemiluminescent technique is recommended for continuous measurement	
O_2	GC (see Table III)	Driscoll (1974)
	Polarographic or paramagnetic techniques are recommended for continuous measurement	Holtzman (1976)

TABLE IV (*Continued*)

Gaseous component(s)	Analytical method	Reference
SO_2	GC (see Table III)	deSouza (1975)
	Titration with barium perchlorate (EPA Method 6). SO_2 is absorbed in impingers containing 3% H_2O_2	
	UV methods are recommended for continuous measurement	Driscoll (1974)
SO_3	Titration with barium perchlorate (EPA Method 6). SO_2 is condensed selectively using a specially designed apparatus (controlled condensation technique)	Driscoll (1974)
H_2O	Condensation–desiccation (EPA Method 4). H_2O is condensed in impingers in an ice bath with a final column of silica gel	Smith *et al.* (1974)
	NDIR methods are recommended for continuous measurement	
Miscellaneous sulfur species (COS, CS_2, etc.)	GC (see Table III)	deSouza (1975)

individual gaseous components and involve the use of special sampling trains. Included within these procedures, for several gaseous components (CO, CO_2, H_2O, O_2, SO_2), are spectroscopic methods (e.g., uv, NDIR) which provide continuous on-line measurements. The level 1 gas chromatographic techniques listed in Table III can be employed for level 2 analysis of CO, CO_2, HCN, H_2S, O_2, SO_2, and miscellaneous sulfur species. In this case, not only must the more stringent accuracy requirements of level 2 be met, but it must be demonstrated that the proposed GC procedures are free of interferences (i.e., identical retention times) from other species.

C. Organic Analysis

1. *Level 1*

The object of level 1 organic analysis is to provide an estimate of the predominant classes of organic compounds present in a given sample. Individual components within these classes will rarely be identified by level 1 procedures alone. Low resolution separation methods (gas and column chromatography) are used in level 1 and thus overlap between classes or organic compounds is to be expected. Sample preparation techniques for various sample types are shown in Fig. 6.

Unfortunately, the SASS train will not capture (and retain for analysis) organic compounds with boiling points in the C_1–C_6 range (≤70°C); volatile materials in the C_7–C_{12} range (100–200°C) are lost to varying degrees in the sample concentration steps required for analysis. Consequently, separate gas chromatography procedures are used for the analysis of these two ranges of materials.

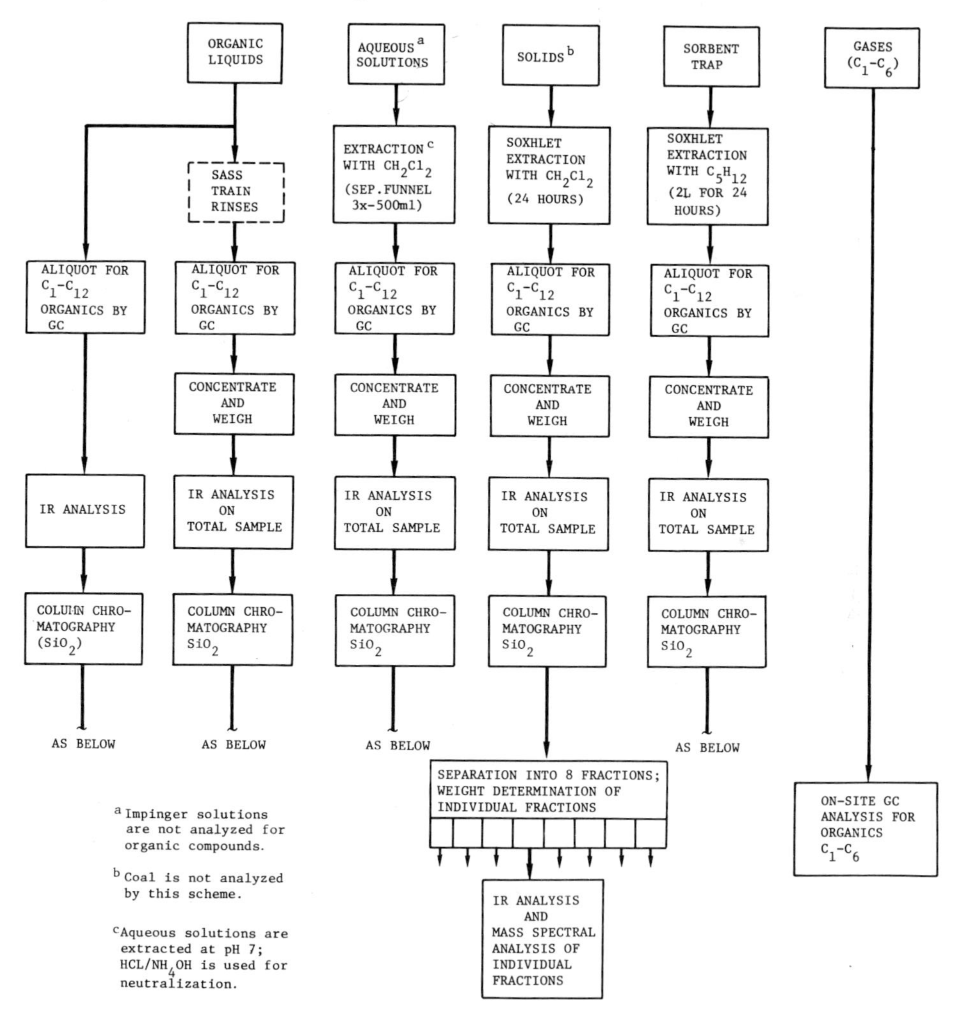

Fig. 6 Level 1 organic analysis scheme.

TABLE V *Recommended Procedures for Organic Analysis[a]: Level 1*

Analysis area	Analytical method
Volatile organic compounds: C_1–C_6 range	Gas chromatography, 6 ft × 0.125 in. ss, Poropak Q (100/120) 50° isothermal, flame ionization detector (FID)
Range b.p. (°C) 1 −160/−100 2 −100/−50 3 −50/0 4 0/30 5 30/60 6 60/90	A temperature calibration curve is prepared from pure samples of C_1–C_6 *n*-paraffins. Peaks observed in the chromatogram will generally represent mixtures of compounds within a given b.p. range rather than individual compounds. Unless additional evidence can be presented as to the specific identity of given peak, material and quantity should be reported as being present within a given range Peak areas may be calculated by a variety of methods: peak height, triangulation, integration, etc.
Organic compounds: C_7–C_{12} range	Gas chromatography, 6 ft × 0.125 in. ss, 1.5% OV 101 (or SE-30) on GasChrom Q (100/120), programmed temperature 50°C isothermal for 5 min then 10°C/min to 150°C final temperature
Range b.p. (°C) 7 90/110 8 110/140 9 140/160 10 160/180 11 180/200 12 200/220	A temperature calibration curve is prepared from pure samples of C_7–C_{12} *n*-paraffins. As the number of possible compounds in these ranges is very large, material and quantity should be reported as being present within a given bp range. Specific compound identification should be supported by additional data
Nonvolatile organic compounds (b.f. > 220°C)	Extraction and fractionation by column chromatography; class identification within each fraction using infrared spectroscopy and low resolution mass spectrometry (LRMS)[b] Nonvolatile organic compounds will be found in organic liquids (including washes), aqueous solutions, solids, and the XAD-2 sorbent trap. Organic liquid separation and analysis may be run neat. Washes (CH_2Cl_2/CH_3OH) are evaporated before separation and analysis. Aqueous solutions (10-liter sample) are extracted three times with 500 ml of CH_2Cl_2, evaporated, and weighed before separation and analysis. Solids are extracted for 24 hr with CH_2Cl_2 in a Soxhlet apparatus. The extract is concentrated and weighed before separation and analysis. The XAD-2 resin (a 2-g sample for inorganic analysis) is extracted with pentane in a large Soxhlet extraction apparatus Samples are prepared for separation by evaporation on a Kuderna–Danish apparatus such that 20–500 mg of sample is obtained (100 mg is optimum) Samples are separated on a 200 × 10.5 mm glass column (Teflon stopcock) packed with 6.0 g of freshly activated (2 hr at 100°C) silica gel (60/100 mesh, grade 950, Fisher Scientific)

TABLE V *(Continued)*

Analysis area	Analytical method
Nonvolatile organic compounds (b.f. > 220°C)	The evaporated sample is weighed in a glass weighing funnel and transferred quantitatively by mixing with 1 g of activated silica gel and placing this onto the column via the weighing funnel. The funnel is rinsed with 5 ml of pentane to complete the transfer. The column is eluted (~1 ml/min) with the following eluents:

Fraction number	Eluent	Volume collected (ml)
1	Pentane	25
2	20% CH_2Cl_2 in pentane	10
3	50% CH_2Cl_2 in pentane	10
4	CH_2Cl_2	10
5	5% CH_3OH in CH_2Cl_2	10
6	20% CH_3OH in CH_2Cl_2	10
7	50% CH_3OH in CH_2Cl	10
8	Conc. $HCl/CH_3OH/CH_2Cl_2$ (5:70:30)	10

After each fraction is collected, it is transferred to a tared aluminum weighing dish (a glass container is used for evaporation of fraction 8), evaporated, and weighed. The total sample and 8 fractions are analyzed by ir spectroscopy for functional groups present

Mass spectra are taken on each fraction which is deemed significant in terms of emissions

Volatile samples may be inserted into the gas inlet; less volatile samples are placed in the solid inlet probe. The probe is programmed to vaporize the sample, and spectra are recorded periodically. Spectra are obtained at 70 eV ionizing voltage, but a lower ionizing voltage (~15 eV) or a CI source, if available, may yield much useful information

[a] Procedures are compiled from Hamersma and Reynolds (1976).
[b] LRMS is defined as a mass spectrum with a resolution ($m/\Delta m$) of 1000.

A gas chromatograph employing a flame ionization detector is used for analysis of both volatile and nonvolatile organic compounds. Using this system, injection of a 1-μl liquid sample should give adequate response for even very dilute extracts. Detailed gas chromatographic requirements for analysis of the C_1–C_6 and C_7–C_{12} organic components are presented in Table V. The GC system should be calibrated for retention time and quantity with C_1–C_{12} n-paraffins. This system will simply be separating and analyzing mixtures of materials within a given boiling point range (and polarity in some cases). Since the chromatogram peaks will represent mixtures of materials present in a certain boiling range, rather than pure, individual compounds, it is recommended that material observed in the chromatogram be reported as being present within boiling point ranges. If, from other information, the identity of specific chromatographic peaks can be ascertained, individual organic compounds may be assigned to these peaks provided the basis of assignment is clearly stated.

While neat organic samples will be taken within the FBC sampling program, the majority of the samples, including the SASS train components, aqueous solutions, and solids, will require extraction with solvent prior to analysis. Both extracted samples and neat organic liquids pose particular problems in regard to solvent interference and specific GC requirements. Solvent interference is, at least in part, minimized by the initial isothermal portion of the temperature program which allows the solvent to elute prior to C_7 hydrocarbons. Neat organic liquids will probably require dilution with an appropriate solvent (e.g., CH_2Cl_2) to prevent overloading and degradation of the column.

Extraction of aqueous solutions should be carried out with methylene chloride using a standard separatory funnel fitted with a Teflon stopcock. Where necessary, ammonium hydroxide or hydrochloric acid is used to adjust the pH of the sample to pH 7 before extraction. Normally, three 50-ml methylene chloride extractions of 10-liter samples are sufficient.

All solid residues, particulates, and feeds, with the exception of coal, should be extracted for 24 hr with methylene chloride in a Soxhlet apparatus. The Soxhlet cup must be previously extracted with methylene chloride in order to avoid contamination. The sample is covered with a plug of glass wool during the extraction to avoid carryover of the sample.

The XAD-2 resin from the SASS trap (400 ml) is extracted with a large Soxhlet extraction apparatus (dumping volume of ~1500 ml). The resin is homogenized, and a 2-g portion set aside for inorganic analysis. The remaining resin is transferred to a previously cleaned extraction thimble and secured with a glass wool plug. Approximately 2 liters of *n*-pentane is added to a 3-liter reflux flask and the resin extracted for 24 hr. (The XAD-2 resin should not be extracted with methylene chloride because the compatibility of this resin with methylene chloride has not been fully evaluated.)

If large quantities of polar materials are extracted, they may precipitate in the reflux flask near the completion of the extraction. Addition of cool methylene chloride to the flask, after extraction is complete, will simplify the subsequent transfer and analysis steps.

Prior to separation by column chromatography, aliquots are taken from organic extracts or liquids for a GC and ir analysis of the C_7–C_{12} hydrocarbons. The ir analysis identifies functional groups in the sample (for an example, see Volume I, Chapter 17, Section II,E). All functional groups identified in the "total sample" must be accounted for in subsequent analyses. Prior to column chromatography, it is recommended that the solvent solutions be concentrated to a volume of 1–2 ml. The Kuderna–Danish apparatus is recommended for sample concentration of volumes less than 1 liter; a rotary evaporator is recommended for volumes which exceed this amount. All sample extracts and neat organic liquids are separated by the following chromatographic procedure if the sample quantity is adequate. A sample of 100 mg is preferred but smaller quantities (>15 mg) can be used. Sample extracts or organic liquids are separated by column chromatography on silica gel, using a solvent gradient series, into eight fractions. A detailed procedure is given in Table V and by Hamersma and Reynolds (1976) (see also Volume I, Chapter 16, Section II,C). Clearly this procedure is not a high resolution technique and consequently there is overlap in class type between many of the fractions. Fraction 1 contains alkanes and possibly some olefins. Fractions 2–4 contain predominately aromatic species. The smaller aromatics (e.g., benzene, naphthalene) will tend to elute in fraction 2 while the larger aromatics (e.g., benzpyrene) will probably elute in fractions 3 and 4. Some low polarity oxygen- and sulfur-containing species may also elute in fraction 4 but most of these will not elute until addition of methanol. Fractions 5–7 will contain polar species including phenols, alcohols, phthalates, amines, ketones, aldehydes, amides, etc. The distribution of class type between these fractions will, by and large, be a function of their polarity and affinity for the silica gel. Some weak acids may elute in fraction 7. Very polar species, primarily carboxylic acids and sulfonic acids, will elute in fraction 8.

After each fraction is collected, it should be transferred to a tared aluminum micro weighing dish for evaporation and gravimetric analysis. Fraction 8 should be dried in a glass container because of its hydrochloric acid content. Each fraction is subsequently analyzed by ir spectrophotometry and, when the quantity is sufficient, low resolution mass spectrometry (LRMS) (Table V). Infrared spectra are preferably obtained with a grating spectrophotometer on KBr salt plates using methylene chloride to transfer the sample to the plates. Sample quantity is adjusted so that the spectra maxima and minima lie between ~10 and 90% transmission, respectively. These spectra are interpreted in terms of functional groups present in each sample. Low resolution mass spectra (LRMS) are obtained on each of the eight fractions deemed to have sufficient quantity, in terms of source concentrations. For the various samples these quantities are gas (SASS train sorbent module), 0.5 mg/m^3; solids, 1.0 mg/kg; and aqueous solutions, 0.1 mg/liter. The mass spectrometer should preferably have a resolution ($m/\Delta m$) of 1000, a batch and direct probe inlet, a variable ionizing voltage source, and electron multiplier detection. Volatile samples are analyzed by insertion in the batch inlet. It is anticipated, however, that most samples will be introduced via the direct insertion probe. A small quantity of sample is placed in the probe capillary and inserted into the cool source. The temperature is then programmed to vaporize the sample. Spectra are recorded periodically throughout this period. Spectra will normally be obtained at 70 eV ionizing voltage, but low voltage (15 eV) or chemical ionization–mass spectroscopy (CI/MS) spectra may yield more useful data in some cases.

Interpretation of the spectra is guided by knowledge of the separation scheme, the ir spectra, and other information about the source. Data are grouped by homologous series based on a most probable structure assignment. Molecular ion series and fragment ions help to identify compound classes (e.g., polynuclear aromatic hydrocarbons are characterized by intense double ionization). Compilations of reference spectra will be useful in spectra interpretation.

2. *Level 2*

Level 2 organic analysis entails the identification and measurement of individual species present in environmentally significant fractions as determined by the level 1 effort. Without the benefit of level 1 results (i.e., a general knowledge of stream composition), it is difficult to specify exact analytical conditions for level 2. In general, level 2 analysis is an extension of level 1 separation techniques using GC and HPLC methods. A greater variety of methods are used for compound identification and quantification and include gas chromatography, infrared spectroscopy, NMR spectroscopy, and GC–mass spectroscopy (GC/MS).

Since level 1 GC procedures for organic compounds are designed only to separate components by boiling point range, a high resolution multipurpose surface coated open tubular (SCOT) column (OV101, 0.4 mm i.d. × 100 m, 2°C/min ambient to 200°C) is used for preliminary screening and identification of volatile species; if specific classes of compounds are suspected, appropriate supports may be chosen from the listing in Table VI.

For all other organic species, HPLC is used for initial separation. Using the information obtained from the level 1 organic analysis, which

TABLE VI *Organic Species Separable by Various GC Columns*

Class of compound	Column
Acids	
C_1–C_9	Chromosorb 101
C_1–C_{18}	FFAP
Alcohols	
C_1–C_5	Poropak Q, Chromosorb 101
C_1–C_{18}	Silar 5CP, Carbowax 20M, FFAP
Polyalcohols	FFAP
Aldehydes	
C_1–C_5	Poropak N, DC-550 Ethofat
C_5–C_{18}	Carbowax 20M, Silar 5CP Poropak Q, Poropak R
Amines	Chromosorb 103, Penwalt 223
Amides	Versamid 900, Igepal CO-630
Esters	Poropak Q, dinonylphthalate Chromosorb 101 or 102
Ethers	Carbowax 20M, Silar 5CP
Glycols	Chromosorb 107
Halides	OV-210, FFAP
Hydrocarbons	
C_5–C_{10}	OV-101, SE-30
Aromatic	Silar %CP, Carbowax 20M
Olefins $>C_6$	DC-550, DC-703
POM	Dexsil 300, OV-101, SE-30
Ketones	Poropak Q, Chromosorb 102, FFAP
Halogenated aromatics	OV-101, OV-225, OV-1, OV-17, SE-30
Phenols	OV-17, Silar 5CP, Carbowax 20M

at minimum provides identification of the major classes of species present, an effective HPLC separation scheme may be designed. Additionally it is possible, indeed likely, that from level 1 results only certain classes of organic compounds [e.g., polycyclic organic materials (POMs)] will be of interest, which further simplifies the separation scheme. For the purposes of this discussion, however, it is assumed that a complete analysis of the complex mixture is desired.

The complex mixture is first fractionated by gel permeation chromatography (GPC) with a previously calibrated column selected for the resolution of the various components by molecular weight. Since the sample is likely to contain relatively small molecules (molecular weights <500), supports with small pore sizes (100–500 Å) should be chosen. The GPC support most often used for sequential analysis is a semirigid styrene divinyl benzene polymer (STYROGEL, BIOBEADS, or equivalent). The column is calibrated by measuring the elution volume of a series of known compounds (preferably compounds of interest). This calibration is only approximate but is sufficient to allow division of the complex multicomponent mixture into manageable fractions. Further information on GPC is given in Volume I, Chapter 16, Section II,D.

These fractions are then separated by reverse phase partition gradient chromatography. Since all components of the GPC fraction are of approximately the same size, this procedure provides a means of separation by functional groups. To maximize column resolution and stability, a microparticle (5–10 μm) chemically bonded silicone polymer packing is recommended. The choice of gradient solvents is detector dependent but typically are water-modified with methanol, isopropanol, acetonitrile, or dioxane. The choice of sorbent and elution solvents is discussed in greater detail by Gurkin (1977).

After this initial reverse phase HPLC, each fraction should be screened to determine whether further separation by HPLC is necessary. When necessary, the same reverse phase column can be used either isocratically or with gradient elution under optimized conditions to further separate the sample. Normal bonded phase chromatography can also be used as an additional separation mode. Once individual compounds are sufficiently well resolved, the following techniques may be used for identification and quantification.

To carry out quantitative analysis by either GC or HPLC, it is necessary that the compounds of interest be reasonably well resolved from other peaks in the chromatogram. Additionally, it is highly desirable that pure samples of the compounds of interest be available for use as standards. (Where these compounds are not available, an approximation may be made by using their homologs.) Calibration is accomplished by preparing a known mixture of internal standard and compounds of interest, and obtaining chromatograms for varying amounts of the mixture. The response factor for each peak is preferably determined using an electronic integrator or computer integration routine. Each compound in the complex mixture is quantified using the previously determined response factors together with known amounts of internal standards. The accuracy and reproducibility of GC and HPLC quantification using these techniques are generally better than $\pm 10\%$. Reproducibility is readily determined at the same time as response ratio calibration.

Under certain circumstances, it is possible to analyze quantitatively for specific components where chromatogram peaks are poorly resolved. If specific detectors such as GC flame photometric detector (fpd) (sulfur specific filter) for sulfur compounds, GC electron capture detector (ecd) for halogenated species, or HPLC-fluorescence for polynuclear aromatic species are used, then detector response is very low for all but the components of interest and an interference-free chromatogram results.

Prior to analysis by GCMS, the complex mixture will have been separated by HPLC into a number of well-resolved fractions. The very nonpolar fractions, consisting of aliphatic hydrocarbons, are most,efficiently analyzed by GC alone. The most polar fractions from the HPLC separation scheme, because of their nonvolatility, are unsuitable for GC separations and should be characterized after further HPLC separation (into individual compounds) with ir, NMR, and high resolution mass spectroscopy. The HPLC fractions between these extremes will be identified by GCMS together with ir and NMR techniques.

Mass spectra may be obtained by electron impact ionization (EI) or by chemical ionization (CI); in the latter mode, sample ionization is accomplished by using an ionized reagent gas such as methane, isobutane, or ammonia. Extensive data files of EI spectra are readily available for spectral matching to aid in identification of unknown compounds. As yet, no comparable files exist for CI spectra and thus EI analysis is better suited to the analysis of complex organic mixtures.

Specific ion current integration is the basis for a very rapid quantification routine. This technique involves determining the ratio of the ion current of the compound of interest to that of an internal standard. Interferences can usually be avoided by using CI and by careful choice of the fragment ions used for quantification. With care, this procedure has an accuracy and reproducibility of better than $\pm 10\%$.

During the separation of complex organic mixtures, ir is used to provide functional group identification in such mixtures, monitor the course of separation, and ultimately identify the separated compounds. However, less information is obtained when very complex mixtures are analyzed as compared to simple mixtures or single compounds. To maximize the information content of an ir analysis of complex mixtures, use of a Fourier transform infrared system (FT-ir) is recommended where such a system is available. FT-ir, as compared to conventional dispersive infrared spectroscopy, has a number of advantages: faster scan speeds (as fast as 0.5 sec), a 30-fold enhancement of signal-to-noise (S/N) ratios, and a 10^2–10^3 increase in sensitivity. Additionally, the FT-ir dedicated computer offers several major data-handling advantages. Not only can absorption bands due to background materials be removed, but spectra can be added, subtracted, multiplied, and divided; thus, spectra may be adjusted in size and unwanted components removed from the spectra without the necessity of chemical separation.

Most experimental details are equally applicable to either FT-ir or conventional dispersive ir; thus, the following comments apply to both types of infrared spectroscopy. Samples will originate from HPLC fractionation as dilute solutions and thus necessitate solvent removal. The same methods of solvent removal and sample preparation used in level 1 are equally applicable to level 2. When possible, the thin film technique is preferred. Occasionally, a highly light scattering solid may be encountered which requires use of the pressed disk (KBr) technique. As separations proceed, fractions may approach a size where microsampling techniques are needed. The most universal method, and thus recommended, is the micropressed disk technique which offers greater sensitivity than the microfilm technique. Where the former is not applicable, the microfilm method is used. It may be emphasized that if FT-ir is used, these micro techniques may be unnecessary. Further details of these techniques are discussed by Potts (1963).

Infrared spectroscopy may also be used to examine fractions separated by gas chromatography. Formerly, GC-ir studies required trapping and collection of each GC peak. Not only was this time-consuming, but often several GC runs were required to collect sufficient material for analysis. With the availability of FT-ir, "on-line" GC-ir became a reality. Sample preparation for GC-ir is the same as for GC/MS and need not be repeated here.

Once an infrared spectrum is obtained, identification is made by matching the unknown spectrum with a reference spectrum from available reference libraries containing up to 150,000 spectra. When an exact match cannot be found, the functional group information together with MS and NMR data can be used to identify the unknown compound.

Nuclear magnetic resonance spectroscopy (NMR) may be used both as a screening technique for complex organic mixtures and as a quantification technique for individual compounds. The presence of various functional groups and identification of types of hydrocarbons are determined from the chemical shift of peaks in the spectrum. A comprehensive listing of chemical shifts for protons as well as a general review of NMR is given by Silverstein and Bassler (1967). NMR studies need not be limited to proton NMR. Using ^{13}C-NMR, functional groups within the carbon backbone may be observed directly. A discussion of ^{13}C-NMR techniques as well as a listing of chemical shifts is given by Strothers (1972). For additional details on ^{1}H- and ^{13}C-NMR, see Chapters 23 and 24.

A primary limitation of conventional continuous wave (cw) NMR is sample size. Milligram quantities are required for proton NMR and natural abundance ^{13}C-NMR is essentially impossible. These difficulties have been largely overcome by Fourier transform NMR. Using this technique, proton NMR requires a minimum of 10 μg of sample, whereas for ^{13}C-NMR milligram quantities are necessary. Another method for increasing sensitivity is using a time-averaging computer (CAT) in the cw mode; however, the time required per sample to achieve an equivalent *S/N* ratio is approximately two orders of magnitude greater than time averaging of FT scans.

Each sample fraction from HPLC is likely to be in solvent unsuitable for NMR (i.e., proton-containing solvents). Thus, the sample must first be separated from the separation and extraction solvents. Methods which may be used include lyophilization, evaporation with a Kuderna–Danish concentrator, evaporation with a stream of inert gas, or removal under vacuum at room temperature. With all of these methods, the principal problem is the possibility of loss of volatile sample components. Once the separation solvents are removed, the sample is redissolved in a suitable NMR solvent. Ideally, for ^{1}H-NMR the solvent should contain no protons; for ^{13}C-NMR it should contain either no carbon or only one type of carbon.

Chemical shifts are reported relative to a standard reference compound. The generally accepted reference for both ^{1}H- and ^{13}C-NMR is tetramethylsilane (TMS). Unfortunately, TMS is not soluble in aqueous solutions. The common reference for aqueous solutions is sodium 2,2-dimethyl-2-silapentane-5-sulfonate (DSS). (Peaks measured with respect to TMS in chloroform will be within a few hundredths of a part per million for the same peaks measured with respect to DSS in water.)

Nuclear magnetic resonance spectroscopy will be used for quantification in only two cases: (1) a single component isolated by the HPLC separation scheme or (2) compounds which are not easily measured by other techniques. Quantification may be accomplished either by using a relative ratio technique or by addition of a known amount of a standard whose resonance will not interfere with those of the sample. These procedures are discussed by Jones *et al.* (1976) and this reference should be consulted for further details.

D. Elemental Analysis

1. *Level 1*

Spark source mass spectroscopy (SSMS) is recommended for elemental analysis at level 1. For details on SSMS, reference should be made to Volume I, Chapter 14. Four specific groups of samples result from the level 1 survey: (1) XAD-2 trap, (2) aqueous samples, (3) organic samples (liquid or solid), and (4) particulate matter, including probe and cyclone washes, spent and raw sorbent, and ash samples. For analysis by SSMS, two general conditions must be met: (1) the sample, if it is not a conductor, must be placed into a conducting medium (graphite), and (2) the sample must be as free as possible from organic matter which can complicate spectra interpretation.

Aqueous samples are prepared by adding a small amount of the sample to powdered graphite and evaporating to dryness. (One milliliter of solution is needed to obtain a 1-$\mu g/l$ sensitivity assuming a basic SSMS sensitivity of 10^{-9} g.) The graphite is then pressed into an electrode.

Particulate matter, ash, and organic samples require oxidation of organic matter by the following procedure: Approximately 1 g of -200 mesh sample is transferred to a clean combustion crucible and weighed to the nearest 0.1 mg. Ten milliliters of 10% HNO_3 is transferred to a Parr bomb, the crucible placed in the electrode support of the bomb, and the fuse wire attached. The bomb is assembled and oxygen added to a pressure of 24 atm (gauge). The bomb is then placed in the calorimeter (cold water in a large stainless steel beaker is also satisfactory) and the sample ignited using safety precautions ordinarily employed in bomb calorimetry work. After combustion, the bomb should be left undisturbed for 10 min to allow temperature equilibration and the absorption of soluble vapors. The pressure is released slowly and the contents transferred to a beaker. Any residue remaining can be brought into solution by fusion techniques. A 1-ml aliquot of this sample is added to graphite, evaporated, and pressed into an electrode.

Analysis of the XAD-2 trap for trace elements is a unique problem because little is known about volatile element retention of this resin. Since adsorption is unlikely to be uniform throughout the length of the trap, the XAD-2 sorbent is first thoroughly mixed to ensure homogeneity, and then a 2-g portion of the sorbent is used for Parr bomb combustion over HNO_3. An aliquot of this sample is then formed into an electrode in the same manner as just described.

Sample homogeneity is of the utmost importance for solid inorganic samples, which are not dissolved prior to analysis. These samples should be reduced to less than 200 mesh in a micromill equipped with Stellite blades. (Alternatively, the sample may be ground with a boron carbide mortar and pestle.) The ground sample is blended with graphite (equal parts) and pressed into electrodes.

SSMS detection systems are of two general types: photographic plate and electrical detection. For level 1 elemental analysis, the photographic system using the "just disappearing line" technique is used. To achieve the highest sensitivity, a series of exposures of the photoplate is made with the sample and is compared to a series of exposures made with a reference sample. Precision and accuracy are highly dependent on spectral line widths and shapes. These parameters define optical densities which are converted to ion densities by means of calibration curves. A number of computer-oriented systems for the derivation and integration of ion intensity profiles have been developed for use in accurate and precise determinations.

While SSMS can, in theory, analyze any element, arsenic, antimony, and mercury are not determined reliably by SSMS; in addition, carbon, hydrogen, nitrogen, and oxygen are not commonly determined by SSMS. Thus, the former are determined by atomic absorption spectroscopy or wet techniques, while the latter group is determined by combustion methods. Table VII summarizes level 1 analytical procedures for all elements not measured by SSMS. Since several additional sample preparation steps are included in this scheme, care must be taken to avoid contamination. Blanks on all solutions, acids, and reagents must be run to ensure accurate and reproducible results.

2. *Level 2*

Level 2 analytical methods are summarized in Tables VIII and IX. Although atomic absorption spectroscopy (either flameless or conventional) is recommended for a majority of elements, it is not unreasonable that, depending on the total number of elements to be analyzed, SSMS might be used at level 2. To achieve the higher accuracy and precision which are required at level 2, a spark source mass spectrometer employing ion-sensitive multiplier phototubes as detectors, is required. With electrical detection, the precision of spark source mass spectroscopy rivals that obtainable by any other analytical method.

A useful modification of atomic absorption spectroscopy (AAS; see Volume I, Chapter 14 for a discussion of this) for the analysis of arsenic,

TABLE VII *Trace Element Analysis: Level 1. Analytical Methods Other than SSMS*

Element	Procedure	Reference
As	Atomic absorption spectroscopy	Pollock (1975)
C	Combustion/gravimetric determination	ASTM (1976)
H	Combustion/gravimetric determination	ASTM (1976)
Hg	Flameless atomic absorption	Pollock (1975)
N	Kjeldahl digestion/titrimetric determination	Dee *et al.* (1973)
O	Combustion/spectrometric	Kuch *et al.* (1967)
Sb	Atomic absorption spectroscopy	Pollock (1975)

TABLE VIII *Trace Element Analysis: Level 2. Atomic Absorption Spectroscopy*[a]

Element	Flame[b]	Analytical wavelength (Å)	Dissolution procedure[c]	Detection limit (μg/ml)	
				Flame	Nonflame
Al	N	3093	W,D	0.1	1×10^{-6}
Sb	A	2175	B	0.03	5×10^{-6}
As	A	1937	B	0.03	8×10^{-6}
Ba	N	5536	W,D	0.02	6×10^{-6}
Be	N	2349	W	0.002	3×10^{-8}
Bi	A	2231	W,D	0.04	4×10^{-6}
B	N	2497	W	3.0	2×10^{-4}
Cd	A	2288	W	0.001	8×10^{-8}
Ca	A	4227	W,D	0.002	4×10^{-2}
Cs	A	8521	D	0.05	4×10^{-7}
Cr	A	3579	W	0.002	2×10^{-6}
Co	A	2407	W,D	0.002	2×10^{-6}
Cu	A	3248	W	0.004	6×10^{-7}
Dy	N	4212	W	0.4	0.007
Er	N	4008	W	0.1	—
Eu	N	4594	W	0.2	0.02
Ga	A	2874	D	0.05	4×10^{-4}
Gd	N	3684	W	4	—
Ge	N	2652	B	0.1	3×10^{-6}
Au	N	2428	W	0.02	1×10^{-6}
Hf	N	3073	W	15	—
Ho	A	4104	W	0.3	—
In	A	3039	W,D	0.03	4×10^{-7}
Ir	N	2640	W	1	—
Fe	A	2483	W,D	0.004	1×10^{-5}
La	N	5501	W	2	—
Pb	A	2833	W	0.01	2×10^{-6}
Li	A	6708	W	0.001	3×10^{-6}
Lu	N	3312	W	3	—
Mg	A	2852	W	0.03	4×10^{-8}
Mn	A	2795	W	0.0008	2×10^{-7}
Hg	—	2537	B	—	2×10^{-5}
Mo	N	3133	W,D	0.03	3×10^{-6}
Nd	N	4634	W	1	—
Ni	A	2320	W,D	0.005	9×10^{-6}
No	N	3344	W	5	—
Os	A	2909	D	1	—
Pd	A	2476	W	0.01	4×10^{-6}
Pt	A	2659	W,D	0.05	1×10^{-5}
K	A	7655	W,D	0.003	4×10^{-5}
Pr	N	4591	W	10	—
Re	N	3460	W	1	—
Rh	A	3435	W	1	—
Rb	A	7800	W,D	0.005	1×10^{-6}

TABLE VIII *(Continued)*

Element	Flame[b]	Analytical wavelength (Å)	Dissolution procedure[c]	Detection limit (μg/ml)	
				Flame	Nonflame
Ru	A	3499	D	0.3	—
Sm	N	4297	D	5	—
Sc	N	3912	D	0.2	—
Se	A	1960	B	0.1	9×10^{-6}
Si	N	2516	W	0.1	5×10^{-6}
Ag	A	3281	W,D	0.001	1×10^{-7}
Na	A	5890	W,D	0.0008	1×10^{-7}
Sr	A	4607	W,D	0.005	1×10^{-6}
Ta	N	2715	W,D	3	—
Te	A	2143	B	0.05	3×10^{-3}
Tb	A	4326	W	0.05	1×10^{-6}
Tl	A	2768	W	0.02	1×10^{-6}
Tn	A	4106	W	1	—
Sn	A	2246	W	0.05	3×10^{-4}
Ti	N	3643	W	0.1	4×10^{-5}
W	N	4009	W	3	—
U	N	3514	W	30	—
V	N	3514	W	0.02	3×10^{-6}
Yb	N	3988	W	0.04	—
Y	N	4077	W	0.3	—
Zn	A	2138	W	0.001	3×10^{-8}
Zr	N	3601	D	5	—

[a] Table compiled from Winefordner (1976), Dean and Rains (1971), Angino and Billings (1967), and Slavin (1968).

[b] Fuel is C_2H_2; oxidant is either N_2O (N) or air (A).

[c] W, Wet ashing; D, dry ashing; B, oxygen bomb dissolution.

TABLE IX *Level 2. Analytical Methods Other Than AAS*

Element	Procedure	Reference
Br	Eschka fusion/spectrometric determination	ASTM (1972) (D 1246)
C	Combustion/gravimetric determination	ASTM (1972) (D 271)
Cl	Eschka fusion/titrimetric determination	ASTM (1972) (D 512)
F	Specific ion electrode determination	Thomas and Gluskoter (1974)
H	Combustion/gravimetric determination	ASTM (1972) (D 271)
I	Combustion/spectrometric determination	ASTM (1972) (D 1246)
N	Kjeldahl digestion/titrimetric determination	ASTM (1972) (D 512)
O	Combustion/spectrometric determination	Kuch *et al.* (1967)
P	Molybdovanado phosphate spectrometric	ASTM (1972) (D 2795)
S	Combustion/gravimetric determination	ASTM (1972) (D 271)

antimony, bismuth, germanium, tellurium, and tin is the hydride evolution technique, where volatile hydrides are generated and analyzed with a conventional AAS system (Pollock, 1975). The advantages of the hydride evolution technique are as follows: (1) Practically all matrix effects are eliminated since matrix materials are left behind; (2) very efficient use is made of the sample since the entire amount of the element being analyzed reaches the flame in a form suitable for efficient atomization; (3) the method is 50–200 times more sensitive than conventional AAS.

The selection of the correct dissolution step is critical to accurate analysis of trace elements. Elements such as sodium, copper, and nickel are easily picked up from the laboratory environment or reagents. Other elements, e.g., mercury and selenium, can be lost in the dissolution step. The dissolution procedure which exposes the sample to the least contamination without potential loss of volatile components should be used for each trace element. (Volatile element dissolution procedures have been discussed at level 1.)

Dry ashing is the simplest prior treatment for samples containing organic material and may be used where high temperature ashing is suitable (as noted in Table VI). A general procedure is as follows (Pollock, 1975): An appropriate amount of sample (1–2 g, −200 mesh) is weighed into a porcelain crucible and placed in a cold vented furnace. The furnace is brought to a temperature of 300°C for 0.5 hr, to 550°C for 0.5 hr, and to 850°C for 1.0 hr. The crucible is removed from the furnace, stirred, and returned to the furnace at 850°C for 1.0 hr with no venting.

The resultant ash is placed in a 100-ml Teflon beaker containing 5 ml of HF (conc.) and 15 ml of HNO_3 (conc.), dissolved by gentle warming, and evaporated until just dry. Distilled water and 1 ml of HNO_3 (conc.) are added to dissolve the salts, and this solution is transferred to a 100-ml volumetric flask. Distilled water is added to adjust the volume to 100 ml. This solution is transferred to a polyethylene bottle and preserved as a stock solution.

Wet digestive procedures are rapid and, in general, less susceptible to volatilization losses; the major disadvantage is the possibility of contamination from the large excess of reagents employed. The most commonly used acids for wet digestion are HNO_3, H_2SO_4, and $HClO_4$. Christian and Feldman (1970) report that a 3 : 1 : 1 mixture, respectively, of these acids dissolve their weight of most organic samples. Many samples, however, will contain large amounts of calcium and because of the dangers of the coprecipitation of trace elements as $CaSO_4$, sulfuric acid should be eliminated from the mixture.

A final method which has been found to be very effective for the dissolution of silicate minerals is described by Bernas (1968): Fifty milligrams of a representative −200 mesh size sample portion is transferred into a Teflon decomposition vessel. Aqua regia (0.5 ml) is added and the sample swirled to ensure thorough wetting. Hydrofluoric acid (3 ml 48%) is added and the vessel sealed. The crucible is placed in a drying oven for 30–40 min at 110°C. After cooling to ambient temperature, the decomposed sample solution is transferred to a polystyrene Spex vial (50 ml). Care should be taken to transfer quantitatively any precipitated metal fluorides which may have formed. The final volume should not exceed 10 ml. Boric acid (1.8 g) is added and stirred with a Teflon stirring bar to hasten the reaction. Upon addition of 5–10 ml of distilled water, any precipitated metal fluorides will dissolve. The solution is transferred to a 100-ml volumetric flask, adjusted to volume, and stored

TABLE X *Anion Analysis*

Species	Analytical method		Reference
	Level 1	Level 2	
NH_3	Reagent test kit	Spectrometric	ASTM (1972) (D 1426)
AsO_4^{3-}/AsO_3^{3-}	SSMS	Spectrometric	Pollock (1975)
Br^-	SSMS	Titrimetric	ASTM (1972) (D 1246)
CO_3^{2-}/HCO_3^-	Reagent test kit	Titrimetric	ASTM (1972) (D 513)
Cl^-	SSMS	Titrimetric	ASTM (1972) (D 512)
CN^-	Reagent test kit	Spectrometric	ASTM (1972) (D 2036)
F^-	SSMS	Specific ion electrode	Thomas and Gluskoter (1974)
I^-	SSMS	Spectrometric	ASTM (1972) (D 1246)
NO_3^-	Reagent test kit	Spectrometric	ASTM (1972) (D 992)
NO_2^-	Reagent test kit	Spectrometric	ASTM (1972) (D 1254)
PO_4^{3-}	Reagent test kit	Spectrometric	ASTM (1972) (D 515)
SO_3^{2-}	Reagent test kit	Titrimetric	ASTM (1972) (D 1339)
SO_4^{2-}	Reagent test kit	Gravimetric	ASTM (1972) (D 516)
S^{2-}	Reagent test kit	Spectrometric	ASTM (1972) (D 2579)
Water quality parameters			
Acidity/alkalinity	Reagent test kit	Titrimetric	ASTM (1972) (D 1007)
Biological oxygen demand (BOD)	Titrimetric	Titrimetric	APHA (1971)
Chemical oxygen demand (COD)	Titrimetric	Titrimetric	ASTM (1972) (D 1252)
Conductivity	Electrometric	Electrometric	APHA (1971)
Dissolved oxygen	Electrometric	Electrometric	ASTM (1972) (D 889, D 1589)
Total dissolved and suspended solids	Gravimetric	Gravimetric	ASTM (1972) (D 1069, D 1888)
pH	Indicator paper	Electrometric	ASTM (1972) (D 1293)

in a polyethylene container. The sample solution should not remain in contact with glass for longer than 2 hr.

E. Anion Analysis

Recommended procedures for anion analysis are presented in Table X. The list within this table does not pretend to cover all possible anions, but rather is intended to present analytical methods for the more common species. Included within this list are seven parameters germane to water quality.

At level 1, anionic species are analyzed using both spark source mass spectroscopy and reagent test kits for specific anions. The former procedure presupposes that each element capable of anion formation is present, *in toto,* in a single anionic form; thus, this can yield only upper limits of anion concentrations. Reagent test kits are used for analysis of the following species and parameters: (1) acidity/alkalinity, (2) ammonia, (3) carbonate, (4) cyanides, (5) nitrate/nitrite, (6) pH, (7) phosphate, (8) sulfate, and (9) sulfite. These kits, manufactured by Hach or Bausch and Lomb, use procedures that usually follow a modified and simplified version of standard methods. The reagents are encapsulated and stored in small plastic pillows in premeasured quantities. Upon addition of the reagent or reagents to the sample, component concentrations are determined colorimetrically or turbidimetrically using reference color disks or portable photometers. In some cases endpoint titrations are used. Although these methods are not as accurate as the standard laboratory procedures, they have sufficient accuracy to satisfy level 1 objectives.

At level 2, all analyses are performed in a laboratory using either standard methods, ASTM, or EPA procedures to provide the increased precision and accuracy required.

REFERENCES

Abelson, H. I., Löwenbach, W. (1977). "Procedures Manual for Environmental Assessment of Fluidized Bed Combustion Processes," EPA Rep. 600/7-77-009. Environ. Prot. Agency, Washington, D.C.

Acurex Corporation (1976). "Source Assessment Sampler: Preliminary Information." Acurex Corporation, Aerotherm Division, Mountain View, California.

American Public Health Association (APHA) (1971). "Standard Methods for the Examination of Waste and Wastewater," 13th Ed. Chicago, Illinois.

American Society for Testing and Materials (ASTM) (1972). "Annual Book of Standards." Philadelphia, Pennsylvania.

American Society for Testing and Materials (ASTM) (1976). "Annual Book of Standards. Philadelphia, Pennsylvania.

Angino, E. E., and Billings, G. K. (1967). "Atomic Absorption Spectroscopy in Geology." Elsevier, Amsterdam.

Bernas, B. (1968). *Anal. Chem.* **40,** 1682–1686.

Christian, G. D., and Feldman, F. J. (1970). "Atomic Absorption Spectroscopy." Wiley, New York.

Dean, J. A., and Rains, T. C. (1971). "Flame Emission and Atomic Absorption Spectrometry, Vol. 2: Components and Techniques." Dekker, New York.

Dee, A., Martens, H. H., Merrill, C. I., and Nakamura, J. J. (1973). *Anal. Chem.* **45,** 1477–1481.

deSouza, T. L. C., Lane, D. C., and Bhatia, S. P. (1975). *Anal. Chem.* **47,** 543–545.

Diehl, H., and Smith, G. F. (1959). *Talanta* **2,** 209–219.

Driscoll, J. N. (1974). "Flue Gas Monitoring Techniques: Manual Determination of Gaseous Pollutants." Ann Arbor Sci. Publ., Ann Arbor, Michigan.

Federal Register (1971). Standards of performance for new stationary sources. *Fed. Regist.* **36,** No. 247.

Federal Register (1974). Determination of CO from stationary sources. *Fed. Regist.* **39,** No. 47.

Federal Register (1975). Emission monitoring standards for new stationary sources. *Fed. Regist.* **40,** No. 194.

Gurkin, M. (1977). *Am. Lab.* **9,** 19–33.

Hamersma, J. W., and Reynolds, S. L. (1976). "Field Test Sampling Analytical Strategies and Implementation Cost Estimates: Coal Gasification and Flue Gas Desulfurization," EPA Rep. 600/2/76-093b. Washington, D.C.

Holtzman, J. L. (1976). Calibration of Oxygen Polography by the Depletion of Oxygen with Hypoxanthine–Xanthine Oxidase–Catalase. *Anal. Chem.* **48**(1), 229–230.

Jones, P. W., Graffeo, A. P., Detrich, R., Clarke, P. A., and Jakobsen, R. J. (1976). "Technical Manual for Analysis of Organic Materials in Process Streams," EPA Rep. 600/2/76-072. Washington, D.C.

Kalb, G. W., and Baldeck, C. (1972). "Development of the Gold Amalgamation Sampling and Analytical Procedures for Investigation of Mercury in Stack Gases." U.S. National Technical Information Service, PB report 210817 Springfield, Va.

Kuch, A. J., Andreatch, A. J., and Mohns, J. P. (1967). *Anal. Chem.* **39,** 1249–1254.

Monsanto Research Corporation (1976). "Technical Manual for Process Sampling Strategies for Organic Materials." U.S. National Technical Information Service PB report 256 696/6BE Springfield, Virginia.

Pollock, E. N. (1975). *Adv. Chem. Ser.* No. 141, 23–34.

Potts, W. J. (1963). "Chemical Infrared Spectroscopy, Vol. 1: Techniques." Wiley, New York.

Ruch, W. E. (1970). "Quantitative Analysis of Gaseous Pollutants." Ann Arbor Sci. Publ., Ann Arbor, Michigan.

Silverstein, R. M., and Bassler, G. C. (1967). "Spectrometric Identification of Organic Compounds." Wiley, New York.

Sittig, M. (1974). "Pollution Detection and Monitoring Handbook." Noyes Data, London.

Slavin, W. (1968). "Atomic Absorption Spectroscopy." Wiley (Interscience), New York.

Smith, F., Wagoner, D. E., and Nelson, A. C., Jr. (1974). "Guidelines for Development of Q.A. V.III—Determination of Moisture in Stack Gases," EPA Rep. 650/4/74-005c. Environ. Prot. Agency, Washington, D.C.

Stothers, J. B. (1972). "Carbon 13 NMR Spectroscopy." Academic Press, New York.

Thomas, J., and Gluskoter, H. J. (1974). *Anal. Chem.* **46,** 1321–1323.

Toggert, A. F. (1945). "Handbook of Mineral Dressing." Wiley, New York.

West, P. W., and Chiang, J. J. (1974). Spectrophotometric Determination of Atmospheric Acidity by means of the Displacement of the Equilibrium of Acid-Base Indicators. *J. Air Pollu. Control Assoc.* **24**(7), 671–673.

Winefordner, J. D., ed. (1976). "Trace Analysis—Spectroscopic Methods for Elements." Wiley, New York.

Chapter 37

Thermal Analysis of Coal and Coal Ashes

N. I. Voina *D. N. Todor*
BUILDING MATERIALS LABORATORY
INSTITUTUL DE CONSTRUCTII
BUCHAREST, ROMANIA

I. INTRODUCTION

Thermal analysis methods, in addition to other instrumental analysis methods, play an important role in the investigation of useful mineral substances. Their application to the study of coals and coal products has increased considerably in the last two decades, and these methods are now being introduced into current laboratory practice. The basis of

ISBN 0-12-399902-2

these methods lies in the physical and chemical transformations which take place in a solid substance when heated. These transformations are closely related to the chemical nature of the substance and to its crystalline structure, and these transformations are recorded as graphs, called thermal curves.

With regard to the application of these methods to the analysis of mineral products, two thermoanalytical methods have been developed since the beginning of the century: first, differential thermal analysis (DTA), and second, thermogravimetry (TG). Both methods have also been applied to coal analysis, but the thermoanalytic results were often treated independently and only rarely were the two methods treated together (Grimshow and Roberts, 1957).

Modern apparatus is often complex enough to embody the two methods together and even include other methods such as derivative thermogravimetry (DTG) in the same instrumental assembly; hence, the applicability of thermal analysis methods has spread considerably. This is due to the fact that the study of one phenomenon by several methods combined can provide, through comparison of the results, much richer information than the analysis of the phenomenon in separate installations. Specialized literature gives more complete details along this line, denoting a larger area of applicability, and includes the works by Bátor and Weltner (1965), Buzágh-Gere and Gál (1974), Heilpern (1974), Hofmann and Garstka (1965), Holowiecki and Chodynski (1974), Todor (1976), and Weltner (1958, 1959a,b, 1961, 1962, 1965, 1966, 1969).

II. BASIC PRINCIPLES OF THERMAL ANALYSIS METHODS: DTA, TG, and DTG

Thermal analyses are those instrumental dynamic analysis methods that monitor the physical and chemical transformations which take place in the structure of a substance being heated or cooled. On this principle, a large array of instrumental methods has been developed based on variations in mass, volume, and temperature between the sample under analysis and a thermally inert substance.

Among previously mentioned methods, those that have given the most encouraging results in compositional analysis are DTA, TG, and DTG. At the present stage, these methods are used either separately or together in a complex instrumental assembly which records the three thermal curves and the oven temperature rise simultaneously for the same sample. Here we shall only present the basic characteristics of the three thermoanalytic methods, and the recommendations on the presentation of thermal curves made by the International Confederation of Thermal Analysis (ICTA).

A. Differential Thermal Analysis

This method covers those techniques which record the temperature difference between a substance and a thermally inert material when the two substances are undergoing identical temperature changes within an environment which is heated or cooled in a controlled ratio. The method was developed following the perfection of thermocouples as precise temperature gauges.

In practice, the differential thermal analysis method involves the simultaneous continuous recording of temperature (T) which exists in the oven, or best in the sample and of the temperature difference (ΔT) which appears between the sample and the thermally inert material. Ideally, the temperature difference ΔT must be recorded with a uniform speed proportional to the temperature of the sample or the inert material. The temperature difference between the sample and the inert material is recorded with a differential thermocouple pair in which one thermocouple is set in the sample and the other in the inert material.

When the heat flux is the same in the oven, the sample, and the inert material, and hence the temperature difference is zero, the recording instrument gives the so-called baseline $\Delta T = 0$. If a phase undergoes change or a decomposition reaction takes place in the sample, with loss or gain of heat, then the temperature gradient with respect to the inert material will change, and the temperature variation will be recorded as a deviation from the baseline. The direction of this deviation is determined by the temperature gradient between the sample and the inert material and indicates the nature of the thermal process. These transformations which take place in the sample, implying an endo- or exothermic process, can lead to negative or positive temperature differences, $\Delta T \neq 0$.

Any physical or chemical reaction caused by temperature gives rise to a maximum on the temperature versus time curve, $\Delta T = f(t)$, and from this maximum it is possible to draw information regarding the temperature and the transformation rate. The curve recorded is called the DTA curve; the temperature difference is given as ΔT on the ordinates, with endothermic phenomena shown downwards and exothermic phenomena upwards, whereas time or temperature is given on the abscissa, increasing from left to right.

The differential thermal analysis method seems at first ideal for the investigation of solids due to its simplicity and rapidity. The method requires only an oven, a fixture for the samples, and a set of thermocouples which allow the measurement of oven temperature and of the differential temperature in the thermoanalytical system. In practice, however, serious complications may arise since the results are affected by some factors related to the apparatus construction, the working schedule, and also the physical and chemical nature of the sample. When literature data are correlated, these factors should necessarily be considered.

B. Thermogravimetry†

This thermoanalytic method covers those techniques in which the weight of the sample is recorded as a function of time or temperature [$m = f(t \text{ or } T)$] while the sample is being heated or cooled at a constant rate in a given environment. This method was developed from the classical method of step-by-step heating and weighing of a solid product. Hence the instrumental setup comprises a set of thermoscales, but the recording of weight variations takes place automatically, thus yielding the curve TG as a function of time [$m = f(t)$]. The TG curve generated by automatic recording has a correct value only when the temperature rise in the oven is the only time constant.

On the basis of these curves one can establish how the sample weight is modified under the action of thermal energy. In general, such a curve has three regions: the region of weight increase; the region of weight decrease; and the region of constant weight (horizontal).

For thermoanalytic practice, especially when monitoring the stability of a compound, the horizontal regions of the TG curve are the most significant and give all the necessary information for a correct thermal treatment. However, the practical use of this method can have major shortcomings, e.g., when two reactions take place closely, overlapping each other over the same temperature range, or when reaction rates are different. In these instances the method becomes uncertain and the interpretation of TG curves is cumbersome and inexact over some temperature ranges. Because of this, the analytical use of the TG curve is generally restricted to the determination of mass variations between initial and final stages.

C. Derivative Thermogravimetry

This thermoanalytic method covers those techniques which record experimentally the first derivative with respect to time of the mass varia-

† Thermogravimetric analysis of chars is described in Chapter 32, Section II,A.

tions [$dm/dt = f'(t)$]. In other words, this method is closely related to thermal gravimetry, since it records the derivative of the curve $m = f(T)$ versus time. The experimental apparatus for obtaining the derivative was devised by Erdey *et al.* (1954) and Waters (1956), and is based on magnetic induction. In general, this is constructed by substituting the other arm of the thermal balance by a coil with a large number of turns. The coil is placed in the homogeneous field of a permanent magnet. When the thermal balance leaves the equilibrium position due to some thermal process taking place inside the specimen, the induction coil crosses the magnetic field lines. The resulting induction current is proportional to the displacement rate of the balance.

The graphical record of this method is the derivative thermogravimetric curve (DTG). The derivative of weight variations must be represented on the ordinates, and time or temperature on the abscissa. Since the method measures weight change rates, the area below the curve represents the total weight change that took place. For analytical interpretations the DTG curve will be the basis for a TG curve, allowing precise determination of mass changes during a thermal process.

D. Multiple and Simultaneous Thermal Analysis (Thermal Derivatography)

Thermal derivatography refers to joining several instrumental techniques for thermal analysis in one self-contained instrumental assembly.

The first instrumental installation of this type was devised by Paulik *et al.* (1958) who called the instrument a *derivatograph* and the method *derivatography*. These definitions were not accepted by the ICTA on the basis of possible confusion with other existing technical terms, so the name of "combined thermal analysis methods" was proposed. At the present time both terms are in use.

Usually such instruments simultaneously measure specimen temperature, the temperature of the thermally inert substance, oven temperature, temperature difference between specimen and the thermally inert substance (DTA curve), specimen weight change function of temperature (TG curve), and the weight change rate (DTG curve), on the same specimen and at the same time. Hence the amount of analytical information is more extensive. In some cases such apparatus is coupled with a gas chromatograph which detects and quantitatively determines the gaseous products resulting from thermal decomposition.

The idea of simultaneous recording, at the same time and on the same specimen, of several thermoanalytical variables was born of necessity.

Thus the three curves are obtained simultaneously and the interpretation of results can be done more conclusively. Whether the DTA curve is identical or different from the DTG and TG curves, there are several ways of interpreting the thermal phenomena. In general, for such an interpretation there are three distinct cases as follows:

(i) When thermal changes take place only on the DTA curve, whereas DTG and TG show no weight change, one can conclude that during heating the specimen suffered only physical changes, or that the chemical reactions took place without weight changes.

(ii) When (a) the DTA curve and (b) the DTG and TG curves are different in shape and development it means that the thermal effects of two or more chemical reactions were recorded, or that some chemical reactions were superimposed on the physical phenomena. In this case the DTG and TG curves record either one or more of these reactions, but only those which took place with weight changes, whereas the DTA curve records the total sum of the chemical reactions and the physical transformations.

(iii) When all three curves indicate chemical changes up to a certain temperature, and thereafter show a constant, i.e., a horizontal plateau, then the products of the chemical reaction are stable.

As a practical example we shall discuss the thermal curves of an ash specimen obtained experimentally in the laboratory through calcination at 1000°C in open air of a sample of carbonated lignite. After calcination the resulting ash was kept in a medium of saturated steam and finally dried at ambient temperature. In Fig. 1 are given the thermal curves of both the ash (denoted by a) and also the curves obtained with lignite in a closed environment (denoted by b). In the experimental case given for the TG curve, the possible total mass loss obtained with ash in a closed medium represents 34.60%. Through graphical interpretation of the DTG and TG curves one can deduce the moisture content, the amount of hydrated calcium oxide, and the amount of carbonate. Thus the total height from A to D corresponds to the total mass loss (34.60%), but each segment represents mass loss for certain components. The segment AB represents the specimen moisture (3.61%) and segment BC represents elimination of the OH groups as water molecules. From hydrated calcium oxide, water loss in this way is 25.46%. Finally, segment CD represents the thermal decomposition of calcium carbonate which takes place through a 5.53% loss of CO_2. From the fraction of the total weight lost through elimination of the OH groups from the calcium hydroxide as water one can deduce the total quantity of hydrated calcium oxide, which is equal to the sum of the amount of dehydrated calcium oxide

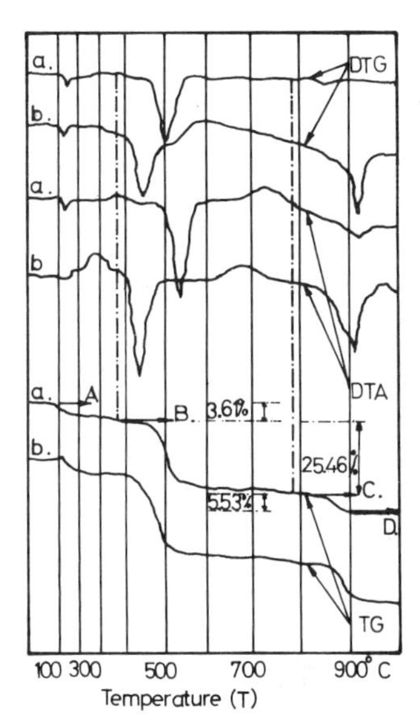

Fig. 1 Thermal analysis curves for (a) ash from lignite and (b) lignite samples.

and the amount of water. [On the same principle one can deduce the quantity of $CaCO_3$ formed in the ash through the action of CO_2 on $Ca(OH)_2$.]

On the DTA curve one can also notice two exothermal effects. The first takes place at about 370°C and represents a polymorphic transformation of the iron oxides; the second takes place between 700 and 800°C and represents a structural reorganization of metasilicates in the presence of calcium oxide.

III. PREPARING THE SPECIMENS FOR ANALYSIS

A. The Usual Methods of Preparation

For the study of coals through thermal methods, an important role is played by the way in which specimens are prepared for analysis. First, we should indicate that preparation of the specimen is closely connected to the aim of the analysis. In general, the specimens need only be ground to a fine granulation after a previous mixing and uniformization which give a representative sample.

In many cases a special preparation of the specimens is used, either through physical or chemical treatments. These special preparations have the aim of better component identification and of revealing some processes that take place during heating, processes which would otherwise be masked.

Coals are natural materials comprising two major substance groups, organic and inorganic, and hence coals raise special problems in thermal analysis. This difficulty manifests itself as interferences due to oxidation of organic substances, on one hand, and structural transformation of inorganic substances, on the other hand. In many cases a special preparation of the specimen is used, in order to lessen these interferences. For a better understanding of such preparations, for the purpose of determining the mineral composition of the specimen, we shall give an example using bituminous coal.

First, the specimen is brought to total uniformity, preparing a representative sample. An amount of the sample is ground to a high degree of fineness, until the granulation cannot be detected by touch. The first determination is performed on this specimen and yields the so-called thermal curves of the untreated specimen (Fig. 2a).

Examination of these thermal curves shows that the thermal phenomena recorded are not too conclusive. In general, one notices that between 50 and 200°C the specimen has undergone the first transformations. They are caused (1) by the elimination of the water due to moisture and the water absorbed in the mineral structure, and (2) by the incipient volatilization and oxidation of several organic compounds.

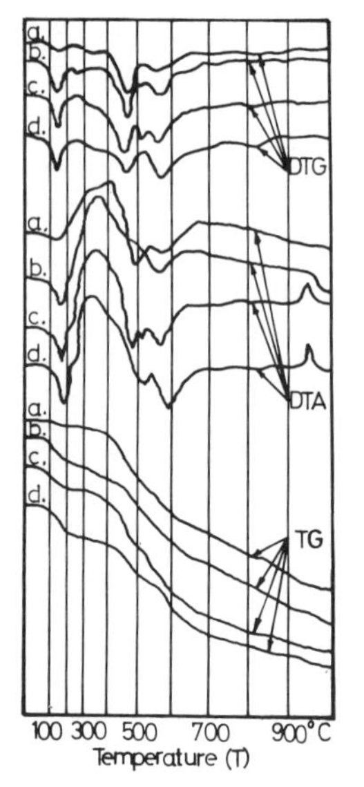

Fig. 2 Thermal analysis curves for bituminous coal samples prepared in different ways: (a) untreated specimen; (b) specimen treated with hydrogen peroxide; (c) specimen after extraction with chloroform; (d) specimen treated with hydrochloric acid solution (2%).

The pyrolysis process is continued and is very pronounced in the range 250–450°C. The thermal curves also show a set of thermal effects, but these are largely masked by the exothermal effect arising from oxidation of organic substances. A conclusive proof of this masking is the absence of the exothermic effect of structural reorganization of kaolinite, which takes place at 950–970°C.

B. Special Preparation of Specimens through Physical and Chemical Treatments

To eliminate the effect of oxidation of organic substances, a part of the initial specimen was treated repeatedly with warm hydrogen peroxide in a water bath. After filtering, washing, and drying at room temperature, parts of the specimen were thermally analyzed (Fig. 2b).

It can be observed on the thermal curves that this treatment has eliminated the exothermic effect of oxidation of organic substances at 100–200°C. In this temperature range one can see a pronounced endothermic effect, specific to clay minerals. The noticeable oxidation of organic substances starts now at ~250°C, at which point on the thermal curve the oxidation is similar to that of the untreated specimen.

Similar and even more conclusive results were obtained when the initial sample was subjected to the extraction of organic substances with heated chloroform in a Soxhlet-type apparatus (Fig. 2c). In this case the thermal effects of elimination of OH groups from illite, kaolinite, and even pyrophyllite appear very well marked, as is the effect of thermal decomposition of siderite and structural reorganization of kaolinite.

Finally, another specimen was analyzed which was treated with weak hydrochloric acid solution (2%), the acid treatment being done on the specimen resulting from extraction with solvent. It was noticed (Fig. 2d) that the endothermal effect completely disappeared from the three d curves in the range 450–600°C, and this indicated the presence of siderite in the specimen of bituminous coal.

The foregoing example is intended to show the reader how varied the methodology of preparing samples for analysis can be and how necessary it is for the method to be followed closely. In general, this methodology is closely related to the analytical system, specifically to whether the analysis is performed in an inert environment (without combustion) or in an oxidizing environment (with combustion). The starting point in choosing the direction of the preparation methodology lies in the aspect of the thermal curves of the untreated sample, i.e., the sample as obtained from the mine.

IV. THERMOANALYTICAL METHODOLOGY FOR COALS

A. Factors Which Influence the Results of Thermal Analysis

As previously shown, in the methods of thermal analysis one monitors both the temperature difference between the specimen and a thermally inert substance and the weight changes which take place with temperature, and their rate. In order that the results be true and reproducible it is necessary to provide identical experimental conditions.

The success and the reproductibility of the thermal curves are closely conditioned by several experimental factors which influence the shape and the temperatures characteristic of the thermal effects that appear during heating. Experimental observation has shown that the temperatures at which certain physical transformations and chemical reactions take place, especially the temperatures at the beginning and ending of these transformations, are very much affected by some experimental factors. Exceptions are the melting points of some pure substances.

Arens (1951) suggested several experimental factors which should be considered in a differential thermal analysis. These factors have been discussed and generalized for all methods of thermal analysis by MacKenzie (1957), Duval (1963), and Todor (1976). Briefly, these factors are the rate of oven temperature rise, the geometric shape of the specimen receptacle, the intimate nature of the specimen, the places at which oven temperature and the differential temperature are picked up, the nature and the properties of the thermocouple, the density and the compactness of the specimen, the effect of the specimen being covered or uncovered, the oven temperature, and others.

For the thermoanalytical results to be reproducible and comparable, these factors must be maintained as constant as possible. Furthermore, when experimental results are given in a report it is necessary that all the experimental data be included to make a critical appraisal possible.

B. Interpreting the Thermal Analysis of Coals in Open and Closed Environments

In the thermal analysis of coals the oven environment has a large influence on the way reactions take place. By oven environment we mean the space inside the oven in which the gases resulting from the chemical processes caused by heating can accumulate. Actually the oven environment varies continuously. Although at the start of the experiment it has the gaseous composition that was prescribed, its composition varies during heating due to the gases produced by the chemical reactions which take place.

If the initial environment in the oven can be controlled and prescribed regarding its composition and pressure, the environment created during heating is difficult to control. The variations in composition and pressure inside the oven have a direct influence on the development of chemical reactions. The thermal decomposition of any product takes place slowly, and at any temperature there is a certain pressure of the gases liberated during the process. This produces a change in the oven environment, and results in changing the decomposition temperature. If the oven accumulates sufficient amounts of gases, we can expect a slowdown and even stoppage of the reaction, which is pushed toward higher temperatures. Because of this, several environments have been tried, such as N_2, O_2, SO_2, air, rare gases, and pressures other than atmospheric.

The gaseous environment which is created inside the specimen due to some thermal process, e.g., dehydration, oxidation, or decomposition, has an influence that is difficult to record. This influence can be explained because the rate of gas formation inside the specimen leads to a rise in pressure with a rise in temperature. If the amount of gases produced is larger than that eliminated through the pores of the specimen, there will take place an accumulation of gases inside the specimen which will slow down the reaction, and hence the decomposition.

The accumulation of gases inside the specimen is also positively influenced by some experimental factors, e.g., surface condition, density, and compactness of the specimen, the amount of substance in the specimen, and the rise rate of the oven temperature. Actually, the gases resulting from decomposition during heating accumulate not only in the space between particles but also in an even larger amount inside the particles. Thus the volatile product must penetrate through a layer of solid material, and hence the rate of thermal decomposition is influenced by the particle size.

Reconsidering the experimental case described in Section III,A, one notices (Fig. 2) that whether or not the specimen is treated, up to 1000°C some amount of organic substances is still present in the specimen, especially carbon, which has not yet been oxidized. This is still present in the specimen for the reasons discussed earlier. To eliminate this impediment, and when knowledge of the ash content of the coal specimen is required, some have used a different type of receptacle in the form of stacked saucers (Fig. 3). On each of these saucers there is a layer of specimen at most 2 mm thick. In this case the organic substances are completely oxidized at 850°C (Fig. 4).

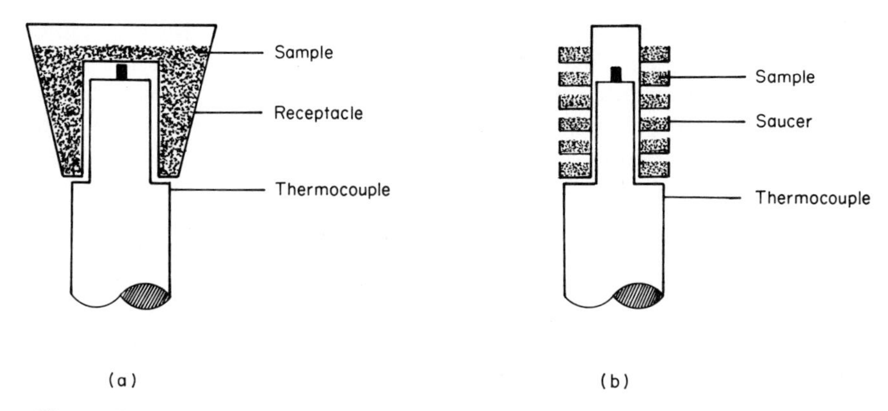

Fig. 3 Comparison of receptacles: (a) classic; (b) in the form of stacked saucers (Paulik system).

In general the thermoanalytical methodology adopted for the analysis is closely related to the purpose in mind. Through this type of analysis in most cases one aims at determining either the mineralogic composition of the compounds which generate the ash or only the ash content. In both cases determinations take place in an open and oxidizing system. In some cases, when the instrumental apparatus allows working in an inert environment, one can determine the cokable content. Two measurements are required, the first in an oxidizing environment to determine the ash content by weight measurements, the second in an inert or closed environment to determine the amount of coke plus ash, also by weight measurements. The coke content is obtained from the difference of the two measurements. Naturally, one can also act inversely .by measuring first in an inert or closed environment, and after the specimen has cooled down, to remeasure in an oxidizing or open environment, using the same specimen. Irregardless of which method is chosen, the measurements must be conducted at the same heating rate, up to 1000–1200°C, maintaining all other experimental parameters constant. In Fig. 5 is given such a measurement for a hard coal specimen of ash content 7.7%.

The thermal curves show that up to 200°C the specimen loses water and possibly a small content of volatile organic substances. The TG curve represents for both measurements in this temperature range, a mass loss of ~22%. As the temperature rises the thermal curves of the two separate measurements start to differ more and more. The thermal curves obtained in the closed system (Fig. 5a) show an exothermal effect at ~420°C which is accompanied by a mass loss, and followed by another effect of identical nature, but of smaller amplitude, at ~700°C.

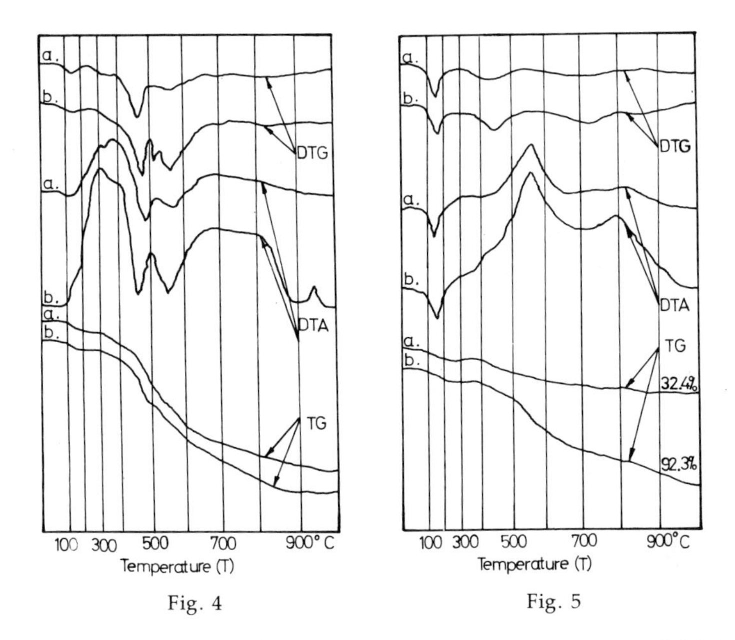

Fig. 4

Fig. 5

Fig. 4 Thermal analysis curves for bituminous coal samples obtained (a) in classic receptacles and (b) in receptacles in the form of stacked saucers.

Fig. 5 Thermal analysis curves for hard coal samples: (a) heated in a closed system; (b) heated in an open system.

The total mass loss at 1000°C is 32.4% of the initial specimen mass, and is caused by eliminating both water and some volatile products from the organic structure. The remaining 77.6% is composed of carbon and ash, representing the actual cokable part of the specimen.

When an open system is used, the oxidation effects of organic substances are very pronounced (Fig. 5b) over the entire temperature range. The mass loss is 92.3%; a reddish residue is left which represents the ash content of the sample (7.7%). The difference between the two measurements allows the calculation of the combustible part of the coke.

From the preceding, one can easily conclude that the thermoanalytical methodology for coal is closely related to the aim in mind. The thermal curves can be interpreted from several viewpoints, but the purely analytical interpretation can be a common starting point. Thus, with the DTA curve one can determine (1) the temperature difference between the sample and the thermally inert substance, which shows if the thermal process took place with heat loss or heat gain, and (2) the characteristic temperature of the thermal process. Additionally, the DTA curve can be used in the analytical interpretation of the TG and DTG curves in order to determine the thermal effects when it is necessary to find the absolute value of the heat change. To calculate the absolute value of the temperature difference ΔT between specimen and thermally inert substance, one can use the area of the thermal effect between its two sides and the baseline for the DTA curve.

Through the analytical interpretation of TG and DTG curves one can determine the mass change during a chemical process. These curves also allow one to establish the reaction rate, the temperature range in which the reaction takes place, and, finally, to determine quantitatively the compounds which generated these reactions.

In the specialized literature numerous procedures are described for the analytical interpretation of the thermal curves, the working methodology being quite varied. Hence we shall only list some bibliographical references: Soule (1952), Barshad (1952), Rey and Kostomarof (1959), MacKenzie (1957), Grimshow and Roberts (1953), Berg and Egunov (1969), Paulik *et al.* (1958), Horowitz and Metzger (1963), and Todor (1972, 1976).

V. THERMAL ANALYSIS OF COAL ASHES

A. Particular Features of Ashes

Ashes are industrial waste which results from coal being burned in thermoelectric stations. The ash is a complex material in texture, microstructure, chemical and mineralogical composition, specific mass, and so on. Hence the disperse heterogeneous ash properties are also a superposition of several factors, among which we mention the nature, composition, and purity of the coal, the type and time of burning, the cooling rate, among others. The noncombustibles present in coal as clay, phosphate, mica, calcite, dolomite, and quartz, undergo physical and chemical changes during the burning process.

In thermoelectric stations coal is burned in powder form in a gas flow. In this situation, and at certain temperatures, the chemical reactions take place on individual granules or on partially agglomerated granules, resulting in fly ash and a mixture of ash and boiler slag. The fly ash is collected by the electrofilters and is directed either toward its utilization or toward a hydraulic outlet to be stored together with the ash and the boiler slag. Structurally, the fly ashes are composed of a vitreous phase (60–85%) and a crystalline phase (15–40%).

According to Voina (1977) the vitreous phase has a heterogeneous quasi-crystalline structure comprising mineral constituents, micro- and cryptocrystalline, representing various degrees of hydraulic activity (pozzolanic and self-hardening).

The crystalline constituents created when the ashes are burned (mullite, quartz, hematite, magnetite, gehlenite, anorthite, etc.) are determined by the phase thermal equilibrium, whereas the vitreous quasi-crystalline phases are formed from liquid phase cooled quickly without crystallization. The heterogeneous quasi-crystalline structure depends on the paragenesis of thermal equilibrium and, accordingly, the self-hardening and hydraulic properties are conditioned for any ash, by the presence of one or more active quasi-crystalline phases.

Ashes containing active quasi-crystalline phases are susceptible to a basic activity whereas ashes composed predominantly of inactive quasi-crystalline phases have reduced hydraulic properties. Depending on the coal quantity and its mineralogical nature, ashes from thermoelectric stations have different compositions (Table I). Since an internationally accepted criterion has not yet been established, we feel that a rational classification can be set up by considering the ratio between SiO_2 and Al_2O_3, as well as the quantities of CaO and SO_3. Hence four classes can be formed:

(i) aluminosiliceous ashes with $SiO_2/Al_2O_3 > 2$ and $CaO < 15\%$;
(ii) silicoaluminous ashes with $SiO_2/Al_2O_3 < 2$ and $CaO < 15\%$;
(iii) sulfocalcic ashes with $CaO > 15\%$ and $SO_3 > 3\%$; and
(iv) calcic ashes with $CaO > 15\%$ and $SO_3 < 3\%$.

With some exceptions the aluminosiliceous and silicoaluminous ashes result from burning of hard coal, whereas the sulfocalcic and calcic ashes result from lignite. In Table I, classifying the ashes exclusively by the

TABLE I *Chemical Composition of Fly Ash Types*

Ash resulting from	Percentage of						Type of ash
	SiO_2	Al_2O_3	Fe_2O_3	CaO	MgO	SO_3	
Hard coal, U.S.A.	55	23	8	6	2	1	Aluminosiliceous
Lignite, U.S.A.	38	17	6	31	9	1	Calcic
Hard coal, France	50	28	9	4	3	1	Silicoaluminous
Lignite, France	25	15	10	35	2	4	Sulfocalcic
Hard coal, Romania	50	27	10	4	3	2	Silicoaluminous
Lignite, Romania	45	21	9	12	4	2	Aluminosiliceous
Hard coal, Hungary	53	28	10	2	2	1	Silicoaluminous
Lignite, Hungary	58	17	10	6	2	3	Aluminosiliceous
Hard coal, Poland	26	11	7	38	4	4	Sulfocalcic

coal type (either hard coal or lignite) corresponds generally to the coals of the United States, Germany, or France. Exceptions to this rule are the hard coal and lignite of Yugoslavia, Poland, Hungary, and Romania. The hard coal and lignite ashes of Romania fall in the silicoaluminous or aluminosiliceous types, with significant variations in the CaO content, whereas sulfocalcic ashes can result from the Polish hard coal. This exception must be taken into account when the ashes are to be used in concrete, concrete products, or stabilizing earth, etc., in order to avoid confusion and an irrational choice of application.

Industrial reclamation of ashes is an important technicoeconomical problem since worldwide there are huge quantities of this material, and results of recent research are extending its areas of application (*International Ash Utilization Symposium*, 1976).

Compatibility should exist between the physicomechanical and chemicomineralogic properties of the ashes (i.e., their type) on one hand and the areas of utilization on the other.

The properties and type of a certain ash can be determined with the thermoanalytical methods. One should attempt to correlate the thermal curves (DTA, DTG, and TG) with the chemicomineralogical compositions, and to determine the efficiency of chemical and chemicothermal activation methods in order to anticipate the industrial use of ashes.

B. Thermal Analysis Applied to Ashes

The thermal analysis of ashes does not generally raise particular problems. To obtain reproducible results the vitreous phase must first be broken up by grinding. To eliminate the effect of grain size and specific surface on the thermal curves, the samples should be ground down to 0.071 mm particle size. The complex apparatus which is used to obtain all four curves (DTA, DTG, TG, and T) simultaneously for coals can also be used for ashes. The 10°C/min heating rate is also adequate.

1. *Some Results and Their Interpretation as a Function of Coal Type, for Dry Fly Ash*

The ash was collected in dry form from the electrofilters of thermoelectric stations which burn hard coal, lignite, or bituminous coal. To establish the long-term effect of moisture, dry ashes, ashes wetted with 30% water for 28 days, as well as wet ashes from stock, were analyzed.

The thermal analysis data obtained on dry ash samples are shown in Fig. 6, and the following endothermic and exothermic effects and their analytical interpretation were noted:

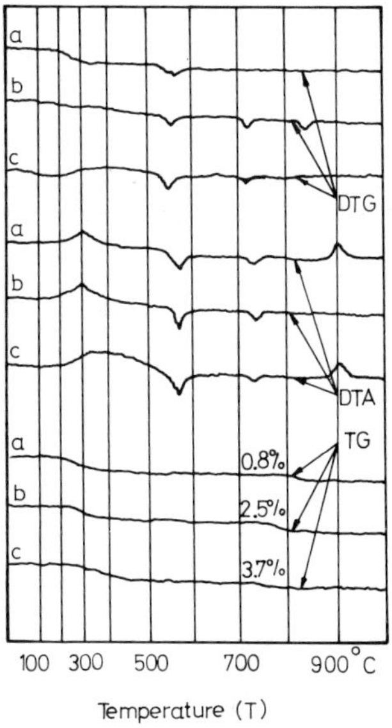

Fig. 6 Thermal analysis curves for dry fly ash samples: (a) silicoaluminous (hard coal); (b) sulfocalcic (lignite); (c) aluminosiliceous (bituminous coal).

(i) There are no endothermic effects in the range 85–140°C, which indicates that there is no water either physically absorbed or fixed in tobormorite gels, e.g., calcium hydrosilicate. In any case, the latter could not be formed without water, whatever the chemicomineralogical composition of the ash sample.

(ii) Exothermic effects (DTA curve) are, however, noticed between 250 and 350°C; they are due to burning of organic substances, and the accompanying weight loss is recorded on the DTA and TG curves. The weight loss depends on the ash type: It was 0.8% for hard coal ash, 2.5% for lignite ash, and 3.7% for bituminous coal ash. This difference in weight loss shows that the lignite and bituminous coal, which contain more noncombustibles than hard coal, did not burn completely during their short time in the burning area of the thermoelectric station. This can signal a defective burning process, either too large a grain size or insufficient gas flow.

(iii) An endothermic effect can be noticed at ~570°C due to the polymorphic transformations of SiO_2, and possibly the loss of the hydroxyl groups (OH) from kaolinite $[Si_2Al_2O_5][OH]_4$. The elimination of water at such a high temperature proves without doubt that the water in the kaolinite structure is bound chemically in OH groups, which are part of the compositional water content.

(iv) Ashes with higher SiO_2 contents, e.g., hard coal or bituminous coal ashes, present stronger endothermic effects as a result of overlapping of the polymorphic transformation of β-quartz into α-quartz. The transformation temperature of 575°C can vary within a few degrees for ashes with fine grains, calcined repeatedly. This indicates that such treatments disturb the quartz network, which allows the polymorphic transformations to take place at a lower temperature. Also, the isomorphic addition of some oxides in the network has a mineralizing effect that activates the transformation.

(v) For sulfocalcic ashes, the CO_2 that results from organic combustion reacts with CaO to form $CaCO_3$. Its presence is noticed on the DTA and TG curves at ~730°C through the endothermic effect of dissociation into CaO and CO_2. Of course, the weight loss at 730°C, as discussed for Fig. 1, allows the determination of the quantity of $CaCO_3$ in the system.

(vi) The DTA curves of the aluminosiliceous and silicoaluminous ashes stress the structural rearrangement at and above 900°C characterized by an exothermic effect with the presence of mullite.

(vii) The endothermic effects at 840°C stress the existence of $CaSO_4$, with the usual dissociation into CaO and SO_3.

For the hard coal used in the United States, France, and Germany, which results in aluminosiliceous or silicoaluminous ashes, and for lignite used in France, which results in sulfocalcic ashes, thermal analysis allows through the interpretation of the thermal effects of the ash, the determination of the exact type of coal that was burned. The silicoaluminous and sulfocalcic ashes resulting from the coals used in Romania, Poland, and Yugoslavia do not give sufficiently conclusive results to deduce the coal type. This statement is made on the basis of the chemical analyses which permitted the classification in Table I.

2. *Some Results and Their Interpretation as a Function of Coal Type, for Wet Fly Ash and for Storage Ash*

Fly ash samples (hard coal, lignite, and bituminous coal) were first wetted with 30% wt/wt water and kept for 28 days in the laboratory at 15 ± 5°C, 60 ± 5% humidity, and atmospheric CO_2. Thermal analysis methods were used to classify the physicochemical processes of the ash–water–CO_2 system. Drawing the DTA, DTG, and TG curves of Fig. 7 allows the following conclusions to be drawn:

(a) *For silicoaluminous and aluminosiliceous ashes* (Fig. 7, a curves):

(i) An endothermic effect which takes place at 102°C. This shows that the water is physically absorbed, and that the concentration of Ca^{2+} and SiO_4^{2-} ions is insufficient to start the reaction to form hydrosilicates of calcium.

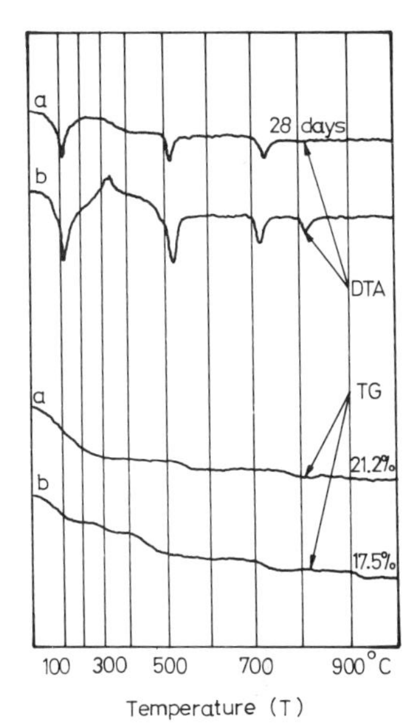

Fig. 7 Thermal analysis curves for wet fly ash samples: (a) silicoaluminous; (b) sulfocalcic.

(ii) The exothermic effect due to burning organic substances is diminished on the DTA curve at 250–350°C, possibly due to water being absorbed in the fine particles and to the thermal inertia of the measuring apparatus.

(iii) The presence of an endothermic effect at 530°C related to OH^- groups being released from the structure of calcium hydroxide formed in small quantities through the long-term action of water on CaO, partly overburned.

(iv) A reduced endothermic effect at 720°C, caused by decarbonation of $CaCO_3$, which was formed through the reaction of $Ca(OH)_2$ and atmospheric CO_2, and CO_2 resulting from burning organic substances.

(v) The weight loss (21.2%) of these samples is larger than before, as can be noticed on the TG curve, due to the initial humidity of 30%.

Thus one can draw the conclusion that these ashes do not possess a self-hardening capacity and that for practical utilization it is required to activate basically. The basic activation is required by the reduced CaO contents, and creates conditions for formation of calcium hydrosilicates with a strong structure.

(b) *For sulfocalcic and calcic ashes* (Fig. 7, b curves): Because of the high contents of CaO, of which ~20% is free, the thermal analysis curves are significantly modified from those of previous ashes (both sulfocalcic, dry, and silicoaluminous, wet). The main effects noticed by thermal differential analysis are as follows:

(i) An endothermic effect which peaks at 140°C and covers a large temperature range. This effect is undoubtedly due to water being released from the calcium hydrosilicate gels. The calcium hydrosilicates were formed by the slow reaction of $Ca(OH)_2$ and vitreous SiO_2, which is dissolved by water with pH greater than 11. Simultaneously, calcium hydrosilicates are also formed by chemical adsorption phenomena.

Tobermoritic gels of calcium hydrosilicate have a variable composition $XCaO \cdot YSiO_2 \cdot ZH_2O$ which can be condensed as $[Ca_xSiO_{2y}O_{(x+y)}]$-$[H_2O]_n$. The fact that the water is contained in a molecular state through capillary absorption forces determines its evaporation at ~140°C. If it were contained in OH^- groups as constitutional water, then its release would take place at higher temperatures, since chemical bonds are stronger than physical bonds.

Thermal analysis shows that for sulfocalcic and calcic ashes the calcium hydrosilicates are formed in a larger percentage than before. This confers on them a self-hardening capacity with high technicoeconomical possibilities of utilization in cement, concrete, soil fixation, and others. However, they are not so highly recommended for the ceramics industry. The self-hardening capacity does not, however, confer the mechanical strength to guarantee total durability, and hence activation methods are also used.

(ii) An exothermic effect at 330°C due to burning organic substances, but less than the preceding since the end of the 140°C endothermic effect overlaps.

(iii) An endothermic effect at 530°C due to OH^- groups being released from the $Ca(OH)_2$ structure, which was formed through the slow hydration of free CaO. This effect is also larger than that for the previous ashes due to the greater percentage of CaO.

(iv) An endothermic effect at 740°C due to the decarbonation reaction of $CaCO_3$.

(v) An endothermic effect at 810°C due to thermal dissociation of $CaSO_4$. As results on the TG curves show, the weight loss is ~17.5%.

Considering the clear differentiation of all endo- and exothermic effects, the thermal analysis curves permit the determination, quantitatively and qualitatively, of chemical compounds such as calcium hydrosilicates, $Ca(OH)_2$, $CaCO_3$, and unburned organic substances. See, for example, the determinations based on Fig. 1.

For sulfocalcic and calcic ashes the exothermic effect of structural reorganization of mullite is less, which is partly explained by the overlapping of the 810°C endothermic effect, and also by the reduced amounts of SiO_2 and Al_2O_3.

The sulfocalcic and calcic ash samples taken from storage, which were wetted naturally, present similar effects with only small, quantitative differences.

3. *Thermal Analysis Applied to Fly Ash Cenospheres*

During the combustion process some ash particles can partially melt, and the surface tension and the simultaneous release of some gases (O_2, N_2, CO_2, etc.) determine the formation of small, empty spheres. These were initially called microballs, and are now termed cenospheres.

Cenospheres with a diameter in the range 20–200 μm and density 0.3–0.6 g/cm^3, can be separated through either dry or wet processes. However, as shown by Pedlow (1973) and Zeeuw and Abresch (1976), the cenospheres present, irrespective of the separation method, are of major interest for industrial use because of their high strength, low density, good thermal and electric capacity, and good tolerance for chemical agents and high temperature.

Thermal analysis with ground cenosphere samples (Fig. 8) shows an endothermic effect at 330°C possibly due to the release of water and some gases from the vitreous mass. These gases were occluded in the vitreous mass at the expansion stage, as, e.g., O_2 and CO_2, among others.

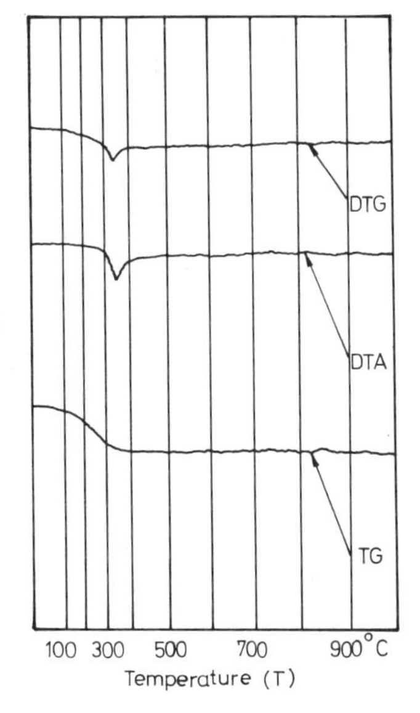

Fig. 8 Thermal analysis curves on ground cenosphere samples.

One should emphasize that at 575°C the thermal effect of polymorphic transformation of quartz is absent, since SiO_2 is present in large quantities in vitreous form. The exothermic effect at 350–450°C, specific to the combustion of organic substances, is also absent. This allows one to conclude that organic substances are absent from the cenosphere structure.

Thermal analysis confirms the vitreous character of the cenospheres with regard to their high stability at elevated temperatures and their melting range (1350–1450°C).

C. Thermal Analysis for Determining the Efficiency of Activating Ashes from Thermoelectric Stations

1. *The Importance of the Problem*

The ashes from thermoelectric stations possess hydraulic characteristics (self-hardening and pozzolanic) of hardening in the presence of some chemical substances, known generally as activators.

Through activation processes, irrespective of their nature, one tries to initiate chemical reactions followed by physical processes of hardening that have the effect of conferring on the structure a strength similar to that of cement. The hydraulic properties are the basis of ash utilization as additive material in the cement industry, concrete, soil stabilization, and so on.

Ash activation for obtaining products with higher strength is done in practice through mixing with aqueous solutions of $Ca(OH)_2$ or donors of Ca^{2+} ions (cement, furnace slags plus cement). The necessary condition for an efficient activation is to ensure a minimum concentration of $Ca(OH)_2$, which in the presence of water and of acid oxides from the ash (SiO_2, Al_2O_3, Fe_2O_3) can initiate the reactions:

$$x\text{CaO} + y\text{SiO}_2 + z\text{H}_2\text{O} \rightarrow x\text{CaO}\cdot y\text{SiO}_2\cdot z\text{H}_2\text{O} \quad (1)$$

$$x'\text{CaO} + y'\text{Al}_2\text{O}_3 + z'\text{H}_2\text{O} \rightarrow x'\text{CaO}\cdot y'\text{Al}_2\text{O}_3\cdot z'\text{H}_2\text{O} \quad (2)$$

The ratios between x, y, z, x', y', z' are dependent on the CaO quantity, the water/ash ratio, the hardening procedure, and other factors.

According to Voina (1975) through convergent ash activations one can form solid solutions from the "hydrogranite" series, for example,

$$\begin{matrix} 3\text{CaO}\cdot\text{Al}_2\text{O}_3\cdot 6\text{H}_2\text{O} & & 3\text{CaO}\cdot\text{Fe}_2\text{O}_3\cdot 6\text{H}_2\text{O} \\ & \times & \\ 3\text{CaO}\cdot\text{Al}_2\text{O}_3\cdot 3\text{SiO}_2 & & 3\text{CaO}\cdot\text{Fe}_2\text{O}_3\cdot 3\text{SiO}_2 \end{matrix} \quad (3)$$

The formation of hydrogranite solid solution series is very important for concrete-plus-ash products since along with a higher strength it gives simultaneously a considerable increase of gel and chemical resistances. These conclusions increase interest in extending the use of ashes in cements, concretes, and concrete products, i.e., obtaining concrete from ash without cement, using the procedure proposed by Voina (1974).

We shall now present the methodology and data interpretation for the thermal analysis of activated silicoaluminous and aluminosiliceous ashes. These ashes were chosen since they are much more abundant than sulfocalcic and calcic ashes, both in the United States and in Europe.

Through the thermal analysis of ashes activated with different chemical substances, with or without thermal treatment, one can evaluate, qualitatively and quantitatively, the $Ca(OH)_2$, the hydrosilicates and calcium hydroaluminates, and the degree of carbonation and other chemical processes. In CaO activation of ashes one can determine the optimum amount for addition by deducing the quantity of $Ca(OH)_2$ not bound in hydrated mineral compounds, knowing that a large quantity of free calcium hydroxide leads to a reduction in mechanical strength and an increase in sensitivity to chemical attack.

2. *Working Methodology*

Samples composed of a well-homogenized mixture of 100 g ash, 15 g $Ca(OH)_2$ powder, and 45 g water were used. In parallel with thermal analysis, these samples were also employed to determine the cumulative effect of a more complex activity using, besides $Ca(OH)_2$, additional quantities of electrolytes as well as hardening through thermal treatments.

The selection criterion for an electrolyte was based on the type of hydrolysis which generated it: e.g., Na_2CO_3, basic hydrolysis; Na_2SO_4, no hydrolysis; $FeCl_3$, acid hydrolysis. The quantities of electrolyte were 2% in all cases.

Assuming that ashes become more reactive with increase in soluble SiO_2 content, Table II shows, besides the samples synthesized for the analysis, the quantity of soluble silica formed by the increase in pH in the solution through lime addition. This allows one to follow the influence of the three types of electrolyte. The salts with basic hydrolysis, as well as those with acid hydrolysis, increase the quantity of soluble SiO_2, facilitating the binding reaction of $Ca(OH)_2$ in hydrated compounds.

The analysis of the complex physicochemical processes which govern formation was done at certain time intervals on the samples, which were kept in a wet environment. For thermal analysis, mean samples were prepared taking 1 g of substance and using a heating rate of 10°C/min.

TABLE II *Composition of Samples and Quantities of Soluble SiO_2*

No.	Composition of samples: Materials	Composition of samples: Quantity (g)	Soluble SiO_2 (%)
1	Ash	100	3.52
2	Ash + lime	100 + 10	5.50
3	Ash + lime + Na_2CO_3	100 + 10 + 2	7.60
4	Ash + lime + Na_2SO_4	100 + 10 + 2	6.90
5	Ash + lime + $FeCl_3$	100 + 10 + 2	5.40

3. *Thermal Analyses of Activated Ashes and Their Interpretation*

In Fig. 9 are shown the thermal analyses of compositions 1 and 2 (Table II), in samples of mechanical mixtures of ash and lime, and in samples which were wetted to trigger the chemical reactions. The analyses were performed 24 hr after mixing.

It can be noticed that the lignite ash (a curves) does not present a thermal behavior different from that of the silicoaluminous or aluminosiliceous ash analyzed in Section V,B,1.

The b curves refer to water cement, and show first an endothermic effect determined by the water present as absorbed moisture, and repre-

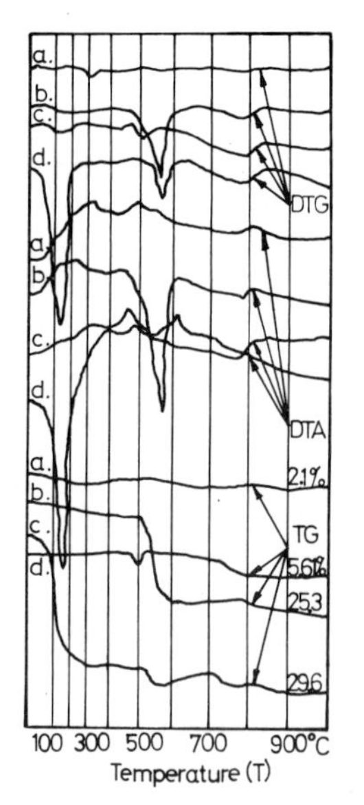

Fig. 9 Thermal analysis curves for (a) ash samples; (b) water cement samples; (c) mixture of ash and hydraulic cement; (d) mixture of ash, lime, and water.

senting 5.4% of the sample weight. In the range 450–580°C the thermal curves indicate an endothermic effect specific for elimination of hydroxyl groups (OH^-) from calcium hydroxide. As we know, the hydroxyl groups are eliminated as water, the total loss being 18.9%. One can also note that the hydrated lime was carbonated to a small degree, as was recorded on the thermal curves in the range 720–780°C, the weight loss being 3%.

From analysis of thermal curves for the mixture of ash and hydraulic cement (c curves) one can observe the expected sequence of effects mentioned for ash and for hydraulic cement.

The endothermic effects for hydraulic cement are reduced and the peaks are 40°C lower, which can be explained if we consider that we worked with a reduced quantity of hydraulic cement in the mixture, i.e., 10%, which means a tenfold reduction. The fact that new supplementary endothermic effects do not appear indicates with clarity that the two components, ash and lime, did not react in the absence of water, which would have led to calcium hydrosilicates and hydroaluminates.

Two supplementary effects should be observed on the thermal curves; first is a slight tendency toward increase in sample weight (TG and DTG curves) between 400 and 550°C, which is explained by the oxidation of Fe^{2+} ions. The second aspect is connected with the increase in the specifically endothermic effect of carbonation, both quantitatively and in a shift toward higher temperatures. Owing to the higher quantity of $CaCO_3$, the peak of the endothermic effect moved toward 800°C. The simultaneous increase in $CaCO_3$ content and decrease in $Ca(OH)_2$ content is determined by the formation of supplementary CO_2, which generated the carbonation of $Ca(OH)_2$ in the sample.

On the thermal d curves the behavior of an ash–lime–water mixture was analyzed after 24 hr of interaction. One can notice, especially on the DTA curves, that the first endothermic effect specific to water loss has increased, and shifted its peak from 105°C to 140°C. This fact indicates that the ash reacted with the lime, yielding calcium hydrosilicates in which water is bound through absorption in tobormoritic gels. Also observed is a partial decrease of the $Ca(OH)_2$ dehydration effect through the consumption of $Ca(OH)_2$ in forming calcium hydrosilicates and calcium carbonates.

Thermal analysis also allows the time-monitoring of kinetic processes of reaction between the oxidic ash compounds and various activating substances, a fact important in elucidating how the structure is formed and how strength and durability are conferred to products and obtained from ash–cements or activated ash. In our case the time evolution up to 90 days of the physicochemical processes in the ash–cement–water sys-

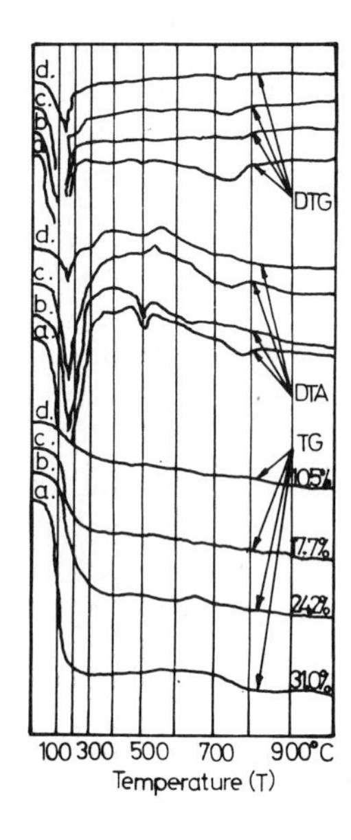

Fig. 10 Thermal analysis curves for ashes activated with lime samples, analyzed at (a) 1, (b) 7, (c) 28, and (d) 90 days.

tem was analyzed through interpretation of the thermal curves in Fig. 10.

It can be observed that in the first 24 hr (a curves), the water is left unbound in the system as absorption water. This determines its release occurring with an endothermic effect, with the peak at ~100°C, and the weight loss reaching 28% (the total up to 1000°C is 31%). The endothermic effect at 500°C due to lime dehydration and that at 780°C due to $CaCO_3$ decarbonation are maintained. One should also consider the possibility that because of the high content of free $Ca(OH)_2$ and through the burning of the organic substances contained in the ash, which releases CO_2, $CaCO_3$ was formed through the interaction of the two components.

From the behavior of the thermal curves for 7, 28, and 90 days (b, c, and d curves), it can be observed that the reactions of formation of calcium hydrosilicate gels are more advanced with time, and the water contained in the gels evaporates at higher temperatures, with peaks toward 180–200°C. This emphasizes the aging phenomenon of gel recrystallization, and bonding water with more intensive forces (the water was initially absorbed and then bound through chemisorption).

On the e and d curves one can notice an endothermic effect with peaks at 260 and 300°C, respectively, related to dehydration of calcium hydroaluminate. The release of water at such high temperatures shows the chemical character of the water bond in the structure to be with OH^-

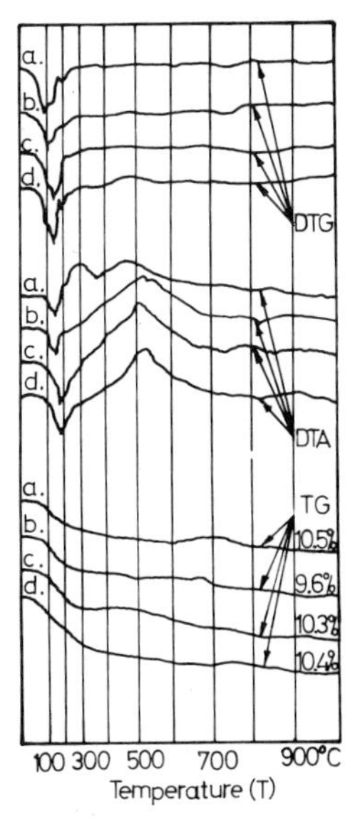

Fig. 11 Thermal analysis curves for (a) ash resulting from lignite, and ash samples activated with (b) lime + Na_2CO_3, (c) lime + Na_2SO_4, and (d) lime + Fe Cl_3.

groups and as crystallization water, respectively. Simultaneously one can deduce the lessening of the endothermic effect characteristic of $Ca(OH)_2$ dehydration, since the content of the hydrated calcium oxide bound in hydrosilicates and hydroaluminates increases with time.

Thermal analyses also permit one to deduce that the specific effects of $Ca(OH)_2$ dehydration are partially modified by the exothermic effects due to burning of organic substances and to polymorphic transformations of SiO_2. The bonding of $Ca(OH)_2$ as hydrosilicates and hydroaluminates also results, as can be seen in the DTA and DTG curves, in a sizable diminishing of the possibility for $CaCO_3$ formation, regardless of the CO_2 present in the atmosphere, or from burning organic substances. The TG curves allow one to deduce that the weight loss diminishes because in time the physical water left in the structure, after water is bound in chemical compounds, can evaporate. The amount of water initially introduced in the mixture is larger than that necessary for the chemical reactions, since one intends to ensure a good workability of the ash–water–activators mixture, which should be easily compactable in various concrete products and prefabricated items.

From the thermal analysis of the ash–lime–Na_2CO_3–water system (Fig. 11, b curves), one can observe that the thermal effects are similar to those of the ash–lime–water system (Fig. 10, c and d curves).

Considering the general character of the basic hydrolysis generated by Na_2CO_3, the resulting effects are analogous to those of lime, with the distinction that in this case the activation effects are convergent [due to the presence of both oxides with basic character, NaOH and $Ca(OH)_2$] and hence appear more pronounced on the DTA curves. This has the following explanation: The solubility of SiO_2 is more pronounced in the presence of NaOH and hence the quantity of calcium hydrosilicates increases. This characteristic is also noticeable on the DTA curve where one can see an obvious enlargement of the range in which water is released from the calcium hydrosilicate gels. The practical use of lime activation, stimulated with other basic hydrolysis salts also, is reasonable since it allows a more efficient utilization of the hydraulic potential of the ashes.

For the mixture ash–lime–Na_2SO_4–water (Fig. 11, c curves), the endothermic effect of water release is slightly shifted toward higher temperatures. This shift is the result of overlapping, in the same temperature range, of two phenomena, the first related to water being released from tobormoritic gels of calcium hydrosilicates, and the second determined by the partial dehydration of the gypsum formed through the double exchange reaction between SO_4^{2-} and Ca^{2+}.

On the thermal analysis curves of the mixture ash–lime–$FeCl_3$–water (Fig. 11, d curves) one can notice some differences. The water release takes place in a narrower temperature range (DTA curve), whereas at ~400°C a bigger exothermic effect takes place, characteristic of the oxidation of Fe ions under the influence of the Cl^- anion and of combustion of the organic substances from the ash (Fig. 11, a, b, and c curves). Simultaneously it can be seen that the samples have a higher weight loss up to 660°C than the samples without $FeCl_3$. This is possibly due to Cl^- ions being released from the system as volatile chlorides.

The thermal analysis of ash–lime–$FeCl_3$–water samples shows that due to the presence of $FeCl_3$ the endothermic effect of calcium carbonate dissociation is absent. This is explained by the decomposition of $FeCl_3$ upon heating and the formation, in the presence of water, of HCl which reacts with calcium carbonate according to

$$2HCl + CaCO_3 \rightarrow CaCl_2 + CO_2 + H_2O \tag{4}$$

4. *Thermal Analysis of Activated Samples Hardened through Thermal Treatments*

Thermal analyses were also used to determine the kinetics of chemical reactions, and the final product of the reaction between ash, lime, and water under the effect of steam and in an autoclave, and to compare it with samples hardened under normal temperatures. A mixture of 1000 g ash, 150 g lime, and 400 cm^3 of water was used. The mixture was homogenized and cast in cubic shapes, divided in three series: The first series was hardened under laboratory conditions for 28 days; the second series was hardened 3 hr after casting with steam at 95°C and 0.5 atm for 12 hr; the third series was hardened 3 hr after casting in an autoclave at 12 atm for 12 hr.

The thermal analyses shown in Fig. 12 are (a) for normally hardened samples, (b) for steamed samples, and (c) for autoclaved samples. For comparison the analysis of a 1 : 3 cement–sand mortar sample hardened for 28 days is also shown (d curves).

The thermal curves for cement (d) and ash–lime, both normally hardened (a) and steamed (b), present very similar endothermic effects due to dehydration of calcium hydrosilicate gels, dehydration of $Ca(OH)_2$, and decarbonation of $CaCO_3$. It is interesting to note that for the samples treated in an autoclave the endothermic effect at 500°C is absent, which demonstrates the absence of $Ca(OH)_2$. This indicates that the whole quantity of $Ca(OH)_2$ was bound in calcium hydrosilicates and hydroaluminates, and that autoclave treatment is more efficient than steaming. The optimum percentage of CaO can be estimated as a function of the hardening process (normal, steam, or autoclave).

The differences in weight loss and the amplification of endothermic effects allow one to deduce the efficiency of hydration processes in the sense that the most intensive hydration was noticed for the cement samples and the ash–lime autoclaved samples, followed by the steamed samples, and finally the samples hardened normally.

The fact that the samples of ash activated with lime present thermal effects largely similar to those of hydrated cement allows us to conclude

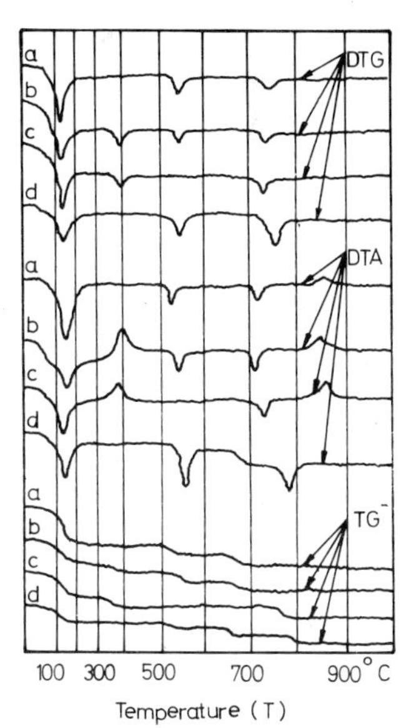

Fig. 12 Thermal analysis curves for ash–lime–water samples: (a) hardened normally, (b) steamed, and (c) autoclaved; (d) 1 : 3 cement–sand mortar, for comparison.

that the reactions between ash and lime give, in the end, the same reaction products as the cement, i.e., calcium hydrosilicates, calcium hydroaluminates, calcium hydroxide, and calcium carbonate. The thermal analyses indicate that the thermal treatments (steaming and autoclaving) present essential advantages for giving strength to products and prefabricates from ash activated with lime.

REFERENCES

Arens, P. L. (1951). "A Study on the Differential Thermal Analysis of Clays and Clay Minerals." Agricultural University, Wageningen, Netherlands.
Barshad, J. (1952). *Am. Mineral.* **37,** 667.
Bátor, B., and Weltner, M. (1965). *Acta Chim. Acad. Sci. Hung.* **43,** 99.
Berg, L. G., and Egunov, V. P. (1969). *J. Therm. Anal.* **1,** 5.
Buzágh-Gere, E., and Gál, S. S. (1974). *Period. Polytech., Eng.* **18,** 33.
Duval, C. (1963). "Inorganic Thermogravimetric Analysis." Elsevier, Amsterdam.
Erdey, L., Paulik, F., and Paulik, J. (1954). *Nature (London)* **174,** 885.
Grimshow, R. W., and Roberts, A. L. (1953). *Trans. Br. Ceram. Soc.* **52**(1), 50.
Grimshow, R. W., and Roberts, A. L. (1957). *In* "The Differential Thermal Investigation of Clays" (R. C. MacKenzie, ed.), Mineral. Soc., London.
Heilpern, S. (1974). *Proc. ICTA Conf., 4th, Budapest* p. 459.
Hofmann, S., and Garstka, H. (1965). *Freiberg. Forschungsh. A* **345,** 143.
Holowiecki, K., and Chodynski, A. (1974). *Proc. ICTA Conf., 4th, Budapest* p. 321.
Horowitz, H. H., and Metzger, G. (1963). *Anal. Chem.* **31,** 1464.
International Ash Utilization Symposium (1976). *4th, St. Louis, Mo.* National Ash Association, Washington, D.C.
MacKenzie, R. C. (1957). "The Differential Thermal Investigation of Clays." Mineral. Soc., London.
Paulik, F., Paulik, J., and Erdey, L. (1958). *Z. Anal. Chem.* **3**(4), 1, 60.
Pedlow, J. W. (1973). *Int. Ash Util. Symp., 3rd, Pittsburgh, Pa.* pp. 33–44.
Rey, M., and Kostomarof, V. (1959). *Silic. Ind.* **24,** 603.
Soule, J. L. (1952). *J. Phys. Radium* **13,** 516.
Todor, D. N. (1972). "Thermal Analysis of Minerals." Editions Tehnica, Bucharest.
Todor, D. N. (1976). "Thermal Analysis of Minerals." Abacus Press, Kent, England.
Voina, N. I. (1974). Ger. Pat. No. 2,336,404.
Voina, N. I. (1975). "I Conferință Privind Valorificarea Cenuşilor de la Centralele Termoelectrice," pp. 54–61. CNIT, Bucharest.
Voina, N. I. (1977). *Rev. Materialelor Constr.* **9,** 15–26.
Waters, P. L. (1956). *Coke Gas* **20,** 256.
Weltner, M. (1958). *Magy. Kem. Lapja* **13,** 200.
Weltner, M. (1959a). *Nature (London)* **183,** 1254.
Weltner, M. (1959b). *Magy. Kem. Lapja* **14,** 192.
Weltner, M. (1961). *Brennst.-Chem.* **42,** 40.
Weltner, M. (1962). *Acta Chim. Acad. Sci. Hung.* **31,** 449.
Weltner, M. (1965). *Acta Chim. Acad. Sci. Hung.* **43,** 89.
Weltner, M. (1966). *Acta Chim. Acad. Sci. Hung.* **47,** 311.
Weltner, M. (1969). *Magy. Kem. Foly.* **75,** 395.
Zeeuw, J. H., and Abresch, R. V. (1976). *Int. Ash Util. Symp., 4th, St. Louis, Mo.* pp. 1–5.

Index

B

C

Q

R

T

W

X

Y

Z